T0227550

Parallel Robots

Mechanics and Control

Parallel Robots

Mechanics and Control

HAMID D. TAGHIRAD

CRC Press
Taylor & Francis Group
Boca Raton London New York

CRC Press is an imprint of the
Taylor & Francis Group, an **informa** business

MATLAB® is a trademark of The MathWorks, Inc. and is used with permission. The MathWorks does not warrant the accuracy of the text or exercises in this book. This book's use or discussion of MATLAB® software or related products does not constitute endorsement or sponsorship by The MathWorks of a particular pedagogical approach or particular use of the MATLAB® software.

CRC Press
Taylor & Francis Group
6000 Broken Sound Parkway NW, Suite 300
Boca Raton, FL 33487-2742

First issued in paperback 2017

© 2013 by Taylor & Francis Group, LLC
CRC Press is an imprint of Taylor & Francis Group, an Informa business

No claim to original U.S. Government works
Version Date: 20121112

ISBN 13: 978-1-138-07738-6 (pbk)
ISBN 13: 978-1-4665-5576-1 (hbk)

This book contains information obtained from authentic and highly regarded sources. Reasonable efforts have been made to publish reliable data and information, but the author and publisher cannot assume responsibility for the validity of all materials or the consequences of their use. The authors and publishers have attempted to trace the copyright holders of all material reproduced in this publication and apologize to copyright holders if permission to publish in this form has not been obtained. If any copyright material has not been acknowledged please write and let us know so we may rectify in any future reprint.

Except as permitted under U.S. Copyright Law, no part of this book may be reprinted, reproduced, transmitted, or utilized in any form by any electronic, mechanical, or other means, now known or hereafter invented, including photocopying, microfilming, and recording, or in any information storage or retrieval system, without written permission from the publishers.

For permission to photocopy or use material electronically from this work, please access www.copyright.com (http://www.copyright.com/) or contact the Copyright Clearance Center, Inc. (CCC), 222 Rosewood Drive, Danvers, MA 01923, 978-750-8400. CCC is a not-for-profit organization that provides licenses and registration for a variety of users. For organizations that have been granted a photocopy license by the CCC, a separate system of payment has been arranged.

Trademark Notice: Product or corporate names may be trademarks or registered trademarks, and are used only for identification and explanation without intent to infringe.

Library of Congress Cataloging-in-Publication Data

Taghirad, Hamid.
Parallel robots : mechanics and control / Hamid Taghirad.
p. cm.
Includes bibliographical references and index.
ISBN 978-1-4665-5576-1 (hardback)
1. Parallel robots. I. Title.

TJ211.4152.T34 2013
629.8'92--dc23 2012040401

Visit the Taylor & Francis Web site at
http://www.taylorandfrancis.com

and the CRC Press Web site at
http://www.crcpress.com

Contents

Preface

Robots have changed the life of human beings in the twenty-first century. In industrial automation, the use of robots is vital to preserve the quantity and quality of production by introducing flexibility to the production line. Industrial robots usually have an articulated structure in which a series of links are connected to each other to provide a large workspace. The motion of the robot is controlled through the disjointed actuators that manipulate individual motion of each link. Although, in such structures, characteristics such as a large workspace and flexibility may be obtained, the accuracy of the last manipulating element is significantly threatened by the serial structure.

For applications in which high precision and low compliance are required or a relatively high load capacity per robot weight is essential, parallel structures are the absolute alternative. A parallel robot has an inherent closed-loop kinematic structure, and its moving platform is linked to the base by several independent kinematic chains. Many industrial applications have adopted parallel structure for their design; however, only a very few textbooks have been published to introduce the analysis of such robots in terms of kinematics, dynamics, and control. This book is intended to give some analysis and design tools for the increasing number of engineers and researchers who are interested in the design and implementation of such robots in industries. In this book, a systematic approach is presented to analyze the kinematics, dynamics, and control of parallel robots.

In order to define the motion characteristics of such robots, it is necessary to represent 3D motion of the robots' moving platform with respect to a fixed-coordinate frame. This naturally leads to the need for a systematic representation of the position, orientation, and location of bodies in space. In Chapter 2, such representations are introduced with an emphasis on screw coordinates, which makes the representation of general motion of the robot much easier to follow. It should be noted that the ideas developed for position and orientation representation will form a basis for linear and angular velocity and acceleration representations, and this is also adopted to represent forces and torques applied in a robotic manipulator.

Kinematic analysis refers to the study of the geometry of motion in a robot without considering the forces and torques that cause the motion. In this analysis, the relation between the geometrical parameters of the manipulator and the final motion of the moving platform is derived and analyzed. A complete treatment of such an analysis is given in Chapter 3, and elaborative case studies are provided for three parallel robots, including a planar cable-driven parallel robot. The analysis of cable-driven parallel robots is formally treated in this book as the promising new generation of parallel structures that provide a very large workspace.

In Chapter 4, kinematic analysis of robot manipulators is further examined beyond static positioning. Differential kinematic analysis plays a vital role in the singular free design of robotic manipulators. Jacobian analysis not only reveals the relation between the joint variable velocities and the moving platform linear and angular velocities, but it also constructs the transformation needed to find the actuator forces from the task space forces and moments acting on the moving platform. A systematic approach to performing Jacobian analysis of parallel manipulators is given in this chapter and the proposed method is examined through the same case studies analyzed in Chapter 3.

The dynamic analysis of parallel manipulators presents an inherent complexity due to the closed-loop structure and kinematic constraints. Nevertheless, the dynamic modeling is quite important for the control, particularly because parallel manipulators are preferred in applications where precise positioning and suitable dynamic performance under high loads are the prime requirements. Although a great deal of research has been presented on the kinematics of parallel manipulators, works on the dynamics and control of parallel manipulators are relatively few, and almost no books cover these issues in detail. These issues are addressed well in this book in Chapter 5, in which dynamic analysis of such robots is examined by three methods, namely the Newton–Euler principle of virtual work and Lagrange formulations. Furthermore, a method is presented in this chapter to formulate the dynamic equation of parallel robots into a closed form, by which the dynamic matrices are more tractable and dynamics verification becomes possible.

The control of a parallel robot is elaborated in the last two chapters of the book, in which both motion and force control schemes are covered. Different model-free and model-based controllers are introduced and robust and adaptive control schemes are elaborated in Chapter 6. The control techniques are applied to two case studies, in which both cable-driven redundant parallel manipulator and fully parallel manipulators are examined through the proposed control schemes. Finally, Chapter 7 covers the force control of parallel robots in detail. In this chapter, stiffness control, direct force control, and impedance control schemes are elaborated and implemented on the same case studies followed in the book.

A key to verify the analysis and the controller performance is computer simulation. Computer simulations are being used for the case studies followed in all chapters throughout the text. Simulations are usually performed by commercially available packages such as MATLAB®, which provides a suitable means to simulate the robot's kinematic or dynamic characteristics and to verify the performance of the control systems. The manuscript was typeset using LaTeX, and the artworks were generated by Smart Draw and Inkscape software.

I am indebted to many people who have supported me either technically or spiritually during the writing of this book. As it involves the knowledge about many disciplines, numerous people have contributed to this work, but a list of the names could not be presented here; however, all of them are acknowledged. I would like to dedicate this book to the late Professor G. Zames and Professor P. R. Bélanger, not just for many things I have learned from them in control theory, but also for the deep influence they have induced in my soul *to make a difference*. I am also indebted to Professors J. Angeles and C. Gosselin who encouraged me to pursue this work. Many of the results presented in this book are mainly the contributions of J. Angeles, C. Gosselin, J.-P. Merlet, L-W. Tsai, and many other prominent researchers in this field. I had the pleasure to organize and further elaborate on these contributions. Any error in the presentation of their work is solely mine.

I acknowledge the enjoyable collaborations I had with Professors M. Nahon and I. Bonev and express my gratitude to them for providing me the visiting opportunities during two critical time periods and allowing me to temporarily escape my regular tight schedule and focus on the book. The content of this book was examined by many students who took the postgraduate course at McGill University and at K. N. Toosi University of Technology, and their comments and corrections have improved the quality of the materials. Among them, I would like to thank Dr. H. Sadjadian who spent a lot of time correcting the manuscript, and R. Oftadeh for his contributions in the dynamic formulation of parallel manipulators. Certainly, the current version of this book is not error-free, and I appreciate any comments and corrections from all respected professional readers. All individuals and institutions who have contributed to graphical materials and artwork are sincerely acknowledged.

I cannot conclude without recalling the support and encouragement I received from my wife, Azam, and my daughter, Matineh, and my deepest regards go to their unlimited support and patience.

Hamid D. Taghirad
Tehran, June 23, 2012

MATLAB® is a registered trademark of The MathWorks, Inc. For product information, please contact:

The MathWorks, Inc.
3 Apple Hill Drive
Natick, MA 01760-2098 USA
Tel: 508-647-7000
Fax: 508-647-7001
E-mail: info@mathworks.com
Web: www.mathworks.com

1

Introduction

Robots are very important assets for today's industry. The use of robots is vital in industrial automation to preserve the quantity and quality of production while introducing flexibility in the manufacturing line. The ever-increasing necessity to introduce new product styles, improve the product quality, and reduce the manufacturing costs has resulted in greater adoption of robotic equipment in various industries. At first, automobile manufacturing companies used robots in their production lines. However, in recent years, other industrial units that produce home appliances, food and pharmaceutical materials, and so on have adopted robotic systems in their production lines. A major reason for the growth in the use of industrial robots in different production lines is their significantly declining cost. In recent years, robot prices have significantly dropped while human labor costs are increasing. Also, robots are becoming more effective, faster, smarter, more accurate, and more flexible.

Industrial robots usually have an articulated structure. In these robotic manipulators, a series of links are connected in order to provide a large workspace. The motion of the robot is controlled through the individual actuators that manipulate the individual motion of each link. Although in such structures, design objectives such as a large workspace and flexibility can be well satisfied, the accuracy of the robot end effector is significantly threatened by its serial structure. For applications in which high precision and stiffness are required or a relatively high load capacity per robot weight is needed, parallel structures are the absolute alternative. Many books have focused on the theoretical and technological advancements of serial robots [5,31,163,168]. However, very few have covered the topics on the analysis, design, and control of parallel robots [105,133]. This book is intended to provide some analysis and design tools for the increasing number of engineers and researchers interested in the design and implementation of parallel robots in industries.

1.1 What Is a Robot?

A robot is a mechanical or virtual artificial agent, usually an electromechanical system, which, by its appearance or movements, conveys the sense that it has intent or agency of its own. While there are still controversies about which machines qualify as robots, a typical robot will have several, although not necessarily all, of the following properties:

- It is not *natural* and has been artificially created
- Can sense its environment
- Can manipulate things in its environment
- Has some degree of intelligence
- Is programmable

- Can move with one or more axes of motion
- Appears to have intent or agency

The last property, the appearance of agency, is important when people are considering whether to call a machine a robot. In general, the more a machine has the appearance of agency, the more it is considered a robot. There is no one definition of robot that satisfies everyone, and many people have written their own. For example, the international standard ISO 8373 defines a robot as

> An automatically controlled, reprogrammable, multipurpose, manipulator, programmable in three or more axes, which may be either fixed in place or mobile for use in industrial automation applications.

Joseph Engelberger, a pioneer of industrial robotics [44], once remarked:

> I can't define a robot, but I know it when I see one.

The *Cambridge Advanced Learner's Dictionary* defines a robot as

> A machine used to perform jobs automatically, which is controlled by a computer.

The Robotics Institute of America used the following definition for a robot:

> A *robot* is a re-programmable multi-functional manipulator designed to move materials, parts, tools, or specialized devices, through variable programmed motions for the performance of a variety of tasks.

This definition includes mechanical manipulators, numerical controlled (NC) machines, walking machines, and humanoids of science fictions. Building a humanoid capable of doing what a human being can do is an ancient dream of humankind, and technologies developments to build machines and mechanisms that can perform like humans may all be seen in the field of robotics research. Hence, robotics is a multidisciplinary engineering field of research. In industry, however, a mechanical manipulator is usually recognized as a robot which resembles the human arm.

The word *robot* entered the vocabulary of English as early as in 1923. This word was first used by Karel Čapek in his book *Rossam's Universal Robots* [183]. Čapek visualized a situation where a bioprocess could create human-like machines devoid of emotions and souls. However, they were very strong and obeyed, and they could be produced quickly and cheaply. Soon, all major countries wanted to equip their armies with hundreds of thousands of slave robotic soldiers, who can fight with dedication but whose loss is not painful. Eventually, the robots decided to become superior to the humans and tried to take over the world. In this story, the word *robota* or worker was coined.

However, the emergence of industrial robots did not occur until after the 1940s. In 1946, George Devol patented a general-purpose playback device for controlling machines using magnetic recording, and in 1954, he designed the first programmable robot and coined the term *universal automation*, planting the seed for the name of his future company—Unimation. In the early 1980s, several robot-producing companies emerged or joined, and the number of industrial robots used in the industries increased significantly. In the second millennium, robotics research was focused more on the technology for building humanoid robots and robotic pets.

1.2 Robot Components

A mechanism or a robotic manipulator is usually built from a number of links connected to each other and to the ground or a movable base by different types of joints. The number of degrees-of-freedom of a robot depends on the number of links and the type of joints used for the construction of the robot. In this section, the definitions of *links, joints, kinematic chains, mechanisms,* and *machines* are given, and then the concept of degrees-of-freedom is described.

The individual rigid bodies that make up a robot are called the *links*. In industrial robots, the rigidity of the links contributes significantly to the precision and performance of the robots, and usually in the design of links, rigidity is a vital requirement. However, in applications such as space robotics or cable-driven manipulators, due to the limitations and type of applications, special designs are adopted in which the links are constructed from flexible elements. Such robots are usually called *flexible link manipulators*. In this book, links are treated as rigid bodies for most of the manipulators which are analyzed in different chapters, unless stated otherwise. The assumption of the rigid bodies makes the analysis of robot manipulators much easier to understand. For cable-driven parallel manipulators, the assumption of rigid bodies for the link is applicable only when the manipulator is operated with high stiffness, and the internal tensions in the cables are relatively high. In such cases, the sagging effect of the cables are negligible, and the assumption of a rigid body for the links gives us good insight into the development of a dynamic analysis and control of such manipulators. From a kinematic point of view, a single link can be defined as an assembly of members connected to each other, such that no relative motion can occur among them. For example, two gears connected by a rigid shaft are treated as a single link.

In robots, the links are connected in pairs, and the connective element between two links is called a *joint*. A joint provides some physical constraints on the relative motion between the two connecting members. Owing to the required relative motion in a kinematic pair, different types of joints may be distinguished.

- A *revolute joint, R,* permits rotation about an axis between two paired elements as shown in Figure 1.1. Hence, a revolute joint imposes five constraints between the connecting links and provides one-degree-of-freedom.
- A *prismatic joint, P,* permits sliding along one axis between two paired elements as shown in Figure 1.1. Hence, a prismatic joint imposes five constraints between the connecting links and provides one-degree-of-freedom.
- A *cylindrical joint, C,* permits rotation about one axis, and independent translation along another axis as shown in Figure 1.2. Hence, a cylindrical joint imposes four constraints between the connecting links and provides two-degrees-of-freedom.
- A *universal joint, U,* permits rotation about two independent axes as shown in Figure 1.2. Hence, a universal joint imposes four constraints between the connecting links and provides two-degrees-of-freedom. A universal joint can be made from two consecutive revolute joints.
- A *spherical joint, S,* permits free rotation of one element with respect to another element about the center of a sphere in all the three directions as shown in Figure 1.3. No translation between the paired element is permitted. Hence, a spherical joint imposes three translational constraints between the connecting links and provides

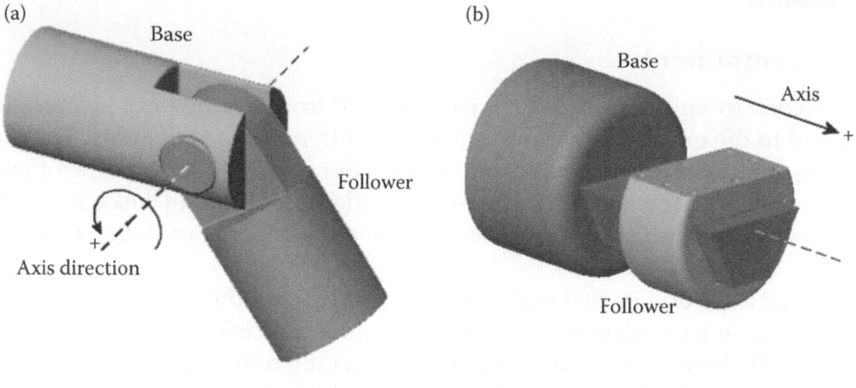

FIGURE 1.1
Schematics of a revolute joint (a) and a prismatic joint (b). (From Mathworks Inc. Schematics of a revolute joint (left) and a prismatic joint (right), 2010. Mathworks. With permission.)

three rotational degrees-of-freedom. As illustrated in Figure 1.3, a ball-and-socket joint has the kinematic structure of a spherical joint.

- A *planar joint*, E, permits two translational degrees-of-freedom along a plane of contact and a rotational degrees-of-freedom about an axis normal to the plane of contact, as shown in Figure 1.4. Hence, it imposes three constraints and provides three-degrees-of-freedom.

A *kinematic chain* is an assembly of links that is connected by joints. When every link in a kinematic chain is connected to other links by at least two distinct paths, then it is called a *closed-loop chain*. On the other hand, if every link is connected to its pair by only one path, the kinematic chain is called an *open-loop chain*. When a mechanism consists of both closed-loop and open-loop kinematic chains, it is called a *hybrid kinematic chain*.

As shown in Figure 1.5, a kinematic chain is called a *mechanism* when one of its links is fixed to the ground, which is called the base. A *machine* is an assembly of one or more mechanisms along with electrical and/or hydraulic components, used to transform external energy into useful work. Although in many texts the terms *mechanism* and *machine* are

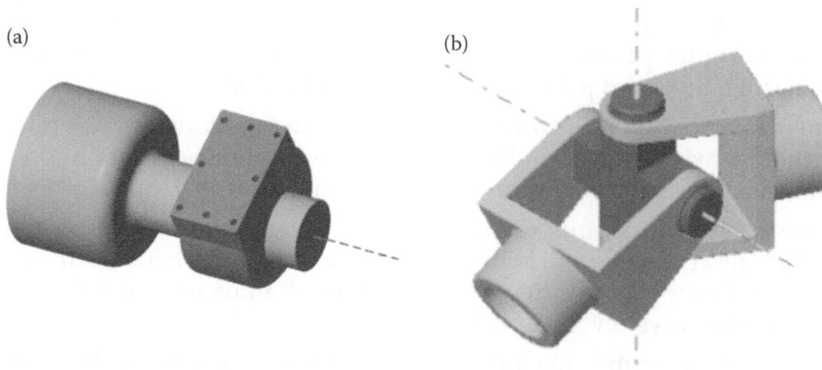

FIGURE 1.2
Schematics of a cylindrical joint (a) and a universal joint (b). (From Mathworks Inc. Schematics of a cylindrical joint (left) and a universal joint (right), 2010. Mathworks. With permission.)

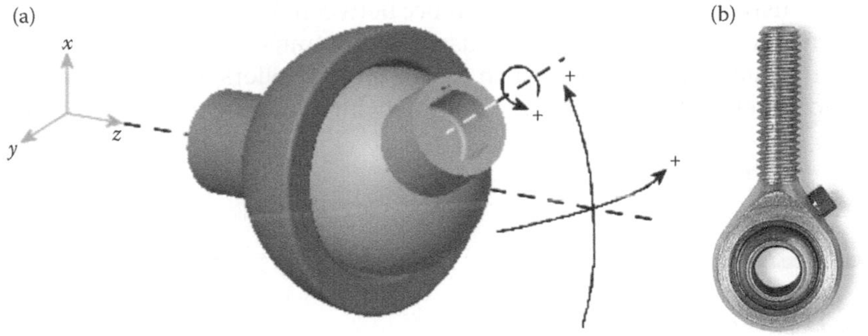

FIGURE 1.3
Schematics of a spherical joint (a) and a ball-and-socket joint (b). (From Mathworks Inc. Schematics of a spherical joint (left) and a ball-and-socket joint (right), 2010. Mathworks. With permission.)

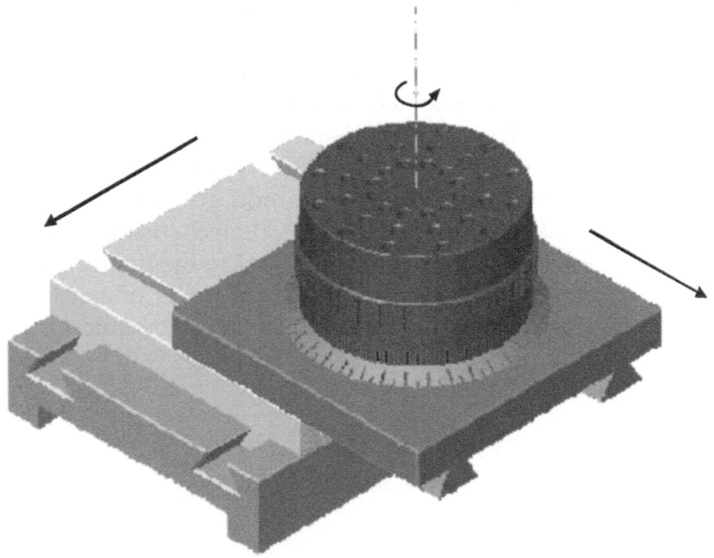

FIGURE 1.4
Schematics of a planar joint: three-degrees-of-freedom.

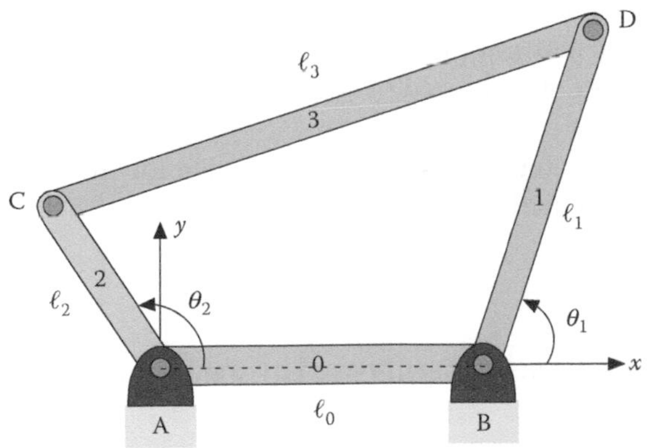

FIGURE 1.5
Schematics of a four bar mechanism.

used synonymously, there is a definite difference between them according to the given definitions. In other words, mechanisms are used for the transmission of motion and can be converted to machines if equipment such as digital controllers, instrumentation systems, actuators, and other accessories are incorporated into their structure to produce useful work.

1.3 Robot Degrees-of-Freedom

The degrees-of-freedom of a mechanism are the number of independent inputs needed to completely specify the configuration of the mechanism. In order to determine the degrees-of-freedom of a mechanism, in most cases a general relation can be used in which the degrees-of-freedom of the mechanism is derived from the number of the links, and the number and type of the joints incorporated into a mechanism. For this, a general joint with n degrees-of-freedom can be interpreted as n binary joints with one independent degrees-of-freedom. For example, a cylindrical joint C can be considered as two independent binary joints, one of revolute type and the other as prismatic. Moreover, a spherical joint S can be viewed as three independent binary joints of revolute type whose axes are intersecting.

Let us define λ as the degrees-of-freedom of the motions that occur in a space, that is, for a planar mechanism $\lambda = 3$ and for a general spatial mechanism $\lambda = 6$. By intuition, the number of degrees-of-freedom of a mechanism is equal to the degrees-of-freedom associated with all the links (except for the base), minus the number of constraints imposed by the joints. Hence, the degrees-of-freedom of a mechanism can be found from the following relation:

$$F = \lambda(n-1) - \sum_{i=1}^{j} c_i \tag{1.1}$$

in which

F = degrees-of-freedom of the mechanism
λ = degrees-of-freedom of the space
n = the number of links in the mechanism including the base
j = the number of binary joints of the mechanism
c_i = the number of constraints imposed by joint i

However, the number of constraints imposed by a joint and the degrees-of-freedom permitted by the joint are equal to λ:

$$\lambda = c_i + f_i. \tag{1.2}$$

Hence, the total number of constraints imposed by the joint is

$$\sum_{i=1}^{j} c_i = \sum_{i=1}^{j} (\lambda - f_i) = j\lambda - \sum_{i=1}^{j} f_i. \tag{1.3}$$

Substituting Equation 1.3 into Equation 1.1 yields the famous Chebyshev–Grübler–Kutzbach (1917) criterion [67]:

$$F = \lambda(n - j - 1) + \sum_{i=1}^{j} f_i. \tag{1.4}$$

The Chebyshev–Grübler–Kutzbach criterion is valid provided that there is no *passive degrees-of-freedom* in the mechanism. To describe passive degrees-of-freedom, consider a link with two spherical joints at both ends. There exists a redundant degree-of-freedom about the axis defined by the two joints, by which the link can freely rotate. Although the link can transmit forces and torques, and thus motion about some other axes, it has no torque transmission capability about this axis of movement. Such degrees-of-freedom are called passive degrees-of-freedom, and although they can be counted as a degree-of-freedom for the motion, they cannot be used for active force or torque transmission. The passive degrees-of-freedom are usually excluded from the degrees-of-freedom of the mechanisms. If f_p is denoted for the total number of such passive degrees-of-freedom, then the remaining degrees-of-freedom are given by

$$F = \lambda(n - j - 1) + \sum_{i=1}^{j} f_i - f_p \tag{1.5}$$

in which

F = degrees-of-freedom of the mechanism
λ = degree-of-freedom of the space
n = the number of links in the mechanism including the base
j = the number of binary joints of the mechanism
f_i = degrees of relative motion permitted by joint i
f_p = the total number of passive degrees-of-freedom

Example 1.1: Piston–Crank Mechanism

Figure 1.6 shows a planar piston–crank mechanism in which the four links generate a single kinematic chain. For this mechanism $\lambda = 3$, since the mechanism is planar, the number of links including the base are four, $n = 4$, and there exist four binary joints

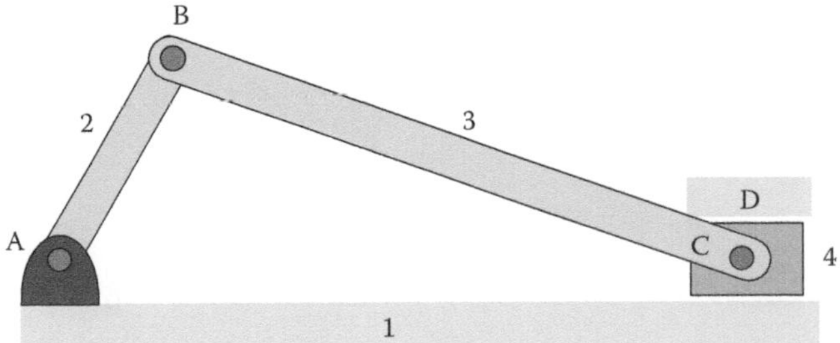

FIGURE 1.6
Schematics of a piston–crank mechanism.

on the mechanism $j = j_1 = 4$, three of them of revolute type and one of them of prismatic type. Moreover, there exists no passive degrees-of-freedom, $f_p = 0$. Applying the Chebyshev–Grübler–Kutzbach criterion (Equation 1.5) yields

$$F = 3 \times (4 - 4 - 1) + (4 \times 1) - 0 = 1.$$

The piston–crank mechanism has one-degree-of-freedom and is used for the transmission of rotational motion from the crank to the cyclic translational motion of the piston. In other applications such as combustion engines, the cyclic force generated in the piston is transmitted on the crankshaft torque.

Example 1.2: Planar 4$R\underline{P}R$ Parallel Mechanism

Figure 1.7 shows a planar 4$R\underline{P}R$ parallel mechanism, in which the motion of the moving platform 10, is generated by four identical piston–cylinder actuators. In this mechanism, the limbs generate multiple closed-loop kinematic chains. For this mechanism $\lambda = 3$, since the mechanism is planar, the number of links including the base is 10, $n = 10$, as indicated in Figure 1.7. There are also eight binary revolute joints A_i's and B_i's and four binary prismatic joints P_i's on the mechanism. Hence, $j = j_1 = 12$. Moreover, there exists no passive degrees-of-freedom, $f_p = 0$. Applying the Chebyshev–Grübler–Kutzbach criterion (Equation 1.5) yields

$$F = 3 \times (10 - 12 - 1) + (12 \times 1) - 0 = 3.$$

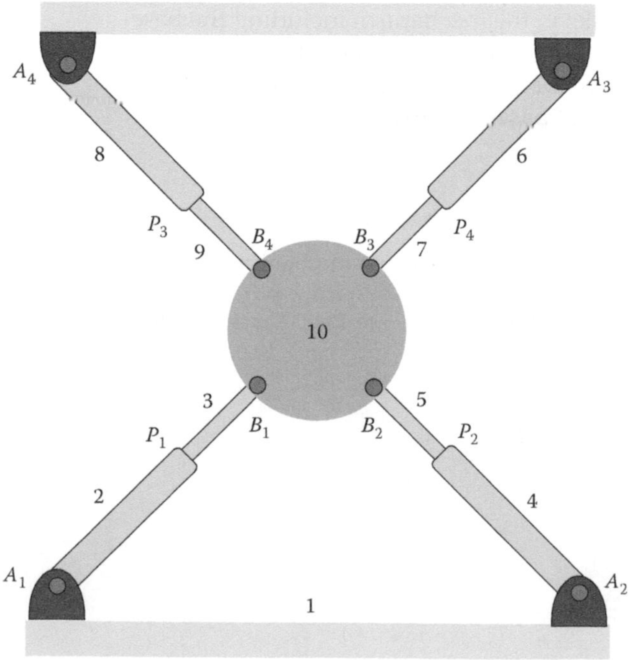

FIGURE 1.7
Schematics of a planar 4$R\underline{P}R$ parallel mechanism.

This planar mechanism has three-degrees-of-freedom. Note that the degrees-of-freedom of the mechanism is equal to λ, and therefore, the mechanism has no kinematic redundancy. However, four actuators are used in the structure, and hence the mechanism has one degree of actuator redundancy.

Example 1.3: The Stewart–Gough Platform

Figure 1.8 shows a spatial parallel manipulator called the Stewart–Gough platform. In this manipulator, the spatial motion of the moving platform is generated by six identical piston–cylinder actuators. For this mechanism $\lambda = 6$, since the mechanism is spatial, the number of links including the base is 14, $n = 6 \times 2 + 2 = 14$. There are also $6 \times 2 = 12$ spherical joints A_i's and B_i's and six binary prismatic joints, P_i's on the mechanism; hence, $j_1 = 6$, $j_3 = 12$, and $j = 18$. Furthermore, for each SPS kinematic structure there exists one passive degrees-of-freedom, and therefore, $f_p = 6$. Applying the Chebyshev–Grübler–Kutzbach criterion (Equation 1.5) yields

$$F = 6 \times (14 - 18 - 1) + (12 \times 3 + 6 \times 1) - 6 = 6.$$

This spatial manipulator has six-degrees-of-freedom and can serve as a general parallel manipulator to produce a complete spatial movement with three-degrees-of-freedom for position and three-degrees-of-freedom for orientation. The number of actuators is also equal to the degrees-of-freedom of the manipulator, and hence, there is no actuator redundancy in this manipulator.

As seen from the examples of parallel mechanisms, the number and location of the actuators must be chosen carefully in order to have complete control on the motion of the

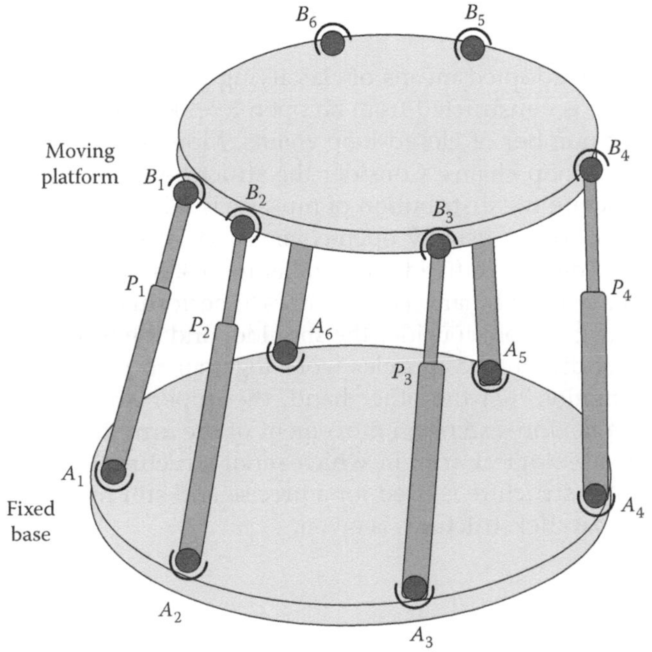

FIGURE 1.8
Schematics of a Stewart–Gough platform.

manipulator. Most robot manipulators have the same number of actuators as the number of degrees-of-freedom required to perform their task. This type of manipulators is called *fully parallel* manipulators. Although this structure is cost efficient, it limits the manipulability of the manipulator to perform the required task. Many biological mechanisms such as the shoulder, arm, and hand in humans benefit from redundancy in actuation. In general, redundant manipulators hire a larger number of actuators than the number of degrees-of-freedom required to perform the required task with some extra desirable user-defined characteristic(s). To control a required motion in this type of manipulators, a suitable actuator effort distribution shall be used. The freedom on how to optimally use the degree(s) of redundancy is usually formulated as an optimal problem to minimize a user-defined performance criterion such as optimal force distribution, singularity avoidance, or other performance indices. Such optimization problems and their corresponding solutions are treated in Section 6.7 of this book in more detail.

1.4 Robot Classification

Classification of robots may be done according to various measures. Obviously, classification based on degrees-of-freedom is one of the first choices. A *general-purpose* robot shall posses six-degrees-of-freedom to freely manipulate an object in space. On the other hand, for some special applications such as an assembly, it is sufficient to have only four-degrees-of-freedom. From this point of view, a *redundant robot* is a robot with more actuators than the required degrees-of-freedom, and a *deficient robot* is one with fewer actuators than the required degrees-of-freedom. A redundant robot provides more freedom to move an object while avoiding the obstacles or singularities that may occur within the workspace of the robot.

Another more widely adopted means of classifying robots is based on their kinematic structure. A *serial robot* is constructed from an open-loop kinematic chain, while a *parallel robot* is made from a number of closed-loop chains. Moreover, a *hybrid robot* consists of both open- and closed-loop chains. Consider the structure of the human arm and wrist. Although in this structure the distribution of muscles is fairly complicated to provide the motion, in mechanical robots usually open-chain or serial structure is used, dedicating one actuator for every degree-of-freedom. Parallel robots, on the contrary, have a closed-chain structure, and use a combination of actuators to perform a multi-degrees-of-freedom motion. To make a comparison, consider the shoulder and the wrist structure of humans, in which a number of distributed muscles work together to perform a three-degrees-of-freedom rotational motion. On the other hand, the upper arm–elbow–lower arm uses direct actuation for a flexion–extension movement of the arm. It is evident that a hybrid structure is used in a biological arm, in which serial structure is used to reach a larger workspace and parallel structure is used for a precise and stiff movement. In this book, a thorough analysis of parallel structures is given.

1.4.1 Serial Robots

Serial robots are the most common robots used in industrial applications. Often they have an anthropomorphic mechanical arm structure, that is, a serial chain of rigid links connected by (mostly revolute) joints, forming a *shoulder*, an *elbow*, and a *wrist*. Their main

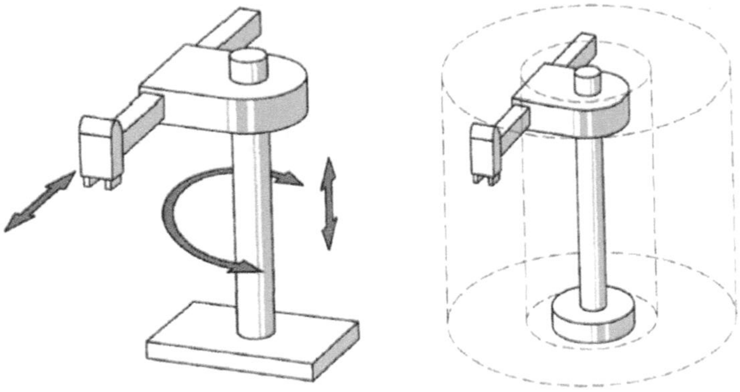

FIGURE 1.9
The schematics and workspace of a cylindrical robot.

advantage is their large workspace with respect to their own volume and occupied floor space. The main disadvantage of serial robots is mainly their low stiffness, inherently caused by their open kinematic structure. Furthermore, their positioning errors are accumulated from link to link. Serial manipulators have a relatively high mass to payload ratio, since they have to carry and move the large weight of most of the actuators. Finally, serial manipulators have a relatively low effective payload capacity. It is known that at least six-degrees-of-freedom are required to place a manipulated object with an arbitrary position and orientation in the workspace of the robot. Hence, many serial robots have six joints. However, the most popular application for serial robots in today's industry is the pick-and-place assembly. Since this only requires four-degrees-of-freedom, special assembly robots of the so-called SCARA* type are being built.

In its most general form, a serial robot consists of a number of rigid links connected with joints. Simplicity considerations in manufacture and control have led to robots with only revolute or prismatic joints that have orthogonal, parallel, and/or intersecting joint axes. Kinematic analysis of a robot is to derive the position and orientation of a robot end effector from the joint positions by means of a geometric model of the robot arm. For serial robots, the mapping from joint positions to end-effector position and orientation is easy, while the inverse mapping is more difficult. Therefore, most industrial robots have special designs that reduce the complexity of the inverse mapping. Donald L. Pieper derived the first practically relevant result in different serial structures [152], and showed that the inverse kinematics of serial manipulators with six revolute joints, and with three consecutive joints intersecting, can be solved analytically. This result had a tremendous influence on the design of industrial robots, where most of them adopt three consecutive intersecting joints in their structure.

Serial robots are themselves classified into different categories. One of these classifications is based on the shape of the robot workspace. As shown in Figures 1.9 through 1.12, different shapes of the robot workspace can be used for pick-and-place robots. Figure 1.9 shows the schematics of a cylindrical robot. The cylindrical shape of the workspace is suitable for pick-and-place applications in circular arrangements. If the workplace and its tools are arranged in different spherical locations, the spherical robot shown in Figure 1.10 may be used. In many applications, a Cartesian placement of the tool in the workspace is

* Selective Compliance Assembly Robot Arm.

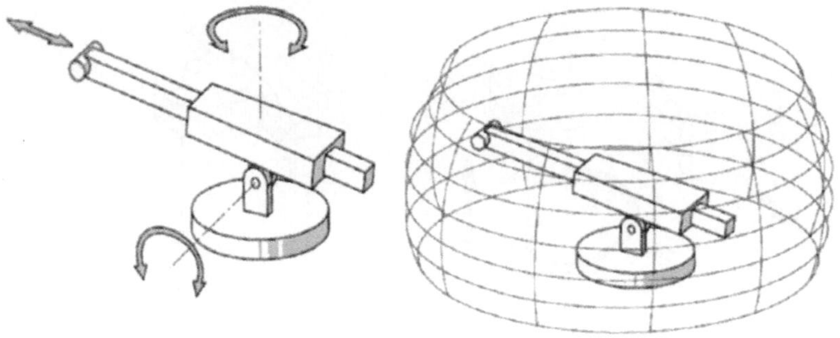

FIGURE 1.10
The schematics and workspace of a spherical robot.

required. In such cases, a Cartesian or gantry robot is recommended. The schematics of a three-degrees-of-freedom gantry robot consisting of three perpendicular prismatic joints is shown in Figure 1.11.

In 1981, a completely new concept for assembly robots was developed called the *Selective Compliance Assembly Robot Arm (SCARA)*. As shown in Figure 1.12, by virtue of the SCARA's parallel-axis joint layout, the arm is slightly compliant in the x–y direction but rigid in the z direction, which allows it to adapt to assembly holes in the xy axes. This is advantageous for many types of assembly operations, that is, inserting a round pin in a round hole without binding. The second attribute of the SCARA robot is the jointed two-link arm layout similar to the articulated human arm. This feature allows the arm to extend into confined areas and then retract or *fold up* out of the way. This is advantageous

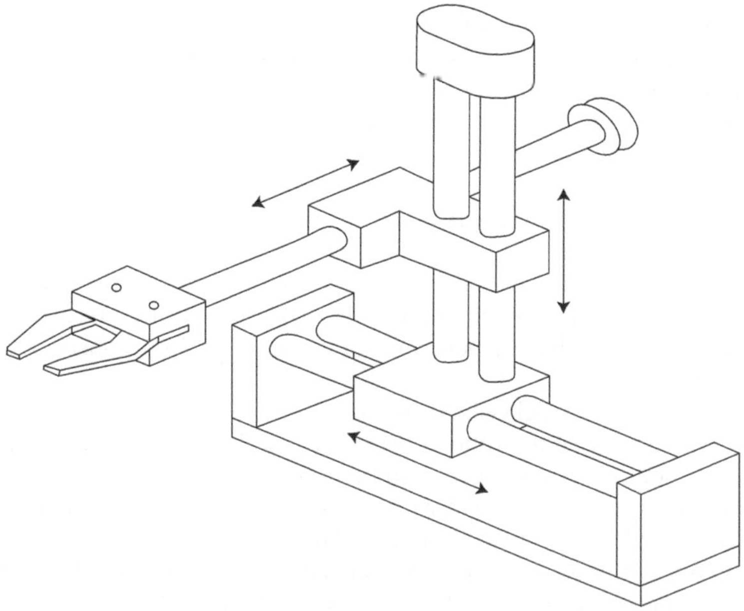

FIGURE 1.11
The schematics of a gantry robot with three-degrees-of-freedom.

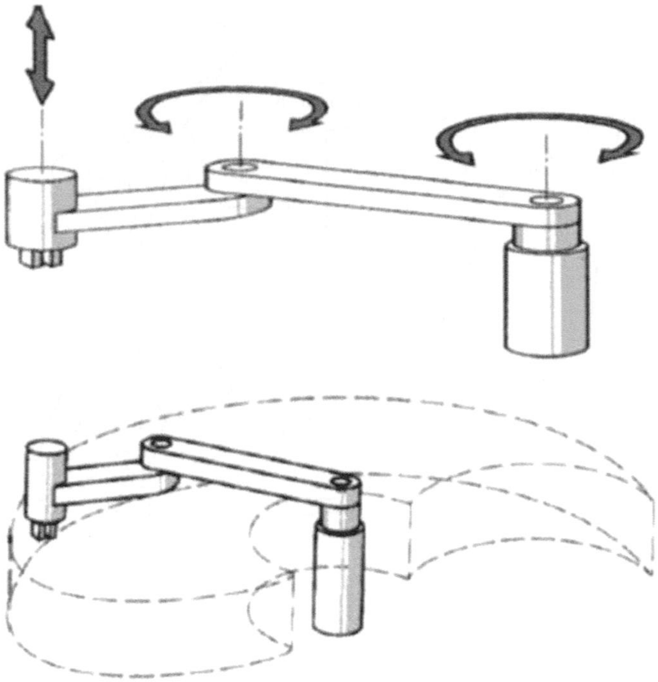

FIGURE 1.12
The schematics and workspace of a SCARA robot.

for transferring parts from one cell to another or for loading/unloading process stations that are enclosed.

The three main drawbacks of serial robots are their poor load-to-mass ratio, low positioning accuracy, and high structural compliance. For a six-degrees-of-freedom general-type serial manipulator, the load-to-mass ratio is less than 0.15, while for the SCARA-type robot, this can increase to a maximum of 0.25 [133]. This means that to have a payload of 100 kg, the weight of a typical serial robot is more than 670 kg or at least 400 kg for a SCARA-type robot. For positioning accuracy, there are two distinct definitions. Absolute accuracy is defined as the absolute positioning error of the end effector to move to a desired pose. Another notion, called repeatability, is defined as the maximum distance between two positions of the end effector reached for the same desired pose from different starting positions. Although users are mostly interested in absolute accuracy, the manufacturers generally indicate repeatability in the robot specifications, which is far better than absolute accuracy. In a serial robot, the positioning accuracy of the robot is usually poor since the errors that occur at each link is accumulated toward the end effector. This is also the reason for the robot structural compliance, which is usually poor for serial manipulators. These drawbacks may be naturally reduced if the structure of the robot is changed. Parallel robots are basically designed based on a different structure, which can significantly reduce these drawbacks.

1.4.2 Parallel Robots

A generalized parallel robot is a closed-loop kinematic chain mechanism whose moving platform is linked to the base by several independent kinematic chains. A parallel

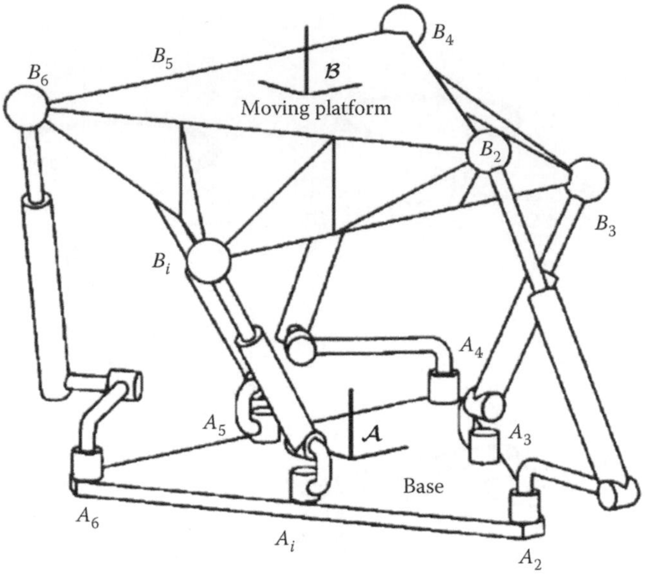

FIGURE 1.13
Schematics of a general parallel robot.

robot consists of a fixed *base* platform connected to a moving platform by means of a number of *limbs*. These limbs often consist of an actuated prismatic joint connected to the platforms through passive spherical and/or universal joints. Hence, the links feel only traction or compression, not bending, which increases their position accuracy and allows a lighter construction. Moreover, in principle, parallel robots have high structural stiffness, since the moving platform is supported by several limbs at the same time. Figure 1.13 shows the schematics of a general parallel robot. All these features result in robots with a wide range of motion capability. Their major drawback is their limited workspace because the limbs can collide and, in addition, each limb has a number of passive joints that have their own mechanical limits. Another drawback of parallel robots is that they may completely lose their stiffness at singular positions, and hence the robot gains extra degrees-of-freedom, which are uncontrollable, and therefore, it becomes shaky or mobile at these configurations.

1.4.2.1 The Stewart–Gough Platform

As historically reviewed by Bonev [14], the most celebrated parallel robot, which is paradoxically referred to in the literature as the Stewart platform was first invented by V. E. Gough in 1947. The idea behind this invention, which is called a universal rig, was borrowed from the design of an octahedral hexapod that was used to determine the properties of tires under combined loads. As shown in Figure 1.14, in a Gough platform there are six independently actuated limbs, where the lengths of the legs are controlled to move the platform to a desired position and orientation. As was reported in Gough's paper in 1956 [63], the octahedral hexapod was not invented from scratch, and at that time, systems with six struts (hexapods) were already in use. These hexapods had usually three vertical struts and three horizontal ones, and have been very popular then and are still used in

FIGURE 1.14
The Gough platform: the moving platform to which a tire is attached and driven by a conveyor belt. The mechanism allows the operator to measure the tire wear and tear under various loading conditions.

different applications [50]. The new idea of the Gough platform was the arrangement of the six struts. Since Gough needed relatively large ranges of motion, he naturally selected a symmetrical arrangement forming an octahedron. The machine was built in the early 1950s and was fully operational in 1954. Although Gough was the first to invent and build the popular octahedral hexapod, Klaus Cappel later designed independently the very same hexapod, patented it [19], and licensed it to the first flight simulator companies. A picture of his first flight simulator is shown in Figure 1.15.

The name of Stewart was attached to this architecture because Gough's earlier work and a photograph of this platform were mentioned in reviewers' remarks on a paper published by Stewart in 1965 [170]. In that paper, Stewart presents another hybrid design, which is shown in Figure 1.16 with three legs having two actuators each. Stewart proposed this six-degrees-of-freedom motion platform for use as a flight simulator. The proposed parallel mechanism, however, is different from the octahedral hexapod invented by Gough that is paradoxically often referred to as the *Stewart platform*. There is no doubt that Stewart's remarkable paper had a great impact on the subsequent development in the field of parallel kinematics. Various suggestions for the use of a hexapod were made, many of which were accurate predictions of the future. In the literature, the famous octahedral hexapod mechanism is mostly referred to as the Stewart platform. In this book, we call variations of such a design as the Stewart–Gough platform to recall the name of its first inventor to the

FIGURE 1.15
A flight simulator designed by Klaus Cappel in the mid-1960s.

reader. Since this structure has six-degrees-of-freedom and is a fully parallel mechanism, many analyses of this structure were reported, which can be used as the basis for a similar parallel mechanism in practice. For this reason, a thorough analysis of this mechanism is described in this book throughout the chapters.

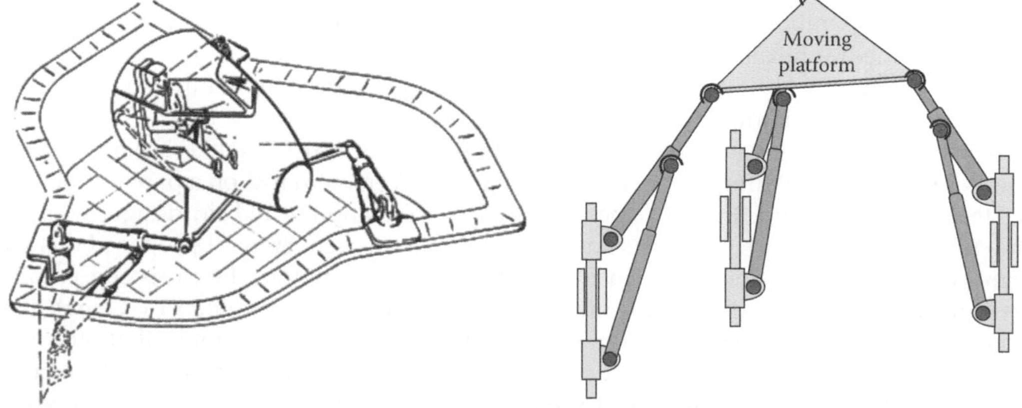

FIGURE 1.16
A schematics of the Stewart platform and its use as flight simulator. (From D. Stewart. A platform with six degrees of freedom. *Proceedings of the UK Institute of Mechanical Engineering*, 180(1): 371–386, 1965.)

1.4.2.2 The Delta Robot

A Delta robot is a type of parallel robot which consists of three arms connected to universal joints at the base (Figure 1.17). The key design feature is the use of parallelograms in the arms, which maintains the orientation of the end effector. Delta robots have popular usage in the picking and packaging factories because they can be quite fast, some executing up to 200 picks per minute [83]. A Delta robot was invented in the early 1980s by Reymond Clavel at the EPFL, Switzerland [28]. The purpose of this new type of robot was to manipulate light and small objects at a very high speed. In 1987, Demaurex company purchased a license for the Delta robot and started the production of Delta robots for packaging industry. In 1991, Reymond Clavel presented his doctoral thesis and received the Golden Robot Award in 1999 for his work and development of the Delta robot. In 1999, ABB Flexible Automation started selling its Delta robot, the FlexPicker.

A Delta robot is a parallel robot, which can also be seen as a spatial generalization of a planar four-bar mechanism. It has four-degrees-of-freedom: three translational and one rotational. The key concept of a Delta robot is the use of parallelograms. A parallelogram allows an output link to remain at a fixed orientation with respect to an input link. The use of three such parallelograms restrains completely the orientation of the mobile platform, which remains only with three purely translational degrees-of-freedom. The robot base is mounted above the workspace. All the actuators are located on this base. From the base, three middle jointed arms are extended. The arms are usually made of lightweight composite material. The ends of the three arms are connected to a small triangular platform. Actuation of the input links will move the triangular platform in the x, y, or z directions. Actuation can be done by linear or rotational actuators. From the base, a fourth leg is used to transmit rotary motion from the base to an end effector mounted on the mobile platform.

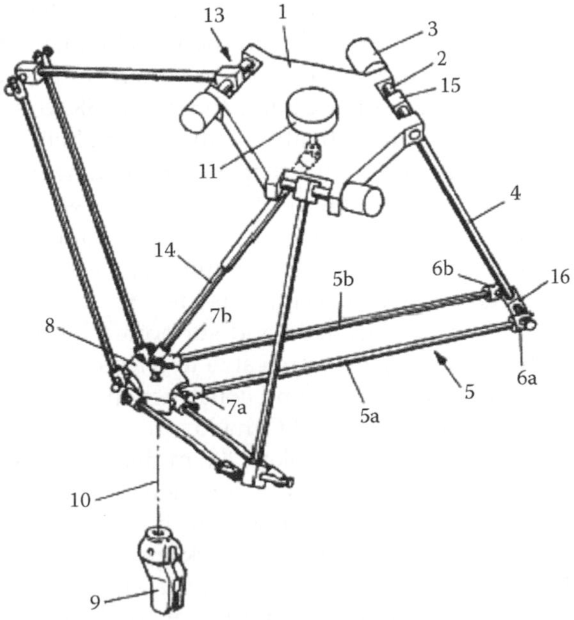

FIGURE 1.17
The schematics of a Delta robot. (From R. Clavel. Device for the movement and positioning of an element in space. US Patent No. 4,976,582, December 1, 1990. With permission.)

Since the actuators are all located on the base, and the arms are made of a composite material, the moving parts of the Delta robot have a small inertia. This allows for very high accelerations; accelerations can be up to 30 g and speeds of 10 m/s may be reached. This makes the Delta robot a perfect candidate for pick-and-place operations of light objects (from 10 g to 1 kg). Delta robots available on the market operate typically in a cylindrical workspace that is 1 m in diameter and 0.2 m high [26]. The industries that take advantage of the high speed of the Delta robot are the packaging, medical, and pharmaceutical industries. Adept has built its delta-type parallel robot called Quattro for packaging industries. This manipulator has one degree of redundancy in actuation and consists of four parallelogram limbs. The structure of the Delta robot has also been used to create commercial haptic devices such as the Novint Falcon.

1.4.3 Cable-Driven Parallel Robots

The increasing performance demands necessitate the design of new types of robots with larger workspace that are capable of performing at higher accelerations. As described in the previous sections, parallel manipulators can generally perform better than serial manipulators in terms of stringent stiffness and acceleration requirements. However, their limited workspace and the existence of singular regions inside their workspace hinder the applications of parallel manipulators for large workspace requirements [79,114]. In a cable-driven parallel manipulator (CDPM), the linear actuators of parallel manipulators are replaced by electrical powered cable drivers, which leads immediately to a larger workspace. As an example, consider the application of video recording a football game using a flying camera over a stadium. In this application, cable-driven parallel structure can serve as an immediate alternative. Such a system has been designed using four pairs of cables suspending a camera package with pan and tilt control. Such systems are commercially available and are called Cablecam, Skycam, or Spidercam.

Another interesting application of cable-driven robots is in the next generation of radio telescopes. An international consortium of radio astronomers and engineers agreed to investigate technologies to build the square kilometer array (SKA), a cm-to-m wave radio telescope for the next generation of investigation into cosmic phenomena [22,87]. A looming *sensitivity barrier* will prevent current telescopes from making much deeper inroads at these wavelengths, particularly in studies of the early universe. The Canadian proposal for the SKA design consists of an array of 30–50 individual antennas whose signals are combined to yield the resolution of a much larger antenna. Each of these antennas would use the large adaptive reflector (LAR) concept. This idea was put forward by a group led by the National Research Council of Canada and supported by university and industry collaborators [20,108]. The LAR design is applicable to telescopes up to several hundreds of meters in diameter. However, the design and construction of a 200 m LAR prototype is pursued by the National Research Council of Canada.

Figure 1.18 is an artist's concept of a complete 200 m diameter LAR installation, which consists of two central components. The first component is a 200 m diameter parabolic reflector with a focal length of 500 m, composed of actuated panels supported at the ground. The second component is the receiver package, which is supported by a tension structure consisting of multiple long tethers and a helium-filled aerostat as shown schematically in Figure 1.19. With funding from the Canada Foundation for Innovation, a one-third scale prototype of the multitethered aerostat subsystem has been designed and implemented in Penticton [108]. It should be noted that even at the 1/3 scale, this system is very large, with a footprint of roughly 1 km^2.

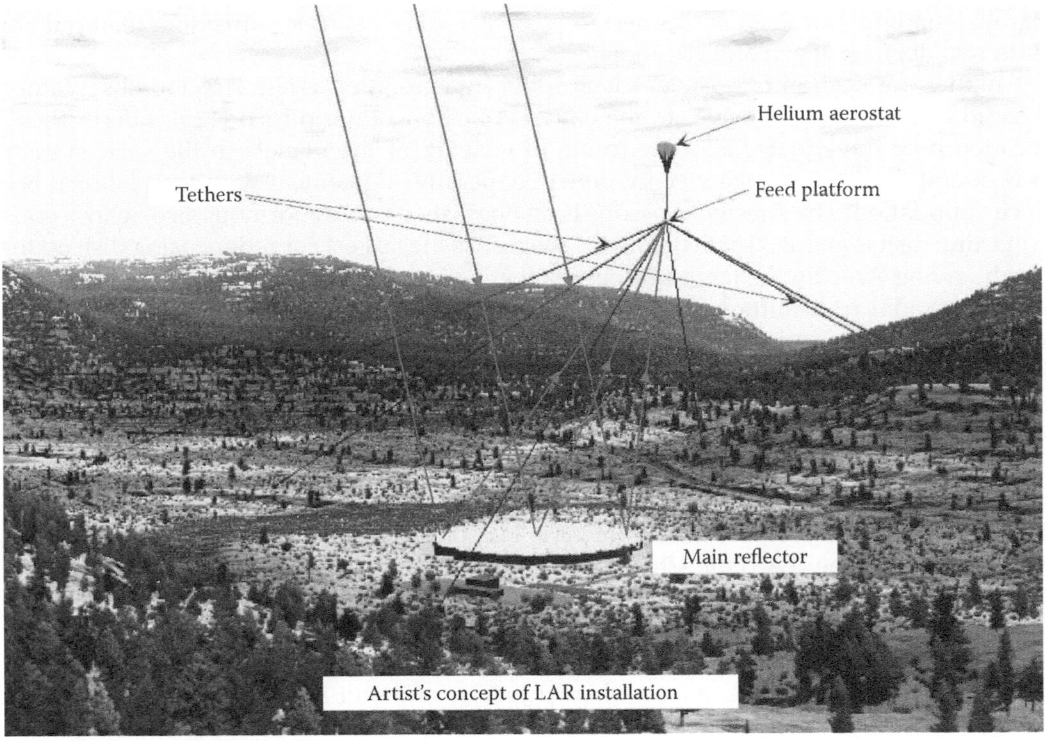

FIGURE 1.18
An artist's concept of a complete 200 m diameter LAR installation.

A challenging problem in this system is accurately positioning the feed (receiver) in the presence of disturbances, such as wind turbulence. As illustrated in Figure 1.19, the receiver is moved to various locations on a circular hemisphere and its positioning is controlled by changing the lengths of eight cables with ground winches. The cable-driven

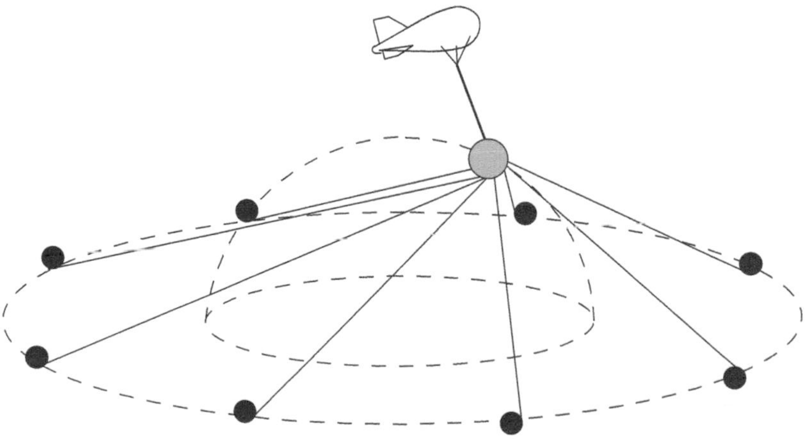

FIGURE 1.19
The schematics of a cable-driven manipulator with eight cables, and the aerostat used in the LAR application.

parallel manipulator used in this design is, in fact, a six-degrees-of-freedom manipulator with two degrees of actuator redundancy.

Another application of a cable-driven robot in astronomy is the Arecibo observatory, a radio telescope located close to the city of Arecibo in Puerto Rico [125]. Puerto Rico's location near the equator allows Arecibo to view all of the planets in the solar system. It is operated by Cornell University under cooperative agreement with the National Science Foundation. The Arecibo telescope is distinguished by its enormous size, whose main collecting dish is about 30 m in diameter. The dish is the largest curved focusing dish on the Earth, giving Arecibo the largest electromagnetic-wave-gathering capacity. The telescope has three radar transmitters and a spherical reflector, as opposed to a common parabolic reflector. This form is due to the method used in the telescope. The telescope's dish is fixed in place, and the receiver is repositioned to intercept signals reflected from different directions by the spherical dish surface. The receiver is located on a 900-ton platform that is suspended 150 m in the air above the dish by 18 cables running from three reinforced concrete towers. The position of the receiver is controlled with a *large cable mechanism* similar to the LAR application [56]. Other similar projects are also being undertaken in other countries. Another project under development is known as the Five-hundred-meter Aperture Spherical Telescope (FAST) in China, which will have a collecting area more than twice that of the Arecibo Observatory [92].

The idea of using cable-driven parallel robots is not limited to applications where a very large workspace is required such as the one mentioned before. This idea has effectively penetrated into applications where a precise and stiff robot is required to operate in high accelerations within a relatively larger workspace than that attainable by conventional parallel robots. One such robot is the WARP virtual acceleration robot [172]. WARP is a six-degrees-of-freedom cable-driven robot having a overconstrained low-mass moving platform. Owing to its light-weight moving structure, a ultra-high speed with more than 40 g acceleration can be attained by this manipulator [95].

As another example the NIST Robocrane can be named, which is a kind of manipulator resembling a Stewart–Gough platform but using an octahedral assembly of cables instead of struts. The Robocrane has six-degrees-of-freedom, and has the capacity to lift and precisely manipulate heavy loads over large volumes with fine control in all six-degrees-of-freedom [4]. The Laboratory Robocrane has demonstrated the ability to manipulate tools such as saws, grinders, and welding torches and to lift and precisely position heavy objects such as steel beams and cast iron pipes. A version of the Robocrane has been commercially developed for the Air Force to enable rapid paint stripping, inspection, and repainting of very large military aircrafts such as the C-5 Galaxy. Potential future applications of the Robocrane include ship building, construction of high-rise buildings, highways, bridges, tunnels, and port facilities cargo handling, ship-to-ship cargo transfer on the sea, radioactive and toxic waste clean-up, and underwater applications such as salvage, drilling, cable maintenance, and undersea waste site management [144].

Although cable-driven robots have distinctive properties that make them attractive for a large number of potential applications, there are many challenging problems related to the design and development of these robots in practice. The most important distinction of cable-driven robots from conventional parallel robots lies in the major property of cables in actuation. Cables work only under tension and can be used only to *pull* and *not to push* any object. Therefore, in the structure of cable-driven robots, special care must be taken to ensure that the cables are operating under tension for all maneuvering tasks throughout their whole workspace. One possible solution is to use an external force acting on the robot moving platform to satisfy this condition. The gravity force is used in the Arecibo

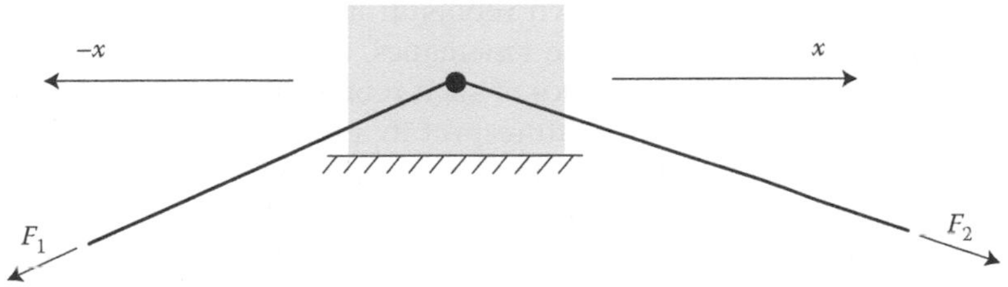

FIGURE 1.20
The concept of using two cables to move an object arbitrarily in one direction.

telescope and Robocrane applications, while helium aerostat lifting force is used in the LAR application. Note that if gravity force is used to ensure tension force in the cables, the robot cannot perform with accelerations more than 1 g in the vertical direction. This limitation is not severe in crane applications where lifting heavy objects is the main task of the robot. However, for a general-purpose six-degrees-of-freedom robot, this method is not recommended. For such applications, a fully constrained structure for the robot with redundancy in actuation may be used as a possible alternative.

To further illustrate the idea behind actuator redundancy, consider the motion of a rigid body by cables along the x-axis on both the positive or negative directions, as illustrated in Figure 1.20. If only one cable is used for actuation, the moving object can be moved only in one direction. However, if a redundant or extra cable is used for actuation, the object can be moved in both directions. Since in a redundant parallel robot, the number of actuators is more than the number of degrees-of-freedom, a redundant planar parallel robot must have at least four actuators, while the general six-degrees-of-freedom redundant robot must have at least seven actuators. In conventional parallel robots, redundancy in actuation is used to avoid singularities within the workspace and/or to provide a more uniform distribution in the actuation efforts. However, in cable-driven robots, redundancy is a necessity to ensure feasible tension forces for all cables within the workspace.

It is clear that by introducing one extra degree of actuation in the robot, the general analysis of the system kinematics, dynamics, and control becomes more sophisticated. Therefore, in the workspace analysis of cable-driven robots, some specific analysis such as a force-feasible workspace or wrench-closure workspace is introduced, to examine the possibility to keep all the cables in tension within a volume of workspace. Furthermore, optimal methods to use the extra degrees of actuation and to ensure force feasibility, in addition to optimally minimizing other cost functions, should be addressed for these robots. Optimal techniques to use these extra degrees-of-freedom open a new area in the analysis of such robots, which is usually named as redundancy resolution. This topic is well addressed in this book in Section 6.7.

1.5 The Aims and Scope of This Book

Many industrial applications have adopted a parallel structure for their design; however, only a very few text books have been published to introduce the analysis of such robots in terms of kinematics, dynamics, and control. This book is intended to give some analysis and design tools for the increasing number of engineers and researchers who are interested in the design and implementation of such robots in industries. In this book, a systematic approach is presented to analyze the kinematics, dynamics, and control of parallel robots. To define the motion characteristics of such robots, it is necessary to represent 3D motion of the robot moving platform with respect to a fixed coordinate. This issue leads to the requirement for 3D representation of position, orientation, and motion of bodies in space. In Chapter 2, such representations are introduced with emphasis on screw coordinates, which makes the representation of the general motion of the robot much easier to follow.

Kinematic analysis refers to the study of robot motion geometry without considering the forces and torques that cause the motion. In this analysis, the relation between the geometrical parameters of the manipulator and the final motion of the moving platform is derived and analyzed. A complete treatment of such analysis is given in Chapter 3, and elaborative case studies are provided for three parallel robots, including a planar cable-driven robot. The analysis of cable-driven robots is formally treated in this book as the promising new generation of parallel robots providing very large workspaces. This analysis may be extended to the linear and angular velocities, which is known as a Jacobian analysis in such robots. In Chapter 4, the kinematic analysis of robot manipulators is further examined beyond static positioning. Jacobian analysis not only reveals the relation between the joint variable velocities of a parallel manipulator and the moving platform linear and angular velocities, but also constructs the transformation needed to find the actuator forces from the forces and moments acting on the moving platform. A systematic means to perform Jacobian analysis of parallel manipulators is given in this chapter, and the proposed method is examined through the same case studies analyzed in Chapter 3.

Dynamic analysis of parallel manipulators presents an inherent complexity due to their closed-loop structure and kinematic constraints. Nevertheless, dynamic modeling is quite important for the control, in particular because parallel manipulators are preferred in applications where precise positioning and suitable dynamic performance under high loads are the prime requirements. Although a great deal of research has been presented on kinematics of parallel manipulators, works on the dynamics and control of parallel manipulators are relatively few. This topic is well addressed in this book, in Chapter 5, in which dynamic analysis of such robots is examined through three methods, namely the Newton–Euler principle of virtual work and Lagrange formulations. Furthermore, a method is presented in this chapter to formulate the dynamic equation of parallel robots into closed form, by which the dynamic matrices are more tractable, and dynamics verification becomes feasible.

The control of parallel robots is elaborated in the last two chapters of the book, in which both the motion and the force control of parallel robots are covered. Different model-free and model-based controllers are introduced, and robust and adaptive control schemes are elaborated in Chapter 6. The control techniques are applied to two case studies in which a cable-driven robot and a fully parallel manipulator are examined by the proposed control schemes. Redundancy resolution schemes are fully elaborated in this chapter, as a required mapping for the implementation of the proposed controller for redundant manipulators. Finally, Chapter 7 covers the force control of parallel robots in detail. In this chapter, stiffness control, direct force control, and impedance control schemes are elaborated, and implemented on the same case studies followed in the book.

2

Motion Representation

In the analysis of robotic manipulators, movement of links, tools, and workpieces in the space is very important. For many other engineering disciplines such as guidance and navigation of mobile robots, airplanes, submarines, missiles, and so on, the representation of spatial location of a rigid body is also of great importance. The ideas developed for this representation originated many years ago [45], when humans were dreaming to fly and many different representations for position and orientation of a rigid body in space were developed since then. In this chapter, some of these representations are introduced with the emphasis on some new tools and descriptions, which are more suitable for robotic manipulators. Moreover, transformation from one representation to the other is also given. It should be noted that the ideas developed for position and orientation representation will form a basis for linear and angular velocity and acceleration representations, and is also adopted to represent forces and torques applied in a robotic manipulator.

In order to define a mathematical representation for the location of a rigid body in the space, the first step is to define a reference coordinate system. Theoretically, it is essential that the reference coordinate system is universally fixed, in order to further develop relations for velocities and accelerations, without adopting the theory of relativity. However, in practice, we may consider a fixed coordinate system attached to the base, and due to its insignificance, neglect the effect of the relative motion of the ground in such analysis. Generally, a *Cartesian* coordinate system is considered for the analysis, although other types of coordinate systems, such as *cylindrical* or *spherical* coordinate systems, may also be adopted.

2.1 Spatial Motion Representation

The *location* or *pose* of a rigid body with respect to a reference coordinate system is known if the *positions* of all its points can be determined. Three independent parameters are sufficient to fully describe the location of a rigid body in a planar motion, and six independent parameters are needed to fully describe the spatial location or pose of that in three-dimensional space. Consider a rigid body in a spatial motion as shown in Figure 2.1. Let us define a fixed reference coordinate system (x, y, z) denoted by frame $\{\boldsymbol{A}\}$, whose origin is located at point O_A. For representation of the rigid body location, a moving coordinate system is attached to the rigid body at point O_B. As shown in Figure 2.1, this Cartesian coordinate system (u, v, w), which is denoted by frame $\{\boldsymbol{B}\}$, has a different position and orientation from that of the fixed frame. The absolute position of a point P of the rigid body can be constructed from the relative position of that point with respect to the moving frame $\{\boldsymbol{B}\}$, and the position and orientation of the moving frame $\{\boldsymbol{B}\}$ with respect to the fixed frame $\{\boldsymbol{A}\}$. Introducing a moving frame attached to the rigid body helps us in constituting the absolute position of an arbitrary point P as by its two components,

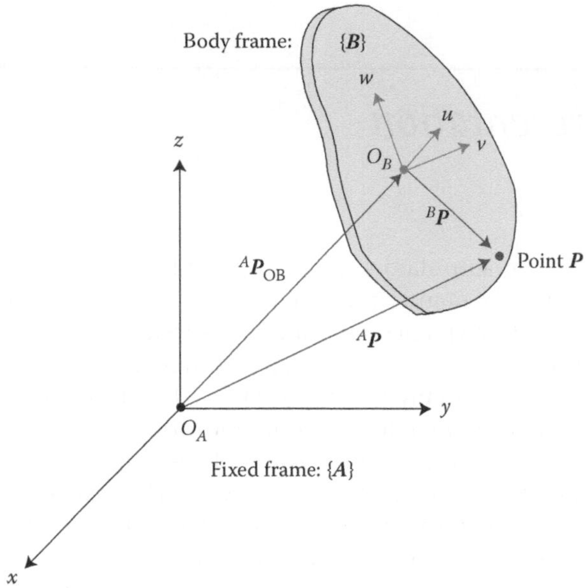

FIGURE 2.1
Representation of a rigid body spatial motion.

namely, relative position of that point with respect to the moving frame $\{B\}$ and absolute position and orientation of frame $\{B\}$ with respect to frame $\{A\}$.

2.1.1 Position of a Point

The position of a point P with respect to a frame $\{A\}$ can be described by a 3×1 *position vector*. If the coordinate frame is a Cartesian coordinate system, the three components of the position vector are the Cartesian components of a vector. If the position is described with respect to a fixed frame, the vector represents an absolute position, and if the frame of reference is moving itself, the representation results in a relative position vector. Because of the significance of the frame of reference to the representation of the position vector, the name of the frame is tagged to the position vector name as a leading superscript as $^A\boldsymbol{P}$, which is read as vector $\boldsymbol{P}$ in frame $\{A\}$. Hence, the position vector can be represented by its three components

$$^A\boldsymbol{P} = \begin{bmatrix} P_x \\ P_y \\ P_z \end{bmatrix}, \tag{2.1}$$

where the subscripts x, y, and z represent the Cartesian components of the position vector. Similarly, the position vector can be represented in other coordinates, such as frame $\{B\}$ in the form $^B\boldsymbol{P}$, as shown in Figure 2.1.

2.1.2 Orientation of a Rigid Body

There is a basic difference between position and orientation representations. For each point of a rigid body, there exists a position vector, but the orientation of the whole rigid body is the same for all its points. Hence, representation of the orientation of a rigid body can be viewed as that for the orientation of a *moving frame attached to the rigid body*. Orientation of a moving coordinate frame with respect to a fixed frame can be represented by several different ways. One general approach to represent orientation is the use of a *rotation matrix*. In this form of representation, a 3×3 matrix is built from the Cartesian components of moving frame unit vectors with respect to the fixed frame. Although this form of representation is universally accepted by all disciplines, the use of nine parameters for this representation is somehow redundant, since from these nine parameters only three are independent. Many other methods try to encapsulate the orientation representation with fewer parameters. In what follows, two major ones, namely the *screw axis representation* and *Euler angles* are described.

2.1.2.1 Rotation Matrix

One convenient method of describing the orientation of a rigid body is to attach a moving frame to the body as shown in Figure 2.2. Since we want to represent orientation in here, consider that the rigid body has been exposed to a pure rotation. As shown in Figure 2.2, consider that the rigid body has changed its orientation from a state shown by dotted line represented by frame $\{A\}$ to its current orientation represented by frame $\{B\}$. A 3×3 rotation matrix ${}^A\boldsymbol{R}_B$ is defined by the following relation:

$$
{}^A\boldsymbol{R}_B = \left[{}^A\hat{\boldsymbol{x}}_B \mid {}^A\hat{\boldsymbol{y}}_B \mid {}^A\hat{\boldsymbol{z}}_B\right] = \begin{bmatrix} r_{11} & r_{12} & r_{13} \\ r_{21} & r_{22} & r_{23} \\ r_{31} & r_{32} & r_{33} \end{bmatrix}, \tag{2.2}
$$

in which ${}^A\hat{\boldsymbol{x}}_B$, ${}^A\hat{\boldsymbol{y}}_B$, and ${}^A\hat{\boldsymbol{z}}_B$ are the Cartesian unit vectors of frame $\{B\}$ represented in frame $\{A\}$. Throughout the book, the superscript $(\hat{\cdot})$ is used to denote a unit vector. In summary, a set of three unit vectors may be used to specify an orientation. If the unit vectors are

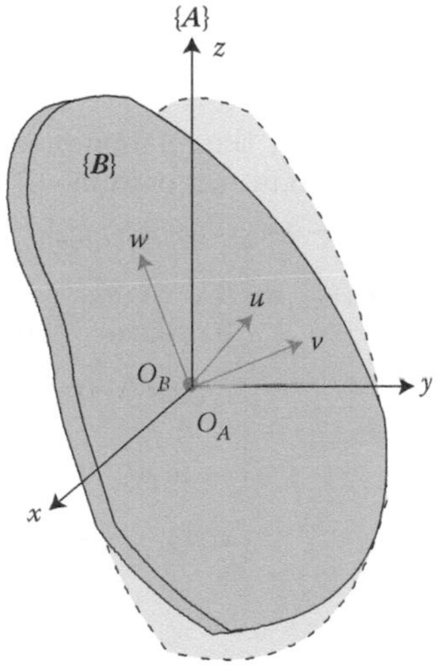

FIGURE 2.2
Pure rotation of a rigid body.

represented componentwise, as follows:

$$^A\hat{x}_B = {^A\hat{u}} = u_x\hat{i} + u_y\hat{j} + u_z\hat{k}, \tag{2.3}$$

$$^A\hat{y}_B = {^A\hat{v}} = v_x\hat{i} + v_y\hat{j} + v_z\hat{k}, \tag{2.4}$$

$$^A\hat{z}_B = {^A\hat{w}} = w_x\hat{i} + w_y\hat{j} + w_z\hat{k}, \tag{2.5}$$

then the rotation matrix can be represented by its nine elements as

$$^A\boldsymbol{R}_B = \left[{^A\hat{u}} \mid {^A\hat{v}} \mid {^A\hat{w}}\right] = \begin{bmatrix} u_x & v_x & w_x \\ u_y & v_y & w_y \\ u_z & v_z & w_z \end{bmatrix}. \tag{2.6}$$

Expressions for these nine elements can be given, noting that the components of any vector in a frame can be simply represented as the projections of that vector onto the unit direction of the reference frame; hence, vector dot product may be used to determine these elements $u_x = \hat{x}_B \cdot \hat{x}_A$. Therefore

$$^A\boldsymbol{R}_B = \left[{^A\hat{x}_B} \mid {^A\hat{y}_B} \mid {^A\hat{z}_B}\right] = \begin{bmatrix} \hat{x}_B \cdot \hat{x}_A & \hat{y}_B \cdot \hat{x}_A & \hat{z}_B \cdot \hat{x}_A \\ \hat{x}_B \cdot \hat{y}_A & \hat{y}_B \cdot \hat{y}_A & \hat{z}_B \cdot \hat{y}_A \\ \hat{x}_B \cdot \hat{z}_A & \hat{y}_B \cdot \hat{z}_A & \hat{z}_B \cdot \hat{z}_A \end{bmatrix}. \tag{2.7}$$

In this representation, the leading superscripts in the rightmost matrix is intentionally omitted, since the choice of frame in which the dot product is performed is arbitrary, as long as it is the same for each dotted pair. The dot product of two unit vectors results in the cosine of the angle between them; therefore, the rotation matrix is sometimes called *direction cosine representation.*

Example 2.1: Simple Axis Rotations

Consider frame $\{\boldsymbol{B}\}$ is rotated about one of the Cartesian axes of frame $\{\boldsymbol{A}\}$ as follows:

a. Rotation about x axis with an angle of α
b. Rotation about y axis with an angle of β
c. Rotation about z axis with an angle of γ

These rotations are shown in Figure 2.3. The rotation matrix $^A\boldsymbol{R}_B$ for these three cases may be found by the projection of the unit axes of frame $\{\boldsymbol{B}\}$ with respect to frame $\{\boldsymbol{A}\}$ as

$$\text{a. } {^A\boldsymbol{R}_B} = \boldsymbol{R_x}(\alpha) = \begin{bmatrix} 1 & 0 & 0 \\ 0 & \cos\alpha & -\sin\alpha \\ 0 & \sin\alpha & \cos\alpha \end{bmatrix},$$

$$\text{b. } {^A\boldsymbol{R}_B} = \boldsymbol{R_y}(\beta) = \begin{bmatrix} \cos\beta & 0 & \sin\beta \\ 0 & 1 & 0 \\ -\sin\beta & 0 & \cos\beta \end{bmatrix},$$

$$\text{c. } {^A\boldsymbol{R}_B} = \boldsymbol{R_z}(\gamma) = \begin{bmatrix} \cos\gamma & -\sin\gamma & 0 \\ \sin\gamma & \cos\gamma & 0 \\ 0 & 0 & 1 \end{bmatrix}.$$

As it is seen in these rotation matrices, the column corresponding to the axis of rotation reads the axis of rotation itself. These simple rotations about one axis is further used to calculate compound rotations.

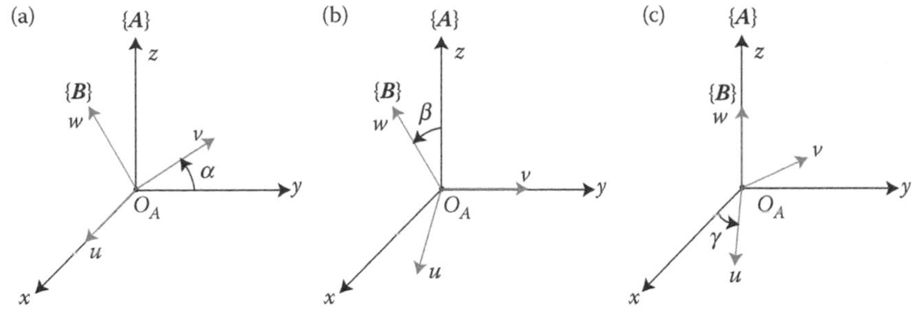

FIGURE 2.3
Simple rotations about major axes. (a) Rotation about x axis with an angle of α. (b) Rotation about y axis with an angle of β. (c) Rotation about z axis with an angle of γ.

2.1.2.2 Rotation Matrix Properties

Property (I)—Orthonormality
Rotation matrix is an orthonormal matrix, meaning that it is composed of column vectors that are unitary

$$\boldsymbol{u}\cdot\boldsymbol{u}=\boldsymbol{v}\cdot\boldsymbol{v}=\boldsymbol{w}\cdot\boldsymbol{w}=1, \tag{2.8}$$

and orthogonal to each other

$$\boldsymbol{u}\cdot\boldsymbol{v}=\boldsymbol{v}\cdot\boldsymbol{w}=\boldsymbol{w}\cdot\boldsymbol{u}=0. \tag{2.9}$$

Because of the orthogonality conditions, cross products of the column vectors satisfy the following equations:

$$\boldsymbol{u}\times\boldsymbol{v}=\boldsymbol{w},\quad \boldsymbol{v}\times\boldsymbol{w}=\boldsymbol{u},\quad \boldsymbol{w}\times\boldsymbol{u}=\boldsymbol{v}. \tag{2.10}$$

Property (II)—Transposition
Further inspection of Equation 2.7 shows that the rows of the rotation matrix ${}^{A}\boldsymbol{R}_{B}$ are the unit vectors of $\{\boldsymbol{A}\}$ expressed in frame $\{\boldsymbol{B}\}$, that is,

$${}^{A}\boldsymbol{R}_{B}=\left[{}^{A}\hat{\boldsymbol{x}}_{B}\mid{}^{A}\hat{\boldsymbol{y}}_{B}\mid{}^{A}\hat{\boldsymbol{z}}_{B}\right]=\begin{bmatrix}{}^{B}\hat{\boldsymbol{x}}_{A}^{T}\\{}^{B}\hat{\boldsymbol{y}}_{A}^{T}\\{}^{B}\hat{\boldsymbol{z}}_{A}^{T}\end{bmatrix}. \tag{2.11}$$

Therefore, ${}^{B}\boldsymbol{R}_{A}$, the orientation description of frame $\{\boldsymbol{A}\}$ with respect to frame $\{\boldsymbol{B}\}$ is given by the transpose of Equation 2.11; that is,

$${}^{B}\boldsymbol{R}_{A}={}^{A}\boldsymbol{R}_{B}^{T}. \tag{2.12}$$

Property (III)—Inverse
Property (II) suggests that the inverse of a rotation matrix is equal to its transpose. Indeed, this property is known from linear algebra [171] that for any orthonormal matrix the

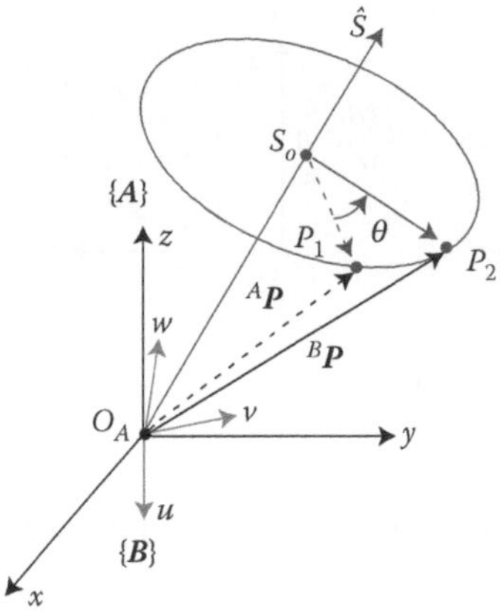

FIGURE 2.4
Pure rotation mapping.

inverse of the matrix is equal to its transpose.* This fact can be easily verified by

$$ {}^A\boldsymbol{R}_B{}^T\,{}^A\boldsymbol{R}_B = \begin{bmatrix} {}^A\hat{\boldsymbol{x}}_B^T \\ {}^A\hat{\boldsymbol{y}}_B^T \\ {}^A\hat{\boldsymbol{z}}_B^T \end{bmatrix} \cdot \left[{}^A\hat{\boldsymbol{x}}_B \mid {}^A\hat{\boldsymbol{y}}_B \mid {}^A\hat{\boldsymbol{z}}_B \right] = \boldsymbol{I}_{3\times 3} \tag{2.13} $$

in which $\boldsymbol{I}_{3\times 3}$ denotes the 3×3 identity matrix. Hence

$$ {}^B\boldsymbol{R}_A = {}^A\boldsymbol{R}_B^{-1} = {}^A\boldsymbol{R}_B^T. \tag{2.14} $$

Property (IV)—Pure Rotation Mapping
In a large number of problems in robotics, we are concerned with the expression of a quantity, such as the position of a point in a rigid body, in terms of various reference frames. The mathematics of mapping for a pure rotation case can be seen as an important property of the rotation matrix. As in Figure 2.4, suppose that the position vector of a point in the rigid body with respect to the moving frame $\{\boldsymbol{B}\}$ is given and is denoted by ${}^B\boldsymbol{P}$. Furthermore, assume that it is necessary to describe this vector in frame $\{\boldsymbol{A}\}$. In this section, consider that the rigid body has been exposed to a *pure rotation*, and therefore, the fixed frame $\{\boldsymbol{A}\}$ and moving frame $\{\boldsymbol{B}\}$ are coincident at their origins. In order to describe ${}^A\boldsymbol{P}$ componentwise with respect to frame $\{\boldsymbol{A}\}$, use the dot product properties of vectors

$$ \begin{aligned} {}^AP_x &= {}^A\hat{\boldsymbol{x}}_A \cdot {}^A\boldsymbol{P} \\ {}^AP_y &= {}^A\hat{\boldsymbol{y}}_A \cdot {}^A\boldsymbol{P} \\ {}^AP_z &= {}^A\hat{\boldsymbol{z}}_A \cdot {}^A\boldsymbol{P}. \end{aligned} \tag{2.15} $$

However, notice that dot product of two vectors can be written in any arbitrary frame, provided both pairs are represented in the same frame. Since the representation of vector

* Refer to Appendix A for more details.

$\boldsymbol{P}$ with respect to frame $\{\boldsymbol{B}\}$ is given, the dot products are written in frame $\{\boldsymbol{B}\}$, as follows:

$$\begin{aligned} {}^{A}P_x &= {}^{B}\hat{\boldsymbol{x}}_A \cdot {}^{B}\boldsymbol{P} = {}^{B}\hat{\boldsymbol{x}}_A^T \; {}^{B}\boldsymbol{P} \\ {}^{A}P_y &= {}^{B}\hat{\boldsymbol{y}}_A \cdot {}^{B}\boldsymbol{P} = {}^{B}\hat{\boldsymbol{y}}_A^T \; {}^{B}\boldsymbol{P} \\ {}^{A}P_z &= {}^{B}\hat{\boldsymbol{z}}_A \cdot {}^{B}\boldsymbol{P} = {}^{B}\hat{\boldsymbol{z}}_A^T \; {}^{B}\boldsymbol{P}. \end{aligned} \tag{2.16}$$

Writing Equation 2.16 in a matrix form yields

$$ {}^{A}\boldsymbol{P} = \begin{bmatrix} {}^{B}\hat{\boldsymbol{x}}_A^T \\ {}^{B}\hat{\boldsymbol{y}}_A^T \\ {}^{B}\hat{\boldsymbol{z}}_A^T \end{bmatrix} {}^{B}\boldsymbol{P}. \tag{2.17}$$

This can be written compactly using the rotation matrix definition in Equation 2.11.

$$ {}^{A}\boldsymbol{P} = {}^{A}\boldsymbol{R}_B \, {}^{B}\boldsymbol{P}. \tag{2.18}$$

Equation 2.18 defines the mapping that transforms the description of a vector from frame $\{\boldsymbol{B}\}$ into frame $\{\boldsymbol{A}\}$. The great help of the notation used for rotation matrices becomes clear here. By this means, you may view that the leading superscript of vector ${}^{B}\boldsymbol{P}$ is canceled with the following subscript of rotation matrix ${}^{A}\boldsymbol{R}_B$.

Example 2.2

Consider the orientation of frame $\{B\}$ with respect to frame $\{A\}$ is given by the following rotation matrix:

$$ {}^{A}\boldsymbol{R}_B = \begin{bmatrix} 0.933 & -0.067 & -0.354 \\ -0.067 & 0.933 & -0.354 \\ 0.354 & 0.354 & 0.866 \end{bmatrix}. $$

Furthermore, consider that the position vector of point P expressed in frame $\{B\}$ is ${}^{B}\boldsymbol{P} = [2 \quad -1 \quad 0]^T$. Position vector ${}^{A}\boldsymbol{P}$ of the same point P in frame $\{A\}$ can be expressed as follows:

$$\begin{aligned} {}^{A}\boldsymbol{P} &= {}^{A}\boldsymbol{R}_B \; {}^{B}\boldsymbol{P} \\ &= \begin{bmatrix} 0.933 & -0.067 & -0.354 \\ -0.067 & 0.933 & -0.354 \\ 0.354 & 0.354 & 0.866 \end{bmatrix} \begin{bmatrix} 2 \\ -1 \\ 0 \end{bmatrix} = \begin{bmatrix} 1.93 \\ -1.07 \\ 0.35 \end{bmatrix}. \end{aligned}$$

Property (V)—Determinant
The determinant of a rotation matrix ${}^{A}\boldsymbol{R}_B$ is equal to 1.

$$ \det({}^{A}\boldsymbol{R}_B) = 1. \tag{2.19}$$

To prove Equation 2.19, consider the rotation matrix ${}^A\boldsymbol{R}_B$ componentwise as given in Equation 2.6, and expand the determinant with respect to its third column coefficients:

$$\begin{aligned}\det({}^A\boldsymbol{R}_B) &= w_x(u_y v_z - v_y u_z) - w_y(u_x v_z - v_x u_z) + w_x(u_x v_y - v_x u_y)\\ &= \boldsymbol{w}\cdot(\boldsymbol{u}\times\boldsymbol{v})\\ &= \boldsymbol{w}\cdot\boldsymbol{w}\\ &= 1.\end{aligned}$$

Property (VI)—Eigenvalues

The eigenvalues of a rotation matrix ${}^A\boldsymbol{R}_B$ are equal to 1, $e^{i\theta}$, and $e^{-i\theta}$, where θ is calculated from

$$\theta = \cos^{-1}\frac{\mathrm{tr}({}^A\boldsymbol{R}_B) - 1}{2}. \tag{2.20}$$

In this equation, $\cos^{-1}$ denotes the inverse cosine function and $\mathrm{tr}({}^A\boldsymbol{R}_B)$ denotes the trace of matrix ${}^A\boldsymbol{R}_B$.

For the proof, consider the rotation matrix ${}^A\boldsymbol{R}_B$ componentwise as given in Equation 2.6, and apply the definition of eigenvalues λ, of the rotation matrix as

$$\det({}^A\boldsymbol{R}_B - \lambda\boldsymbol{I}) = \begin{vmatrix} u_x-\lambda & v_x & w_x\\ u_y & v_y-\lambda & w_y\\ u_z & v_z & w_z-\lambda\end{vmatrix} = 0. \tag{2.21}$$

Let us expand Equation 2.21 along the first column and consider Equations 2.10 and 2.19 to simplify it as follows:

$$\begin{aligned}\det({}^A\boldsymbol{R}_B - \lambda\boldsymbol{I}) &= (u_x-\lambda)[(v_y-\lambda)(w_z-\lambda) - v_z w_y]\\ &\quad - u_y[v_x(w_z-\lambda) - v_z w_x] + u_z[v_x w_y - w_x(v_y-\lambda)]\\ &= -\lambda^3 + (u_x+v_y+w_z)\lambda^2\\ &\quad - [(v_y w_z - v_z w_y) + (u_x w_z - u_z w_x) + (u_x v_y - u_y v_x)]\lambda + \det({}^A\boldsymbol{R}_B)\\ &= -\lambda^3 + \mathrm{tr}({}^A\boldsymbol{R}_B)\lambda^2 - [(\boldsymbol{v}\times\boldsymbol{w})_x + (\boldsymbol{w}\times\boldsymbol{u})_y + (\boldsymbol{u}\times\boldsymbol{v})_z]\lambda + 1\\ &= -\lambda^3 + \mathrm{tr}({}^A\boldsymbol{R}_B)\lambda^2 - (u_x+v_y+w_z)\lambda + 1.\end{aligned}$$

Hence, the characteristic equation simplifies to

$$\lambda^3 - \mathrm{tr}({}^A\boldsymbol{R}_B)\lambda^2 + \mathrm{tr}({}^A\boldsymbol{R}_B)\lambda - 1 = 0. \tag{2.22}$$

Note that Equation 2.22 contains $(\lambda - 1)$ as a factor, and can be simplified as follows:

$$(\lambda-1)(\lambda^2 + (1-\mathrm{tr}({}^A\boldsymbol{R}_B))\lambda + 1) = 0. \tag{2.23}$$

Hence, the roots of characteristic equation are

$$\lambda = 1, e^{i\theta}, \text{ and } e^{-i\theta}, \tag{2.24}$$

in which $i = \sqrt{-1}$, and $e^{\pm i\theta}$ is defined as

$$e^{\pm i\theta} \triangleq \cos\theta \pm i \sin\theta.$$

Hence,

$$\cos\theta = \frac{\operatorname{tr}({}^{A}\boldsymbol{R}_{B}) - 1}{2}; \quad \sin\theta = \sqrt{1 - \cos^2\theta}.$$

Example 2.3

For the rotation matrix given in Example 2.2, find the corresponding eigenvalues and eigenvectors, and verify Property (IV).

The eigenvalues and eigenvectors of the given rotation matrix are determined as follows:

$$\lambda_1 = 1; \; v_1 = \begin{bmatrix} -0.7071 \\ +0.7071 \\ 0 \end{bmatrix},$$

$$\lambda_{2,3} = 0.866 \pm 0.5i = e^{\pm\pi/6}; \; v_{2,3} = \begin{bmatrix} \pm 0.5i \\ \pm 0.5i \\ 0.7071 \end{bmatrix}.$$

It can be easily verified from Equation 2.20 that $\theta = \pi/6$.

The angle θ is an invariant parameter of a rotation matrix, which can be used to represent the orientation of the frame $\{\boldsymbol{B}\}$ with respect to frame $\{\boldsymbol{A}\}$ by another means of representation. This is called screw axis representation, which is described in the next section.

2.1.2.3 Screw Axis Representation

As seen in the properties of the rotation matrix, there exists an invariant angle θ corresponding to the rotation matrix. This angle is indeed an equivalent angle of rotation, which can represent the orientation of frame $\{\boldsymbol{B}\}$ with respect to frame $\{\boldsymbol{A}\}$. The rotation is a spatial change of orientation about an axis, which is called the *screw axis*. It can be shown that the screw axis is also an invariant parameter of the rotation matrix, and is indeed the real eigenvector of the rotation matrix corresponding to the eigenvalue $\lambda = 1$. If rotation matrix is considered as a map that converts frame $\{\boldsymbol{A}\}$ into frame $\{\boldsymbol{B}\}$, the magnitude of the eigenvalues of this map designates the gain of the map. For the rotation matrix, the magnitude of all its eigenvalues and the determinant are both equal to one. This indicates the fact that this mapping preserves the magnitudes, and what it does is just a rotation in the space. The angle θ, which is given in Equation 2.20, indicates the amount of rotation, and the real eigenvector of the rotation matrix describes a real axis in space, about which the rotation takes place. Use of the term *screw axis* for this axis of rotation has the benefit that a general motion of a rigid body, which is composed as a pure translation and a pure rotation, can be further represented by the same axis of rotation. This generalization is described in Section 2.3.3.

Screw axis representation of orientation has the benefit of using only four parameters to describe a pure rotation. These parameters are the angle of rotation θ, and the axis of rotation, or the screw axis, which is a unit vector denoted by ${}^{A}\hat{\mathbf{s}} = [s_x, s_y, s_z]^T$. Since a unit vector is used to represent the axis of rotation in space, the components of the screw axis must satisfy an algebraic constraint, namely

$$\hat{\mathbf{s}} \cdot \hat{\mathbf{s}} = 1 \text{ or } s_x^2 + s_y^2 + s_z^2 = 1. \tag{2.25}$$

As it is shown before, the angle of rotation θ may be uniquely derived from the rotation matrix ${}^{A}\boldsymbol{R}_B$ by Equation 2.20. In order to derive expressions for screw axis components ${}^{A}\hat{\mathbf{s}} = [s_x, s_y, s_z]^T$, property (IV) of rotation matrix may be used. Consider a pure rotation of rigid body and the relation of position vector of a point P with respect to the fixed and moving frame, as stated in Equation 2.18. To represent this relation using screw axis representation, consider the pure rotation of rigid body about an axis ${}^{A}\hat{\mathbf{s}}$, with an angle θ as shown in Figure 2.5. As shown in this figure, the point P of rigid body is in state $\boldsymbol{P}_1$ before rotation, and it is in state $\boldsymbol{P}_2$ when the rotation is completed. The moving frame was coincident to the fixed frame at state 1, and is rotated about the screw axis ${}^{A}\hat{\mathbf{s}}$, with an angle of θ. In this figure, the moving frame $\{\boldsymbol{B}\}$ is shown after the rotation. From the geometry shown in Figure 2.5, one may obtain

$$\overrightarrow{S_o P_1} = \boldsymbol{P}_1 - \overrightarrow{O_A S_o} = \boldsymbol{P}_1 - (\boldsymbol{P}_1 \cdot \hat{\mathbf{s}})\hat{\mathbf{s}} \tag{2.26}$$

$$\overrightarrow{S_o P_2} = \boldsymbol{P}_2 - \overrightarrow{O_A S_o} = \boldsymbol{P}_2 - (\boldsymbol{P}_1 \cdot \hat{\mathbf{s}})\hat{\mathbf{s}} \tag{2.27}$$

As it is seen in this equation, $\overrightarrow{O_A S_o}$ is determined as the projection of vector $\boldsymbol{P}_1$ onto the screw axis $\hat{\mathbf{s}}$ using dot product. Furthermore, Figure 2.5b shows the normal plane to the screw axis, and by means of geometric interpretations, two relations for $\cos\theta$ and $\sin\theta$ can be derived as follows:

$$\overrightarrow{S_o N} = \overrightarrow{S_o P_1} \cos\theta. \tag{2.28}$$

$$\overrightarrow{N P_2} = (\hat{\mathbf{s}} \times \overrightarrow{S_o P_1}) \sin\theta = (\hat{\mathbf{s}} \times \boldsymbol{P}_1) \sin\theta \tag{2.29}$$

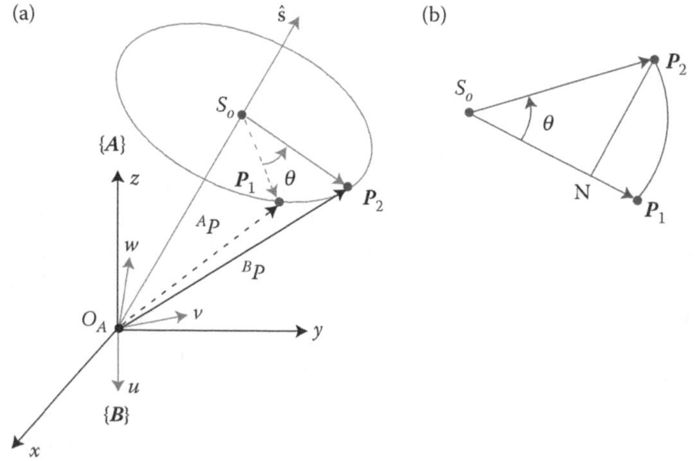

FIGURE 2.5
Pure rotation about a screw axis: a vector diagram. (a) Screw axis rotation. (b) Normal plane to the screw axis.

in which $\times$ denotes vector cross product. In order to relate $\boldsymbol{P}_1$ and $\boldsymbol{P}_2$, a vector relation for $\overrightarrow{S_o P_2}$ as shown in Figure 2.5b is as follows:

$$\overrightarrow{S_o P_2} = \overrightarrow{S_o N} + \overrightarrow{N P_2}. \tag{2.30}$$

Substitute Equations 2.26 through 2.29 into 2.30

$$\boldsymbol{P}_2 - (\boldsymbol{P}_1 \cdot \hat{\mathbf{s}})\hat{\mathbf{s}} = \overrightarrow{S_o P_1} \cos\theta + (\hat{\mathbf{s}} \times \boldsymbol{P}_1) \sin\theta \tag{2.31}$$

$$\boldsymbol{P}_2 - (\boldsymbol{P}_1 \cdot \hat{\mathbf{s}})\hat{\mathbf{s}} = (\boldsymbol{P}_1 - (\boldsymbol{P}_1 \cdot \hat{\mathbf{s}})\hat{\mathbf{s}}) \cos\theta + (\hat{\mathbf{s}} \times \boldsymbol{P}_1) \sin\theta \tag{2.32}$$

Rearranging this equation yields a formula that is known as Rodrigues's rotation formula for spatial rotation of a rigid body:

$$\boldsymbol{P}_2 = \boldsymbol{P}_1 \cos\theta + (\hat{\mathbf{s}} \times \boldsymbol{P}_1) \sin\theta + (\boldsymbol{P}_1 \cdot \hat{\mathbf{s}})(1 - \cos\theta)\hat{\mathbf{s}} \tag{2.33}$$

This formula is in vector form, and may be rewritten in a matrix form using Property VI of rotation matrices. By this means, rotation matrix can be stated as a function of the screw axis components ${}^A\hat{\mathbf{s}} = [s_x, s_y, s_z]^T$ and the angle of rotation θ. Notice that the pure rotation map of a vector can be stated as

$$\boldsymbol{P}_2 = {}^A\boldsymbol{R}_B \boldsymbol{P}_1, \tag{2.34}$$

in which the rotation matrix components can be derived using Rodrigues rotation formula 2.33 as follows:

$${}^A\boldsymbol{R}_B = \begin{bmatrix} s_x^2 v\theta + c\theta & s_x s_y v\theta - s_z s\theta & s_x s_z v\theta + s_y s\theta \\ s_y s_x v\theta + s_z s\theta & s_y^2 v\theta + c\theta & s_y s_z v\theta - s_x s\theta \\ s_z s_x v\theta - s_y s\theta & s_z s_y v\theta + s_x s\theta & s_z^2 v\theta + c\theta \end{bmatrix}, \tag{2.35}$$

in which $s\theta = \sin\theta, c\theta = \cos\theta$, and $v\theta = 1 - \cos\theta$ are used here and further in this book as shorthand notations. Equation 2.35 is called the *screw axis representation* of the orientation of a rigid body. Given the screw axis and angle of rotation, the rotation matrix may be derived from Equation 2.35. The inverse map can be obtained easily as well by the following equations. The rotation angle is derived from Equation 2.20, and the screw axis components are obtained by taking the differences between each pair of two opposing off-diagonal elements as follows:

$$\begin{aligned} s_x &= \frac{r_{32} - r_{23}}{2s\theta}, \\ s_y &= \frac{r_{13} - r_{31}}{2s\theta}, \\ s_z &= \frac{r_{21} - r_{12}}{2s\theta}. \end{aligned} \tag{2.36}$$

From Equations 2.20 and 2.36, it seems that there exist two solutions for the screw axis and rotation angle, one being the negative of the other. In fact, these two solutions represent the same screw, since a $(-\theta)$ rotation about the $-\hat{\mathbf{s}}$ axis produce the same result as a $(+\theta)$ rotation about the $+\hat{\mathbf{s}}$ axis.

Example 2.4: Simple Axis Rotations

Consider the following simple rotations as in Example 2.1:

a. Rotation about x axis with an angle of α
b. Rotation about y axis with an angle of β
c. Rotation about z axis with an angle of γ

The rotations are shown in Figure 2.3. The screw axis representation for these three rotations can be found using Equations 2.20 and 2.36 as follows:

a. For the first rotation matrix

$$\theta = \cos^{-1}((1 + 2\cos\alpha - 1)/2) = \alpha,$$
$$s_x = (s\alpha - (-s\alpha))/2s\theta = 1,$$
$$s_y = 0,$$
$$s_z = 0.$$

Hence

$$\theta = \alpha;\ \hat{s} = \begin{bmatrix} 1 & 0 & 0 \end{bmatrix}.$$

b. Similarly

$$\theta = \beta;\ \hat{s} = \begin{bmatrix} 0 & 1 & 0 \end{bmatrix}.$$

c. And

$$\theta = \gamma;\ \hat{s} = \begin{bmatrix} 0 & 0 & 1 \end{bmatrix}.$$

As it is seen for these simple rotations, the calculated rotation angles and screw axis components can be verified by inspection.

Example 2.5

Consider rotation about $\hat{s} = \frac{1}{\sqrt{3}}[1 \quad 1 \quad 1]^T$ axis with an angle of $\theta = \pi/6$. The resulting rotation matrix is as follows:

$$^{A}\boldsymbol{R}_{B} = \begin{bmatrix} 0.9107 & -0.244 & +0.333 \\ +0.333 & 0.9107 & -0.244 \\ -0.244 & +0.333 & 0.9107 \end{bmatrix}.$$

Note that the eigenvalues of $^{A}\boldsymbol{R}_{B}$ are $\lambda_{1,2,3} = 1, e^{i\pi/6}, e^{-i\pi/6}$ and the eigenvector corresponding to eigenvalue $\lambda_1 = 1$ is

$$v_1 = [0.5774 \quad 0.5774 \quad 0.5774]^T,$$

which is equal to $\hat{s}$.

Example 2.6

Consider the rotation matrix introduced in Example 2.2 and further analyzed in Example 2.3. This rotation matrix can be represented by the screw axis representation $\{\theta = \pi/6, \hat{s} = [-0.7071 \quad 0.7071 \quad 0]^T\}$.

2.1.2.4 Euler Angles

The representation of orientation with rotation matrix requires nine parameters and the screw axis representation requires four. Since rotation in space is a motion with three-degrees-of-freedom, a set of three independent parameters are sufficient to represent the orientation. Perhaps the simplest way for a human operator to assign the orientation of a robot end effector in space is by using only three independent parameters, without considering the inherent orthonormal constraints on the columns of a rotation matrix, or unitary constraints of the screw axis representation. Several sets of three-parameter representations have been reported in the literature throughout the years, originating by the way an orientation in space can be measured. The most commonly used sets are called the Euler angles. In the following sections, several such representations are introduced.

In an Euler angle representation, three successive rotations about the coordinate system of either fixed or moving frame are used to describe the orientation of the rigid body. It is thus convenient to use rotations about the principle axes of the coordinate frames for such descriptions. These rotations are introduced as $\boldsymbol{R}_x(\alpha), \boldsymbol{R}_y(\beta)$, and $\boldsymbol{R}_z(\gamma)$ in Examples 2.2 and 2.3 by rotation matrix, and screw axis representations, respectively. In what follows, one type of Euler angle in which the rotations are considered with respect to the fixed frame is described. This representation is called *pitch–roll–yaw* or *fixed* X–Y–Z Euler angles. Three other types of Euler angles, in which the rotations are considered with respect to a moving frame, comes later. Those representations are denoted by *w–v–u*, *w–v–w*, and *w–u–w* Euler angles.

Pitch–Roll–Yaw Euler Angles

The pitch, roll, and yaw angles are defined for a moving object in space as the rotations along the lateral, longitudinal, and vertical axes attached to the moving object, respectively. These angles are shown on an airplane in Figure 2.6, in which the pitch angle represents the orientation of the airplane about the lateral axis, the roll angle represents the rotation about the longitudinal axis, and the yaw angle is the rotation about the vertical axis.

In order to find the relation between theses angles and the rotation matrix, three successive rotations about x, y, and z axes of the fixed frame $\{A\}$ are defined as follows:

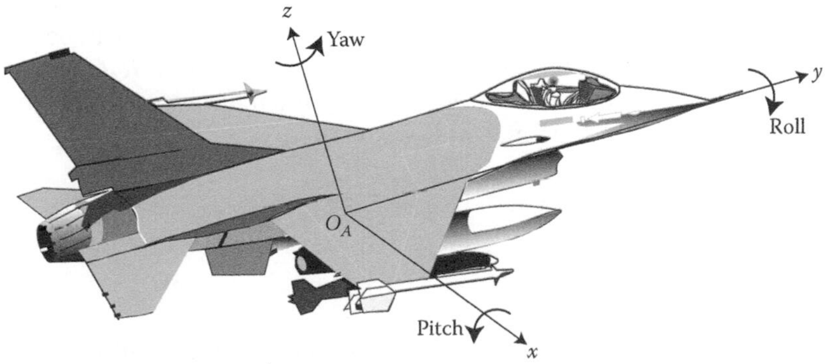

FIGURE 2.6
Definition of pitch, roll, and yaw angles on an airplane.

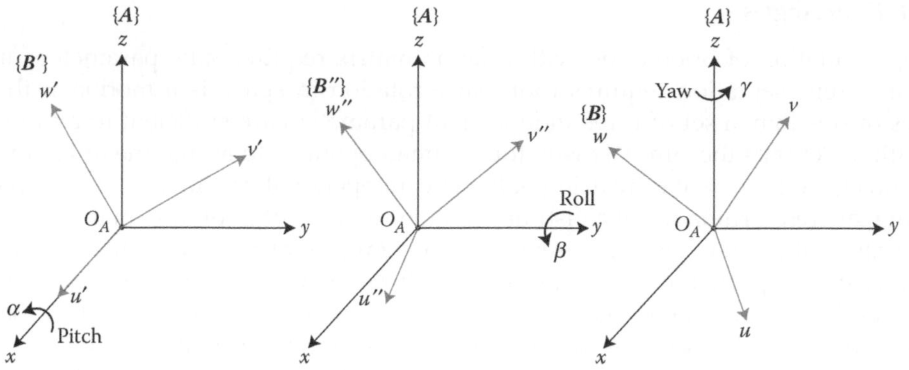

FIGURE 2.7
Successive rotations about the fixed coordinate axes; pitch angle α about x axis, roll angle β about y axis, and yaw angle γ about z axis.

> As shown in Figure 2.7, start with the state that the moving frame $\{\boldsymbol{B}\}$ coincides with the fixed frame $\{\boldsymbol{A}\}$. Rotate the moving frame about the x axis of the fixed frame by an angle α and denote the moving frame in this state as $\{\boldsymbol{B}'\}$. Then rotate the moving frame $\{\boldsymbol{B}'\}$ about the y axis of the fixed frame by an angle β and denote the moving frame in this state as $\{\boldsymbol{B}''\}$. Finally, rotate the moving frame $\{\boldsymbol{B}''\}$ about the z axis of the fixed frame by an angle γ and denote the final moving frame as $\{\boldsymbol{B}\}$.

Since all three rotations take place about the axes of a fixed coordinate frame, the resulting rotation matrix is obtained by premultiplying the three basic rotation matrices as follows:

$$\begin{aligned}\boldsymbol{R}_{PRY}(\alpha,\beta,\gamma) &= \boldsymbol{R}_z(\gamma)\boldsymbol{R}_y(\beta)\boldsymbol{R}_x(\alpha)\\ &= \begin{bmatrix} c\gamma & -s\gamma & 0\\ s\gamma & c\gamma & 0\\ 0 & 0 & 1\end{bmatrix}\begin{bmatrix} c\beta & 0 & s\beta\\ 0 & 1 & 0\\ -s\beta & 0 & c\beta\end{bmatrix}\begin{bmatrix} 1 & 0 & 0\\ 0 & c\alpha & -s\alpha\\ 0 & s\alpha & c\alpha\end{bmatrix},\end{aligned} \tag{2.37}$$

where $c\alpha$ is shorthand for $\cos\alpha$, $s\alpha$ for $\sin\alpha$, and so on. Note that the successive rotations about the fixed coordinate axes result in *premultiplication* of the matrices, and since the matrix multiplication does not commute, the order of rotations cannot be exchanged. Considering the rotations as operators in space, the sequence of rotations can be considered as the matrix multiplications from the right. Hence, the first operation $\boldsymbol{R}_x(\alpha)$ is considered as the right-most matrix for multiplication. Multiplying the matrices in Equation 2.37 results in the final pitch–roll–yaw rotation matrix $\boldsymbol{R}_{PRY}$ as follows:

$$\boldsymbol{R}_{PRY}(\alpha,\beta,\gamma) = \begin{bmatrix} c\beta c\gamma & s\alpha s\beta c\gamma - c\alpha s\gamma & c\alpha s\beta c\gamma + s\alpha s\gamma\\ c\beta s\gamma & s\alpha s\beta s\gamma + c\alpha c\gamma & c\alpha s\beta s\gamma - s\alpha c\gamma\\ -s\beta & s\alpha c\beta & c\alpha c\beta\end{bmatrix}. \tag{2.38}$$

The inverse problem is also often of interest, in which, given a rotation matrix representation for the orientation, the pitch, roll, and yaw angles must be computed. The solution involves nine trigonometric equations, with six dependencies and three unknowns, resulting in three independent equations and three unknowns. Consider the first column of the rotation matrix. The sum of squares of matrix elements r_{11} and r_{21} results in $c^2\beta(c^2\gamma + s^2\gamma) = c^2\beta$. Therefore, β can be uniquely computed using $\sqrt{r_{11}^2 + r_{21}^2}$ and r_{31}. Furthermore, γ is solved from r_{11} and r_{21}, and α from r_{32} and r_{33}, provided $c\beta \neq 0$ as follows:

$$\begin{aligned} \beta &= \text{Atan2}\,(-r_{31}, \pm\sqrt{r_{11}^2 + r_{21}^2}), \\ \gamma &= \text{Atan2}\,(r_{21}/c\beta, r_{11}/c\beta), \\ \alpha &= \text{Atan2}\,(r_{32}/c\beta, r_{33}/c\beta). \end{aligned} \tag{2.39}$$

Atan2 (y, x) is a four-quadrant arc tangent function that computes $\tan^{-1}(y/x)$ but uses the sign of both x and y to identify the quadrant in which the resulting angle lies. For example, $\text{Atan}(-1.0/-1.0) = \pi/4$, whereas $\text{Atan2}\,(-1.0, -1.0) = -3\pi/4$. This provides an important distinction that will be lost with a two-quadrant arc tangent. Therefore, the resulting angle from Atan2 function may lie in all four quadrants, and range over the full $[-\pi, \pi]$ region. This function is repeatedly used in this book to replace Atan and many programming language libraries have it predefined.

Equation 2.39 gives two solutions for pitch–roll–yaw angles, for which one solution lies in $-\pi/2 \leq \beta \leq \pi/2$, and the second solution that is found by using the negative square root formula for β lies in the other two quadrants. The inverse solution 2.39 degenerates if $c\beta = 0$ or $\beta = \pm\pi/2$. In such cases, only the sum or the difference of α and γ can be computed as follows:

$$\begin{cases} \beta = \pi/2 \\ \alpha - \gamma = \text{Atan2}\,(r_{12}, r_{22}) \end{cases} \quad \text{or} \quad \begin{cases} \beta = -\pi/2 \\ \alpha + \gamma = \text{Atan2}\,(-r_{12}, r_{22}) \end{cases} \tag{2.40}$$

Example 2.7

Consider the rotation matrix introduced in Example 2.2 and further analyzed in Examples 2.3 and 2.6. This rotation can be represented by the following pitch–roll–yaw Euler angles using Equation 2.39:

$$\alpha = 22.23^\circ; \quad \beta = -20.73^\circ; \quad \gamma = -4.11^\circ.$$

Example 2.8

Consider a rotation represented by the following rotation matrix:

$$\boldsymbol{R} = \begin{bmatrix} 0 & 0.5 & 0.866 \\ 0 & 0.866 & -0.5 \\ -1 & 0 & 0 \end{bmatrix}.$$

The solution to the pitch–roll–yaw Euler angles for this specific rotation degenerates since $r_{11} = r_{21} = 0$. This means that for this rotation, $c\beta = 0$ or $\beta = \pm\pi/2$. Hence, in this case, only the sum or the difference of α and γ can be computed as follows:

$$\begin{cases} \beta = \pi/2 \\ \alpha - \gamma = \pi/6 \end{cases} \quad \text{or} \quad \begin{cases} \beta = -\pi/2 \\ \alpha + \gamma = -\pi/6. \end{cases}$$

u–v–w Euler Angles

Another way to describe the orientation of a moving object is to consider three successive rotations about the coordinate axes of the moving frame as follows:

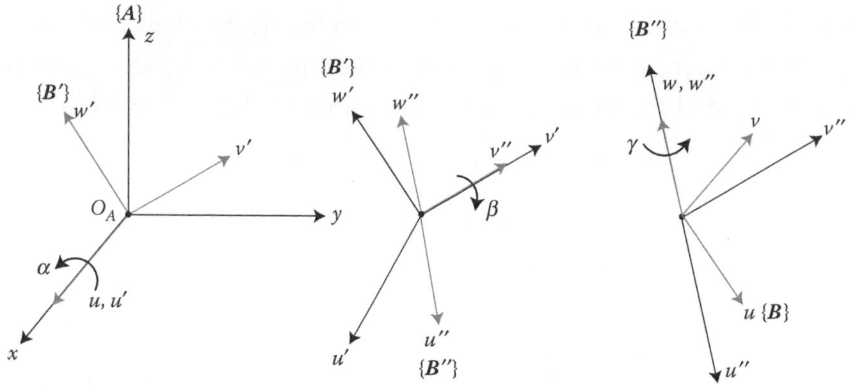

FIGURE 2.8
Successive rotations about the moving coordinate axes the first Euler angle α about u axis, the second Euler angle β about v' axis, and the third Euler angle γ about w'' axis.

As shown in Figure 2.8, start with the state that the moving frame $\{\boldsymbol{B}\}$ coincides with the fixed frame $\{\boldsymbol{A}\}$. Rotate the moving frame about u axis of the moving frame by an angle α and denote the moving frame in this state as $\{\boldsymbol{B}'\}$. Then rotate the moving frame $\{\boldsymbol{B}'\}$ about its v' axis by an angle β and denote the moving frame in this state as $\{\boldsymbol{B}''\}$. Finally, rotate the moving frame $\{\boldsymbol{B}''\}$ about its w'' axis by an angle γ and denote the final moving frame as $\{\boldsymbol{B}\}$.

Note that each rotation occurs about an axis the location of which depends on the preceding rotations. Since the rotations do not occur about fixed axes, premultiplication of the individual rotation matrices fails to give the correct solution. It can be shown that the resulting matrix can be derived by *postmultiplication* of the individual rotation matrices for the cases in which rotations are about the moving frame axes. This can be verified by kinematic inversion [181]. Consider the orientation of frame $\{\boldsymbol{A}\}$ with respect to frame $\{\boldsymbol{B}\}$. This inverse problem can be stated as rotation of frame $\{\boldsymbol{A}\}$ about the w axis by angle $-\gamma$, followed by a second rotation of $-\beta$ about the v axis, and by a third rotation of $-\alpha$ about the u axis. Under the kinematic inversion, it can be considered that the coordinate axes of frame $\{\boldsymbol{B}\}$ are fixed, and therefore the premultiplication rule can be used in this case meaning that

$$^{B}\boldsymbol{R}_{A}(-\alpha, -\beta, -\gamma) = \boldsymbol{R}_{w}(-\gamma)\boldsymbol{R}_{v}(-\beta)\boldsymbol{R}_{u}(-\alpha). \tag{2.41}$$

Since $^{A}\boldsymbol{R}_{B} = {}^{B}\boldsymbol{R}_{A}^{-1}$, and furthermore, $\boldsymbol{R}_{w}^{-1}(-\gamma) = \boldsymbol{R}_{w}(\gamma)$, $\boldsymbol{R}_{v}^{-1}(-\beta) = \boldsymbol{R}_{v}(\beta)$, and $\boldsymbol{R}_{u}^{-1}(-\alpha) = \boldsymbol{R}_{u}(\alpha)$, Equation 2.41 can be expanded as follows:

$$\begin{aligned} ^{A}\boldsymbol{R}_{B}(\alpha, \beta, \gamma) &= {}^{B}\boldsymbol{R}_{A}^{-1}(-\alpha, -\beta, -\gamma) \\ &= [\boldsymbol{R}_{w}(-\gamma)\boldsymbol{R}_{v}(-\beta)\boldsymbol{R}_{u}(-\alpha)]^{-1} \\ &= \boldsymbol{R}_{u}^{-1}(-\alpha)\boldsymbol{R}_{v}^{-1}(-\beta)\boldsymbol{R}_{w}^{-1}(-\gamma). \end{aligned}$$

Hence,

$$^{A}\boldsymbol{R}_{B}(\alpha, \beta, \gamma) = \boldsymbol{R}_{u}(\alpha)\boldsymbol{R}_{v}(\beta)\boldsymbol{R}_{w}(\gamma). \tag{2.42}$$

Equation 2.42 gives a general formula for deriving the resulting rotation matrix in case the rotations are about the coordinate axes of moving frames. Comparing this formula to that of rotations about fixed frame axes, Equation 2.37, it is clear that when rotations are about the moving frame axes *postmultiplication*, and when rotations are about the fixed frame axes *premultiplication* of individual rotation matrices should be used. Expanding the matrices in Equation 2.42 results in

$$\begin{aligned} \boldsymbol{R}_{uvw}(\alpha,\beta,\gamma) &= \begin{bmatrix} 1 & 0 & 0 \\ 0 & c\alpha & -s\alpha \\ 0 & s\alpha & c\alpha \end{bmatrix} \begin{bmatrix} c\beta & 0 & s\beta \\ 0 & 1 & 0 \\ -s\beta & 0 & c\beta \end{bmatrix} \begin{bmatrix} c\gamma & -s\gamma & 0 \\ s\gamma & c\gamma & 0 \\ 0 & 0 & 1 \end{bmatrix} \\ &= \begin{bmatrix} c\beta c\gamma & -c\beta s\gamma & s\beta \\ s\alpha s\beta c\gamma + c\alpha s\gamma & -s\alpha s\beta s\gamma + c\alpha c\gamma & -s\alpha c\beta \\ -c\alpha s\beta c\gamma + s\alpha s\gamma & c\alpha s\beta s\gamma + s\alpha c\gamma & c\alpha c\beta \end{bmatrix}. \end{aligned} \tag{2.43}$$

The inverse solution for the *u–v–w* Euler angles can be obtained similar to that of the pitch–roll–yaw Euler angles as follows. For $c\beta \neq 0$

$$\begin{aligned} \beta &= \text{Atan2}\ (r_{13}, \pm\sqrt{r_{11}^2 + r_{12}^2}), \\ \alpha &= \text{Atan2}\ (-r_{23}/c\beta, r_{33}/c\beta), \\ \gamma &= \text{Atan2}\ (-r_{12}/c\beta, r_{11}/c\beta). \end{aligned} \tag{2.44}$$

If $c\beta = 0$, then following two cases may occur:

$$\begin{cases} \beta = \pi/2 \\ \alpha + \gamma = \text{Atan2}\ (r_{32}, r_{22}) \end{cases} \quad \text{or} \quad \begin{cases} \beta = -\pi/2 \\ \alpha - \gamma = \text{Atan2}\ (r_{32}, r_{22}) \end{cases} \tag{2.45}$$

Example 2.9

Consider the rotation matrix introduced in Example 2.2 and further analyzed in Examples 2.3, 2.5, and 2.7. This rotation can be represented by the following *u–v–w* Euler angles using Equation 2.44.

$$\alpha = 22.23^o; \quad \beta = -20.73^o; \quad \gamma = 4.11^o.$$

Example 2.10

Consider a rotation represented by the following rotation matrix:

$$\boldsymbol{R} = \begin{bmatrix} 0 & 0 & 1.0000 \\ 0.9658 & -0.2588 & 0 \\ 0.2588 & 0.9658 & 0 \end{bmatrix}.$$

The solution to the *u–v–w* Euler angles for this specific rotation degenerates since $r_{11} = r_{12} = 0$. This means that for this rotation, $c\beta = 0$ or $\beta = \pm\pi/2$. Hence, in this case, only the sum or the difference of α and γ can be computed as

$$\begin{cases} \beta = \pi/2 \\ \alpha + \gamma = 5\pi/12 \end{cases} \quad \text{or} \quad \begin{cases} \beta = -\pi/2 \\ \alpha - \gamma = 5\pi/12. \end{cases}$$

There exist 12 possible cases for Euler angle representations; two of which are described as follows:

w–v–w Euler Angles

Another type of Euler angle representation consists of a rotation of angle α about the w axis, followed by a second rotation of β about the rotated v axis, and then by a third rotation of γ about the rotated w axis. Using the postmultiplication rule, the resulting rotation matrix is obtained as follows:

$$\begin{aligned} \boldsymbol{R}_{wvw}(\alpha,\beta,\gamma) &= \boldsymbol{R}_w(\alpha)\boldsymbol{R}_v(\beta)\boldsymbol{R}_w(\gamma) \\ &= \begin{bmatrix} c\alpha c\beta c\gamma - s\alpha s\gamma & -c\alpha c\beta s\gamma - s\alpha c\gamma & c\alpha s\beta \\ s\alpha c\beta c\gamma + c\alpha s\gamma & -s\alpha c\beta s\gamma + c\alpha c\gamma & s\alpha s\beta \\ -s\beta c\gamma & s\beta s\gamma & c\beta \end{bmatrix}. \end{aligned} \tag{2.46}$$

Similar to what has been done for pitch–roll–yaw angles, the inverse map can be obtained as follows. If $s\beta \neq 0$, then

$$\begin{aligned} \beta &= \text{Atan2}\,(\pm\sqrt{r_{31}^2 + r_{32}^2}, r_{33}), \\ \alpha &= \text{Atan2}\,(r_{23}/s\beta, r_{13}/s\beta), \\ \gamma &= \text{Atan2}\,(r_{32}/s\beta, -r_{31}/s\beta). \end{aligned} \tag{2.47}$$

If $s\beta = 0$, then the following two cases may occur:

$$\begin{cases} \beta = 0 \\ \gamma + \alpha = \text{Atan2}\,(-r_{12}, r_{11}) \end{cases} \quad \text{or} \quad \begin{cases} \beta = \pi \\ \gamma - \alpha = \text{Atan2}\,(r_{12}, -r_{11}). \end{cases} \tag{2.48}$$

Example 2.11

Consider the rotation matrix introduced in Example 2.2. This rotation can be represented by the following w–v–w Euler angles using Equation 2.47.

$$\alpha = -135^{\circ}; \quad \beta = 30^{\circ}; \quad \gamma = 135^{\circ}.$$

w–u–w Euler Angles

The last example of Euler angles described here is similar to the previous case, with the difference that the second rotation is about the rotated u axis (instead of v axis in w–v–w Euler angles). The resulting rotation matrix is obtained as follows:

$$\begin{aligned} \boldsymbol{R}_{wuw}(\alpha,\beta,\gamma) &= \boldsymbol{R}_w(\alpha)\boldsymbol{R}_u(\beta)\boldsymbol{R}_w(\gamma) \\ &= \begin{bmatrix} c\alpha c\gamma - s\alpha c\beta s\gamma & -c\alpha s\gamma - s\alpha c\beta c\gamma & s\alpha s\beta \\ s\alpha c\gamma + c\alpha c\beta s\gamma & -s\alpha s\gamma + c\alpha c\beta c\gamma & -c\alpha s\beta \\ s\beta s\gamma & s\beta c\gamma & c\beta \end{bmatrix}. \end{aligned} \tag{2.49}$$

The inverse map may be obtained as follows. If $s\beta \neq 0$, then

$$\begin{aligned} \beta &= \text{Atan2}\,(\pm\sqrt{r_{31}^2 + r_{32}^2}, r_{33}), \\ \alpha &= \text{Atan2}\,(r_{13}/s\beta, -r_{23}/s\beta), \\ \gamma &= \text{Atan2}\,(r_{31}/s\beta, r_{32}/s\beta). \end{aligned} \tag{2.50}$$

If $s\beta = 0$, then the following two cases may occur:

$$\begin{cases} \beta = 0 \\ \alpha + \gamma = \text{Atan2}\,(-r_{12}, r_{11}) \end{cases} \quad \text{or} \quad \begin{cases} \beta = \pi \\ \alpha - \gamma = \text{Atan2}\,(r_{12}, r_{11}) \end{cases} \tag{2.51}$$

Example 2.12

Consider the rotation matrix introduced in Example 2.2. This rotation can be represented by the following *w–u–w* Euler angles using Equation 2.50.

$$\alpha = -45^\circ; \quad \beta = -30^\circ; \quad \gamma = 45^\circ.$$

Example 2.13

Consider rotation of a rigid body as shown in Figure 2.9. The initial frame is denoted by frame $\{\boldsymbol{I}\}$, and the final state is represented by frame $\{\boldsymbol{F}\}$. The rotation matrix can be derived from definition 2.2 as

$${}^{I}\boldsymbol{R}_F = \begin{bmatrix} -1 & 0 & 0 \\ 0 & 0 & 1 \\ 0 & 1 & 0 \end{bmatrix}.$$

This rotation can be represented by the following *w–u–w* Euler angles using Equation 2.50.

$$\alpha = \pi; \quad \beta = \pi/2; \quad \gamma = 0.$$

Note that there exist only two rotations in this example, and they can be easily verified by inspection from Figure 2.10. As shown in this figure, by rotation of the rigid body about w axis with an angle of π, the rigid body can be represented by frame $\{\boldsymbol{X}\}$, and further rotation about u axis with an angle of $\pi/2$ moves the rigid body to the final state represented by frame $\{\boldsymbol{F}\}$.

Notice that regardless of the type of Euler angle description used, if the Euler angle is given, a unique rotation matrix is determined for the orientation of the rigid body. However, the inverse map is not one–to–one, and at least two Euler angle sets can be found for each orientation. Moreover, the choice of Euler angles becomes multiple in a special orientation where in the inverse solution degenerates. Therefore, special care must

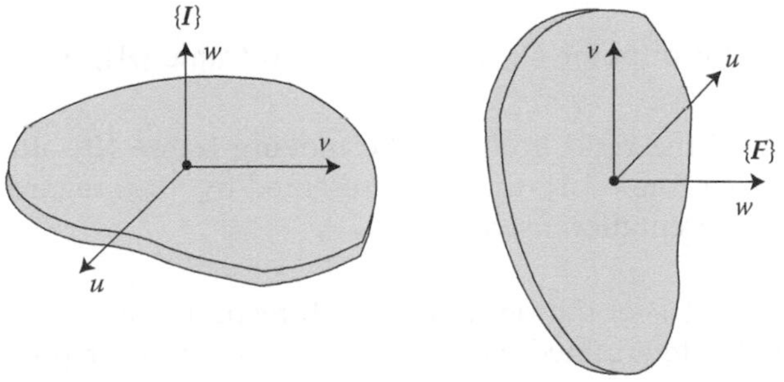

FIGURE 2.9
Example of a rigid body pure rotation.

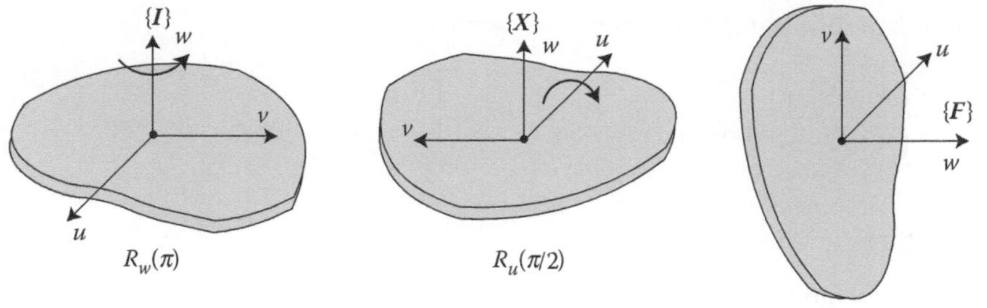

FIGURE 2.10
Representation of the rigid body rotation by two w–u–w Euler angles.

be taken if Euler angles are used for representation of the orientation. From the continuity of the motion a suitable solution may be chosen, such that no abrupt changes are seen in the variation of the Euler angles in a typical maneuver. Moreover, cases in which the inverse solution degenerates, the type of Euler angle chosen becomes singular and will not be capable of giving independent Euler angles for the motion. In such cases, usually two sets of different Euler angles are redundantly used, such that in case one of them becomes singular the second set describes the orientation of the rigid body. Noting these complications, the use of rotation matrix to represent the orientation of a rigid body is generally preferred, although there are nine parameters for that description.

2.2 Motion of a Rigid Body

In the previous sections, different representations for the orientation of a rigid body and the position of a point are described. Refer to Figure 2.1. Since the relative positions of a rigid body with respect to the moving frame $\{B\}$ attached to it is fixed for all time, it is sufficient to know the position of the origin of the moving frame O_B and the orientation of the moving frame $\{B\}$ with respect to the fixed frame $\{A\}$, to represent the position of any point P in the space. Representation of the position of O_B is uniquely given by the position vector, while orientation of the rigid body is represented in different forms. As it is detailed in Section 2.1.2, for all possible orientation representations, a rotation matrix ${}^{A}\boldsymbol{R}_B$ can be derived. Therefore, the location or pose of a rigid body, shown in Figure 2.1, can be fully determined by

1. The position vector of point O_B with respect to frame $\{A\}$, which is denoted by ${}^{A}\boldsymbol{P}_{O_B}$.
2. The orientation of the rigid body, or the moving frame $\{B\}$ attached to it with respect to the fixed frame $\{A\}$, that is represented by ${}^{A}\boldsymbol{R}_B$, regardless of any type of orientation representation used.

The position of any point P of the rigid body with respect to a fixed frame $\{A\}$, which is denoted as ${}^{A}\boldsymbol{P}$, may be determined as follows, if the location or pose of the rigid body $\{{}^{A}\boldsymbol{R}_B, {}^{A}\boldsymbol{P}_{O_B}\}$ is given:

$$ {}^{A}\boldsymbol{P} = {}^{A}\boldsymbol{R}_B\,{}^{B}\boldsymbol{P} + {}^{A}\boldsymbol{P}_{O_B}. \tag{2.52} $$

Equation 2.52 describes the absolute position of any point P of the rigid body, ${}^{A}\boldsymbol{P}$, by the relative position of that point with respect to the moving frame, ${}^{B}\boldsymbol{P}$, and the location or pose of the rigid body represented by the set $\{{}^{A}\boldsymbol{R}_B, {}^{A}\boldsymbol{P}_{O_B}\}$. In this equation, the leading superscript of ${}^{B}\boldsymbol{P}$ cancels with the trailing subscript of ${}^{A}\boldsymbol{R}_B$, leaving all quantities as vectors expressed in the fixed frame $\{\boldsymbol{A}\}$.

When the motion of a rigid body is expressed in this way, since it is assumed that initially the moving frame coincides with the fixed frame, ${}^{B}\boldsymbol{P}$ may be considered as the position of a point P before the motion, while ${}^{A}\boldsymbol{P}$ is the position of the same point after the motion. Therefore, as suggested in Equation 2.52, the motion of the rigid body is divided into two parts. The first term in the right-hand side of Equation 2.52, represents the contribution due to a pure rotation of the rigid body, while the second term represents the contribution due to the translation along the vector ${}^{A}\boldsymbol{P}_{O_B}$. This relation is well known as *Chasles' theorem*, that general spatial displacement of a rigid body can be considered as a rotation plus a translation.

Example 2.14

Consider a rigid body in an initial state represented by frame {0}, and a point P on the rigid body with the coordinates $\boldsymbol{P} = [0.1 \;\; 0.1 \;\; 0]^T$, as shown in Figure 2.11. Furthermore, consider the rigid body is moved to a location represented by frame {1}. The absolute position vector of point P is denoted by ${}^{0}\boldsymbol{P}$, and can be calculated by Equation 2.52 as

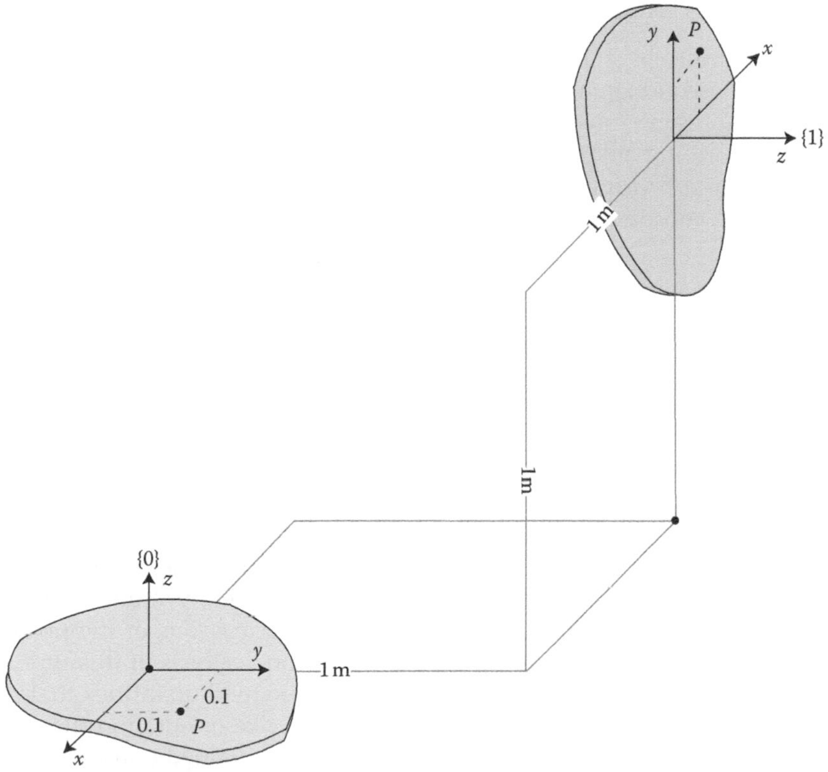

FIGURE 2.11
Example of a rigid body motion.

follows. The motion of the rigid body can be divided into a translation represented by the vector ${}^0\boldsymbol{P}_{O_1}$, and a rotation represented by ${}^0\boldsymbol{R}_1$ as

$$
{}^0\boldsymbol{P}_{O_1} = \begin{bmatrix} -1 \\ 1 \\ 1 \end{bmatrix}; \quad {}^0\boldsymbol{R}_1 = \begin{bmatrix} -1 & 0 & 0 \\ 0 & 0 & 1 \\ 0 & 1 & 0 \end{bmatrix}.
$$

Therefore, absolute position of point P with respect to the fixed frame $\{0\}$ is

$$
\begin{aligned}
{}^0\boldsymbol{P} &= {}^0\boldsymbol{R}_1\, {}^1\boldsymbol{P} + {}^0\boldsymbol{P}_{O_1} \\
&= \begin{bmatrix} -1 & 0 & 0 \\ 0 & 0 & 1 \\ 0 & 1 & 0 \end{bmatrix} \begin{bmatrix} 0.1 \\ 0.1 \\ 0 \end{bmatrix} + \begin{bmatrix} -1 \\ 1 \\ 1 \end{bmatrix} \\
&= \begin{bmatrix} -1.1 \\ 1.0 \\ 1.1 \end{bmatrix},
\end{aligned}
$$

which can be easily verified by inspection from Figure 2.11.

2.3 Homogeneous Transformations

Equation 2.52 describes a general transformation of a vector from its description in one frame to that in a second frame. This equation is not in a compact form, since the 3×3 rotation matrix represents only the orientation, it is good for pure rotations, and it is not capable of describing any translational motion. To write the mapping given in Equation 2.52 in a better-appearing form, the 3×1 Cartesian coordinates are generalized to a 4×1 homogeneous coordinates, and Equation 2.52 is generalized to a conceptual form of

$$
{}^A\boldsymbol{P} = {}^A\boldsymbol{T}_B\, {}^B\boldsymbol{P}, \tag{2.53}
$$

in which ${}^A\boldsymbol{T}_B$ is a 4×4 homogeneous transformation matrix.

2.3.1 Homogeneous Coordinates

In this chapter, we have primarily focused on position vectors. However, other quantities that are represented by vectors such as linear and angular velocities, linear and angular accelerations, forces, and moments have been discussed in the following chapters. There are two basic classes of vector quantities, the generalization to homogeneous coordinates of which are different. The first type is called *line vector*. Line vectors refer to a vector-valued quantity the amount of which depends on the *line of action*, or the position where the vector is applied. As seen in this chapter, the position vector is of this type. Similarly, linear velocity, linear acceleration, and force are vector-valued quantities that depend on the line of action. On the contrary, there exist quantities like orientation that hold for the whole rigid body, and do not correspond to a point. Vector-valued quantities like angular velocity, angular accelerations, and moments are such quantities that hold for the whole rigid body, and can be positioned freely throughout the whole rigid body, without any change in their quantity. These types of vectors are called *free vectors*.

For line vectors, both orientation and translation of the moving frame contribute to their value. Hence, in order to use Equation 2.53, the 4×1 homogeneous coordinate of such vectors is generated by appending 1 to the three components of that vector as follows:

$$\text{For line vectors } \boldsymbol{V} = \left[\begin{array}{c} v_x \\ v_y \\ v_z \\ \hline 1 \end{array}\right]. \tag{2.54}$$

For free vectors, only the orientation of the moving frame contributes to their value. Hence, in order to use Equation 2.53, the 4×1 homogeneous coordinate of such vectors is generated by appending 0 to the three components of that vector as follows:

$$\text{For free vectors } \boldsymbol{\omega} = \left[\begin{array}{c} \omega_x \\ \omega_y \\ \omega_z \\ \hline 0 \end{array}\right]. \tag{2.55}$$

The convention adopted in here is that a vector-valued quantity is represented by a 3×1 or 4×1 vector, depending on whether it is multiplied by a 3×3 rotation matrix or a 4×4 homogeneous transformation matrix.

2.3.2 Homogeneous Transformation Matrix

The *homogeneous transformation matrix* is a 4×4 matrix, defined for the purpose of transformation mapping of a vector in a homogeneous coordinate from one frame to another in a compact form as depicted in Equation 2.53. The matrix is composed of the rotation matrix ${}^{A}\boldsymbol{R}_{B}$, representing the orientation, and the position vector ${}^{A}\boldsymbol{P}_{O_B}$, representing the translation partitioned as follows:

$${}^{A}\boldsymbol{T}_{B} = \left[\begin{array}{ccc|c} & {}^{A}\boldsymbol{R}_{B} & & {}^{A}\boldsymbol{P}_{O_B} \\ \hline 0 & 0 & 0 & 1 \end{array}\right]. \tag{2.56}$$

The homogeneous transformation matrix ${}^{A}\boldsymbol{T}_{B}$ is a 4×4 matrix operator mapping vector-valued quantities represented by 4×1 homogeneous coordinates. Using the homogeneous coordinates for a line vector like position, the transformation Equation 2.53 yields

$$\left[\begin{array}{c} {}^{A}\boldsymbol{P} \\ \hline 1 \end{array}\right] = \left[\begin{array}{ccc|c} & {}^{A}\boldsymbol{R}_{B} & & {}^{A}\boldsymbol{P}_{O_B} \\ \hline 0 & 0 & 0 & 1 \end{array}\right] \left[\begin{array}{c} {}^{B}\boldsymbol{P} \\ \hline 1 \end{array}\right] \tag{2.57}$$

$${}^{A}\boldsymbol{P} = {}^{A}\boldsymbol{R}_{B}\,{}^{B}\boldsymbol{P} + {}^{A}\boldsymbol{P}_{O_B} \tag{2.58}$$

$$1 = 1.$$

As seen, using homogeneous transformation matrix, the compact form of mapping Equation 2.53 leads to the same result obtained in Equation 2.52.

Using homogeneous coordinates for a free vector like angular velocity of a rigid body, the transformation Equation 2.53 yields

$$\left[\begin{array}{c} {}^{A}\boldsymbol{\omega} \\ \hline 0 \end{array}\right] = \left[\begin{array}{ccc|c} & {}^{A}\boldsymbol{R}_{B} & & {}^{A}\boldsymbol{P}_{O_B} \\ \hline 0 & 0 & 0 & 1 \end{array}\right] \left[\begin{array}{c} {}^{B}\boldsymbol{\omega} \\ \hline 0 \end{array}\right] \tag{2.59}$$

$$\begin{aligned} {}^{A}\boldsymbol{\omega} &= {}^{A}\boldsymbol{R}_{B}\,{}^{B}\boldsymbol{\omega} \\ 0 &= 0, \end{aligned} \tag{2.60}$$

which confirms the property of free vectors in general motions that only change in orientation affects their representation, while translation of the frame does not.

Example 2.15

Consider the general motion of the rigid body shown in Figure 2.11. Furthermore, assume that the position of point P with respect to the moving frame {1}, denoted by the vector ${}^{1}\boldsymbol{P}$, and the angular velocity of the rigid body with respect to frame {1}, denoted by ${}^{1}\boldsymbol{\omega}$ are given as

$${}^{1}\boldsymbol{P} = \begin{bmatrix} 0.1 \\ 0.1 \\ 0 \end{bmatrix}; \quad {}^{1}\boldsymbol{\omega} = \begin{bmatrix} 1 \\ 0 \\ 1.2 \end{bmatrix}.$$

In order to obtain the absolute position vector ${}^{0}\boldsymbol{P}$ and the angular velocity of the rigid body represented in frame {0}, namely ${}^{0}\boldsymbol{\omega}$, the homogeneous transformation can be used by Equation 2.53. For this example, the homogeneous transformation matrix is simplified to

$${}^{0}\boldsymbol{T}_{1} = \left[\begin{array}{ccc|c} & {}^{0}\boldsymbol{R}_{1} & & {}^{0}\boldsymbol{P}_{O_1} \\ \hline 0 & 0 & 0 & 1 \end{array}\right] = \left[\begin{array}{ccc|c} -1 & 0 & 0 & 1 \\ 0 & 0 & 1 & 1 \\ 0 & 1 & 0 & 1 \\ \hline 0 & 0 & 0 & 1 \end{array}\right].$$

Furthermore, since the position vector is a line vector while angular velocity is a free vector, their corresponding homogeneous coordinates are given as

$${}^{1}\boldsymbol{P} = \left[\begin{array}{c} 0.1 \\ 0.1 \\ 0 \\ \hline 1 \end{array}\right]; \quad {}^{1}\boldsymbol{\omega} = \left[\begin{array}{c} 1 \\ 0 \\ 1.2 \\ \hline 0 \end{array}\right].$$

Therefore, ${}^{0}\boldsymbol{P}$ and ${}^{0}\boldsymbol{\omega}$ are calculated as follows:

$${}^{0}\boldsymbol{P} = {}^{0}\boldsymbol{T}_{1}\,{}^{1}\boldsymbol{P} = \left[\begin{array}{c} -1.1 \\ 1.0 \\ 1.1 \\ \hline 1 \end{array}\right]; \quad {}^{0}\boldsymbol{\omega} = {}^{0}\boldsymbol{T}_{1}\,{}^{1}\boldsymbol{\omega} = \left[\begin{array}{c} -1 \\ 1.2 \\ 0 \\ \hline 0 \end{array}\right].$$

2.3.3 Screw Displacement

As discussed earlier, *Chasles' theorem* states that the most general rigid body displacement can be produced by a translation along a line followed by a rotation about the same line. Because this displacement is reminiscent of the displacement of a screw, it is called a *screw displacement*, and the line or axis is called the *screw axis*. Earlier in this chapter, we have represented pure rotation by an equivalent angle of rotation about the screw axis. Now consider a general motion of the rigid body, wherein a combination of translation and rotation has occurred. Figure 2.12 shows such a general motion in which the rigid body has moved from an initial state represented by frame $\{A\}$ to a final state which is represented by frame $\{B\}$. An intermediate frame $\{A'\}$ is considered in this figure in order to separate the translational motion of the rigid body from the general motion. As depicted in Figure 2.12, frame $\{A'\}$ coincides with the origin of frame $\{B\}$, but it is parallel to $\{A\}$. By this means, the translational motion of the rigid body is represented by frame $\{A'\}$, which represents the motion with the translation vector of ${}^{A}\boldsymbol{P}_{O_B}$.

Rotation of the rigid body can be considered as the relative motion of the two frames $\{A'\}$ and $\{B\}$. As studied in Section 2.1.2.3, the pure rotation of the rigid body can be represented by a screw axis of rotation $\hat{\mathbf{s}}$, and an equivalent angle of rotation θ, shown in Figure 2.12. The screw axis coordinates $\hat{\mathbf{s}} = [s_x, s_y, s_z]^T$, and the equivalent angle of rotation θ may be derived using Equations 2.36 and 2.20, respectively. This representation can be generalized to incorporate the translational motion as well. According to the Chasles' theorem, the general rigid body displacement can be represented by a translation along the same screw axis. However, the translation of frame $\{A'\}$, with respect to frame $\{A\}$, has been earlier represented by the vector ${}^{A}\boldsymbol{P}_{O_B}$, which is not necessarily along the screw axis. In order to

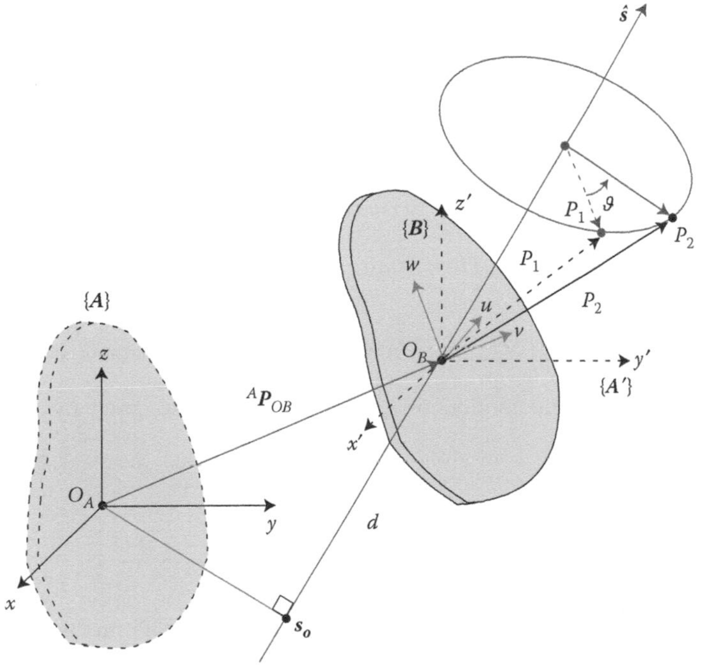

FIGURE 2.12
General motion of a rigid body represented by screw displacement.

represent the translational part of the general motion along the screw axis, it is sufficient to use the projection of vector ${}^A\mathbf{P}_{O_B}$ on the screw axis $\hat{\mathbf{s}}$, denoted by a scalar d (Figure 2.12). Furthermore, contrary to the pure rotation case, depicted in Figure 2.5, the screw axis does not pass through the origin of the fixed frame $\{\boldsymbol{A}\}$. Any point $\mathbf{s}_o$ on the screw axis can be chosen to represent its distance to the frame $\{\boldsymbol{A}\}$, whereas we may choose $\mathbf{s}_o$ coincident with O_B, the origin of the frame $\{\boldsymbol{B}\}$. Hence, the general motion of the rigid body can be represented by two scalars, namely θ and d, and two vectors, namely $\hat{\mathbf{s}}$ and $\mathbf{s}_o$, where

$$\hat{\mathbf{s}}^T\hat{\mathbf{s}} = 1; \quad \mathbf{s}_o^T\hat{\mathbf{s}} = d. \tag{2.61}$$

If the transformation matrix is given to represent the general motion of a rigid body, the screw displacement parameters can be easily obtained as follows. The rotation parameters, namely θ and $\hat{\mathbf{s}}$, are found by Equations 2.20 and 2.36, respectively. The translation parameters are derived by

$$\mathbf{s}_o = {}^A\mathbf{P}_{O_B}; \quad d = \mathbf{s}_o^T\hat{\mathbf{s}}. \tag{2.62}$$

Otherwise, if the screw displacement parameters

$$\{\theta, \hat{\mathbf{s}} = [s_x, s_y, s_z]^T; \quad d, \mathbf{s}_o = [s_{o_x}, s_{o_y}, s_{o_z}]^T\} \tag{2.63}$$

are given, the transformation matrix is determined by

$$ {}^A\boldsymbol{T}_B = \left[\begin{array}{ccc|c} s_x^2 v\theta + c\theta & s_x s_y v\theta - s_z s\theta & s_x s_z v\theta + s_y s\theta & p_x \\ s_y s_x v\theta + s_z s\theta & s_y^2 v\theta + c\theta & s_y s_z v\theta - s_x s\theta & p_y \\ s_z s_x v\theta - s_y s\theta & s_z s_y v\theta + s_x s\theta & s_z^2 v\theta + c\theta & p_z \\ \hline 0 & 0 & 0 & 1 \end{array}\right] \tag{2.64}$$

in which

$$\begin{aligned} p_x &= d\, s_x - s_{o_x}(s_x^2 - 1)v\theta - s_{o_y}(s_x s_y v\theta - s_z s\theta) - s_{o_z}(s_x s_z v\theta + s_y s\theta) \\ p_y &= d\, s_y - s_{o_x}(s_y s_x v\theta + s_z s\theta) - s_{o_y}(s_y^2 - 1)v\theta - s_{o_z}(s_y s_z v\theta - s_x s\theta) \\ p_z &= d\, s_z - s_{o_x}(s_z s_x v\theta - s_y s\theta) - s_{o_y}(s_z s_y v\theta + s_x s\theta) - s_{o_z}(s_z^2 - 1)v\theta \end{aligned} \tag{2.65}$$

and as herein before, $s\theta = \sin\theta, c\theta = \cos\theta$, and $v\theta = 1 - \cos\theta$ are used as shorthand notations.

Example 2.16: Screw Displacement

Consider that the general motion of a rigid body is represented by the following transformation matrix:

$$ {}^A\boldsymbol{T}_B = \left[\begin{array}{ccc|c} +0.7500 & 0.6124 & -0.2500 & +1.4142 \\ -0.6124 & 0.5000 & -0.6124 & -1.5764 \\ -0.2500 & 0.6124 & +0.7500 & -1.4142 \\ \hline 0 & 0 & 0 & 1 \end{array}\right].$$

The screw displacement parameters are determined as follows. From the rotation matrix ${}^A\boldsymbol{R}_B$, θ and $\hat{\mathbf{s}}$ are derived by using Equations 2.20 and 2.36, respectively.

$$\left\{\theta = \frac{\pi}{3}; \quad \hat{\mathbf{s}} = \frac{1}{\sqrt{2}}[1, 0, -1]^T\right\}.$$

Furthermore, using Equation 2.62, d and s_O are determined as follows:

$$\left\{d = 2; \quad s_O = [1.4142, -1.5764, -1.4142]^T\right\}.$$

Example 2.17

Consider that the general motion of a rigid body is represented by the following screw displacement parameters:

$$\left\{\theta = \frac{\pi}{4}, \hat{s} = \frac{1}{\sqrt{6}}[1, -2, -1]^T; \quad d = 0.5, s_O = [0, 0, 1]^T\right\}.$$

Note that, in this example, s_O is considered as an arbitrary point in the space. For this case, Equation 2.64 is used to determine the transformation matrix as follows:

$$^{A}\boldsymbol{T}_B = \left[\begin{array}{ccc|c} 0.7559 & 0.1910 & -0.6262 & 0.8303 \\ -0.3863 & 0.9024 & -0.1910 & -0.2172 \\ 0.5285 & 0.3863 & 0.7559 & 0.0400 \\ \hline 0 & 0 & 0 & 1 \end{array}\right].$$

2.3.4 Transformation Arithmetics

In this section, the required arithmetic to generate the total transformation matrix representing a number of consecutive transformations is described. Furthermore, rules to invert the transformations will be given.

2.3.4.1 Consecutive Transformations

As shown in Figure 2.13, consider the motion of a rigid body described at three locations. Frame $\{A\}$ represents the initial location, frame $\{B\}$ is an intermediate location, and frame

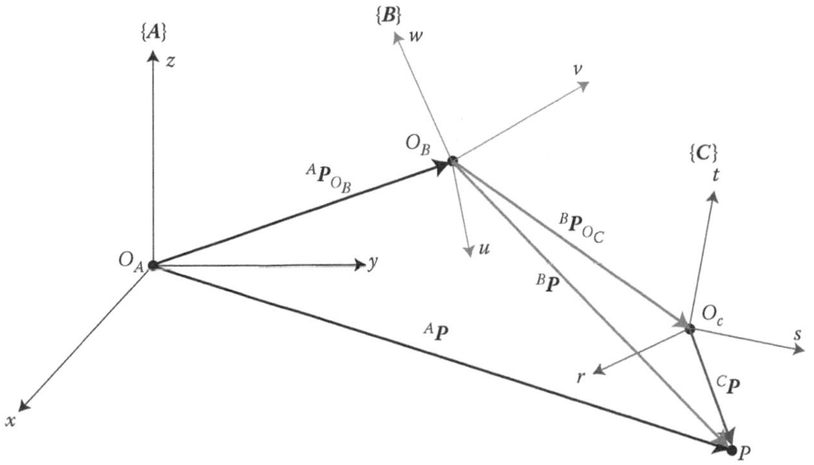

FIGURE 2.13
Motion of a rigid body represented at three locations by frames $\{A\}$, $\{B\}$, and $\{C\}$.

$\{C\}$ represents the rigid body at its final location. Furthermore, suppose the position vector of a point P of the rigid body is given in the final location, that is ${}^C\boldsymbol{P}$ is given, and the position of this point is to be found in the fixed frame $\{A\}$, that is ${}^A\boldsymbol{P}$. Since the locations of the rigid body is known relative to each other, ${}^C\boldsymbol{P}$ can be transformed to ${}^B\boldsymbol{P}$ using transformation matrix ${}^B\boldsymbol{T}_C$ as

$$ {}^B\boldsymbol{P} = {}^B\boldsymbol{T}_C\,{}^C\boldsymbol{P}. \tag{2.66}$$

Now, ${}^B\boldsymbol{P}$ can be transformed into ${}^A\boldsymbol{P}$, using transformation matrix ${}^A\boldsymbol{T}_B$ as

$$ {}^A\boldsymbol{P} = {}^A\boldsymbol{T}_B\,{}^B\boldsymbol{P}. \tag{2.67}$$

Combining Equations 2.66 and 2.67 yields

$$ {}^A\boldsymbol{P} = {}^A\boldsymbol{T}_B\,{}^B\boldsymbol{T}_C\,{}^C\boldsymbol{P}. \tag{2.68}$$

From which, the consecutive transformation can be defined as follows:

$$ {}^A\boldsymbol{T}_C = {}^A\boldsymbol{T}_B\,{}^B\boldsymbol{T}_C. \tag{2.69}$$

The notation used for transformation matrices makes such manipulations simple. If ${}^A\boldsymbol{T}_C$ is to be derived from the known descriptions of the frames $\{B\}$ and $\{C\}$, the following relation can be used:

$$ {}^A\boldsymbol{T}_C = \left[\begin{array}{ccc|c} & {}^A\boldsymbol{R}_B\,{}^B\boldsymbol{R}_C & & {}^A\boldsymbol{R}_B\,{}^B\boldsymbol{P}_{O_C} + {}^A\boldsymbol{P}_{O_B} \\ \hline 0 & 0 & 0 & 1 \end{array}\right]. \tag{2.70}$$

Furthermore, if more than two consecutive transformations exist, relation 2.69 can be generalized to

$$ {}^A\boldsymbol{T}_Z = {}^A\boldsymbol{T}_B\,{}^B\boldsymbol{T}_C\,{}^C\boldsymbol{T}_D \cdots {}^Y\boldsymbol{T}_Z. \tag{2.71}$$

Example 2.18: Consecutive Transformations

Consider three locations of a rigid body shown in Figure 2.14, and represented by frames $\{0\}$, $\{1\}$, and $\{2\}$. The homogeneous transfer matrices ${}^0\boldsymbol{T}_1$, ${}^1\boldsymbol{T}_2$, and ${}^0\boldsymbol{T}_2$ can be determined by Equation 2.56 as follows:

$$ {}^0\boldsymbol{T}_1 = \left[\begin{array}{ccc|c} 0 & -1 & 0 & 0 \\ -1 & 0 & 0 & 1 \\ 0 & 0 & -1 & 1 \\ \hline 0 & 0 & 0 & 1 \end{array}\right], $$

$$ {}^1\boldsymbol{T}_2 = \left[\begin{array}{ccc|c} 1 & 0 & 0 & 0 \\ 0 & -1 & 0 & 1 \\ 0 & 0 & -1 & 1 \\ \hline 0 & 0 & 0 & 1 \end{array}\right], $$

$$ {}^0\boldsymbol{T}_2 = \left[\begin{array}{ccc|c} 0 & 1 & 0 & -1 \\ -1 & 0 & 0 & 1 \\ 0 & 0 & 1 & 0 \\ \hline 0 & 0 & 0 & 1 \end{array}\right]. $$

Multiply ${}^0\boldsymbol{T}_1$ by ${}^1\boldsymbol{T}_2$, and show that ${}^0\boldsymbol{T}_2 = {}^0\boldsymbol{T}_1\,{}^1\boldsymbol{T}_2$ as given in Equation 2.71.

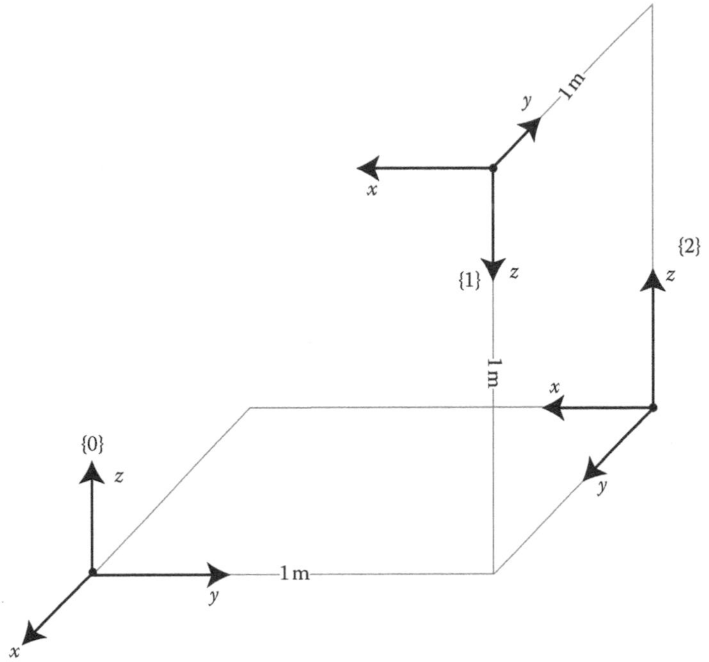

FIGURE 2.14
Motion of a rigid body represented in three locations by frames {0}, {1}, and {2}.

Example 2.19: Finite Angle Rotations

In this example, it is shown that finite angle rotations cannot be represented by vectors since they do not generally commute. Moreover, it is shown that in contrast to finite rotations, infinitesimal rotations can be represented by a vector $\delta\boldsymbol{\theta}$ since they commute.

Consider two consecutive rotations, first about $\hat{x}$ with a finite angle θ_1 and then about $\hat{z}$ with another finite angle θ_2. The final rotation matrix is derived by using consecutive transformation formula 2.71 as follows:

$$\boldsymbol{R}_x(\theta_1) = \begin{bmatrix} 1 & 0 & 0 \\ 0 & c_1 & -s_1 \\ 0 & s_1 & c_1 \end{bmatrix},$$

$$\boldsymbol{R}_z(\theta_2) = \begin{bmatrix} c_2 & -s_2 & 0 \\ s_2 & c_2 & 0 \\ 0 & 0 & 1 \end{bmatrix},$$

$$\boldsymbol{R}_1 = \boldsymbol{R}_x(\theta_1)\,\boldsymbol{R}_z(\theta_2) = \begin{bmatrix} c_2 & -s_2 & 0 \\ c_1s_2 & c_1c_2 & -s_1 \\ s_1s_2 & s_1c_2 & c_1 \end{bmatrix},$$

in which $c_1 = \cos(\theta_1), s_1 = \sin(\theta_1), c_2 = \cos(\theta_2)$, and $s_2 = \sin(\theta_2)$ are used as shorthand notations. Since finite rotations are generally represented by a rotation matrix, and matrix multiplication does not commute, if the order of rotations is changed, the resulting orientation is represented by

$$\boldsymbol{R}_2 = \boldsymbol{R}_z(\theta_2)\,\boldsymbol{R}_x(\theta_1) = \begin{bmatrix} c_2 & -c_1s_2 & s_1s_2 \\ s_2 & c_1c_2 & -s_1c_2 \\ 0 & s_1 & c_1 \end{bmatrix}.$$

Therefore, generally, $R_1 \neq R_2$. In other words, if the order of finite rotations changes, the resulting motion is different. This case can be used as a counterexample that finite rotations cannot be represented by vectors. However, if rotations become infinitesimal, $\theta_1 = \delta\theta_1$, and $\theta_2 = \delta\theta_2$, then the approximations $s_1 \simeq \delta\theta_1, c_1 \simeq 1, s_2 \simeq \delta\theta_2$, and $c_2 \simeq 1$ can be used in corresponding rotation matrices. Therefore, for infinitesimal rotations, we have

$$R_1 = R_x(\theta_1)\, R_z(\theta_2) = \begin{bmatrix} 1 & -\delta\theta_2 & 0 \\ \delta\theta_2 & 1 & -\delta\theta_1 \\ \delta\theta_1\delta\theta_2 & \delta\theta_1 & 1 \end{bmatrix},$$

$$R_2 = R_z(\theta_2)\, R_x(\theta_1) = \begin{bmatrix} 1 & -\delta\theta_2 & \delta\theta_1\delta\theta_2 \\ \delta\theta_2 & 1 & -\delta\theta_1 \\ 0 & \delta\theta_1 & 1 \end{bmatrix},$$

where for small $\delta\theta_i$'s, the higher-order terms can be neglected, that is, $\delta\theta_1\delta\theta_2 \simeq 0$. Therefore, in this case (and in general) $R_1 = R_2$. Hence, infinitesimal rotation angles commute, and therefore, can be represented by vectors.

2.3.4.2 Inverse Transformation

Direct inversion of a 4×4 homogeneous transfer matrix ${}^A T_B$, to obtain ${}^B T_A$, might be computationally intensive. Since the homogeneous transfer matrix is composed of a rotation matrix, a position vector describing the origin of frame $\{B\}$ with respect to frame $\{A\}$, and a row with constant values, it is much easier to use this structure for inversion. In order to obtain ${}^B T_A$ using this structure, it is necessary to compute ${}^B R_A$ and ${}^B P_{OA}$ from the known ${}^A R_B$ and ${}^A P_{OB}$ and generate ${}^B T_A$ as follows:

$${}^B T_A = \left[\begin{array}{ccc|c} & {}^B R_A & & {}^B P_{OA} \\ \hline 0 & 0 & 0 & 1 \end{array}\right]. \tag{2.72}$$

First recall that ${}^B R_A = {}^A R_B^T$. Next, describe the position vector of the origin of frame $\{A\}$ in $\{B\}$ as follows:

$$\begin{aligned} {}^A P_{O_A} &= -{}^A P_{O_B} \\ {}^B P_{O_A} &= {}^B R_A\, {}^A P_{O_A} = -{}^B R_A\, {}^A P_{O_B} \\ &= -{}^A R_B^T\, {}^A P_{O_B} \end{aligned} \tag{2.73}$$

Hence, the inverse of the transformation matrix can be obtained by

$${}^B T_A = {}^A T_B^{-1} = \left[\begin{array}{ccc|c} & {}^A R_B^T & & -{}^A R_B^T\, {}^A P_{O_B} \\ \hline 0 & 0 & 0 & 1 \end{array}\right]. \tag{2.74}$$

Equation 2.74 is a general and computationally efficient way of computing the inverse of a homogeneous transformation matrix.

Example 2.20

The orientation of an object frame {1} with respect to frame {0} is given by the following rotation matrix:

$$^{0}\mathbf{R}_1 = \begin{bmatrix} 0.933 & 0.167 & 0.354 \\ 0.067 & 0.933 & -0.354 \\ -0.354 & 0.354 & 0.866 \end{bmatrix}.$$

We suspect that the elements of $^{0}\mathbf{R}_1$ may be wrong, and check the orthonormality of its column vectors. It can be easily verified that the norm of the first and the third column vectors are equal to one, but the norm of the second column vector is greater than one. By inspection, it can be seen that r_{12} is mistyped and should be equal to 0.067.

Furthermore, consider that the origin of frame {1} is represented by the position vector $^{0}\mathbf{P}_{O_1} = \begin{bmatrix} 2 & -1 & 0 \end{bmatrix}^T$. If the position vector of point P expressed in frame {0} is $^{0}\mathbf{P} = \begin{bmatrix} 0.25 & 0.43 & 0.86 \end{bmatrix}^T$, the position vector $^{1}\mathbf{P}$ of the same point P in frame {1} can be obtained as follows:

$$^{1}\mathbf{P} = {}^{1}\mathbf{T}_0\, {}^{0}\mathbf{P} = {}^{0}\mathbf{T}_1^{-1}\, {}^{0}\mathbf{P}$$

In order to complete the calculation, it is necessary to determine the inverse of the homogeneous transfer matrix $^{0}\mathbf{T}_1$. The inversion can be performed using Equation 2.74 as follows:

$$^{1}\mathbf{T}_0 = {}^{0}\mathbf{T}_1^{-1} = \left[\begin{array}{ccc|c} & {}^{0}\mathbf{R}_1^T & & -{}^{0}\mathbf{R}_1^T\, {}^{0}\mathbf{P}_{O_1} \\ \hline 0 & 0 & 0 & 1 \end{array}\right]$$

$$= \left[\begin{array}{ccc|c} 0.933 & 0.067 & -0.354 & -1.80 \\ 0.067 & 0.933 & 0.354 & 0.80 \\ 0.354 & -0.354 & 0.866 & -1.06 \\ \hline 0 & 0 & 0 & 1 \end{array}\right].$$

Hence, $^{1}\mathbf{P} = \begin{bmatrix} -1.84 & 1.52 & -0.38 \end{bmatrix}^T$.

PROBLEMS

1. Consider the frames $\{A\}$, $\{B\}$, and $\{C\}$ as depicted in Figure 2.15. Give the values of $^{A}\mathbf{R}_B, {}^{A}\mathbf{R}_C$, and $^{B}\mathbf{R}_C$.
2. For the frames defined in Figure 2.15, find the following rotation matrices $^{B}\mathbf{R}_A, {}^{C}\mathbf{R}_A$, and $^{C}\mathbf{R}_B$ by definition. Verify the results by numerical inversion of the rotation matrices found in Problem 1.
3. Repeat Problem 1 for the frames given in Figure 2.16.
4. Repeat Problem 2 for the frames given in Figure 2.16.
5. Repeat Problem 1 for the frames given in Figure 2.17.
6. Repeat Problem 2 for the frames given in Figure 2.17.
7. The rotation of a rigid body is represented by the following rotation matrix:

$$^{A}\mathbf{R}_B = \begin{bmatrix} 0.5000 & 0.0795 & 0.8624 \\ 0.5000 & 0.7866 & -0.3624 \\ -0.7071 & 0.6124 & 0.3536 \end{bmatrix}$$

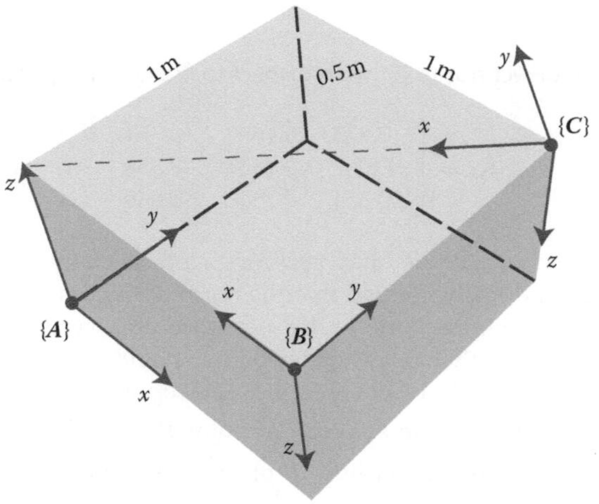

FIGURE 2.15
Frames $\{A\}$, $\{B\}$, and $\{C\}$ at the corners of a rectangular box.

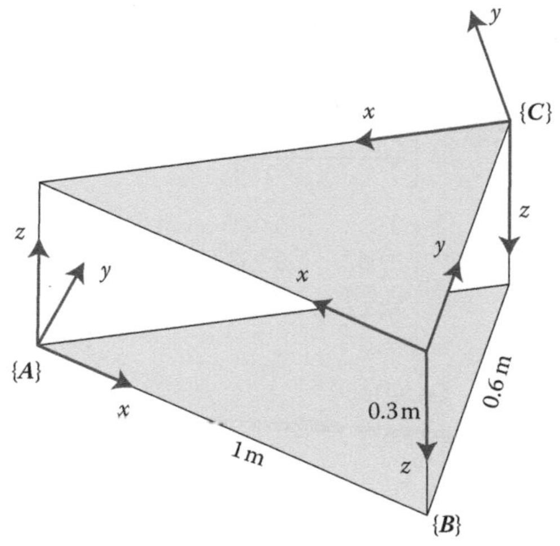

FIGURE 2.16
Frames $\{A\}$, $\{B\}$, and $\{C\}$ at the corners of a wedge.

a. If the position vector ${}^{B}\boldsymbol{P} = [0.7071 \quad -0.7071 \quad 0]^T$ is given, find ${}^{A}\boldsymbol{P}$.
b. If the position vector ${}^{A}\boldsymbol{P} = [0.5774 \quad -0.5774 \quad -0.5774]^T$ is given, find ${}^{B}\boldsymbol{P}$.
c. Find the eigenvalues and eigenvectors of ${}^{A}\boldsymbol{R}_{B}$.
d. Give the screw axis representation of the rotation, namely the equivalent angle of rotation θ, and the components of the screw axis ${}^{A}\hat{\mathbf{s}} = [s_x \quad s_y \quad s_z]^T$. Verify the results by the eigenvalues and eigenvectors of the rotation matrix obtained in part (c).
e. Give the equivalent pitch–roll–yaw Euler angles for this rotation.
f. Give the equivalent w–v–u Euler angles that can represent this rotation.

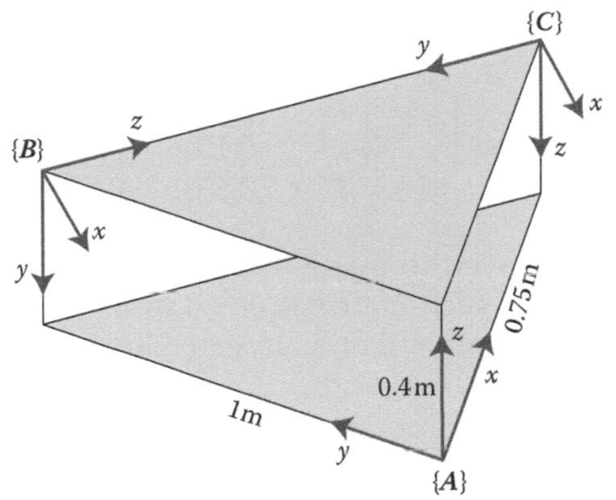

FIGURE 2.17
Frames $\{A\}$, $\{B\}$, and $\{C\}$ at the corners of a wedge.

8. Repeat Problem 7 for the following rotation matrix:

$$^{A}\mathbf{R}_B = \begin{bmatrix} 0.5000 & 0.0795 & 0.8624 \\ 0.5000 & 0.7866 & -0.3624 \\ -0.7071 & 0.6124 & 0.3536 \end{bmatrix}.$$

9. The rotation of a rigid body is represented by the following rotation matrix:

$$^{A}\mathbf{R}_B = \begin{bmatrix} -0.1464 & -0.8536 & 0.5000 \\ 0.8536 & 0.1464 & 0.5000 \\ -0.5000 & 0.5000 & 0.7071 \end{bmatrix}.$$

a. Find the eigenvalues and eigenvectors of $^{A}\mathbf{R}_B$.
b. Give the screw axis representation of the rotation, namely the equivalent angle of rotation θ and the components of the screw axis $^{A}\hat{\mathbf{s}} = [s_x \quad s_y \quad s_z]^T$. Verify the results by the eigenvalues and eigenvectors of the rotation matrix obtained in part (c).
c. Give the equivalent w–v–w Euler angles for this rotation.
d. Give the equivalent w–v–u Euler angles that can represent this rotation.

10. Repeat Problem 9 for the following rotation matrix:

$$^{A}\mathbf{R}_B = \begin{bmatrix} 0 & 0.2588 & 0.9659 \\ 0 & 0.9659 & -0.2588 \\ -1.0000 & 0 & 0 \end{bmatrix}.$$

11. A rotation matrix is given as follows. We suspect that there are few errors in logging this rotation. Using the properties of a rotation matrix, find the errors and correct them:

$$^{A}\mathbf{R}_B = \begin{bmatrix} 0.0990 & -0.3696 & 0.9239 \\ 0.9839 & -0.1124 & -0.1464 \\ 0.1487 & 0.9235 & 0.3536 \end{bmatrix}$$

12. Repeat Problem 11 for the following rotation matrix:

$$^{A}\boldsymbol{R}_B = \begin{bmatrix} -0.8001 & -0.4975 & 0.3351 \\ 0.4619 & 0.1546 & 0.8733 \\ -0.3827 & 0.6536 & 0.3536 \end{bmatrix}$$

13. In this problem, it is intended to prove that a rotation matrix is an orthonormal matrix, and the inverse of such matrix is equal to its transpose. For this reason, suppose that two unit vectors $\hat{s}$ and $\hat{r}$ are embedded in a rigid body, and note that the geometric angle between such vectors is preserved for any rotations. Use this fact to provide a concise proof for the orthonormality condition of a rotation matrix.
14. Suppose the rotation of a rigid body is represented by the following pitch–roll–yaw Euler angles $\boldsymbol{R}_x(\pi/6), \boldsymbol{R}_y(\pi/4)$, and $\boldsymbol{R}_z(\pi/4)$.
    a. Find the corresponding rotation matrix $^{A}\boldsymbol{R}_B$.
    b. Give the screw axis representation of the rotation, namely the equivalent angle of rotation θ and the components of the screw axis $^{A}\hat{\mathbf{s}} = [s_x \quad s_y \quad s_z]^T$.
    c. Give the equivalent w–u–w Euler angles for this rotation.
    d. Give the equivalent w–v–u Euler angles that can represent this rotation.
15. Suppose the rotation of a rigid body is represented by the following three rotations about fixed axes, $\boldsymbol{R}_z(\alpha), \boldsymbol{R}_y(\beta)$, and $\boldsymbol{R}_z(\gamma)$:
    a. Find the corresponding rotation matrix $^{A}\boldsymbol{R}_B$.
    b. Evaluate the rotation matrix for $\alpha = \pi/2, \beta = -\pi/4$, and $\gamma = \pi/6$.
    c. Give the screw axis representation of the rotation, namely the equivalent angle of rotation θ and the components of the screw axis $^{A}\hat{\mathbf{s}} = [s_x \quad s_y \quad s_z]^T$.
    d. Give the equivalent w–v–w Euler angles for this rotation.
    e. Give the equivalent w–v–u Euler angles that can represent this rotation.
16. Suppose the rotation of a rigid body is represented by the following three rotations about moving axes $\boldsymbol{R}_v(\alpha), \boldsymbol{R}_w(\beta)$, and $\boldsymbol{R}_u(\gamma)$:
    a. Find the corresponding rotation matrix $^{A}\boldsymbol{R}_B$.
    b. Evaluate the rotation matrix for $\alpha = \pi/4, \beta = -\pi/2$, and $\gamma = \pi/3$.
    c. Give the screw axis representation of the rotation, namely the equivalent angle of rotation θ, and the components of the screw axis $^{A}\hat{\mathbf{s}} = [s_x \quad s_y \quad s_z]^T$.
    d. Give the equivalent w–v–w Euler angles for this rotation.
    e. Give the equivalent w–u–w Euler angles that can represent this rotation.
17. Suppose the rotation of a rigid body is given by the following screw axis representation:

$$\left\{\theta = \pi/2, \hat{\mathbf{s}} = [-0.5774 \quad 0.5774 \quad 0.5774]^T\right\}.$$

    a. Find the corresponding rotation matrix $^{A}\boldsymbol{R}_B$.
    b. Give the equivalent pitch–roll–yaw Euler angles that represent this rotation.
    c. Give the equivalent w–v–w Euler angles for this rotation.
    d. Give the equivalent w–u–w Euler angles that can represent this rotation.

18. Consider the pure rotation of a rigid body that is represented by the screw axis parameters $\{\theta, \hat{\mathbf{s}} = [s_x \quad s_y \quad s_z]^T\}$. Show that the corresponding rotation matrix can be evaluated by the following matrix exponential function:

$$^{A}\mathbf{R}_B = e^{\hat{\mathbf{s}}_\times \theta}, \tag{2.75}$$

in which the matrix $\hat{\mathbf{s}}_\times$ is defined as

$$\hat{\mathbf{s}}_\times = \begin{bmatrix} 0 & -s_z & s_y \\ s_z & 0 & -s_x \\ -s_y & s_x & 0 \end{bmatrix}. \tag{2.76}$$

19. In this problem, *Euler parameters* or *unit quaternion* are introduced. Consider the pure rotation of a rigid body that is represented by the screw axis parameters $\{\theta, \hat{\mathbf{s}} = [s_x \quad s_y \quad s_z]^T\}$. Four Euler parameters are defined as

$$\epsilon_1 = s_x \sin(\theta/2); \quad \epsilon_2 = s_y \sin(\theta/2); \quad \epsilon_3 = s_z \sin(\theta/2); \quad \epsilon_4 = \cos(\theta/2). \tag{2.77}$$

Euler parameters can be put together into a 4×1 tuple

$$\boldsymbol{\epsilon} = \begin{bmatrix} \epsilon_1 & \epsilon_2 & \epsilon_3 & \epsilon_4 \end{bmatrix}^T,$$

and are also known as a *unit quaternion*. Since

$$\|\boldsymbol{\epsilon}\|^2 = \epsilon_1^2 + \epsilon_2^2 + \epsilon_3^2 + \epsilon_4^2 = 1$$

a. Show that the rotation matrix can be represented by Euler parameters as follows:

$$^{A}\mathbf{R}_B = \begin{bmatrix} 1 - 2\epsilon_2^2 - 2\epsilon_3^2 & 2(\epsilon_1\epsilon_2 - \epsilon_3\epsilon_4) & 2(\epsilon_1\epsilon_3 + \epsilon_2\epsilon_4) \\ 2(\epsilon_1\epsilon_2 + \epsilon_3\epsilon_4) & 1 - 2\epsilon_1^2 - 2\epsilon_3^2 & 2(\epsilon_2\epsilon_3 - \epsilon_1\epsilon_4) \\ 2(\epsilon_1\epsilon_3 - \epsilon_2\epsilon_4) & 2(\epsilon_2\epsilon_3 + \epsilon_1\epsilon_4) & 1 - 2\epsilon_1^2 - 2\epsilon_2^2 \end{bmatrix}. \tag{2.78}$$

b. Show that given a rotation matrix, the inverse solution for the Euler parameters are

$$\epsilon_1 = \frac{(r_{32} - r_{23})}{4\epsilon_4};$$
$$\epsilon_2 = \frac{(r_{13} - r_{31})}{4\epsilon_4};$$
$$\epsilon_3 = \frac{(r_{21} - r_{12})}{4\epsilon_4};$$
$$\epsilon_4 = \frac{1}{2}\sqrt{1 + r_{11} + r_{22} + r_{33}}.$$

20. Consider the frames $\{A\}$, $\{B\}$, and $\{C\}$ as depicted in Figure 2.15. Give the values of $^{A}\mathbf{T}_B$, $^{A}\mathbf{T}_C$, and $^{B}\mathbf{T}_C$.

21. For the frames defined in Figure 2.15, find the following homogeneous transformation matrices ${}^{B}\boldsymbol{T}_{A}$, ${}^{C}\boldsymbol{T}_{A}$, and ${}^{C}\boldsymbol{T}_{B}$, by definition. Verify the results through inversion of the homogeneous transformation matrices found in Problem 20, using the inversion formula 2.74.
22. Repeat Problem 20 for the frames given in Figure 2.16.
23. Repeat Problem 21 for the frames given in Figure 2.16.
24. Repeat Problem 20 for the frames given in Figure 2.17.
25. Repeat Problem 21 for the frames given in Figure 2.17.
26. For a general motion of rigid body depicted in Figure 2.12, show that the corresponding general *Rodrigues formula* similar to that given in Equation 2.33 is as follows:

$$ {}^{A}\boldsymbol{P} = \boldsymbol{s}_o + d\,\hat{\boldsymbol{s}} + ({}^{B}\boldsymbol{P} - \boldsymbol{s}_o)c\theta + \hat{\boldsymbol{s}} \times ({}^{B}\boldsymbol{P} - \boldsymbol{s}_o)s\theta + [({}^{B}\boldsymbol{P} - \boldsymbol{s}_o)^T\hat{\boldsymbol{s}}]v\theta\,\hat{\boldsymbol{s}}, $$

in which $c\theta = \cos\theta, s\theta = \sin\theta$, and $v\theta = 1 - \cos\theta$. Furthermore, prove Equation 2.65.

27. Given ${}^{A}\boldsymbol{T}_{B}$ as follows, find ${}^{B}\boldsymbol{T}_{A}$.

$$ {}^{A}\boldsymbol{T}_{B} = \left[\begin{array}{ccc|c} 0.2500 & 0.6124 & 0.7500 & 0.4223 \\ 0.6124 & 0.5000 & -0.6124 & 1.6717 \\ -0.7500 & 0.6124 & -0.2500 & -0.1723 \\ \hline 0 & 0 & 0 & 1 \end{array}\right]. $$

28. Consider the frames $\{A\}, \{B\}$, and $\{C\}$ as depicted in Figure 2.15. Furthermore, assume that the position vector of a point of the rigid body and the angular velocity of the rigid body with respect to frame $\{C\}$ are given as follows:

$$ {}^{C}\boldsymbol{P} = \begin{bmatrix} 1 \\ -1 \\ 1 \end{bmatrix}; \quad {}^{C}\boldsymbol{\omega} = \begin{bmatrix} 1 \\ 1.75 \\ -1.2 \end{bmatrix}. $$

Using the proper homogeneous transformations, find ${}^{B}\boldsymbol{P}$ and ${}^{B}\boldsymbol{\omega}$. Using these vectors and proper homogeneous transformations, find ${}^{A}\boldsymbol{P}$ and ${}^{A}\boldsymbol{\omega}$. Finally, evaluate ${}^{A}\boldsymbol{P}$ and ${}^{A}\boldsymbol{\omega}$ directly from the first vectors, using proper transformations, and verify your results.

29. Repeat Problem 28 using the frames given in Figure 2.16.
30. Repeat Problem 28 using the frames given in Figure 2.17.

3

Kinematics

3.1 Introduction

In this chapter kinematic analysis of parallel robots is discussed in detail. Kinematic analysis refers to the study of the geometry of motion of a robot, without considering the forces and torques that cause the motion. In this analysis, the relation between the geometrical parameters of the manipulator with the final motion of the moving platform is derived and analyzed.

A parallel robot is a mechanism with a number of closed kinematic chains, and its moving platform is linked to the base by several independent kinematic chains. This definition of parallel robots is very extensive, and it includes redundant parallel manipulators with more actuators than the number of controlled degrees-of-freedom as well as under constrained parallel manipulators. Parallel robots for which the number of kinematic chains is equal to the number of degrees-of-freedom of the moving platform are called fully parallel robots [58,133]. If in addition to this condition, the type and number of joints at each limb, and the number and location of the actuated joints are identical in all the limbs, such a parallel robot is called *symmetric*. Since kinematic analysis of symmetrical parallel manipulators is more convenient, this structure is used in many industrial robots, such as Stewart–Gough manipulator, or Delta robot.

There are three main cases for fully parallel manipulators. Planar robots with two translation and one rotational degree-of-freedom in the plane; spatial orientation manipulators with three rotational degrees-of-freedom in space, and a general spatial robot with three translational and three rotational degrees-of-freedom in space. Figure 3.1 illustrates a symmetrical planar parallel robot the limbs of which consist of an identical structure of three revolute joints, which can be named as $3\underline{R}RR$ manipulator. The underlined joint in this convention refers to the actuated one. Figure 3.32 shows a redundant $4\underline{R}RR$ similar to this manipulator, in which the first revolute joint is actuated. As another planar robot, consider the manipulator shown in Figure 1.7. This manipulator is a redundant manipulator with the structure of $4R\underline{P}R$, which means a symmetrical limb structure of $R\underline{P}R$ is used and the prismatic joint is actuated.

Examples of spatial orientation manipulators are shown in Figures 3.33 and 3.10, in which the first manipulator is a fully parallel three-degrees-of-freedom manipulator, while the latter is a redundant parallel manipulator. Both manipulators are designed to provide a pure orientation motion for their moving platform. The first manipulator shown in Figure 3.33, consists of two tetrahedrals connected to each other by a spherical joint, limiting the motion of the moving platform into three rotational degrees-of-freedom. The limbs in this robot have an $S\underline{P}S$ structure, in which their prismatic joint is actuated. Likewise, in the redundant manipulator shown in Figure 3.10, which is called hydraulic shoulder

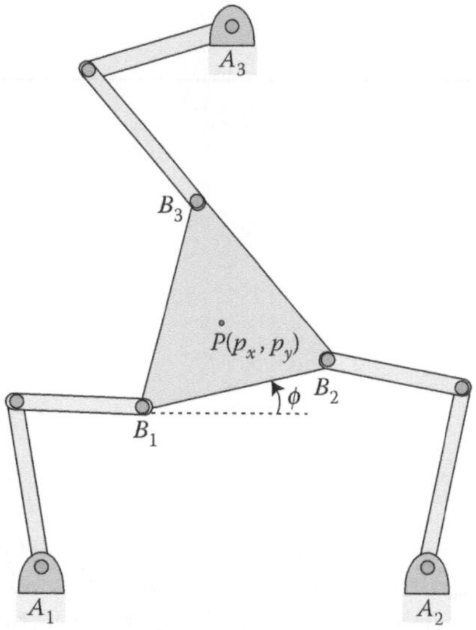

FIGURE 3.1
A symmetrical planar parallel robot.

manipulator, the passive link which is connected to the base with a spherical joint ($S6$), limits the motion of the moving platform to only three rotational degrees-of-freedom. However, in this manipulator, four limbs with identical structures of $U\underline{P}U$ are used to generate a redundant parallel robot. In this book, thorough kinematic analyses of both the redundant planar manipulator as well as the redundant spatial orientation manipulator are given, while the analysis of the fully parallel manipulators is left to the reader to be examined in the problems.

The Stewart–Gough platform (SGP) is also adopted for the analysis of fully parallel six-degrees-of-freedom manipulators. This manipulator consists of six identical limbs with an $S\underline{P}S$ structure. Although kinematic analysis of the SGP in its most general form has been considered by many researchers and fascinating results have been obtained [52,190], in this book, the closed-form solution of a special geometry of the SGP, which is called $6\text{–}6^P$ SGP is chosen. This is mainly due to the reason that the analytical forward kinematic solution of this manipulator is tractable and can be more easily followed by the reader than the other geometries. The $6\text{–}6^P$ SGP is shown in Figure 3.18.

It is known that unlike serial manipulators, inverse kinematic analysis of parallel robots is usually simple and straightforward. In most cases, limb variables may be computed independently using the given pose of the moving platform, and the solution in most cases even for redundant manipulators is uniquely determined. However, forward kinematics of parallel manipulators is generally very complicated, and its solution usually involves systems of nonlinear equations, which are highly coupled and in general have no closed form and unique solution [123,124]. Different approaches are given in the literature to solve this problem either in general [131] or in special cases [15,90]. In general, different solutions to the forward kinematic problem of parallel manipulators can be found by using numerical [131] or analytical approaches [90]. In what follows, a systematic means to perform kinematic analysis of parallel manipulators is given and this method is examined through the above-mentioned three selected case studies.

3.2 Loop Closure Method

As illustrated in Figure 3.2, a typical parallel manipulator consists of two main bodies. Body A is arbitrarily designated as fixed and is called the *base*, while body B is designed to be movable and is called the *moving platform*, or sometimes the *end effector*. These two bodies are coupled via n limbs, each attached to points A_i and B_i, and called fixed- and moving-attachment points of the limb i, respectively. At the displacement level, the forward kinematic problem pertains to the determination of the actual location or pose of the moving platform relative to the base from a set of joint–position readouts. At the velocity level, the forward kinematic problem refers to the determination of the actual twist, that is, the translational and angular velocities of the moving platform relative to the base, from a set of joint-velocity readouts and for a known configuration. Since the properties of parallel manipulators are usually complementary to those of serial architectures, they can be used in situations where the properties of the latter do not meet the application requirements. However, the higher degree of complexity of parallel mechanisms leads to more challenging forward kinematic problems, at both the displacement and velocity levels. The complexity of the forward kinematic problems depends extensively on the architecture, geometry, and joint–sensor layouts.

To describe the motion of the moving platform relative to the base, attach frame $\{A\}$ to body A and frame $\{B\}$ to body B, as shown in Figure 3.2. The pose of the moving platform relative to the base is thus defined by a position vector $\boldsymbol{p}$ in addition to a rotation matrix $\boldsymbol{R}$, in which $\boldsymbol{p}$ denotes the position vector of the origin of $\{B\}$ with respect to frame $\{A\}$, and furthermore, the orientation of $\{B\}$ with respect to $\{A\}$ is represented by a 3×3 rotation matrix $\boldsymbol{R}$. Each limb of a parallel manipulator defines a *kinematic loop* passing through the origins of frames $\{A\}$ and $\{B\}$, and through the two limb attachment points A_i and B_i.

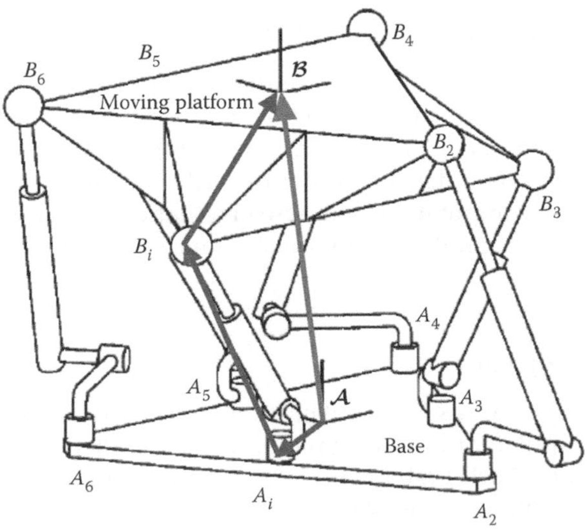

FIGURE 3.2
Vector-loop closures in a general parallel manipulator.

At the displacement level, the closure of each kinematic loop can be expressed in the vector form as

$$\overrightarrow{AB} = \overrightarrow{AA_i} + \overrightarrow{A_iB_i} - \overrightarrow{BB_i} \quad \text{for } i = 1, 2, \ldots, n, \tag{3.1}$$

in which $\overrightarrow{AA_i}$ and $\overrightarrow{BB_i}$ can be easily obtained from the geometry of the attachment points in the base and in the moving platform. Let us define vector $\boldsymbol{a}_i = \overrightarrow{AA_i}$ in the fixed frame $\{A\}$, and $\boldsymbol{b}_i = \overrightarrow{BB_i}$ in the moving frame $\{B\}$. Furthermore, $\boldsymbol{q}_i = \overrightarrow{A_iB_i}$ is defined as the limb variable, which indicates the geometry of the limb, and generally includes the active and passive limb segments. Hence, the loop closure can be written as the unknown pose variables $\boldsymbol{p}$, and $\boldsymbol{R}$, the position vectors describing the known geometry of the base and the moving platform, $\boldsymbol{a}_i$ and $\boldsymbol{b}_i$. Furthermore, the limb vector $\boldsymbol{q}_i$ is only known from the kinematic equation of the limb together with the readouts of the position sensors located at the different joints of that limb. Therefore, one may write the loop closure equations as follows:

$$\boldsymbol{p} = \boldsymbol{a}_i + \boldsymbol{q}_i - \boldsymbol{R}\boldsymbol{b}_i \quad \text{for } i = 1, 2, \ldots, n. \tag{3.2}$$

These equations are the main body of kinematic analysis. For an inverse kinematic problem, it is assumed that the moving platform position $\boldsymbol{p}$ and orientation $\boldsymbol{R}$ are given and the problem is to solve the active limb variables. Hence, from the above equations the passive limb variables must be eliminated to solve the inverse kinematic problem. This analysis is usually straightforward and results in a unique solution for the limb variables, even for redundant parallel manipulators. However, the inverse solution is not straightforward, and usually numerical methods are used for forward kinematic solution. In the proceeding sections, the inverse and forward kinematic analysis of three case studies are elaborated in detail.

3.3 Kinematic Analysis of a Planar Manipulator

In this section, the kinematics of a planar parallel manipulator is discussed in detail. In this analysis, the mechanism kinematic structure is described first and then the geometry of the manipulator is elaborated. Next, the inverse and forward kinematic analyses of the manipulator are worked out in detail. In order to verify the formulations, in Section 3.3.5, the simulation results for computation of inverse kinematics and forward kinematics of the manipulator are presented, and the accuracy of a numerical solution to forward kinematic problem is verified.

3.3.1 Mechanism Description

The architecture of a planar $4R\underline{P}R$ parallel manipulator considered here is shown in Figure 3.3. In this manipulator the moving platform is supported by four limbs of an identical kinematic structure. Each limb connects the fixed base to the manipulator moving platform by a revolute joint (R), followed by an actuated prismatic joint ($\underline{P}$), and another revolute joint (R). Hence, the total structure of the manipulator is $4R\underline{P}R$. The kinematic structure of a prismatic joint is used to model either a piston-cylinder actuator at each limb, or a cable-driven one. In order to avoid singularities at the central position of the

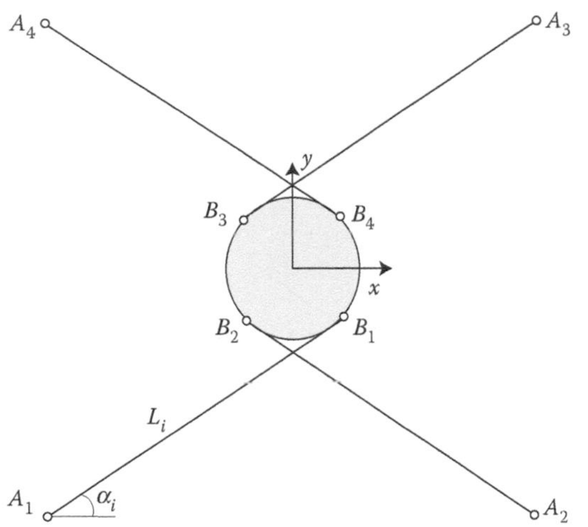

FIGURE 3.3
The schematics of a $4R\underline{P}R$ parallel manipulator.

manipulator, the limbs are considered to be crossed. Complete singularity analysis of the mechanism is presented in Section 4.6.2.

As shown in Figure 3.3, A_i denotes the fixed base points of the limbs, B_i denotes point of connection of the limbs to the moving platform, L_i denotes the limbs' lengths, and α_i denotes the limbs' angles. The position of the center of the moving platform G is denoted by $G = [x_G, y_G]$, and orientation of the manipulator moving platform is denoted by ϕ with respect to the fixed coordinate frame. Hence, the manipulator possesses three-degrees-of-freedom $\boldsymbol{\mathcal{X}} = [x_G, y_G, \phi]$, with one-degree-of-redundancy in the actuators.

3.3.2 Geometry of the Manipulator

For the purpose of analysis and as illustrated in Figure 3.4, a fixed frame O: xy is attached to the fixed base at point O, the center of the base point circle which passes through A_i's. Moreover, another moving coordinate frame G: UV is attached to the manipulator moving platform at point G. Furthermore, assume that point A_i lies at a radial distance of R_A from point O, and point B_i lies at a radial distance of R_B from point G in the xy plane, when the manipulator is at a central location.

In order to specify the geometry of the manipulator, let us define θ_{Ai} and θ_{Bi} as the absolute angles of points A_i and B_i at the central configuration of the manipulator with respect to the fixed frame O. The instantaneous orientation angle of B_i's is defined as

$$\phi_i = \phi + \theta_{B_i}. \tag{3.3}$$

Therefore, for each limb, $i = 1, 2, 3, 4$, the positions of the base points A_i are given by

$$A_i = [R_A \cos(\theta_{A_i}), R_A \sin(\theta_{A_i})]^T. \tag{3.4}$$

3.3.3 Inverse Kinematics

For inverse kinematic analysis, it is assumed that the position and orientation of the moving platform $\boldsymbol{\mathcal{X}} = [x_G, y_G, \phi]^T$ is given and the problem is to find the joint variables of the

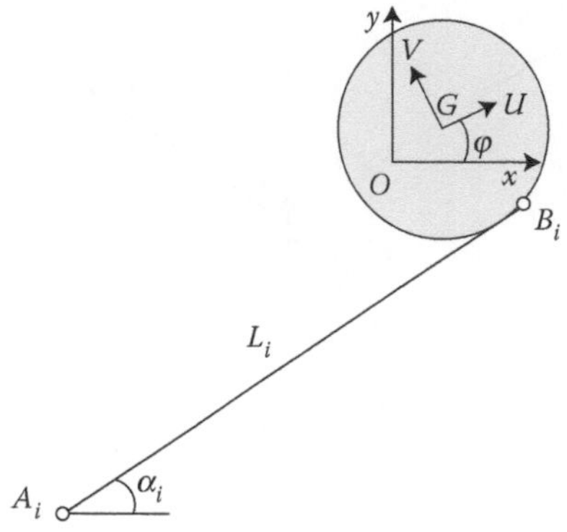

FIGURE 3.4
Kinematic configuration of the $4R\underline{P}R$ planar manipulator.

manipulator, $\boldsymbol{L} = [L_1, L_2, L_3, L_4]^T$. From the geometry of the manipulator, as illustrated in Figure 3.4, the loop closure equation for each limb, $i = 1, 2, 3, 4$, may be written as,

$$\overrightarrow{A_i G} = \overrightarrow{A_i B_i} - \overrightarrow{G B_i}. \tag{3.5}$$

Rewriting the vector-loop closure componentwise,

$$x_G - x_{Ai} = L_i \cos(\alpha_i) - R_B \cos(\phi_i), \tag{3.6}$$

$$y_G - y_{Ai} = L_i \sin(\alpha_i) - R_B \sin(\phi_i), \tag{3.7}$$

in which α_i's are the absolute limb angles. To solve the inverse kinematic problem, it is required to eliminate α_i's from the above equations and solve L_i's. This can be accomplished by reordering the above equations as,

$$L_i \cos(\alpha_i) = x_G - x_{Ai} + R_B \cos(\phi_i) \tag{3.8}$$

$$L_i \sin(\alpha_i) = y_G - y_{Ai} + R_B \sin(\phi_i). \tag{3.9}$$

By adding the square of both the sides of Equations 3.8 and 3.9, the limb lengths are uniquely determined as follows:

$$L_i = \left[(x_G - x_{Ai} + R_B \cos(\phi_i))^2 + (y_G - y_{Ai} + R_B \sin(\phi_i))^2 \right]^{1/2}. \tag{3.10}$$

Furthermore, the limb angles α_i's can be determined from the following equation:

$$\alpha_i = \text{Atan2}\left[(y_G - y_{Ai} + R_B \sin(\phi_i)), (x_G - x_{Ai} + R_B \cos(\phi_i)) \right]. \tag{3.11}$$

Hence, corresponding to each given manipulator location $\mathcal{X} = [x_G, y_G, \phi]^T$, there is a unique solution for the limb lengths L_i's, and limb angles α_i's.

3.3.4 Forward Kinematics

In the forward kinematic problem, the joint variable L_i's are given, and the position and orientation of the moving platform $\boldsymbol{\mathcal{X}} = [x_G, y_G, \phi]^T$ are to be found. This can be accomplished by eliminating x_G and y_G from Equations 3.6 and 3.7 as follows. Let us define two intermediate variables x_i and y_i as follows:

$$\begin{cases} x_i = -x_{Ai} + R_B \cos(\phi_i) \\ y_i = -y_{Ai} + R_B \sin(\phi_i). \end{cases} \tag{3.12}$$

In order to simplify the calculations, consider the square of Equation 3.10:

$$L_i^2 = (x_G + x_i)^2 + (y_G + y_i)^2. \tag{3.13}$$

We first try to solve x_G and y_G. This can be accomplished by reordering Equation 3.13 into:

$$x_G^2 + y_G^2 + r_i x_G + s_i y_G + u_i = 0, \tag{3.14}$$

in which $i = 1, 2, 3, 4$,

$$r_i = 2x_i, \quad s_i = 2y_i, \quad u_i = x_i^2 + y_i^2 - L_i^2. \tag{3.15}$$

Equation 3.14 provides four quadratic relations for x_G and y_G for $i = 1, 2, 3, 4$. Subtracting each of the two equations from each other results in linear equations in terms of x_G and y_G.

$$\boldsymbol{A} \cdot \begin{bmatrix} x_G \\ y_G \end{bmatrix} = \boldsymbol{b}, \tag{3.16}$$

in which

$$\boldsymbol{A} = \begin{bmatrix} r_1 - r_2 & s_1 - s_2 \\ r_2 - r_3 & s_2 - s_3 \\ r_3 - r_4 & s_3 - s_4 \\ r_4 - r_1 & s_4 - s_1 \end{bmatrix}, \quad \boldsymbol{b} = \begin{bmatrix} u_2 - u_1 \\ u_3 - u_2 \\ u_4 - u_3 \\ u_1 - u_4 \end{bmatrix}. \tag{3.17}$$

The components of $\boldsymbol{A}$ and $\boldsymbol{b}$ are all functions of ϕ. It appears that only two equations of the above equations of all are sufficient to evaluate x_G and y_G in terms of ϕ, but all four equations have been used here to have tractable solutions even at singular configurations. As shown in Appendix A, overdetermined Equations 3.16 can be solved using left pseudo-inverse,

$$\begin{bmatrix} x_G \\ y_G \end{bmatrix} = \boldsymbol{A}^{\dagger} \cdot \boldsymbol{b}, \tag{3.18}$$

in which $A^{\dagger}$ denotes the left pseudo-inverse solution of A, and for an overdetermined set of equations this is calculated from,

$$\boldsymbol{A}^{\dagger} = (\boldsymbol{A}^T \boldsymbol{A})^{-1} \boldsymbol{A}^T. \tag{3.19}$$

There exist tractable numerical methods such as Householder reflection, or Cholesky decomposition* to evaluate the pseudo-inverse. Note that the components of $\boldsymbol{A}$ and $\boldsymbol{b}$ are all functions of ϕ, and for a given ϕ Equation 3.18 gives the solution with least-squares error for x_G and y_G. In order to solve ϕ, this solution can be substituted in Equation 3.14, and a function of only one unknown variable ϕ is obtained:

$$f_i(\phi) = x_G^2 + y_G^2 + r_i x_G + s_i y_G + u_i. \tag{3.20}$$

Consider

$$f(\phi) = \sum_{i=1}^{4} f_i(\phi), \tag{3.21}$$

and use numerical methods that use iterative search routines† to find the final solution of $f(\phi) = 0$. It appears that any function $f_i(\phi) = 0$ could be used to obtain the solution of forward kinematics ϕ. However, the summation of all four equations is used here to have a tractable solution even at singular configurations.

The flowchart given in Figure 3.5 reveals the details of the iterative method of the forward kinematic solution. As seen in this flowchart, for a given ϕ the values of x_i, y_i, and ϕ_i are calculated using Equations 3.12 and 3.3, respectively. Then r_i, s_i, and u_i are calculated using Equation 3.15, and the values of matrices $\boldsymbol{A}$ and $\boldsymbol{b}$ are derived from Equation 3.17. For the given values of $\boldsymbol{A}$ and $\boldsymbol{b}$, the least-squares solutions for x_i and y_i are derived from Equation 3.19. Then the value of $f(\phi)$ is calculated from Equation 3.21, and if it is not very close to zero, an effective search routine is used to recalculate a new value for ϕ. This iteration is followed to obtain a solution for $f(\phi) = 0$ with an accuracy of $\varepsilon \ll 1$.

Multiple solutions may exist for equation $f(\phi) = 0$, and in order to avoid jumps in the forward kinematic solutions, in the numerical routine the solution at the previous iteration is used for the search of the next solution. Simulation results, detailed in Section 3.3.5, illustrate the integrity and accuracy of the numerical routines used to solve forward kinematics. If the limb angles α_i's need to be derived, Equation 3.11 can be used directly by substituting x_G, y_G, and ϕ.

3.3.5 Simulations

Geometric parameters used in the simulations are adopted from a cable-driven parallel manipulator design [177], and are given in Table 3.1. This manipulator design is adopted from the Canadian concept of next-generation telescopes, [178,179]. For this special application, the limb lengths are much greater than those of conventional manipulators, resulting in a larger workspace compared with that of conventional manipulators. For simulations, a typical trajectory for the $4R\underline{P}R$ manipulator $\boldsymbol{\mathcal{X}}_d$ is considered and illustrated in Figure 3.6. As seen in this figure, 200 m motion in the x direction, and 100 m movement in the y direction is considered, while a 45° rotation is considered for the orientation of the moving platform ϕ, in 10 s. For the given trajectory, inverse kinematic solution of the manipulator is obtained from Equation 3.10, and the manipulator limbs lengths are uniquely determined and have been illustrated in Figure 3.7. As seen in this figure,

---

* Use `pinv` function in MATLAB.

† Use `fzero` function in MATLAB.

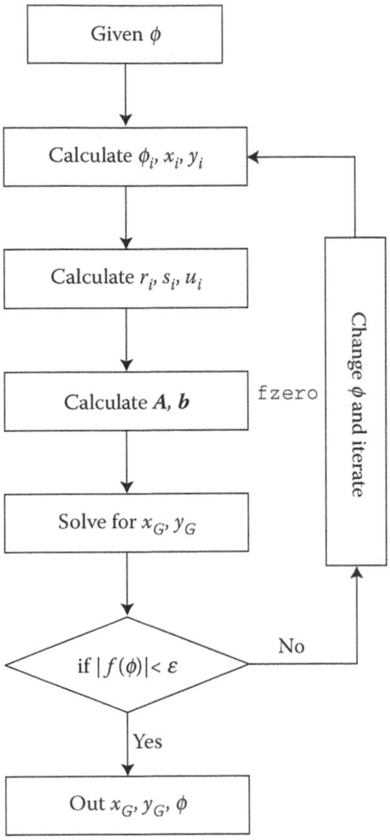

FIGURE 3.5
Flowchart of iterative routine used to solve the forward kinematics of the $4R\underline{P}R$ manipulator.

TABLE 3.1
Geometric Parameters of the $4R\underline{P}R$ Parallel Manipulator

Description	Quantity
R_A: Radius of the fixed base points A_i's	900 m
R_B: Radius of the moving platform points B_i's	10 m
θ_{A_i}: Angles of the fixed base points A_i's	$[-\frac{3\pi}{4}, -\frac{\pi}{4}, \frac{\pi}{4}, \frac{3\pi}{4}]$
θ_{B_i}: Angles of the moving platform points B_i's	$[-\frac{\pi}{4}, -\frac{3\pi}{4}, \frac{3\pi}{4}, \frac{\pi}{4}]$

manipulator limb lengths change in the order of hundred meters, in accordance with the desired trajectory variations.

As shown in Figure 3.8, to verify the accuracy and integrity of the numerical solutions, the forward kinematic solutions for the manipulator are derived for the given limbs lengths. The difference between the numerical solutions $\mathcal{X}_c$ and the desired motion $\mathcal{X}_d$ are illustrated in Figure 3.9. It is observed that the calculated numerical solutions from the forward kinematic problem are completely identical to the desired trajectories, with an accuracy of 10^{-13}. This confirms the accuracy and integrity of the numerical method to

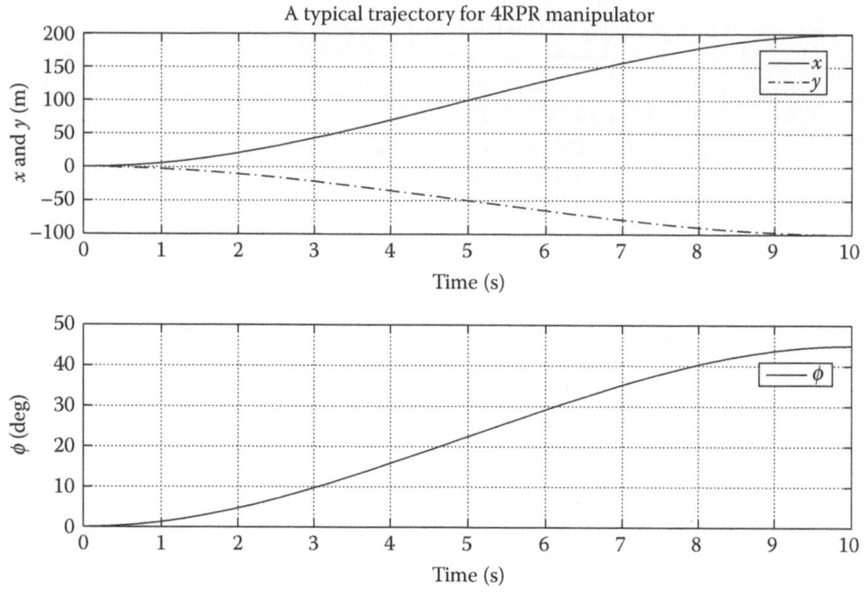

FIGURE 3.6
A typical trajectory of a $4R\underline{P}R$ parallel manipulator.

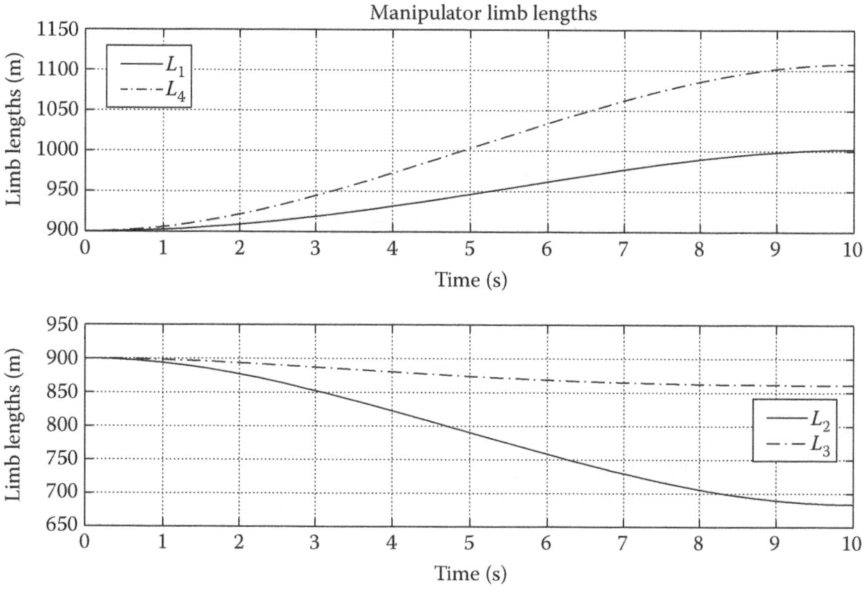

FIGURE 3.7
The inverse kinematic solution for the given trajectory.

obtain the forward kinematic solution. It should be noted that forward kinematic solutions are not unique, and to avoid converging towards other solutions at each step, the last step solution is used as the initial guess for the next step of iterations. By this means, the numerical solution converges toward the right solution in all the examined points.

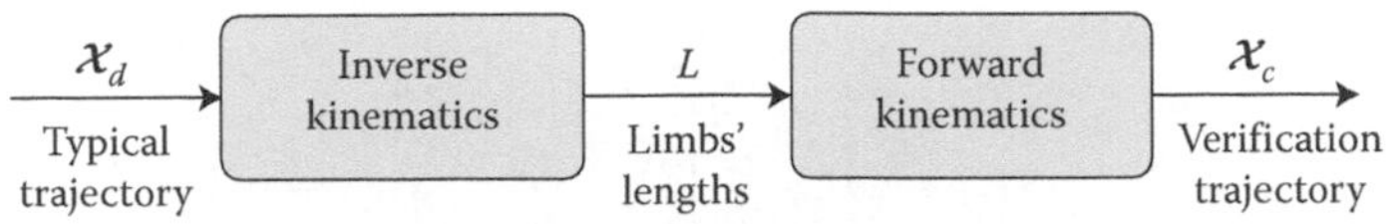

FIGURE 3.8
Mechanism of the forward kinematics verification for a given trajectory.

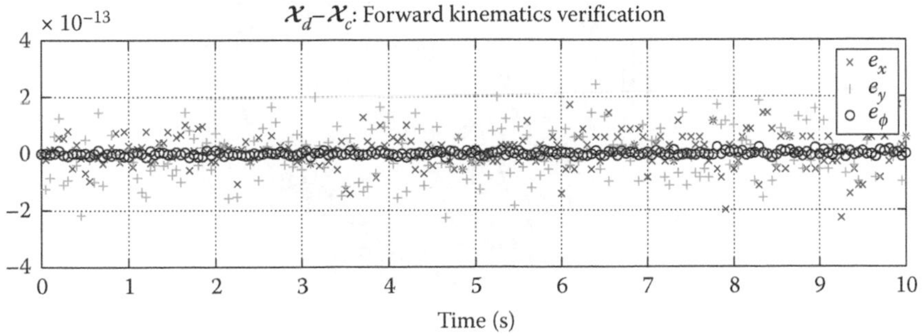

FIGURE 3.9
The accuracy and integrity of the forward and inverse kinematic solutions.

3.4 Kinematic Analysis of Shoulder Manipulator

In this section, kinematic analysis of a three-degrees-of-freedom spatial orientation manipulator is studied in detail. In this analysis, first the mechanism kinematic structure is described and then the geometry of the system is worked out. Next, the inverse and a *closed-form* forward kinematic solution of the manipulator are analyzed in detail. In order to verify the formulations, in Section 3.4.5, simulation results for computation of the inverse and forward kinematics of the manipulator are presented, and the accuracy of the forward kinematic solution is verified.

3.4.1 Mechanism Description

In this section, a complete kinematic analysis for a three-degrees-of-freedom parallel manipulator with one degree of actuator redundancy is reported. The mechanism, which is called *the shoulder manipulator*, is designed by Vincent Hayward [73,74] who did so by borrowing design ideas from biological manipulators particularly the biological shoulder. The interesting features of this mechanism and its similarity to the human shoulder have made its design unique and appropriate to serve as the basis of a proper experimental setup for parallel robotics research. In the anatomy of a human shoulder, there are at least four groups of muscles identified to generate a pure rotational motion for the shoulder [64]. These muscle groups keep the ball and socket kinematic structure of the shoulder in place, while providing the joint with three independent rotational movements [147]. The use of at least one redundant muscle group in human shoulder inspires the engineers to design an artificial shoulder manipulator with one degree of actuation redundancy [74]. This redundancy increases the manipulator stiffness, and moreover it enlarges the singular free workspace of the manipulator [72].

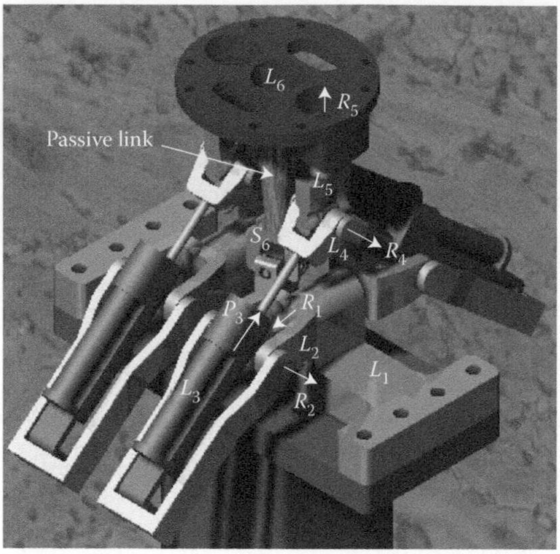

FIGURE 3.10
A CAD drawing of the shoulder manipulator.

The shoulder manipulator was designed in 1992 [71], by borrowing design ideas from the human biological shoulder. A CAD model of the mechanism is shown in Figure 3.10, and a picture of the manipulator in a twisted configuration is shown in Figure 3.11. As shown in these figures, the moving platform is constrained to a three-degrees-of-freedom purely orientation motion. Four high-performance hydraulic piston actuators are used to generate three-degrees-of-freedom motion in the moving platform, and hence, like the human shoulder structure, one degree of actuator redundancy is considered in this design. Each actuator includes a position sensor of LVDT-type and an embedded Hall effect force sensor. Simple revolute joints are used in the manipulator to build spherical and universal

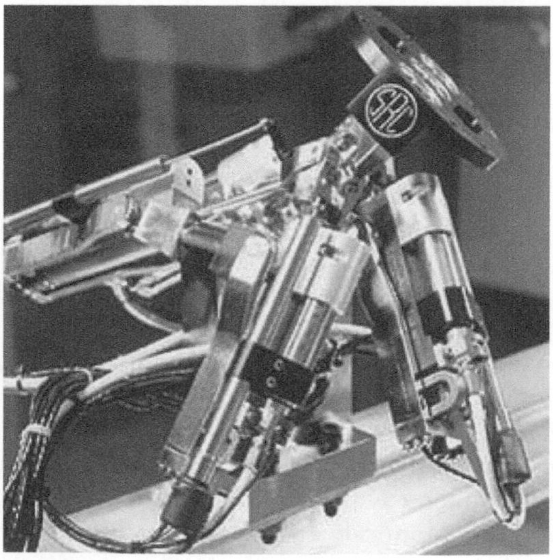

FIGURE 3.11
A picture of the shoulder manipulator in a twisted configuration.

joints, and all the four limbs share an identical kinematic structure. The four active limbs share a $U\underline{P}U$ kinematic structure in which each of the two limbs are connected to the moving platform through a single connecting link.

As shown in Figure 3.10, the base link L_1 is connected to the cylinder link L_3 via a connecting link L_2. Therefore, the universal joint used in the manipulator can be decomposed into two binary revolute joints R_1 and R_2. The prismatic joint is denoted by P_3 in this figure, and the universal joint connecting the piston link L_4 to the moving platform L_6 is also decomposed into two binary revolute joints R_4 and R_5. It is important to note that the universal joints of each of the two pairs of piston actuators share one connecting link denoted by L_5 in the figure. Finally, a passive limb connects the moving platform to the fixed base by a spherical joint S_6, by which the pure translations of the moving platform are suppressed. As seen in this figure, the spherical joint S_6 is also built from three perpendicular revolute joints.

From the structural point of view, the shoulder manipulator can be considered as a shoulder for a light-weight seven-degrees-of-freedom robotic arm, which can carry loads several times its own weight. The workspace of such a mechanism can be considered as part of a spherical surface, while the orientation angles are limited to vary between $-\pi/6$ and $\pi/6$. No sensors are available for measuring the orientation angles of the moving platform. This fact justifies the importance of the forward kinematic analysis as a key element in the position control of the shoulder manipulator, using the LVDT position sensors as the only source of measurement in such a control scheme. A complete kinematic analysis of such a manipulator will provide the required means to better understand the characteristics of the structure and to design suitable control laws to accomplish the required tracking performance.

3.4.2 Geometry of the Manipulator

Figure 3.12 depicts a geometric model for the shoulder manipulator. The parameters used in the kinematics can be defined as

$$l_b = \|\overrightarrow{CA_i}\|, \quad l_p = \|\overrightarrow{CP}\|, \quad l_d = \|\overrightarrow{PP_i}\|_v, \quad l_k = \|\overrightarrow{PP_i}\|_w, \tag{3.22}$$

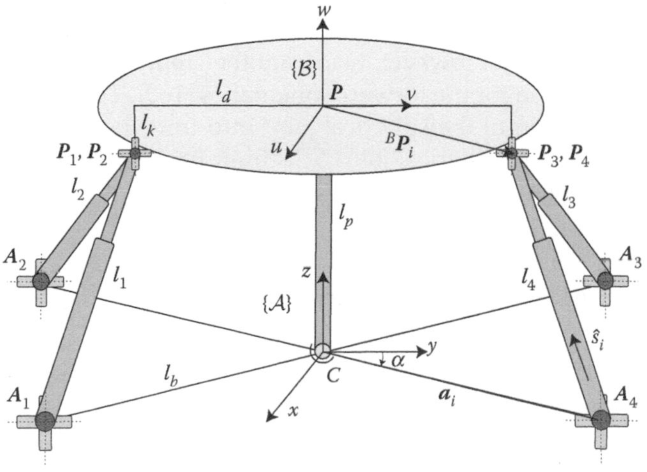

FIGURE 3.12
Geometry of the shoulder manipulator.

in which the latter two norms denote the component of $\|\overrightarrow{PP_i}\|$ in v and w directions, respectively. Furthermore, α is the angle between CA_4 and y; C is the center of the reference frame; P is the center of the moving platform; l_i are the actuator lengths for $i = 1, 2, 3, 4$; P_i's denote the moving endpoints of the actuators; and A_i's are the fixed endpoints of the actuators. Two coordinate frames are defined for the purpose of analysis. The base coordinate frame $\{A\} : xyz$ is attached to the fixed base at point C (rotation center) with its z-axis perpendicular to the plane defined by the actuator base points A_1, A_2, A_3, A_4, and its x-axis parallel to the bisector of angle $\angle A_1 C A_4$. The second coordinate frame $\{B\} : uvw$ is attached to the center of the moving platform P with its w-axis perpendicular to the line defined by the actuators' moving end points $P_1 P_3$ along the passive limb. It is assumed that the actuators' fixed end points lie on the same plane as the rotation center C. The position of the moving platform center P is defined by

$$^{A}\boldsymbol{P} = \begin{bmatrix} p_x & p_y & p_z \end{bmatrix}^T. \tag{3.23}$$

Moreover, a rotation matrix $^{A}\boldsymbol{R}_B$ is used to define the orientation of the moving platform with respect to the base frame, using a pitch–roll–yaw representation:

$$\begin{aligned} ^{A}\boldsymbol{R}_B &= \boldsymbol{R}_z(\theta_z)\boldsymbol{R}_y(\theta_y)\boldsymbol{R}_x(\theta_x) \\ &= \begin{bmatrix} c\theta_z c\theta_y & c\theta_z s\theta_y s\theta_x - s\theta_z c\theta_x & c\theta_z s\theta_y c\theta_x + s\theta_z s\theta_x \\ s\theta_z c\theta_y & s\theta_z s\theta_y s\theta_x + c\theta_z c\theta_x & s\theta_z s\theta_y c\theta_x - c\theta_z s\theta_x \\ -s\theta_y & c\theta_y s\theta_x & c\theta_y c\theta_x \end{bmatrix}, \end{aligned} \tag{3.24}$$

where θ_x, θ_y, and θ_z are the Euler angles of the moving platform denoting rotations of the moving frame about the fixed x, y, and z axes, respectively. Furthermore, $c\theta$ and $s\theta$ are short-hand notations representing $\cos(\theta)$ and $\sin(\theta)$, respectively. Although pitch–roll–yaw Euler angles are a convenient way to represent the orientation of the moving platform, it is not very convenient for the forward kinematic solution. A thorough kinematic analysis of this manipulator is reported in [159–161]. Using a pitch–roll–yaw representation for the orientation yields a complex forward kinematic formulation for the system, which can only be solved numerically, using neural networks [161], Taylor series expansion, or quasi-closed solution [159]. However, using another representation for the orientation yields a closed-form solution for the forward kinematics [160]. This representation is based on kinematic decomposition of the spherical joint into three perpendicular binary joints. As shown in Figure 3.10, the spherical joint S_6 is built from three perpendicular revolute joints intersecting at a point. These rotations can be represented by u–v–w Euler angles as follows:

$$\begin{aligned} \boldsymbol{R}_{uvw}(\theta_1, \theta_2, \theta_3) &= \boldsymbol{R}_u(\theta_1)\ \boldsymbol{R}_v(\theta_2)\ \boldsymbol{R}_w(\theta_3) \\ &= \begin{bmatrix} c_3c_2 & -s_3c_2 & s_2 \\ s_3c_1 + c_3s_2s_1 & c_3c_1 - s_3s_2s_1 & -c_2s_1 \\ s_3s_1 - c_3s_2c_1 & c_3s_1 + s_3s_2c_1 & c_2c_1 \end{bmatrix}. \end{aligned} \tag{3.25}$$

Hence, the position and orientation of the moving platform are completely defined by six variables, from which only three orientation angles θ_1, θ_2, and θ_3 are independently specified as independent variables of the shoulder manipulator.

3.4.3 Inverse Kinematics

Although the shoulder manipulator is a redundantly actuated manipulator, given the orientation of the moving platform its inverse kinematics is easily solved. In an inverse kinematic solution the actuator lengths l_i's must be determined as a function of the manipulator orientation variables θ_1, θ_2, and θ_3. In order to find the actuator lengths l_i's, the loop-closure equation for each actuated limb is written as follows:

$$\boldsymbol{L}_i = l_i\,\hat{\boldsymbol{s}}_i = {}^A\boldsymbol{P} + {}^A\boldsymbol{R}_B\,{}^B\boldsymbol{P}_i - \boldsymbol{a}_i, \tag{3.26}$$

in which l_i is the length of the ith actuated limb and $\hat{\boldsymbol{s}}_i$ is the unit vector pointing along the direction of the ith actuated limb. Moreover, ${}^A\boldsymbol{R}_B$ is the rotation matrix of the moving platform, and ${}^A\boldsymbol{P}$ is its center position vector:

$${}^A\boldsymbol{P} = l_p\begin{bmatrix} s_2 & -c_2 s_1 & c_2 c_1\end{bmatrix}^T. \tag{3.27}$$

Furthermore, as shown in Figure 3.12, vectors $\boldsymbol{a}_i$ and ${}^B\boldsymbol{P}_i$ denote the fixed end points of the actuators (A_i) in the base frame and the moving attachment points of the actuators, respectively. Hence, for the given geometry of the manipulator we have

$$\begin{aligned}
\boldsymbol{a}_1 &= l_b\begin{bmatrix} +\sin\alpha & -\cos\alpha & 0\end{bmatrix}^T,\\
\boldsymbol{a}_2 &= l_b\begin{bmatrix} -\sin\alpha & -\cos\alpha & 0\end{bmatrix}^T,\\
\boldsymbol{a}_3 &= l_b\begin{bmatrix} -\sin\alpha & +\cos\alpha & 0\end{bmatrix}^T,\\
\boldsymbol{a}_4 &= l_b\begin{bmatrix} +\sin\alpha & +\cos\alpha & 0\end{bmatrix}^T,
\end{aligned} \tag{3.28}$$

and

$$\begin{aligned}
{}^B\boldsymbol{P}_1 = {}^B\boldsymbol{P}_2 &= \begin{bmatrix} 0 & -l_d & l_p - l_k\end{bmatrix}^T,\\
{}^B\boldsymbol{P}_3 = {}^B\boldsymbol{P}_4 &= \begin{bmatrix} 0 & +l_d & l_p - l_k\end{bmatrix}^T.
\end{aligned} \tag{3.29}$$

The actuator lengths, l_i, can be easily computed by dot multiplying Equation 3.26 to itself:

$$\boldsymbol{L}_i^T\boldsymbol{L}_i = l_i^2 = \left[{}^A\boldsymbol{P} + {}^A\boldsymbol{R}_B\,{}^B\boldsymbol{P}_i - \boldsymbol{a}_i\right]^T\left[{}^A\boldsymbol{P} + {}^A\boldsymbol{R}_B\,{}^B\boldsymbol{P}_i - \boldsymbol{a}_i\right]. \tag{3.30}$$

Writing Equation 3.30 four times with the corresponding parameters given in Equations 3.25, 3.27 through 3.29, and simplifying the results yields

$$l_1^2 = k_1 + k_2 s_2 - k_3 c_2 s_1 + k_4 s_3 c_2 + k_5(c_3 c_1 - s_3 s_2 s_1), \tag{3.31}$$

$$l_2^2 = k_1 - k_2 s_2 - k_3 c_2 s_1 - k_4 s_3 c_2 + k_5(c_3 c_1 - s_3 s_2 s_1), \tag{3.32}$$

$$l_3^2 = k_1 - k_2 s_2 + k_3 c_2 s_1 + k_4 s_3 c_2 + k_5(c_3 c_1 - s_3 s_2 s_1), \tag{3.33}$$

$$l_4^2 = k_1 + k_2 s_2 + k_3 c_2 s_1 - k_4 s_3 c_2 + k_5(c_3 c_1 - s_3 s_2 s_1), \tag{3.34}$$

in which

$$\begin{aligned} k_1 &= 4l_p^2 + l_d^2 + l_k^2 + l_b^2 - 4l_p\, l_k, \\ k_2 &= 2l_b\,(l_k - 2l_p)\sin\alpha, \\ k_3 &= 2l_b\,(2l_p - l_k)\cos\alpha, \\ k_4 &= -2l_d\, l_b \sin\alpha, \\ k_5 &= -2l_d\, l_b \cos\alpha. \end{aligned} \tag{3.35}$$

Therefore, the actuator lengths are uniquely determined by Equations 3.31 through 3.34 as a function of the moving platform orientation variables.

3.4.4 Forward Kinematics

In the forward kinematic analysis of the shoulder manipulator, we shall find all the possible orientations of the moving platform for a given set of actuated limb lengths. Equations 3.31 through 3.34 may be used to obtain the forward kinematic solution of the hydraulic shoulder with the actuator lengths as the input variables. In fact, since the manipulator is redundantly actuated we have four nonlinear equations to solve for three unknowns. First, let us try to express the moving platform position and orientation in terms of the joint variables θ_1, θ_2, and θ_3 using Equations 3.25 and 3.27. Subtract Equation 3.32 from Equation 3.31 and similarly Equation 3.34 from Equation 3.33 to simplify the equations to

$$l_1^2 - l_2^2 = +2k_2\, s_2 + 2k_4\, s_3\, c_2, \tag{3.36}$$

$$l_3^2 - l_4^2 = -2k_2\, s_2 + 2k_4\, s_3\, c_2. \tag{3.37}$$

Subtract Equation 3.37 from Equation 3.36, and solve s_2 as follows:

$$s_2 = \frac{l_1^2 - l_2^2 - l_3^2 + l_4^2}{4k_2}. \tag{3.38}$$

Substituting Equation 3.38 into the trigonometric identity $s_2^2 + c_2^2 = 1$ results in two solutions for c_2.

$$c_2 = \pm\sqrt{1 - s_2^2}. \tag{3.39}$$

Having s_2 and c_2 in hand, we can solve s_3 from Equation 3.36 as

$$s_3 = \frac{l_1^2 - l_2^2 - 2k_2\, s_2}{2k_4\, c_2} \quad \text{or} \quad \frac{l_3^2 - l_4^2 + 2k_2\, s_2}{2k_4\, c_2}. \tag{3.40}$$

Similarly,

$$c_3 = \pm\sqrt{1 - s_3^2}. \tag{3.41}$$

Up to here, four possible solutions for the forward kinematics have been generated. In order to solve the remaining unknowns, c_1 and s_1, subtract Equation 3.31 from Equation 3.33 and similarly subtract Equation 3.32 from Equation 3.34:

$$l_3^2 - l_1^2 = 2k_3\, c_2\, s_1 - 2k_2\, s_2, \tag{3.42}$$

$$l_4^2 - l_2^2 = 2k_3\, c_2\, s_1 + 2k_2\, s_2. \tag{3.43}$$

Hence,

$$s_1 = \frac{l_3^2 - l_1^2 + 2k_2\, s_2}{2k_3\, c_2} \quad \text{or} \quad \frac{l_4^2 - l_2^2 - 2k_2\, s_2}{2k_3\, c_2}, \tag{3.44}$$

and

$$c_1 = \pm\sqrt{1 - s_1^2}. \tag{3.45}$$

Therefore, totally eight solutions for the forward kinematics can be summarized as

$$\theta_1 = \text{A}\tan 2(s_1, c_1), \quad \theta_2 = \text{A}\tan 2(s_2, c_2), \quad \theta_3 = \text{A}\tan 2(s_3, c_3). \tag{3.46}$$

The solutions degenerate if $c_2 = 0$. In this case two independent values for θ_1 and θ_3 cannot be obtained. If $c_2 = 0$ and $s_2 = 1$, then

$$\theta_3 + \theta_1 = \cos^{-1}\frac{l_1^2 - k_1 - k_2}{k_5} \quad \text{or} \quad \cos^{-1}\frac{l_2^2 - k_1 + k_2}{k_5}. \tag{3.47}$$

Otherwise, if $c_2 = 0$ and $s_2 = -1$, then

$$\theta_3 - \theta_1 = \cos^{-1}\frac{l_1^2 - k_1 + k_2}{k_5} \quad \text{or} \quad \cos^{-1}\frac{l_2^2 - k_1 - k_2}{k_5}. \tag{3.48}$$

Furthermore, the moving platform position ${}^A\boldsymbol{P}$ and orientation ${}^A\boldsymbol{R}_B$ can be obtained by Equations 3.27 and 3.25. If there is any need to represent the orientation with the pitch–roll–yaw Euler angles, we can solve θ_x, θ_y, and θ_z using Equation 3.24 which completes the solution process to the forward kinematics of the shoulder manipulator. It should be noted that, as stated before, there are eight solutions for the forward kinematics due to several square roots involved in the process, where the erroneous solutions must be identified and omitted. However, for the limited workspace of the shoulder manipulator, in which the orientation angles are limited to $\pm\pi/6$, the positive square roots in Equations 3.45, 3.39, and 3.41 yield the right solution.

3.4.5 Simulations

To verify the results of the forward kinematic solution, a simulation study is performed in which a sample trajectory is considered within the reachable workspace of the manipulator. The provided inverse and forward solutions are tested along such a trajectory and the results are compared. Consider a smooth motion specified in terms of a desired orientation of the moving platform. The sample trajectory is generated by a cubic function given the

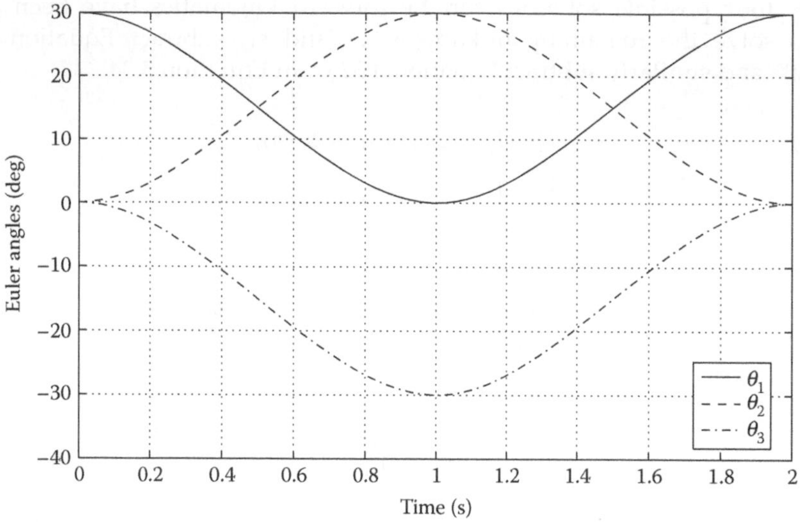

FIGURE 3.13
A typical trajectory of the shoulder manipulator depicted for the u–v–w Euler angles.

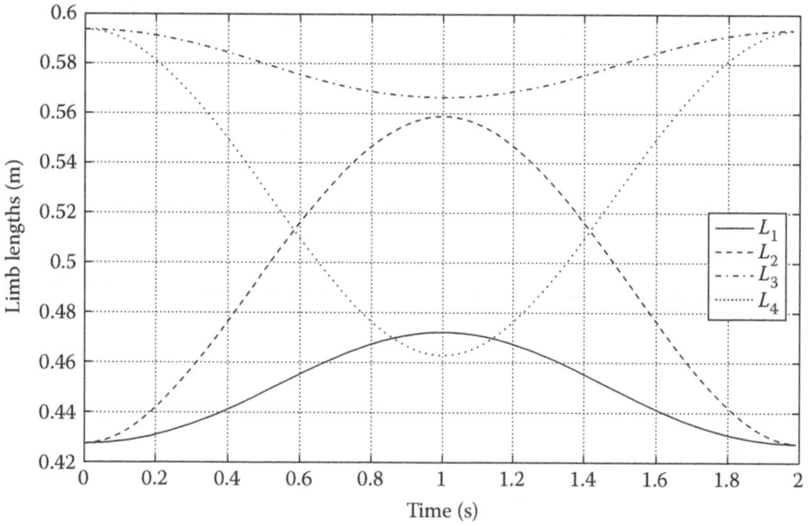

FIGURE 3.14
Inverse kinematic solution of the shoulder manipulator for the given trajectory.

initial and final Euler angles and the elapsed time.* Figure 3.13 shows the sample trajectory for the shoulder manipulator by orientation angles represented by u–v–w Euler angles. As seen in this figure, the Euler angles vary between ±30° within the reachable workspace of the hydraulic shoulder manipulator. For the given trajectory, inverse kinematic solution of the manipulator is found from Equations 3.31 through 3.34, and the manipulator limbs' lengths are uniquely determined and illustrated in Figure 3.14. As seen in this figure,

* Refer to Appendix B for more details on cubic trajectory planning.

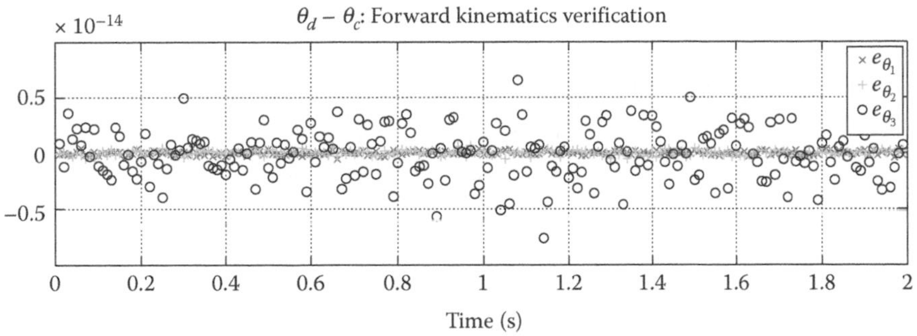

FIGURE 3.15
The accuracy and integrity of the forward and inverse kinematic solutions for the shoulder manipulator.

manipulator limbs' lengths change a few centimeters, according to the desired trajectory variations.

To verify the accuracy and integrity of the formulations, the forward kinematic solution of the manipulator is obtained for the given limbs' lengths using Equation 3.46. The difference between the forward kinematic solutions θ_c and the desired motion θ_d is illustrated in Figure 3.15. It is observed that the calculated forward kinematic solution is completely identical to the desired trajectories with an accuracy of 10^{-14}. These simulations verify the integrity of the forward kinematic formulations.

3.5 Kinematic Analysis of Stewart–Gough Platform

In this section, the kinematic analysis of the Stewart–Gough platform (SGP) is studied in detail. In this analysis, first the manipulator kinematic structure is explained and then the geometry of the system is given. Next, the inverse and forward kinematic analyses of the manipulator are worked out.

3.5.1 Mechanism Description

Figure 3.16 shows the schematic of an SGP. In this manipulator, the spatial motion of the moving platform is generated by six identical piston–cylinder actuators. Each piston–cylinder actuator consists of two parts connected with a prismatic joint. In practice, either hydraulic actuators or electric motors with ball screw generate the prismatic motion in the limbs. The actuators connect the fixed base to the moving platform by spherical joints at points A_i and B_i, $i = 1, 2, \ldots, 6$. Note that in Figure 3.16 all the attachment points A_i's lie in the base plane, and all B_i's lie in the moving platform plane. However, in a general SGP the attachment points are not necessarily confined to lie in a plane.

As shown in Figure 3.16, and in order to analyze the kinematics of the manipulator, frame $\{\boldsymbol{A}\}$ is attached to the fixed base and frame $\{\boldsymbol{B}\}$ is attached to the moving platform at points O_A and O_B, respectively. The kinematic structure of each limb is an $S\underline{P}S$ arrangement. As analyzed in Example 1.3 in Chapter 1, in this manipulator there exist 14 links connected by 6 prismatic joints and 12 spherical joints. Moreover, for each $S\underline{P}S$ kinematic structure, there exists one passive degree-of-freedom. Hence, this spatial

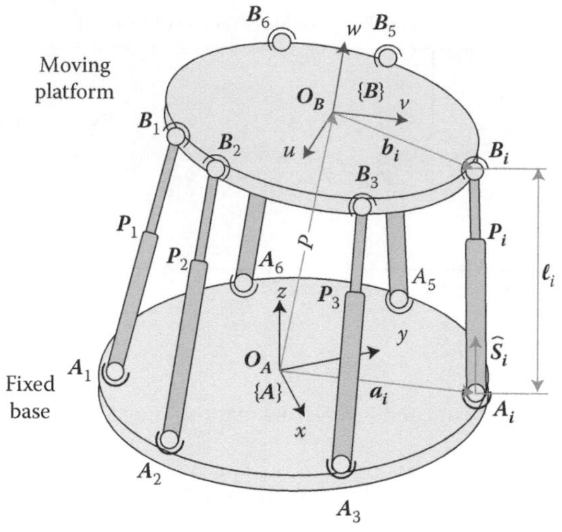

FIGURE 3.16
Schematic of a Stewart–Gough platform and the frames attached to the fixed base and to the moving platform.

manipulator has six-degrees-of-freedom and can serve as a general parallel manipulator to produce a complete spatial movement with three-degrees-of-freedom in position and three-degrees-of-freedom in orientation. The number of actuators are also equal to the degrees-of-freedom of the manipulator and hence the manipulator is fully parallel. Since all the limbs are connected to the moving platform and to the base by spherical joints, no twisting torques can be transmitted through the limbs. Therefore, the transmitting force acting on each limb is directed along the longitudinal axis of the limb.

3.5.2 Geometry of the Manipulator

As shown in Figure 3.16, and for the purpose of kinematic analysis, the geometry of the manipulator is described by two coordinate frames attached to the fixed base and the moving platform. The fixed frame $\{\boldsymbol{A}\}$ is attached to the base at point O_A, and the position of attachment points on the fixed base is denoted by the vectors $\boldsymbol{a}_i$'s. Similarly, the moving frame $\{\boldsymbol{B}\}$ is attached to the moving platform at point O_B, and the positions of moving attachment points are denoted by the vectors $\boldsymbol{b}_i$'s. The geometry of each limb is described by its length ℓ_i, and its direction is denoted by a unit vector $\hat{\boldsymbol{s}}_i$. Position of the point O_B of the moving platform is described by the position vector ${}^A\boldsymbol{P} = \begin{bmatrix} p_x & p_y & p_z \end{bmatrix}^T$, and orientation of the moving platform is described by the rotation matrix ${}^A\boldsymbol{R}_B$. The rotation matrix can be represented componentwise by the components of the unit vectors $\hat{\boldsymbol{u}}, \hat{\boldsymbol{v}}$, and $\hat{\boldsymbol{w}}$, as follows:

$$ {}^A\boldsymbol{R}_B = \begin{bmatrix} u_x & v_x & w_x \\ u_y & v_y & w_y \\ u_z & v_z & w_z \end{bmatrix}. \tag{3.49} $$

Note that the rotation matrix is orthonormal and satisfies the orthonormality conditions detailed in Equations 2.8 through 2.10. The orientation of the moving platform can also be represented by screw axis representation or different Euler angles. In the forward kinematic analysis of the SGP, which will be presented in Section 3.5.4, the following screw axis representation is used:

$$\left\{\hat{\mathbf{s}} = \begin{bmatrix} s_x & s_y & s_z \end{bmatrix}^T, \theta\right\}. \tag{3.50}$$

In terms of the geometry of the moving platform and the base, the SGP can be classified into many cases. A class of SGPs for which both attachment points of six limbs, A_i's and B_i's are coplanar, has been extensively investigated in the literature. For that class of SGP, the centers of the ball-joints generally form two hexagons on the top platform and the base. When a ball-joint is concentrated on the other, the number of joining points of the six limbs in the moving and the base platforms is respectively reduced. If the number of the joining points in the moving platform is n, and that in base is N, then the SGP is named as type $N-n$ SGP. The fewer the number, the easier to solve the forward kinematics; however, concentric ball-joints may cause design problems. To avoid the use of concentric ball-joints, the ball-joints of limbs can be separated to become a higher type while retaining the shape.

3.5.3 Inverse Kinematics

For inverse kinematic analysis, it is assumed that the position ${}^A\mathbf{P}$ and orientation of the moving platform ${}^A\mathbf{R}_B$ are given and the problem is to obtain the joint variables, namely, $\mathbf{L} = [\ell_1, \ell_2, \ell_3, \ell_4, \ell_5, \ell_6]^T$. From the geometry of the manipulator, as illustrated in Figure 3.16, the loop closure equation for each limb, $i = 1, 2, \ldots, 6$, can be written as

$$\begin{aligned} \ell_i\, {}^A\hat{\mathbf{s}}_i &= {}^A\mathbf{P} + {}^A\mathbf{b}_i - {}^A\mathbf{a}_i \\ &= {}^A\mathbf{P} + {}^A\mathbf{R}_B\, {}^B\mathbf{b}_i - {}^A\mathbf{a}_i. \end{aligned} \tag{3.51}$$

To obtain the length of each actuator and eliminate $\hat{\mathbf{s}}_i$, it is sufficient to dot multiply each side by itself:

$$\ell_i^2 \left[{}^A\hat{\mathbf{s}}_i{}^T\, {}^A\hat{\mathbf{s}}_i\right] = \left[{}^A\mathbf{P} + {}^A\mathbf{R}_B\, {}^B\mathbf{b}_i - {}^A\mathbf{a}_i\right]^T \left[{}^A\mathbf{P} + {}^A\mathbf{R}_B\, {}^B\mathbf{b}_i - {}^A\mathbf{a}_i\right] \tag{3.52}$$

or

$$\begin{aligned} \ell_i^2 = {}& {}^A\mathbf{P}^T\, {}^A\mathbf{P} + {}^B\mathbf{b}_i^T\, {}^B\mathbf{b}_i + {}^A\mathbf{a}_i^T\, {}^A\mathbf{a}_i - 2\, {}^A\mathbf{P}^T\, {}^A\mathbf{a}_i \\ & + 2\, {}^A\mathbf{P}^T \left[{}^A\mathbf{R}_B\, {}^B\mathbf{b}_i\right] - 2\left[{}^A\mathbf{R}_B\, {}^B\mathbf{b}_i\right]^T {}^A\mathbf{a}_i. \end{aligned} \tag{3.53}$$

Hence, for $i = 1, 2, \ldots, 6$, each limb length can be uniquely determined by

$$\begin{aligned} \ell_i = {}& \Big[{}^A\mathbf{P}^T\, {}^A\mathbf{P} + {}^B\mathbf{b}_i^T\, {}^B\mathbf{b}_i + {}^A\mathbf{a}_i^T\, {}^A\mathbf{a}_i - 2\, {}^A\mathbf{P}^T\, {}^A\mathbf{a}_i \\ & + 2\, {}^A\mathbf{P}^T \left[{}^A\mathbf{R}_B\, {}^B\mathbf{b}_i\right] - 2\left[{}^A\mathbf{R}_B\, {}^B\mathbf{b}_i\right]^T {}^A\mathbf{a}_i\Big]^{1/2}. \end{aligned} \tag{3.54}$$

If the position and orientation of the moving platform lie in the feasible workspace of the manipulator, one unique solution to the limb length is determined by Equation 3.54. Otherwise, when the limbs' lengths derived from Equation 3.54 yield complex numbers, then the position or orientation of the moving platform is not reachable.

3.5.4 Forward Kinematics

In forward kinematic analysis, it is assumed that the vector of limb lengths $\boldsymbol{L} = [\ell_1, \ell_2, \ell_3, \ell_4, \ell_5, \ell_6]^T$ are given and the problem is to find the position ${}^A\boldsymbol{P}$, and the orientation of the moving platform ${}^A\boldsymbol{R}_B$. The size of the problem depends on the representation used for orientation. If a rotation matrix is used for this representation, the number of unknowns in the forward kinematic problem are 12. Position vector ${}^A\boldsymbol{P}$ introduced three unknowns p_x, p_y, p_z, and nine other unknowns are the components of the rotation matrix. On the other hand, Equation 3.53 provides six nonlinear equations, for $i = 1, 2, \ldots, 6$, and the orthonormality conditions 2.8 through 2.10 provide the other six equations to be solved simultaneously. Therefore, the forward kinematic problem is recast into 12 nonlinear equations with 12 unknowns. This problem is highly nonlinear and is extremely difficult to solve, and only few practical closed-form solutions have been obtained for the general SGP. Moreover, even a numerical iterative procedure is not suitable for an SGP, because it leads to a heavy computational burden and depends highly on a good initial guess in order to converge toward the right solution. In case of using screw axis or unit quaternion representation for the orientation, the number of equations and the unknowns reduce to seven and in case of using any Euler angle representations, the number of equations reduces to six. However, the complexity of the equations significantly increases, resulting in a more expensive computational cost.

3.5.4.1 Background Literature

Although the forward kinematic problem has been addressed in numerous works, a majority of them focus on finding all the possible solutions to the forward kinematics of certain kinds of parallel manipulators [37,82,89,132]. These approaches usually use algebraic formulations to generate a high degree of polynomial or set of nonlinear equations. Then, methods such as algebraic elimination [85,110], interval analysis [132], and continuation method [155,185] are used to obtain the roots of the polynomial. The forward kinematic problem is not fully solved simply by finding all the possible solutions. Schemes are further needed to find a unique actual pose of the platform among all the possible solutions. Use of an iterative numerical procedure [84,106,131,188] and auxiliary sensors [10,24,68,150] are the two commonly adopted schemes to yield a unique solution. Numerical iteration is usually sensitive to the choice of initial values and the nature of the resulting constraint equations. The auxiliary sensors approach has practical limitations, such as cost and measurement errors. Some researchers have also tried using neural networks for solving the forward kinematic problem [53,151,195]. No matter how the forward kinematic problem is solved, direct determination of a unique solution is still a challenging problem. The complexity of the forward kinematic problem depends widely on the manipulator architecture, geometry, and joint sensor layouts. Though promising results have been achieved for certain simplified configurations, the pursuit of practical algorithms for the general SGP continues [189].

As explained earlier, the majority of reported research results focus on finding the solutions to the forward kinematics of certain kinds of parallel manipulators. Figure 3.17 shows a 3-3 SGP whose six limbs meet in a pairwise fashion at three points in the moving platform and the base. This type of special construction makes closed-form forward kinematic solution feasible. This structure is adopted in many flight training simulators. Forward kinematics of such a structure are considered in [66,141], and [3]. It is shown in [66] and [141] that the 3–3 SGP is kinematically equivalent to a spatial $3R\underline{P}S$ mechanism, and

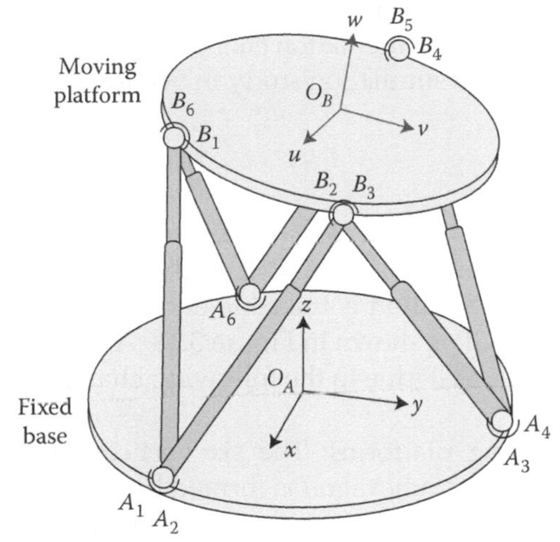

FIGURE 3.17
Schematic of a 3–3 Stewart–Gough platform.

therefore, the forward kinematics can be reduced to a 16th-order polynomial equation. It has been further shown that in order to obtain the set of 16 solutions, it is sufficient to solve only an 8th-order polynomial.

A more general case of a 6–3 SGP is analyzed in [118]. In this structure the six limbs meet in a pairwise fashion at three points on the moving platform. However, they are arbitrarily located in a plane at the base. In the algorithm proposed in [118], the forward kinematic solution is found through a numerical solution of three simultaneous nonlinear algebraic equations. No specified initial guess is needed for convergence toward the right solution. Once the three equations are solved, the unique kinematic solutions are obtained by some algebraic operations.

In the literature, References 52, 189, and 190 can be named as representatives of the analysis performed on a nearly general 6–6 SGP. The analysis given in [190] leads to a 20th-order polynomial equation in one unknown, from which with other equations 40 different locations of the platform can be derived. A generalization of the SGP is introduced in [52], by considering all possible geometric constraints between six pairs of geometric primitives on the base and the platform, respectively. This gives 3850 types of generalized SGPs with the original SGP as one of the cases. Furthermore, closed-form solutions to general SGPs are also given. It is proved that the original SGP could have 40 real solutions.

All analytical solutions like the one given in [52,190] are much involved and fairly complex for implementation. Wang has presented a numerical method for forward kinematics of nearly general SGPs, which can directly generate a unique solution [189]. This method utilizes the trivial nature of inverse kinematics of parallel manipulators, and derives a straightforward linear relationship between the small change in joint variables (limbs' lengths) and the resulting small motion of the platform. The solution to the forward kinematics is then obtained through a series of small changes in joint variables.

In fact, mathematical complexity of the forward kinematics of SGPs has become a serious deficiency that prevents it from being used in many high-speed, real-time, and online implementations. In the following sections, an analytical solution to a special structure of an SGP, and a numerical method to find the forward kinematics for a general Stewart platform structure are given. The main purpose of introducing these two methods as representative of the vast research work in this area is to give a scientific judgment to the performance, limitations, and implementation characteristics of these methods in practice. The methods are compared by a simulation study in Section 3.5.5.

3.5.4.2 Analytical Solution

In this section, the closed-form solution of a special geometry of an SGP is elaborated. The special geometry adopted is called $6\text{–}6^p$ SGP. It is chosen for the analysis given in this section, because its analytical solution is tractable and can be followed more easily than other geometries. The $6\text{–}6^P$ SGP is shown in Figure 3.18. As defined in [169], this platform is different from a general 6–6 SGP due to the following characteristics.

1. Planar base and moving platforms: The six vertices of the base platform are required to be in a plane, so a hexagon is formed by the six vertices. Similarly, the six vertices of the moving platform are also in a plane to form another hexagon.
2. Similar hexagons: The two hexagons are required to be similar. In other words, if the six vertices of the base plate are given, the corresponding vertices of the moving platform will be determined simply by multiplying the positions of the base plate by a scalar factor.

This special $6\text{–}6^p$ SGP has been studied by many researchers. It is well-known that a closed-form solution in the forward kinematics usually introduces unnecessary complex roots [18]. Yang and Geng analyzed the condition when complex roots would vanish, and the condition was used such that the unnecessary complex roots were avoided and only four real roots were found from a 4th-order polynomial with eight solutions to the kinematics of the platform [194]. In this section, a simpler closed-form forward kinematic

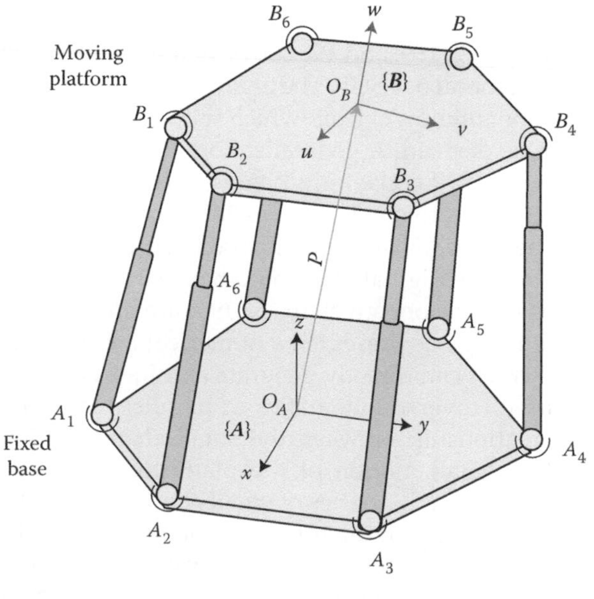

FIGURE 3.18
Schematic of a $6\text{–}6^P$ Stewart–Gough platform.

solution to the 6–6^p SGP will be given from a set of quadratic equations. This set of quadratic equations is obtained by separating orientation from position equations [89]. Besides, the screw axis representation is introduced to get the orientation of the moving platform. Obviously, by this means, solving a degree four polynomial equation is avoided. Another advantage is that unnecessary complex roots vanish automatically.

As shown in Figure 3.18, the two coordinate systems $\{A\}$ and $\{B\}$ are attached to the base and the moving platform, respectively. As shown in Figure 3.16, the platform geometry can be described by vectors $\boldsymbol{a}_i$ and $\boldsymbol{b}_i$ for $i = 1, 2, \ldots, 6$, which represent the vertices of the base platform in frame $\{A\}$ and the vertices of the moving platform in frame $\{B\}$, respectively. The six limbs are denoted by vectors $\ell_i \hat{\boldsymbol{s}}_i$ that connect vertices A_i and B_i for $i = 1, 2, \ldots, 6$. In this 6–6^p SGP, since two hexagons are similar, the following condition exists:

$$\boldsymbol{b}_i = \mu\, \boldsymbol{a}_i; \qquad i = 1, 2, \ldots, 6, \tag{3.55}$$

where μ is a designated constant called the scaling factor [169]. In general

$$\boldsymbol{a}_i = a_{x_i}\hat{\boldsymbol{x}} + a_{y_i}\hat{\boldsymbol{y}} + a_{z_i}\hat{\boldsymbol{z}}, \tag{3.56}$$

where a_{x_i}, a_{y_i}, and a_{z_i} are three components of $\boldsymbol{a}_i$, and $\hat{\boldsymbol{x}}, \hat{\boldsymbol{y}}$, and $\hat{\boldsymbol{z}}$ are three unit vectors of frame $\{A\}$. In the 6–6^p SGP, $a_{z_i} = 0$, therefore

$$\boldsymbol{a}_i = a_{x_i}\hat{\boldsymbol{x}} + a_{y_i}, \hat{\boldsymbol{y}} \tag{3.57}$$

$$\boldsymbol{b}_i = \mu(a_{x_i}\hat{\boldsymbol{x}} + a_{y_i}\hat{\boldsymbol{y}}). \tag{3.58}$$

Recall the kinematic loop closure Equation 3.51 and substitute Equation 3.56. Hence, for $i = 1, 2, \ldots, 6$

$$\ell_i\, {}^A\hat{\boldsymbol{s}}_i = {}^A\boldsymbol{P} + {}^A\boldsymbol{R}_B\, {}^B\boldsymbol{b}_i - {}^A\boldsymbol{a}_i,$$

$$\ell_i\, \hat{\boldsymbol{s}}_i = \boldsymbol{P} + (\mu\boldsymbol{R} - \boldsymbol{I})\, \boldsymbol{a}_i, \tag{3.59}$$

in which $\boldsymbol{I}$ is a 3×3 identity matrix, $\boldsymbol{R} = {}^A\boldsymbol{R}_B$, and for simplicity, the superscript ^{A} is omitted from the notations. Therefore, for $i = 1, 2, \ldots, 6$, the limbs' lengths can be obtained from

$$\begin{aligned} \ell_i^2 &= [(\mu R - \boldsymbol{I})\, \boldsymbol{a}_i + \boldsymbol{P}]^T\, [(\mu R - \boldsymbol{I})\, \boldsymbol{a}_i + \boldsymbol{P}] \\ &= \boldsymbol{P}^T\boldsymbol{P} + \boldsymbol{a}_i^T \left[(\mu\boldsymbol{R}^T - \boldsymbol{I})(\mu\boldsymbol{R} - \boldsymbol{I})\right] \boldsymbol{a}_i + 2\boldsymbol{a}_i^T(\mu\boldsymbol{R}^T - \boldsymbol{I})\,\boldsymbol{P}. \end{aligned} \tag{3.60}$$

By substituting Equation 3.57 into Equation 3.60, and for $i = 1, 2, \ldots, 6$, we have

$$\begin{aligned} \ell_i^2 &= \boldsymbol{P}^T\boldsymbol{P} + a_{x_i}\left[\hat{\boldsymbol{x}}^T(\mu\boldsymbol{R}^T\boldsymbol{P} - \boldsymbol{P})\right] + 2a_{y_i}\left[\hat{\boldsymbol{y}}^T(\mu\boldsymbol{R}^T\boldsymbol{P} - \boldsymbol{P})\right] \\ &\quad + 2\mu\left\{a_{x_i}^2(\hat{\boldsymbol{x}}^T\boldsymbol{R}\hat{\boldsymbol{x}}) + a_{x_i}a_{y_i}(\hat{\boldsymbol{x}}^T\boldsymbol{R}\hat{\boldsymbol{y}} + \hat{\boldsymbol{y}}^T\boldsymbol{R}\hat{\boldsymbol{x}}) + a_{y_i}^2(\hat{\boldsymbol{y}}^T\boldsymbol{R}\hat{\boldsymbol{y}})\right\} \\ &\quad + (1 + \mu^2)(a_{x_i}^2 + a_{y_i}^2). \end{aligned} \tag{3.61}$$

By defining $\boldsymbol{W} = [W_1, W_2, W_3, W_4, W_5, W_6]^T$ with

$$\begin{aligned} W_1 &= \boldsymbol{P}^T\boldsymbol{P}, \\ W_2 &= \hat{\boldsymbol{x}}^T(\mu\boldsymbol{R}^T\boldsymbol{P} - \boldsymbol{P}), \\ W_3 &= \hat{\boldsymbol{y}}^T(\mu\boldsymbol{R}^T\boldsymbol{P} - \boldsymbol{P}), \end{aligned} \tag{3.62}$$

and

$$W_4 = \hat{\boldsymbol{x}}^T \boldsymbol{R}\hat{\boldsymbol{x}},$$
$$W_5 = \hat{\boldsymbol{x}}^T \boldsymbol{R}\hat{\boldsymbol{y}} + \hat{\boldsymbol{y}}^T \boldsymbol{R}\hat{\boldsymbol{x}}, \tag{3.63}$$
$$W_6 = \hat{\boldsymbol{y}}^T \boldsymbol{R}\hat{\boldsymbol{y}}.$$

Equation 3.61 becomes a linear algebraic set of equations represented by

$$\boldsymbol{Q}\,\boldsymbol{W} = \boldsymbol{d}. \tag{3.64}$$

Matrix $\boldsymbol{Q} = [Q_1, Q_2, Q_3, Q_4, Q_5, Q_6]^T$ may be derived componentwise by

$$Q_i = \left[1, 2a_{x_i}, 2a_{y_i}, -2\mu a_{x_i}^2, -2\mu a_{x_i} a_{y_i}, -2\mu a_{x_i}^2\right], \tag{3.65}$$

and vector $\boldsymbol{d} = [d_1, d_2, d_3, d_4, d_5, d_6]^T$ may be obtained by

$$d_i = \ell_i^2 - (1+\mu^2)(a_{x_i}^2 + a_{y_i}^2). \tag{3.66}$$

It is worth noting that, by this formulation, the position of the moving platform is solely a function of W_1, W_2, and W_3, and is separated from the orientation of the moving platform, which is a function of W_4, W_5, and W_6. If matrix $\boldsymbol{Q}$ is not singular, that is, the six vertices of the base platform or the moving platform are not in a quadratic curve, $\boldsymbol{W}$ can be obtained by

$$\boldsymbol{W} = \boldsymbol{Q}^{-1}\,\boldsymbol{d}. \tag{3.67}$$

Since $\boldsymbol{W}$ is now determined, Equation 3.63 can be used to analyze the orientation, and Equation 3.62 can be used to derive the position of the moving platform as follows. Let us represent the orientation of the moving platform by the screw axis $\left\{\hat{s} = \begin{bmatrix} s_x & s_y & s_z \end{bmatrix}^T, \theta\right\}$. Hence, the rotation matrix can be written as

$$\boldsymbol{R} = \begin{bmatrix} s_x^2 v\theta + c\theta & s_x s_y v\theta - s_z s\theta & s_x s_z v\theta + s_y s\theta \\ s_y s_x v\theta + s_z s\theta & s_y^2 v\theta + c\theta & s_y s_z v\theta - s_x s\theta \\ s_z s_x v\theta - s_y s\theta & s_z s_y v\theta + s_x s\theta & s_z^2 v\theta + c\theta \end{bmatrix}. \tag{3.68}$$

Note that

$$W_4 = \hat{\boldsymbol{x}}^T \boldsymbol{R}\hat{\boldsymbol{x}} = r_{11},$$
$$W_5 = \hat{\boldsymbol{x}}^T \boldsymbol{R}\hat{\boldsymbol{y}} + \hat{\boldsymbol{y}}^T \boldsymbol{R}\hat{\boldsymbol{x}} = r_{12} + r_{21}, \tag{3.69}$$
$$W_6 = \hat{\boldsymbol{y}}^T \boldsymbol{R}\hat{\boldsymbol{y}} = r_{22},$$

in which r_{ij} denotes the components of the rotation matrix $\boldsymbol{R}$, and $s\theta = \sin\theta, c\theta = \cos\theta$, and $v\theta = 1 - \cos\theta$ are used as the shorthand notations. Equation 3.69 provides three nonlinear equations for the screw parameters. Considering the unitary property of the screw axis, the following four equations must be solved to find the screw parameters:

$$W_4 = s_x^2 v\theta + c\theta, \tag{3.70}$$

$$W_5 = 2s_x s_y v\theta, \tag{3.71}$$

$$W_6 = s_y^2 v\theta + c\theta, \tag{3.72}$$

$$1 = s_x^2 + s_y^2 + s_z^2. \tag{3.73}$$

From Equations 3.70 and 3.72, s_x^2 and s_y^2 can be determined as follows:

$$s_x^2 = \frac{W_4 - c\theta}{v\theta}; \quad s_y^2 = \frac{W_6 - c\theta}{v\theta}. \tag{3.74}$$

By substituting Equation 3.74 into Equation 3.71, and with some manipulation, a quadratic equation in $\cos\theta$ is obtained,

$$\cos^2\theta - B\cos\theta + C = 0, \tag{3.75}$$

in which

$$B = W_4 + W_6; \quad C = W_4 W_6 - W_5^2/4. \tag{3.76}$$

Therefore, the screw parameters can be determined from

$$\cos\theta = \frac{1}{2}\left(B - \sqrt{B^2 - 4C}\right), \tag{3.77}$$

and if $W_4 \neq c\theta$, then

$$s_x = \sqrt{\frac{W_4 - c\theta}{v\theta}}; \quad s_y = \frac{W_5}{2s_x v\theta}; \quad s_z = \sqrt{1 - s_x^2 - s_y^2}; \tag{3.78}$$

otherwise, if $W_6 \neq c\theta$, then

$$s_y = \sqrt{\frac{W_6 - c\theta}{v\theta}}; \quad s_x = \frac{W_5}{2s_y v\theta}; \quad s_z = \sqrt{1 - s_x^2 - s_y^2}; \tag{3.79}$$

and in case $W_4 = W_6 = c\theta$, then

$$s_x = s_y = 0; \quad s_z = 1. \tag{3.80}$$

Here, the positive root of $B^2 - 4C$ is omitted, such that unnecessary complex roots of θ are avoided. Furthermore, the negative roots of s_x, s_z provide four different screw representations for the rotation of the moving platform. These four rotations are

$$\mathbf{s} = \begin{bmatrix} s_x & s_y & s_z \end{bmatrix}^T, \quad \mathbf{s} = \begin{bmatrix} s_x & s_y & -s_z \end{bmatrix}^T,$$
$$\mathbf{s} = \begin{bmatrix} -s_x & -s_y & s_z \end{bmatrix}^T, \quad \mathbf{s} = \begin{bmatrix} -s_x & -s_y & -s_z \end{bmatrix}^T.$$

By substituting the screw parameters into Equation 3.68, four different transformation matrices may be obtained as $\boldsymbol{R}_i, i = 1, 2, 3, 4$, correspondingly.

Now that the orientation of the moving platform is determined, the position of its center point is found using W_1, W_2, and W_3. Let us define

$$\boldsymbol{U}^T = \hat{x}^T\,(\mu \boldsymbol{R}^T - \boldsymbol{I}), \tag{3.81}$$

$$\boldsymbol{V}^T = \hat{y}^T\,(\mu \boldsymbol{R}^T - \boldsymbol{I}). \tag{3.82}$$

Vectors $\boldsymbol{U}$ and $\boldsymbol{V}$ are determined from each transformation matrix $\boldsymbol{R}_i$, $i = 1, 2, 3, 4$, therefore, $\boldsymbol{U}$ and $\boldsymbol{V}$ have four cases as well. Now Equation 3.62 becomes

$$W_1 = \boldsymbol{P}^T \boldsymbol{P}, \tag{3.83}$$

$$W_2 = \boldsymbol{U}^T \boldsymbol{P}, \tag{3.84}$$

$$W_3 = \boldsymbol{V}^T \boldsymbol{P}. \tag{3.85}$$

Obviously, Equations 3.84 and 3.85 represent two planes and their intersection is a line represented as

$$\boldsymbol{P} = \boldsymbol{r}_0 + t\boldsymbol{r}_1, \tag{3.86}$$

where t is the parameter of the line, r_0 is a line in the plane spanned by vectors $\boldsymbol{U}$ and $\boldsymbol{V}$.

$$\boldsymbol{r}_0 = \frac{(\boldsymbol{V}^T\boldsymbol{V})W_2 - (\boldsymbol{U}^T\boldsymbol{V})W_3}{(\boldsymbol{U}^T\boldsymbol{U})(\boldsymbol{V}^T\boldsymbol{V}) - (\boldsymbol{U}^T\boldsymbol{V})^2}\boldsymbol{U} + \frac{(\boldsymbol{U}^T\boldsymbol{V})W_2 - (\boldsymbol{U}^T\boldsymbol{U})W_3}{(\boldsymbol{U}^T\boldsymbol{U})(\boldsymbol{V}^T\boldsymbol{V}) - (\boldsymbol{U}^T\boldsymbol{V})^2}\boldsymbol{V} \tag{3.87}$$

and $\boldsymbol{r}_1$ is a unit vector perpendicular to both vectors $\boldsymbol{U}$ and $\boldsymbol{V}$:

$$\boldsymbol{r}_1 = \frac{\boldsymbol{U} \times \boldsymbol{V}}{\|\boldsymbol{U} \times \boldsymbol{V}\|} \tag{3.88}$$

in which $\times$ denotes the vector cross product. Obviously

$$\boldsymbol{r}_0^T \boldsymbol{r}_1 = \boldsymbol{r}_1^T \boldsymbol{r}_0 = 0. \tag{3.89}$$

The line Equation 3.86, intersects the sphere Equation 3.83 at two points $\boldsymbol{P}_1$ and $\boldsymbol{P}_2$, which are the solutions to Equations 3.83 through 3.85. Here

$$\boldsymbol{P}_1 = \boldsymbol{r}_0 + t\boldsymbol{r}_1, \tag{3.90}$$

$$\boldsymbol{P}_2 = \boldsymbol{r}_0 - t\boldsymbol{r}_1, \tag{3.91}$$

and

$$t = \sqrt{W_1 - \boldsymbol{r}_0^T \boldsymbol{r}_0} \tag{3.92}$$

considering Equation 3.89. So far, both orientation matrix $\boldsymbol{R}$ and position vector $\boldsymbol{P}$ have been obtained. The forward kinematics for the 6–6^p Stewart platform have eight solutions if the equations are not degenerated.

The analytical solution presented here for the 6–6^P SGP, has major advantages to other solutions presented in the literature for a nearly general SGP, such as [190,194]. The presented analytical solution is computationally more efficient, and more tractable than the others. However, there are specific deficiencies in any analytical solution, which is reported less in the literature. The analytical solution of the 6–6^P SGP is not excluded from this rule, and two main deficiencies have been reported in the simulations carried out on this solution.

The first deficiency is the dependency of the orientation parameters of the solution, namely W_4, W_5, and W_6 on the value of $c\theta$. Reconsider Equations 3.70 through 3.72 for the case where $\theta = 0$ or $\theta = \pi$, by which $c\theta = 1$ and $v\theta = 0$. In this case, $W_4 = W_6 = 1$, and W_5 becomes zero, and these equations fail to generate any solution for s_x, s_y, and s_z. Unfortunately, there is no means to overcome this problem in the solution, and this is a singular orientation for the given analytical solution.

The second deficiency is more general and it applies to all analytical solutions. As seen in the given analysis, eight possible solutions for a given set of limbs' lengths are produced. Schemes are further needed to find a unique actual position, and orientation of the platform among all the possible solutions. Use of an iterative numerical procedure [84,106,131,188] and auxiliary sensors [10,24,68,150] are the two commonly adopted schemes to further lead to a unique solution. Numerical iteration is usually sensitive to the choice of initial values and the nature of the resulting constraint equations. The auxiliary sensors approach has practical limitations, such as cost and measurement errors. No matter how the forward kinematic problem has been solved, the direct determination of a unique solution is still a challenging problem.

As detailed in Section 3.5.5, the scheme used here to determine a unique solution of all the possible solutions, is based on the fact that at each step, the solutions are compared to those of the previous step. The solution whose error norm to the previous solution is smaller than a threshold is selected as the right solution. This causes two deficiencies in the analytical solution. The first deficiency is that although the analytical method does not inherently need any initial guess to determine the possible solutions, in order to determine a unique solution at initial time, an initial guess of position and orientation is needed, and converging towards the desired solution depends on the accuracy of the initial guess. The second deficiency observed in the simulation is for the cases where either one of s_x, s_y or especially s_z approaches zero. Although the solution at the time when either of the screw axes components are zero is correctly obtained by the proposed method, right after this point in time, the proposed method cannot distinguish between the positive or negative values of the solutions, and may fail in providing the right solution. The problem is exaggerated if two of the screw axes components approach zero simultaneously.

From the above-mentioned observation, it can be concluded that for the cases where none of the screw axes parameters, namely s_x, s_y, s_z, and θ, are zero or approaching zero, the analytical solution leads to the right solution. But otherwise, the solution may degenerate, or may diverge from the right solution. In the next section, a numerical solution to the problem is given, in which although the computational time is 10 times greater than that of the analytical solution, it is not much sensitive to the special cases of the orientation to converge towards the right solution.

3.5.4.3 Numerical Solution

In order to generate a numerical solution to the forward kinematic problem, assume that the limbs' lengths $\boldsymbol{L} = [\ell_1, \ell_2, \ell_3, \ell_4, \ell_5, \ell_6]^T$ are given and the problem is to obtain the position ${}^{A}\boldsymbol{P} = [p_x, p_y, p_z]^T$, and orientation of the moving platform represented by screw parameters $\{\hat{\mathbf{s}} = [s_x, s_y, s_z]^T, \theta\}$. Therefore, seven unknown variables can be encapsulated by the vector $\boldsymbol{x}$ as

$$\boldsymbol{x} = \begin{bmatrix} p_x & p_y & p_z & s_x & s_y & s_z & \theta \end{bmatrix}^T \tag{3.93}$$

$$= \begin{bmatrix} x_1 & x_2 & x_3 & x_4 & x_5 & x_6 & x_7 \end{bmatrix}^T. \tag{3.94}$$

The unknown position vector $\boldsymbol{P}$, and orientation matrix $\boldsymbol{R}(\boldsymbol{x})$ can be determined as a function of $\boldsymbol{x}$ by the following equations:

$$\boldsymbol{P}(\boldsymbol{x}) = \begin{bmatrix} x_1 & x_2 & x_3 \end{bmatrix}^T \tag{3.95}$$

and

$$\boldsymbol{R}(\boldsymbol{x}) = \begin{bmatrix} x_4^2\, v x_7 + c x_7 & x_4 x_5\, v x_7 - x_6\, s x_7 & x_4 x_6\, v x_7 + x_5\, s x_7 \\ x_5 x_4\, v x_7 + x_6\, s x_7 & x_5^2\, v x_7 + c x_7 & x_5 x_6\, v x_7 - x_4\, s x_7 \\ x_6 x_4\, v x_7 - x_5\, s x_7 & x_6 x_5\, v x_7 + x_4\, s x_7 & x_6^2\, v x_7 + c x_7 \end{bmatrix}. \tag{3.96}$$

Assume that the geometry of the base and moving platforms' attachment points A_i and B_i are given by vectors $\boldsymbol{a}_i$, and $\boldsymbol{b}_i$, respectively. The numerical solution involves iteratively finding the solution to the following seven nonlinear equations:

$$\begin{cases} F_i(\boldsymbol{x}) = -\ell_i^2 + [\boldsymbol{P}(\boldsymbol{x}) + \boldsymbol{R}(\boldsymbol{x})\,\boldsymbol{b}_i - \boldsymbol{a}_i]^T\,[\boldsymbol{P}(\boldsymbol{x}) + \boldsymbol{R}(\boldsymbol{x})\,\boldsymbol{b}_i - \boldsymbol{a}_i] \\ \qquad \text{for } i = 1, 2, \ldots, 6 \\ F_7(\boldsymbol{x}) = x_4^2 + x_5^2 + x_6^2 - 1 \end{cases} \tag{3.97}$$

Numerical methods using nonlinear least-squares optimization routines [138], can be used to obtain a solution for $\boldsymbol{F}(\boldsymbol{x}) = \boldsymbol{0}$. In a least-squares problem, the functional $f(\boldsymbol{x}) = \frac{1}{2}\sum_i F_i(\boldsymbol{x})^2$ is minimized over $\boldsymbol{x} \in \mathbb{R}^n$ [34]. The Gauss–Newton and Levenberg–Marquardt methods [112,122] are two main search routines used to solve the nonlinear least-squares problem.* The flowchart given in Figure 3.19 reveals the details of the iterative method used to obtain the forward kinematic solution. As seen in this flowchart, for a given $\boldsymbol{x}$, the values of $\boldsymbol{P}(\boldsymbol{x})$ and $\boldsymbol{R}(\boldsymbol{x})$ are calculated using Equations 3.95 and 3.96, respectively. Next, $F_i(\boldsymbol{x})$ for $i = 1, 2, \ldots, 7$ are calculated using Equation 3.97. Then, the value of $f(\boldsymbol{x}) = \frac{1}{2}\sum_i F_i(\boldsymbol{x})^2$ is calculated, and if it is not close to zero with the required accuracy, an optimal search routine is used to recalculate a new value for $\boldsymbol{x}$. This iteration is followed to obtain a solution for $f(\boldsymbol{x}) = 0$ with an accuracy of $\epsilon \ll 1$. Multiple solutions may exist for the equation $\boldsymbol{F}(\boldsymbol{x}) = \boldsymbol{0}$, and in order to avoid abrupt changes in the forward kinematic solutions, the solution of the previous iteration is used to search for the next step solution. Simulation results detailed in Section 3.5.5 illustrate the effectiveness, and accuracy of the numerical routines used to solve the forward kinematics.

3.5.5 Simulations

To verify the accuracy and integrity of the solutions, a simulation study is performed for a typical 6–6^p SGP in this section. The forward kinematic solution results are given for

* Use `fsolve` function in MATLAB.

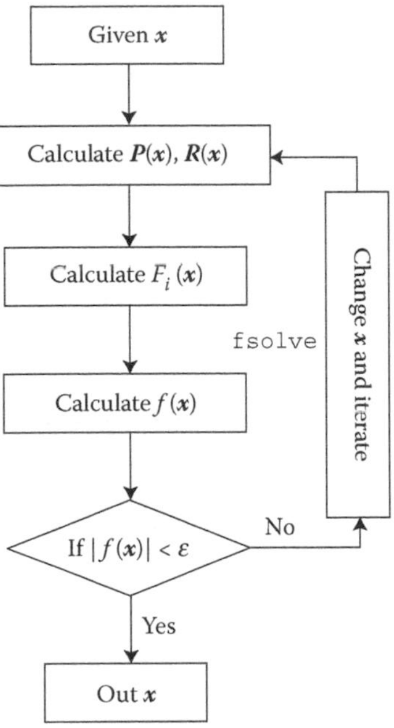

FIGURE 3.19
Flowchart of the iterative routine used to solve the forward kinematics of Stewart–Gough platform.

TABLE 3.2
Geometric Parameters of the 6–6^P Stewart–Gough Platform

Description	Quantity	Unit
a_1: Position vector of point A_1	$[+0.8, +0.0, 0]^T$	m
a_2: Position vector of point A_2	$[+0.5, +1.2, 0]^T$	m
a_3: Position vector of point A_3	$[-0.2, +1.0, 0]^T$	m
a_4: Position vector of point A_4	$[-0.6, +0.4, 0]^T$	m
a_5: Position vector of point A_5	$[-0.4, -0.8, 0]^T$	m
a_6: Position vector of point A_6	$[+0.4, -1.2, 0]^T$	m
μ: The moving platform size ratio $b_i = \mu a_i$	0.6	–

the analytical solution and they are compared to those of the numerical solution. The geometric parameters used in the simulations are given in Table 3.2, and the base platform geometry is shown in Figure 3.20.

3.5.5.1 *Analytical Solution*

For simulations, a typical spatial trajectory for the 6–6^p SGP is considered. The trajectories consist of three position trajectories and three orientation trajectories represented by the screw parameter illustrated in Figure 3.21. As seen in this figure, in the x and y directions 1 and 1.5 m movements are considered, respectively, while for the z direction, the initial height is 1 m, and the final height is considered 1.5 m. The orientation of the moving platform is considered as a spatial rotation about a constant screw axis $\hat{s} = [1, 1, 1]^T/\sqrt{3}$, with $\pi/6$ initial angle and $\pi/3$ maximum rotation.

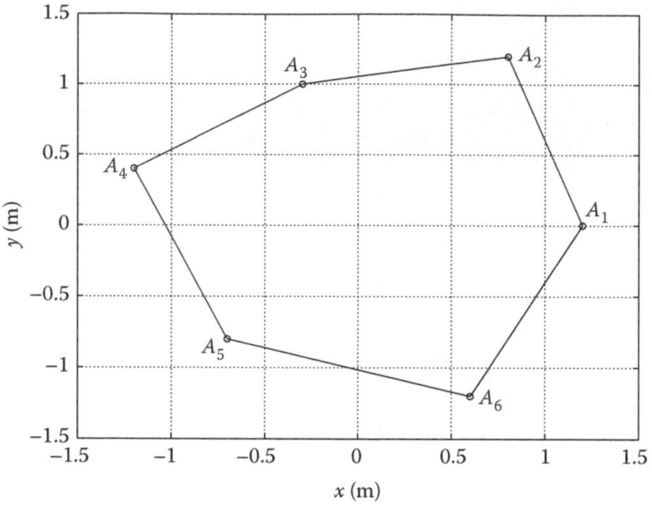

FIGURE 3.20
The base geometry of the examined 6–6^P Stewart–Gough platform.

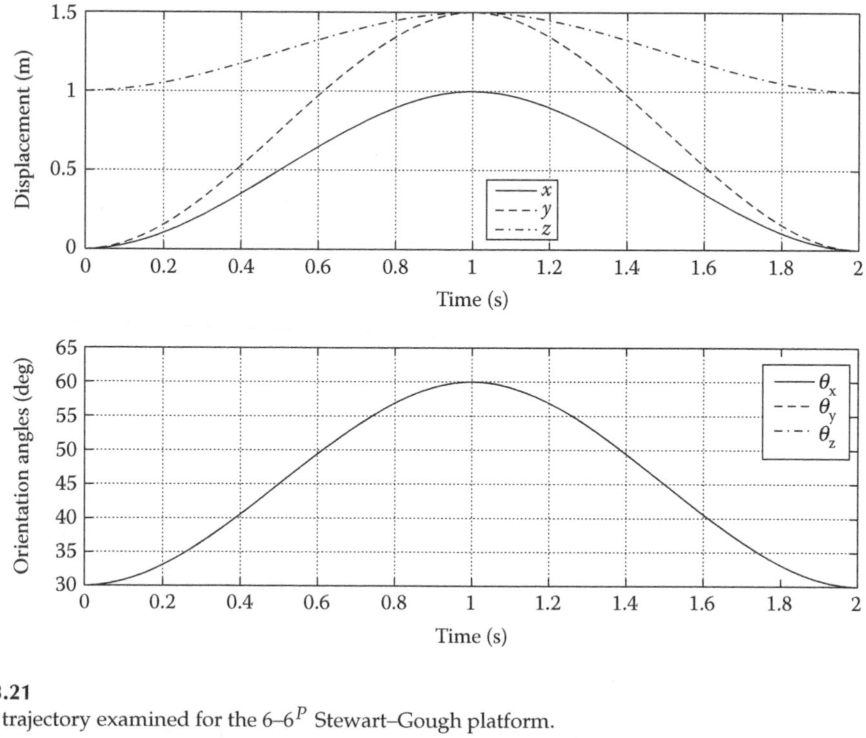

FIGURE 3.21
A typical trajectory examined for the 6–6^P Stewart–Gough platform.

For the given trajectory, inverse kinematic solution of the manipulator is found from Equation 3.54, and manipulator limbs' lengths are uniquely determined and illustrated in Figure 3.22. As seen in this figure, manipulator limbs' lengths vary from 1 m to about 3 m, according to the desired trajectory variations. As shown in Figure 3.23, and to verify the accuracy and integrity of the solutions, the forward kinematic solution for the manipulator

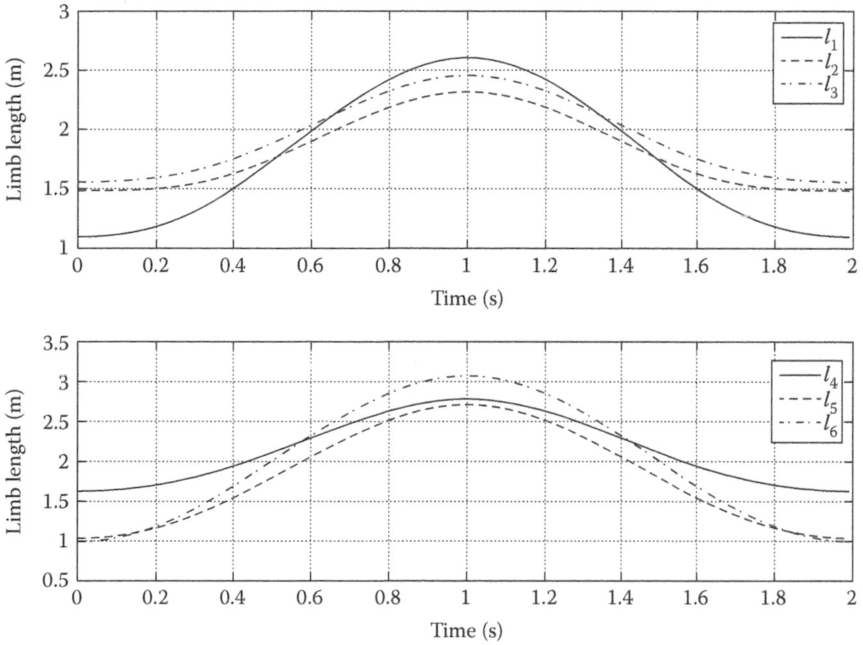

FIGURE 3.22
Inverse kinematic solution of a 6–6^P Stewart–Gough platform for the given trajectory.

is derived for the determined limbs' lengths of the inverse kinematics, which are illustrated in Figure 3.22. The difference between the forward kinematic solution $\mathcal{X}_c$ and the desired motion $\mathcal{X}_d$ is illustrated in Figure 3.23. It is observed that the calculated analytical solution to the forward kinematic problem is completely identical to the desired trajectories to an accuracy of 10^{-14}. It should be noted that the forward kinematic solutions are not unique, and to avoid converging toward other solutions at each step, previous step solution is used as the initial guess for the next step iteration. By this means, the numerical solution converges toward the right solution at all the examined points. Similar results are observed for other trajectories, satisfying the conditions which would avoid degeneration of the analytical solutions. As explained earlier, in cases where $\theta_d = 0$ or $s_z = 0$ are considered in the desired trajectories, the analytical solution fails to provide the right solution.

3.5.5.2 Numerical Solution

In order to compare the accuracy and the computational effort of the analytical solution to that of the numerical solution, first the same trajectory depicted in Figure 3.21 is considered for the simulation. This time, numerical methods are used to obtain the forward kinematic solution and the difference between the forward kinematic solution $\mathcal{X}_c$ and the desired motion $\mathcal{X}_d$ are illustrated in Figure 3.24. It is observed that the numerical solutions are identical to the desired trajectories to an accuracy of 10^{-8} in positions, and 10^{-10} in orientation parameters. Although the obtained numerical solutions are less accurate than those of analytical solutions, they are quite acceptable and can meet the required precision.

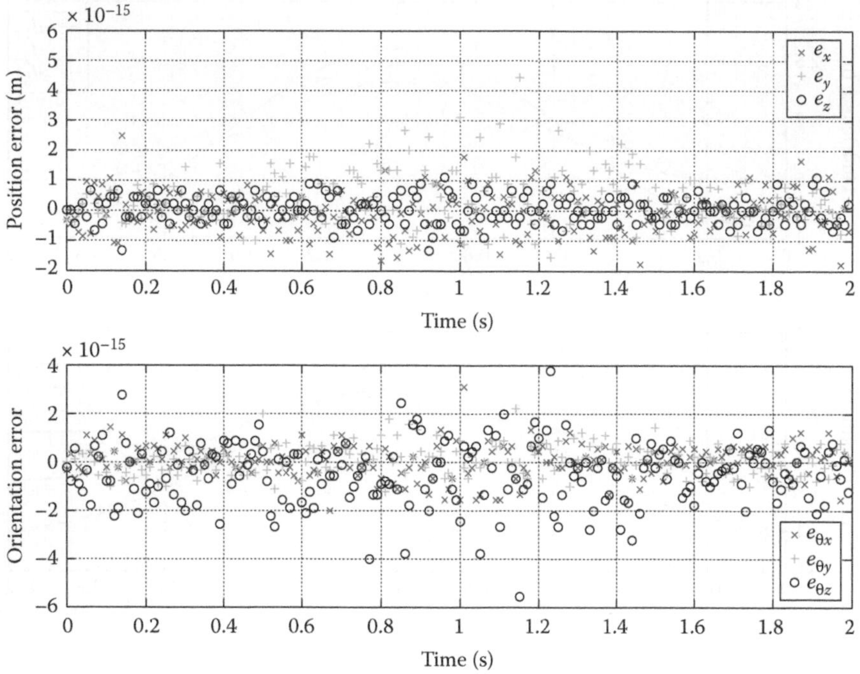

FIGURE 3.23
The accuracy and integrity of the forward and inverse kinematic solutions; analytical approach.

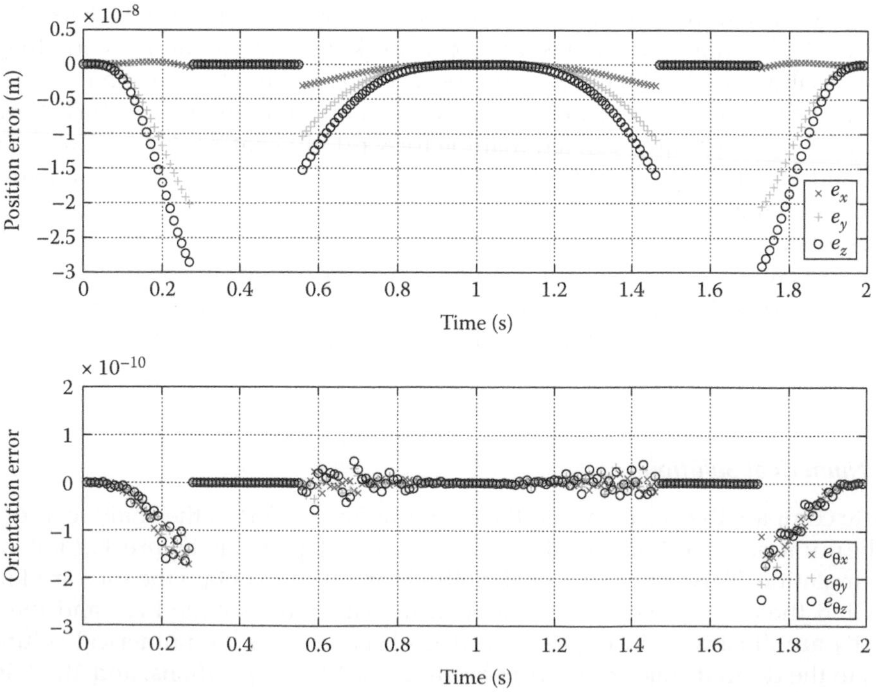

FIGURE 3.24
The accuracy and integrity of the forward and inverse kinematic solutions; numerical approach.

In order to compare the computation time of each method, the simulation has been performed on a Sony laptop, VGN-TX2 Series, with a 1.20 GHz Intel Pentium processor in 200 step times. For the total simulation time of 2 s, the elapsed time for numerical solutions is about 5.89 s, while that for analytical solutions is about 0.65 s. Furthermore, the average computation time of the forward kinematic solution at each iteration step is about 26.33 ms for the numerical method, and about 0.86 ms for the analytical method. It is evident that the computational speed of the numerical solution is much lower than that of the analytical solution.

However, the numerical solution is not sensitive to the orientation parameters, and works well in cases where analytical solution degenerates. Moreover, it can solve forward kinematics of a general SGP and there is no restriction to use a special structure. As an example, consider a general SGP, whose geometric parameters are given in Table 3.3. In this structure, the six vertices of the base and moving platforms do not lie in a plane, and they are not similar, and therefore, this example may be considered as a general SGP. The moving platform position trajectory is as before. However, the orientation of the moving platform is considered such that the screw parameters become zero at a certain instant of time as shown in Figure 3.25. Both the structure and the type of trajectory cause the forward kinematic problem to be unsolvable for the analytical method; however, the problem is still tractable by numerical method.

For the given trajectory, inverse kinematic solution of the manipulator is found from Equation 3.54, and the manipulator limbs' lengths are uniquely determined and illustrated in Figure 3.26. Although the position trajectory in this case is the same as that for the $6\text{–}6^P$ manipulator, the variation of the limbs' lengths in these two cases is different due to different geometries of the two manipulators. As shown in Figure 3.8, the forward kinematic solutions for the manipulator are derived for the determined limbs' lengths illustrated in Figure 3.26. The difference between the forward kinematic solution $\mathcal{X}_c$ and the desired trajectory $\mathcal{X}_d$ is illustrated in Figure 3.27. It is observed that the numerical forward kinematic solutions are identical to the desired trajectories, to an accuracy of 10^{-8}. The accuracies obtained here are almost the same as those in Figure 3.24 for various positions; however,

TABLE 3.3

Geometric Parameters of a General Stewart–Gough Platform

Description	Quantity	Unit
a_1: Position vector of point A_1	$[+0.8, +0.0, +0.1]^T$	m
a_2: Position vector of point A_2	$[+0.5, +1.2, -0.2]^T$	m
a_3: Position vector of point A_3	$[-0.2, +1.0, -0.1]^T$	m
a_4: Position vector of point A_4	$[-0.6, +0.4, +0.2]^T$	m
a_5: Position vector of point A_5	$[-0.4, -0.8, -0.1]^T$	m
a_6: Position vector of point A_6	$[+0.4, -1.2, +0.1]^T$	m
b_1: Position vector of point B_1	$[+0.6, +0.2, -0.1]^T$	m
b_2: Position vector of point B_2	$[+0.3, +1.0, +0.2]^T$	m
b_3: Position vector of point B_3	$[-0.1, +1.2, -0.2]^T$	m
b_4: Position vector of point B_4	$[-0.4, +0.2, -0.1]^T$	m
b_5: Position vector of point B_5	$[-0.2, -1.0, +0.1]^T$	m
b_6: Position vector of point B_6	$[+0.3, -1.0, -0.1]^T$	m

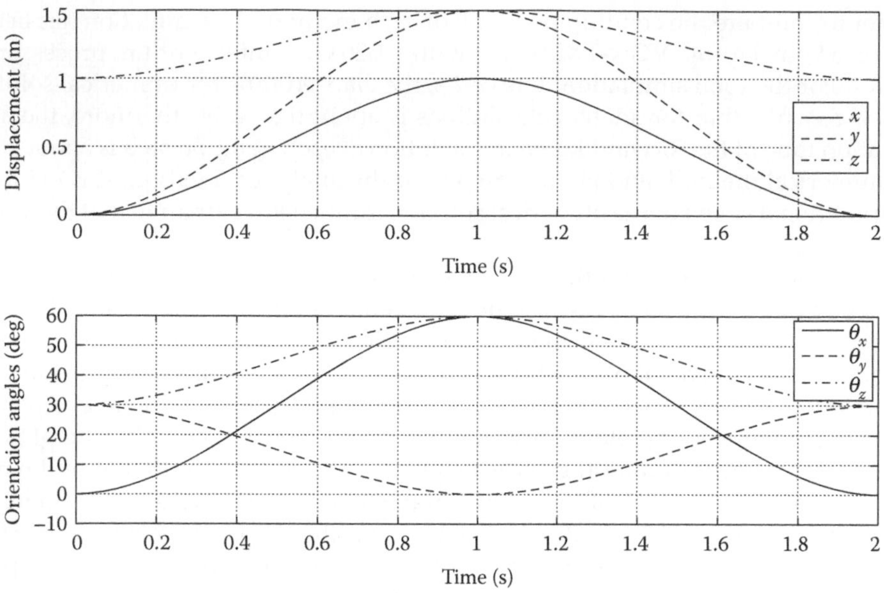

FIGURE 3.25
A typical trajectory examined for the general Stewart–Gough platform.

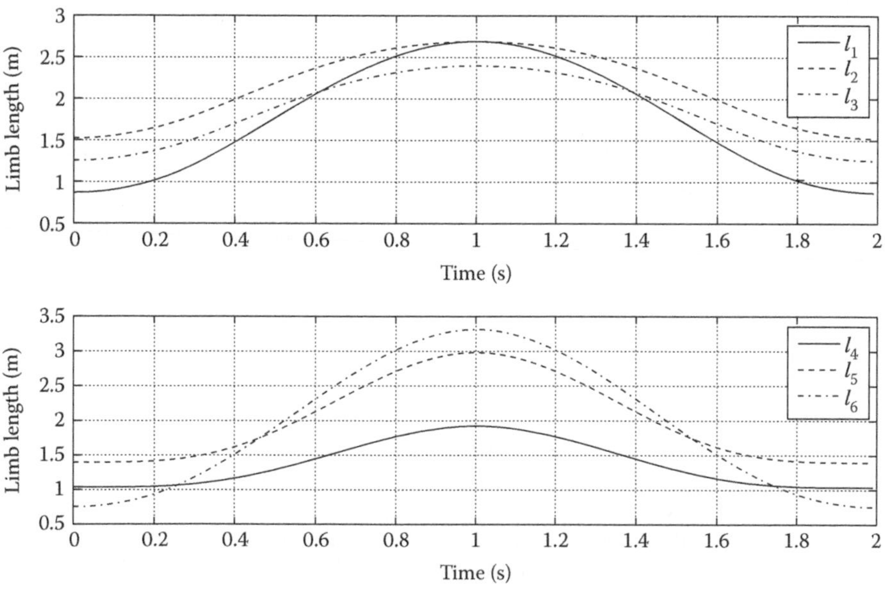

FIGURE 3.26
Inverse kinematic solution for the given trajectory for general Stewart–Gough platform.

they are less than those of the analytical solutions given in Figure 3.23. Nevertheless, the accuracies still satisfy practical requirements. Finally, it should be mentioned that the numerical solution is sensitive to the initial guess, and may diverge from the right solutions if it is not well assigned.

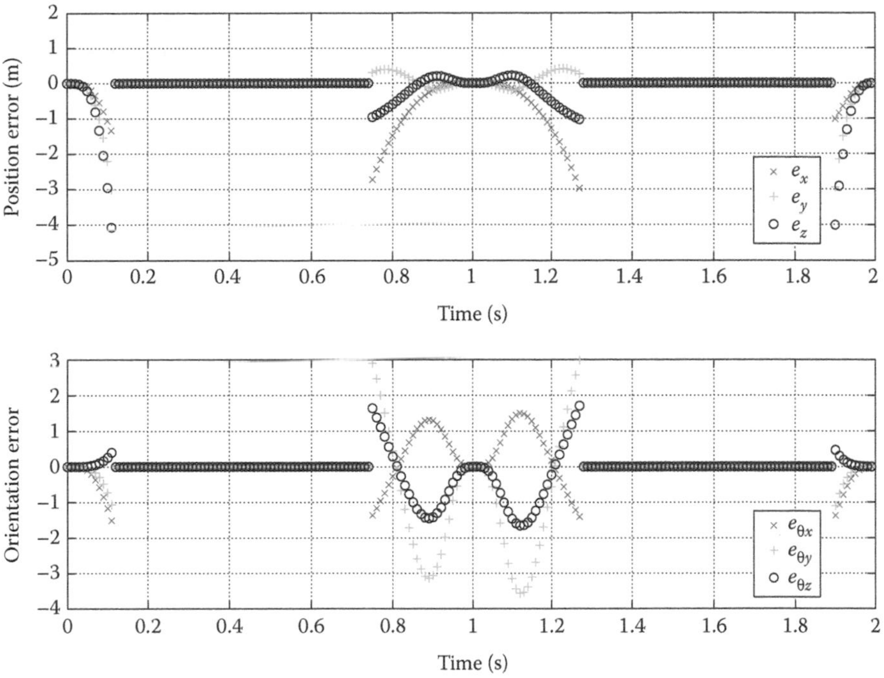

FIGURE 3.27
The accuracy and integrity of the forward and inverse kinematic solutions for the general Stewart–Gough platform.

PROBLEMS

1. Consider the five-bar mechanism shown in Figure 3.28. How many degrees-of-freedom exist in this mechanism? Find the end-effector position vector $\boldsymbol{P} = [p_x, p_y]^T$ as a function of the vector of two input angles $\boldsymbol{\theta} = [\theta_1, \theta_2]^T$.

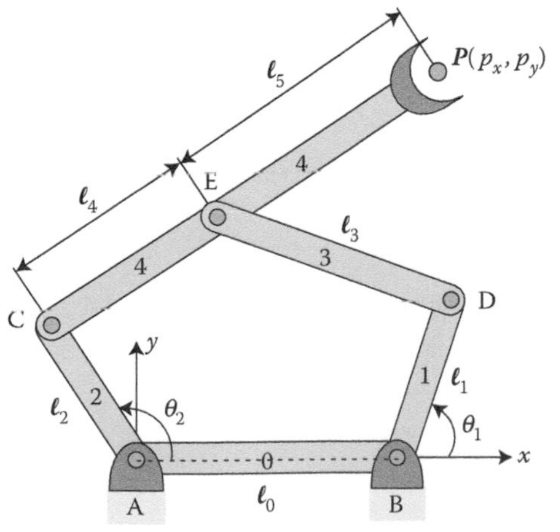

FIGURE 3.28
Five bar mechanism.

2. Consider a five-bar mechanism shown in Figure 3.29 which is constructed from four revolute joints and one prismatic joint. Find the end-effector position vector $\boldsymbol{P} = [p_x, p_y]^T$ as a function of two input variables ℓ_1, θ_2.
3. Figure 3.30 shows a planar pantograph mechanism in which the ground-connected joints are prismatic, and the remaining joints are revolute. The structure of this mechanism is based on a parallelogram, in which links 3 and 5 and links 4 and 6

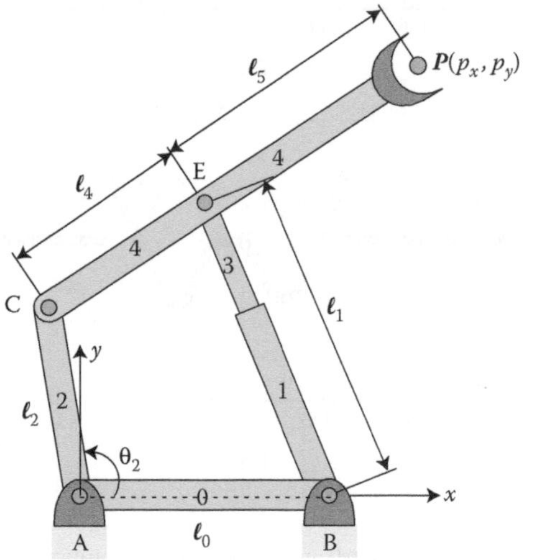

FIGURE 3.29
Five-bar, 4P1R mechanism.

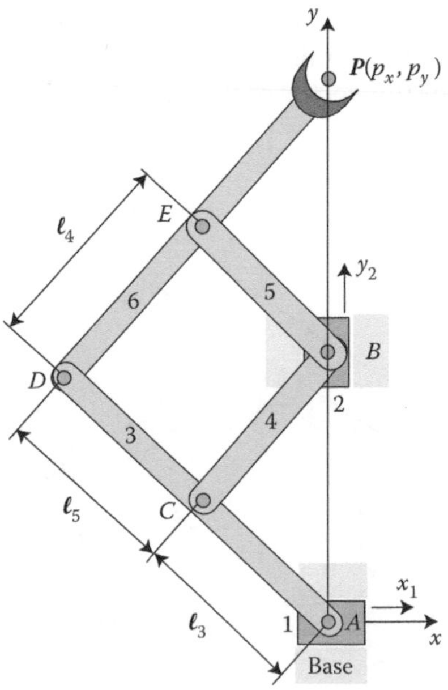

FIGURE 3.30
Pantograph mechanism.

are parallel, respectively. For this mechanism, show that the number of degrees-of-freedom is 2. Moreover, find the position vector of the end effector $\boldsymbol{P} = [p_x, p_y]^T$, as a function of the two input links linear displacements x_1, y_2.

4. A planar three-degrees-of-freedom $3\underline{R}RR$ parallel manipulator is shown in Figure 3.31. In this manipulator, three fixed pivots, A_1, A_2, and A_3 define the geometry of the base, and three moving pivots B_1, B_2, and B_3 define the geometry of the moving platform. A fixed frame $\{\mathcal{A} : xy\}$ is defined at the central position of the moving plate $B_1B_2B_3$, with its x direction along the line connecting A_1 to A_2, while a moving frame $\{\mathcal{B} : uv\}$ is attached to the moving plate at point $\boldsymbol{P} = [p_x, p_y]^T$, the center of the equilateral triangle $\triangle B_1B_2B_3$ with the link length of ℓ. The fixed base attachment points A_1, A_2, A_3 generate another equilateral triangle with link length L. The positions of the fixed base attachment points A_i's are denoted by $\boldsymbol{A}_i = [A_{x_i}, A_{y_i}]^T$, and that for the moving platform are denoted by $\boldsymbol{B}_i = [B_{x_i}, B_{y_i}]^T$. Frame $\{\boldsymbol{B}\}$ coincides with frame $\{\boldsymbol{A}\}$, when the moving platform is at the central position, and the orientation angle ϕ denotes the angle of the moving platform with respect to a fixed frame. The limbs' lengths of the manipulator are denoted by a_i's and b_i's for $i = 1, 2, 3$. The input limb angles are denoted by the vector $\boldsymbol{\alpha} = [\alpha_1, \alpha_2, \alpha_3]^T$, while the position and orientation of the moving platform are represented by $\boldsymbol{P} = [p_x, p_y]^T$, and ϕ, respectively.
   a. Using the Chebyshev–Grübler–Kutzbach criteria, show that the degrees-of-freedom for this manipulator is three.
   b. From the geometry of the moving platform, find the position of point $\boldsymbol{P}$ as a function of position vectors of the moving platform vertices $\boldsymbol{B}_i = [B_{i_x}, B_{i_y}]^T$.
   c. Form the vector closure loops, and by assuming that the moving platform position and orientation $\boldsymbol{\mathcal{X}} = [p_x, p_y, \phi]^T$ are given, find expressions for the limbs angles $\boldsymbol{\alpha}$ as a function of $\boldsymbol{x}$. Eliminate the unactuated limbs' angles β_i's

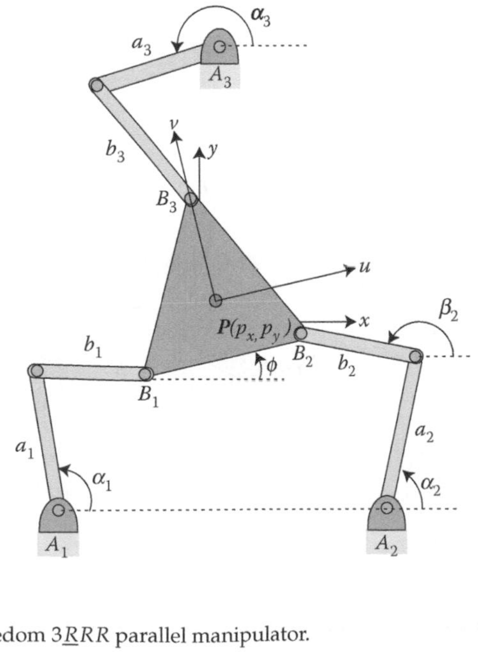

FIGURE 3.31
A planar three-degrees-of-freedom $3\underline{R}RR$ parallel manipulator.

from these equations, and give expressions for the inverse kinematics of the manipulator.

d. Assume that the limbs' angles $\boldsymbol{\alpha} = [\alpha_1, \alpha_2, \alpha_3]^T$ are given, solve the forward kinematics of the manipulator, and find expressions for the position and orientation of the moving platform $\boldsymbol{\mathcal{X}} = [p_x, p_y, \phi]^T$.

e. Develop a program in MATLAB with the following numerical values for the manipulator parameters, in which for a given moving platform trajectory $\boldsymbol{x}(t)$, the limbs' angles $\boldsymbol{\alpha}(t)$ are found. Then, for the given time trajectories of the limbs' angles $\boldsymbol{\alpha}(t)$ determined from the inverse kinematic simulations, solve the forward kinematics of the manipulator, and compare the calculated $\boldsymbol{\mathcal{X}}(t)$ with the original one. Note that due to multiple solutions in inverse kinematics, the solution must be found in such a way that no jumps are observed in the manipulator trajectories. Provide illustrative plots of $\alpha_i(t)$s, and possibly $\boldsymbol{\mathcal{X}}_d(t) - \boldsymbol{\mathcal{X}}(t)$, for verification purpose.

$$a_i = b_i = 1(m); \quad L = 5(m); \quad \ell = 1.5(m).$$

The trajectories are cubic polynomials with:

$$\boldsymbol{x}_o = [0, 0, 0]^T; \quad \boldsymbol{x}_f = [0.2, -0.1, \pi/3]^T; \quad t_o = 0; \quad t_f = 2.$$

5. A planar three-degrees-of-freedom redundant parallel manipulator is shown in Figure 3.32. In this manipulator, the four fixed pivots, A_1, A_2, A_3, and A_4 define the geometry of the base, and the four moving pivots B_1, B_2, B_4, and B_4 define the geometry of the moving platform. The fixed frame $\{\boldsymbol{\mathcal{A}} : xy\}$ is defined at the central position of the moving plate $B_1 B_2 B_3 B_4$, with its x direction along the line

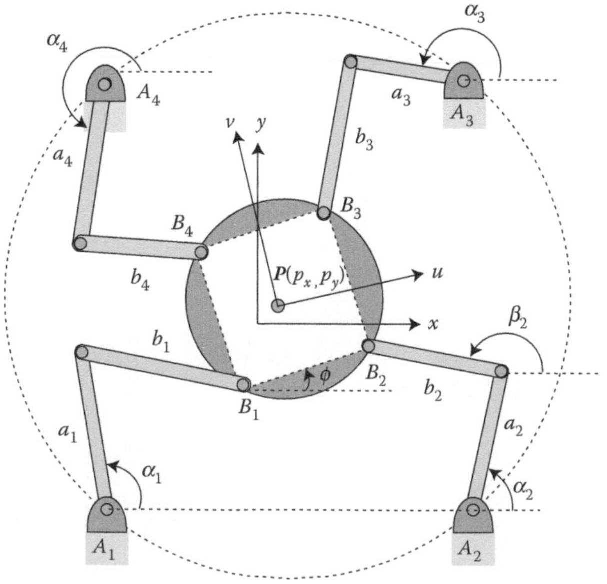

FIGURE 3.32
A planar three-degrees-of-freedom $4\underline{R}RR$ redundant parallel manipulator.

connecting A_1 to A_2, while a moving frame $\{\mathcal{B}: uv\}$ is attached to the moving plate at point $\boldsymbol{P} = [p_x, p_y]^T$, and the center of the circle passing through the moving platform attachment points B_i's. The positions of the fixed base attachment points A_i's are denoted by $\boldsymbol{A}_i = [A_{x_i}, A_{y_i}]^T$, and that for the moving platform are denoted by $\boldsymbol{B}_i = [B_{x_i}, B_{y_i}]^T$. Frame $\{B\}$ coincides with frame $\{A\}$ when the moving plate is at the central position and the moving platform orientation angle ϕ is defined as the rotation angle between the moving frame with respect to the fixed frame. Furthermore, the manipulator limbs' lengths are denoted by a_i's and b_i's for $i = 1, 2, 3, 4$. The input limb angles are denoted by the vector $\boldsymbol{\alpha} = [\alpha_1, \alpha_2, \alpha_3, \alpha_4]^T$, while the position and the orientation of the moving platform are represented by $\boldsymbol{P} = [p_x, p_y]^T$, and ϕ, respectively.

a. Using the Chebyshev–Grübler–Kutzbach criteria, show that the degrees-of-freedom of this redundant manipulator is three.
b. From the geometry of the moving platform, find position of point $\boldsymbol{P}$ as a function of position vectors of the moving platform vertices $\boldsymbol{B}_i = [B_{i_x}, B_{i_y}]^T$.
c. From the vector-loop equations, and assuming that the moving platform position and orientation $\boldsymbol{\mathcal{X}} = [p_x, p_y, \phi]^T$ is given, find expressions for the limbs' angles $\boldsymbol{\alpha}$ as a function of $\boldsymbol{\mathcal{X}}$. Eliminate the unactuated limbs angle β_i's from these equations, and give expressions for the inverse kinematics of the manipulator.
d. Assume that the limbs' angles $\boldsymbol{\alpha} = [\alpha_1, \alpha_2, \alpha_3, \alpha_4]^T$ are given, solve the forward kinematics and find expressions for position and orientation of the moving platform $\boldsymbol{\mathcal{X}} = [p_x, p_y, \phi]^T$.
e. Develop a program in MATLAB with the following numerical values for the manipulator parameters, in which for a given moving platform trajectory $\boldsymbol{\mathcal{X}}(t)$, the limbs' angles $\boldsymbol{\alpha}(t)$ are found. Then, for the given limbs' angles trajectories $\boldsymbol{\alpha}(t)$ determined from the inverse kinematic simulations, solve the forward kinematics of the manipulator, and compare the calculated $\boldsymbol{\mathcal{X}}(t)$ with the original one, $\boldsymbol{\mathcal{X}}_d(t)$. Note that due to multiple solutions in the inverse kinematics, the solution must be found in such a way that no jumps are observed in the manipulator trajectories. Provide illustrative plots of $\alpha_i(t)$s, and possibly $\boldsymbol{x}_d(t) - \boldsymbol{x}(t)$ for verification purpose.

   For simulations, consider the base and moving platform attachment points lie on a circle with radius $R_A = 3$(m) and $R_B = 1$(m), respectively. In order to assign the positions of the attachment points, consider the angle of their placement on the circle, as the following:

$$\theta_{A_i} = \theta_{B_i} = [-3\pi/4, -\pi/4, \pi/4, 3\pi/4]^T;$$
$$a_i = b_i = 1.5(\text{m}).$$

   Finally, consider the trajectories are cubic polynomials with:

$$\boldsymbol{x}_o = [0, 0, 0]^T; \ \ \boldsymbol{x}_f = [0.2, -0.1, \pi/3]^T; \ \ t_o = 0; \ \ t_f = 2.$$

6. Figure 3.33 shows a spatial orientation manipulator that is made up of two tetrahedrons that are pivoted at point O with a spherical joint. The fixed base tetrahedron is denoted by $OA_1A_2A_3$, and the moving platform is shown by the $OB_1B_2B_3$ tetrahedron. Three extensible limbs connect the moving platform at points B_i to the

fixed base at points A_i by spherical joints. Each limb consists of a cylinder–piston pair connected to each other by a prismatic joint. This manipulator can be used for generating a pure orientation for the moving platform with respect to the base. For analysis purpose, consider a fixed frame $\{A\}$ attached to the base at point O. The z-axis is considered along OA_1, while the x-axis is parallel to A_1A_2. Similarly, a moving frame $\{B\}$ is attached to the moving platform at point O, with w-axis along OB_1, and the v-axis parallel to B_1B_2. The positions of the fixed-base attachment points A_i's are denoted by ${}^A\boldsymbol{A}_i = [A_{x_i}, A_{y_i}, A_{z_i}]^T$, and that for the moving platform are denoted by ${}^B\boldsymbol{B}_i = [B_{u_i}, B_{v_i}, B_{w_i}]^T$. The limbs' lengths of the manipulator are denoted by ℓ_i's for $i = 1, 2, 3$. The input limbs' lengths are denoted by vector $\boldsymbol{L} = [\ell_1, \ell_2, \ell_3]^T$, while the orientation of the moving platform is represented by the resulting rotation matrix ${}^A\boldsymbol{R}_B$.

a. Using the Chebyshev–Grübler–Kutzbach criteria, show that the degrees-of-freedom for this spatial manipulator is three.
b. Use a w–u–w Euler angle representation for orientation, and represent the position vector ${}^A\boldsymbol{B}$ as a function of corresponding. Euler angles $[\alpha, \beta, \gamma]^T$.
c. Assume that the orientation of the moving platform is given. From the vector–loop equations solve the inverse kinematics, and give expressions for the limbs' lengths of the manipulator.
d. Assume that the manipulator limbs' lengths $\boldsymbol{L} = [\ell_1, \ell_2, \ell_3]^T$, are given. Solve the forward kinematics of the manipulator, and find the orientation of the moving platform represented by w–u–w Euler angles $[\alpha, \beta, \gamma]^T$, [88].
e. Give some parametric values to the geometry of the manipulator, and develop a program in MATLAB in which for a given moving platform orientation trajectory $[\alpha(t), \beta(t), \gamma(t)]^T$, the limbs' lengths $\boldsymbol{L}(t) = [\ell_1(t), \ell_2(t), \ell_3(t)]^T$ are found.

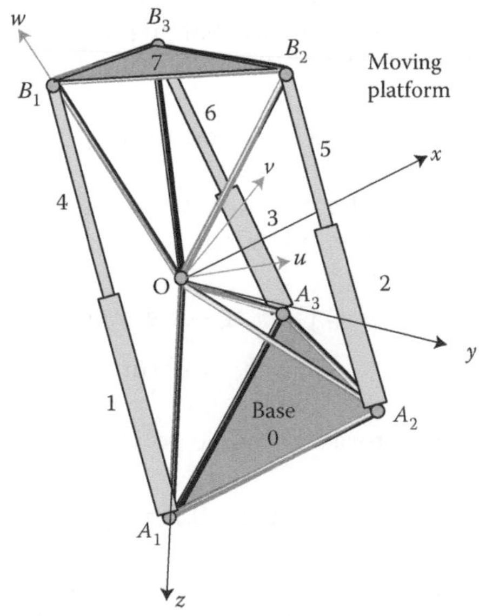

FIGURE 3.33
Two tetrahedral orientation manipulator.

Then, for the given limbs' lengths trajectories solve forward kinematics and calculate Euler angles with respect to time. Note that due to multiple solutions in the forward kinematics, a solution must be found such that no jumps are observed in manipulator trajectories. Then compare the calculated Euler angles with the original ones to verify your programs. Provide illustrative plots of $\ell_i(t)$s, and possibly the difference of the calculated and given Euler angles with respect to time, for verification purpose.

7. The architecture of a $3U\underline{P}U$ parallel manipulator is shown in Figure 3.34. This manipulator consists of a moving platform, a fixed base, and three supporting limbs of identical kinematic structure [93]. Each limb connects the fixed base to the moving platform by a universal joint followed by an actuated prismatic joint and another universal joint. A linear actuator drives each prismatic joint. Since the joint degrees-of-freedom of each limb are equal to five, each limb provides one constraint to the moving platform. It has been shown that the universal joints in each limb can be arranged such that it provides one rotational constraint to the moving platform. Hence, if the constraints are independent of one another, the moving platform possesses pure translational motion [182].

   For inverse kinematics, the position vector $\boldsymbol{P} = [p_x, p_y, p_z]^T$ of a reference point P in the moving platform is given and the problem is to find the input joint variables. Assume that a fixed coordinate frame is attached to the fixed base at point O, and another coordinate frame is attached to the moving platform at point P. The ith actuated limb is connected to the moving platform at point B_i and to the fixed base at point A_i. Furthermore, assume that points B_1, B_2, B_3 lie on the u–v plane at a radial distance of r_b from point P. Points A_1, A_2, A_3 lie on the x–y plane at a radial distance of r_a from point O. Since the orientation of the moving platform

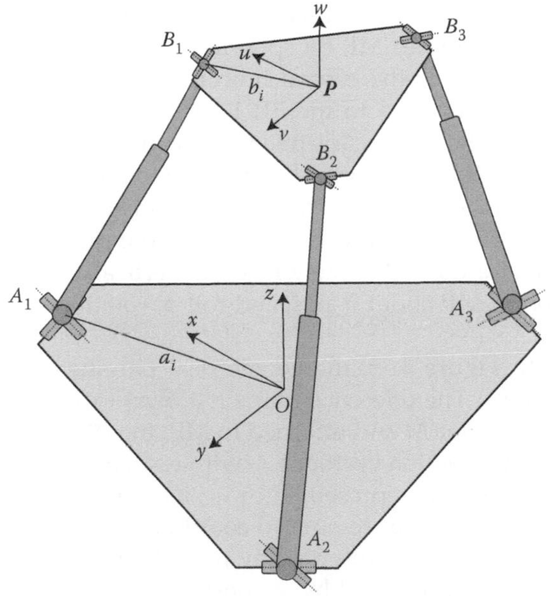

FIGURE 3.34
A three-degrees-of-freedom $3U\underline{P}U$ parallel manipulator. (Adapted from L. W. Tsai and S. A. Joshi. *ASME Journal of Mechanical Design*, 122:439–446, 2000.)

with respect to the fixed base remains constant in all positions the rotation matrix ${}^{A}R_{B} = I$, identically, and hence, ${}^{A}b_i = {}^{B}b_i$. Angle θ is measured from line PB_1 to the u axis.

a. Using the Chebyshev–Grübler–Kutzbach criteria show that the degrees-of-freedom of this manipulator is three.
b. Given p, solve the inverse kinematics and determine the limbs' lengths, q.
c. Given the limbs' positions q, solve the forward kinematics and determine the position vector P of the moving platform.
d. The geometric properties of the manipulator are as follows: The moving and base platforms are equilateral with sides of 1 and 2 m, respectively. For these parametric values, develop a program in MATLAB and find the limbs' lengths $q(t) = [q_1(t), q_2(t), q_3(t)]^T$ for the following moving platform trajectories. Consider cubic polynomial trajectories with:

$$p_o = [0, 0, 1]^T; \;\; p_f = [0.2, -0.3, 1.5]^T; \;\; t_o = 0; \;\; t_f = 2.$$

e. For the determined limbs' length trajectories in part (d), solve the forward kinematics and calculate the position and orientation of the moving platform with respect to time. Note that due to multiple solutions in the forward kinematics, the solution must be found in such a way that no jumps are observed in the manipulator trajectories. Then, compare the calculated position and orientation of the moving platform with the original one to verify your programs. Provide illustrative plots of $q_i(t)$s, and possibly the difference of the calculated and given position components with respect to time for verification purpose.

8. Consider a $3U\underline{P}S$ parallel manipulator introduced in [47], composed of a moving platform (MP), a base platform (BP), and four limbs, as depicted in Figure 3.35. Three limbs are connected to MP by spherical joints and coupled to BP by universal joints. The fourth or the central limb is connected to the centroid of MP by universal joint and is fixed to the BP. Each limb contains two links coupled by a prismatic joint. The degrees-of-freedom for this manipulator is 3 using the Chebyshev–Grübler–Kutzbach formula. Thus, three linear hydraulic actuators are used to derive the prismatic joints of three limbs and an idle prismatic joint is used for the central limb. This design with the central limb provides three independent degrees-of-freedom for the MP, namely, heave h, vertical displacement of MP along z; pitch ψ, rotation of MP about v axis; and roll ϕ, rotation of MP about u axis, as shown in Figure 3.35.

   With reference to Figure 3.35, the geometrical parameters of the manipulator are defined as follows: The reference frame $\{\mathcal{A} : xyz\}$ located at O; the coordinate frame $\{\mathcal{B} : uvw\}$ located at M and attached to MP; the ith limb is connected to MP by B_i and to BP by A_i; position vector of A_i with respect to $\{A\}$ is represented by a_i; the length of the ith limb is represented by q_i; its unit vector is represented by $\hat{e}_i$ and position vector of B_i with respect to coordinate frame $\{\mathcal{B}\}$ is represented by b_i, angle of limb i with z is represented by β_i and angle of the projection of limb i on BP with x-axis is represented by α_i, position vector of A_i with respect to M expressed in reference frame and frame $\{B\}$ are represented by b_i and ${}^{B}b_i$, respectively ($i = 1, 2, 3$). Moreover, position vector of M with respect to O is defined by p and rotation matrix of MP with respect to the reference frame is defined

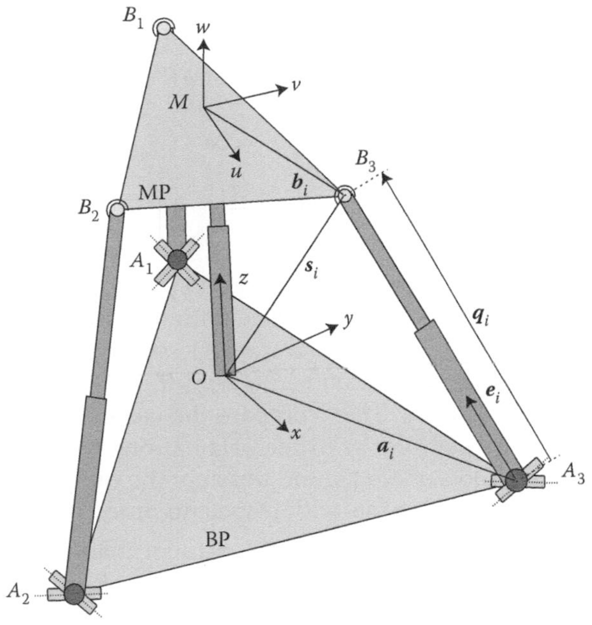

FIGURE 3.35
A four-legged three-degrees-of-freedom 3*U*<u>*P*</u>*S* parallel manipulator. (Adapted from A. Fattah and G. Kasaei. *Robotica*, 18(5):535–543, 2000.)

by $\boldsymbol{R}$. The rotation matrix $\boldsymbol{R}$ can be readily determined having the roll and pitch angles of MP.

a. Using the Chebyshev–Grübler–Kutzbach criteria, show that the degrees-of-freedom of this manipulator is three.

b. Given $\boldsymbol{p}$ and $\boldsymbol{R}$, determine the motion of actuators, q_i and $\boldsymbol{e}_i$.

c. Given the limb positions q_i, determine the motion of the moving platform, that is, the position vector $\boldsymbol{P}$ and the orientation of MP denoted by $\boldsymbol{R}$.

d. The geometric properties of the manipulator are as follows: The moving and base platforms are equilaterals with sides of 1 and 2 m, respectively. The minimum length of the central limb is 1 m and its stroke, heave of MP, is 0.4 m. The pitch and roll angles of MP are varied from $-\pi/6$ to $\pi/6$. For these parametric values, develop a program in MATLAB in which for the following given moving platform position and orientation, the limbs' lengths $\boldsymbol{q}(t) = [q_1(t), q_2(t), q_3(t)]^T$ are found. To solve inverse position kinematics, position vector of point M of MP, $\boldsymbol{p}$, with respect to the reference frame is defined by a prescribed cycloidal maneuver as

$$\boldsymbol{p} = [0 \quad 0 \quad h(t)]^T;$$

$$h(t) = 1 + 0.4\left[\frac{t}{T} - \frac{1}{2\pi}\sin\left(\frac{2\pi t}{T}\right)\right], (m) \quad 0 \le t \le T,$$

where $h(t)$ shows the heave of MP and T is the period of the maneuver in terms of seconds. The rotation matrix $\boldsymbol{R}$ of MP with respect to the reference frame can be defined as

$$R = \begin{bmatrix} \cos\psi(t) & 0 & \sin\psi(t) \\ 0 & 1 & 0 \\ -\sin\psi(t) & 0 & \cos\psi(t) \end{bmatrix};$$

$$\psi(t) = \frac{\pi}{6}\left[\frac{2t}{T} - \frac{1}{\pi}\sin(\frac{2\pi t}{T}) - 1\right], \quad 0 \le t \le T.$$

e. For the determined limbs' length trajectories in part (d), solve the forward kinematics and calculate the position and orientation of the moving platform with respect to the time. Note that due to multiple solutions in the forward kinematics, the solution must be found in a way that no jumps are observed in the manipulator trajectories. Then compare the calculated position and orientation of the moving platform with the original one to verify your programs. Provide illustrative plots of $q_i(t)$'s, and possibly the difference of the calculated and given trajectory components with respect to time, for verification purpose.

9. Consider a three-degrees-of-freedom $3R\underline{P}S$ parallel manipulator represented in Figure 3.36. The manipulator consists of the moving $(B_1\, B_2\, B_3)$ and base $(A_1\, A_2\, A_3)$ platforms that are equilateral triangles with ℓ and $k\ell$ as the lengths of the sides, respectively. Three extensible prismatic limbs connect the moving platform by spherical joints and the base by means of rotational ones; at each joint of the base the revolute axis is parallel to the opposite edge. The $3R\underline{P}S$ joint structure corresponds to the revolute joints between the limbs and the base, the actuated prismatic joints in the limbs and the spherical joints between the limbs and the moving platform. The actuated prismatic joints of the limbs drive the manipulator.

   Assume that the end effector of the manipulator coincides with the center point E of the moving platform. Denote its coordinates in the base coordinate system attached at the center point O of the base platform by frame $\{\mathcal{A} : (x, y, z)\}$, with the plane x–y coinciding with the base platform. The joint coordinates are the lengths ℓ_1, ℓ_2, and ℓ_3 of the platform limbs as shown in Figure 3.36. The frame $\{\mathcal{B} : (u, v, w)\}$ is the coordinate system attached to the moving platform with origin at point E. Consider the coordinates of the vertices B_1, B_2, B_3 of the moving platform. From the constraints applied by the rotational joints at points A_1, A_2, A_3, the coordinates of the moving platform points are obtained as follows:

$$y_{B_1} = x_{B_1/\sqrt{3}}; \quad y_{B_3} = -x_{B_3/\sqrt{3}}; \quad x_{B_2} = 0.$$

   From the equilateral triangular form of the upper platform it is clear, that

$$x = (x_{B_1} + x_{B_2} + x_{B_3})/3; \quad y = (y_{B_1} + y_{B_2} + y_{B_3})/3\sqrt{3};$$
$$z = (z_{B_1} + z_{B_2} + z_{B_3})/3 = 0.$$

   Hence by substitution, and some manipulations, reach the following constraint to the positions of the end effector.

$$x_{B_1} = (3x + 3\sqrt{3}y - y_{B_2}\sqrt{3})/2; \quad x_{B_3} = (3x - 3\sqrt{3}y + y_{B_2}\sqrt{3})/2.$$

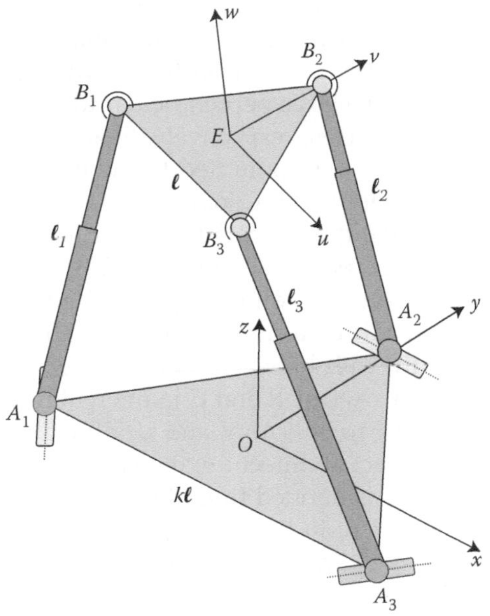

FIGURE 3.36
A three-degrees-of-freedom $3R\underline{P}S$ structure parallel manipulator. (Adapted from A. Sokolov and P. Xirouchakis. *Robotica*, 23(2):207–217, 2005.)

a. Using the Chebyshev–Grübler–Kutzbach criteria show that the degrees-of-freedom of this manipulator is three.
b. From the system of four equations of the distances:

$$B_1E^2 = \ell^2/3; \quad B_1B_2^2 = \ell^2; \quad B_2B_3^2 = \ell^2; \quad B_1B_3^2 = \ell^2.$$

Find the coordinates of the points B_i as functions of the end effector E coordinates $[x, y, z]^T$ and length ℓ.
c. These equations show that the inverse kinematic problem has several solutions. Obtain the correct solutions with real values for the coordinates of the B_i points and continuous curves for the joint coordinates as functions of the trajectories of the end effector [167].

10. The schematic diagram of a $3\underline{P}RC$ translational parallel manipulator is shown in Figure 3.37. It consists of a moving platform, a fixed base, and three limbs with identical kinematic structure. Each limb connects the fixed base to the moving platform by an actuated P joint, an R joint, and a C joint in sequence. Thus, the moving platform is attached to the base by three identical $\underline{P}RC$ linkages. The mobility analysis of this manipulator [115], shows that in order to avoid any change of orientation in the moving platform, it is sufficient for the three axes of joints within the same limb to satisfy certain geometric conditions. Referring to Figure 3.37, these conditions are
    - The R joint axis ($\boldsymbol{r}_i$) and the C joint axis ($\boldsymbol{c}_i$) within the ith limb, for $i = 1, 2, 3$ are parallel to the same unit vector $\boldsymbol{s}_{i_0}$.
    - The limbs are arranged such that $\boldsymbol{s}_{i_0} \neq \boldsymbol{s}_{j_0}$ for $i \neq j$, and $i, j = 1, 2, 3$.

To facilitate the kinematic analysis, as shown in Figures 3.37 and 3.38, assign a fixed Cartesian frame $\{\mathcal{A}:(x,y,z)\}$ at the center point O of the fixed base, and a moving Cartesian frame $\{\mathcal{B}:(u,v,w)\}$ on the triangle moving platform at the center point P, with the z and w axes perpendicular to the platform, and the x and y axes parallel to the u and v axes, respectively. In addition, the ith limb $C_i B_i$ for $(i=1,2,3)$ with the length of ℓ is connected to the moving platform at point B_i which is a point on the axis of the i'th C joint.

B_i' denotes the point on the moving platform that is coincident with the initial position of B_i, and the three points B_i' for $i=1,2,3$ lie on a circle of radius b. The three rails $M_i N_i$ intersect each other at point D and intersect the x–y plane at points A_1, A_2, and A_3 that lie on a circle of radius a. The sliders of P joints C_i are restricted to move along the rails between M_i and N_i. Moreover, the axis of the P joint is perpendicular to the axes of R and C joints within the ith limb. Angle α is measured from the fixed base to rails $M_i N_i$ and is defined as the layout angle of the actuators. To obtain a compact architecture, the value of α is designed within the range $0 \le \alpha \le \pi/2$. Angle ϕ_i is defined from the x-axis to OA_i, in the fixed frame, and also from the u-axis to PB_i' in the moving frame. Without loss of generality, let the x-axis point along OA_1, and the u-axis direct along PB_1'. Then, we have $\phi_1 = 0$.

a. Show by using the Chebyshev–Grübler–Kutzbach criteria that the degrees-of-freedom of this manipulator in general form is zero, and due to the specific geometric design of this manipulator stated above, its moving platform possesses three-degrees-of-freedom.

b. The purpose of the inverse kinematics is to solve the actuated variables from a given position of the moving platform. Write the vector-loop closures, and assuming that the moving platform position $\boldsymbol{x}=[p_x, p_y, \phi]^T$ is given, while its orientation is ${}^A\boldsymbol{R}_B = \boldsymbol{I}$, give expressions for the inverse kinematics of the manipulator.

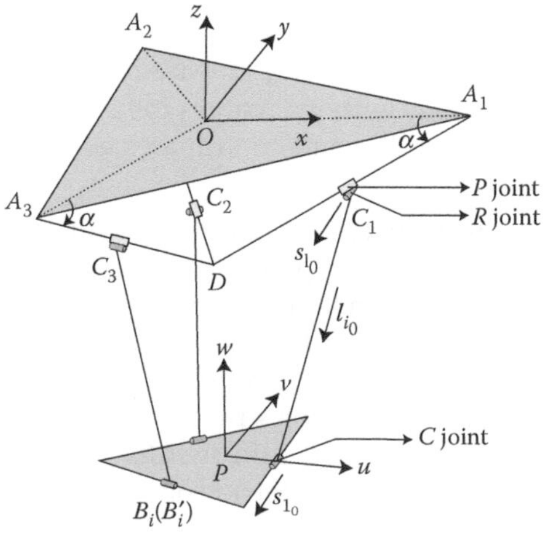

FIGURE 3.37
The schematics of a three-degrees-of-freedom $3\underline{P}RC$ translational parallel manipulator. (Adapted from Y. Y. Li and Q. Xu. *Journal of Mechanical Design*, 128(4):729–737, 2006.)

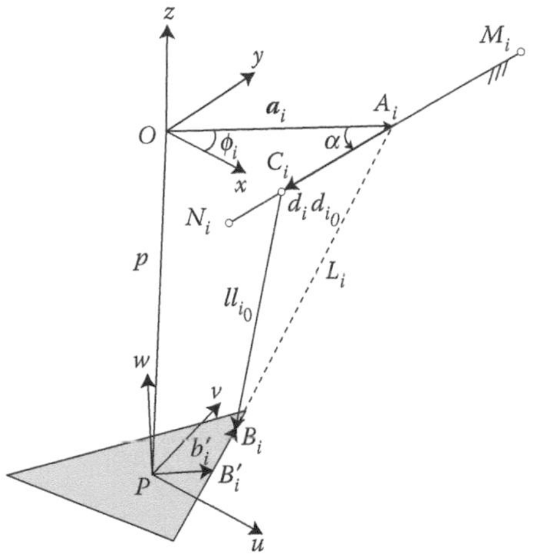

FIGURE 3.38
A schematic for the 3<u>P</u>RC translational parallel manipulator used for kinematic analysis of the manipulator.

c. Given a set of actuated inputs, the position of the moving platform can be solved by forward kinematic analysis. Using the inverse kinematic equations, find three second-degree algebraic equations in the unknowns of p_x, p_y, and p_z.

d. Use the Sylvester dialytic elimination method [115] to reduce the above system of equations to an eighth-degree polynomial in only one variable.

e. Give some parametric values to the geometry of the manipulator, and develop a program in MATLAB in which for a given moving platform position trajectory $[p_x(t), p_y(t), p_z(t)]^T$ (e.g., that given in Problem 7), the actuator inputs d_i's are found. Then, for the determined actuator input trajectories, solve the forward kinematics and calculate the position vector of the moving platform with respect to time. In the forward kinematic simulations use two methods. First use numerical solution of part (c), and next use numerical solutions of part (d). Note that due to multiple solutions in forward kinematics, the solution must be found in such a way that no jumps are observed in the manipulator trajectories. Then, compare the calculated position vectors in each solution to original one to verify your programs. Provide illustrative plots for actuator inputs, and possibly the difference of the calculated and given position vectors with respect to time, for verification purpose. Give your analysis, on the computation effort, and the convergence of the forward kinematic solutions in each method.

11. The Delta robot was invented in the early 1980s by Reymond Clavel [26] at EPFL, Switzerland. The purpose of this new type of robot was to manipulate light and small objects at a very high speed. As shown in Figure 3.39, this parallel robot consists of a fixed base, a moving platform and three arms. Each arm consists of two parts, the actuated lower arm and the upper arm. The lower arm is connected to the base by revolute joints. The key design feature in the upper arm is the use of parallelograms, which maintains the orientation of the end effector. These parallelograms are connected to the lower arm and the moving platform by spherical joints. Hence, the moving platform has three translational degrees-of-freedom.

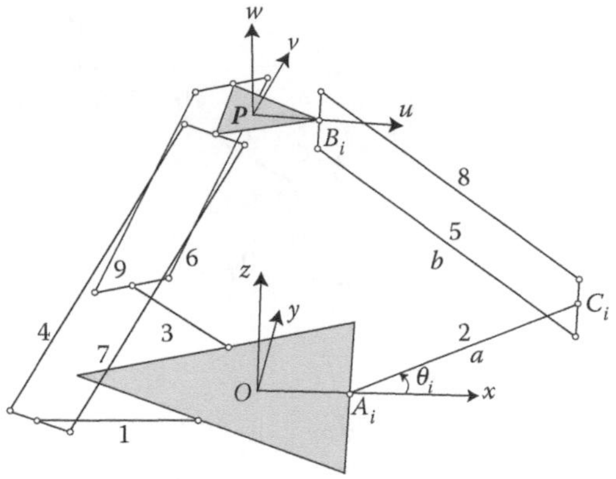

FIGURE 3.39
The schematic of a three-degrees-of-freedom Delta robot.

For the kinematic analysis, consider the delta robot without the gripper actuator, as shown in Figure 3.39. In this figure, a fixed Cartesian frame $\{\mathcal{A} : (x, y, z)\}$ is attached to the center point O of the fixed base, and a moving Cartesian frame $\{\mathcal{B} : (u, v, w)\}$ on the triangle moving platform at the center point P, with the x-axis along OA_1 direction and the u-axis along PB_1, respectively. The lengths of the lower arms are considered to be equal and are denoted by a, while the lengths of the upper arms are also considered to be equal and denoted by b.

a. Show by using the Chebyshev–Grübler–Kutzbach criteria, that the degrees-of-freedom of this manipulator is three.
b. Solve the inverse kinematics of this manipulator and derive expressions for the lower arm angles $\boldsymbol{\theta} = [\theta_1, \theta_2, \theta_3]^T$ as a function of given moving platform position $\boldsymbol{P} = [p_x, p_y, p_z]^T$.
c. Use the inverse kinematic formulations and find a forward kinematic solution for the manipulator using numerical methods.
d. Give some parametric values to the geometry of the manipulator, and develop a program in MATLAB in which for a given moving platform position trajectory $[p_x(t), p_y(t), p_z(t)]^T$ (e.g., that given in Problem 7), the actuator inputs θ_i's are found. Then, for the determined actuator input trajectories, solve the forward kinematics and calculate the position vector of the moving platform with respect to time. Note that due to multiple solutions in forward kinematics, the right solution must be found in a way that no jumps are observed in the manipulator trajectories. Then, compare the calculated position vectors to the original ones to verify your programs. Provide illustrative plots of the actuator inputs, and possibly the difference of the calculated and given position vectors with respect to time, for verification purpose.

12. A picture of a Quadrupteron robot, and the schematic of one of its limb's kinematics is shown in Figure 3.40 [157]. It consists of a moving platform connected to a fixed base by four limbs each having five joints namely, from base to moving platform, a P joint and four R joints which are labeled from 1 to 5 in sequence. For practical reasons, the axes of the last two R joints within a limb are arranged to have at least one common point. The P joints, mounted directly on the base, are

the only actuated ones. In order for the architecture to produce three translational and one rotational motion and to be drivable by the four actuated P joints, the following geometric constraints must be satisfied:

a. The axes of the four R joints attached to the moving platform are all parallel.
b. The axes of joints 2, 3, and 4 within a same limb are all parallel.
c. The axes of joints 1 and 2 within the same limb are not orthogonal to each other.
d. The axes of joints 2 of all four limbs are not all parallel to a plane that is parallel to the axes of the R joints attached to the moving platform.
e. In at the most one of the four limbs, all the axes of joints 2–5 within the same limb are parallel. In this case, joints 4 and 5 degenerate into one joint. Such a limb, if it exists, is numbered as limb 1 for convenience.
f. The axes of joints 2 of all four limbs, are not all parallel to a plane. This requires that the axes of joints 4 and 5 within the same limb are not orthogonal to each other in at least one of the four limbs.

Conditions (a)–(c) guarantee that the moving platform can undergo at least four-degrees-of-freedom 3T1R motion. Condition (d), together with conditions (a)–(c), guarantee that the degrees-of-freedom of the moving platform is 4. Conditions (e) and (f) further guarantee that the four P joints can be used to control the motion of the moving platform. As shown in Figure 3.40, two unit vectors $\boldsymbol{c}_i$ and $\boldsymbol{e}_i$ are used to represent, respectively, the direction of joint 5 as well as the direction of joints 2, 3, 4 of the ith limb. Furthermore, it is assumed that $i = 1, 2, 3, 4$. A $\underline{P}4R$ limb kinematics exerts one constraint on the moving platform which prevents it from rotating about any axis parallel to $\boldsymbol{c}_i \times \boldsymbol{e}_i$ if $\boldsymbol{c}_i \neq \boldsymbol{e}_i$ or exerts two constraints on the moving platform which prevents it from rotating about any axis perpendicular to $\boldsymbol{c}_1$ if $\boldsymbol{c}_1 = \boldsymbol{e}_1$. Since all $\boldsymbol{c}_i$ vectors are parallel, the directions in which rotations are prevented are all parallel to a plane and the moving platform can rotate only in a direction orthogonal to that plane, that is, in the direction of $\boldsymbol{c}_i$, thereby leading to the 3T1R motions.

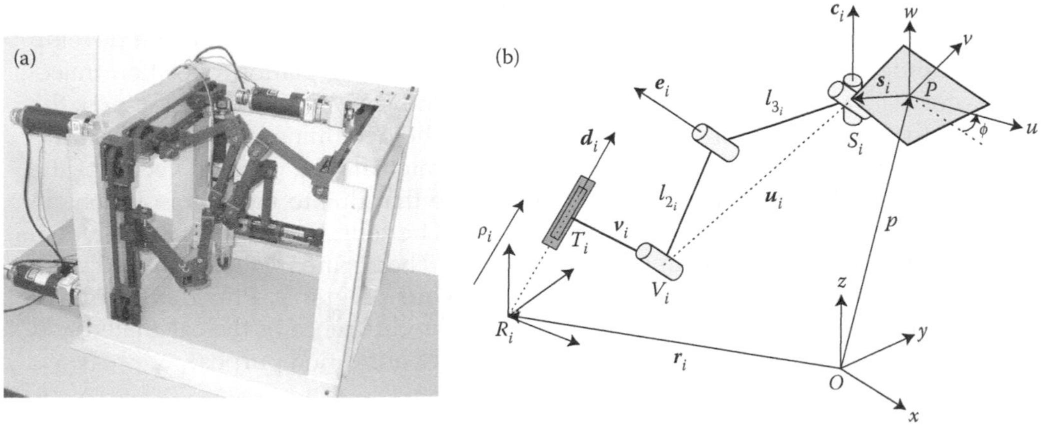

FIGURE 3.40
The Quadrupteron: a parallel manipulator with 3T1R degrees-of-freedom (Courtesy of Clément Gosselin, Laval University, Canada.) (a) and a schematic of the kinematic structure of one of its limbs (b) (Adapted from P. L. Richard and C. M. Gosselin. *ASME Journal of Mechanical Design*, 129(6):611–616, 2007.)

Referring to Figure 3.40, the following quantities are used to describe the geometry of the manipulator. A fixed reference frame $\{O - xyz\}$ is defined on the base and a moving reference frame $\{P - uvw\}$ is attached to the moving platform. The position of the moving platform is given by the vector connecting O to P, denoted by $\boldsymbol{P} = [p_x, p_y, p_z]^T$. The orientation of the moving platform with respect to the base is given by matrix $\boldsymbol{R}$. Without loss of generality, both the z and w axes are chosen to be parallel to $\boldsymbol{c}_i$. Hence, we have $\boldsymbol{c}_i = [0, 0, 1]^T$ and the rotation matrix $\boldsymbol{R}$ can be written as

$$\boldsymbol{R} = \begin{bmatrix} \cos\phi & -\sin\phi & 0 \\ \sin\phi & \cos\phi & 0 \\ 0 & 0 & 1 \end{bmatrix}.$$

The direction of the ith fixed P joint is denoted by unit vector $\boldsymbol{d}_i$. A reference point R_i is defined on the axis of the ith fixed P joint, as shown in Figure 3.40, and the motion of the P joint is measured with respect to the latter reference and denoted by the signed distance ρ_i between R_i and a point T_i defined on the sliding body. Point S_i is defined as the intersection of the axes of the last two joints in limb i. Furthermore, the position vector of point R_i (the vector connecting O to R_i) in the fixed frame is denoted by $\boldsymbol{r}_i = [r_{x_i}, r_{y_i}, r_{z_i}]^T$, while the position vector of point S_i (the vector connecting P to S_i) is denoted by ${}^P\boldsymbol{s}_i = [s_{x_i}, s_{y_i}, s_{z_i}]^T$ in the moving frame and $\boldsymbol{s}_i$ in the fixed frame, in which $\boldsymbol{s}_i = \boldsymbol{R}\,{}^P\boldsymbol{s}_i$. Since the axes of joints 2, 3, and 4 of a given limb are parallel, it is possible to define a point V_i on the axis of joint 2 such that the vector connecting V_i to S_i, denoted by $\boldsymbol{u}_i$ is orthogonal to the unit vector $\boldsymbol{e}_i = [e_{x_i}, e_{y_i}, e_{z_i}]^T$.

a. Show that the degrees-of-freedom of this manipulator is four, using the Chebyshev–Grübler–Kutzbach criteria.

b. Solve the inverse kinematics of this manipulator and derive expressions for the input limbs' lengths ρ_i as a function of given moving platform position $\boldsymbol{P} = [p_x, p_y, p_z]^T$ and orientation ϕ.

c. Use the inverse kinematic formulations, and find a forward kinematic solution for the manipulator using numerical methods.

d. Give some parametric values to the geometry of the manipulator, and develop a program in MATLAB in which for a given moving platform position trajectory $[p_x(t), p_y(t), p_z(t)]^T$ (e.g., that given in Problem 7) and orientation $\phi(t)$, the actuator inputs $\rho_i(t)$'s are found. Then, for the determined actuator input trajectories solve the forward kinematics and calculate the position vector of the moving platform with respect to time. Note that due to multiple solutions in forward kinematics, the right solution must be found in a way that no jumps are observed in the manipulator trajectories. Then compare the calculated position vectors to the original ones to verify your programs. Provide illustrative plots of the actuator inputs, and possibly the difference of the calculated and given position vector with respect to time for verification purpose.

4

Jacobians: Velocities and Static Forces

4.1 Introduction

In this chapter, kinematic analysis of robot manipulators is further examined beyond static positioning. The notions of linear velocity of a point and angular velocity of a rigid body are defined in the same way as position of a point and orientation of a rigid body are defined in Chapter 2. Furthermore, these concepts are used to analyze the motion of parallel manipulators. Differential kinematic analysis plays a vital role in the study of robotic manipulators. It turns out that the study of velocities in a parallel manipulator leads to the definition of the Jacobian matrix. The Jacobian matrix not only reveals the relation between the joint variable velocities of a parallel manipulator to the moving platform linear and angular velocities, it also constructs the transformation needed to find the actuator forces from the forces and moments acting on the moving platform.

The optimal kinematic design of manipulators is an important and challenging topic in parallel robotics research [6]. Most of the work that has been performed on the subject has been directed toward the optimization of dexterity or manipulability indices [103,196], which are in fact related to the kinematic accuracy and to the Jacobian analysis of such systems [62,78]. In the context of parallel manipulators, kinematic accuracy is a very important issue. Indeed, it is well known that a special type of local degeneracy can occur in the motion of these manipulators [61]. Physically, these configurations lead to

a. An instantaneous change in the degrees-of-freedom of the system and hence a loss of controllability.
b. An important degradation of the natural stiffness that may lead to very high joint forces or torques.

Therefore, it is very important to identify singular configurations at the design stage to improve the performance. The singularities of parallel manipulators have been studied by several researchers [29,101,120]. In the Jacobian analysis of the parallel manipulator, Gosselin and Angeles [61] maybe the first who suggest the use of two Jacobian matrices to specify inverse and forward kinematics singularities. In [61], it has been shown that by writing the velocity equations of closed chains in terms of two Jacobian matrices, it is possible to classify singularities in three different types that have different physical interpretations. In [134], Merlet used Grassmann geometry to study the singularities of spatial six-degrees-of-freedom parallel manipulators. He obtained an exhaustive list of possible singular configurations. The latter method is general and can be applied to various mechanisms with any degrees-of-freedom. The screw theory has also been used extensively to derive Jacobian matrices in parallel structures [101,135]. In what follows, a systematic approach to analyze the differential kinematics of parallel robots is given.

4.2 Angular and Linear Velocities

In this section, differential kinematics of a rigid body have been studied in detail. In differential kinematics, we study the variations of the position of a point and the orientation of a rigid body with respect to time. Basically, derivative of the position vector with respect to time leads to linear velocity of a point. However, to determine the absolute linear velocity of a point, the derivative must be calculated relative to a fixed frame. Differentiation of a position vector with respect to a moving frame results in a relative velocity. In the study of robotic manipulators, usually multiple moving frames are defined, to analyze the motion of the moving platform. Therefore, it is necessary to define the required arithmetics to transform the relative velocities to absolute ones. Furthermore, the moving frames are usually attached to the rigid bodies to define their orientation in space. Change of orientation of these frames is defined as the angular velocity of the rigid body.

4.2.1 Angular Velocity of a Rigid Body

To define angular velocity of a rigid body, consider that a moving frame $\{B\}$ is attached to the rigid body, and motion is analyzed with respect to a fixed frame as shown in Figure 4.1. Angular velocity is an attribute of a rigid body and describes the rotational motion of the frame $\{B\}$ that is attached to the rigid body. The *angular velocity vector*, denoted by the symbol $\boldsymbol{\Omega}$, describes the instantaneous rotation of frame $\{B\}$ with respect to the fixed frame $\{A\}$. At any time instant, the direction of $\boldsymbol{\Omega}$ indicates the instantaneous axis of rotation, which is coincident with the previously defined screw axis $\hat{\mathbf{s}}$. Furthermore, the magnitude of $\boldsymbol{\Omega}$ indicates the speed of rotation that is equivalent to the rate of change of the screw

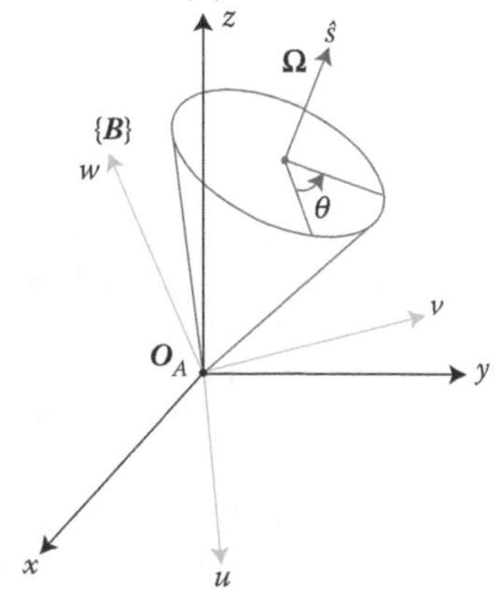

FIGURE 4.1
Instantaneous rotation of a rigid body.

parameter θ. Hence,

$$\boldsymbol{\Omega} \doteq \dot{\theta}\,\hat{\boldsymbol{s}}. \tag{4.1}$$

Note that angular velocity is a vector and can be represented in any frame. The leading superscript is used as before, to denote the frame that angular velocity is represented in. For example, ${}^{B}\boldsymbol{\Omega}$ denotes that angular velocity of the rigid body is expressed in frame $\{\boldsymbol{B}\}$. Therefore, angular velocity of a rigid body can be expressed componentwise in any frame:

$$\begin{aligned} {}^{A}\boldsymbol{\Omega} &= \Omega_x\hat{\boldsymbol{x}} + \Omega_y\hat{\boldsymbol{y}} + \Omega_z\hat{\boldsymbol{z}} \\ &= \dot{\theta}\,(s_x\hat{\boldsymbol{x}} + s_y\hat{\boldsymbol{y}} + s_z\hat{\boldsymbol{z}}) \end{aligned} \tag{4.2}$$

in which Ω_x, Ω_y, and Ω_z are the three components of angular velocity of a rigid body expressed in frame $\{\boldsymbol{A}\}$.

4.2.1.1 Angular Velocity and Rotation Matrix Rate

As described in Chapter 2, orientation of a rigid body can be represented by rotation matrix, screw axis, and Euler angles in different forms. Equation 4.1 defines angular velocity using a screw–axis representation. However, there are certain relationships between the defined angular velocity and the rate of change of the rotation matrix, and the rate of change of the Euler angles. In this section, the relation between angular velocity and rate of change of the rotation matrix is examined.

As seen hereinbefore, rotation matrix is an orthonormal matrix, and inverse of a rotation matrix is equal to its transpose. Hence,

$${}^{A}\boldsymbol{R}_B\,{}^{A}\boldsymbol{R}_B^T = \boldsymbol{I}, \tag{4.3}$$

in which $\boldsymbol{I}$ is a 3×3 identity matrix. Differentiate both sides of Equation 4.3, with respect to time. This leads to

$${}^{A}\dot{\boldsymbol{R}}_B\,{}^{A}\boldsymbol{R}_B^T + {}^{A}\boldsymbol{R}_B\,{}^{A}\dot{\boldsymbol{R}}_B^T = \boldsymbol{0}. \tag{4.4}$$

Substitute ${}^{A}\boldsymbol{R}_B^T = {}^{A}\boldsymbol{R}_B^{-1}$ and ${}^{A}\boldsymbol{R}_B = \left({}^{A}\boldsymbol{R}_B^{-1}\right)^T$ into Equation 4.4 to obtain

$$\left({}^{A}\dot{\boldsymbol{R}}_B\,{}^{A}\boldsymbol{R}_B^{-1}\right) + \left({}^{A}\dot{\boldsymbol{R}}_B\,{}^{A}\boldsymbol{R}_B^{-1}\right)^T = \boldsymbol{0}. \tag{4.5}$$

This means that the resulting matrix ${}^{A}\dot{\boldsymbol{R}}_B\,{}^{A}\boldsymbol{R}_B^{-1}$ is a 3×3 skew-symmetric matrix, and hence can be represented by a general skew-symmetric matrix $\boldsymbol{\Omega}^{\times}$ as

$$\boldsymbol{\Omega}^{\times} \equiv {}^{A}\dot{\boldsymbol{R}}_B\,{}^{A}\boldsymbol{R}_B^{-1} = \begin{bmatrix} 0 & -\Omega_z & \Omega_y \\ \Omega_z & 0 & -\Omega_x \\ -\Omega_y & \Omega_x & 0 \end{bmatrix}. \tag{4.6}$$

It can be shown that the parameters Ω_x, Ω_y, and Ω_z used here to define the rate of change of the rotation matrix are in fact the same three components of the angular velocity of the rigid body given in Equation 4.2.

4.2.1.2 Angular Velocity and Euler Angles Rate

In this section, the relation between angular velocity and the rate of change of the Euler angles is examined. Note that for any type of Euler angles $[\alpha, \beta, \gamma]^T$, the angular velocity is not equal to the rate of change of the Euler angles, that is,

$$\boldsymbol{\Omega} \neq \begin{bmatrix} \dot{\alpha} \\ \dot{\beta} \\ \dot{\gamma} \end{bmatrix}. \tag{4.7}$$

To derive the relation between the angular velocity and the rate of Euler angles, Equation 4.6 is used. From this matrix equation, the following three independent equations are extracted.

$$\begin{aligned} \Omega_x &= \dot{r}_{31}r_{21} + \dot{r}_{32}r_{22} + \dot{r}_{33}r_{23}, \\ \Omega_y &= \dot{r}_{11}r_{31} + \dot{r}_{12}r_{32} + \dot{r}_{13}r_{33}, \\ \Omega_z &= \dot{r}_{21}r_{11} + \dot{r}_{22}r_{12} + \dot{r}_{23}r_{13}, \end{aligned} \tag{4.8}$$

in which r_{ij} denotes the (i, j) components of the corresponding rotation matrix. These equations can be written in a matrix form

$$\boldsymbol{\Omega} = \boldsymbol{E}(\alpha, \beta, \gamma) \begin{bmatrix} \dot{\alpha} \\ \dot{\beta} \\ \dot{\gamma} \end{bmatrix}, \tag{4.9}$$

in which $\boldsymbol{E}(\alpha, \beta, \gamma)$ is the matrix that relates the rate of Euler angles to the angular velocity. This matrix is a function of the three Euler angles. For example, consider w–v–w Euler angles. For this representation, the rotation matrix is given in Equation 2.46. Differentiating $\boldsymbol{R}_{wvw}$ with respect to time and substituting into Equation 4.8 yields

$$\boldsymbol{E}_{wvw} = \begin{bmatrix} 0 & -s\alpha & c\alpha s\beta \\ 0 & c\alpha & s\alpha s\beta \\ 1 & 0 & c\beta \end{bmatrix}. \tag{4.10}$$

4.2.2 Linear Velocity of a Point

Linear velocity of a point P can be easily determined by the time derivative of the position vector $\boldsymbol{p}$ of that point with respect to a fixed frame.

$$v_p = \dot{\boldsymbol{p}} = \left(\frac{\mathrm{d}\boldsymbol{p}}{\mathrm{d}t}\right)_{fix}. \tag{4.11}$$

Note that this is correct only if the derivative is taken with respect to a fixed frame, as denoted by $(\mathrm{d}(\cdot)/\mathrm{d}t)_{fix}$ in Equation 4.11. If variation of the position vector is determined with respect to a moving fame, *relative* velocity is obtained. Relative velocity is denoted by the following notation:

$$v_{rel} = \left(\frac{\partial \boldsymbol{p}}{\partial t}\right)_{mov}, \tag{4.12}$$

in which v_{rel} denotes the relative velocity and $(\partial(\cdot)/\partial t)_{mov}$, denotes the time derivative with respect to a moving frame. For all points of a rigid body, relative velocity with respect to the moving frame attached to the body is zero, while absolute velocity may be determined from the velocity of the origin of the moving frame. In classical mechanics, it is shown that the relation between absolute derivative of any vector to its relative derivative is given by the following equation [130]:

$$\left(\frac{d(\cdot)}{dt}\right)_{fix} = \left(\frac{\partial(\cdot)}{\partial t}\right)_{mov} + \boldsymbol{\Omega} \times (\cdot), \tag{4.13}$$

in which $\boldsymbol{\Omega}$ denotes the angular velocity of the moving frame with respect to the fixed frame. The term $\boldsymbol{\Omega} \times (\cdot)$ denotes the cross product of two vectors $\boldsymbol{\Omega}$ and vector $(\cdot)$. To write this equation in matrix form, the matrix $\boldsymbol{\Omega}^{\times}$ can be used as the following:

$$\left(\frac{d(\cdot)}{dt}\right)_{fix} = \left(\frac{\partial(\cdot)}{\partial t}\right)_{mov} + \boldsymbol{\Omega}^{\times}(\cdot), \tag{4.14}$$

in which matrix $\boldsymbol{\Omega}^{\times}$ denotes a skew-symmetric matrix defined by the components of angular velocity vector Ω_x, Ω_y, and Ω_z as given in Equation 4.6. Using this notation, makes the relative derivative formula applicable even to matrices. For example, consider the problem of finding the derivative of rotation matrix with respect to time, ${}^{A}\dot{\boldsymbol{R}}_{B}$. Applying this formulation results in

$$\left(\frac{d({}^{A}\boldsymbol{R}_{B})}{dt}\right)_{fix} = \left(\frac{\partial({}^{A}\boldsymbol{R}_{B})}{\partial t}\right)_{mov} + \boldsymbol{\Omega}^{\times}({}^{A}\boldsymbol{R}_{B}), \tag{4.15}$$

while

$$\left(\frac{\partial({}^{A}\boldsymbol{R}_{B})}{\partial t}\right)_{mov} = 0.$$

Hence

$${}^{A}\dot{\boldsymbol{R}}_{B} = \boldsymbol{\Omega}^{\times}\, {}^{A}\boldsymbol{R}_{B}. \tag{4.16}$$

This is one way of verification of Equation 4.6, in which $\boldsymbol{\Omega}^{\times}$ is defined as the multiplication of ${}^{A}\dot{\boldsymbol{R}}_{B}\, {}^{A}\boldsymbol{R}_{B}^{-1}$.

Now, consider the general motion of a rigid body shown in Figure 4.2, in which a moving frame $\{\boldsymbol{B}\}$ is attached to the rigid body and the problem is to find the absolute velocity of point P with respect to a fixed frame $\{\boldsymbol{A}\}$. The rigid body perform a general motion, which is a combination of a translation, denoted by the vector ${}^{A}\boldsymbol{P}_{O_B}$, and an instantaneous rotation shown in Figure 4.2.

To determine the velocity of point P, start with the relation between absolute and relative position vectors of point P as the following. Rewrite Equation 2.52,

$${}^{A}\boldsymbol{P} = {}^{A}\boldsymbol{P}_{O_B} + {}^{A}\boldsymbol{R}_{B}\, {}^{B}\boldsymbol{P}, \tag{4.17}$$

in which ${}^{A}\boldsymbol{P}$ and ${}^{A}\boldsymbol{P}_{O_B}$ are absolute position vectors of point P and absolute position vectors of the origin of frame $\{\boldsymbol{B}\}$, respectively. Furthermore, ${}^{B}\boldsymbol{P}$ is the relative position vector

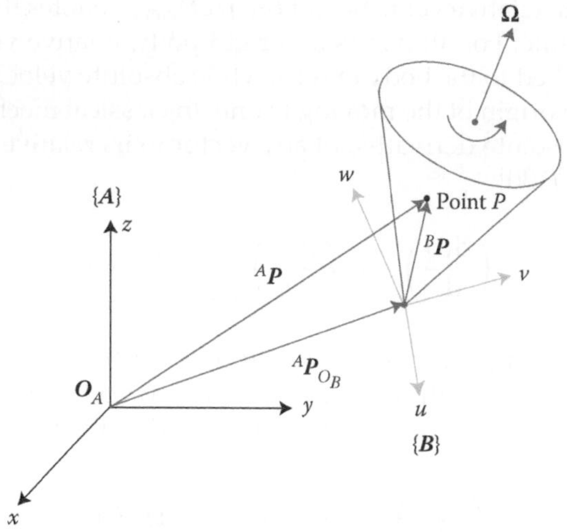

FIGURE 4.2
Instantaneous velocity of a point P with respect to a moving frame $\{\boldsymbol{B}\}$.

of point P with respect to frame $\{\boldsymbol{B}\}$. To derive the velocity of point P, differentiate both sides with respect to time as

$$\begin{aligned} {}^A\dot{\boldsymbol{P}} &= {}^A\dot{\boldsymbol{P}}_{O_B} + {}^A\dot{\boldsymbol{R}}_B\, {}^B\boldsymbol{P} + {}^A\boldsymbol{R}_B\, {}^B\dot{\boldsymbol{P}} \\ {}^A\boldsymbol{v}_p &= {}^A\boldsymbol{v}_{O_B} + {}^A\dot{\boldsymbol{R}}_B\, {}^B\boldsymbol{P} + {}^A\boldsymbol{R}_B\, {}^B\boldsymbol{v}_p, \end{aligned} \tag{4.18}$$

in which ${}^B\boldsymbol{v}_p = v_{rel}$ is the relative velocity of point P with respect to frame $\{\boldsymbol{B}\}$. The time derivative of rotation matrix ${}^A\dot{\boldsymbol{R}}_B$ can be derived from Equation 4.6, which is verified by Equation 4.16 as

$${}^A\dot{\boldsymbol{R}}_B = {}^A\boldsymbol{\Omega}^\times\, {}^A\boldsymbol{R}_B. \tag{4.19}$$

In this relation, matrix ${}^A\boldsymbol{\Omega}^\times$ is represented in Equation 4.6 by the components of angular velocity of frame $\{\boldsymbol{B}\}$ with respect to frame $\{\boldsymbol{A}\}$. Hence, the velocity of a point in a general motion represented by Equation 4.18 is simplified to

$${}^A\boldsymbol{v}_p = {}^A\boldsymbol{v}_{O_B} + {}^A\boldsymbol{R}_B\, {}^B\boldsymbol{v}_p + {}^A\boldsymbol{\Omega}^\times\, {}^A\boldsymbol{R}_B\, {}^B\boldsymbol{P}. \tag{4.20}$$

For the case where P is a point embedded in the rigid body, the relative velocity ${}^B\boldsymbol{v}_p$ is equal to zero, thus Equation 4.20 simplifies to

$${}^A\boldsymbol{v}_p = {}^A\boldsymbol{v}_{O_B} + {}^A\boldsymbol{\Omega}^\times\, {}^A\boldsymbol{R}_B\, {}^B\boldsymbol{P}. \tag{4.21}$$

4.2.3 Screw Coordinates

So far, we have shown that finite rotation of a rigid body can be expressed as a rotation θ about a screw axis $\hat{\mathbf{s}}$. Furthermore, it is shown that the angular velocity of a rigid body is also defined as the rate of instantaneous rotation angle $\dot{\theta}$ about the same screw axis $\hat{\mathbf{s}}$. As discussed earlier, the *Chasles' theorem* states that the most general rigid-body displacement can be produced by a translation along a line followed by a rotation about the same line. Since this displacement is reminiscent of the displacement of a screw, it is called a *screw displacement*, and the line or axis is called the *screw axis*. *Screw coordinates* is defined in this section to encapsulate the general motion of a rigid body using screw representation. Assume the general motion of a rigid body as a combination of translation along and rotation about the screw axis. Furthermore, assume that the ratio of the instantaneous translation d to the instantaneous rotation θ is called the pitch, λ, similar to the pitch defined for an ordinary screw. Therefore, for finite motions, $\lambda = d/\theta$ is expressed in m/rad unit. Furthermore, for infinitesimal motions λ may be defined by the following relation:

$$\lambda = \frac{\dot{d}}{\dot{\theta}}. \tag{4.22}$$

Hence, general motion can be expressed conveniently by a 6×1 coordinates system called *screw coordinates* as the following [9,158]. Define the coordinate of a *unit screw* $\hat{\$}$, by a pair of the following two vectors:

$$\hat{\$} = \begin{bmatrix} \hat{s} \\ s_o \times \hat{s} + \lambda \hat{s} \end{bmatrix} = \begin{bmatrix} \$_1 \\ \$_2 \\ \$_3 \\ \$_4 \\ \$_5 \\ \$_6 \end{bmatrix}, \tag{4.23}$$

in which $\hat{s}$ is the unit vector along the screw axis and s_o is the distance of the screw axis to the origin of the fixed frame $\{A\}$, as shown in Figure 2.12. Note that $s_o \times \hat{s}$ is perpendicular to both s_o and $\hat{s}$, and defines the moment of the screw axis about the origin of frame $\{A\}$. Furthermore, since the screw axis is a unit vector, even for the case where s_o is selected as an arbitrary point on the screw axis $\hat{s}$, the magnitude of the resulting cross product of the two vectors s_o and $\hat{s}$ is equal to the Euclidean distance of the screw axis to the origin of the fixed frame $\{A\}$.

The general displacement of a point in a rigid body is completely determined if the *amplitude* or the *intensity* of screw coordinates is specified. Let $\dot{q}$ be the intensity of the screw coordinate at velocity level, then the resulting *twist* is defined by the following screw coordinate:

$$\$ = \dot{q}\, \hat{\$} \tag{4.24}$$

in which, for a general motion, $\dot{q}$ is equal to the magnitude of the angular velocity of the rigid body, $\dot{q} = \dot{\theta}$. Note that the first three components of the screw coordinates are equal to the vector of angular velocity of the rigid body

$$\hat{s}\,\dot{\theta} = {}^A\boldsymbol{\Omega}. \tag{4.25}$$

The next three components denote the linear velocity vector of a point of rigid body coincident with the origin of frame $\{A\}$, as seen by the following manipulations:

$$\begin{aligned} (s_o \times \hat{s} + \lambda \hat{s})\dot{\theta} &= s_o \times \dot{\theta}\hat{s} + \lambda \dot{\theta}\hat{s} \\ &= s_o \times \boldsymbol{\Omega} + \lambda \dot{\theta}\hat{s} \\ &= \boldsymbol{\Omega} \times (-s_o) + \dot{d}\hat{s} \\ &= \boldsymbol{\Omega} \times {}^B\boldsymbol{P}_{O_A} + \dot{d}\hat{s}. \end{aligned} \tag{4.26}$$

The last equation is derived from Figure 2.12, wherein ${}^{A}\boldsymbol{P}_{O_B} = \boldsymbol{s}_o$, therefore ${}^{B}\boldsymbol{P}_{O_A} = -\boldsymbol{s}_o$. Comparing the resulting Equation 4.26 to Equation 4.21, the second term in Equation 4.26 is the linear velocity of the rigid body due to translational motion denoted by ${}^{A}\boldsymbol{v}_{O_B}$ in Equation 4.21. Furthermore, the first term in Equation 4.26 is due to the rotational motion of the rigid body and denotes the velocity of a point in the rigid body that coincides with O_A.

Hence, the last three components of the screw coordinates in Equation 4.24 denote *the velocity of a point of the rigid body that coincides instantaneously with the origin of the frame* $\{\boldsymbol{A}\}$. Therefore, to represent angular velocity of a rigid body and linear velocity of a point on it, it is possible to use a screw coordinate representation, in which the origin of the reference frame $\{\boldsymbol{A}\}$ is instantaneously attached to the point of interest.

For special cases in which only translational, or pure rotational motion occurs, the screw coordinates can be conveniently used as the following. For pure rotation, $\lambda = 0$, and $\dot{q} = \dot{\theta}$ as before, and the twist represented in screw coordinates reduces to

$$\$ = \begin{bmatrix} \hat{\boldsymbol{s}} \\ \boldsymbol{s}_o \times \hat{\boldsymbol{s}} \end{bmatrix} \dot{\theta}. \tag{4.27}$$

For a pure translational motion, $\lambda = \infty$ and $\dot{q} = \dot{d}$, and the twist represented in screw coordinates is defined as

$$\$ = \begin{bmatrix} \boldsymbol{0} \\ \hat{\boldsymbol{s}} \end{bmatrix} \dot{d}. \tag{4.28}$$

Note that the Chebyshev–Grübler–Kutzbach criteria may be conveniently used to specify the degrees-of-freedom of a mechanism; however, it cannot be used to specify the nature and characteristics of the degrees-of-freedom. On the other hand, screw coordinates may be conveniently used to provide an analytical tool to specify the characteristics of the degrees-of-freedom of a parallel robot [105].

4.3 Jacobian Matrices of a Parallel Manipulator

A general parallel manipulator such as a typical one, shown in Figure 3.2, consists of a moving platform that is connected to a fixed base by several limbs. For a fully parallel manipulator, the number of limbs are equal to the number of degrees-of-freedom of the moving platform. Each limb consists of a number of joints, in which only one of them is actuated. Thus, some of the joints are driven by actuators, whereas others are passive. In general, the number of actuated joints are greater than or equal to the number of degrees-of-freedom. In such cases, where the number of actuators are greater than the number of degrees-of-freedom, the manipulator is considered to be redundantly actuated. Let $\boldsymbol{q}$ denote the vector of actuated joint coordinates and $\boldsymbol{\mathcal{X}}$ denote the vector of moving platform motion variables, consisting of position and orientation variables. In general, we have

$$\boldsymbol{q} = [q_1, q_2, \ldots, q_m]^T \tag{4.29}$$

and

$$\boldsymbol{\mathcal{X}} = [x_1, x_2, \ldots, x_n]^T, \tag{4.30}$$

where q_i represents either the angular rotation of an actuated revolute joint or the linear displacement of an actuated prismatic joint and x_i represents the position or orientation variable of the moving platform. Furthermore, m denotes the number of actuated joints in the manipulator, while n denotes the number of degrees-of-freedom of the manipulator. Generally $m \geq n$, in which for fully parallel manipulators, $m = n$, whereas for redundant manipulators $m > n$. We also refer to vectors $\mathbf{q}$ and $\boldsymbol{\mathcal{X}}$ as the (active) joint and moving platform variables, respectively. These two vectors are related through a system of nonlinear algebraic equations representing the kinematic constraints imposed by the limbs, which can be generally written as

$$f(q, \boldsymbol{\mathcal{X}}) = \mathbf{0}. \tag{4.31}$$

This equation is a set of m nonlinear algebraic equations, in which $\mathbf{0}$ denotes an m-dimensional zero vector. Differentiate Equation 4.31 with respect to time to find a relationship between the input joint rates and the moving platform output twist as follows:

$$J_x \dot{\boldsymbol{\mathcal{X}}} = J_q \dot{q}, \tag{4.32}$$

where

$$J_x = \frac{\partial f}{\partial \boldsymbol{\mathcal{X}}} \quad \text{and} \quad J_q = -\frac{\partial f}{\partial \mathbf{q}}. \tag{4.33}$$

These two matrices are called the Jacobian matrices. The above derivation leads to two separate Jacobian matrices as proposed in [61], where $\boldsymbol{J}_q$ is a square matrix of dimension m and $\boldsymbol{J}_x$ is an $m \times n$ matrix. In the case where $m = n$, both Jacobian matrices become a square. The general Jacobian matrix, $\boldsymbol{J}$, of a parallel manipulator is defined as

$$\dot{q} = J\dot{\boldsymbol{\mathcal{X}}}, \tag{4.34}$$

where, for nonsingular J_q,

$$J = J_q^{-1} J_x. \tag{4.35}$$

Note that the general Jacobian of a parallel manipulator $\boldsymbol{J}$ defined in Equation 4.34 corresponds to the inverse Jacobian definition of a serial manipulator.

4.4 Velocity Loop Closure

To derive the Jacobian matrices of a parallel manipulator and to determine the relation between the moving platform twist, $\dot{\boldsymbol{\mathcal{X}}}$ and the joint rates $\dot{\mathbf{q}}$, two main methods are introduced in the literature. In one approach, the screw coordinates are used to represent the moving platform twist and the joint rates. This method of obtaining the Jacobian matrices of a parallel manipulator is called the *screw-based Jacobian*. In the other more common method, the conventional *velocity loop closures* are used. This method appears to be more straightforward, and therefore, it is examined in this section.

Generally, velocity loop closures are derived by direct differentiation of kinematic loop closures. As elaborated in Section 3.2, at a displacement level, closure of each kinematic loop can be expressed in vector form as in Equation 3.2. Let us rewrite this equation for further analysis for the case where the number of limbs, and therefore, the number of loop closures are m, as

$$\boldsymbol{p} = \boldsymbol{a}_i + \boldsymbol{d}_i - \boldsymbol{R}\,\boldsymbol{b}_i \quad \text{for } i = 1, \ldots, m. \tag{4.36}$$

This loop closure is written as a function of the position and orientation variables of the moving platform $\boldsymbol{p}$, and $\boldsymbol{R}$, the position vectors describing the known geometry of the base and the moving platform denoted by $\boldsymbol{a}_i$ and $\boldsymbol{b}_i$, respectively, and the limb vector denoted by $\boldsymbol{d}_i$. The limb vector $\boldsymbol{d}_i$ is a function of the active joint variables denoted by q_i, and possibly some other passive joint variables, corresponding to the unactuated joints. The velocity of the moving platform relative to the base can be obtained upon taking the time derivative of both sides of Equation 4.36, which yields to

$$\dot{\boldsymbol{p}} = \dot{\boldsymbol{d}}_i - \boldsymbol{\omega} \times \boldsymbol{R}\,\boldsymbol{b}_i \quad \text{for } i = 1, \ldots, m. \tag{4.37}$$

Note that the time derivatives of $\boldsymbol{a}_i$ vectors with respect to the fixed frame $\{\boldsymbol{A}\}$, and that of $\boldsymbol{b}_i$ vectors with respect to the moving frame $\{\boldsymbol{B}\}$ are all zero. This equation relates the active and passive joint rates $\dot{\boldsymbol{d}}_i$, to the moving platform linear and angular velocities $\dot{\boldsymbol{p}}$ and $\boldsymbol{\omega}$, respectively.

To eliminate the unactuated joint rates in each limb, the velocity loop closures are dot multiplied to an appropriate vector that is normal to all vectors of passive joint rates. By this means only the active joint rates $\dot{q}_i$ remain in the vector loop closures. Although this part of the Jacobian derivation is repeatedly carried out for all parallel manipulators, obtaining the right vector for the dot product depends on the structure and geometry of each manipulator. The details of this derivation are illustrated by different examples in the following sections. Finally, the resulting equations are written in a matrix form and Jacobian matrices of the manipulator are extracted. For consistency, the moving platform twist vector $\dot{\mathcal{X}}$ for the conventional Jacobian is defined as a six-dimensional vector consisting of the linear velocity of a point of the moving platform followed by the angular velocity of the moving platform, as follows:

$$\dot{\mathcal{X}} = \begin{bmatrix} v_p \\ \omega \end{bmatrix}. \tag{4.38}$$

This method can be best illustrated by examples, which will be detailed in Sections 4.6 through 4.8.

4.5 Singularity Analysis of Parallel Manipulators

Singularities of parallel mechanisms have been examined by many researchers. Gosselin and Angeles [61] defined three types of singularities for closed-loop mechanisms that were based on the roots of the determinant of the Jacobian matrices. With the same definitions, Tsai [181] terms the first, second, and third kinds of singularities as inverse, forward, and combined singularities, respectively. These classifications have been further refined by Zlatanov, Fenton, and Benhabib [199], where six types of singular configurations are introduced with detailed physical interpretations. Later, Zlatanov, Bonev, and Gosselin [198], discovered another type of singularity, which is termed as constraint singularity. In this book, we adopt the definition given in [61], and further used in [181], for singularity classification. These singularities occur, respectively, when

1. Matrix J_q is rank deficient, or
2. Matrix J_x is rank deficient, or
3. The positioning equations degenerate.

As pointed out in [61], only the first type of singularity is possible for serial manipulators. The physical interpretation of these types of singularities are briefly represented in the sequel.

4.5.1 Inverse Kinematic Singularity

Referring to Equation 4.32, inverse kinematic singularity occurs when matrix J_q is rank deficient. Since for all parallel manipulators, J_q is an $m \times m$ square matrix; this can happen when

$$\det(J_q) = 0. \tag{4.39}$$

The corresponding configurations are located at the boundary of the manipulator workspace or on the internal boundaries between subregions of the workspace where the number of solutions of inverse kinematic problem is not the same. In such cases, since the null space of J_q is not empty, there exist nonzero vectors $\dot{q}$, which correspond to a vanishing Cartesian twist vector $\dot{\mathcal{X}}$. In other words, infinitesimal motion of the moving platform along certain directions cannot be accomplished. Hence, the manipulator looses one or more degrees-of-freedom when inverse kinematic singularity happens. Moreover, in virtue of the kinematic–static duality [184], there exist wrenches that will not affect the actuators when applied to the moving platform. In other words, at an inverse kinematic singular configuration, the manipulator can resist forces or moments in some directions with zero actuator forces or torques. Inverse kinematic singularity is similar to serial manipulator singularity.

4.5.2 Forward Kinematic Singularity

Referring to Equation 4.32, forward kinematic singularity occurs when matrix J_x is rank deficient. Since for parallel manipulators, J_x is an $m \times n$ matrix with $m \geq n$, this can happen when

$$\det(J_x^T J_x) = 0. \tag{4.40}$$

If the manipulator is not redundantly actuated, then $m = n$, and the Jacobian matrix J_x becomes a square. Hence, in this case, forward kinematic singularity happens when

$$\det(J_x) = 0. \tag{4.41}$$

As opposed to inverse kinematic singularity, this type of degeneracy can occur inside the manipulator's Cartesian workspace and corresponds to the set of configurations for which two different branches of forward kinematic problem meet. In fact, this is why this type of singularity cannot occur for serial manipulators, where forward kinematic problem always leads to a unique solution.

From Equations 4.32 and 4.40, since the null space of matrix $\boldsymbol{J}_x$ is not empty, there exist nonzero Cartesian twist vectors $\dot{\boldsymbol{\mathcal{X}}}$ that are mapped into a vanishing actuator velocity vector $\dot{q}$. The corresponding configuration will be one in which an infinitesimal motion of the end effector is possible even if the actuators are locked. In other words, the end effector gains one or several degrees-of-freedom and the manipulator's stiffness vanishes in the corresponding direction(s).

4.5.3 Combined Singularity

This type of singularity is of a slightly different nature from the first two. A combined singularity occurs when $\boldsymbol{J}_x$ and $\boldsymbol{J}_q$ both are rank deficient. It corresponds to a degeneracy of the position or orientation equations. Some conditions on the geometric parameters of the robot are required for this type of singularity to occur. Indeed, only certain special architectures lead to this type of singularity that justifies the term architecture singularity used in [120] to designate them. In fact, the geometric conditions under which such manipulators exhibit combined singularities are given in [61] and can easily be avoided. Such singularities will lead to configurations where a finite motion of the end effector is possible even if the actuators are locked, or in situations where a finite motion of the actuators produces no motion of the end effector. In both cases, the manipulator cannot be controlled.

4.6 Jacobian Analysis of a Planar Manipulator

Jacobian analysis plays a vital role in the study of robotic manipulators. The Jacobian matrix not only reveals the relation between the joint variable velocities $\dot{\boldsymbol{L}}$ and the moving platform velocities $\dot{\boldsymbol{\mathcal{X}}}$, it also constructs the transformation needed to find the actuator forces $\boldsymbol{\tau}$ from the forces and moments acting on the moving platform $\boldsymbol{\mathcal{F}}$. In this section, Jacobian analysis for the planar $4R\underline{P}R$ manipulator is performed. The manipulator Jacobian is then used for singularity and sensitivity analysis in Sections 4.6.2 and 4.6.3, respectively.

4.6.1 Velocity Loop Closure

The architecture of a planar $4R\underline{P}R$ parallel manipulator considered here is described in detail in Section 3.3.1 and shown in Figure 1.7. As explained hereinbefore, in this manipulator, the moving platform is supported by four limbs of an identical kinematic structure. Each limb connects the fixed base to the manipulator moving platform by a revolute joint (R) followed by an actuated prismatic joint (P̲) and another revolute joint (R). The kinematic structure of a prismatic joint is used to model either a piston–cylinder actuator at each limb or a cable driven one. To avoid singularities at the central position of the manipulator at each level, the limbs are considered to be crossed. Complete singularity analysis of the mechanism is analyzed and presented in Section 4.6.2.

As shown in Figure 1.7, A_i denote the fixed base points of the limbs, B_i denote points of connection of the limbs on the moving platform, L_i denote the limbs' lengths, and α_i denote the limbs' angles. The position of the center of the moving platform G, is denoted by $G = [x_G, y_G]$, and the orientation of the manipulator moving platform is denoted by ϕ with respect to the fixed coordinate frame. Hence, the manipulator possesses three-degrees-of-freedom represented by $\boldsymbol{\mathcal{X}} = [x_G, y_G, \phi]$, and one degree of actuation redundancy. For the purpose of analysis and as illustrated in Figure 3.4, a fixed frame $O : xy$ is attached to the fixed base at point O, the center of the base point circle that passes through $A_i's$. Moreover, another moving coordinate frame $G : UV$ is attached to the manipulator moving platform at point G. The geometrical parameters of the manipulator are given in Section 3.3.2.

Contrary to serial manipulators, the Jacobian matrix of a parallel manipulator is defined as the transformation matrix that converts the moving platform twist to joint variable velocities, that is,

$$\dot{\boldsymbol{L}} = \boldsymbol{J} \cdot \dot{\boldsymbol{\mathcal{X}}}, \tag{4.42}$$

in which for the planar $4R\underline{P}R$ manipulator, $\dot{\boldsymbol{L}} = [\dot{L}_1, \dot{L}_2, \dot{L}_3, \dot{L}_4]$ is the 4×1 limb velocity vector, and $\dot{\boldsymbol{\mathcal{X}}} = [\dot{x}_G, \dot{y}_G, \dot{\phi}]$, the 3×1 moving platform twist vector. Since the manipulator is redundantly actuated, the Jacobian matrix $\boldsymbol{J}$ is a nonsquare 4×3 matrix. From the geometry of the manipulator as illustrated in Figure 3.4, the loop closure Equation 3.5 for each limb, is rewritten here for further analysis:

$$\overrightarrow{A_iG} = \overrightarrow{A_iB_i} + \overrightarrow{B_iG}. \tag{4.43}$$

Considering the vector definitions $\hat{\boldsymbol{S}}_i$ and $\boldsymbol{E}_i$ illustrated in Figure 4.3, the vector loop closure is simplified to the following equation

$$\boldsymbol{G} = L_i\hat{\boldsymbol{S}}_i - \boldsymbol{E}_i \quad \text{or} \quad \boldsymbol{G} + \boldsymbol{E}_i = L_i\hat{\boldsymbol{S}}_i, \tag{4.44}$$

in which $\boldsymbol{G}$ is the position vector of point G. To obtain the Jacobian matrix, let us differentiate the vector loop Equation 4.44 with respect to time. The time derivative of position vector $\boldsymbol{G}$ is denoted by the velocity vector v_G. The time derivative of vector $\boldsymbol{E}_i$ is derived from the relative differentiation formula 4.13 as follows:

$$\left(\frac{\mathrm{d}(\boldsymbol{E}_i)}{\mathrm{d}t}\right)_{fix} = \left(\frac{\partial(\boldsymbol{E}_i)}{\partial t}\right)_{mov} + \boldsymbol{\Omega} \times (\boldsymbol{E}_i), \tag{4.45}$$

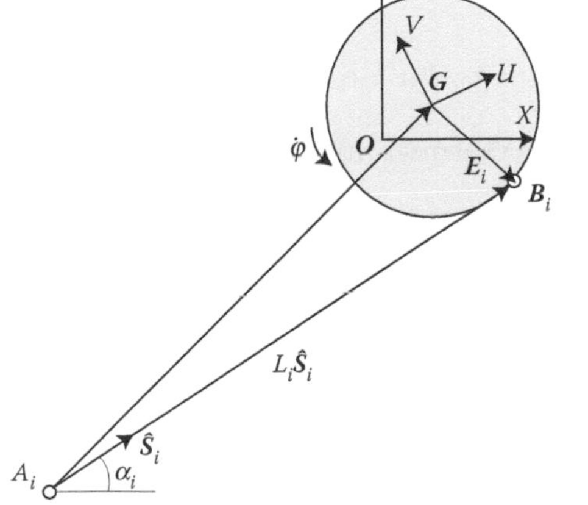

FIGURE 4.3
Vectors' definitions for Jacobian derivation of the planar $4R\underline{P}R$ manipulator.

where the time derivative of vector $\boldsymbol{E}_i$ with respect to the moving frame is zero, and the angular velocity of the moving frame is $\boldsymbol{\Omega} = \dot{\phi}\hat{\boldsymbol{K}}$. Hence,

$$\dot{\boldsymbol{E}}_i = \dot{\phi}(\hat{\boldsymbol{K}} \times \boldsymbol{E}_i). \tag{4.46}$$

Finally, the time derivative of the last term $L_i\hat{\boldsymbol{S}}_i$ is determined by the following equation:

$$\frac{\mathrm{d}}{\mathrm{d}t}(L_i\hat{\boldsymbol{S}}_i) = \dot{L}_i\hat{\boldsymbol{S}}_i + L_i\dot{\hat{\boldsymbol{S}}}_i, \tag{4.47}$$

in which the time derivative of the second term $L_i\dot{\hat{\boldsymbol{S}}}_i$ is determined by the relative differentiation rule similar to that for $\dot{\boldsymbol{E}}_i$, noting that the angular velocity of this vector is $\dot{\alpha}\boldsymbol{K}$. Hence,

$$\frac{d}{dt}(L_i\hat{\boldsymbol{S}}_i) = \dot{L}_i\hat{\boldsymbol{S}}_i + \dot{\alpha}_i L_i(\hat{\boldsymbol{K}} \times \hat{\boldsymbol{S}}_i). \tag{4.48}$$

Thus, for each limb $i = 1, 2, 3, 4$, the velocity loop closure results in

$$\boldsymbol{v}_G + \dot{\phi}(\hat{\boldsymbol{K}} \times \boldsymbol{E}_i) = \dot{L}_i\hat{\boldsymbol{S}}_i + \dot{\alpha}_i L_i(\hat{\boldsymbol{K}} \times \hat{\boldsymbol{S}}_i), \tag{4.49}$$

in which $\boldsymbol{v}_G = [\dot{x_G}, \dot{y_G}]^T$ is the velocity of the moving platform at point G and $\hat{\boldsymbol{K}}$ is the unit vector in z direction of the fixed coordinate frame $\{\boldsymbol{A}\}$. To eliminate the passive joint variable $\dot{\alpha_i}$, dot multiply both sides of Equation 4.49 by $\hat{\boldsymbol{S}}_i$.

$$\hat{\boldsymbol{S}}_i \cdot \boldsymbol{v}_G + \hat{\boldsymbol{S}}_i \cdot (\hat{\boldsymbol{K}} \times \boldsymbol{E}_i)\,\dot{\phi} = \dot{L}_i\hat{\boldsymbol{S}}_i \cdot \hat{\boldsymbol{S}}_i + \dot{\alpha}_i\, L_i\,(\hat{\boldsymbol{K}} \times \hat{\boldsymbol{S}}_i) \cdot \hat{\boldsymbol{S}}_i. \tag{4.50}$$

Note that $\hat{\boldsymbol{S}}_i$ is a unit vector, and hence, $\dot{L}_i\hat{\boldsymbol{S}}_i \cdot \hat{\boldsymbol{S}}_i = \dot{L}_i$. Furthermore, the vector $\hat{\boldsymbol{K}} \times \hat{\boldsymbol{S}}_i$ is orthogonal to $\hat{\boldsymbol{S}}_i$, and hence, $(\hat{\boldsymbol{K}} \times \hat{\boldsymbol{S}}_i) \cdot \hat{\boldsymbol{S}}_i = 0$. Using the dot and cross multiplication interchanging rule

$$\boldsymbol{a} \cdot (\boldsymbol{b} \times \boldsymbol{c}) = \boldsymbol{b} \cdot (\boldsymbol{c} \times \boldsymbol{a}). \tag{4.51}$$

Equation 4.50 simplifies to

$$\hat{\boldsymbol{S}}_i \cdot \boldsymbol{v}_G + \hat{\boldsymbol{K}} \cdot (\boldsymbol{E}_i \times \hat{\boldsymbol{S}}_i)\,\dot{\phi} = \dot{L}_i. \tag{4.52}$$

Rewriting Equation 4.52 in a matrix form results in

$$\dot{L_i} = \left[S_{i_x} \mid S_{i_y} \mid E_{i_x}S_{i_y} - E_{i_y}S_{i_x}\right] \cdot \begin{bmatrix} v_{G_x} \\ v_{G_y} \\ \dot{\phi} \end{bmatrix}. \tag{4.53}$$

Using Equation 4.53 for $i = 1, 2, 3, 4$, the manipulator Jacobian matrix $\boldsymbol{J}$ is derived.

$$\boldsymbol{J} = \left[\begin{array}{c|c|c} S_{1_x} & S_{1_y} & E_{1_x}S_{1_y} - E_{1_y}S_{1_x} \\ S_{2_x} & S_{2_y} & E_{2_x}S_{2_y} - E_{2_y}S_{2_x} \\ S_{3_x} & S_{3_y} & E_{3_x}S_{3_y} - E_{3_y}S_{3_x} \\ S_{4_x} & S_{4_y} & E_{4_x}S_{4_y} - E_{4_y}S_{4_x} \end{array}\right]. \tag{4.54}$$

Note that Jacobian matrix J is a nonsquare 4×3 matrix, since the manipulator is redundantly actuated.

To get an expression for $\dot{\alpha}_i$, cross-multiply both sides of Equation 4.49 by $\hat{S}_i$:

$$\hat{S}_i \times v_G + \dot{\phi}\,(E_i \cdot \hat{S}_i)\,\hat{K} = \dot{\alpha}_i\, L_i\, \hat{K}. \tag{4.55}$$

Rewriting Equation 4.55 in a matrix form as

$$\dot{\alpha}_i = \frac{1}{L_i} \cdot \left[-S_{i_y} \mid S_{i_x} \mid E_{i_x} S_{i_x} + E_{i_y} S_{i_y} \right] \cdot \begin{bmatrix} v_{Gx} \\ v_{Gy} \\ \dot{\phi} \end{bmatrix}. \tag{4.56}$$

Therefore, J_α is defined as the matrix relating the vector of the moving platform velocities, $\dot{\mathcal{X}} = [\dot{x}_G, \dot{y}_G, \dot{\phi}]$, to the vector of angular velocities of the limbs $\dot{\boldsymbol{\alpha}} = [\dot{\alpha}_1, \dot{\alpha}_2, \dot{\alpha}_3, \dot{\alpha}_4]$ as

$$\dot{\boldsymbol{\alpha}} = J_\alpha \cdot \dot{\mathcal{X}}, \tag{4.57}$$

in which

$$J_\alpha = \frac{1}{L_i} \cdot \left[\begin{array}{c|c|c} -S_{1_y} & S_{1_x} & E_{1_x} S_{1_x} + E_{1_y} S_{1_y} \\ -S_{2_y} & S_{2_x} & E_{2_x} S_{2_x} + E_{2_y} S_{2_y} \\ -S_{3_y} & S_{3_x} & E_{3_x} S_{3_x} + E_{3_y} S_{3_y} \\ -S_{4_y} & S_{4_x} & E_{4_x} S_{4_x} + E_{4_y} S_{4_y} \end{array} \right]. \tag{4.58}$$

4.6.2 Singularity Analysis

This section is devoted to singularity analysis of the planar $4R\underline{P}R$ parallel manipulator. Jacobian analysis of a parallel manipulator is generally much more comprehensive than that of a serial manipulator. An important limitation of the parallel manipulator is that singular configurations may exist within the workspace, where the manipulator can gain one or more degrees-of-freedom, and therefore, loses its stiffness. Moreover, as suggested in [61], singularities of closed loop mechanisms are separated by the analysis of two Jacobian matrices.

To perform singularity analysis of the manipulator in hand, consider the Jacobian matrix of the $4R\underline{P}R$ manipulator as given in Equation 4.54. Considering the general Jacobian definition in 4.34 and by comparing it with Equation 4.33, it is clear that for the planar $4R\underline{P}R$ manipulator $J_q = I$ and $J_x = J$. Hence, the manipulator has no inverse kinematic singularity, but may possess forward kinematic singularity when J becomes rank deficient. Note that since the manipulator is redundantly actuated and J is a 4×3 matrix, the robot experiences a singular configuration when

$$\det(J^T J) = 0. \tag{4.59}$$

Singular configuration can be derived either by calculation of Equation 4.59 at grid points in the entire manipulator workspace, or by geometrical examination of Jacobian matrix properties. Following the latter, since J consists of three columns of $\hat{S}_x$, $\hat{S}_y$, and $E \times \hat{S}$, the matrix may become singular if either of the columns becomes zero, or two columns become linearly dependent. From the geometrical representation of vectors E and $\hat{S}$, depicted in Figure 4.3, it may be concluded that $E \times \hat{S} = 0$ only if the two vectors become parallel. This

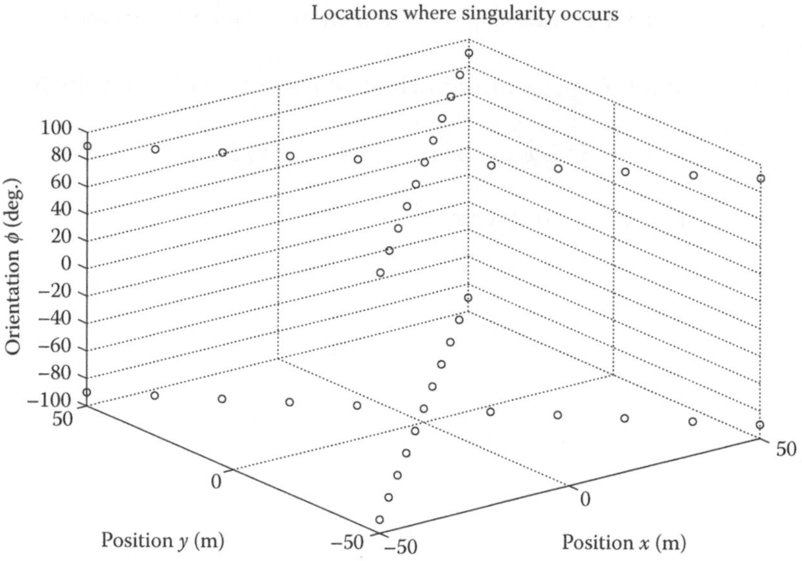

FIGURE 4.4
Singular configurations of the planar $4R\underline{P}R$ manipulator.

can happen at configurations like the central position when $x_G = y_G = 0$, if the moving platform angle is $\phi = \pm 90°$.

When the moving platform moves from the central position, geometric interpretation becomes nontrivial. Therefore, complete singular configurations for the $4R\underline{P}R$ manipulator are derived from the calculation of Equation 4.59 for a number of grid points in the workspace of the manipulator. This calculation has been done for 1089 grid points within the ranges $-50 \le x_G \le 50$, $-50 \le y_G \le 50$, and $-\pi \le \phi \le \pi$, and singular configurations are given in the three-dimensional (3D) plot of Figure 4.4. As seen in this figure, the geometrical interpretation of the singular configurations is confirmed and is extended to points in two planes where

$$x_G = \pm y_G \quad \text{and} \quad \phi = \pm 90°. \tag{4.60}$$

Note that in the design of the manipulator, the cable crossed configuration is considered. The main reason for this choice is to avoid having a singular configuration at a central position and orientation $x_G = y_G = \phi = 0$. In this proposed design, the manipulator is nonsingular for the entire workspace, provided the required orientation workspace is smaller than $\pm 90°$.

4.6.3 Sensitivity Analysis

As seen in Section 4.6.2, singular points of the manipulator are limited to a few configurations where $\phi = \pm 90°$. In this section, the conditioning of the Jacobian matrix is analyzed in configurations close to singular points. The measure proposed to analyze the good conditioning of the manipulator Jacobian matrices, is the inverse of the Jacobian matrix condition number. In fact, for a nonsquare matrix, like the redundant manipulator Jacobian matrices, this measure quantifies the ratio of the smallest to the largest singular value of the matrix. If the system is close to singular configurations, the smallest singular value

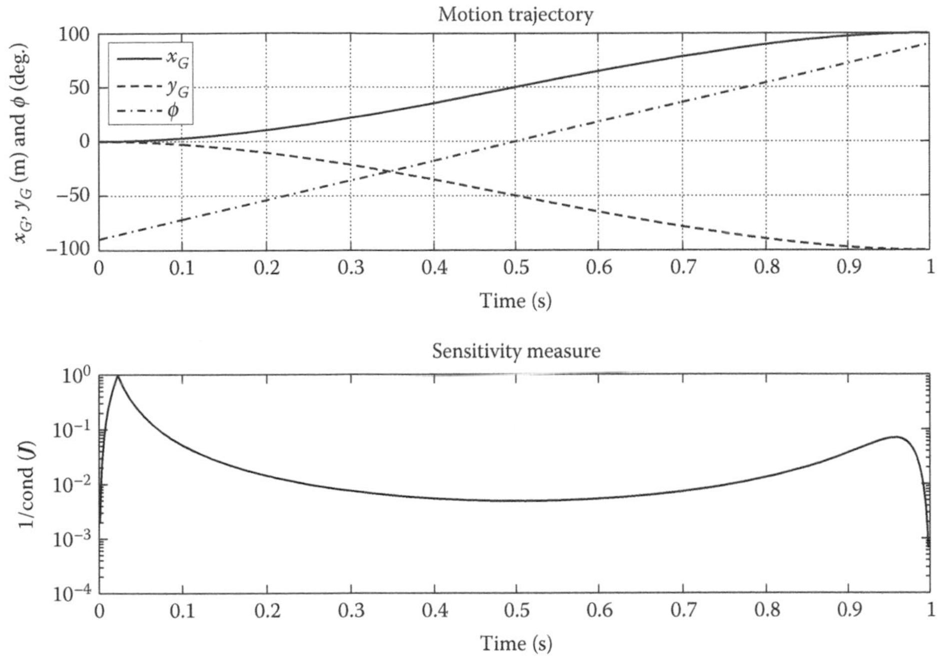

FIGURE 4.5
Sensitivity of the planar $4R\underline{P}R$ manipulator Jacobian for a typical trajectory.

tends to be zero, resulting in having a small value for the sensitivity measure. On the contrary, values close to 1 for the sensitivity measure depict poses where the singular values are close to each other, or the manipulator is near an isotropic configuration.

Figure 4.5 illustrates the sensitivity measure of the manipulator Jacobian matrix for a typical trajectory. As illustrated in the top figure, a typical trajectory is considered for the moving platform, in which the horizontal and vertical motions are about 100 m, while the orientation of the moving platform is about 180°. The initial orientation is very close to −90° singular configuration and the final orientation approached +90°. The bottom figure gives a logarithmic scale plot of the sensitivity measure of the manipulator Jacobian J for the typical trajectory. As seen in this figure, the sensitivity measure of the Jacobian matrix is close to zero at initial and final configurations, which are close to singular configuration. However, the Jacobian matrix becomes more and more well conditioned, as the trajectory moves away from the singular configuration. Since the sensitivity measures are given in a logarithmic scale, it is clear that for the configurations not very close to the singular configurations, the Jacobian matrix of the manipulator is well conditioned and can be tractably used in relating numerical computations.

4.7 Jacobian Analysis of Shoulder Manipulator

In this section, Jacobian analysis of the three-degrees-of-freedom, spatial orientation shoulder manipulator is studied in detail. In this analysis, first the manipulator Jacobian matrices are determined, and then the singularity of the manipulator is analyzed.

4.7.1 Velocity Loop Closure

As shown in Figure 3.10, in the shoulder manipulator, the moving platform is constrained to a three-degrees-of-freedom of purely orientation motion. Four high-performance hydraulic piston actuators are used to provide three-degrees-of-freedom for the moving platform, and hence one degree of actuator redundancy is adopted in this design. Simple revolute joints are used in the manipulator to build spherical and universal joints and all the limbs share an identical kinematic structure. The four active limbs share a $U\underline{P}U$ kinematic structure, in which each of the two limbs are connected to the moving platform through a single connecting link. Figure 3.12 depicts a geometric model for the mechanism and a detailed geometrical description of the manipulator is given in Section 3.4.2.

In this section, we derive Jacobian matrices of the shoulder manipulator as a basic requirement for singularity and stiffness analysis for this manipulator. The Jacobian matrix of the shoulder manipulator relates the angular velocity of the moving platform $\boldsymbol{\omega}$ to the vector of the actuated joint rates, $\dot{\boldsymbol{L}} = [\dot{l}_1, \dot{l}_2, \dot{l}_3, \dot{l}_4]^T$ by

$$\dot{\boldsymbol{L}} = \boldsymbol{J}\boldsymbol{\omega}. \tag{4.61}$$

From the above definition, it is easily observed that the Jacobian of this manipulator is a 4×3 matrix since the mechanism is a redundant manipulator. Using the same idea of mapping between joint and task space variables, it is clear that the Jacobian matrix depends on the actuated limbs as well as the passive supporting limb. Therefore, one may first derive a 4×6 Jacobian, $\boldsymbol{J}_v$, relating the six-dimensional (6D) twist vector of the moving platform, $\dot{\boldsymbol{\mathcal{X}}}$, to the vector of actuated limbs' rates, $\dot{\boldsymbol{L}}$. Then, we find the 6×3 Jacobian of the passive supporting limb, $\boldsymbol{J}_p$. The Jacobian of the shoulder manipulator will be finally derived as

$$\boldsymbol{J} = \boldsymbol{J}_v \boldsymbol{J}_p. \tag{4.62}$$

4.7.1.1 *Jacobian of the Actuated Limbs*

Jacobian of the actuated limbs, $\boldsymbol{J}_v$, relates the 6D twist vector of the moving platform, $\dot{\boldsymbol{\mathcal{X}}}$, to the vector of actuated limb rates, $\dot{\boldsymbol{L}}$, such that

$$\dot{\boldsymbol{L}} = \boldsymbol{J}_v \dot{\boldsymbol{\mathcal{X}}}, \tag{4.63}$$

in which the moving platform twist can be written as

$$\dot{\boldsymbol{\mathcal{X}}} = \begin{bmatrix} {}^A\boldsymbol{v}_p \\ \boldsymbol{\omega} \end{bmatrix} = \begin{bmatrix} v_{p_x} & v_{p_y} & v_{p_z} & \omega_x & \omega_y & \omega_z \end{bmatrix}^T. \tag{4.64}$$

In Equation 4.64, ${}^A\boldsymbol{v}_p$ denotes the velocity of the moving platform center point and $\boldsymbol{\omega}$ denotes the angular velocity of the moving platform. Let us rewrite the kinematic vector-loop equation for each actuated limb, Equation 3.26, here for further analysis.

$$\boldsymbol{L}_i = l_i\,\hat{\boldsymbol{s}}_i = {}^A\boldsymbol{P} + {}^A\boldsymbol{R}_B\,{}^B\boldsymbol{P}_i - \boldsymbol{a}_i, \tag{4.65}$$

where l_i is the length of the i'th actuated limb, $\hat{\boldsymbol{s}}_i$ is the unit vector pointing along the direction of the i'th actuated limb, ${}^A\boldsymbol{R}_B$ is the rotation matrix of the moving platform, and

${}^{A}\boldsymbol{P}$ is its center position vector. To derive the velocity loop closure, differentiate kinematic vector loop closure 4.65 with respect to time

$$\dot{l}_i\hat{\boldsymbol{s}}_i + (\boldsymbol{\omega}_i \times \hat{\boldsymbol{s}}_i)\, l_i = {}^{A}\boldsymbol{v}_p + \boldsymbol{\omega} \times {}^{A}\boldsymbol{R}_B\, {}^{B}\boldsymbol{P}_i, \tag{4.66}$$

where $\boldsymbol{\omega}_i$ is the angular velocity of the i'th limb written in the base frame. Dot multiply Equation 4.66 by $\hat{\boldsymbol{s}}_i$

$$\dot{l}_i = \hat{\boldsymbol{s}}_i^T\, {}^{A}\boldsymbol{v}_p + \left[({}^{A}\boldsymbol{R}_B\, {}^{B}\boldsymbol{P}_i) \times \hat{\boldsymbol{s}}_i\right]^T \boldsymbol{\omega}. \tag{4.67}$$

Writing the above equation four times for each actuated limb in a vector form as that in Equation 4.63 results in the Jacobian matrix of the actuated limbs, $\boldsymbol{J}_v$.

$$\boldsymbol{J}_v = \begin{bmatrix} {}^{A}\hat{\boldsymbol{s}}_1^T & ({}^{A}\boldsymbol{P}_1 \times {}^{A}\hat{\boldsymbol{s}}_1)^T \\ {}^{A}\hat{\boldsymbol{s}}_2^T & ({}^{A}\boldsymbol{P}_2 \times {}^{A}\hat{\boldsymbol{s}}_2)^T \\ {}^{A}\hat{\boldsymbol{s}}_3^T & ({}^{A}\boldsymbol{P}_3 \times {}^{A}\hat{\boldsymbol{s}}_3)^T \\ {}^{A}\hat{\boldsymbol{s}}_4^T & ({}^{A}\boldsymbol{P}_4 \times {}^{A}\hat{\boldsymbol{s}}_4)^T \end{bmatrix}_{4\times 6}. \tag{4.68}$$

Note that $\boldsymbol{P}_1 = \boldsymbol{P}_2$ is the connecting point of the limbs 1 and 2 to the moving platform, while $\boldsymbol{P}_3 = \boldsymbol{P}_4$ is that of the limbs 3 and 4.

4.7.1.2 Jacobian of the Passive Limb

To find the manipulator Jacobian, we need to find a relationship between the 6D twist vector of the moving platform, $\dot{\boldsymbol{\mathcal{X}}}$, and the angular velocity of the moving platform, $\boldsymbol{\omega}$. This relation is defined by the Jacobian of the passive limb $\boldsymbol{J}_p$ as

$$\dot{\boldsymbol{\mathcal{X}}} = \begin{bmatrix} {}^{A}\boldsymbol{v}_p \\ \boldsymbol{\omega} \end{bmatrix} = \boldsymbol{J}_p \boldsymbol{\omega}. \tag{4.69}$$

Let us first rewrite the relation of position vector ${}^{A}\boldsymbol{P}$ with respect to u–v–w Euler angles of Equation 3.27 to derive ${}^{A}\boldsymbol{v}_p$

$${}^{A}\boldsymbol{P} = l_p \begin{bmatrix} s_2 & -c_2 s_1 & c_2 c_1 \end{bmatrix}^T. \tag{4.70}$$

Differentiate Equation 4.70 with respect to time

$${}^{A}\boldsymbol{v}_p = \begin{bmatrix} 0 & c_2 & 0 \\ -c_2 c_1 & s_2 s_1 & 0 \\ -c_2 s_1 & -s_2 c_1 & 0 \end{bmatrix} l_p \dot{\boldsymbol{\theta}}, \tag{4.71}$$

where $\dot{\boldsymbol{\theta}} = [\dot{\theta}_1, \dot{\theta}_2, \dot{\theta}_3]^T$ is the vector of Euler angle rates. The angular velocity of the moving platform can also be expressed as a function of Euler angles rate using Equation 4.8, in which the rotation matrix ${}^{A}\boldsymbol{R}_B$ is given in Equation 3.25.

$$\boldsymbol{\omega} = \begin{bmatrix} 1 & 0 & s_2 \\ 0 & c_1 & -c_2 s_1 \\ 0 & s_1 & c_2 c_1 \end{bmatrix} \dot{\boldsymbol{\theta}} \tag{4.72}$$

Solving Equation 4.72 for $\dot{\theta}$

$$\dot{\theta} = \frac{1}{c_2}\begin{bmatrix} c_2 & s_2 s_1 & -s_2 c_1 \\ 0 & c_2 c_1 & c_2 s_1 \\ 0 & -s_1 & c_1 \end{bmatrix} \omega, \tag{4.73}$$

and substituting Equation 4.73 into Equation 4.71

$$^{A}v_p = \begin{bmatrix} 0 & c_2 c_1 & c_2 s_1 \\ -c_2 c_1 & 0 & s_2 \\ -c_2 s_1 & -s_2 & 0 \end{bmatrix} l_p \omega. \tag{4.74}$$

Finally, the Jacobian matrix of the passive limb is obtained by augmenting the identity map $\omega = I_{3\times 3}\omega$ with Equation 4.74, referring to Equation 4.69.

$$J_p = \begin{bmatrix} 0 & l_p c_2 c_1 & l_p c_2 s_1 \\ -l_p c_2 c_1 & 0 & l_p s_2 \\ -l_p c_2 s_1 & -l_p s_2 & 0 \\ 1 & 0 & 0 \\ 0 & 1 & 0 \\ 0 & 0 & 1 \end{bmatrix}. \tag{4.75}$$

Having J_v and J_p in hand, the shoulder manipulator Jacobian $J_{4\times 3}$ will be easily found using Equation 4.62

$$J_{4\times 3} = J_v J_p = \begin{bmatrix} (^{A}P_{i_z} - l_p c_2 c_1)^{A}s_{i_y} - (^{A}P_{i_y} + l_p c_2 s_1)^{A}s_{i_z} \\ (^{A}P_{i_x} - l_p s_2)^{A}s_{i_z} - (^{A}P_{i_z} - l_p c_2 c_1)^{A}s_{i_x} \\ (^{A}P_{i_y} + l_p c_2 s_1)^{A}s_{i_x} - (^{A}P_{i_x} - l_p s_2)^{A}s_{i_y} \end{bmatrix}^{T}_{i=1,\ldots,4}. \tag{4.76}$$

4.7.2 Singularity Analysis

In this section, singularity analysis of the shoulder manipulator is performed. To analyze singular configurations of the shoulder manipulator, consider its Jacobian matrix J as given in Equation 4.76. Considering the general Jacobian definition in 4.34 and by comparing it with Equation 4.61, it is clear that for the shoulder manipulator $J_q = I$ and $J_x = J$. Hence the manipulator has no inverse kinematic singularity, but may possess forward kinematic singularity when J becomes rank deficient. Note that since the manipulator is redundantly actuated and J is a 4×3 matrix, the system experiences a singular configuration when

$$\det(J^T J) = 0. \tag{4.77}$$

The singular configuration can be derived either by direct calculation of Equation 4.77, or by examining the determinants of all 3×3 minors of J. Both the methods are presented here. First, the complete singularity configurations for the shoulder manipulator are derived from calculation of Equation 4.77, for a number of grid points in the whole workspace. This calculation has been carried out for 9261 grid points within the ranges

$-\pi/3 \leq \theta_1, \theta_2, \theta_3 \leq \pi/3$. It is observed that in all the examined grid points, there exists no configuration that Equation 4.77 will hold. Therefore, there exists no singular configuration throughout the entire workspace of the shoulder manipulator. This has been claimed in the original work of the manipulator design [74].

For a better understanding of the effect of actuator redundancy to produce a singularity-free workspace, the same analysis is performed using the determinants of all 3×3 minors of $\boldsymbol{J}$. It is well understood that for a redundant parallel manipulator, in general, singularities will occur if the rank of the Jacobian is lower than n, the number of degrees-of-freedom of the moving platform, or equivalently if

$$\det(\boldsymbol{M}) = 0, \quad \forall \boldsymbol{M} \in \{n \times n \text{ submatrices of } \boldsymbol{J}\}. \tag{4.78}$$

It should be noted that only $m - n + 1$ equations from Equation 4.78 are independent, in which m is the number of actuated joints. Hence, singularities are found at the intersection of $m - n + 1$ hypersurfaces resulting in a lower dimensional manifold with a dimension of $n - (m - n + 1) = n - 1 - n_r$ in the task space, where n_r is the number of redundant actuators. Thus, actuator redundancy can be effectively used to reduce or even eliminate the singularities in the workspace.

For the shoulder manipulator, as shown in the previous section, the linear velocities of the actuators $\dot{\boldsymbol{L}}$ are related to the angular velocity of the moving platform $\boldsymbol{\omega}$ by Equation 4.61, in which $\boldsymbol{J}$ is given in 4.76. Thus, the singularities are characterized by rank deficiency of $\boldsymbol{J}$, which occurs only if the determinants of all 3×3 minors of $\boldsymbol{J}$ are identically zero. These square minors correspond to the Jacobian matrices of the shoulder manipulator with one of the actuating limbs removed. Therefore, the redundant manipulator will be in a singular configuration only if all the nonredundant structures resulted by suppressing one of the actuating limbs are in a singular configuration. Such a case does not occur in the workspace of the shoulder manipulator as shown hereinbefore. In fact, one of the remarkable features of adding the fourth actuator is elimination of the loci of singularities through the whole workspace.

Figure 4.6 shows the determinants of the four minor Jacobian matrices computed in the workspace of the manipulator denoted by DM1 through DM4. It should be noted that in computing DM1, DM2, DM3, and DM4, limbs 4, 3, 2, and 1 were removed, respectively, to obtain the corresponding nonredundant structure. As shown in Figure 4.6, the values of the minor Jacobian matrices vary throughout the spherical surface of the workspace of the manipulator. However, they do not *simultaneously* reach a minimum at a particular configuration. To see this more clearly in Figure 4.6, note that, for example, at the configurations where DM2 becomes minimum, DM1 and DM3 get their midvalue, and DM4 possesses its high value. Furthermore, similar patterns can be observed at configurations where the other Jacobian minors become minimum. Therefore, there exists no configuration throughout the entire workspace of the manipulator where all minor Jacobian matrices reach their minimum simultaneously. The possibility of getting into a singular configuration is significantly increased when one of the redundant actuators is removed.

Note that although this analysis is performed for a specific parallel manipulator in hand, minor Jacobian matrices can be used in a similar way to analyze the singularity of other parallel manipulators. In cases where there is no redundancy in actuation, a similar analysis can be performed to obtain singular configurations. Moreover, in the case where one of the actuators malfunctions, its effect on the singularity of the manipulator can be examined in detail.

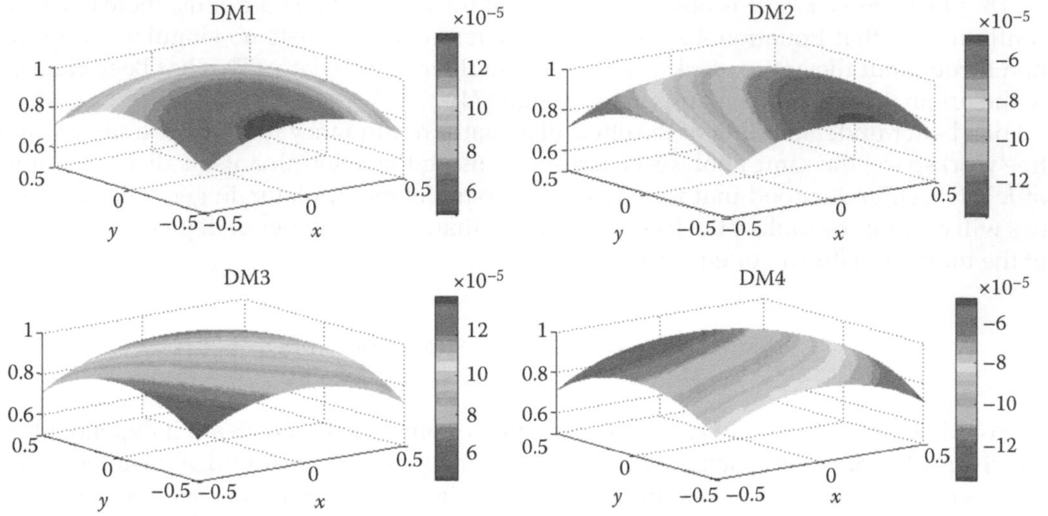

FIGURE 4.6
Minor determinants for the nonredundant structures for the shoulder manipulator; in computing DM1 through DM4, limbs 4, 3, 2, and 1 were removed, respectively.

4.8 Jacobian Analysis of the Stewart–Gough Platform

In this section, Jacobian analysis of the Stewart–Gough platform is studied in detail. In this analysis, first the manipulator Jacobian matrices are determined, and then the singularity analysis of the manipulator is performed on its entire workspace.

4.8.1 Velocity Loop Closure

As shown in Figure 3.16, spatial motion of the Stewart–Gough moving platform is generated by six identical piston–cylinder actuators. Each piston–cylinder actuator consists of two parts connected by a prismatic joint. The actuators connect the fixed base to the moving platform by spherical joints at points A_i and B_i, $i = 1, 2, \ldots, 6$, respectively. Note that in Figure 3.16, the attachment points A_i's lie in the base plane and B_i's lie in the moving platform plane. However, in a general Stewart–Gough platform, the attachment points do not necessarily lie in a plane.

As shown in Figure 3.16, and to analyze the kinematics of the manipulator, a fixed frame $\{\boldsymbol{A}\}$ is attached to the fixed base at point O_A, and attachment points on the fixed base are denoted by $\boldsymbol{a}_i$ vectors. Similarly, the moving frame $\{\boldsymbol{B}\}$ is attached to the moving platform at point O_B, and the attachment points at the moving platform are denoted by $\boldsymbol{b}_i$ vectors. The geometry of each limb is described by its length ℓ_i, and its direction by a unit vector $\hat{\boldsymbol{s}}_i$. The position of point O_B of the moving platform is described by the position vector ${}^A\boldsymbol{P} = \begin{bmatrix} p_x & p_y & p_z \end{bmatrix}^T$, and the orientation of the moving platform is represented by the rotation matrix ${}^A\boldsymbol{R}_B$.

Kinematic structure of each limb is an $S\underline{P}S$ arrangement. As analyzed in Example 1.3 in Chapter 1, this manipulator has six-degrees-of-freedom and can serve as a general parallel manipulator to produce complete spatial motion. The number of actuators are also equal to the number of degrees-of-freedom, and hence, there is no actuator redundancy in this manipulator.

From the geometry of the manipulator as illustrated in Figure 3.16, the input joint rate is denoted by $\dot{\boldsymbol{L}} = [\dot{\ell}_1, \dot{\ell}_2, \dot{\ell}_3, \dot{\ell}_4, \dot{\ell}_5, \dot{\ell}_6]^T$, and the output twist vector can be described by the velocity of point P, and the angular velocity of the moving platform, denoted by $\dot{\boldsymbol{\mathcal{X}}} = [{}^A\boldsymbol{v}_p, {}^A\boldsymbol{\omega}]^T$. The Jacobian matrix can be derived by formulating a velocity loop closure equation for each limb. The loop closure equations for each limb, $i = 1, 2, \ldots, 6$, are derived in Equation 3.51, and rewritten here for further analysis

$$ {}^A\boldsymbol{P} + {}^A\boldsymbol{R}_B\, {}^B\boldsymbol{b}_i = \ell_i\, {}^A\hat{\boldsymbol{s}}_i + {}^A\boldsymbol{a}_i. \tag{4.79}$$

Differentiate Equation 4.79 with respect to time

$$ {}^A\boldsymbol{v}_p + {}^A\dot{\boldsymbol{R}}_B\, {}^B\boldsymbol{b}_i + {}^A\boldsymbol{R}_B\, {}^B\dot{\boldsymbol{b}}_i = \dot{\ell}_i\, {}^A\hat{\boldsymbol{s}}_i + \ell_i\, {}^A\dot{\hat{\boldsymbol{s}}}_i + {}^A\dot{\boldsymbol{a}}_i. \tag{4.80}$$

Note that the time derivative of vector $\boldsymbol{b}_i$ with respect to the moving frame $\{\boldsymbol{B}\}$ is zero, ${}^B\dot{\boldsymbol{b}}_i = 0$, and similarly, ${}^A\dot{\boldsymbol{a}}_i = 0$. Furthermore

$$ {}^A\dot{\boldsymbol{R}}_B\, {}^B\boldsymbol{b}_i = {}^A\boldsymbol{\omega} \times ({}^A\boldsymbol{R}_B\, {}^B\boldsymbol{b}_i) = {}^A\boldsymbol{\omega} \times {}^A\boldsymbol{b}_i,$$

in which ${}^A\boldsymbol{\omega}$ denotes the angular velocity of the moving platform expressed in the fixed frame $\{\boldsymbol{A}\}$. Similarly, the time derivative of the unit vector $\hat{\boldsymbol{s}}_i$ can be determined from relative differentiation rule Equation 4.13 as

$$ \ell_i\, {}^A\dot{\hat{\boldsymbol{s}}}_i = \ell_i\, ({}^A\boldsymbol{\omega}_i \times \hat{\boldsymbol{s}}_i),$$

in which ${}^A\boldsymbol{\omega}_i$ is the angular velocity of limb i expressed in fixed frame $\{\boldsymbol{A}\}$. Hence, the velocity loop closure 4.80 simplifies to

$$ {}^A\boldsymbol{v}_p + {}^A\boldsymbol{\omega} \times {}^A\boldsymbol{b}_i = \dot{\ell}_i\, {}^A\hat{\boldsymbol{s}}_i + \ell_i\, ({}^A\boldsymbol{\omega}_i \times \hat{\boldsymbol{s}}_i), \tag{4.81}$$

in which ${}^A\boldsymbol{v}_p = [\dot{x}_p, \dot{y}_p, \dot{z}_p]^T$ is the velocity of the moving platform at point P, ${}^A\boldsymbol{\omega}$ denotes the angular velocity of the moving platform expressed in the fixed frame $\{\boldsymbol{A}\}$, and $\hat{\boldsymbol{s}}_i$ is the unit vector in the direction of limb i. To eliminate passive joint rate ${}^A\boldsymbol{\omega}_i$, dot multiply both sides of Equation 4.81 by $\hat{\boldsymbol{s}}_i$, and use dot and cross multiplication interchange rules in Equation 4.51 to

$$ \hat{\boldsymbol{s}}_i \cdot \boldsymbol{v}_p + (\boldsymbol{b}_i \times \hat{\boldsymbol{s}}_i)\, \boldsymbol{\omega} = \dot{\ell}_i, \tag{4.82}$$

in which the leading superscript A is omitted for all vectors for short hand. Write Equation 4.82 six times for $i = 1, 2, \ldots, 6$ and rearrange it into a matrix form

$$ \dot{\boldsymbol{L}} = \boldsymbol{J} \dot{\boldsymbol{\mathcal{X}}}, \tag{4.83}$$

in which the input joint rate is denoted by $\dot{\boldsymbol{L}} = [\dot{\ell}_1, \dot{\ell}_2, \dot{\ell}_3, \dot{\ell}_4, \dot{\ell}_5, \dot{\ell}_6]^T$, the output twist vector is denoted by $\dot{\boldsymbol{\mathcal{X}}} = [\boldsymbol{v}_p, \boldsymbol{\omega}]^T$, and the Jacobian matrix $\boldsymbol{J}$ is a 6×6 square matrix, determined as

$$ \boldsymbol{J} = \begin{bmatrix} \hat{\boldsymbol{s}}_1^T & (\boldsymbol{b}_1 \times \hat{\boldsymbol{s}}_1)^T \\ \hat{\boldsymbol{s}}_2^T & (\boldsymbol{b}_2 \times \hat{\boldsymbol{s}}_2)^T \\ \vdots & \vdots \\ \hat{\boldsymbol{s}}_6^T & (\boldsymbol{b}_6 \times \hat{\boldsymbol{s}}_6)^T \end{bmatrix}. \tag{4.84}$$

4.8.2 Singularity Analysis

This section is devoted to the singularity analysis of the Stewart–Gough platform. An important limitation of the parallel manipulator is that singular configurations may exist within the workspace, where the manipulator can gain one or more degrees-of-freedom, and therefore becomes uncontrollable. Furthermore, the actuator forces may become very large and may result in a breakdown of the mechanism. Therefore, it is of primary importance to avoid singularities in a given workspace.

To study the singularity configurations of the Stewart–Gough platform, consider the Jacobian matrix of the manipulator given in Equation 4.84. Considering the general Jacobian definition in 4.34 and comparing this with Equation 4.33, it is clear that for the Stewart–Gough platform, $J_q = I$ and $J_x = J$. Hence, the manipulator has no inverse kinematic singularity within the manipulator workspace, but may possess forward kinematic singularity when J becomes rank deficient. Note that inverse kinematic singularities can occur at the workspace boundary, where one or more limbs are in fully stretched or retracted positions. Forward kinematic singularity can occur within the workspace of the manipulator, when J becomes rank deficient. Since the manipulator Jacobian is a square 6×6 matrix, this may occur when

$$\det(J) = 0. \tag{4.85}$$

4.8.2.1 Background Literature

Derivation of the exact loci where this type of singularity occurs within the workspace of the manipulator is a comprehensive task, and many researchers have focused on this issue. Hunt [81] found that a singularity can occur when all the lines associated with the prismatic actuators intersect a common line. Then Fichter [48] studied the special structure of the Gough–Stewart platform, that is, triangle simplified symmetric manipulator (TSSM), and found that another singular configuration occurs when the platform is rotated about an axis perpendicular to the base plane by $\pm 90°$. Huang [80] studied the singularity loci and distribution characteristics of the TSSM architecture as shown in Figure 4.7e, where all singularities have been classified into three different linear–complex singularities. In [35], 3–6 fully parallel manipulator (Figure 4.7e, TSSM architecture) was studied and a singularity locus equation was obtained using the Rodrigues parameters. Later, Di Gregorio [36] presented the singularity locus expression of a 6–6 fully parallel manipulator (Figure 4.7d, simplified symmetric manipulator (SSM) architecture) using the mixed products of vectors. Wolf and Shoham [192] used the line geometry and screw theory to determine the singular configurations of parallel mechanisms and their behavior at these points, in which examples of Hunt's and Fichter's and a few selected examples from Merlet's work [134] have been analyzed by this method.

Merlet [134] studied the simplified Gough–Stewart platforms using Grassmann geometry, namely, the SSM (Figure 4.7d), the TSSM (Figure 4.7e), and the minimal simplified symmetric manipulator (MSSM) (Figure 4.7f). For each of these mechanisms, the number of geometric conditions leading to singular configurations is different. The method is general and can be applied to any parallel mechanism. However, it is sometimes difficult to express the conditions mathematically. Only in 1998, Kim [102] obtained an analytical expression of the singularity locus equation for the general Stewart–Gough platform with constant orientation using the concept of a local structurization method with the extra sensors, which is a third degree polynomial in three position variables.

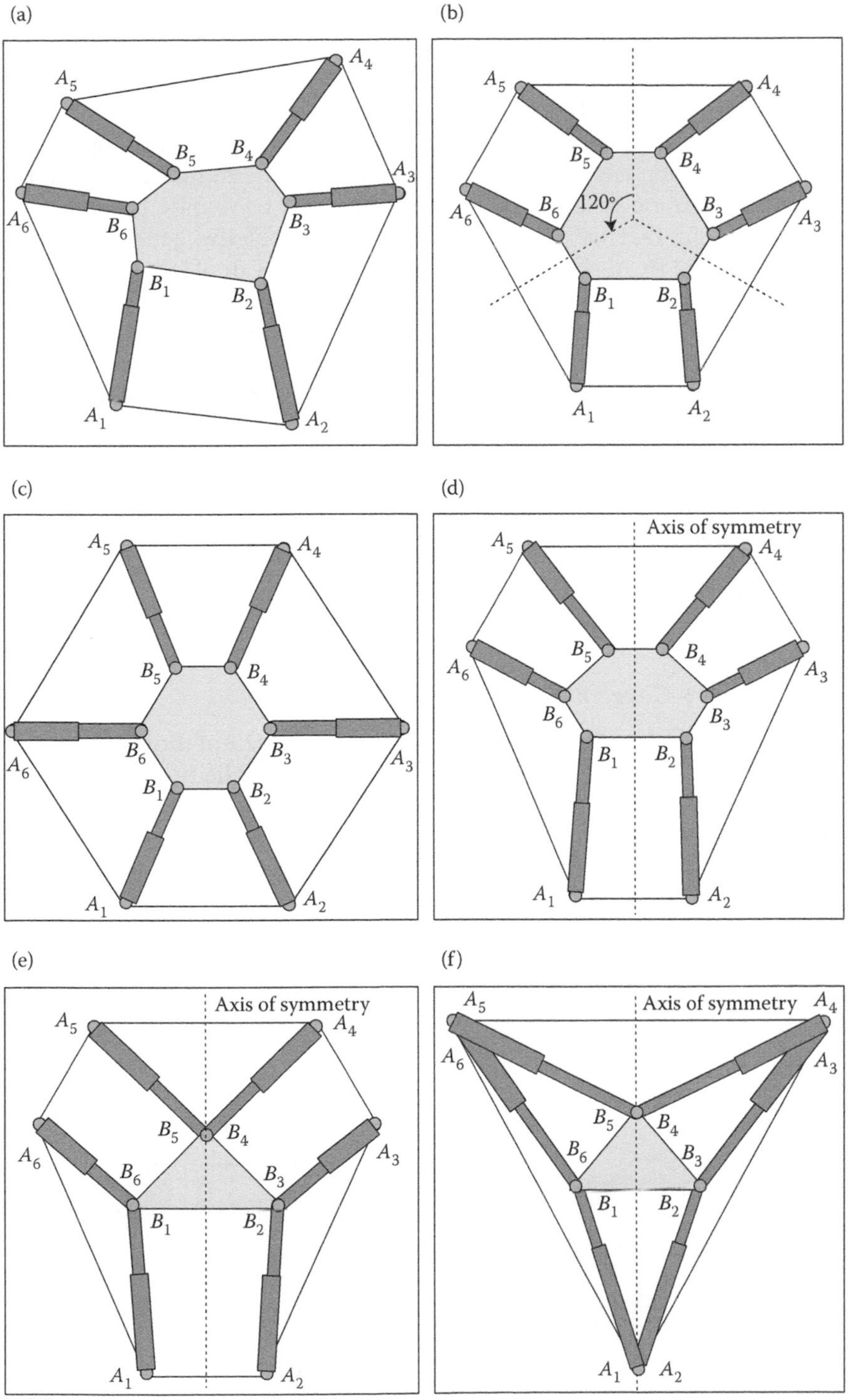

FIGURE 4.7
Top view of some Stewart–Gough architectures. (a) Platform with irregular hexagons, (b) platform with semi-regular hexagons, (c) platform with regular hexagons, (d) SSM architecture, (e) TSSM architecture, and (f) MSSM architecture.

Later, Mayer St-Onge and Gosselin [127] found the same result by expanding the Jacobian matrix of the mechanism with constant orientation using linear decomposition and cofactor expansion. From the design point of view, it is desirable to obtain an analytical expression of the singularity locus of a mechanism. Then, with a given set of design parameters, the designers can obtain a graphical representation of the singularity locus of the mechanism. It is then easy to identify the locations of singularities within the given workspace and determine whether the singularities can be avoided. Later, in [114], the same authors showed that using the velocity equation of the general Stewart–Gough platform, an analytical expression for the singularity locus in 6D Cartesian space can be obtained, that is, a polynomial in six variables (three position variables x, y, z, and three orientation variables, e.g., three Euler angles α, β, γ). It is shown that although the expression is very complex, it can be handled both numerically and analytically. Since it is impossible to illustrate the singularity locus in 6D space, a numerical procedure is introduced in [114] to represent the singularity locus graphically within a given 3D workspace for either constant orientation or constant position or the cases in which any three of the six variables are fixed.

It is important to note that since the Stewart–Gough platform is not redundantly actuated, there may exist many possible singularity loci within its workspace. To show that in detail, and as a representative, the singular configuration of a 3–6 Stewart–Gough platform is examined in the following subsection.

4.8.2.2 A 3–6 Stewart–Gough Platform

The forward kinematic singularities for a 3–6 Stewart–Gough platform for general orientation are studied in detail in [79], and due to its relative simplicity, it has been adopted in this book. On the basis of the singularity kinematics principle, a planar singularity equivalent mechanism has been proposed in [79], by which the complicated singularity analysis of that parallel mechanism is transformed into a simpler position analysis of the planar mechanism.

A typical 3–6 Stewart–Gough platform is represented schematically in Figure 4.8. It consists of an equilateral triangle moving platform $B_1B_3B_5$, and a semiregular hexagon base $C_1, C_2, \ldots, C_6$. A fixed frame O: xyz is attached to the base platform, and a moving frame P: $x'y'z'$ is attached to the moving platform. Point P represented by the position

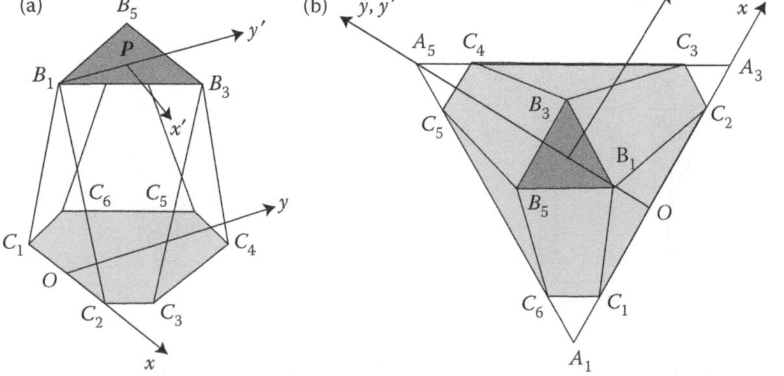

FIGURE 4.8
(a) Schematic of a 3–6 Stewart–Gough platform and (b) its top view.

vector $\boldsymbol{P} = [x, y, z]^T$ in the fixed frame is the geometric center of the moving platform. The orientation of the moving platform is represented by the z–y–z Euler angles (α, β, γ). The structure of the singularity surface of 3–6 Stewart–Gough platform for all different orientations (α, β, γ) is classified in [79], as follows:

1. When $\beta = 0$, the moving and the base plates are parallel to each other. The special configuration of the mechanism is $(90°, 0, 0)$ for which the mechanism is singular for all positions of the moving platform [48].
2. When $\beta = 0$ and $Z = 0$, the moving platform and the base are coincident. The mechanism is singular for all the other two Euler angles.
3. When $\beta \neq 0$, $\alpha = (\pm 30°, \pm 90°, \pm 150°)$, and $\gamma \neq (\pm 30°, \pm 90°, \pm 150°)$, the singularity loci for this orientation include a plane (Hunt plane) and a hyperbolic paraboloid [80]. The intersecting line between the moving platform and the base is parallel to one side of the triangle $A_1 A_3 A_5$.
4. When $\beta \neq 0$, $\alpha = (\pm 30°, \pm 90°, \pm 150°)$, and γ is one of the angles, $\pm 30°$, $\pm 90°$, or $\pm 150°$, the singularity loci for this orientation are three intersecting planes [80].
5. When $\beta \neq 0$, $\alpha \neq (\pm 30°, \pm 90°, \pm 150°)$, and $\gamma = (\pm 30°, \pm 90°, \pm 150°)$, the singularity loci for this orientation also include a plane and a hyperbolic paraboloid.
6. When $\beta \neq 0$ and neither α nor γ is equal to any one of the angles, $\pm 30°$, $\pm 90°$, $\pm 150°$, it is the most general case. The singularity equation for this orientation is a special irresolvable polynomial expression of degree three, and the structure of the singularity loci in infinite parallel principal sections includes a parabola, four pairs of intersecting straight lines, and an infinity of hyperbolas.

To visualize the singularity locations within the workspace of a 3–6 Stewart–Gough platform, singular configurations are derived from calculation of Equation 4.41, for a number of grid points within the workspace of the manipulator. Note that the workspace of this manipulator has six dimensions and is not possible to visualize the whole workspace in a 3D plot. Therefore, only variation of three coordinates is examined while the other three coordinates are considered to be fixed. This calculation has been done for 4410 grid points within the ranges $-1 \leq x \leq 1$, $-1 \leq y \leq 1$, and $0 \leq \theta \leq \pi/2$, while the manipulator height is considered fixed at $z = 1.0$. Three cases are considered for this analysis, in which the rotation is considered for x, y, and z directions, and singular configurations are given within the constant orientation workspace of the manipulator in three 3D plots of Figures 4.9 through 4.11, respectively. As seen in Figures 4.9 and 4.10, it is observed that as the rotation of the moving platform is raised to more than $\pi/4$, a singular configuration appears and grows in volume within the constant orientation workspace; while for $\theta_x = \pi/2$ or $\theta_y = \pi/2$, the whole workspace planes become singular. This is different for θ_z, the rotation about z-axis. A singular configuration for this case occurs only when $\theta_z = \pi/2$, as shown in Figure 4.11.

These figures confirm the intuitive remark mentioned hereinbefore that since the Stewart–Gough platform is not redundantly actuated, there may exist many possible singularity loci within its workspace. Singular configuration intensity increases generally, if the manipulator design is more symmetrical, and by introducing asymmetry into the manipulator design singular configuration may be significantly reduced. In any case, singular configuration analysis plays an important role in the optimal design of parallel manipulators, and special care must be taken in the trajectory planning and control of such manipulators to avoid singular configurations.

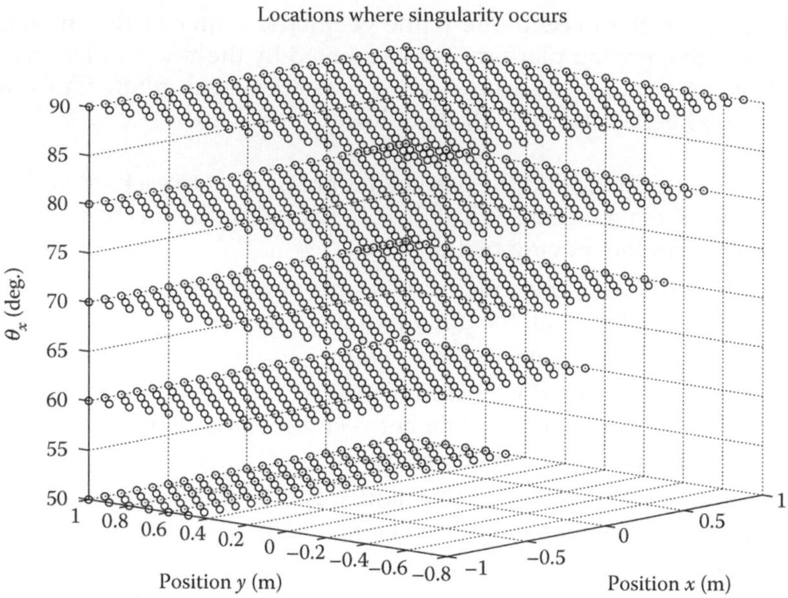

FIGURE 4.9
Singular configurations of a 3–6 Stewart–Gough platform for fixed $z = 1.0$, and θ_x rotation about x-axis.

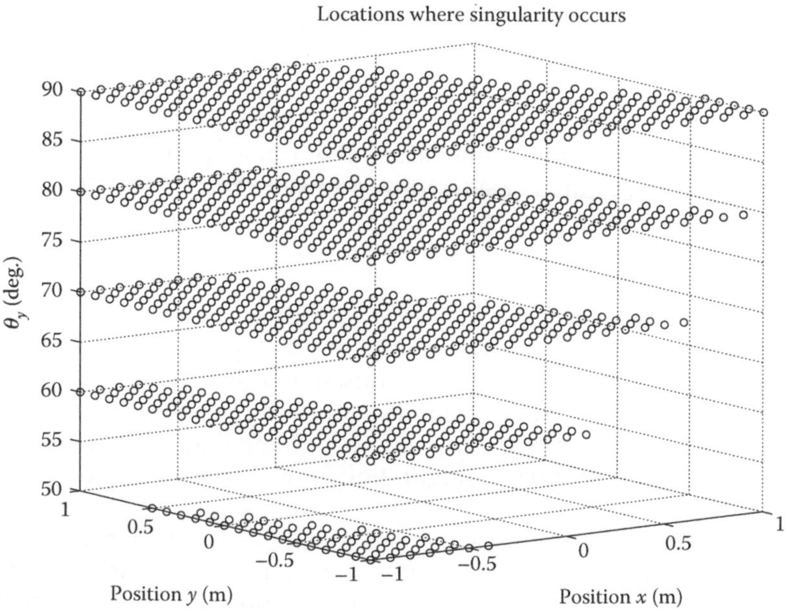

FIGURE 4.10
Singular configurations of a 3–6 Stewart–Gough platform for fixed $z = 1.0$, and θ_y rotation about y-axis.

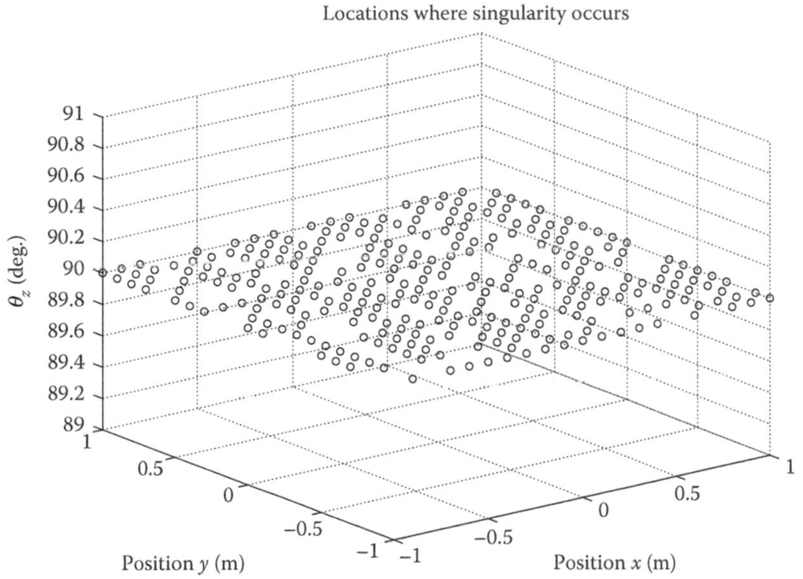

FIGURE 4.11
Singular configurations of a 3–6 Stewart–Gough platform for fixed $z = 1.0$, and θ_z rotation about z-axis.

4.9 Static Forces in Parallel Manipulators

When a manipulator is used in a task like grinding where its moving platform is in contact with a stiff environment, it is necessary to analyze the forces and moments applied to the environment at the point of contact. These forces and moments are generated by the actuator forces of the manipulator. The relation between the forces and moments exerted at the point of contact to the actuator forces/torques is determined and analyzed in the study of static force analysis.

In this study it is assumed that the manipulator is at a static equilibrium, and actuator forces required to produce the desired contact forces are determined. Owing to the existence of closed loops in the kinematic structure, the study of static forces in a parallel manipulator is more comprehensive than that in serial manipulators, and contrary to serial manipulators, in parallel manipulators the recursive methods to determine the static forces are not applicable. In parallel manipulators, usually either the free-body diagram approach is used to determine the static forces, or the principle of virtual work is applied. In the free-body diagram approach, the actuator forces are determined to produce desired contact forces/moments as well as the internal and interacting forces/torques applied at the limbs. The analysis of such forces/torques are essential in the design of a manipulator to determine the stresses and deflections of each link and joint. However, if only the actuator forces are desired to be determined, the principle of virtual work is more efficient and computationally less expensive. In what follows, these two approaches are introduced and illustrated by some examples.

4.9.1 Free-Body Diagram Approach

In the static force analysis, the free-body diagram approach is investigated in this section. A parallel manipulator consists of a moving platform that is connected to a fixed base by several limbs. The number of limbs are at least equal to the number of degrees-of-freedom of the moving platform such that each limb is driven by no more than one actuator. Because of the closed-loop structure, the joints cannot be controlled independently. Thus some of the joints are driven by actuators, whereas others are passive. The static analysis of such manipulator is greatly simplified, if it is assumed that the limbs consist of binary links that can bear only tension or compression forces. This assumption is generally applicable, if no external forces other than the actuator forces are applied to the links, and it is assumed that all the external forces and disturbances are applied to the moving platform. Therefore, if no external force is applied on a binary link in a planar manipulator, the static force can be assumed to be along the link axis, and the link is subject to a pure tension or compression force. Similarly, if no external force is applied to a spherical–spherical binary link chain in a spatial manipulator, the link will be subject to tension or compression force along a line passing through the two spherical joints.

In a free-body diagram approach, the free-body diagram of all forces and moments exerted on each link of the manipulator, and the moving platform is drawn separately. The external forces/torques acting on the moving platform is assumed to be given and the actuator forces/torques are to be found. By writing the static balance of forces and moments for all the free-body diagrams, and eliminating the internal forces acting between the base and links, links to each other, and the links and moving platform, the relation between the external forces/torques and the actuator forces are derived. It can be shown that for a general parallel manipulator, this relation can be written in a matrix form. This method is applied in some case studies in the following subsections.

4.9.2 Virtual Work Approach

In this section, the principle of virtual work is applied for the analysis of the static forces in parallel manipulators. A virtual displacement for a parallel manipulator refers to an infinitesimal change in the general displacement of the moving platform as a result of any arbitrary infinitesimal changes in the active and passive joint variables at a given instant of time. The term *virtual displacement* is used to distinguish it from an *actual displacement*, for which the active and passive joint forces/torques may be changing at the instant. The notation $\delta\boldsymbol{\mathcal{X}}$ is used to denote a virtual displacement, as opposed to $d\boldsymbol{\mathcal{X}}$ for an actual displacement. The virtual displacement at the actuated joints of a parallel manipulator can be written as $\delta\boldsymbol{q} = [\delta q_1, \delta q_2, \ldots, \delta q_m]^T$ for an m-actuated manipulator, and $\delta\boldsymbol{\mathcal{X}} = [\delta x, \delta y, \delta z, \delta\theta_x, \delta\theta_y, \delta\theta_z]^T$, denotes the virtual displacement of a contacting point of the moving platform, in which x, y, z are the position variables, while

$$\begin{bmatrix} \delta\theta_x \\ \delta\theta_y \\ \delta\theta_z \end{bmatrix} = \delta\theta\,\hat{\mathbf{s}} \tag{4.86}$$

are the orientation variables represented by screw coordinates. Furthermore, as shown in Figure 4.12, let the vector of actuator forces/torques be denoted by $\boldsymbol{\tau} = [\tau_1, \tau_2, \ldots, \tau_m]^T$, and the external force/torque acting on the contact point of the moving platform denoted by a wrench, in a screw coordinate as

$$\boldsymbol{\mathcal{F}} = \begin{bmatrix} f \\ n \end{bmatrix}, \tag{4.87}$$

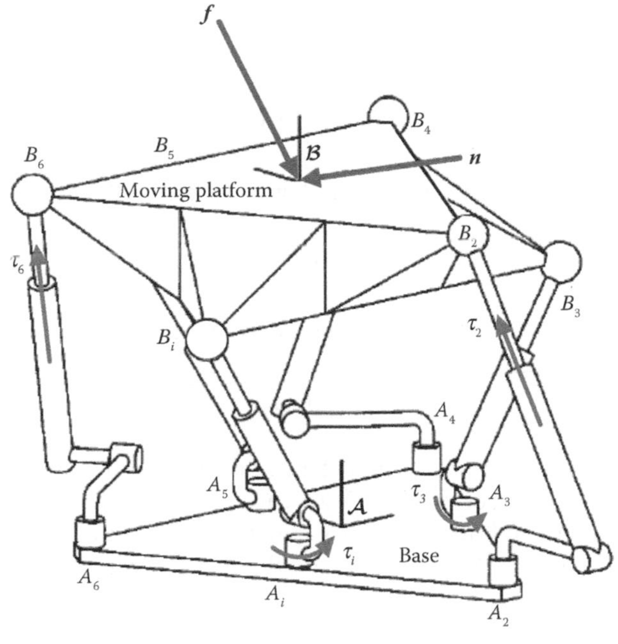

FIGURE 4.12
Schematic of a general parallel manipulator on which the external wrench and the actuator forces/torques are indicated.

in which $\boldsymbol{f} = [f_x, f_y, f_z]^T$ denotes the external force, and $\boldsymbol{n} = [n_x, n_y, n_z]^T$ denotes the external torque acting on the moving platform point of contact to the environment.

Assume that the frictional forces acting on the joints are negligible, and also that the gravitational forces of the limb links are much smaller than the interacting force of the moving platform to the environment. The principle of virtual work states that the total virtual work, δW, done by all actuators and external forces is equal to zero. The total virtual work consists of the virtual work done by the actuators and the virtual work done by external wrench, namely

$$\delta W = \boldsymbol{\tau}^T \delta \boldsymbol{q} - \boldsymbol{\mathcal{F}}^T \delta \boldsymbol{\mathcal{X}} = 0. \tag{4.88}$$

In this equation, the virtual work done by the actuators can be directly determined by $\boldsymbol{\tau}^T \delta \boldsymbol{q}$, since the virtual displacement $\delta \boldsymbol{q}$ is defined along the active torque exerted by the actuators. Furthermore, the virtual work done by the external wrench on the contact point is determined by $-\boldsymbol{\mathcal{F}}^T \delta \boldsymbol{\mathcal{X}}$. Note that the negative sign of this term is due to the fact that $\boldsymbol{\mathcal{F}}$ denotes the external wrench exerted on the moving platform from the environment, while the wrench applied by the manipulator to the environment is equal to $-\boldsymbol{\mathcal{F}}$. Furthermore, from the definition of the manipulator Jacobian matrix given in Equation 4.34, the virtual displacements, $\delta \boldsymbol{q}$ and $\delta \boldsymbol{\mathcal{X}}$ are related to each other by the Jacobian matrix*

$$\delta \boldsymbol{q} = \boldsymbol{J} \delta \boldsymbol{\mathcal{X}}, \tag{4.89}$$

* Note that Jacobian relation does not hold for finite orientation angles, but it holds at velocity level, and for infinitesimal orientation angles as for the virtual displacements.

in which for a general parallel manipulator $J = J_q^{-1} J_x$ as defined in Equation 4.35. Substituting Equation 4.89 into Equation 4.88 yields

$$\left(\boldsymbol{\tau}^T J - \mathcal{F}^T\right) \delta \mathcal{X} = 0. \tag{4.90}$$

This equation holds for any arbitrary virtual displacement, $\delta\mathcal{X}$, hence

$$\boldsymbol{\tau}^T J - \mathcal{F}^T = 0. \tag{4.91}$$

Transposition of Equation 4.91 results in the following equation:

$$\mathcal{F} = J^T \boldsymbol{\tau}. \tag{4.92}$$

Equation 4.92 indicates an important characteristic of the Jacobian matrix. Jacobian matrix not only reveals the relation between the joint variables velocities $\dot{q}$ and the moving platform velocities $\dot{\mathcal{X}}$ as given in Equation 4.34, it constructs the transformation needed to find the actuator forces $\boldsymbol{\tau}$ from the wrench acting on the moving platform $\mathcal{F}$ by Equation 4.92.

4.9.3 Static Forces of a Planar Manipulator

We first examine the free-body diagram approach for static forces analysis. Reconsider the architecture of the planar $4R\underline{P}R$ parallel manipulator as described in detail in Section 3.3.1 and shown in Figure 1.7. As explained earlier, in this manipulator, the moving platform is supported by the four limbs of the identical kinematic structure. Each limb connects the fixed base to the manipulator moving platform by a revolute joint (R) followed by an actuated prismatic joint (P) and another revolute joint (R). The kinematic structure of a prismatic joint is used to model either a piston–cylinder actuator at each limb or a cable-driven one. As it is shown in Figure 1.7, A_i denote the fixed base points of the limbs, B_i denote point of connection of the limbs to the moving platform, L_i denote the limbs' lengths, and α_i denote the limbs' angles. The position of the moving platform center point G is denoted by $G = [x_G, y_G]$, and orientation of the manipulator moving platform is denoted by ϕ with respect to the fixed coordinate frame. Hence the manipulator has three-degrees-of-freedom $\mathcal{X} = [x_G, y_G, \phi]$, with one-degree-of-actuation redundancy.

For the purpose of static forces analysis and as is illustrated in Figure 4.13, a fixed frame O: xy is considered for the system, while the free-body diagrams of the moving platform and the limbs are drawn separately. Consider the actuator forces denoted by $\boldsymbol{\tau} = [f_1, f_2, f_3, f_4]^T$, and force and moment applied to the environment by the moving platform at the center point G, denoted by $\boldsymbol{f} = \left[f_x, f_y, 0\right]^T$ and $\boldsymbol{n} = [0, 0, n_z]^T$, respectively. Notice that since the manipulator is performing in a plane, only three components of the wrench $\mathcal{F}$ is applied to the environment, and therefore, $\mathcal{F} = \left[f_x, f_y, n_z\right]^T$ is considered for the analysis. For the sake of simplicity, and because of the similarity of the applied forces by the limbs to the moving platform, only one limb is illustrated in Figure 4.13.

First, consider the external forces applied to each limb. It is assumed that no external forces are applied directly to the limbs except the actuator forces. Therefore, the static force can be assumed to be along the limb axis $\hat{\mathbf{S}}_i$, and the limb is subject to a tension force $-f_i$. Next, consider the free-body diagram of forces and moments applied on the moving platform. The summation of all acting forces on the moving platform shall be zero.

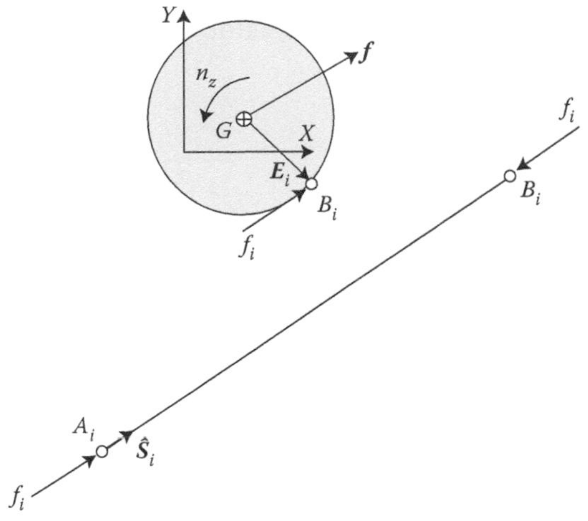

FIGURE 4.13
Free-body diagram of forces and moments acting on the moving platform and each limb of planar $4R\underline{P}R$ manipulator.

Therefore

$$-\boldsymbol{f} + \sum_{i=1}^{4} f_i \hat{\mathbf{S}}_i = \mathbf{0}, \tag{4.93}$$

in which $-\boldsymbol{f}$ is the external force applied to the moving platform by the environment. Furthermore, the summation of moments contributed by all forces acting on the moving platform about G is as follows:

$$-n_z\hat{\mathbf{Z}} + \sum_{i=1}^{4} \boldsymbol{E}_i \times f_i \hat{\mathbf{S}}_i = \mathbf{0}, \tag{4.94}$$

in which $-n_z\hat{\mathbf{Z}}$ is the external moment applied to the moving platform by the environment, and $\hat{\mathbf{Z}}$ is the unit vector along the Z-axis. By eliminating the trivial equations from Equations 4.93 and 4.94, the remaining three equations can be written in a matrix form as follows:

$$\begin{bmatrix} S_{i_x} \\ S_{i_y} \\ E_{i_x} S_{i_y} - E_{i_y} S_{i_x} \end{bmatrix}_{3\times 4} \cdot \begin{bmatrix} f_1 \\ f_2 \\ f_3 \\ f_4 \end{bmatrix} = \begin{bmatrix} f_x \\ f_y \\ n \end{bmatrix}, \tag{4.95}$$

in which S_{i_x} and S_{i_y} are the x and y components of the vectors $\hat{\mathbf{S}}_i$, and E_{i_x} and E_{i_y} are the x and y components of the vector $\boldsymbol{E}_i$, respectively. The left matrix in Equation 4.95 is a 3×4 matrix whose ith column is given in this equation for the sake of brevity. Comparing this matrix to Jacobian matrix of $4R\underline{P}R$ manipulator given in Equation 4.54, it is clearly seen that this matrix is equivalent to the transpose of Jacobian matrix of the manipulator. Therefore, Equation 4.95 can be written in the form given by virtual work approach 4.92 as

$$\mathcal{F} = J^T \boldsymbol{\tau}, \tag{4.96}$$

in which $\boldsymbol{\tau} = [f_1, f_2, f_3, f_4]^T$ is the vector of actuator forces, and $\mathcal{F} = \left[f_x, f_y, n_z\right]^T$ is the 3D wrench generated by the planar manipulator.

4.9.4 Static Forces of Shoulder Manipulator

For the shoulder manipulator introduced in Section 3.4.1, the free-body diagram approach is used to perform the static forces analysis, and the result will be verified by virtual work approach. Reconsider the architecture of the shoulder manipulator as shown in Figure 3.10. As explained earlier, in this spatial orientation manipulator the moving platform is supported by four limbs of the identical kinematic structure. Figure 4.14 depicts a geometric model for the shoulder manipulator, in which the moving platform center point is denoted by P; l_i are the actuator lengths for $i = 1, 2, 3, 4$; P_i are the moving attachment points of the actuators; and A_i are the fixed attachment points. Two coordinate frames are defined for the purpose of analysis. The base coordinate frame $\{\boldsymbol{A}\} : xyz$ is attached to the fixed base at point C, while the second coordinate frame $\{\boldsymbol{B}\} : uvw$ is attached to the center of the moving platform P.

Consider the actuator forces denoted by $\boldsymbol{\tau} = [f_1, f_2, f_3, f_4]^T$, the force acting on the passive limb denoted by $\boldsymbol{f}_p$, and the external forces and moments applied to the moving platform at point P by $\boldsymbol{f}$ and $\boldsymbol{n}$, respectively. Since this manipulator is a spatial orientation manipulator with three rotational degrees-of-freedom, the external wrench applied to the moving platform is a 3D vector $\mathcal{F}$ equal to the external moment $\boldsymbol{n}$ in this case. The external forces applied to the moving platform are completely supported by the force that is generated in the passive limb. It is further assumed that no external forces other than the actuator forces are applied to the limbs. Therefore, considering the free-body diagram of the limbs shown in Figure 4.14, the static force can be assumed to be along the limb axis $\hat{\boldsymbol{s}}_i$, and the limb is subject to a pure tension/compression force f_i. Similarly, assuming that no external forces are applied to the passive limb, the static force acting on this limb is also along the limb axis $\hat{\boldsymbol{s}}_p$.

Next, consider the free-body diagram of forces and moments applied to the moving platform as shown in Figure 4.14. At static equilibrium, the summation of all acting forces

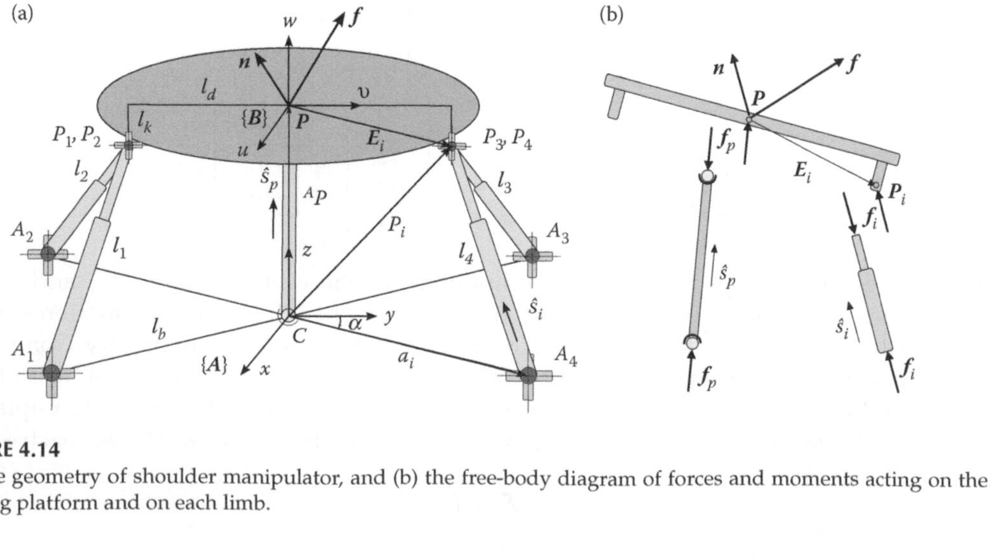

FIGURE 4.14
(a) The geometry of shoulder manipulator, and (b) the free-body diagram of forces and moments acting on the moving platform and on each limb.

on the moving platform should be zero. Thus

$$f + \sum_{i=1}^{4} f_i \hat{s}_i + f_p \hat{s}_p = 0, \tag{4.97}$$

in which f is the external force applied to the moving platform by the environment. Note that this equation can be used to determine the contribution of external force that is supported by passive force, $f_p = f_p \hat{s}_p$, and may be used for design purposes. However, to determine the actuator forces the summation of moments contributed by all forces acting on the moving platform about point P must be used as follows:

$$n + \sum_{i=1}^{4} E_i \times f_i \hat{s}_i = 0, \tag{4.98}$$

in which n is the external moment applied to the moving platform by the environment, and as illustrated in Figure 4.14, E_i is the distance of the applied actuator force with respect to the point P. Note that the passive force $f_p = f_p \hat{s}_p$ does not contribute in the moments applied to the moving platform about point P, and hence it is missing in Equation 4.98. Therefore, this equation may be directly used to obtain a relation between the actuator forces required to produce an external moment n. As seen in Figure 4.14, E_i vector can be derived from the two previously defined vectors ${}^{A}P$ and ${}^{A}P_i$ as follows:

$${}^{A}E_i = {}^{A}P_i - {}^{A}P. \tag{4.99}$$

Hence

$$\mathcal{F} = n = -\sum_{i=1}^{4} f_i \left[({}^{A}P_i - {}^{A}P) \times {}^{A}\hat{s}_i \right]. \tag{4.100}$$

Substituting ${}^{A}P = l_p \begin{bmatrix} s_2 & -c_2 s_1 & c_2 c_1 \end{bmatrix}^T$ from Equation 3.27 and using the components of the vector ${}^{A}P_i = {}^{A}R_B {}^{B}P_i = \begin{bmatrix} {}^{A}P_{i_x} & {}^{A}P_{i_y} & {}^{A}P_{i_z} \end{bmatrix}^T$ into Equation 4.99, this equation is written componentwise as follows:

$${}^{A}P_i - {}^{A}P = \begin{bmatrix} {}^{A}P_{i_x} - l_p s_2 \\ {}^{A}P_{i_y} + l_p c_2 s_1 \\ {}^{A}P_{i_z} - l_p c_2 c_1 \end{bmatrix}. \tag{4.101}$$

Hence, Equation 4.100 may be written in a matrix form as

$$\begin{bmatrix} ({}^{A}P_{i_z} - l_p c_2 c_1){}^{A}s_{i_y} - ({}^{A}P_{i_y} + l_p c_2 s_1){}^{A}s_{i_z} \\ ({}^{A}P_{i_x} - l_p s_2){}^{A}s_{i_z} - ({}^{A}P_{i_z} - l_p c_2 c_1){}^{A}s_{i_x} \\ ({}^{A}P_{i_y} + l_p c_2 s_1){}^{A}s_{i_x} - ({}^{A}P_{i_x} - l_p s_2){}^{A}s_{i_y} \end{bmatrix}_{3\times 4} \cdot \begin{bmatrix} f_1 \\ f_2 \\ f_3 \\ f_4 \end{bmatrix} = \mathcal{F} \tag{4.102}$$

in which the leftmost matrix is the Jacobian transpose of shoulder manipulator given in Equation 4.76. Note that Jacobian transpose matrix is a 3×4 matrix, and only its i'th column is written in detail in Equations 4.102 and 4.76 for the sake of brevity. Therefore, Equation 4.102 can be written in the form given by virtual work approach Equation 4.92 as

$$\mathcal{F} = J^T \tau, \tag{4.103}$$

in which $\boldsymbol{\tau} = [f_1, f_2, f_3, f_4]^T$ is the vector of actuator forces, and $\boldsymbol{\mathcal{F}} = \boldsymbol{n}$ is the 3D wrench applied to the spatial orientation manipulator.

4.9.5 Static Forces of the Stewart–Gough Platform

Consider the Stewart–Gough platform introduced in Section 3.5.1, and shown in Figure 3.16. In this manipulator, the spatial motion of the moving platform is generated by six identical piston–cylinder actuators. The actuators connected the fixed base to the moving platform by spherical joints at points A_i and B_i, $i = 1, 2, \ldots, 6$, respectively. As shown in Figure 3.16, a fixed frame $\{\boldsymbol{A}\}$ is attached to the fixed base at point O_A, and the fixed attachment points are denoted by $\boldsymbol{a}_i$ vectors. Similarly, a moving frame $\{\boldsymbol{B}\}$ is attached to the moving platform at point O_B, and moving attachment points are denoted by $\boldsymbol{b}_i$ vectors. The geometry of each limb is described by its length ℓ_i, and its direction through a unit vector $\hat{\boldsymbol{s}}_i$.

As shown in Figure 4.15, the twist of moving platform is described by a 6D vector denoted by $\dot{\boldsymbol{\mathcal{X}}} = \left[{}^A\boldsymbol{v}_p \quad {}^A\boldsymbol{\omega}\right]^T$, in which ${}^A\boldsymbol{v}_p$ is the velocity of point O_B, and ${}^A\omega$ is the angular velocity of moving platform. Furthermore, consider an external wrench generated by the manipulator and applied to the environment at the point O_B denoted by $\boldsymbol{\mathcal{F}} = \left[\boldsymbol{f} \quad \boldsymbol{n}\right]^T$. The vector $\boldsymbol{f}$ denotes the force generated by the moving platform and applied to the environment at point O_B, and vector $\boldsymbol{n}$ is the moment applied to the environment. For the sake of brevity, and because of the similarity of the limbs, only one generic limb is illustrated in Figure 4.15.

First consider the external forces applied to each limb. It is assumed that no external forces are applied to the limbs except the actuator forces. Therefore, the static force can be assumed to be along the limb axis $\hat{\boldsymbol{s}}_i$, and the limb is subject to a tension/compression force f_i. Next, consider the free-body diagram of forces and moments applied to the moving platform. At static equilibrium, the summation of all acting forces on the moving platform

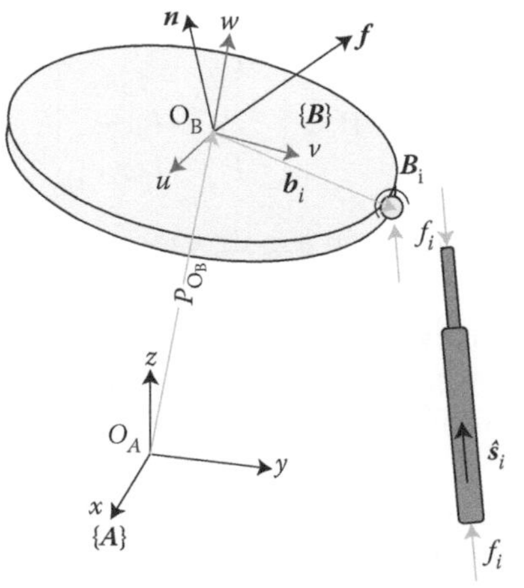

FIGURE 4.15
Free-body diagram of forces and moments acting on the moving platform and each limb of the Stewart–Gough platform.

shall be zero. Therefore

$$-\boldsymbol{f} + \sum_{i=1}^{6} f_i \hat{\boldsymbol{s}}_i = \boldsymbol{0}, \tag{4.104}$$

in which $-\boldsymbol{f}$ is the external force applied to the moving platform from the environment. Furthermore, the summation of moments contributed by all forces acting on the moving platform about O_B is as follows:

$$-\boldsymbol{n} + \sum_{i=1}^{6} \boldsymbol{E}_i \times f_i \hat{\boldsymbol{s}}_i = \boldsymbol{0}, \tag{4.105}$$

in which $-\boldsymbol{n}$ is the external moment applied to the moving platform by the environment, and $\boldsymbol{b}_i$ is the position vector from the point O_B to the attached point $\boldsymbol{B}_i$ on the moving platform. Writing Equations 4.104 and 4.105 together and in a matrix form results in

$$\begin{bmatrix} \hat{\boldsymbol{s}}_1 & \hat{\boldsymbol{s}}_2 & \cdots & \hat{\boldsymbol{s}}_6 \\ \boldsymbol{b}_1 \times \hat{\boldsymbol{s}}_1 & \boldsymbol{b}_2 \times \hat{\boldsymbol{s}}_2 & \cdots & \boldsymbol{b}_6 \times \hat{\boldsymbol{s}}_6 \end{bmatrix} \cdot \begin{bmatrix} f_1 \\ f_2 \\ \vdots \\ f_6 \end{bmatrix} = \begin{bmatrix} \boldsymbol{f} \\ \boldsymbol{n} \end{bmatrix}. \tag{4.106}$$

Comparing the leftmost matrix of Equation 4.106 to that of the Jacobian matrix of the Stewart–Gough platform given in Equation 4.84, it is clearly seen that this matrix is equivalent to the Jacobian transpose of the manipulator. Therefore, Equation 4.106 can be written in the form given by the virtual work approach Equation 4.92 as

$$\boldsymbol{\mathcal{F}} = \boldsymbol{J}^T \boldsymbol{\tau}, \tag{4.107}$$

in which $\boldsymbol{\tau} = [f_1, f_2, \ldots, f_6]^T$ is the vector of actuator forces, and $\boldsymbol{\mathcal{F}} = \left[\boldsymbol{f}, \boldsymbol{n}\right]^T$ is the 6D wrench applied by the manipulator to the environment.

4.10 Stiffness Analysis of Parallel Manipulators

In many applications, the moving platform of a parallel manipulator is in contact with a stiff environment, and applies wrenches to the environment. In the previous section, it is shown that Jacobian transpose is a projection map between the applied wrench to the environment and the actuator forces causing this wrench. In this section, we focus our attention on the deflections of the manipulator moving platform that are the result of the applied wrench to the environment. The amount of these deflections are a function of the applied wrench as well as the manipulator structural stiffness. Thus, the stiffness of a manipulator has a direct impact on its overall positioning accuracy if the manipulator moving platform is in contact with a stiff environment.

The stiffness of a manipulator depends on several factors [43]. The most important factor is the structure of the manipulator. Using closed-kinematic chains in the structure of the robot contributes significantly to higher stiffness and better positioning accuracy, and this is one of the leading advantages of parallel robots over their serial counterparts. Other factors such as the size and materials of the links, the type of actuators and transmission systems, and the control systems also contribute to the overall stiffness of a manipulator. As the links become longer and more slender, they become a major source of manipulator compliance. This is particularly true for space robots and cable-driven manipulators, in which long and lightweight links are used in the structure of the robot. However, in a general manipulator, the main source of compliance is usually the actuators, transmissions, and possibly flexible elements used for force and torque transducers. In what follows, we assume that only the actuators and their transmission systems are the main sources of manipulator compliance.

4.10.1 Stiffness and Compliance Matrices

The actuators of a typical parallel manipulator consist of either hydraulic piston–cylinders or electrically driven linear/rotary servo systems with geared transmissions and possibly force/torque transducers. Each element of such actuator may contribute in the compliance of the limb. Applying large amount of forces through the actuators to generate the required wrench on the manipulator moving platform causes the actuator to deflect along the applied force axis. The relation between the applied actuator force f_i and the corresponding small deflection Δq_i along the applied force axis can be approximated as a linear function

$$f_i = k_i \cdot \Delta q_i, \tag{4.108}$$

in which k_i denotes the *stiffness constant* of the actuator caused by all its compliant components. Rewriting Equation 4.108 for all limbs into a matrix form result in

$$\boldsymbol{\tau} = \mathcal{K} \cdot \Delta \boldsymbol{q}, \tag{4.109}$$

in which $\boldsymbol{\tau} = \begin{bmatrix} f_1 & f_2 & \cdots & f_m \end{bmatrix}^T$ is the vector of actuator forces, the vector of the corresponding actuator deflections is denoted by $\Delta \boldsymbol{q} = [\Delta q_1 \;\; \Delta q_2 \;\; \cdots \;\; \Delta q_m]^T$, and $\mathcal{K} = \mathrm{diag}\begin{bmatrix} k_1 & k_2 & \cdots & k_m \end{bmatrix}$ is an $m \times m$ diagonal matrix composed of the actuator stiffness constants.

In parallel manipulators, Jacobian matrix not only reveals the relation between infinitesimal deflection of the actuators $\Delta \boldsymbol{q}$ to that of the moving platform $\Delta \boldsymbol{\mathcal{X}}$, it also constructs the transformation map between the actuator forces $\boldsymbol{\tau}$ to the wrench exerted by the moving platform $\boldsymbol{\mathcal{F}}$. Writing the Jacobian relation given in Equation 4.34 for infinitesimal deflections reads

$$\Delta \boldsymbol{q} = \boldsymbol{J} \Delta \boldsymbol{\mathcal{X}}, \tag{4.110}$$

in which $\Delta \boldsymbol{\mathcal{X}} = \begin{bmatrix} \Delta x & \Delta y & \Delta z & \Delta \theta_x & \Delta \theta_y & \Delta \theta_z \end{bmatrix}$ is the infinitesimal linear and angular deflection of the moving platform. Furthermore, rewriting the Jacobian relation for parallel manipulators as the projection of actuator forces to the moving platform wrench 4.92 gives

$$\boldsymbol{\mathcal{F}} = \boldsymbol{J}^T \boldsymbol{\tau}. \tag{4.111}$$

Hence, by substitution of Equations 4.110 and 4.111 into Equation 4.109, this equation may be written in task space

$$\boldsymbol{\mathcal{F}} = \boldsymbol{J}^T \mathcal{K} \boldsymbol{J} \cdot \Delta \boldsymbol{\mathcal{X}}. \tag{4.112}$$

The general stiffness relation of a manipulator relates the infinitesimal deflection of the moving platform to the applied wrench as

$$\mathcal{F} = K \cdot \Delta\mathcal{X}, \tag{4.113}$$

in which K is the manipulator *stiffness matrix* computed from the following equation:

$$K = J^T \mathcal{K} J. \tag{4.114}$$

Equation 4.114 implies that the moving platform output wrench is related to its deflection by the stiffness matrix K. The stiffness matrix has desirable characteristics for analysis. It is a symmetric positive definite matrix, however, it is configuration dependent. Furthermore, if the manipulator actuators are all of the same type and have the same stiffness constants, that is, $k_1 = k_2 = \cdots = k_m = k$, the stiffness matrix is reduced to the simple form of $K = kJ^T J$. Note that the stiffness matrix is invertible if

$$\det(J^T J) \neq 0. \tag{4.115}$$

Referring to the singularity conditions elaborated in Section 4.5.2, the condition given in Equation 4.115 means that the parallel manipulator has no forward kinematics singularity. From a physical point of view, at a forward kinematic singularity configuration, the manipulator gains one or more uncontrollable degrees-of-freedom, and therefore loses its stiffness in some directions.

If condition 4.115 holds, the *compliance matrix* of the manipulator denoted by C is defined as

$$C = K^{-1} = (J^T \mathcal{K} J)^{-1}. \tag{4.116}$$

The compliance matrix of a manipulator shows the mapping of the moving platform wrench to its deflection by

$$\Delta\mathcal{X} = C \cdot \mathcal{F}. \tag{4.117}$$

If parallel manipulator is fully actuated and is not redundant, that is, $m = n$, then, Jacobian matrix is squared and compliance matrix may be derived from

$$C = J^{-1} \mathcal{K}^{-1} J^{-T}. \tag{4.118}$$

Note that the compliance matrix of a manipulator, like the stiffness matrix, is a function of not only the compliance of each limb but also the Jacobian matrix.

4.10.2 Transformation Ellipsoid

As seen in Section 4.3, Jacobian matrix J in parallel manipulators transforms n-dimensional moving platform velocity vector $\dot{\mathcal{X}}$ into m-dimensional actuated joint velocity $\dot{q}$. Furthermore, it is seen in Section 4.9.2 that Jacobian transpose J^T maps m-dimensional actuated joint forces $\boldsymbol{\tau}$ into n-dimensional applied wrench $\mathcal{F}$. There are many measures defined in the literature to describe the characteristics of these transformations. One way to characterize the transformation is to compare the amplitude and direction of the moving platform velocity generated by a unit actuator joint velocity. To achieve this goal, confine the actuator joint velocity vector on an m-dimensional unit sphere

$$\dot{q}^T \dot{q} = 1, \tag{4.119}$$

and compare the resulting moving platform velocity in n-dimensional space. Substitute Equation 4.34 into Equation 4.119

$$\dot{\mathcal{X}}^T J^T J \dot{\mathcal{X}} = 1. \tag{4.120}$$

Similarly, confine the exerted moving platform wrench by an n-dimensional unit sphere

$$\mathcal{F}^T \mathcal{F} = 1, \tag{4.121}$$

and compare the required actuator forces in m-dimensional space. Substitute Equation 4.92 into Equation 4.121

$$\boldsymbol{\tau}^T J J^T \boldsymbol{\tau} = 1. \tag{4.122}$$

Note that in general, the dimension of matrix JJ^T is $m \times m$, in which m is the number of actuated limbs, while the dimension for $J^T J$ is $n \times n$, where n is the number of degrees-of-freedom of the manipulator. Consider the case of fully parallel manipulators with no actuator redundancy. In this case, $m = n$, and all the J, J^T, JJ^T, and $J^T J$ transformations are represented by $n \times n$ matrices. Moreover, both the transformations indicated by Equations 4.120 and 4.122 generate symmetric and positive semidefinite matrices the eigenvectors of which are orthogonal, and the eigenvalues of both the transformations are the same.* Hence, geometrically, these transformations represent a hyper-ellipsoid in n-dimensional space [197], whose principal axes coincide with the eigenvectors of $J^T J$ and JJ^T, respectively. Furthermore, the lengths of the principal axes are equal to the reciprocals of the square roots of the eigenvalues of JJ^T and $J^T J$, which in fact is equal to the reciprocals of the singular values of J.

The shape of this hyper-ellipsoid in space indicates the characteristics of the transformation. As this hyper-ellipsoid is closer to a hyper-sphere, the transformation becomes more uniform in different directions. This transformation is called *isotropic*, when the principal axes of the hyper-ellipsoid are all of equal lengths. Since Jacobian matrix is configuration dependent, the shape of the hyper-ellipsoid is also configuration dependent, and as the moving platform moves from one pose to the other, the shape of the hyper-ellipsoid changes accordingly. Hence, at an isotropic configuration and at velocity level, a unit sphere of actuator joint velocities is mapped into a unit sphere of moving platform velocity. In other words, unit change of displacement in one actuator is mapped in a one-to-one manner to a unit change of displacement in one direction in task space. On the other hand, in the vicinity of singular configurations, one or more eigenvalues tend to zero and the shape of hyperellipsoid degenerates to a hypercylinder. This means that physically at these configurations, an infinitesimal motion of moving platform is possible even if the actuators are locked. In other words, the manipulator's stiffness vanishes in some directions.

As another means to compare the characteristics of the twist or wrench mappings of the manipulators, and for those manipulators with only one type of task, namely, either pure

* In fact the eigenvalues of both $J^T J$ and JJ^T matrices are equal to the square of Jacobian matrix singular values.

translational motion or pure rotational movement, but not both, the reciprocal of Jacobian matrix condition number κ can be used. This measure is defined as the square root of the ratio of the smallest eigenvalue to the largest one [58], as follows:

$$\frac{1}{\kappa} = \sqrt{\frac{\lambda_{min}}{\lambda_{max}}}, \tag{4.123}$$

in which λ_{min} and λ_{max} are the smallest and the largest eigenvalues of Jacobian matrix, respectively. Note that in general, when Jacobian matrix is not square, this measure can be defined as

$$\frac{1}{\kappa} = \frac{\sigma_{min}}{\sigma_{max}}, \tag{4.124}$$

in which σ_{min} and σ_{max} are the smallest and the largest singular values of Jacobian matrix, respectively. This measure is used to characterize the dexterity of the manipulator [103,162]. Note that this dexterity measure is configuration dependent, and as the moving platform moves from one pose to another, the Jacobian matrix varies, and therefore, this dexterity measure changes accordingly. Furthermore, dexterity measure defined based on the condition number κ varies between zero and one, and configurations of the manipulator at which this measure is equal to one corresponds to the *isotropic points* [162]. These configurations correspond to the situations in which the hyperellipsoid is turned into a hypersphere, as discussed hereinbefore.

To summarize this section, and to give further reading references, it should be noted that the concept of manipulability ellipsoids was first introduced by Yoshikawa as a measure of the capability of a manipulator to execute a specific task in a given configuration [197]. Since then, a number of interesting extensions and applications of manipulability ellipsoids have appeared in the literature. As a representative of these vast researches, Klein and Blaho [103] review and apply several dexterity measures to find *optimal* configurations for a given end effector pose and related problems. Possible inconsistencies are derived from the improper use of the Euclidean metric in the 6D force (wrench) and velocity (twist) vector spaces commonly encountered in robotics [41]. As a consequence, the various dexterity measures depend on the singular values or the reciprocal of the condition number of the manipulator Jacobian J. It has been shown that these quantities have no physical consistency and change with changes of scale, physical units, and coordinate frame. As a good review paper [39] has reviewed the basic definitions of different manipulability ellipsoids and demonstrated the difficulties in the previously defined manipulability measures. It has been shown that all manipulability measures in space, which include both linear and angular terms together, regardless of the robot degrees-of-freedom, may lead to such an inconsistency. Even planar manipulators with at least one revolute joint generate both linear and angular motions, so that vectors with noncommensurate components cannot be correctly avoided in even the simplest situations. An alternative formulation of the manipulability ellipsoid is given based on a tuple $(n, \overline{\omega}_n)$. The degree-of-manipulability n equals the rank of the ellipsoid matrix, and $\overline{\omega}_n$ equals the ellipsoid volume in n-space. The volume of the ellipsoid matrix has been redefined as the product of the nonzero diagonal terms, by which, as long as the robot has motion capability it has some degree-of-manipulability.

4.10.3 Stiffness Analysis of a Planar Manipulator

The architecture of a planar $4R\underline{P}R$ parallel manipulator considered here is described in detail in Section 3.3.1 and shown in Figure 3.3. To perform the stiffness analysis for the manipulator, assume that the vector of actuated joint forces is denoted by $\boldsymbol{\tau} = \begin{bmatrix} f_1 & f_2 & f_3 & f_4 \end{bmatrix}$, and the corresponding vector of infinitesimal displacements of the actuated limbs is denoted by $\Delta \boldsymbol{L} = \begin{bmatrix} \Delta l_1 & \Delta l_2 & \Delta l_3 & \Delta l_4 \end{bmatrix}$. The relation between the joint displacement $\Delta \boldsymbol{L}$ to the actuator force $\boldsymbol{\tau}$ is described by a diagonal 4×4 matrix of joint stiffness constants $\mathcal{K} = \text{diag}\begin{bmatrix} k_1 & k_2 & k_3 & k_4 \end{bmatrix}$, as follows:

$$\boldsymbol{\tau} = \mathcal{K} \cdot \Delta \boldsymbol{L}. \tag{4.125}$$

From the definition of the Jacobian in Equation 4.34, we have

$$\Delta \boldsymbol{L} = \boldsymbol{J} \cdot \Delta \boldsymbol{\mathcal{X}}, \tag{4.126}$$

in which $\Delta \boldsymbol{\mathcal{X}} = \begin{bmatrix} \Delta x_G & \Delta y_G & \Delta \phi \end{bmatrix}$ is the vector of infinitesimal linear and angular motions of the moving platform. Also, the 3D vector of the moving platform output wrench, which in this case, is $\boldsymbol{\mathcal{F}} = \begin{bmatrix} f_x & f_y & n_z \end{bmatrix}$, is related to the vector of actuated joint forces $\boldsymbol{\tau}$ by the principle of virtual work as

$$\boldsymbol{\mathcal{F}} = \boldsymbol{J}^T \cdot \boldsymbol{\tau}. \tag{4.127}$$

Substituting Equations 4.125 and 4.126 into Equation 4.127, yields

$$\boldsymbol{\mathcal{F}} = \boldsymbol{K} \cdot \Delta \boldsymbol{\mathcal{X}}, \tag{4.128}$$

in which

$$\boldsymbol{K} = \boldsymbol{J}^T \mathcal{K} \boldsymbol{J}. \tag{4.129}$$

$\boldsymbol{K}$ is called the stiffness matrix of the planar $4R\underline{P}R$ manipulator, which is obviously configuration dependent.

For a given configuration of the moving platform, the eigenvalue of stiffness matrix represents the stiffness of the manipulator in the corresponding eigenvector direction. These directions are in fact represented by twist vectors. Furthermore, the reciprocal of the condition number of the stiffness matrix can be used to represent the dexterity of the manipulator. This measure is plotted in Figure 4.16, assuming equal stiffness constants for all the actuators. Since the dexterity measure is configuration dependent, Figure 4.16 illustrates the dexterity measure of the planar manipulator within the whole translational workspace, with a constant orientation of the moving platform. In this figure, the dexterity measure for four different orientations, namely $\phi = 0°, 45°, 60°$, and $75°$, is given. As the dexterity measure becomes closer to one, the twist and wrench transformations are conditioned better, and on the other hand, as the dexterity measure becomes smaller, the transformation characteristics get worse. It is seen in Figure 4.16 that generally the dexterity measure is not close to one, and the manipulator is not isotropic within its whole workspace. However, the dexterity measure of the manipulator is conditioned better when close to its central position, and as the moving platform moves toward the edges of the workspace it becomes more ill-conditioned. Comparing different constant orientation workspaces to each other, it is seen that at the central position, the dexterity measure becomes better conditioned as the orientation of the moving platform increases. However, at the corners, having larger orientation in the moving platform makes the dexterity measure worse.

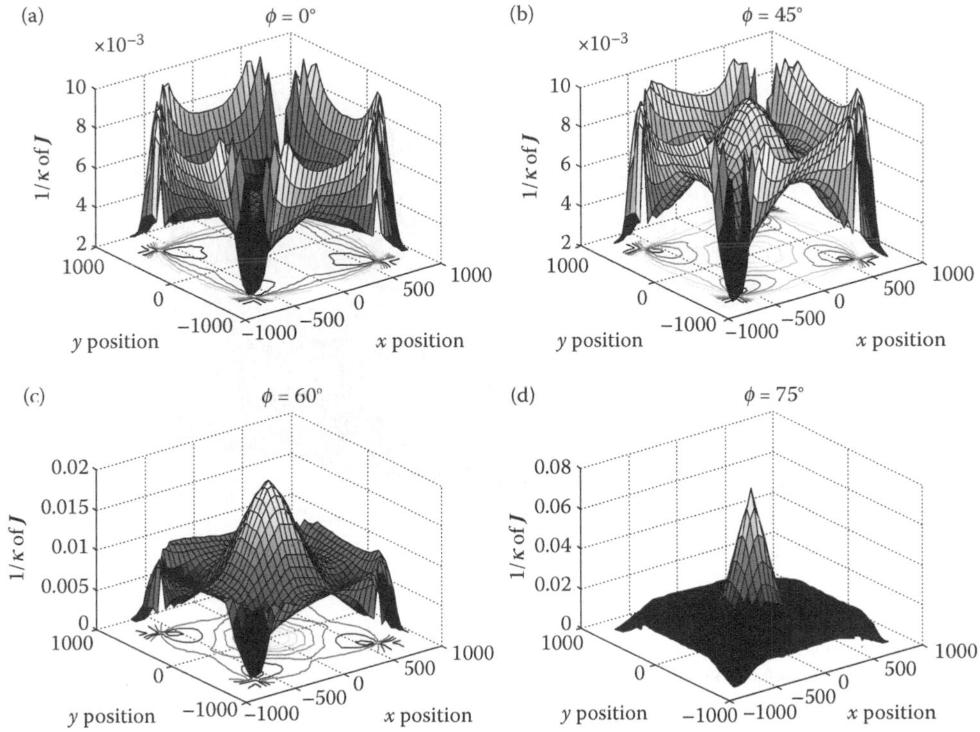

FIGURE 4.16
Reciprocal of the condition number of planar manipulator stiffness matrix over the constant orientation workspace: (a) $\phi = 0°$, (b) $\phi = 45°$, (c) $\phi = 60°$, and (d) $\phi = 75°$.

Furthermore, it can be analyzed how the actuator redundancy can improve the total stiffness of the planar manipulator. This is clearly shown in Figure 4.17, where the dexterity measure of $1/\kappa$ is plotted for the redundant and nonredundant manipulators by suppressing one of the redundant limbs over the constant orientation workspace of the manipulator with $\phi = 0°$. It is clearly observed in Figure 4.17 that losing one actuator significantly reduces the dexterity of the manipulator, especially at the positions where the suppressed actuator is located.

4.10.4 Stiffness Analysis of Shoulder Manipulator

As explained in Section 4.10.1, the stiffness of a parallel manipulator at a given point of its workspace can be characterized by its stiffness matrix, which relates the wrench exerted by the moving platform in the Cartesian space to the corresponding linear and angular Cartesian deflections. In this section, a detail stiffness analysis is performed for the shoulder manipulator. In fact, this analysis is a critical initiating point for further research of this manipulator in fields such as accurate position control schemes or the problem of controlling the manipulator in an environment with kinematic and force constraints, in which a desired and appropriate stiffness should be synthesized through various stiffness control schemes. Furthermore, the stiffness matrix gives information on some important kinematic properties of manipulator such as dexterity or manipulability [59].

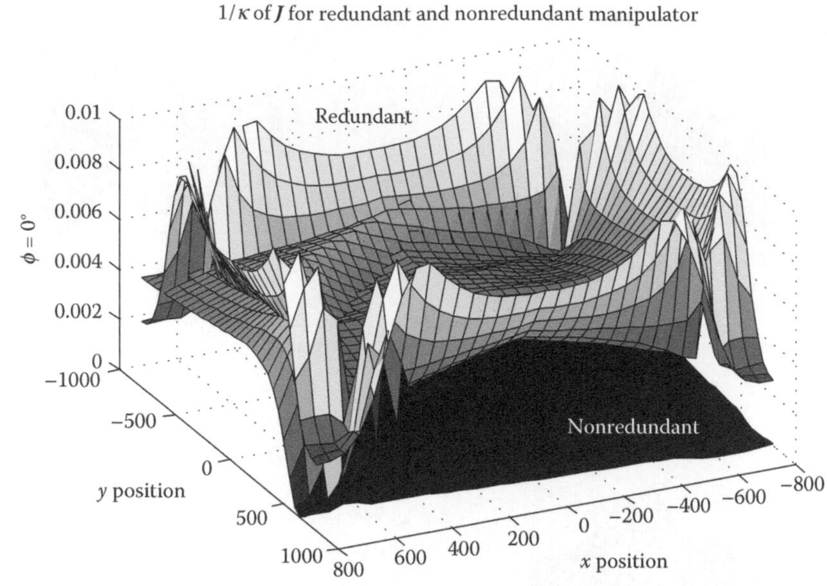

FIGURE 4.17
Comparison of stiffness values for the redundant and nonredundant structures of planar manipulator within its constant orientation workspace with $\phi = 0°$.

To perform the stiffness analysis for shoulder manipulator, assume that the vector of actuated joint forces are denoted by $\boldsymbol{\tau} = \begin{bmatrix} f_1 & f_2 & f_3 & f_4 \end{bmatrix}$, and the corresponding vector of infinitesimal displacements of the actuated limbs are denoted by $\Delta \boldsymbol{L} = \begin{bmatrix} \Delta l_1 & \Delta l_2 & \Delta l_3 & \Delta l_4 \end{bmatrix}$. The relation between joint displacement $\Delta \boldsymbol{L}$ to actuator force $\boldsymbol{\tau}$ is described by a diagonal 4×4 matrix of joint stiffness constants $\mathcal{K} = \text{diag}\begin{bmatrix} k_1 & k_2 & k_3 & k_4 \end{bmatrix}$ as follows:

$$\boldsymbol{\tau} = \mathcal{K} \cdot \Delta \boldsymbol{L}. \tag{4.130}$$

From the definition of the Jacobian in Equation 4.61, we have

$$\Delta \boldsymbol{L} = \boldsymbol{J} \cdot \Delta \theta \tag{4.131}$$

in which $\Delta \theta = \begin{bmatrix} \Delta \theta_x & \Delta \theta_y & \Delta \theta_z \end{bmatrix}$ is the vector of infinitesimal angular displacement of the moving platform. Also, 3D vector of the moving platform output wrench, which is, in this case as $\boldsymbol{\mathcal{F}} = \begin{bmatrix} n_x & n_y & n_z \end{bmatrix}$, is related to the vector of actuated joint forces $\boldsymbol{\tau}$ by

$$\boldsymbol{\tau} = \boldsymbol{J}^T \cdot \boldsymbol{\mathcal{F}}. \tag{4.132}$$

Substituting Equations 4.131 and 4.132 into Equation 4.130, yields

$$\boldsymbol{\mathcal{F}} = \boldsymbol{K} \cdot \Delta \theta, \tag{4.133}$$

in which

$$\boldsymbol{K} = \boldsymbol{J}^T \mathcal{K} \boldsymbol{J}. \tag{4.134}$$

$\boldsymbol{K}$ is called the stiffness matrix of the shoulder manipulator, which is obviously configuration dependent.

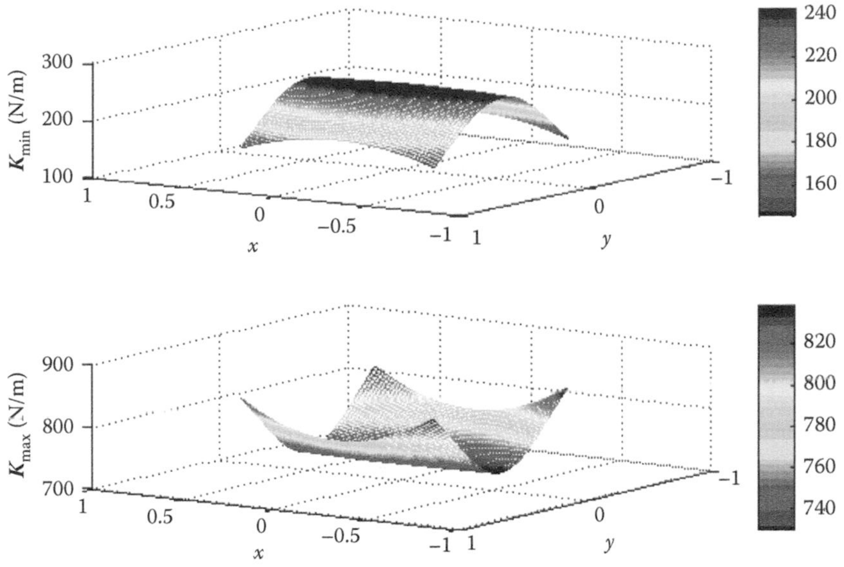

FIGURE 4.18
Minimum and maximum stiffness mappings of the shoulder manipulator in its whole workspace.

For a given configuration, the eigenvalue of stiffness matrix represents the stiffness of shoulder manipulator in the corresponding eigenvector direction. These directions are in fact represented by twist vectors. Similarly, minimum and maximum stiffness mappings may be obtained using the corresponding eigenvalues and eigenvectors. These maps are shown in Figure 4.18, assuming equal stiffness constants of 10^5 N/m for all the actuators. Moreover, the reciprocal of the manipulator Jacobian condition number may be used as a measure of dexterity of the manipulator [160]. Using this measure, one may analyze how actuator redundancy has improved the total stiffness of the shoulder manipulator. This is shown in Figure 4.19, where the Euclidean norms of stiffness matrix is computed for the redundant and nonredundant manipulator by suppressing one of the redundant legs. It is clearly observed in Figure 4.19, that loosing one actuator significantly reduces the stiffness of the manipulator.

4.10.5 Stiffness Analysis of the Stewart–Gough Platform

The architecture of the Stewart–Gough platform is described in detail in Section 3.5.1 and shown in Figure 3.16. To perform the stiffness analysis for this manipulator, different structures of this manipulator may be examined. In this section, we restrict our analysis to a 3–6 structure of this manipulator, in which, as shown in Figure 4.20, there exist six distinct attachment points A_i's on the fixed base. The moving platform is connected through six actuators to only three moving attachment point B_i's. Denote the vector of actuated joint forces by $\boldsymbol{\tau} = \begin{bmatrix} f_1 & f_2 & f_3 & f_4 & f_5 & f_6 \end{bmatrix}$, and the corresponding vector of infinitesimal displacements of actuated limbs denoted by $\Delta \boldsymbol{L} = \begin{bmatrix} \Delta l_1 & \Delta l_2 & \Delta l_3 & \Delta l_4 & \Delta l_5 & \Delta l_6 \end{bmatrix}$. The relation between the joint displacement $\Delta \boldsymbol{L}$ to actuator force $\boldsymbol{\tau}$ is described by a diagonal 6×6 matrix of joint stiffness constants $\mathcal{K} = \text{diag}\begin{bmatrix} k_1 & k_2 & k_3 & k_4 & k_5 & k_6 \end{bmatrix}$ as follows:

$$\boldsymbol{\tau} = \mathcal{K} \cdot \Delta \boldsymbol{L}. \tag{4.135}$$

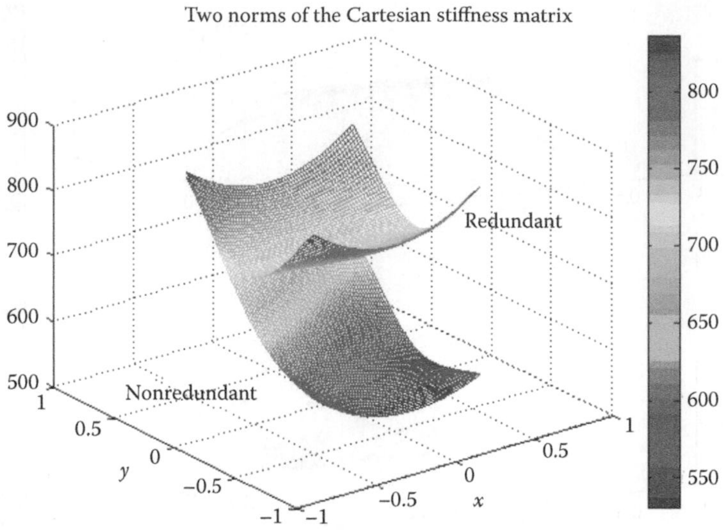

FIGURE 4.19
Comparison of the stiffness values for the redundant and nonredundant structures of the shoulder manipulator within its whole workspace.

From the definition of the Jacobian in Equation 4.34, we have

$$\Delta \boldsymbol{L} = \boldsymbol{J} \cdot \Delta \boldsymbol{\mathcal{X}}, \tag{4.136}$$

in which $\Delta \boldsymbol{\mathcal{X}} = \begin{bmatrix} \Delta x & \Delta y & \Delta z & \Delta\theta_x & \Delta\theta_y & \Delta\theta_z \end{bmatrix}$ is the vector of infinitesimal linear and angular motions of the moving platform. Also, a 3D vector of the moving platform output wrench denoted by $\boldsymbol{\mathcal{F}} = \begin{bmatrix} f_x & f_y & f_x & n_x & n_y & n_z \end{bmatrix}$ is related to the vector of

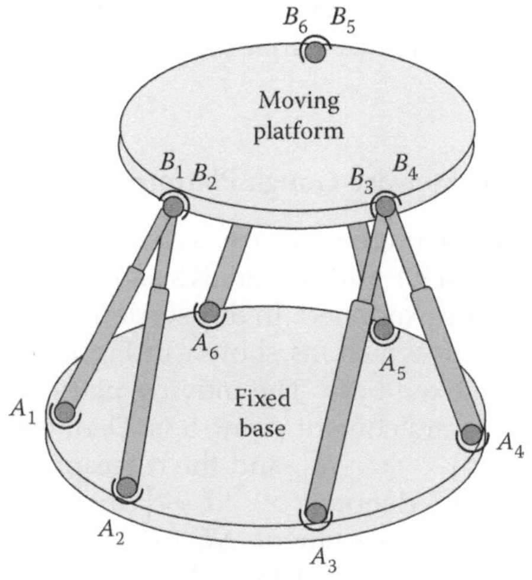

FIGURE 4.20
Schematic of a 3–6 Stewart–Gough platform.

actuated joint forces $\boldsymbol{\tau}$ by

$$\mathcal{F} = J^T \cdot \boldsymbol{\tau}. \tag{4.137}$$

Substituting Equations 4.135 and 4.136 into Equations 4.137, yields

$$\mathcal{F} = K \cdot \Delta \mathcal{X}, \tag{4.138}$$

in which

$$K = J^T \mathcal{K} J. \tag{4.139}$$

K is called the stiffness matrix of the Stewart–Gough manipulator.

For a given configuration of the moving platform, the eigenvalue of the stiffness matrix represents the stiffness of the manipulator in the corresponding eigenvector direction. These directions are in fact represented by twist vectors. Furthermore, the reciprocal of the stiffness matrix condition number may be used to represent the dexterity of the manipulator. This measure is plotted in Figure 4.21, assuming equal stiffness constants for all the actuators. Since dexterity measure is configuration dependent, Figure 4.21 illustrates the dexterity measure of the 3–6 Stewart–Gough manipulator within the whole translational workspace, with a constant heave $z = 1$, and constant orientation of the moving

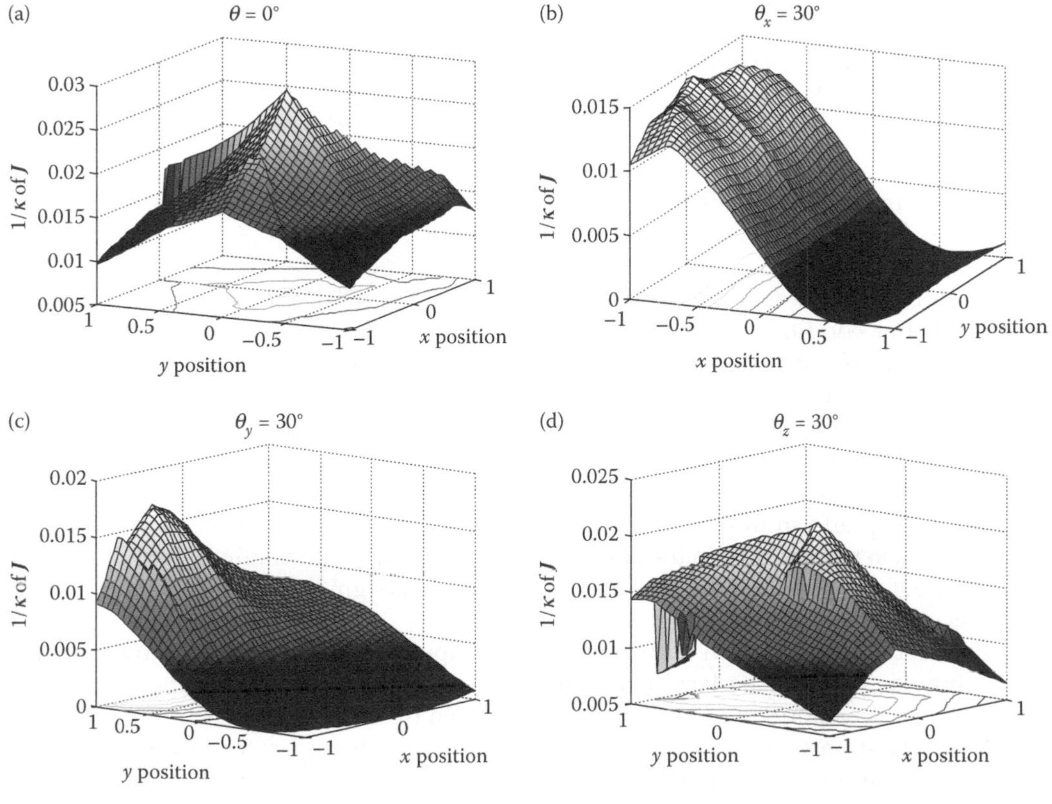

FIGURE 4.21
Reciprocal of the condition number of a 3–6 Stewart–Gough platform stiffness matrix over the constant orientation workspace with varying x and y and fixed $z = 0$ and (a) $\boldsymbol{\theta}_x = 0$, (b) $\theta_x = 30°$, (c) $\theta_y = 30°$, and (d) $\theta_z = 30°$.

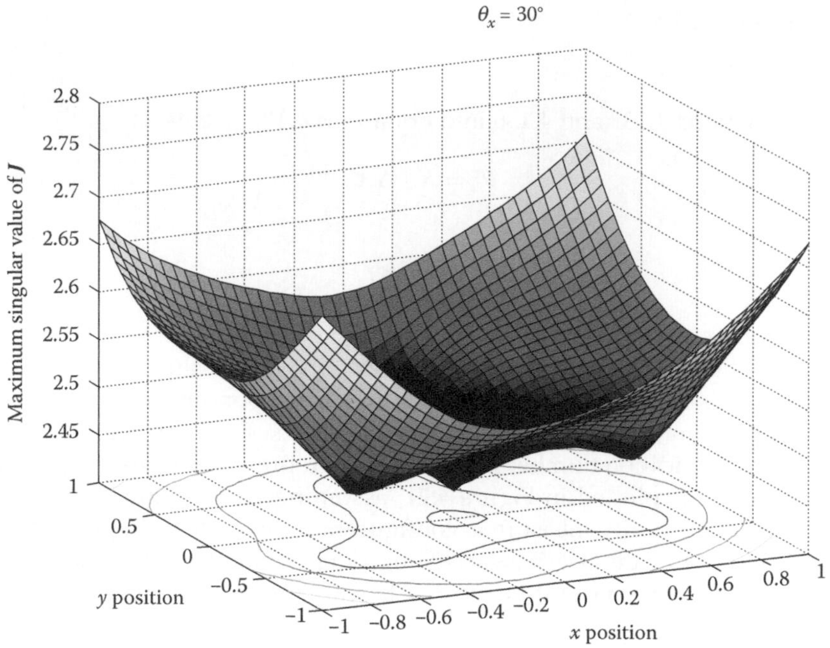

FIGURE 4.22
Maximum singular value of the Jacobian matrix over the constant orientation workspace with fixed $z = 0$ and $\theta_x = 0, \theta_y = 0$, and $\theta_z = 0$.

platform. In this figure, the dexterity measure for four different orientations, namely (a) $\theta_x = 0, \theta_y = 0, \theta_z = 0$, (b) $\theta_x = 30°, \theta_y = 0, \theta_z = 0$, (c) $\theta_x = 0, \theta_y = 30^0, \theta_z = 0$, and (d) $\theta_x = 0, \theta_y = 0, \theta_z = 30°$, is given.

As the dexterity measure becomes closer to one, the twist and wrench transformations are better conditioned, and on the other hand, as the dexterity measure becomes smaller, the transformation characteristics get worse. It is seen in Figure 4.21 that generally the dexterity measure is not close to one, and the manipulator is not isotropic within its whole workspace. However, the dexterity measure of the manipulator is conditioned better in cases (a) and (d) in which the rotation of the manipulator about x- and y-axes are zero. In these two cases, best dexterity is obtained close to its central position, and as the moving platform moves toward the edges of the workspace, it becomes more ill-conditioned. On the contrary, in the cases (b) and (c), in which, the manipulator has angular rotation about either x- or y-axes, the dexterity measure is significantly reduced toward the workspace edges.

Furthermore, the maximum stiffness of the manipulator can be analyzed by the maximum singular value of the Jacobian matrix. As explained hereinbefore, the largest axis of the stiffness transformation hyperellipsoid is given by this value at each configuration. Figure 4.22 illustrates the maximum singular value of the manipulator Jacobian over the constant orientation workspace with varying x and y and fixed $z = 0$ and $\boldsymbol{\theta} = 0$. It is clearly observed that as the moving platform moves toward the edges, the largest singular value of the Jacobian matrix becomes larger, and therefore, the transformation hyperellipsoid deviates from a hypersphere. Hence, from this point of view, the result obtained from the dexterity measure is verified, which indicates less dexterity at the edges of the constant orientation workspace. Similar conclusions can be drawn by such analysis at different other orientations of the manipulator.

PROBLEMS

1. Consider the five-bar mechanism shown in Figure 3.28 and described in Problem 3.1. Find the end-effector velocity vector $\boldsymbol{v}_p = [v_x, v_y]^T$ as a function of two input angular velocities $\dot{\boldsymbol{\theta}} = [\dot{\theta}_1, \dot{\theta}_2]^T$. Derive the Jacobian matrix of the manipulator and identify singular configurations.
2. Consider the five-bar mechanism shown in Figure 3.29 and described in Problem 3.2. Find the end-effector velocity vector $\boldsymbol{v}_p = [v_x, v_y]^T$ as a function of the two input velocity variables $\dot{\ell}_1, \dot{\theta}_2$. Derive the Jacobian matrix of the manipulator and identify singular configurations.
3. Consider the planar pantograph mechanism shown in Figure 3.30 and described in Problem 3.3. For this mechanism, find the end-effector velocity vector $\boldsymbol{v}_p = [v_x, v_y]^T$, as a function of the two input links linear velocities $\dot{x}_1, \dot{y}_2$. Derive the Jacobian matrix of the manipulator, and identify singular configurations.
4. Consider the planar three-degrees-of-freedom $3\underline{R}RR$ parallel manipulator shown in Figure 3.31 and described in Problem 3.4.
   a. For this manipulator, form the vector loops closure, and find relations between the moving platform twist vector $\dot{\boldsymbol{\mathcal{X}}} = [v_x, v_y, \dot{\phi}]^T$ as a function of angular velocity of the actuated limbs $\dot{\alpha}_i$'s and passive limbs $\dot{\beta}_i$'s.
   b. Eliminate the passive limb angular velocities $\dot{\beta}_i$'s from the equations obtained in part (a), and find expressions for Jacobian matrices $\boldsymbol{J}_x$ and $\boldsymbol{J}_q$ of the manipulator.
   c. Identify inverse and forward singularity configurations of the manipulator from the structure of Jacobian matrices by inspection.
   d. Develop a program in MATLAB with the following numerical values for the manipulator parameters. Generate a grid space for the entire workspace of the manipulator, calculate the numerical values of Jacobian matrices at each grid point, and identify singular configurations. Illustrate singular configurations of the manipulator in a 3D plot, and compare the result to the one obtained in part (c).

   $$a_i = b_i = 1(m), \quad L = 5(m), \quad \ell = 1.5(m).$$

   Consider the manipulator workspace within the following limits:

   $$-2(m) \leq x \leq 2(m), \quad -2(m) \leq y \leq 2(m), \quad -90^\circ \leq \phi \leq 90^\circ.$$

   e. Identify any configuration in which the manipulator has combined singularity, if any.
5. Consider the planar three-degrees-of-freedom $4\underline{R}RR$ parallel manipulator shown in Figure 3.32 and described in Problem 3.5.
   a. For this manipulator, form the vector loops closure, and find a relation between the moving platform twist vector $\dot{\boldsymbol{\mathcal{X}}} = [v_x, v_y, \dot{\phi}]^T$ to the angular velocity of the actuated limbs $\dot{\alpha}_i$'s and passive limbs $\dot{\beta}_i$'s.

b. Eliminate passive limb angular velocities $\dot{\beta}_i$'s from the equation obtained in part (a), and find expressions for Jacobian matrices $\boldsymbol{J}_x$ and $\boldsymbol{J}_q$ of the manipulator.

c. Identify inverse and forward singularity configurations of the manipulator from the structure of Jacobian matrices by inspection.

d. Develop a program in MATLAB with the following numerical values for the manipulator parameters. Generate a grid space for the entire workspace of the manipulator, calculate the numerical values of Jacobian matrices at each grid point, and identify singular configurations. Illustrate singular configuration of the manipulator in a 3D plot, and compare the result to the one obtained in part (c).

$$R_A = 3(m), R_B = 1(m), a_i = b_i = 1.5(m)$$

$$\theta_{A_i} = \theta_{B_i} = [-135°, -45°, 45°, 135°]^T.$$

Consider the manipulator workspace within the following limits

$$-2(m) \leq x \leq 2(m), \quad -2(m) \leq y \leq 2(m), \quad -90° \leq \phi \leq 90°.$$

e. Identify any configuration in which the manipulator has combined singularity, if any.

6. Consider the spatial parallel manipulator shown in Figure 3.33 and described in Problem 3.6.

a. For this manipulator, form the vector loops closure, and find a relation between the moving platform angular velocity $\boldsymbol{\omega}$ to the linear velocity vector of the actuated limbs $\dot{\ell}_i$'s and their angular velocity ω_i's.

b. Eliminate ω_i's from the equations obtained in part (a), and find expressions for Jacobian matrices $\boldsymbol{J}_x$ and $\boldsymbol{J}_q$ of the manipulator.

c. Identify inverse and forward singularity configurations of the manipulator from the structure of the Jacobian matrices by inspection.

d. Give some parametric values to the geometry of the manipulator, and develop a program in MATLAB to generate a grid space for entire workspace of the manipulator, calculate the numerical values of Jacobian matrices at each grid point, and identify singular configurations. Illustrate singular configurations of the manipulator in a 3D plot, and compare the result to the one obtained in part (c). Consider the manipulator workspace within the following limits:

$$-30° \leq \alpha, \beta, \gamma \leq 30°.$$

7. Consider the $3U\underline{P}U$ parallel manipulator shown in Figure 3.34 and described in Problem 3.7.

a. For this manipulator, form the vector loops closure, and find a relation between the moving platform velocity v_P to the linear velocity vector of the actuated limbs $\dot{q}_i$'s, and their angular velocities of the unactuated joints ω_i's.

b. Eliminate ω_i's from the equations obtained in part (a), and find expressions for Jacobian matrices $\boldsymbol{J}_x$ and $\boldsymbol{J}_q$ of the manipulator.

c. Identify inverse and forward singularity configurations of the manipulator from the structure of Jacobian matrices by inspection.

d. For the parametric values given in Problem 3.7, develop a program in MATLAB to generate a grid space for the entire workspace of the manipulator, calculate numerical values for Jacobian matrices at each grid point, and identify singular configurations. Illustrate singular configurations of the manipulator in a 3D plot, and compare the result to the one obtained in part (c). Consider manipulator workspace within the following limits:

$$-1(m) \leq x, y, z \leq 1(m).$$

8. Consider a $3U\underline{P}S$ parallel manipulator introduced in [47], shown in Figure 3.35 and described in Problem 3.8.

   a. For this manipulator, form the velocity loop closure, and find a relation between the 6D twist of the moving platform, $\dot{\mathcal{X}}$, to the vector of actuated limb velocities, $\dot{L}$, by $\dot{L} = J_v \dot{\mathcal{X}}$. Then, find the Jacobian of the passive supporting limb, J_p, relating the 6D twist of the moving platform, $\dot{\mathcal{X}}$, to the vector of linear and angular velocity of the passive limb, $v_p = [\dot{h}, \omega_{p_x}, \omega_{p_y}]^T$, by $\dot{\mathcal{X}} = J_p v_p$.

   b. Find the Jacobian matrix of the manipulator relating actuator velocities $\dot{L}$, to the moving platform generalized velocity v_p, by $\dot{L} = J v_p$, using the active and passive limb Jacobian matrices $J = J_v J_p$.

   c. For the parametric values given in Problem 3.8, develop a program in MATLAB to generate a grid space for entire workspace of the manipulator, calculate the numerical values of Jacobian matrix at each grid point, and identify singular configurations. Illustrate singular configurations of the manipulator in a 3D plot. Consider the manipulator workspace within the following limits:

$$0 \leq h \leq 0.4(m), \quad -30^\circ \leq \psi, \phi \leq 30^\circ.$$

9. Consider the three-degrees-of-freedom $3R\underline{P}S$ manipulator represented in Figure 3.36 and described in Problem 3.9.

   a. For this manipulator, form the velocity loop closure, and find a relation between the moving platform linear velocity v_E to the linear velocity vector of the actuated limbs, $\dot{\ell}_i$'s, and their angular velocities ω_i's.

   b. Eliminate ω_i's from the equations obtained in part (a), and find expressions for Jacobian matrices J_x and J_q of the manipulator.

   c. Identify inverse and forward singularity configurations of the manipulator from the structure of Jacobian matrices by inspection.

   d. Give some parametric values to the geometry of the manipulator, and develop a program in MATLAB to generate a grid space for entire workspace of the manipulator, calculate numerical values of Jacobian matrices at each grid point, and identify singular configurations. Illustrate singular configurations of the manipulator in a 3D plot, and compare the result to the one obtained in part (c).

10. Consider the three-degrees-of-freedom $3\underline{P}RC$ translational parallel manipulator represented in Figure 3.36 and described in Problem 3.10.

    a. For this manipulator, form the velocity loop closure, and find a relation between the moving platform linear velocity v_P to the linear velocity vector of the actuated limbs $\dot{d}_i$'s, and the limbs' angular velocities ω_i's.

b. Eliminate ω_i's from the equations obtained in part (a), and find expressions for Jacobian matrices J_x and J_q of the manipulator.

c. Identify inverse and forward singularity configurations of the manipulator from the structure of Jacobian matrices by inspection.

d. Give some parametric values to the geometry of the manipulator, and develop a program in MATLAB to generate a grid space for entire workspace of the manipulator, calculate numerical values of Jacobian matrices at each grid point, and identify singular configurations. Illustrate singular configurations of the manipulator in a 3D plot, and compare the result to the one obtained in part (c).

11. Consider the Delta robot represented in Figure 3.39 and described in Problem 3.11.

a. For this manipulator, form the velocity loop closure, and find the relation between the moving platform linear velocity $\boldsymbol{v}_P$ to the angular velocity vector of actuated limbs $\dot{\theta}_i$'s, and that of passive limbs' angular velocities $\dot{\beta}_i$'s.

b. Eliminate β_i's from the equations obtained in part (a), and find expressions for Jacobian matrices J_x and J_q of the manipulator.

c. Identify inverse and forward singularity configurations of the manipulator from the structure of Jacobian matrices by inspection.

d. Give some parametric values to the geometry of the manipulator, and develop a program in MATLAB to generate a grid space for entire workspace of the manipulator, calculate numerical values of Jacobian matrices at each grid point, and identify singular configurations. Illustrate singular configurations of the manipulator in a 3D plot, and compare the result to the one obtained in part (c).

12. Consider the Quadrupteron robot represented in Figure 3.40 and described in Problem 3.12 [157].

a. For this manipulator, form the velocity loop closure, and find a relation between the moving platform linear velocity $\boldsymbol{v}_P$ and its angular velocity $\dot{\phi}$ to the linear velocity vector of the actuated limbs $\dot{\rho}_i$'s, and that of passive limbs' angular velocities.

b. Eliminate passive limbs' angular velocities from the equations obtained in part (a), and find expressions for Jacobian matrices J_x and J_q of the manipulator.

c. Identify inverse and forward singularity configurations of the manipulator from the structure of Jacobian matrices by inspection.

d. Give some parametric values to the geometry of the manipulator, and develop a program in MATLAB to generate a grid space for the entire constant orientation workspace of the manipulator, calculate numerical values of Jacobian matrices at each grid point, and identify singular configurations. Illustrate the singular configurations of the manipulator in a 3D plot, and compare the result to the one obtained in part (c).

13. Reconsider the five-bar mechanism shown in Figure 3.28 and described in Problems 3.1 and 4.1.

a. Apply the free-body diagram approach and calculate the actuator torques $\boldsymbol{\tau} = [\tau_a, \tau_2]^T$, required to generate an output force of $\boldsymbol{F} = [f_x, f_y]^T$ at point P.

b. Show that the relation obtained in part (a), can be written in the form of $F = J^T \tau$, and compare the obtained Jacobian matrix with the one obtained in Problem 4.1.

c. Investigate the conditions that would result in unbounded input torques.

14. Reconsider the five-bar mechanism shown in Figure 3.29 and described in Problems 3.2 and 4.2.

a. Apply the free-body diagram approach and calculate the actuator torques $\tau = [f_1, \tau_2]^T$, required to generate an output force of $F = [f_x, f_y]^T$ at point P.

b. Show that the relation obtained in part (a) can be written in the form of $F = J^T \tau$, and verify the obtained Jacobian matrix with the one obtained in Problem 4.2.

c. Investigate the conditions that would result in unbounded input torques.

15. Reconsider the planar pantograph mechanism shown in Figure 3.30 and described in Problems 3.3 and 4.3.

a. Apply the free-body diagram approach and calculate the actuator torques $\tau = [f_1, f_2]^T$, required to generate an output force of $F = [f_x, f_y]^T$ at point P.

b. Show that the relation obtained in part (a) can be written in the form of $F = J^T \tau$, and verify the obtained Jacobian matrix with the one obtained in Problem 4.3.

c. Investigate the conditions that will result in unbounded input torques.

16. Reconsider the planar three-degrees-of-freedom 3$\underline{R}RR$ parallel manipulator shown in Figure 3.31 and described in Problems 3.4 and 4.4.

a. For this manipulator, apply the free-body diagram approach and calculate the actuator torques $\tau = [\tau_1, \tau_2, \tau_3]^T$, required to generate an output wrench of $\mathcal{F} = [f_x, f_y, n_z]^T$ at point P.

b. Show that the relation obtained in part (a) can be written in the form of $\mathcal{F} = J^T \tau$, and verify the obtained Jacobian matrix with the one obtained in Problem 4.4.

c. Consider unit stiffness values for all manipulator actuators, and derive an expression for stiffness matrix of the manipulator.

d. Develop a program in MATLAB with the numerical values given in Problem 4.4, and calculate the dexterity measure $1/\kappa$ of the manipulator stiffness matrix over the entire constant orientation workspace of the manipulator. Use four different orientations $\phi = 0, 30^\circ, 60^\circ$, and 75°, and plot the dexterity measures in four different 3D plots. Analyze the dexterity plots and comment on the design of the manipulator.

e. Further, develop the program written for part (d), and calculate and plot the maximum singular value, and the two norms of the stiffness matrix over the entire constant orientation workspace with $\phi = 0$. Analyze the plots and comment on the stiffness characteristics of the manipulator.

17. Reconsider the planar redundant 4$\underline{R}RR$ parallel manipulator shown in Figure 3.32 and described in Problems 3.5 and 4.5.

a. For this manipulator, apply the free-body diagram approach and calculate the actuator torques $\tau = [\tau_1, \tau_2, \tau_3, \tau_4]^T$ required to generate an output wrench of $\mathcal{F} = [f_x, f_y, n_z]^T$ at point P.

b. Show that the relation obtained in part (a) can be written in the form of $\mathcal{F} = J^T \tau$, and verify the obtained Jacobian matrix with the one obtained in Problem 4.5.

c. Consider unit stiffness values for all manipulator actuators, and derive an expression for stiffness matrix of the manipulator.

d. Develop a program in MATLAB with the numerical values given in Problem 4.5, and calculate the dexterity measure $1/\kappa$ of the manipulator stiffness matrix over the entire constant orientation workspace of the manipulator. Use four different orientations $\phi = 0, 30°, 60°$, and $75°$, and plot the dexterity measures in four different 3D plots. Analyze the dexterity plots and comment on the design of the manipulator.

e. Further, develop the program written for part (d), and calculate and plot the maximum singular value, and the two norms of the stiffness matrix over the entire constant orientation workspace with $\phi = 0$. Analyze the plots and comment on the stiffness characteristics of the manipulator.

f. Suppress one of the actuators of redundant manipulator, and repeat part (d) and (e). Give comparison plots for the redundant and nonredundant structures, and comment on the effect of the redundancy in stiffness characteristics and dexterity of the manipulator.

18. Reconsider the spatial parallel manipulator shown in Figure 3.33 and described in Problems 3.6 and 4.6.

a. For this manipulator, apply the free-body diagram approach and calculate the actuator torques $\boldsymbol{\tau} = [f_1, f_2, f_3]^T$ required to generate an output wrench of $\mathcal{F} = [n_x, n_y, n_z]^T$ in the moving platform.

b. Show that the relation obtained in part (a) can be written in the form of $\mathcal{F} = J^T \boldsymbol{\tau}$, and verify the obtained Jacobian matrix with the one obtained in Problem 4.6.

c. Consider unit stiffness values for all manipulator actuators, and derive an expression for stiffness matrix of the manipulator.

d. Develop a program in MATLAB with the numerical values used in Problem 4.6, and calculate the dexterity measure $1/\kappa$ of the manipulator stiffness matrix over the entire constant orientation workspace of the manipulator. Use four different orientations $\gamma = 0, 30°, 60°$, and $75°$, and plot the dexterity measures in four different 3D plots. Analyze the dexterity plots and comment on the design of the manipulator.

e. Further develop the program written for part (d), and calculate and plot the maximum singular value, and the two norms of the stiffness matrix over the entire constant orientation workspace with $\gamma = 0$. Analyze the plots and comment on the stiffness characteristics of the manipulator.

19. Reconsider the $3U\underline{P}U$ parallel manipulator shown in Figure 3.34 and described in Problems 3.7 and 4.7.

a. For this manipulator, apply the free-body diagram approach and calculate the actuator torques $\boldsymbol{\tau} = [f_1, f_2, f_3]^T$ required to generate an output wrench of $\mathcal{F} = [f_x, f_y, f_z]^T$ at point P.

b. Show that the relation obtained in part (a) can be written in the form of $\mathcal{F} = J^T \boldsymbol{\tau}$, and verify the obtained Jacobian matrix with the one obtained in Problem 4.7.

c. Consider unit stiffness values for all manipulator actuators, and derive an expression for stiffness matrix of the manipulator.

d. Develop a program in MATLAB with the numerical values considered in Problem 4.7, and calculate the dexterity measure $1/\kappa$ of the manipulator stiffness matrix over the entire workspace of the manipulator. Use four different heave values $z = 0, 0.2, 0.4, 0.6$, and plot the dexterity measures in four different 3D plots. Analyze the dexterity plots and comment on the design of the manipulator.

e. Further develop the program written for part (d), and calculate and plot the maximum singular value, and the two norms of the stiffness matrix over the entire workspace with $z = 0$. Analyze the plots and comment on the stiffness characteristics of the manipulator.

20. Reconsider a $3U\underline{P}S$ parallel manipulator introduced in [47], shown in Figure 3.35 and described in Problems 3.8 and 4.8.

a. For this manipulator, apply the free-body diagram approach and calculate the actuator torques $\boldsymbol{\tau} = [f_1, f_2, f_3]^T$ required to generate an output wrench of $\mathcal{F} = [f_x, n_x, n_y]^T$ at point M.

b. Show that the relation obtained in part (a) can be written in the form of $\mathcal{F} = J^T\boldsymbol{\tau}$, and verify the obtained Jacobian matrix with the one obtained in Problem 4.8.

c. Consider unit stiffness values for all manipulator actuators, and derive an expression for stiffness matrix of the manipulator.

d. Develop a program in MATLAB with the numerical values considered in Problem 4.8, and calculate the dexterity measure $1/\kappa$ of the manipulator stiffness matrix over the entire workspace of the manipulator. Use four different heave values $h = 0, 0.1, 0.2$, and 0.4, and plot the dexterity measures in four different 3D plots. Analyze the dexterity plots and comment on the design of the manipulator.

e. Further, develop the program written for part (d), and calculate and plot the maximum singular value, and the two norms of the stiffness matrix over the entire workspace with $h = 0.4$. Analyze the plots and comment on the stiffness characteristics of the manipulator.

21. Reconsider the three-degrees-of-freedom $3R\underline{P}S$ manipulator represented in Figure 3.36 and described in Problems 3.9 and 4.9.

a. For this manipulator, apply the free-body diagram approach and calculate the actuator torques $\boldsymbol{\tau} = [f_1, f_2, f_3]^T$ required to generate an output wrench of $\mathcal{F} = [f_x, f_y, f_z]^T$ at point P.

b. Show that the relation obtained in part (a) can be written in the form of $\mathcal{F} = J^T\boldsymbol{\tau}$, and verify the obtained Jacobian matrix with the one obtained in Problem 4.9.

c. Consider unit stiffness values for all manipulator actuators, and derive an expression for stiffness matrix of the manipulator.

d. Develop a program in MATLAB with the numerical values considered in Problem 4.9, and calculate the dexterity measure $1/\kappa$ of the manipulator stiffness matrix over the entire workspace of the manipulator. Use four different heave values and plot the dexterity measures in four different 3D plots. Analyze the dexterity plots and comment on the design of the manipulator.

e. Further, develop the program written for part (d), and calculate and plot the maximum singular value, and the two norms of the stiffness matrix over the entire workspace with $z = 0$. Analyze the plots and comment on the stiffness characteristics of the manipulator.

22. Reconsider the three-degrees-of-freedom $3\underline{P}RC$ translational parallel manipulator represented in Figure 3.36 and described in Problems 3.10 and 4.10.
    a. For this manipulator, apply the free-body diagram approach and calculate the actuator torques $\boldsymbol{\tau} = [f_1, f_2, f_3]^T$ required to generate an output wrench of $\mathcal{F} = [f_x, f_y, f_z]^T$ at point P.
    b. Show that the relation obtained in part (a) can be written in the form of $\mathcal{F} = J^T \boldsymbol{\tau}$, and verify the obtained Jacobian matrix with the one obtained in Problem 4.10.
    c. Consider unit stiffness values for all manipulator actuators, and derive an expression for stiffness matrix of the manipulator.
    d. Develop a program in MATLAB with the numerical values considered in Problem 4.10, and calculate the dexterity measure $1/\kappa$ of the manipulator stiffness matrix over the entire workspace of the manipulator. Use four different heave values and plot the dexterity measures in four different 3D plots. Analyze the dexterity plots and comment on the design of the manipulator.
    e. Further, develop the program written for part (d), and calculate and plot the maximum singular value, and the two norms of the stiffness matrix over the entire workspace with $z = 0$. Analyze the plots and comment on the stiffness characteristics of the manipulator.

23. Reconsider the Delta robot represented in Figure 3.39 and described in Problems 3.11 and 4.11.
    a. For this manipulator, apply the free-body diagram approach and calculate the actuator torques $\boldsymbol{\tau} = [\tau_1, \tau_2, \tau_3]^T$ required to generate an output wrench of $\mathcal{F} = [f_x, f_y, f_z]^T$ at point P.
    b. Show that the relation obtained in part (a) can be written in the form of $\mathcal{F} = J^T \boldsymbol{\tau}$ and verify the obtained Jacobian matrix with the one obtained in Problem 4.11.
    c. Consider unit stiffness values for all manipulator actuators, and derive an expression for stiffness matrix of the manipulator.
    d. Develop a program in MATLAB with the numerical values considered in Problem 4.11, and calculate the dexterity measure $1/\kappa$ of the manipulator stiffness matrix over the entire workspace of the manipulator. Use four different heave values and plot the dexterity measures in four different 3D plots. Analyze the dexterity plots and comment on the design of the manipulator.
    e. Further, develop the program written for part (d), and calculate and plot the maximum singular value, and the two norms of the stiffness matrix over the entire workspace with $z = 0$. Analyze the plots and comment on the stiffness characteristics of the manipulator.

5

Dynamics

5.1 Introduction

In contrast to the open-chain serial manipulators, dynamic analysis of parallel manipulators presents an inherent complexity due to their closed-loop structure. Nevertheless, dynamic modeling is quite important for control, particularly for parallel manipulators used in applications where precise positioning and desirable dynamic performance are the prime requirements. Several approaches have been proposed for the dynamic analysis of parallel manipulators. Traditional Newton–Euler formulation is used for dynamic analysis of general parallel manipulators [60], and also for the Stewart–Gough platform [33]. In this formulation, the equations of motion for each limb and the moving platform must be derived, which inevitably leads to a large number of equations and less computational efficiency. On the other hand, all the reaction forces can be computed, which is very useful in the design of a parallel manipulator.

The Lagrangian formulation eliminates all the unwanted reaction forces at the outset, and therefore, it is quite efficient [142]. However, because of the constraints imposed by the closed-loop structure of the parallel manipulator, deriving explicit equations of motion in terms of a set of independent generalized coordinates becomes a prohibitive task [109]. A third approach is to use the principle of virtual work, in which the computation of the constraint forces are bypassed [186]. In this method, inertial forces and moments are computed using linear and angular accelerations of the bodies. Then, the whole manipulator is considered to be in static equilibrium by using the d'Alembert's principle, and the principle of virtual work is applied to derive the input forces or torques [186]. Since constraint forces and moments do not need to be computed for the purpose of dynamics simulations and control, this approach leads to faster computational algorithms for a parallel manipulator [128].

Different objectives require different forms of formulations, and among these methods, there are three key issues pursued to derive dynamic formulation of parallel manipulators

- Calculation of internal forces either active or passive for the design process of the manipulator [33,91].
- Study on dynamical properties of the manipulator for controller design [180,193].
- Utilization of dynamic specifications in an inverse dynamics controller (IDC) or any model-based control topology, which usually requires online computations [148,187].

As explained hereinbefore, and widely reported in the literature, the first item is categorized as the main advantage of the Newton–Euler formulation, while the second and third items are reported as the benefits of using the Lagrange or virtual work approaches. By using traditional Newton–Euler formulation all the internal forces can be computed, which is very useful for the design of the manipulator. On the other hand, in this formulation the equation of motion for each limb and the moving platform must be derived, which inevitably leads to a large number of equations and less computational efficiency.

Along with complexity, closed-loop constrains in parallel manipulators bring up other issues. A more preferable task in attaining the dynamic formulation is to select a set of independent generalized coordinates, and then derive the dynamic equations in terms of these generalized coordinates and their time derivatives. These parameters are usually the coordinates of the moving platform. In order to achieve this kind of formulation, the internal forces and other passive joint variables should be eliminated. In addition, kinematic variables of internal parts should be derived as a function of these desired coordinates. If this elimination process can be perfectly done, the dynamic equations can be written in an *explicit* form of

$$M(\mathcal{X})\ddot{\mathcal{X}} + C(\mathcal{X}, \dot{\mathcal{X}})\dot{\mathcal{X}} + G(\mathcal{X}) = \mathcal{F}, \tag{5.1}$$

in which $\mathcal{X}$ is a vector of the generalized coordinates, $M(\mathcal{X})$ denotes the system mass matrix, $C(\mathcal{X}, \dot{\mathcal{X}})$ denotes the Coriolis and centrifugal matrix, $G(\mathcal{X})$ denotes the gravity vector, and $\mathcal{F}$ denotes the generalized force.

Deriving dynamic formulation in the form of Equation 5.1 is well studied for serial manipulators [113,168]. However, for parallel manipulators, because of the constraints imposed by the closed-loop nature of the manipulator, deriving explicit equations of motion in terms of independent generalized coordinates becomes a prohibitive task [109]. In terms of taking advantage of dynamic formulation in the design and implementation of control systems, deriving the equations of motion in this special form is usually very assisting. Since the Lagrange formulation eliminates all the unwanted reaction forces at the outset, this form is frequently considered to be attainable by the Lagrange method [113,168]. Nevertheless, by means of Lie algebra, Reference 153 proposed a procedure that utilizes the recursive Newton–Euler method to attain this explicit form of equations for general multibody systems.

In contrast to the dynamics of open-chain serial manipulators, although some excellent results have been accomplished for parallel manipulators, there is no generally approved procedure given to derive the dynamic equations of a general parallel manipulator. This is because of the inherent complexity of closed-loop structures and their kinematic constraints. There are a number of methods reported in the literature to attain dynamic equations for a general parallel manipulator. Perusing those general solutions for parallel manipulators, although the methods and formulations are totaly varied, some common grounds can be detected among them. Almost all of these methods use disconnecting the limbs from the moving platform, deriving the dynamics of each limb separately, and then combining the local models and coordinates to get the final dynamic formulation for the whole manipulator. As another common ground, it is seen that local coordinates are considered for each limb and for the moving platform. For the limbs, position vectors of the moving attachment points may be used to represent the generalized coordinates. For the moving platform, position of its center of mass and its orientation angle is a good nominee for generalized coordinate. A dynamic model of each part is derived locally based on the selected generalized coordinates, and a coordinate transformation scheme is used in some papers to transform the local models to global ones, based on the generalized coordinates of the whole manipulator. Furthermore, these dynamic models are combined to attain a global dynamic model of the whole manipulator. These ideas are followed and extended in this book in the derivation of the closed-form dynamic formulation of the case studies examined in this chapter.

In this area, Reference 60 has introduced an approach based on the Newton–Euler method for inverse dynamics of parallel manipulators. Furthermore, Reference 51 has used screw theory along with the principal of virtual work to propose a method for dynamic analysis of such manipulators. Khalil and Ibrahim [97] have proposed a closed-form method based on different Jacobian matrices of the manipulator and joint variables in dynamic formulation. Dasgupta and Choudhury [32] have established a general procedure based on the Newton–Euler method to derive inverse dynamic formulation for parallel manipulators, and converting them into a closed-form formulation.

Abdellatif and Heimann [1] have introduced another set of formulations to derive a dynamic model of parallel manipulators based on the Lagrange method. It is stated that sensible selection of local generalized coordinates for each limb may significantly affect the computational efficiency and simplicity of the eventuated dynamic model. Although in this Lagrange formulation the necessity of evaluating Lagrange multipliers for closed-chain constrains are avoided, componentwise derivation is still used to derive dynamic equations of the limbs and hence result in a very complex model for a manipulator such as the Stewart–Gough platform. In this book, a matrix-based manipulation technique is introduced by which the componentwise manipulation is avoided and this method considerably reduces the complexity of the equations while keeping them manageable without need to any symbolic software packages.

In this chapter, first a review on the basic definitions and algorithms for dynamic analysis of a rigid body is presented, and then the aforementioned three methods for dynamic analysis are presented and examined by a few case studies.

5.2 Dynamics of Rigid Bodies: A Review

In this section, basic definitions and important laws in the dynamic analysis of rigid bodies is reviewed. The first step in dynamic analysis is to derive angular acceleration of a rigid body, and furthermore, to calculate linear acceleration of a point P of the rigid body. This analysis is directly used in the Newton–Euler formulation and the principle of virtual work approach to derive dynamic equations for a parallel manipulator.

5.2.1 Acceleration of Rigid Bodies

In this section, differential kinematics of a rigid body at the acceleration level has been studied in detail. By acceleration analysis, we study the variations of linear velocity of a point and angular velocity of a rigid body with respect to time. Basically, direct differentiation of these vectors with respect to time in a fixed frame leads to linear velocity of a point and angular velocity of a rigid body, respectively. Note that, to determine the absolute linear velocity of a point, the derivative must be calculated relative to a fixed frame. Differentiation of a velocity vector with respect to a moving frame results into relative acceleration. In the study of robotic manipulators, usually multiple moving frames are defined to carefully determine the motion of the moving platform. Therefore, it is necessary to define the required arithmetics to transform the relative accelerations into absolute ones. This issue is carefully studied in the following subsections.

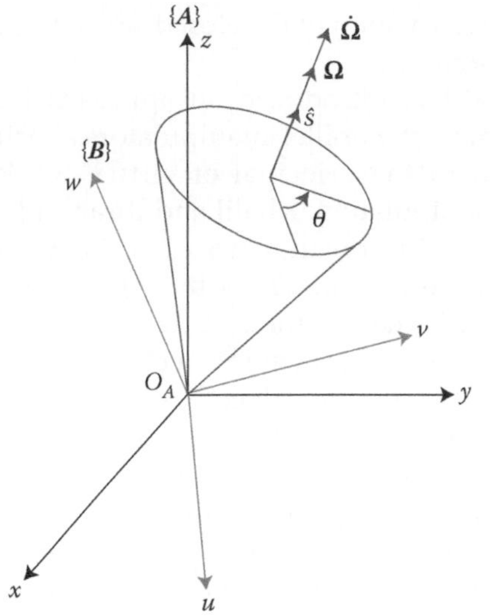

FIGURE 5.1
Angular acceleration of a rigid body in a pure rotational motion.

5.2.1.1 Angular Acceleration of a Rigid Body

To define angular acceleration of a rigid body, consider a moving frame $\{\boldsymbol{B}\}$ attached to the rigid body, and the motion analyzed with respect to a fixed frame as shown in Figure 5.1. Angular acceleration is an attribute of a rigid body and describes the variation of angular velocity of frame $\{\boldsymbol{B}\}$, attached to the rigid body, with respect to time. *Angular acceleration vector*, denoted by the symbol $\dot{\boldsymbol{\Omega}}$, describes the instantaneous change of the angular velocity of frame $\{\boldsymbol{B}\}$, denoted by $\boldsymbol{\Omega}$, with respect to the fixed frame $\{\boldsymbol{A}\}$. Consider the definition of angular velocity given in Equation 4.1, angular acceleration is derived as

$$\begin{aligned}\dot{\boldsymbol{\Omega}} = \frac{\mathrm{d}\boldsymbol{\Omega}}{\mathrm{d}t} &= \ddot{\theta}\hat{\mathbf{s}} + \dot{\theta}\dot{\hat{\mathbf{s}}} \\ &= \ddot{\theta}\hat{\mathbf{s}} + \dot{\theta}(\boldsymbol{\Omega} \times \mathbf{s}) \\ &= \ddot{\theta}\hat{\mathbf{s}}. \end{aligned} \tag{5.2}$$

In Equation 5.2, rotation of the rigid body is represented by screw parameters, $\{\theta, \hat{\mathbf{s}}\}$. Furthermore, $\ddot{\theta}$ represents the magnitude of angular acceleration, and as seen in Figure 5.1, since $\boldsymbol{\Omega}$ is along $\hat{\mathbf{s}}$, their cross product is zero $\boldsymbol{\Omega} \times \mathbf{s} = \mathbf{0}$. Therefore, angular acceleration of a rigid body is also along the screw axis $\hat{\mathbf{s}}$, as shown in Figure 5.1.

5.2.1.2 Linear Acceleration of a Point

Linear acceleration of a point P can be easily determined by time derivative of the velocity vector $\boldsymbol{v}_p$ of that point with respect to a fixed frame.

$$\boldsymbol{a}_p = \dot{\boldsymbol{v}}_p = \left(\frac{\mathrm{d}\boldsymbol{v}_p}{\mathrm{d}t}\right)_{\text{fix}}. \tag{5.3}$$

Note that this is correct only if the derivative is taken with respect to a fixed frame, as denoted by $(\mathrm{d}(\cdot)/\mathrm{d}t)_{\text{fix}}$ in Equation 5.3. If variation of linear velocity is determined with respect to a moving frame, the *relative* acceleration is obtained, which is denoted by the following equation

$$\boldsymbol{a}_{\text{rel}} = \left(\frac{\partial \boldsymbol{v}_p}{\partial t}\right)_{\text{mov}}, \tag{5.4}$$

in which $\boldsymbol{a}_{\text{rel}}$ denotes relative acceleration, and $(\partial(\cdot)/\partial t)_{\text{mov}}$ denotes differentiation with respect to a moving frame. Relation between absolute derivative of any vector to its relative derivative is given in Equations 4.13 and 4.14.

Now consider general motion of a rigid body shown in Figure 4.2, in which a moving frame $\{\boldsymbol{B}\}$ is attached to a rigid body and the problem is to find absolute acceleration of point P with respect to the fixed frame $\{\boldsymbol{A}\}$. The rigid body performs a general motion, which is a combination of a translation, denoted by the velocity vector ${}^{A}\boldsymbol{v}_{O_B}$, and an instantaneous angular rotation denoted by $\boldsymbol{\Omega}$ as shown in Figure 4.2. To determine acceleration of point P, start with the relation between absolute and relative velocities of point P as follows. Rewrite Equation 4.20 as

$${}^{A}\boldsymbol{v}_p = {}^{A}\boldsymbol{v}_{O_B} + {}^{A}\boldsymbol{R}_B\,{}^{B}\boldsymbol{v}_p + {}^{A}\boldsymbol{\Omega}^{\times}\,{}^{A}\boldsymbol{R}_B\,{}^{B}\boldsymbol{P}. \tag{5.5}$$

In order to derive acceleration of point $\boldsymbol{P}$, differentiate both sides of Equation 5.5 with respect to time.

$${}^{A}\dot{\boldsymbol{v}}_p = {}^{A}\dot{\boldsymbol{v}}_{O_B} + \frac{\mathrm{d}}{\mathrm{d}t}\left({}^{A}\boldsymbol{R}_B\,{}^{B}\boldsymbol{v}_p\right) + \frac{\mathrm{d}}{\mathrm{d}t}\left({}^{A}\boldsymbol{\Omega}^{\times}\,{}^{A}\boldsymbol{R}_B\,{}^{B}\boldsymbol{P}\right) \tag{5.6}$$

$$\begin{aligned}{}^{A}\boldsymbol{a}_p &= {}^{A}\boldsymbol{a}_{O_B} + \left({}^{A}\boldsymbol{R}_B\,{}^{B}\boldsymbol{a}_p + {}^{A}\boldsymbol{\Omega}^{\times}\,{}^{A}\boldsymbol{R}_B\,{}^{B}\boldsymbol{v}_P\right)\\ &\quad + \left({}^{A}\dot{\boldsymbol{\Omega}}^{\times}\,{}^{A}\boldsymbol{R}_B\,{}^{B}\boldsymbol{P} + {}^{A}\boldsymbol{\Omega}^{\times}\left({}^{A}\boldsymbol{\Omega}^{\times}\,{}^{A}\boldsymbol{R}_B\,{}^{B}\boldsymbol{P}\right) + {}^{A}\boldsymbol{\Omega}^{\times}\,{}^{A}\boldsymbol{R}_B\,{}^{B}\boldsymbol{v}_P\right).\end{aligned} \tag{5.7}$$

Hence, acceleration of a point $\boldsymbol{P}$ in its most general form is derived from

$$\begin{aligned}{}^{A}\boldsymbol{a}_p &= {}^{A}\boldsymbol{a}_{O_B} + {}^{A}\boldsymbol{R}_B\,{}^{B}\boldsymbol{a}_p\\ &\quad + {}^{A}\dot{\boldsymbol{\Omega}}^{\times}\,{}^{A}\boldsymbol{R}_B\,{}^{B}\boldsymbol{P} + {}^{A}\boldsymbol{\Omega}^{\times}\left({}^{A}\boldsymbol{\Omega}^{\times}\,{}^{A}\boldsymbol{R}_B\,{}^{B}\boldsymbol{P}\right) + 2\,{}^{A}\boldsymbol{\Omega}^{\times}\,{}^{A}\boldsymbol{R}_B\,{}^{B}\boldsymbol{v}_P\end{aligned} \tag{5.8}$$

In this relation, ${}^{A}\boldsymbol{a}_{O_B}$ represents linear acceleration of the origin of frame $\{\boldsymbol{B}\}$, ${}^{A}\boldsymbol{R}_B\,{}^{B}\boldsymbol{a}_p$ represents relative acceleration of point $\boldsymbol{P}$ with respect to the origin of frame $\{\boldsymbol{B}\}$, ${}^{A}\dot{\boldsymbol{\Omega}}^{\times}\,{}^{A}\boldsymbol{R}_B\,{}^{B}\boldsymbol{P}$ represents the contribution of angular acceleration of frame $\{\boldsymbol{B}\}$ to the linear acceleration of point $\boldsymbol{P}$, and the last two terms are the *centrifugal* and *Coriolis* acceleration terms, respectively. For the case where P is a point embedded in the rigid body, relative velocity ${}^{B}\boldsymbol{v}_p$ and relative acceleration ${}^{B}\boldsymbol{a}_p$ are equal to zero, and Equation 5.8 simplifies to

$${}^{A}\boldsymbol{a}_p = {}^{A}\boldsymbol{a}_{O_B} + {}^{A}\dot{\boldsymbol{\Omega}}^{\times}\,{}^{A}\boldsymbol{R}_B\,{}^{B}\boldsymbol{P} + {}^{A}\boldsymbol{\Omega}^{\times}\left({}^{A}\boldsymbol{\Omega}^{\times}\,{}^{A}\boldsymbol{R}_B\,{}^{B}\boldsymbol{P}\right). \tag{5.9}$$

5.2.2 Mass Properties

In this section, the properties of mass, namely center of mass, moments of inertia and its characteristics, and the required transformations are described.

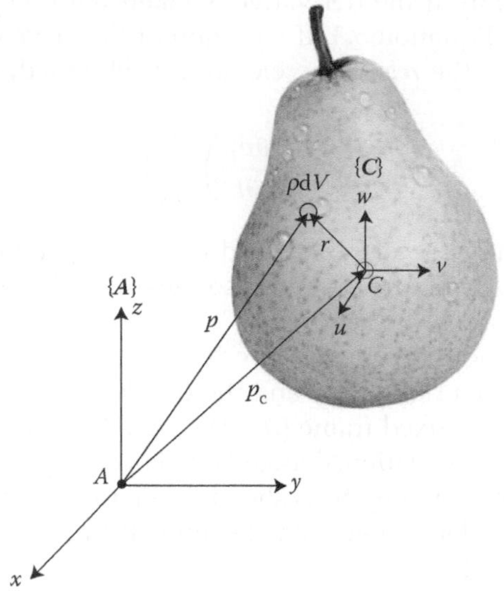

FIGURE 5.2
Mass properties of a rigid body.

5.2.2.1 *Center of Mass*

We are all familiar with the concept of *mass*, from our early studies in physics. Mass can be defined as the property of a body that causes it to have weight in a gravitational field. Referring to Figure 5.2, the distribution of mass in a rigid body forms the size, shape, and the weight of it. Consider a reference frame $\{A\}$, in which the mass distribution of a material body is measured, and let $\boldsymbol{p}$ denote the position vector of a differential mass $\rho \mathrm{d}V$ with respect to a reference frame. The *center of mass* of a rigid body is defined as the point C, the position of which satisfies the following condition:

$$\boldsymbol{p}_c = \frac{1}{m} \int_V \boldsymbol{p}\, \rho \mathrm{d}V, \tag{5.10}$$

in which the mass of the material body $\{\boldsymbol{B}\}$ with density ρ and volume V is defined as

$$m = \int_V \rho \mathrm{d}V. \tag{5.11}$$

5.2.2.2 *Moments of Inertia*

As opposed to the mass, which introduces inertia to linear accelerations, moment of inertia is the property of mass which introduces inertia to angular accelerations. Basically, for rotational motion, the distribution of mass with respect to the axis of rotation introduces resistance to the angular acceleration. Consider the material body illustrated in Figure 5.2 having a pure rotational motion about the origin of reference frame $\{A\}$. Moments of inertia matrix $\boldsymbol{I}$ about A is defined by the second moment of the mass with respect to a reference frame of rotation as

$$^{A}I = \begin{bmatrix} I_{xx} & I_{xy} & I_{xz} \\ I_{yx} & I_{yy} & I_{yz} \\ I_{zx} & I_{zy} & I_{zz} \end{bmatrix}, \tag{5.12}$$

in which

$$I_{xx} = \int_V (y^2 + z^2)\rho \mathrm{d}V, \quad I_{xy} = I_{yx} = -\int_V xy\, \rho \mathrm{d}V,$$

$$I_{yy} = \int_V (x^2 + z^2)\rho \mathrm{d}V, \quad I_{yz} = I_{zy} = -\int_V yz\, \rho \mathrm{d}V,$$

$$I_{zz} = \int_V (x^2 + y^2)\rho \mathrm{d}V, \quad I_{xz} = I_{zx} = -\int_V xz\, \rho \mathrm{d}V.$$

5.2.2.3 Principal Axes

As seen in Equation 5.12, the inertia matrix elements are a function of mass distribution of the rigid body with respect to the frame $\{A\}$. Hence, it is possible to find orientations of frame $\{A\}$ in which the product-of-inertia terms vanish and inertia matrix becomes diagonal

$$^{A}I = \begin{bmatrix} I_{xx} & 0 & 0 \\ 0 & I_{yy} & 0 \\ 0 & 0 & I_{zz} \end{bmatrix}. \tag{5.13}$$

Such axes are called the principal axes of inertia, and diagonal terms I_{xx}, I_{yy}, and I_{zz} are called the principal moments of inertia, which represent the maximum, minimum, and intermediate values of the moments of inertia, respectively, for a particular chosen origin A. It can be shown that the principal moments of inertia are invariant parameters, and can be determined from the eigenvalues of inertia matrix in any configuration of the reference frame $\{A\}$, while the principal axes are the corresponding eigenvectors of inertia matrix. From a physical point of view, principal axes correspond to the axes of symmetry for the rigid body.

5.2.2.4 Inertia Matrix Transformations

Calculation of moment of inertia from the Definition 5.12 is quite cumbersome and therefore, usually CAD software outputs or standard tables are used to determine these values [130]. The center of mass and moment of inertia of many homogeneous objects are given in such tables for a set of predetermined frames. These frames usually pass through the center of mass or the corners of the objects. The inertia matrix changes under pure translation of the reference frame, and the *parallel axis theorem* is used to relate the inertia matrix with respect to the center of mass to any other arbitrary point [130].

Consider frame $\{C\}$ attached to the center of mass of a rigid body as shown in Figure 5.2, to be parallel to the reference frame $\{A\}$, and let $\boldsymbol{p}_c = [x_c, y_c, z_c]^T$ denote the vector of the position of the center of mass with respect to frame $\{A\}$. The relation between the inertia matrix about A and that about C is given by the following relation:

$$^{A}I = {}^{C}I + m\left(\boldsymbol{p}_c^T \boldsymbol{p}_c I_{3\times 3} - \boldsymbol{p}_c \boldsymbol{p}_c^T\right), \tag{5.14}$$

in which m denotes the mass of rigid body, and $I_{3\times3}$ denotes the 3×3 identity matrix. This relation can be written componentwise as follows:

$$\begin{aligned} {}^{A}I_{xx} &= {}^{C}I_{xx} + m\,(y_c^2 + z_c^2), & {}^{A}I_{xy} &= {}^{C}I_{xy} + m\ x_c y_c, \\ {}^{A}I_{yy} &= {}^{C}I_{yy} + m\,(x_c^2 + z_c^2), & {}^{A}I_{yz} &= {}^{C}I_{yz} + m\ y_c z_c, \\ {}^{A}I_{zz} &= {}^{C}I_{zz} + m\,(x_c^2 + y_c^2), & {}^{A}I_{xz} &= {}^{C}I_{xz} + m\ x_c z_c. \end{aligned} \tag{5.15}$$

On the other hand, if the reference frame has *pure rotation* with respect to the frame attached to the center of mass, rotation matrices can be used by the following relation to determine the required transformation

$${}^{A}I = {}^{A}\boldsymbol{R}_C\,{}^{C}I\,{}^{A}\boldsymbol{R}_C^T. \tag{5.16}$$

5.2.3 Momentum and Kinetic Energy

In this section, linear and angular momenta and kinetic energy of a rigid body are defined.

5.2.3.1 Linear Momentum

Linear momentum of a material body, shown in Figure 5.2, with respect to a reference frame $\{A\}$ is defined as

$${}^{A}\boldsymbol{G} = \int_V \frac{\mathrm{d}\boldsymbol{p}}{\mathrm{d}t}\,\rho \mathrm{d}V. \tag{5.17}$$

The importance of the center of mass can be seen in the following simplification of Equation 5.17. As shown in Figure 5.2, for any mass element $\rho \mathrm{d}V$ the position vector $\boldsymbol{p}$ can be written as

$$\boldsymbol{p} = \boldsymbol{p}_c + \boldsymbol{r}. \tag{5.18}$$

Substitute Equation 5.18 into Equation 5.17

$${}^{A}\boldsymbol{G} = \int_V \frac{\mathrm{d}\boldsymbol{p}_c}{\mathrm{d}t}\,\rho \mathrm{d}V + \frac{\mathrm{d}}{\mathrm{d}t}\left(\int_V \boldsymbol{r}\,\rho \mathrm{d}V\right). \tag{5.19}$$

Consider the definition of center of mass for a rigid body given in Equation 5.10 written with respect to frame $\{C\}$. In this frame, the center of mass is located at the origin, that is, $m \cdot {}^{C}\boldsymbol{p}_c = 0$, therefore

$$\int_V {}^{C}\boldsymbol{p}_c\,\rho \mathrm{d}V = \int_V \boldsymbol{r}\,\rho \mathrm{d}V = 0. \tag{5.20}$$

Hence, the second term in the right-hand side of Equation 5.19 vanishes, and the definition of linear momentum of a rigid body is reduced to the following equation:

$${}^{A}\boldsymbol{G} = \int_V \frac{\mathrm{d}\boldsymbol{p}_c}{\mathrm{d}t}\,\rho \mathrm{d}V = \frac{\mathrm{d}\boldsymbol{p}_c}{\mathrm{d}t}\int_V \rho \mathrm{d}V. \tag{5.21}$$

Therefore

$$^{A}G = m \cdot {}^{A}v_{c}, \tag{5.22}$$

in which $^{A}v_{c}$ denotes the velocity of the center of mass with respect to the frame $\{A\}$. This result implies that the total linear momentum of differential masses is equal to the linear momentum of a point mass m located at the center of mass. Equation 5.22 highlights the importance of the center of mass in dynamic formulation of rigid bodies.

5.2.3.2 Angular Momentum

Consider the material body represented in Figure 5.2. Angular momentum of the differential masses ρdV about a reference point A, expressed in the reference frame $\{A\}$ is defined as

$$^{A}\boldsymbol{H} = \int_{V} \left(\boldsymbol{p} \times \frac{d\boldsymbol{p}}{dt} \right) \rho dV, \tag{5.23}$$

in which $d\boldsymbol{p}/dt$ denotes the velocity of differential mass with respect to the reference frame $\{A\}$. Substitute Equation 5.18 into Equation 5.23

$$^{A}\boldsymbol{H} = (\boldsymbol{p}_c \times \boldsymbol{v}_c) \int_{V} \rho dV + \int_{V} \left(\boldsymbol{r} \times \frac{d\boldsymbol{r}}{dt} \right) \rho dV$$

$$+ \boldsymbol{p}_c \times \left(\frac{d}{dt} \int_{V} \boldsymbol{r} \rho dV \right) + \left(\int_{V} \boldsymbol{r} \rho dV \right) \times \boldsymbol{v}_c. \tag{5.24}$$

From the definition of the center of mass and referring to Equation 5.20, the last two terms in Equation 5.24 vanish. Hence

$$^{A}\boldsymbol{H} = \boldsymbol{p}_c \times m \boldsymbol{v}_c + \int_{V} \boldsymbol{r} \times (\boldsymbol{\Omega} \times \boldsymbol{r}) \rho dV. \tag{5.25}$$

Therefore, angular momentum of the rigid body about point A is reduced to

$$^{A}\boldsymbol{H} = \boldsymbol{p}_c \times \boldsymbol{G}_c + {}^{C}\boldsymbol{H}, \tag{5.26}$$

in which

$$^{C}\boldsymbol{H} = \int_{V} \boldsymbol{r} \times (\boldsymbol{\Omega} \times \boldsymbol{r}) \rho dV = {}^{C}\boldsymbol{I} \cdot \boldsymbol{\Omega}. \tag{5.27}$$

Note that by writing vector $\boldsymbol{r} = [r_x, r_y, r_z]^T$ componentwise, and using the definition of moment of inertia matrix given in Equation 5.12, about center of mass C, the integrand in Equation 5.27 will be simplified to $^{C}\boldsymbol{I} \cdot \boldsymbol{\Omega}$.

As shown in Figure 5.3, Equation 5.26 reveals an important fact that angular momentum of a rigid body about a point A can be written as $\boldsymbol{p}_c \times \boldsymbol{G}_c$, which is the contribution of linear momentum of the rigid body about point A, and $^{C}\boldsymbol{H}$ which is the angular momentum of the rigid body about the center of mass. This equation also highlights the importance of the center of mass in the dynamic analysis of rigid bodies, and illustrates that if the center of mass is taken as the reference point, the relation describing angular momentum of a rigid body (Equation 5.27) is very analogous to that of linear momentum (Equation 5.22).

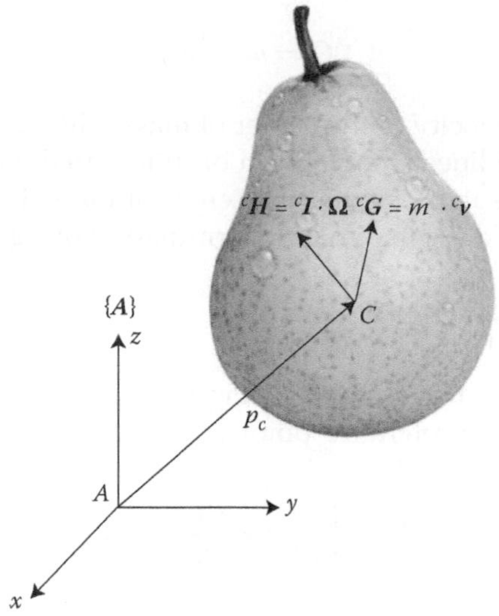

FIGURE 5.3
The components of the angular momentum of a rigid body about A.

For the sake of simplicity, in this book, unless otherwise is specified, the center of mass is taken as the reference point for calculation of angular momentum.

5.2.3.3 Kinetic Energy

Kinetic energy of a rigid body is defined as

$$K = \frac{1}{2}\int_V v \cdot v \rho \mathrm{d}V. \tag{5.28}$$

Similar to what followed in the previous section, velocity of a differential mass $\rho\,\mathrm{d}V$ can be represented by linear velocity of the center of mass and angular velocity of the rigid body as

$$v = v_p + \boldsymbol{\Omega} \times \boldsymbol{r}. \tag{5.29}$$

Substitute Equation 5.29 into Equation 5.28, and use the property of the center of mass (Equation 5.20). By some manipulation the kinetic energy of the rigid body may be obtained by

$$k = \frac{1}{2} v_C \cdot G_c + \frac{1}{2}\boldsymbol{\Omega} \cdot {}^C\boldsymbol{H}, \tag{5.30}$$

in which G_c is linear momentum of the rigid body given in Equation 5.22, and ${}^C\boldsymbol{H}$ is the angular momentum of the rigid body about the center of mass, given in Equation 5.27. This equation reveals that kinetic energy of a moving body can be represented as the kinetic energy of a point mass located at the center of mass, in addition to the kinetic energy of the body rotating about the center of mass.

5.2.4 Newton–Euler Laws

In this section, Newton and Euler laws are reviewed. These laws can be written for three different cases in which angular motion is about a fixed point in space, or is represented about the center of mass, or about an arbitrary moving point in space. For the sake of brevity, and as noted in last few sections, we examine only the case here, in which all rotations are represented about the center of mass. Without loss of generality this case can be used for the other two cases, although, in some cases using other representation may simplify the formulations.

Consider a rigid body under general motion, that is, a combination of translation and rotation. The Newton's law relates the change of linear momentum of the rigid body to the resulting external forces applied to it.

$$\sum \boldsymbol{f}_{\text{ext}} = \frac{\mathrm{d}\boldsymbol{G}_c}{\mathrm{d}t}. \tag{5.31}$$

For the case of constant mass rigid body, this law is reduced to

$$\sum \boldsymbol{f}_{\text{ext}} = m\,\frac{\mathrm{d}\boldsymbol{v}_c}{\mathrm{d}t} = m\,\boldsymbol{a}_c, \tag{5.32}$$

in which $\boldsymbol{a}_c$ is linear acceleration of the center of mass. Note that the above equations are vector equations that may be written componentwise. Furthermore, the time derivatives of vectors are with respect to a fixed frame and formulations relating the relative accelerations to the absolute one, namely Equations 5.8 and 5.9, may be used for simplification. Finally, note that $\Sigma\boldsymbol{f}_{\text{ext}}$ includes *all* external forces, and hence the free-body diagram of forces applied to individual bodies may be used to identify them.

The Euler law relates the change of angular momentum of a rigid body about the center of mass, to the summation of all external moments applied to the rigid body about center of mass

$$\sum {}^{c}\boldsymbol{n}_{\text{ext}} = \frac{\mathrm{d}}{\mathrm{d}t}({}^{c}\boldsymbol{H}). \tag{5.33}$$

For the case of constant mass rigid body, this law is reduced to

$$\sum {}^{c}\boldsymbol{n}_{\text{ext}} = \frac{\mathrm{d}}{\mathrm{d}t}({}^{c}\boldsymbol{I}\,\boldsymbol{\Omega}) = {}^{c}\boldsymbol{I}\,\dot{\boldsymbol{\Omega}} + \boldsymbol{\Omega} \times ({}^{c}\boldsymbol{I}\,\boldsymbol{\Omega}), \tag{5.34}$$

in which $\Sigma{}^{c}\boldsymbol{n}_{\text{ext}}$ is the summation of *all* external moments applied to the rigid body about center of mass, ${}^{c}\boldsymbol{I}$ is the moment of inertia about the center of mass, and $\boldsymbol{\Omega}$ and $\dot{\boldsymbol{\Omega}}$ are angular velocity and angular acceleration of the rigid body, respectively. For a special case, in which frame $\{C\}$ coincides with the principal axes of the rigid body, Equation 5.34 can be written componentwise as

$$\begin{aligned}
\sum {}^{c}n_x &= I_{xx}\dot{\Omega}_x - \Omega_y\Omega_z(I_{yy} - I_{zz}),\\
\sum {}^{c}n_y &= I_{yy}\dot{\Omega}_y - \Omega_z\Omega_x(I_{zz} - I_{xx}),\\
\sum {}^{c}n_z &= I_{zz}\dot{\Omega}_z - \Omega_x\Omega_y(I_{xx} - I_{yy}),
\end{aligned} \tag{5.35}$$

where I_{xx}, I_{yy}, and I_{zz} are the principal moments of inertia about the center of mass.

5.2.5 Variable-Mass Systems

The Newton–Euler laws given in Equations 5.31 and 5.33 are derived for a constant mass system. These equations can be extended for the case in which the masses of the system components are varying. The general treatment for such a case is more complex than what is given in Equations 5.31 and 5.33, and has been given in this section. Note that for special cases, in which the system gains or loses mass by overtaking or expelling a stream of mass at a velocity $v_0 \neq v_c$, the resulting Newton's second law for such systems becomes identical to Equation 5.33 as follows:

$$\sum f_{\text{ext}} = \frac{\mathrm{d}G_c}{\mathrm{d}t} = \frac{\mathrm{d}m\, v_c}{\mathrm{d}t} = m\,\dot{v}_c + \dot{m}\,v_c. \tag{5.36}$$

In order to derive this equation, it is necessary to consider the force required to accelerate or decelerate the stream of mass overtaken into, or expelled from the system, and therefore, direct differentiation from the linear momentum is not applicable. Furthermore, this formulation is correct *only* when the body picks up mass initially at rest or when it expels mass at a zero absolute velocity [129]. Equation 5.36 may also be used by direct differentiation of the momentum, provided a proper system of constant total mass is chosen.

For the general case of variable mass a control volume treatment for the system is advised. Consider the system in Figure 5.4 defined by a closed control volume whose shape and position may vary with time, and the quantity and distribution of mass within the system are also time-varying. Furthermore, it is assumed that a mass stream continuously enters the system through section E. The total mass of the system at time t is m, while the mass entering the system is denoted by Δm. Instantaneous position of the center of mass G of the varying mass system is denoted by the position vector $\bar{\boldsymbol{r}}$, and a representative particle of mass m_i of the system is located from O and G by the respective position vectors $\boldsymbol{r}_i$ and $\boldsymbol{\rho}_i$. The position of Δm measured from a fixed frame $\{A\}$ is denoted by vector $\boldsymbol{r}_0$, and from the center of mass G by vector $\boldsymbol{\rho}_0$. The position of the mass entrance section E from these two reference points are given by $\boldsymbol{r}_e$ and $\boldsymbol{\rho}_e$, respectively. It should be noted that although $\boldsymbol{r}_0 = \boldsymbol{r}_e$ and $\boldsymbol{\rho}_e = \boldsymbol{\rho}_0$, their time derivatives represent different velocities, and therefore, $\dot{\boldsymbol{r}}_0 \neq \dot{\boldsymbol{r}}_e$ and $\dot{\boldsymbol{\rho}}_e \neq \dot{\boldsymbol{\rho}}_0$.

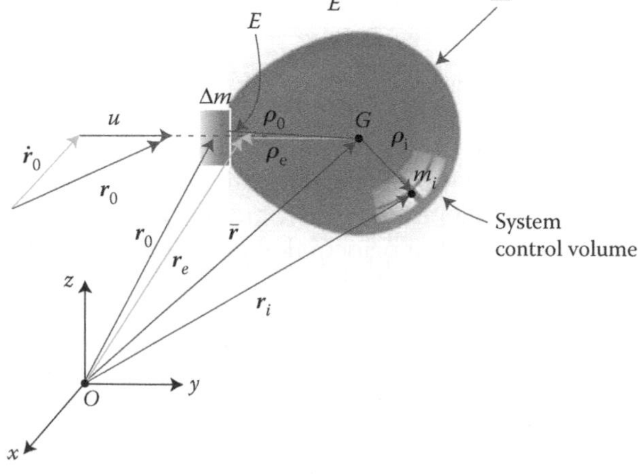

FIGURE 5.4
A control volume treatment of the variable-mass system.

The linear momentum of the system at time t can be derived by the sum of linear momentum of all particles within the control volume

$$G = \sum m_i \dot{\boldsymbol{r}}_i. \tag{5.37}$$

At time $t + \Delta t$ the linear momentum of the added mass is also included in the linear momentum of the system

$$G + \Delta G = \sum m_i (\dot{\boldsymbol{r}}_i + \Delta \dot{\boldsymbol{r}}_i) + \Delta m \, (\dot{\boldsymbol{r}}_0 + \Delta \dot{\boldsymbol{r}}_0). \tag{5.38}$$

Subtract the above two equations, divide by Δt, take the limit by $\Delta t \to 0$, and neglect higher-order terms to simplify the result into

$$\dot{G} = \sum m_i \ddot{\boldsymbol{r}}_i + \dot{m} \, \dot{\boldsymbol{r}}_0. \tag{5.39}$$

From the second Newton's law, $\Sigma \boldsymbol{F}_{\text{ext}} = \Sigma m_i \ddot{\boldsymbol{r}}_i$, hence

$$\sum \boldsymbol{F}_{\text{ext}} = \dot{G} - \dot{m} \dot{\boldsymbol{r}}_0. \tag{5.40}$$

Equation 5.40 is the generalization of the force–momentum relation (Equation 5.31) for a time-varying mass. This equation states that the resultant forces acting on a time-varying mass system equals the rate of change of linear momentum of the system minus the rate at which linear momentum is added to the system by incoming mass.

Although Equation 5.40 is a general formulation for varying mass systems, it is more convenient to express $\boldsymbol{G}$ in terms of some measurable quantities. Similar to the derivation of Equation 5.39, it can be shown that

$$m\bar{\boldsymbol{r}} = \sum m_i \boldsymbol{r}_i,$$

$$\frac{\mathrm{d}}{\mathrm{d}t}(m\bar{\boldsymbol{r}}) = \sum m_i \dot{\boldsymbol{r}}_i + \dot{m}\boldsymbol{r}_0.$$

Hence

$$G = \frac{\mathrm{d}}{\mathrm{d}t}(m\bar{\boldsymbol{r}}) - \dot{m}\boldsymbol{r}_0. \tag{5.41}$$

Substitute $\boldsymbol{r}_0 - \bar{\boldsymbol{r}} = \boldsymbol{\rho}_0 = \boldsymbol{\rho}_e$, into Equation 5.41

$$G = m\dot{\bar{\boldsymbol{r}}} - \dot{m}\boldsymbol{\rho}_e. \tag{5.42}$$

From this equation, it can be concluded that the linear momentum of a time-varying mass is a function of the location where mass is added to the system. With the substitutions $\dot{\bar{\boldsymbol{r}}} = \dot{\boldsymbol{r}}_e - \dot{\boldsymbol{\rho}}_e$ and $\boldsymbol{u} = \dot{\boldsymbol{r}}_0 - \dot{\boldsymbol{r}}_e$ into Equation 5.40, and some manipulations, the final formula for force balance of the time-varying system is obtained as follows:

$$\sum \boldsymbol{F}_{\text{ext}} = m\ddot{\boldsymbol{r}}_e - \dot{m}\boldsymbol{u} - \frac{\mathrm{d}^2}{\mathrm{d}t^2}(m\boldsymbol{\rho}_e). \tag{5.43}$$

This equation is more convenient for use, compared with Equation 5.40, since it is expressed by acceleration $\ddot{r}_e$, rather than acceleration of the center of mass of the varying-mass system, which may change its position with respect to the control volume. If the mass is leaving the system, $\dot{m}$ is negative and u will be in the reverse direction. Furthermore, if masses enter and leave the control volume through more than one opening, the last term of Equation 5.40 may be replaced by corresponding terms for each mass stream.

Angular momentum relations can be treated in a similar way. It can be shown that resultant moments about the center of mass of the varying mass system equals the rate of change of angular momentum about the center of mass minus the rate at which the added mass increases the angular momentum relative to G

$$\sum^{c} n_{\text{ext}} = {}^{c}\dot{H} - \rho_e \times \dot{m}\dot{\rho}_0. \tag{5.44}$$

The term $\dot{\rho}_0$ is velocity of the entering mass relative to G, and equals the relative velocity u only if $\dot{\rho}_e$ is zero. A similar moment equation may be written for a fixed point O, as follows:

$$\sum^{o} n_{\text{ext}} = {}^{o}\dot{H} - r_e \times \dot{m}\dot{r}_0. \tag{5.45}$$

5.3 Newton–Euler Formulation

The most popular approach used in robotics to derive the dynamic equation of motion of a parallel manipulator is the Newton–Euler formulation. A parallel manipulator such as the one shown in Figure 3.2, typically consists of a moving platform that is connected to a fixed base by several limbs. Because of the closed-loop structure, all the joints are not independently controlled. Thus, some of the joints are driven by actuators, whereas others are passive. In the Newton–Euler formulation, the free-body diagrams of all the limbs and moving platform are considered and the Newton–Euler laws are applied to each isolated body. To apply the laws to each body, it is necessary to derive linear acceleration of links, center of mass, as well as angular acceleration of the links. Hence, acceleration analysis would be performed on all the links of the manipulator and the moving platform.

Furthermore, all the external forces and moments applied to the links and to the moving platform must be carefully determined. Gravitational forces, acting on the center of masses, frictional forces and moments acting on the joints, and any possible disturbance force or moment applied to the links and to the moving platform would be identified. The most important external forces or moments applied on the manipulator are the one applied by the actuators, denoted by $\tau = [\tau_1, \tau_2, \ldots, \tau_m]^T$. Moreover, the internal forces acting between base and links, links to each other, and links and the moving platform shall be considered in the free-body diagram. The forces and moments shall be derived from the set of Newton–Euler laws, which are written separately for each link and the moving platform. Finally, by elimination of these constraints forces and moments on the Newton–Euler equations written for the moving platform, the dynamic equations relating the actuator forces and moments τ to the motion variables of the moving platform $\mathcal{X}, \dot{\mathcal{X}}$, and $\ddot{\mathcal{X}}$ are derived. This method is demonstrated by examples in the following subsections.

5.3.1 Dynamic Formulation of a Planar Manipulator: Constant Mass Treatment

In this section, dynamic formulation of a planar 4$R\underline{P}R$ manipulator is studied in detail. The architecture of this manipulator is described in detail in Section 3.3.1 and shown in Figure 1.7. As explained hereinbefore, in this manipulator the moving platform is supported by four limbs of identical $R\underline{P}R$ kinematic structure. To avoid singularities at the central position of the manipulator at each level, the limbs are considered to be crossed. A complete Jacobian and singularity analyses of this manipulator is presented in Section 4.6.

5.3.1.1 Acceleration Analysis

Acceleration analysis of the limbs and the moving platform is needed for the Newton–Euler formulation of any parallel manipulator. In this section, acceleration analysis for planar 4$R\underline{P}R$ manipulator has been carried out. In acceleration analysis, it is intended to derive expressions for linear and angular accelerations of the limbs, namely $\ddot{L}_i$ and $\ddot{\alpha}_i$, as a function of moving platform acceleration vector $\ddot{\boldsymbol{\mathcal{X}}} = [\ddot{x}_G, \ddot{y}_G, \ddot{\phi}]^T$. To obtain such a relation, rewrite velocity loop closure (4.49) here, for further manipulations.

$$\boldsymbol{v}_G + \dot{\phi}\,(\hat{\boldsymbol{K}} \times \boldsymbol{E}_i) = \dot{L}_i\hat{\boldsymbol{S}}_i + \dot{\alpha}_i\,L_i\,(\hat{\boldsymbol{K}} \times \hat{\boldsymbol{S}}_i). \tag{5.46}$$

Differentiate Equation 5.46 with respect to time, and consider vector definitions $\hat{\boldsymbol{S}}_i$ and $\boldsymbol{E}_i$ illustrated in Figure 4.3, and note that $\dot{\hat{\boldsymbol{S}}}_i = \dot{\alpha}_i(\hat{\boldsymbol{K}} \times \hat{\boldsymbol{S}}_i)$ and $\dot{\boldsymbol{E}}_i = \dot{\phi}(\hat{\boldsymbol{K}} \times \boldsymbol{E}_i)$. With some manipulations, acceleration loop closure equations can be derived for each loop $i = 1, 2, 3, 4$, as

$$\boldsymbol{a}_G + \ddot{\phi}\,(\hat{\boldsymbol{K}} \times \boldsymbol{E}_i) - \dot{\phi}^2\boldsymbol{E}_i = \ddot{L}_i\,\hat{\boldsymbol{S}}_i + 2\,\dot{L}_i\,\dot{\alpha}_i\,(\hat{\boldsymbol{K}} \times \hat{\boldsymbol{S}}_i) + \ddot{\alpha}_i\,L_i\,(\hat{\boldsymbol{K}} \times \hat{\boldsymbol{S}}_i) - \dot{\alpha}_i^2\,L_i\,\hat{\boldsymbol{S}}_i. \tag{5.47}$$

To eliminate $\ddot{\alpha}_i$ and get an expression for $\ddot{L}_i$, dot multiply both sides by $\hat{\boldsymbol{S}}_i$ and reorder into

$$\ddot{L}_i = \boldsymbol{a}_G \cdot \hat{\boldsymbol{S}}_i + \ddot{\phi}\,\hat{\boldsymbol{K}}(\boldsymbol{E}_i \times \hat{\boldsymbol{S}}_i) - \dot{\phi}^2(\boldsymbol{E}_i \cdot \hat{\boldsymbol{S}}_i) + \dot{\alpha}_i^2\,L_i. \tag{5.48}$$

In order to eliminate $\ddot{L}_i$ and get an expression for $\ddot{\alpha}_i$, cross multiply both sides of Equation 5.47 by $\hat{\boldsymbol{S}}_i$

$$\hat{\boldsymbol{S}}_i \times \boldsymbol{a}_G + \ddot{\phi}\,(\boldsymbol{E}_i \cdot \hat{\boldsymbol{S}}_i)\,\hat{\boldsymbol{K}} - \dot{\phi}^2(\hat{\boldsymbol{S}}_i \times \boldsymbol{E}_i) = (2\,\dot{L}_i\,\dot{\alpha}_i + \ddot{\alpha}_i\,L_i)\,\hat{\boldsymbol{K}}. \tag{5.49}$$

This simplifies to

$$\begin{aligned}\ddot{\alpha}_i &= \frac{1}{L_i}[-S_{i_y}|S_{i_x}|E_{i_x}S_{i_x} + E_{i_y}S_{i_y}]\begin{bmatrix} a_{Gx} \\ a_{Gy} \\ \ddot{\phi} \end{bmatrix} \\ &\quad - \frac{1}{L_i}\left((E_{i_y}S_{i_x} - E_{i_x}S_{i_y})\dot{\phi}^2 + 2\,\dot{L}_i\,\dot{\alpha}_i\right).\end{aligned} \tag{5.50}$$

Note that if this equation is written for all the four limbs, the first term constitutes $\boldsymbol{J}_\alpha$, as defined in Equation 4.58.

To complete the manipulator acceleration analysis it is necessary to derive expressions for linear accelerations of center of masses of each limb. Since that manipulator is cable-driven, the center of mass of each limb is located at the center of the limb. Denote the velocity and acceleration of the limbs' center of mass as v_{c_i} and a_{c_i}, respectively. The velocity of the center of mass is composed as the tangential and normal components. Note that tangential component of the velocity of the cables at all points along their lengths is $\dot{L}_i\hat{S}_i$, while normal component varies along the lengths. Therefore

$$v_{c_i} = \dot{L}_i\hat{S}_i + \frac{1}{2}L_i\,\dot{\alpha}_i\,(\hat{K}\times\hat{S}_i). \tag{5.51}$$

To obtain the relation for the center of mass acceleration of each limb, differentiate Equation 5.51 with respect to time

$$a_{c_i} = \frac{1}{2}\left((2\ddot{L}_i - L_i\,\dot{\alpha}^2)\,\hat{S}_i + (L_i\,\ddot{\alpha}_i + 3\,\dot{L}_i\,\dot{\alpha}_i)(\hat{K}\times\hat{S}_i)\right). \tag{5.52}$$

Note that velocity and acceleration of the center of mass of the limbs v_{c_i} and a_{c_i} are functions of $\dot{L}_i, \dot{\alpha}_i, \ddot{L}_i$ and $\ddot{\alpha}_i$, the relation of which to the manipulator velocity and acceleration $\dot{\mathcal{X}}$ and $\ddot{\mathcal{X}}$ are given in Equations 4.53, 4.56, 5.48, and 5.50, respectively.

5.3.1.2 Dynamic Formulation of the Limbs

To derive dynamic formulation of the planar $4R\underline{P}R$ manipulator, assume that the moving platform center of mass is located at the center point G and has a mass of M and its moment of inertia about the center of mass in z-axis is denoted by I_{c_i}. Furthermore, it is assumed that the cables are homogeneous, with a circular cross section, and have density per unit length of ρ_m. The cables are considered to be in a straight line and modeled as rigid bodies, with a mass of $m_i = \rho_m\, L_i$. The moment of inertia of the cables are determined by assuming that they are slender bars with length L_i. Therefore, the mass and the moment of inertia of the cables about their centers of mass c_i in z-axis are given by

$$m_i = \rho_m\, L_i; \quad I_{c_i} = \frac{m_i}{12}\,L_i^2 = \frac{\rho_m}{12}\,L_i^3. \tag{5.53}$$

Furthermore, assume that the planar $4R\underline{P}R$ manipulator does not lie necessarily in the horizontal plane, and the gravity forces are given by $m_i\,g$ vector as shown in Figure 5.5. With these assumptions, consider the free-body diagrams of the limbs and the moving platform as illustrated in Figure 5.5. The reaction forces at fixed points A_i's are illustrated componentwise and are denoted by $F_{A_i}^S$ and $F_{A_i}^N$, in which $\hat{S}_i$ is along the limb direction and $\hat{N}_i$ is perpendicular to the limb direction. Similarly, internal force at points B_i's are denoted componentwise by $F_{B_i}^S$ and $F_{B_i}^N$, respectively. The velocity and acceleration of the limb center of mass, v_{c_i} and a_{c_i}, are also shown in this figure. Assume that the only external disturbance wrench existing, acts on the moving platform, and is denoted by $\mathcal{F}_d = [f_{d_x}, f_{d_y}, \tau_d]^T$.

Let us first derive the equations of motion of the limbs. Referring to Equations 5.31 and 5.33, the Newton–Euler equations for a constant mass system can be written as

$$\sum F_{\text{ext}} = m_i\, a_{c_i} \tag{5.54}$$

$$\sum {}^{c_i}n_{\text{ext}} = I_{c_i}\,\ddot{\alpha}_i \tag{5.55}$$

in which ΣF_{ext} is the summation of all external forces acting on each limb and $\Sigma^{c_i}n_{\text{ext}}$ denotes the resulting external moments in the z-direction about the center of mass of each

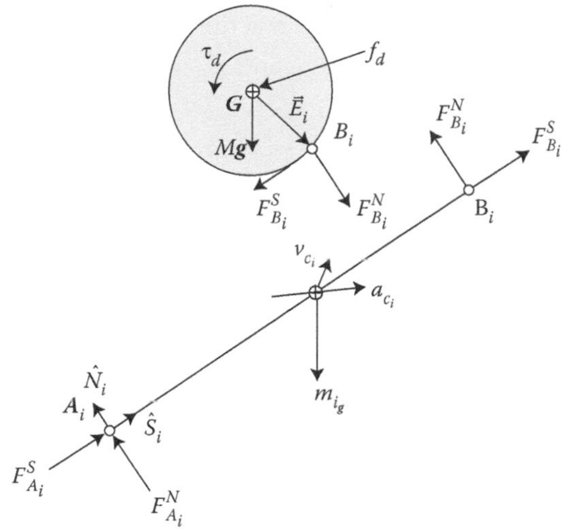

FIGURE 5.5
Free-body diagram of the planar $4R\underline{P}R$ manipulator's limb and the moving platform.

limb c_i. The linear velocity and acceleration of each limb at the center of mass, $\boldsymbol{v}_{c_i}, \boldsymbol{a}_{c_i}$, are given in Equations 5.51 and 5.52, respectively. Considering the free-body diagram of the limb in Figure 5.5, the resulting forces and moments can be determined in a vector form as

$$\left(F^S_{B_i} + F^S_{A_i}\right)\hat{\mathbf{S}}_i + \left(F^N_{B_i} + F^N_{A_i}\right)\hat{\mathbf{N}}_i + m_i \boldsymbol{g} = m_i\, \boldsymbol{a}_{c_i} \tag{5.56}$$

$$\left(F^N_{B_i} - F^N_{A_i}\right)\frac{L_i}{2} = I_{c_i}\,\ddot{\alpha}_i. \tag{5.57}$$

Substituting $\boldsymbol{v}_{c_i}$ and $\boldsymbol{a}_{c_i}$ from Equations 5.51 and 5.52, and writing Equation 5.56 componentwise in $\hat{\mathbf{S}}_i$ and $\hat{\mathbf{N}}_i$ directions results in

$$F^S_{B_i} + F^S_{A_i} = \frac{\rho_m}{2}\left(2L_i\ddot{L}_i - (L_i\dot{\alpha}_i)^2 - 2L_i g^S\right) \tag{5.58}$$

$$F^N_{B_i} + F^N_{A_i} = \frac{\rho_m}{2}\left(L_i^2\ddot{\alpha}_i + 3L_i\dot{L}_i\dot{\alpha}_i - 2L_i g^N\right) \tag{5.59}$$

$$F^N_{B_i} - F^N_{A_i} = \frac{\rho_m}{6}L_i^2\ddot{\alpha}_i, \tag{5.60}$$

in which the gravitational acceleration is considered componentwise as $\boldsymbol{g} = g^S\hat{\mathbf{S}}_i + g^N\hat{\mathbf{N}}_i + g^K\hat{\boldsymbol{K}}$, where g^S and g^N represent the components of $\boldsymbol{g}$ along $\hat{\mathbf{S}}$, and $\hat{\boldsymbol{N}}$ directions, respectively. These components may be derived from the following equations:

$$g^S = \boldsymbol{g}\cdot\hat{\mathbf{S}}_i; \quad g^N = \boldsymbol{g}\cdot\hat{\mathbf{N}}_i; \quad g^K = \boldsymbol{g}\cdot\hat{\mathbf{K}}. \tag{5.61}$$

Note that $F^N_{A_i}$ denotes the pivot reaction force and $F^N_{B_i}$ is the internal force between the limbs and moving platform, and can be determined from Equations 5.60 and 5.59. Furthermore, note that $F^S_{A_i}$ are the actuator forces acting on the limbs that may be denoted by $F^S_{A_i} = -\tau_i$ to consider positive tension forces in the actuators. Therefore, from Equations 5.58 through 5.60, the pivot reaction force $F_{A_i}^N$, and the interacting forces between the limbs and moving platform, $F_{B_i}^N$ and $F_{B_i}^S$ are determined as follows:

$$F_{A_i}^N = \rho_m \left(\frac{1}{6} L_i^2 \ddot{\alpha}_i + \frac{3}{4} L_i \dot{L}_i \dot{\alpha}_i - \frac{1}{2} L_i g^N \right) \tag{5.62}$$

$$F_{B_i}^N = \rho_m \left(\frac{1}{3} L_i^2 \ddot{\alpha}_i + \frac{3}{4} L_i \dot{L}_i \dot{\alpha}_i - \frac{1}{2} L_i g^N \right) \tag{5.63}$$

$$F_{B_i}^S = \tau_i + \rho_m \left(L_i \ddot{L}_i - \frac{1}{2} (L_i \dot{\alpha}_i)^2 - L_i g^S \right). \tag{5.64}$$

These relations would be used in the dynamic equation of the moving platform.

5.3.1.3 Dynamic Formulation of the Moving Platform

Now the dynamic analysis of the moving platform is carried out using the free-body diagram depicted in Figure 5.5. The Newton–Euler equations of the moving platform is as follows:

$$\sum \boldsymbol{F}_{\text{ext}} = M\boldsymbol{g} + \boldsymbol{f}_d - \sum_{i=1}^{4} \left(F_{B_i}^S \hat{\boldsymbol{S}}_i + F_{B_i}^N \hat{\boldsymbol{N}}_i \right) = M\boldsymbol{a}_G \tag{5.65}$$

$$\sum {}^G\boldsymbol{n}_{\text{ext}} = \tau_d \hat{\boldsymbol{K}} - \sum_{i=1}^{4} \boldsymbol{E}_i \times \left(F_{B_i}^S \hat{\boldsymbol{S}}_i + F_{B_i}^N \hat{\boldsymbol{N}}_i \right) = I_m \ddot{\phi} \hat{\boldsymbol{K}}. \tag{5.66}$$

Write the force Equation 5.65 componentwise. With some manipulation, this leads to the following set of equations:

$$M\ddot{x}_G - M(\boldsymbol{g} \cdot \hat{\boldsymbol{x}}) - f_{d_x} + \sum_{i=1}^{4} \left(F_{B_i}^S S_{i_x} - F_{B_i}^N S_{i_y} \right) = 0 \tag{5.67}$$

$$M\ddot{y}_G - M(\boldsymbol{g} \cdot \hat{\boldsymbol{y}}) - f_{d_y} + \sum_{i=1}^{4} \left(F_{B_i}^S S_{i_y} + F_{B_i}^N S_{i_x} \right) = 0 \tag{5.68}$$

$$I_m \ddot{\phi} - \tau_d - \sum_{i=1}^{4} \left(F_{B_i}^S (E_{i_x} S_{i_y} - E_{i_y} S_{i_x}) + F_{B_i}^N (\boldsymbol{E}_i \cdot \hat{\boldsymbol{S}}_i) \right) = 0. \tag{5.69}$$

Note that in these manipulations the unit vectors are written componentwise as $\hat{\boldsymbol{S}}_i = [S_{i_x},\ S_{i_y}]^T$ and hence, $\hat{\boldsymbol{N}}_i = [N_{ix},\ N_{iy}]^T = [-S_{i_y},\ S_{i_x}]^T$.

Equations 5.67 through 5.69 are the governing equations of the motion of the manipulator, in which $\boldsymbol{\mathcal{F}}_d = [f_{Dx}, f_{Dy}, \tau_d]^T$ is the disturbance wrench applied to the moving platform, and the interaction forces between the limbs and moving platform $F_{B_i}^S$ and $F_{B_i}^N$ are derived from the limb dynamics in Equations 5.64 and 5.63, respectively. Furthermore, vectors $\boldsymbol{E}_i$ and $\hat{\boldsymbol{S}}_i$ can be determined from

$$\hat{\boldsymbol{S}}_i = \begin{bmatrix} \cos(\alpha_i) & \sin(\alpha_i) \end{bmatrix}^T \tag{5.70}$$

$$\boldsymbol{E}_i = \begin{bmatrix} R_B \cos(\theta_{B_i} + \phi) & R_B \sin(\theta_{B_i} + \phi) \end{bmatrix}^T. \tag{5.71}$$

Equations 5.67 through 5.69 can be viewed in an implicit vector form of

$$f(\ddot{\mathcal{X}}, \dot{\mathcal{X}}, \mathcal{X}, g, \mathcal{F}_d, \boldsymbol{\tau}) = \mathbf{0}. \tag{5.72}$$

The dynamic formulation of the manipulator may be used in two ways. The first use of dynamic formulation is its implementation in dynamics simulation of the robot. In dynamics simulation, it is assumed that the actuator forces $\boldsymbol{\tau}(t)$ are given and the manipulator motion trajectory $\mathcal{X}(t)$ is needed to be determined. This form of using the dynamic equations is called the forward dynamic analysis, which is elaborated hereafter. Owing to the implicit nature of the dynamic formulation, special integration routine capable of integrating implicit differential equations are used for simulations. On the other hand, dynamic equations may be used to evaluate the actuator forces $\boldsymbol{\tau}$ needed to produce a prescribed trajectory $\mathcal{X}(t)$ in presence of any disturbance wrench $\mathcal{F}_d$ applied to the robot. This form of dynamic formulation implementation is called inverse dynamics, the details of which have been given later in this section. The most important use of inverse dynamic formulation though is its direct use in the structure of control laws, which are elaborated in Chapter 6.

5.3.1.4 Forward Dynamics Simulations

As explained hereinbefore, the dynamic formulation of planar $4R\underline{P}R$ manipulator can be used for forward dynamics simulations. As shown in Figure 5.6, in forward dynamics, it is assumed that the actuator forces and the external disturbances applied to the manipulator are given, and the resulting trajectory of the moving platform is to be determined. The dynamic formulation, namely Equations 5.67 through 5.69, are the main equations to be integrated using an appropriate integration routine. Owing to the implicit nature of the dynamic formulation of this manipulator, usual numerical integration routines such as the Runge–Kutta methods [166] cannot be used to solve the problem. However, a special integration routine* capable of integrating implicit functions may be used for dynamics simulations. In the first step, the initial condition for accelerations $\ddot{\mathcal{X}}(0)$ has to be determined by solving the implicit equation at initial time. The dynamic formulation written in vector form in Equation 5.72 is then integrated iteratively with respect to time.

A number of simulations are examined to verify the accuracy and integrity of dynamic formulation. In all the simulations it is assumed that the manipulator motion is in a horizontal plane and the gravitational force acting in $-\hat{K}$ direction. In the first simulation, as illustrated in Figure 5.7, the actuator forces are considered as $\boldsymbol{\tau} = [1, 1, 0, 0]^T kN$ to have the resulting force of the actuators in $-y$ direction. By inspection, we expect to have the resulting motion of the manipulator for this simulation only along the $-y$-axis, and have

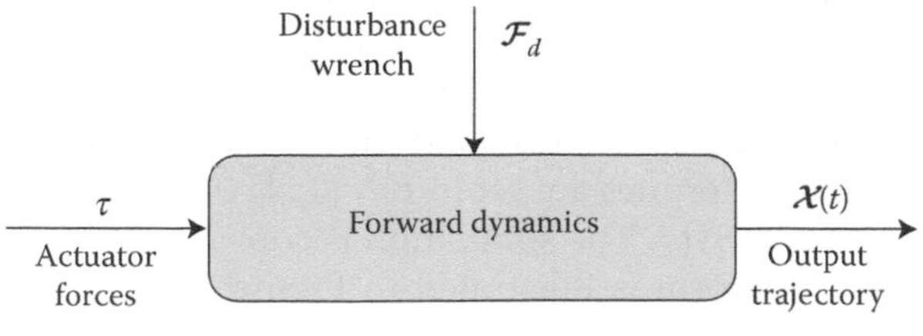

FIGURE 5.6
Block diagram of forward dynamics simulations.

* `ode15i` function of MATLAB.

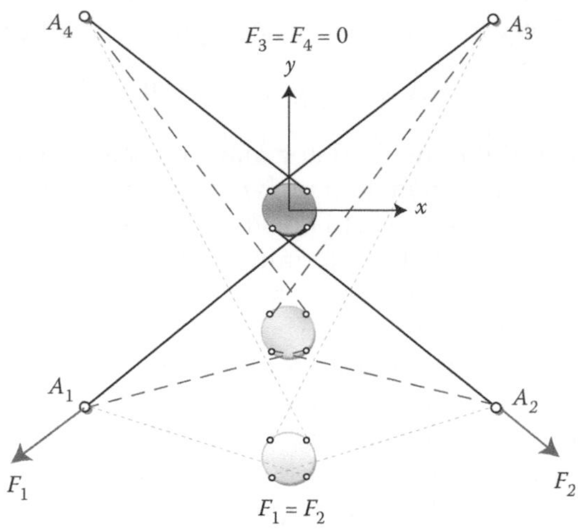

FIGURE 5.7
First simulation scenario for the planar $4R\underline{P}R$ manipulator with $\boldsymbol{\tau} = [1, 1, 0, 0]^T kN$.

TABLE 5.1
Geometric and Inertial Parameters of the $4R\underline{P}R$ Parallel Manipulator

Description	Quantity
R_A: Radius of the fixed base points A_i's	900 m
R_B: Radius of the moving platform points B_i's	10 m
θ_{A_i}: Angle of the fixed base points A_i's	[−135°, −45°, 45°, 135°]
θ_{B_i}: Angle of the moving platform points B_i's	[−45°, −135°, 135°, 45°]
M: The moving platform mass	2500 kg
I_m: The moving platform moment of inertia	3.5×10^5 kg · m^2
ρ_m: The limb density per length	0.215 kg/m

no motion in x and ϕ directions. Furthermore, as illustrated in Figure 5.7, the motion in the $-y$ direction continues even after the moving platform center of mass passes the line A_1A_2. This is because at the time of crossing this line, the moving platform has a positive acceleration, and after that the motion becomes oscillatory.

The dynamic equations are integrated in a forward dynamics simulation for the planar $4R\underline{P}R$ manipulator, the geometrical and inertial parameters of which are given in Table 5.1. For this simulation the disturbance wrench, and the initial conditions for all the states are set to zero, while the actuator torques are set to $\boldsymbol{\tau} = [1,\ 1,\ 0,\ 0]^T kN$, and the gravity vector is considered as $\boldsymbol{g} = [0,\ 0,\ -9.81]^T$. The simulation is considered for 350 s, and the output trajectory of the moving platform is illustrated in Figure 5.8. As seen in this figure, the position $x(t)$ and the orientation $\phi(t)$ of the moving platform are equal to zero, while $y(t)$ has an oscillatory motion. The amplitude of the oscillation is such that the moving platform center of mass passes the line A_1A_2 as expected, and furthermore, the moving platform returns to its initial position in about 220 s. The output trajectories completely agree with the expected behavior of the system.

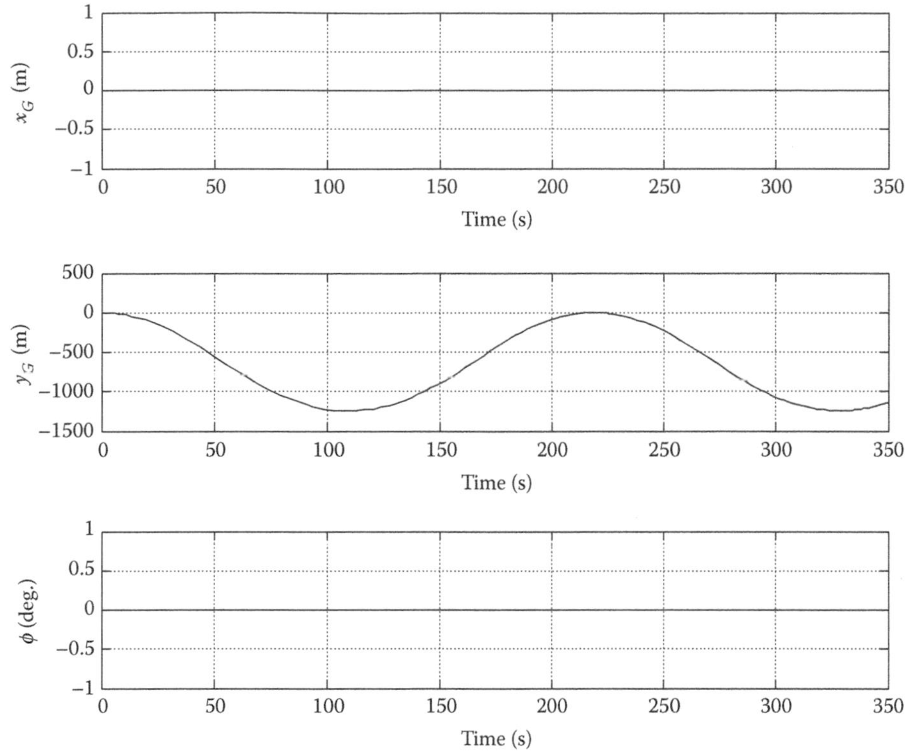

FIGURE 5.8
The output trajectory of the first simulation scenario for the planar $4R\underline{P}R$ manipulator.

In the second simulation, as illustrated in Figure 5.9, the actuator forces are considered as $\boldsymbol{\tau} = [1, 1, 1, 1]^T kN$ and the gravity vector is set at $g = [0, 0, -9.81]^T$, to have zero resulting force acting on the moving platform in its planar motion. If the initial states are all set at zero, no motion is expected to be generated at the moving platform. However, consider initial states as $\boldsymbol{\mathcal{X}}(0) = [1, 1, 0]^T$, in which no orientation is given to the moving platform, while x and y initial movements are set equally at one. By inspection, we expect to have no angular motion for the moving platform for this situation, and have an oscillatory motion in both x and y directions. Furthermore, as it is illustrated in Figure 5.9, the amplitude of oscillation is expected to be one, that is, the same as the initial states in x and y directions. The dynamic equations are integrated in a forward dynamics simulation for the planar $4R\underline{P}R$ manipulator, with zero disturbance wrench, and the assigned initial conditions and actuator forces. The simulation is considered for 350 s, and the output trajectory of the moving platform is illustrated in Figure 5.10. As seen in this figure, the orientation $\phi(t)$ of the moving platform is equal to zero, while the positions $x(t)$ and $y(t)$ both show oscillatory motions with amplitude one. Furthermore, it is seen that the oscillations have a sustained amplitude of one, and the moving platform returns to its initial position in about 220 s. These simulation results completely agree with the expected behavior of the system as well.

In the third simulation, forward dynamics is implemented in a closed-loop block diagram as illustrated in Figure 5.11. Suppose it is intended to have a desired trajectory for the motion of the manipulator moving platform as illustrated in Figure 5.12. To produce the actuator forces required for such a motion, a closed-loop control system can be

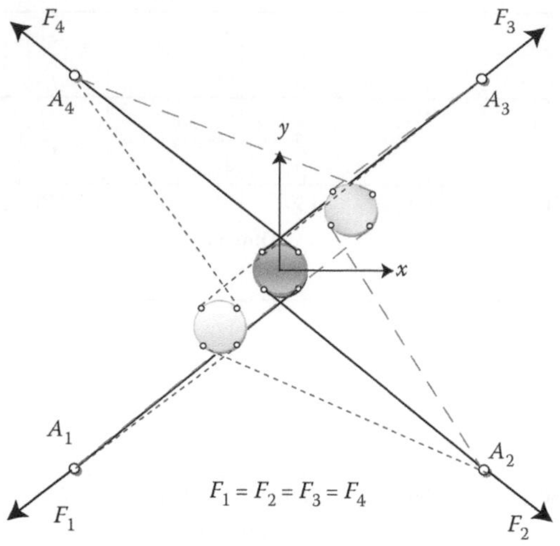

FIGURE 5.9
Second simulation scenario for the planar 4$R\underline{P}R$ manipulator with $\boldsymbol{\tau} = [1, 1, 1, 1]^T\, kN$, and $\mathbf{X}(0) = [1, 1, 0]^T$.

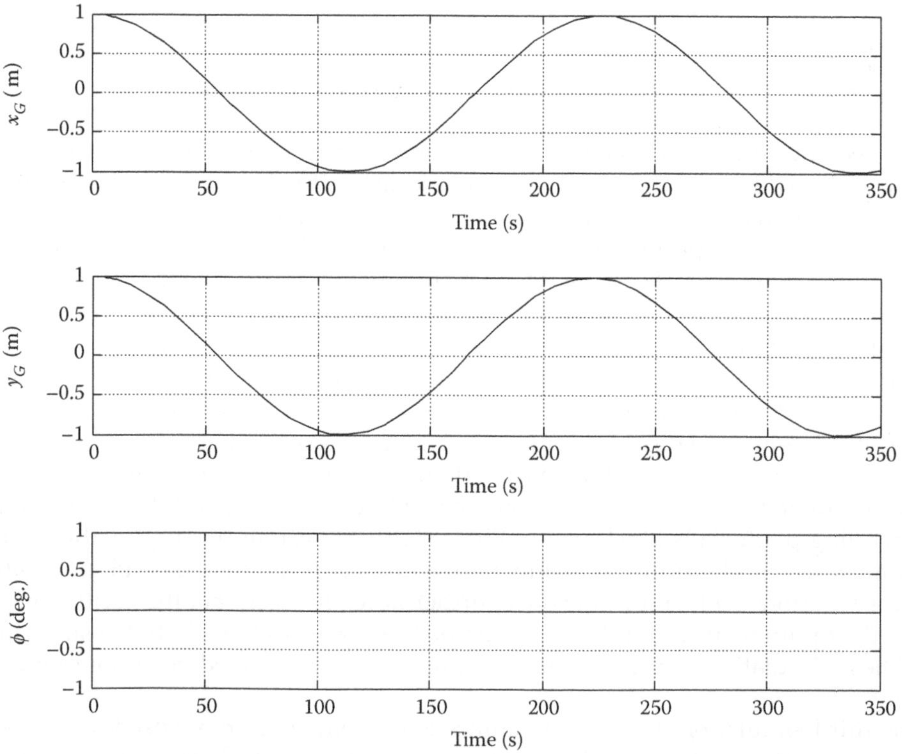

FIGURE 5.10
The output trajectory of the second simulation scenario for the planar 4$R\underline{P}R$ manipulator.

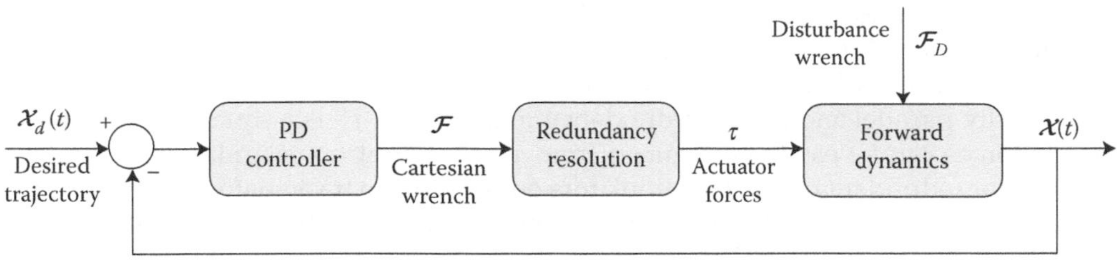

FIGURE 5.11
Block diagram of the forward dynamics implementation in closed loop.

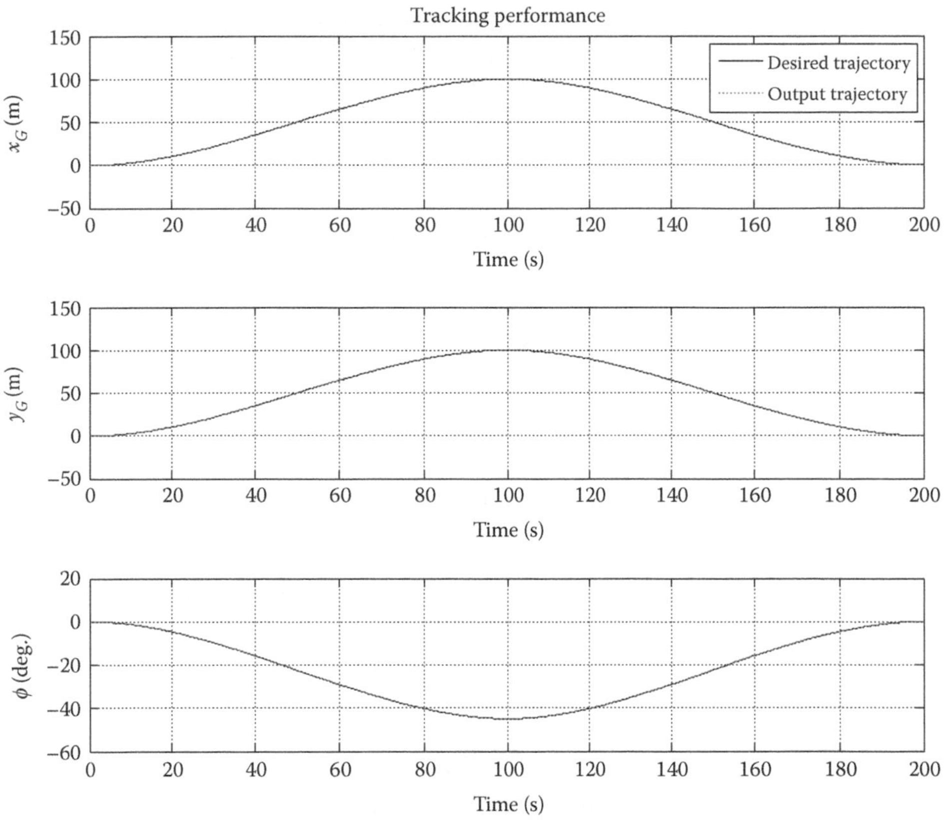

FIGURE 5.12
The desired and output trajectories of the planar $4R\underline{P}R$ manipulator in closed loop.

designed for the system. In this structure, the actual trajectory of the moving platform $\mathcal{X}(t)$ is measured and compared to the desired trajectory $\mathcal{X}_d(t)$. Furthermore, the tracking error $e(t) = \mathcal{X}_d(t) - \mathcal{X}(t)$ is fed to a controller to determine the required wrench needed to be applied to the moving platform. Different control topologies can be used to define such a wrench, which is elaborated in detail in Chapter 6. Here for the sake of simplicity, a simple PD controller is applied for each component of the tracking error $e(t)$ to generate the required wrench $\mathcal{F}$ on the manipulator moving platform.

The actuator forces required to accomplish the task are in fact the projection of the computed wrench $\mathcal{F}$ from task space to joint space. As explained in Section 4.9.2, this projection is accomplished by $\boldsymbol{J}^T$ as given in Equation 4.92, that is, $\mathcal{F} = \boldsymbol{J}^T \boldsymbol{\tau}$. If the manipulator is fully parallel and has no redundancy in actuation, $\boldsymbol{J}^T$ is a square matrix and the actuator forces can be easily determined from $\boldsymbol{\tau} = \boldsymbol{J}^{-T} \mathcal{F}$ at nonsingular configurations. However, for redundant parallel manipulators one or more extra actuator(s) exist(s) to provide better performance for the manipulator. Hence, infinite solutions for actuator efforts $\boldsymbol{\tau}$ exist to generate the required Cartesian wrench $\mathcal{F}$. There exist many different algorithms developed to use the extra degree of actuation optimally, which is elaborated in detail in Section 6.7; however, for the sake of simplicity, the Moore–Penrose pseudo inverse of Jacobian transpose, $\boldsymbol{J}^{T^\dagger}$ is used here as the simplest way of redundancy resolution. In this method, the minimum norm solution for actuator forces $\boldsymbol{\tau}_{\min}$ is found to satisfy the wrench projection $\mathcal{F} = \boldsymbol{J}^T \boldsymbol{\tau}_{\min}$. Although this method is very popular for many applications, it is not recommended for a cable-driven manipulator, because it may result in compression forces for the actuators. Complete analysis of this case is given in Chapter 6, and in this section the intention is to see how forward dynamics is used in a closed loop structure.

For this simulation, the geometric and inertial parameters are given in Table 5.1, and the disturbance wrench, and initial conditions for the states are all set at zero. Furthermore, the gravity vector is considered as $\boldsymbol{g} = [0, 0, -9.81]^T$. The desired trajectory of the moving platform is considered to be a smooth and cubic polynomial function for x_d, y_d, and ϕ_d, as shown in Figure 5.12. The time of simulation is considered to be 200 s, and the PD gains used in the simulations are $K_P = K_D = 10^4 \cdot \mathrm{diag}[1, 1, 100]$. The closed-loop tracking performance of planar $4R\underline{P}R$ manipulator is illustrated in Figures 5.12 and 5.13.

As seen in Figure 5.12 the difference between the desired and final closed-loop motions of the system are not distinguishable. As seen in greater detail in Figure 5.13, a PD controller for the manipulator is capable of reducing the tracking errors to less than 0.02 m in position and to less than 0.012° in orientation. Note that the controller used in this simulation is very simple, and the intention of this example is to see how forward dynamics can be implemented in closed-loop simulations. This controller is very sensitive to the external disturbances and is not generally suitable for closed-loop applications.

5.3.1.5 Inverse Dynamics Simulation

In inverse dynamics simulations, it is assumed that trajectory of the manipulator is given, and actuator forces required to generate such trajectories are to be determined. As illustrated in Figure 5.14, inverse dynamic formulation is implemented by the following sequence. The first step is trajectory generation for the manipulator moving platform. Many different algorithms have been developed for smooth trajectory generation,* from these methods a cubic polynomial trajectory planner is used in the simulations, as shown in Figure 5.12. For such a trajectory $\mathcal{X}_d(t)$, the next step is to solve inverse kinematics of the manipulator and to find $\boldsymbol{L}(t)$ and $\boldsymbol{\alpha}(t)$, as a function of the manipulator trajectory using Equations 3.10 and 3.11, respectively. Then, manipulator Jacobian matrices $\boldsymbol{J}$ and $\boldsymbol{J}_\alpha$ are calculated through Equations 4.54 and 4.58, respectively. In this step, $\dot{\boldsymbol{L}}(t)$ and $\dot{\boldsymbol{\alpha}}(t)$ are calculated as well. Then the accelerations are evaluated using acceleration analysis Equations 5.48 and 5.50. Finally, the interaction forces between the limbs and moving platform

* See Appendix B.

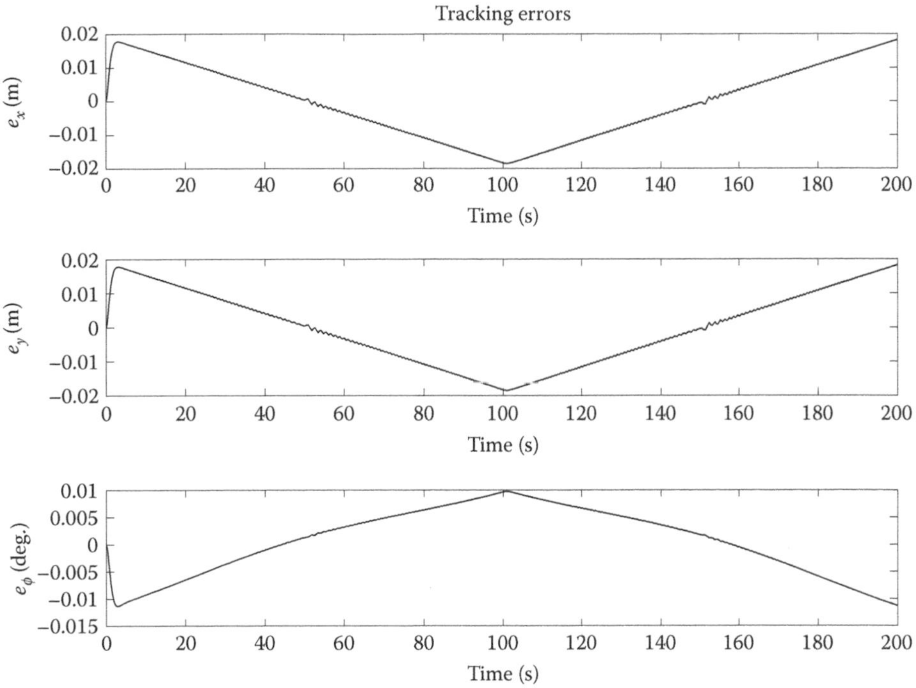

FIGURE 5.13
The closed-loop tracking performance of the planar $4R\underline{P}R$ manipulator.

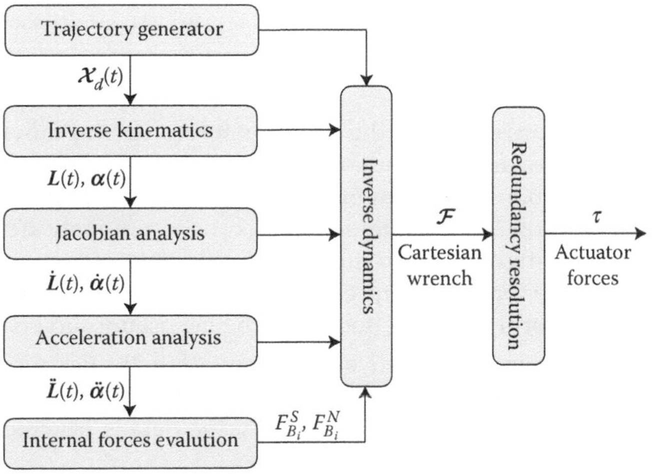

FIGURE 5.14
Flowchart of inverse dynamics implementation sequence.

$F^S_{B_i}$, $F^N_{B_i}$ are computed from Equations 5.64 and 5.63, respectively and are substituted in the dynamic formulation of the manipulator.

Let us define $\mathcal{F}$ as the resulting Cartesian wrench applied to the moving platform, which can be calculated from the summation of all inertial and external forces *excluding the actuator torques* $\boldsymbol{\tau}$ in dynamic Equations 5.67 through 5.69. Hence, by equation $\mathcal{F} = J^T\boldsymbol{\tau}$,

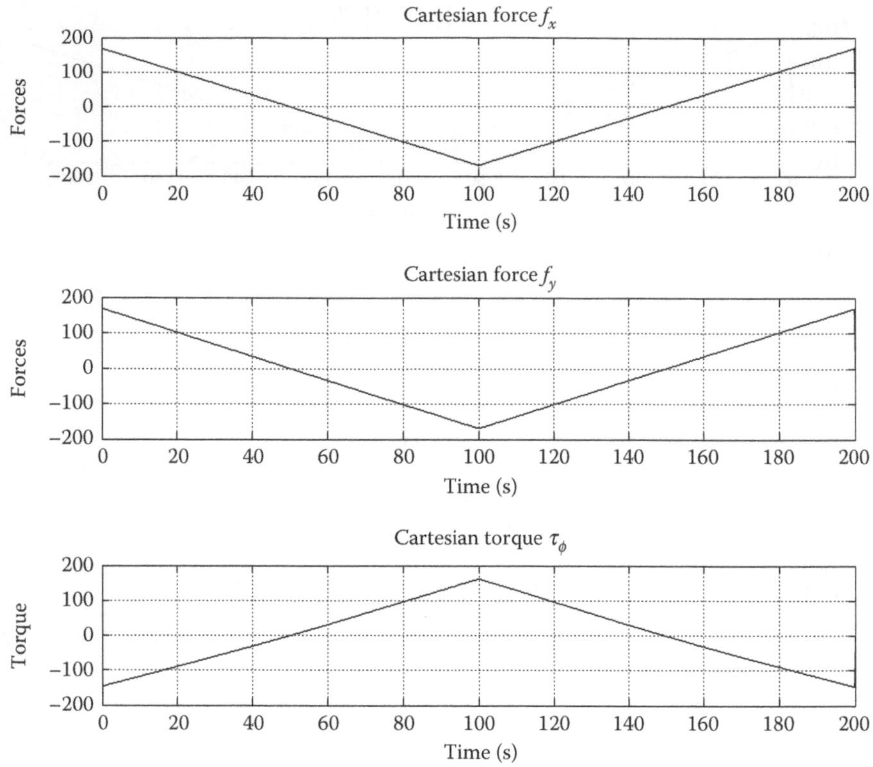

FIGURE 5.15
The Cartesian wrench of the planar 4$R\underline{P}R$ manipulator required to generate trajectory shown in Figure 5.12.

$\mathcal{F}$ can be viewed as the projection of the actuator forces on the moving platform and can be uniquely determined from dynamic equations by exclusion of the actuator forces from these dynamic equations. If the manipulator has no redundancy in actuation the Jacobian matrix J is squared, and actuator forces can be uniquely determined by $\boldsymbol{\tau} = J^{-T}\mathcal{F}$, provided J is nonsingular. For redundant manipulators, however, there are infinite solutions for $\boldsymbol{\tau}$ to be projected into $\mathcal{F}$. As explained hereinbefore, the simplest solution is a minimum norm solution, found from the pseudo-inverse of J^T, given as $\boldsymbol{\tau} = J^{T^{\dagger}}\mathcal{F}$. This solution is implemented in the simulation studies reported in this section. Other optimization techniques can be used to find actuator forces projected from $\mathcal{F}$ subject to more detailed manipulator constraints, the details of which are reported in Section 6.7.

Inverse dynamics of the manipulator is simulated in the absence of any disturbing forces $\mathcal{F}_d = \mathbf{0}$. Simulation results are illustrated in Figures 5.15 and 5.16. Typical third-order polynomial trajectories for the manipulator are considered in this simulation, depicted in Figure 5.12. Cartesian wrench at the moving platform, $\mathcal{F} = [f_x, f_y, \tau_\phi]^T$, are illustrated in Figure 5.15. As seen, Cartesian forces have similar pattern to the desired trajectory accelerations, which are linear for cubic trajectories. Similarly, actuator forces of the manipulator are illustrated in Figure 5.16. It is observed that since the manipulator moves in positive x and y directions, the actuator forces of the first and third limbs are dominant.

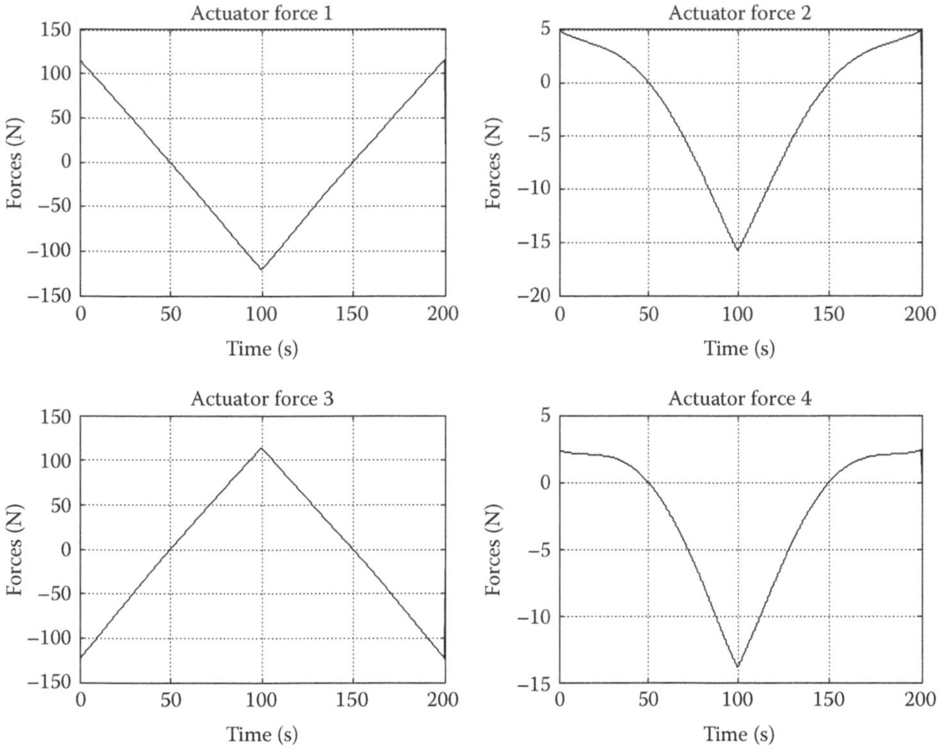

FIGURE 5.16
The actuator forces of the planar $4R\underline{P}R$ manipulator required to generate trajectory shown in Figure 5.12.

5.3.2 Dynamic Formulation of a Planar Manipulator: Variable-Mass Treatment

In this section, dynamic analysis of a planar $4R\underline{P}R$ manipulator in its most general form has been studied in detail. Contrary to the analysis given in Section 5.3.1, in which the effect of mass variation in the cables are neglected, here the dynamic formulation is derived for the system considering variable-mass effects. The architecture of this manipulator is considered as in the previous subsection and is shown in Figure 1.7. The kinematic structure of the system is also as the earlier one, but lengthening or shortening of the cables are considered as a mass stream entering or leaving the control volume containing the limbs and the moving platform.

5.3.2.1 Acceleration Analysis

In acceleration analysis, it is intended to derive expressions for the linear and angular accelerations of the limbs, namely $\ddot{L}_i$ and $\dot{\omega}_i$, as a function of the moving platform acceleration. In the following analysis, consider a vector formulation for linear and angular velocities, and denote the moving platform twist as $\dot{\boldsymbol{\mathcal{X}}} = [v_G, \boldsymbol{\omega}]^T$, in which v_G denotes the moving platform center of mass linear velocity, and $\boldsymbol{\omega}$ denotes the moving platform angular velocity. To obtain such a relation, let us rewrite the velocity loop closure Equation 4.49 in a vector form as

$$v_G + \boldsymbol{\omega} \times \boldsymbol{b}_i = \dot{L}_i\, \hat{\mathbf{S}}_i + L_i\, (\boldsymbol{\omega}_i \times \hat{\mathbf{S}}_i) \tag{5.73}$$

in which ${}^{A}\boldsymbol{b}_i$ denotes the position vector of points B_i, $\boldsymbol{\omega} = \dot{\phi}\hat{\boldsymbol{K}}$ is the moving platform angular velocity, and $\boldsymbol{\omega}_i = \dot{\alpha}_i\hat{\boldsymbol{K}}$ is the limb angular velocity. Rewriting the velocity loop closure in this format enables us to have a general vector base treatment for any cable-driven parallel manipulator, not necessarily being planar. Let us first define an intermediate variable $\boldsymbol{v}_{b_i}$ as the velocity of point $\boldsymbol{b}_i$

$$\boldsymbol{v}_{b_i} = \boldsymbol{v}_G + \boldsymbol{\omega} \times \boldsymbol{b}_i \tag{5.74}$$

$$= \dot{L}_i\,\hat{\boldsymbol{S}}_i + L_i\,(\boldsymbol{\omega}_i \times \hat{\boldsymbol{S}}_i). \tag{5.75}$$

To obtain the angular velocity of the limbs $\boldsymbol{\omega}_i$, cross multiply $\hat{\boldsymbol{S}}_i$ to both sides of Equation 5.73.

$$\hat{\boldsymbol{S}}_i \times \boldsymbol{v}_G + \hat{\boldsymbol{S}}_i \times (\boldsymbol{\omega} \times \boldsymbol{b}_i) = L_i\left(\hat{\boldsymbol{S}}_i \times (\boldsymbol{\omega}_i \times \hat{\boldsymbol{S}}_i)\right). \tag{5.76}$$

Note that for planar motion of the limbs, the axial spin of the limbs is zero. Hence

$$\boldsymbol{\omega}_i \cdot \hat{\boldsymbol{S}}_i = \dot{\boldsymbol{\omega}}_i \cdot \hat{\boldsymbol{S}}_i = 0. \tag{5.77}$$

This means that vector $\boldsymbol{\omega}_i$ is perpendicular to the limb axis $\hat{\boldsymbol{S}}_i$, and therefore

$$\left(\hat{\boldsymbol{S}}_i \times (\boldsymbol{\omega}_i \times \hat{\boldsymbol{S}}_i)\right) = \boldsymbol{\omega}_i. \tag{5.78}$$

Considering Equation 5.78, the angular velocity of the limb is determined by the following equation:

$$\boldsymbol{\omega}_i = \frac{1}{L_i}\left(\hat{\boldsymbol{S}}_i \times \boldsymbol{v}_{b_i}\right)$$

$$= \frac{1}{L_i}\hat{\boldsymbol{S}}_i \times (\boldsymbol{v}_G + \boldsymbol{\omega} \times \boldsymbol{b}_i). \tag{5.79}$$

Similar to Equation 5.74, linear acceleration of $\boldsymbol{b}_i$ points can be defined by differentiation of Equation 5.74

$$\boldsymbol{a}_{b_i} = \boldsymbol{a}_G + \dot{\boldsymbol{\omega}} \times \boldsymbol{b}_i + \boldsymbol{\omega} \times (\boldsymbol{\omega} \times \boldsymbol{b}_i) \tag{5.80}$$

$$= \ddot{L}_i\hat{\boldsymbol{S}}_i + 2\dot{L}_i\boldsymbol{\omega}_i \times \hat{\boldsymbol{S}}_i + L_i\boldsymbol{\omega}_i \times (\boldsymbol{\omega}_i \times \hat{\boldsymbol{S}}_i). \tag{5.81}$$

To eliminate $\dot{\boldsymbol{\omega}}_i$ and get an expression for $\ddot{L}_i$, dot multiply both sides of Equation 5.81 by $\hat{s}_i$, considering the fact that $\hat{\boldsymbol{S}}_i \cdot (\boldsymbol{\omega}_i \times \hat{\boldsymbol{S}}_i) = 0$ and $\boldsymbol{\omega}_i \cdot \hat{\boldsymbol{S}}_i = 0$, then reorder the same into

$$\ddot{L}_i = \boldsymbol{a}_{b_i} \cdot \hat{\boldsymbol{S}}_i + L_i(\boldsymbol{\omega}_i \cdot \boldsymbol{\omega}_i)$$

$$= (\boldsymbol{a}_G + \dot{\boldsymbol{\omega}} \times \boldsymbol{b}_i + \boldsymbol{\omega} \times (\boldsymbol{\omega} \times \boldsymbol{b}_i)) \cdot \hat{\boldsymbol{S}}_i + L_i(\boldsymbol{\omega}_i \cdot \boldsymbol{\omega}_i). \tag{5.82}$$

To eliminate $\ddot{L}_i$ and get an expression for $\dot{\boldsymbol{\omega}}_i$, cross multiply both sides of Equation 5.81 by $\hat{s}_i$, and note that

$$\boldsymbol{\omega}_i \times (\boldsymbol{\omega}_i \times \hat{\boldsymbol{S}}) = \boldsymbol{0}; \quad \hat{\boldsymbol{S}}_i \times (\dot{\boldsymbol{\omega}}_i \times \hat{\boldsymbol{S}}) = \dot{\boldsymbol{\omega}}_i. \tag{5.83}$$

This can be simplified to

$$\dot{\omega}_i = \frac{1}{L_i}\hat{S}_i \times a_{b_i} - 2\dot{L}_i\omega_i$$
$$= \frac{1}{L_i}(a_G + \dot{\omega} \times b_i + \omega \times (\omega \times b_i)). \tag{5.84}$$

To complete the manipulator acceleration analysis, since the variable-mass approach is followed in this section, the velocity and acceleration terms of the stream mass entering the system, and the acceleration terms of the limb center of mass is needed to be determined. Recall the right-hand-side terms in Equation 5.43, which reads $m\ddot{r}_e - \dot{m}\dot{u} - \frac{d^2}{dt^2}(m\rho_e)$. To calculate these terms, note that in this formulation, and as shown in Figure 5.4, the total mass of the system at time t is denoted by m, the instantaneous position of the center of mass of the varying mass system is denoted by $\bar{r}$, the position of the entering mass is denoted by vector r_0, and from the center of mass by vector ρ_0. The position of the mass entrance section E from these two reference points are given by r_e and ρ_e, respectively. It should be noted that although $r_0 = r_e$ and $\rho_e = \rho_0$, their time derivatives represent different velocities such that $\dot{r}_0 \neq \dot{r}_e$ and $\dot{\rho}_e \neq \dot{\rho}_0$.

Consider the reference frame for evaluation of these terms to be $\{A_i\}$. This frame is considered to be parallel to the fixed frame $\{A\}$, but located at point A_i. With respect to this frame, the position vector r_e is equal to the position vector r_0, both equal to zero. The time derivative of r_e is also equal to zero, $\dot{r}_e = \ddot{r}_e = 0$, whereas the time derivative of r_0 is not zero and is equal to the velocity of the cable in $\hat{S}_i$ direction, therefore

$$\dot{r}_0 = \dot{L}_i\hat{S}_i$$
$$u = \dot{r}_0 - \dot{r}_e = \dot{L}_i\hat{S}_i. \tag{5.85}$$

Furthermore, mass of the ith cable at time t is equal to $m(t) = \rho_m L_i$, in which ρ_m is the cable density; therefore, $\dot{m} = \rho_m \dot{L}_i$. The position vector of the mass entrance with respect to the cable center of mass is derived as $\rho_e = -\frac{1}{2}L_i\hat{S}_i$. Finally, the velocity of a point instantaneously coinciding with the cable center of mass is denoted by $\dot{\bar{r}}$, and is calculated from

$$\dot{\bar{r}} = \dot{L}_i\hat{S}_i + \frac{L_i}{2}\omega_i \times \hat{S}_i; \tag{5.86}$$

hence

$$\dot{\rho}_0 = \dot{r}_0 - \dot{\bar{r}}$$
$$= -\frac{L_i}{2}\omega_i \times \hat{S}_i. \tag{5.87}$$

Therefore, the right-hand-side terms in Equation 5.43 can be reduced to

$$m\ddot{r}_e - \dot{m}\dot{u} - \frac{d^2}{dt^2}(m\rho_e) = 0 - \rho_m\dot{L}_i^2\hat{S}_i - \frac{d^2}{dt^2}\left(-\frac{1}{2}\rho_m L_i^2\hat{S}_i\right)$$
$$= \rho_m\left(-\dot{L}_i^2\hat{S}_i + L_i\ddot{L}_i\hat{S}_i + \dot{L}_i^2\hat{S}_i + 2L_i\dot{L}_i(\omega_i \times \hat{S}_i)\right)$$
$$+ \frac{\rho_m}{2}\left(L_i^2(\dot{\omega}_i \times \hat{S}_i) + L_i^2\omega_i \times (\omega_i \times \hat{S}_i)\right)$$
$$= \rho_m\left(L_i\ddot{L}_i\hat{S}_i + 2L_i\dot{L}_i(\omega_i \times \hat{S}_i) + \frac{1}{2}(L_i^2(\dot{\omega}_i \times \hat{S}_i) + L_i^2\omega_i \times (\omega_i \times \hat{S}_i))\right). \tag{5.88}$$

5.3.2.2 Dynamic Analysis of the Limbs

Consider free-body diagrams of the limbs and the moving platform of $4R\underline{P}R$ manipulator as what is considered in the previous section, and illustrated in Figure 5.5. The general assumptions considered in the previous section is adopted here, except for the assumption of considering the limbs having constant mass. As hereinbefore, consider the reaction forces at fixed points A_i's illustrated componentwise and denoted by $F^N_{A_i}$ and $F^S_{A_i}$, in which $\hat{\mathbf{S}}_i$ is along the limb direction and $\hat{\mathbf{N}}_i$ is perpendicular to the limb. Similarly, the internal force at points B_i's are denoted componentwise by $F^N_{B_i}$ and $F^S_{B_i}$. Furthermore, assume that the only external disturbance wrench existing acts on the moving platform and is denoted by $\boldsymbol{\mathcal{F}}_d = [f_{d_x}, f_{d_y}, \tau_d]^T$.

Let us first derive the Newton–Euler formulation for the limbs with a varying-mass approach. Referring to Equations 5.43 and 5.44, Newton–Euler formulations for variable-mass system can be written as

$$\sum \boldsymbol{F}_{\text{ext}} = m\,\ddot{\boldsymbol{r}}_e - \dot{m}\,\dot{\boldsymbol{u}} - \frac{\mathrm{d}^2}{\mathrm{d}t^2}(m\,\boldsymbol{\rho}_e) \tag{5.89}$$

$$\sum {}^{c_i}\boldsymbol{n}_{\text{ext}} = {}^{c_i}\dot{\boldsymbol{H}} - \boldsymbol{\rho}_e \times \dot{m}\,\dot{\boldsymbol{\rho}}_0, \tag{5.90}$$

in which $\Sigma \boldsymbol{F}_{\text{ext}}$ is the summation of all external forces acting on each limb and $\Sigma^{c_i}\boldsymbol{n}_{\text{ext}}$ denotes the resulting external moments about the center of mass of each limb c_i, and the right-hand-side terms of Equation 5.89 is given in Equation 5.88. Consider the free-body diagram of the limb in Figure 5.5, and determine the resulting forces and moments in a vector form as

$$\sum \boldsymbol{F}_{\text{ext}} = \left(F^S_{B_i} + F^S_{A_i}\right)\hat{\mathbf{S}}_i + \left(F^N_{B_i} + F^N_{A_i}\right)\hat{\mathbf{N}}_i + m_i\boldsymbol{g} \tag{5.91}$$

$$\sum {}^{c_i}n_{\text{ext}} = \frac{L_i}{2}\left(F^N_{B_i} - F^N_{A_i}\right)(\hat{\mathbf{S}}_i \times \hat{\mathbf{N}}_i). \tag{5.92}$$

To calculate the right-hand side of the Euler Equation 5.92, note that for planar manipulator the left-hand side is a vector in $\hat{\boldsymbol{K}}_i = \hat{\mathbf{S}}_i \times \hat{\mathbf{N}}_i$ direction; therefore, the z component of ${}^{c_i}\dot{\boldsymbol{H}}$ can be determined as follows, assuming the cables to be slender rods:

$$\begin{aligned}
{}^{c_i}\boldsymbol{H} &= I_{c_i}\boldsymbol{\omega}_i = \frac{\rho_m}{12}L_i^3\boldsymbol{\omega}_i \\
{}^{c_i}\dot{\boldsymbol{H}} &= \frac{\rho_m}{12}L_i^3\dot{\boldsymbol{\omega}}_i + \frac{\rho_m}{4}L_i^2\dot{L}_i\boldsymbol{\omega}_i.
\end{aligned} \tag{5.93}$$

Furthermore, the other term can be determined through the following manipulation:

$$\begin{aligned}
\boldsymbol{\rho}_e \times \dot{m}\,\dot{\boldsymbol{\rho}}_0 &= \left(-\frac{1}{2}L_i\hat{\mathbf{S}}_i\right) \times (\rho_m\,\dot{L}_i)\left(-\frac{1}{2}L_i\boldsymbol{\omega}_i \times \hat{\mathbf{S}}_i\right) \\
&= \frac{\rho_m}{4}L_i^2\dot{L}_i\left(\hat{\mathbf{S}}_i \times (\boldsymbol{\omega}_i \times \hat{\mathbf{S}}_i)\right) \\
&= \frac{\rho_m}{4}L_i^2\dot{L}_i\boldsymbol{\omega}_i.
\end{aligned} \tag{5.94}$$

It is notable that this term will cancel the second term in Equation 5.93, and therefore, the right-hand side of the Euler equation 5.92 simplifies to

$$ {}^{c_i}\dot{\boldsymbol{H}} - \boldsymbol{\rho}_e \times \dot{m}\,\boldsymbol{\rho}_0 = \frac{\rho_m}{12} L_i^3 \dot{\boldsymbol{\omega}}_i. \tag{5.95} $$

Substitute the above-derived equations in the Newton–Euler formulation for the limbs, namely Equations 5.89 and 5.90. This leads to

$$ \left(F_{B_i}^S + F_{A_i}^S\right)\hat{\boldsymbol{S}}_i + \left(F_{B_i}^N + F_{A_i}^N\right)\hat{\boldsymbol{N}}_i + m_i\boldsymbol{g} \tag{5.96} $$

$$ = \rho_m\left(L_i\ddot{L}_i\hat{\boldsymbol{S}}_i + 2L_i\dot{L}_i(\boldsymbol{\omega}_i \times \hat{\boldsymbol{S}}_i) + \frac{1}{2}(L_i^2(\dot{\boldsymbol{\omega}}_i \times \hat{\boldsymbol{S}}_i) + L_i^2\boldsymbol{\omega}_i \times (\boldsymbol{\omega}_i \times \hat{\boldsymbol{S}}_i))\right) $$

$$ \times \frac{L_i}{2}\left(F_{B_i}^N - F_{A_i}^N\right)\hat{\boldsymbol{K}}_i = \frac{\rho_m}{12}L_i^3\dot{\boldsymbol{\omega}}_i. \tag{5.97} $$

To write the Newton equation componentwise in $\hat{\boldsymbol{S}}_i$ and $\hat{\boldsymbol{N}}_i$ directions, first dot multiply both sides of Equation 5.96 to $\hat{\boldsymbol{S}}_i$, and then cross multiply it to $\hat{\boldsymbol{S}}_i$, and note that the Euler equation has only one component in $\hat{\boldsymbol{K}}$ direction. Hence, Equations 5.96 and 5.97 simplify into the following three equations

$$ (F_{B_i}^S + F_{A_i}^S)\hat{\boldsymbol{S}}_i = \rho_m\left(L_i\ddot{L}_i - L_i^2(\boldsymbol{\omega}_i \cdot \boldsymbol{\omega}_i) - L_i g^S\right)\hat{\boldsymbol{S}}_i \tag{5.98} $$

$$ (F_{B_i}^N + F_{A_i}^N)\hat{\boldsymbol{N}}_i = \rho_m\left(\frac{1}{2}L_i^2(\dot{\boldsymbol{\omega}}_i \times \hat{\boldsymbol{S}}_i) + 2L_i\dot{L}_i(\boldsymbol{\omega}_i \times \hat{\boldsymbol{S}}_i) - L_i g^N\hat{\boldsymbol{N}}_i\right) \tag{5.99} $$

$$ (F_{B_i}^N - F_{A_i}^N)\hat{\boldsymbol{K}} = \frac{\rho_m}{6}L_i^2\dot{\boldsymbol{\omega}}_i, \tag{5.100} $$

in which g^S and g^N represent the components of vector $\boldsymbol{g}$ in $\hat{\boldsymbol{S}}$ and $\hat{\boldsymbol{N}}$ directions, respectively, and may be derived from Equation 5.61. Substitute $\boldsymbol{\omega}_i = \dot{\alpha}\hat{\boldsymbol{K}}$, $\dot{\boldsymbol{\omega}}_i = \ddot{\alpha}\hat{\boldsymbol{K}}$ and simplify as

$$ F_{B_i}^S + F_{A_i}^S = \rho_m\left(L_i\ddot{L}_i - (L_i\dot{\alpha}_i)^2 - L_i g^S\right) \tag{5.101} $$

$$ F_{B_i}^N + F_{A_i}^N = \rho_m\left(\frac{1}{2}L_i^2\ddot{\alpha}_i + 2L_i\dot{L}_i\dot{\alpha}_i - L_i g^N\right) \tag{5.102} $$

$$ F_{B_i}^N - F_{A_i}^N = \frac{\rho_m}{6}L_i^2\ddot{\alpha}_i. \tag{5.103} $$

Note that $F_{A_i}^S$ is the actuator forces acting on the limbs. Let us denote it by $F_{A_i}^S = -\tau_i$ to consider positive tension forces in the actuators. Therefore, from Equations 5.101 through 5.103 the pivot reaction force $F_{A_i}^N$, and the interacting forces between limbs and moving platform, namely $F_{B_i}^N$ and $F_{B_i}^S$ are determined as follows:

$$ F_{A_i}^N = \rho_m\left(\frac{1}{6}L_i^2\ddot{\alpha}_i + L_i\dot{L}_i\dot{\alpha}_i - \frac{1}{2}L_i g^N\right) \tag{5.104} $$

$$ F_{B_i}^N = \rho_m\left(\frac{1}{3}L_i^2\ddot{\alpha}_i + L_i\dot{L}_i\dot{\alpha}_i - \frac{1}{2}L_i g^N\right) \tag{5.105} $$

$$ F_{B_i}^S = \tau_i + \rho_m\left(L_i\ddot{L}_i - (L_i\dot{\alpha}_i)^2 - L_i g^S\right). \tag{5.106} $$

These equations would be used in the dynamic equation of the moving platform.

There is no difference in the dynamic analysis of the moving platform considering the variable mass treatment to that of constant mass treatment given in Section 5.3.1. Therefore, Equations 5.67 through 5.69 are the governing dynamic formulations of the manipulator, in which the interaction forces between the limbs and the moving platform $F_{B_i}^S$ and $F_{B_i}^N$ are derived from newly derived limb dynamics, given in Equations 5.106 and 5.105, respectively. As before, Equations 5.67 through 5.69 can be viewed in an implicit vector form given in Equation 5.72 and may be simulated in forward or inverse dynamics forms.

To compare the effect of variable mass treatment in the dynamic formulation, first compare Equations 5.101 through 5.103 to the corresponding Equations 5.58 through 5.60. It is notable that in Equation 5.101 both the inertial terms, $L_i \ddot{L}_i$, $(L_i \dot{\alpha}_i)^2$ are doubled considering the variable mass treatment, while in Equation 5.102 only the Coriolis term $L_i \dot{L}_i \dot{\alpha}_i$ is doubled. Therefore, we expect a minor difference in the overall dynamics of the cable-driven manipulators, if the cables are not very bulky and heavy compared to the weight of the moving platform. To examine this difference quantitatively, the inverse dynamics of the manipulator, the parameters of which are given in Table 5.1 is compared for the two cases for a typical third-order trajectory already examined in Figure 5.12. The Cartesian wrench of the moving platform, $\mathcal{F} = [f_x, f_y, \tau_\phi]^T$, in both the cases are illustrated in Figure 5.17.

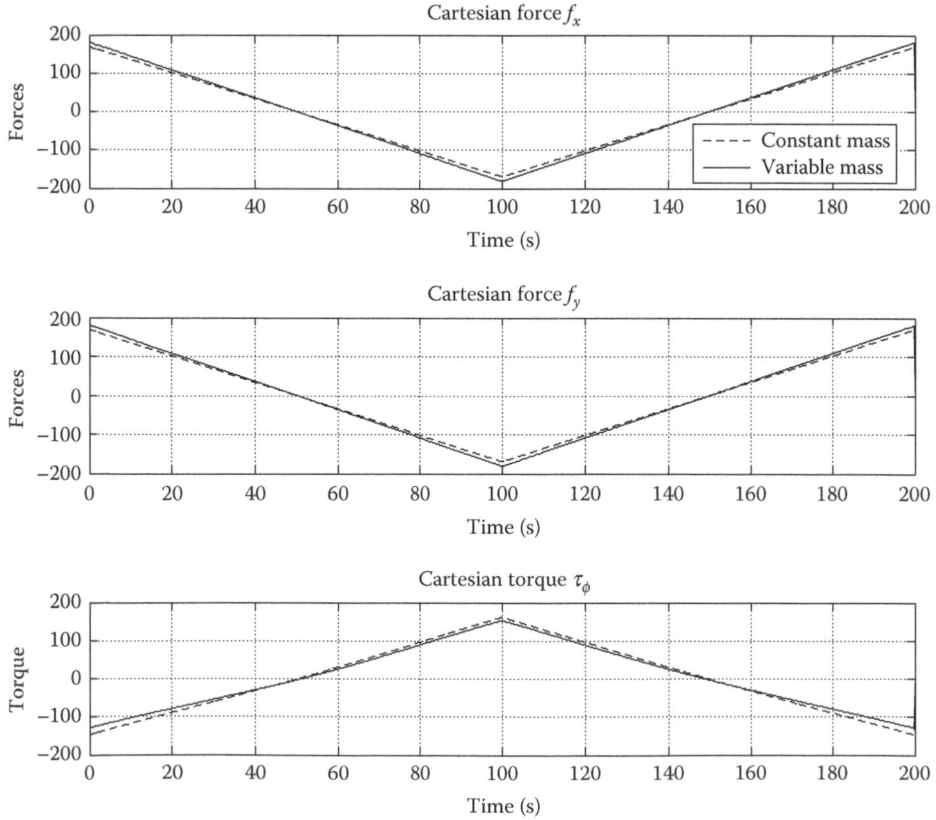

FIGURE 5.17
Comparison of the Cartesian wrench of the planar 4$R\underline{P}R$ manipulator; variable mass treatment (solid); constant mass treatment (dashed).

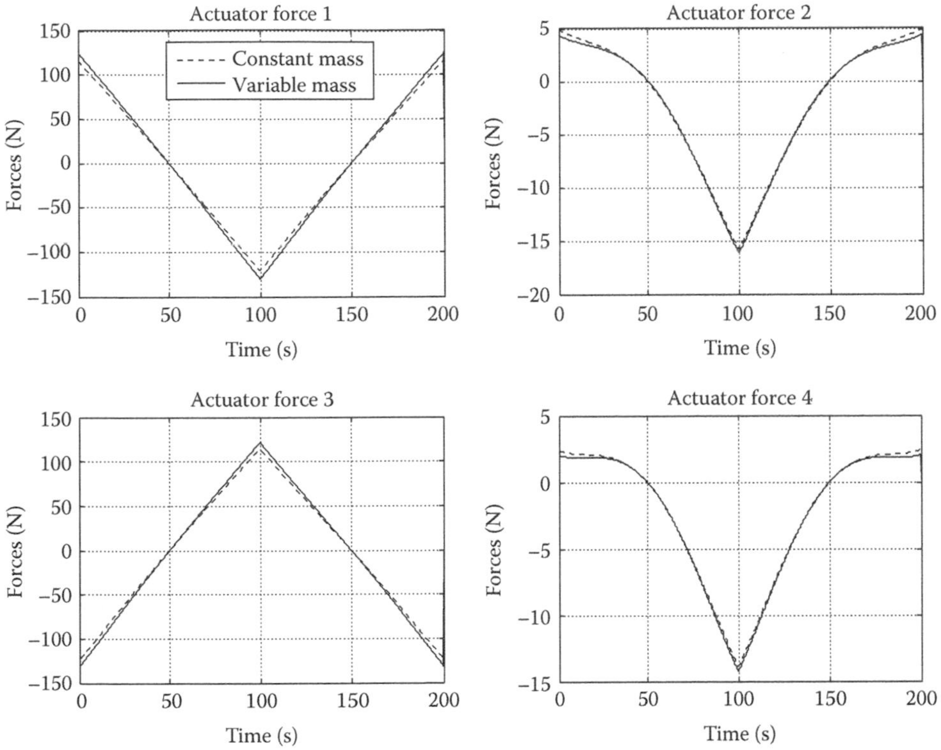

FIGURE 5.18
Comparison of the actuator forces of the planar $4R\underline{P}R$ manipulator; variable mass treatment (solid); constant mass treatment (dashed).

As seen in this figure, to generate this trajectory the Cartesian forces f_x and f_y show about 7% increase in their amount considering the variable mass treatment, while for Cartesian torque τ_ϕ this amount goes up to 12%. Therefore, the variable mass treatment contributes about 10% more Cartesian wrench required to generate such typical trajectory. Similarly, the actuator forces of the manipulator in both cases are illustrated in Figure 5.18. Similar to Cartesian wrench comparison, the required forces to generate a typical trajectory such as the one illustrated in Figure 5.12 shows an amount of increase between 7% and 15% among the actuators, considering variable mass treatment.

5.3.3 Dynamic Formulation of the Stewart–Gough Platform

In this section, dynamic analysis of the Stewart–Gough platform has been studied in detail. The architecture of this manipulator is described in Section 3.5.1 and shown in Figure 3.16. As explained earlier, in this manipulator spatial motion of the moving platform is generated by six identical piston–cylinder actuators. A complete kinematic analysis of this manipulator is given in Section 3.5 and its Jacobian analysis is presented in Section 4.8.

5.3.3.1 Acceleration Analysis

Acceleration analysis of the limbs and the moving platform has been carried out here to be employed in the Newton–Euler formulation. In acceleration analysis, it is intended to derive expressions for linear and angular accelerations of the limbs, namely $\ddot{\ell}_i$ and $\dot{\omega}_i$ as a function of the moving platform acceleration $\ddot{\mathcal{X}} = [\dot{v}_p, \dot{\omega}]^T$. To obtain such a relation, let us rewrite the velocity loop closure Equation 4.81 as

$$v_p + \omega \times b_i = \dot{\ell}_i \, \hat{s}_i + \ell_i \, (\omega_i \times \hat{s}_i). \tag{5.107}$$

Let us first define an intermediate variable v_{b_i} as the velocity of point b_i.

$$v_{b_i} = v_p + \omega \times b_i. \tag{5.108}$$

To obtain angular velocity of the limbs ω_i, cross multiply $\hat{s}_i$ to both sides of Equation 5.107.

$$\hat{s}_i \times v_p + \hat{s}_i \times (\omega \times b_i) = \ell_i \, \left(\hat{s}_i \times (\omega_i \times \hat{s}_i)\right). \tag{5.109}$$

Note that since there is no actuation torque about $\hat{s}_i$, the limb angular velocity and acceleration vectors are normal to $\hat{s}_i$ provided the following assumptions are considered for the platform:

- Both end joints of the limb are spherical. It is interesting to note that if the Stewart–Gough platform structure consists of universal joints at the base, the angular velocity and acceleration vectors of the limbs have a component along $\hat{s}_i$ [21].
- The limbs are symmetric with respect to their axes.
- The effects of friction in spherical joints are neglected.

Considering these assumptions, it can be concluded that the limbs cannot spin about their axes. Hence

$$\omega_i \cdot \hat{s}_i = \dot{\omega}_i \cdot \hat{s}_i = 0. \tag{5.110}$$

This means that vector ω_i is perpendicular to the limb axis $\hat{s}_i$, and therefore

$$\left(\hat{s}_i \times (\omega_i \times \hat{s}_i)\right) = \omega_i. \tag{5.111}$$

Taking Equations 5.108 and 5.111 into account, the angular velocity of the limb is determined by the following equation:

$$\begin{aligned} \omega_i &= \frac{1}{\ell_i} \left(\hat{s}_i \times v_p + \hat{s}_i \times (\omega \times b_i)\right) \\ &= \frac{1}{\ell_i} (\hat{s}_i \times v_{b_i}). \end{aligned} \tag{5.112}$$

Furthermore, as illustrated in Figure 5.19, the piston–cylinder structure of the limbs is decomposed into two separate parts, the masses of which are denoted by m_{i_1} and m_{i_2} as shown in this figure. Position vector of these two center of masses can be determined by the following equations:

$$p_{i_1} = a_i + c_{i_1} \hat{s}_i \tag{5.113}$$

$$p_{i_2} = a_i + (\ell_i - c_{i_2}) \hat{s}_i. \tag{5.114}$$

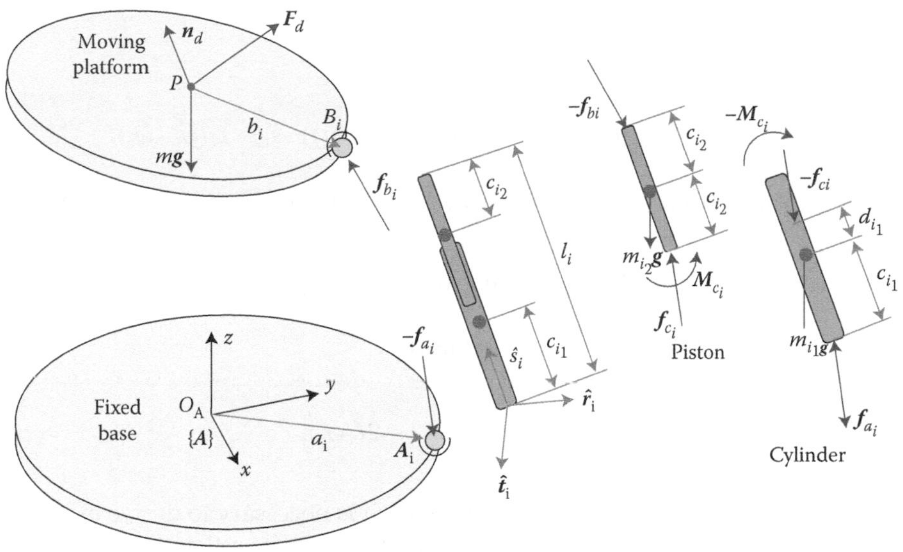

FIGURE 5.19
Free-body diagram of the limbs and the moving platform of a general Stewart–Gough manipulator.

Differentiate these two equations to find the velocities

$$v_{i_1} = c_{i_1}(\omega_i \times \hat{s}_i) \tag{5.115}$$

$$v_{i_2} = (\ell_i - c_{i_2})(\omega_i \times \hat{s}_i) + \dot{\ell}_i \hat{s}_i. \tag{5.116}$$

Now that linear velocities of the limbs' centers of masses and limbs' angular velocities are determined, acceleration terms are derived as follows. First, differentiate Equation 5.107 with respect to time

$$\begin{aligned} a_p + \dot{\omega} \times b_i + \omega \times (\omega \times b_i) &= \ddot{\ell}_i \hat{s}_i + 2\dot{\ell}_i(\omega_i \times \hat{s}_i) + \ell_i(\dot{\omega}_i \times \hat{s}_i) \\ &\quad + \ell_i \left(\omega_i \times (\omega_i \times \hat{s}_i)\right). \end{aligned} \tag{5.117}$$

Similar to Equation 5.108, linear acceleration of b_i points can be defined as the left-hand side of Equation 5.117 by

$$a_{b_i} = a_p + \dot{\omega} \times b_i + \omega \times (\omega \times b_i). \tag{5.118}$$

To eliminate $\dot{\omega}_i$ and get an expression for $\ddot{\ell}_i$, dot multiply both sides of Equation 5.117 by $\hat{s}_i$ and reorder it into

$$\begin{aligned} \ddot{\ell}_i &= a_p \cdot \hat{s}_i + (b_i \times \hat{s}_i) \cdot \dot{\omega} \\ &\quad + (\omega \times (\omega \times \hat{b}_i)) \cdot \hat{s}_i - \ell_i(\omega_i \times (\omega_i \times \hat{s}_i))) \cdot \hat{s}_i. \end{aligned} \tag{5.119}$$

Note that as indicated in Equation 5.110, the vector ω_i is perpendicular to the limb axis $\hat{s}_i$, and therefore

$$(\omega_i \times (\omega_i \times \hat{s}_i)) \cdot \hat{s}_i = -\omega_i \cdot \omega_i. \tag{5.120}$$

Hence, Equation 5.119 can be simplified into

$$\begin{aligned}\ddot{\ell}_i &= \boldsymbol{a}_{b_i} \times \hat{\boldsymbol{s}}_i + \ell_i(\boldsymbol{\omega}_i \cdot \boldsymbol{\omega}_i) \\ &= \boldsymbol{a}_p \cdot \hat{\boldsymbol{s}}_i + (\boldsymbol{b}_i \times \hat{\boldsymbol{s}}_i) \cdot \dot{\boldsymbol{\omega}} + (\boldsymbol{\omega} \times (\boldsymbol{\omega} \times \hat{\boldsymbol{b}}_i)) \cdot \hat{\boldsymbol{s}}_i + \ell_i(\boldsymbol{\omega}_i \cdot \boldsymbol{\omega}_i). \end{aligned} \tag{5.121}$$

To eliminate $\ddot{\ell}_i$ and get an expression for $\dot{\boldsymbol{\omega}}_i$, cross multiply both sides of Equation 5.117 by $\hat{\boldsymbol{s}}_i$

$$\hat{\boldsymbol{s}}_i \times (\boldsymbol{a}_p + \dot{\boldsymbol{\omega}} \times \boldsymbol{b}_i + \boldsymbol{\omega} \times (\boldsymbol{\omega} \times \boldsymbol{b}_i)) = \ell_i \dot{\boldsymbol{\omega}}_i + 2\dot{\ell}_i \boldsymbol{\omega}_i. \tag{5.122}$$

Using Equation 5.118, this can be simplified into

$$\dot{\boldsymbol{\omega}}_i = \frac{1}{\ell_i}(\hat{\boldsymbol{s}}_i \times \boldsymbol{a}_{b_i} - 2\dot{\ell}_i \boldsymbol{\omega}_i). \tag{5.123}$$

To complete the manipulator acceleration analysis, it is necessary to derive expressions for linear accelerations of the center of mass of each limb. Differentiate Equations 5.115 and 5.116 with respect to time to find the following expressions for accelerations:

$$\boldsymbol{a}_{i_1} = c_{i_1}(\dot{\boldsymbol{\omega}}_i \times \hat{\boldsymbol{s}}_i + \boldsymbol{\omega}_i \times (\boldsymbol{\omega}_i \times \hat{\boldsymbol{s}}_i)), \tag{5.124}$$

$$\boldsymbol{a}_{i_2} = (\ell_i - c_{i_2})(\dot{\boldsymbol{\omega}}_i \times \hat{\boldsymbol{s}}_i - (\boldsymbol{\omega}_i \cdot \boldsymbol{\omega}_i)\hat{\boldsymbol{s}}_i) + 2\dot{\ell}_i(\boldsymbol{\omega}_i \times \hat{\boldsymbol{s}}_i) + \ddot{\ell}_i \hat{\boldsymbol{s}}_i. \tag{5.125}$$

5.3.3.2 Dynamic Formulation of the Limbs

To derive the dynamic formulation of the Stewart–Gough platform, and as shown in Figure 5.19, decompose manipulator into the moving platform and six identical limbs. Furthermore, assume that each limb consists of two parts, the cylinder and the piston, where the velocities and accelerations of their centers of masses are determined. Assume that the centers of masses of the cylinder and the piston are located at a distance of c_{i_1} and c_{i_2} above their foot points, and their masses are denoted by m_{i_1} and m_{i_2}, respectively. Moreover, consider that the pistons are symmetric about their axes, and their centers of masses lie at their midlengths.

The free-body diagrams of the limbs and the moving platform is given in Figure 5.19. The reaction forces at fixed points A_i's are denoted by $\boldsymbol{f}_{a_i}$, the internal force at moving points B_i's are denoted by $\boldsymbol{f}_{b_i}$, and the internal forces and moments between cylinders and pistons are denoted by $\boldsymbol{f}_{c_i}$ and $\boldsymbol{M}_{c_i}$, respectively. The gravitational forces of the limb segments and the moving platform are also shown in this figure. Assume that the only existing external disturbance wrench is applied on the moving platform and is denoted by $\boldsymbol{\mathcal{F}}_d = [\boldsymbol{F}_d, \boldsymbol{n}_d]^T$. To define internal and interaction forces applied on the limb segments componentwise, define unit vectors $\hat{\boldsymbol{t}}_i$ and $\hat{\boldsymbol{r}}_i$ as follows:

$$\hat{\boldsymbol{t}}_i = \frac{\hat{\boldsymbol{s}}_i \times \boldsymbol{a}_i}{\|\hat{\boldsymbol{s}}_i \times \boldsymbol{a}_i\|}, \tag{5.126}$$

$$\hat{\boldsymbol{r}}_i = \hat{\boldsymbol{s}}_i \times \hat{\boldsymbol{t}}_i. \tag{5.127}$$

Considering these assumptions and definitions, one can derive each limb segment equations of motion as following. Referring to Equations 5.31 and 5.33, the Newton–Euler equations for cylinder segments are given as

$$\sum F_{\text{ext}} = m_{i_1} a_{i_1}$$
$$= f_{a_i} - f_{c_i} + m_{i_1} g \tag{5.128}$$
$$\sum {}^{c_{i_1}} n_{\text{ext}} = {}^{A}I_{c_{i_1}}\, \dot{\omega}_i + \omega_i \times {}^{A}I_{c_{i_1}} \cdot \omega_i$$
$$= c_{i_1}(-\hat{s}_i \times f_{a_i}) + d_{i_1}(\hat{s}_i \times -f_{c_i}) - M_{c_i} \tag{5.129}$$

in which

$$d_{i_1} = \ell_i - c_{i_1} - 2c_{i_2}. \tag{5.130}$$

Note that in this formulation, ${}^{A}I_{c_{i_1}}$ denotes the inertia matrix of the cylinder evaluated at a frame parallel to $\{A\}$ located at point c_{i_1}. Since the leg is moving with respect to this frame, this matrix can be evaluated by rotation rule given in Equation 5.16 from the inertia matrix easily obtained about $\{A_i\}$ attached to the cylinder at point c_{i_1}. This can be done through the following transformation:

$${}^{A}I_{c_{i_1}} = {}^{A}R_{A_i}\, {}^{A_i}I_{c_{i_1}}\, {}^{A}R_{A_i}^{T}. \tag{5.131}$$

Furthermore, ΣF_{ext} denotes the summation of all external forces acting on each cylinder and $\Sigma^{c_{i_1}} n_{\text{ext}}$ denotes the resulting external moments about the center of mass c_{i_1} of each cylinder. Linear acceleration at the center of mass, a_{i_1}, limb angular velocity ω_i, and angular acceleration $\dot{\omega}_i$ are given in Equations 5.124, 5.112, and 5.123, respectively. In a similar way, Newton–Euler equations for the piston segments are derived as follows:

$$\sum F_{\text{ext}} = m_{i_2} a_{i_2}$$
$$= f_{c_i} - f_{b_i} + m_{i_2} g \tag{5.132}$$
$$\sum {}^{c_{i_2}} n_{\text{ext}} = {}^{A}I_{c_{i_2}}\, \dot{\omega}_i + \omega_i \times {}^{A}I_{c_{i_2}} \cdot \omega_i$$
$$= c_{i_2}(-\hat{s}_i \times f_{c_i}) + c_{i_2}(\hat{s}_i \times -f_{b_i}) + M_{c_i} \tag{5.133}$$

in which

$${}^{A}I_{c_{i_2}} = {}^{A}R_{A_i}\, {}^{A_i}I_{c_{i_2}}\, {}^{A}R_{A_i}^{T} \tag{5.134}$$

and acceleration of the piston center of mass a_{i_2} is given in Equation 5.125. Since moving frame $\{A_i\}$ is attached to the limb i, by using unit vector $\hat{s}_i$, $\hat{t}_i$, and $\hat{r}_i$ defined in Equations 5.126 and 5.127, the rotation matrix ${}^{A}R_{A_i}$ can be represented by

$${}^{A}R_{A_i} = \begin{bmatrix} \hat{s}_i \cdot \hat{x} & \hat{t}_i \cdot \hat{x} & \hat{r}_i \cdot \hat{x} \\ \hat{s}_i \cdot \hat{y} & \hat{t}_i \cdot \hat{y} & \hat{r}_i \cdot \hat{y} \\ \hat{s}_i \cdot \hat{z} & \hat{t}_i \cdot \hat{z} & \hat{r}_i \cdot \hat{z} \end{bmatrix}, \tag{5.135}$$

in which $\hat{x}$, $\hat{y}$, and $\hat{z}$ are unit vectors of the fixed frame A. Moreover, as described hereinbefore, if the limbs are considered to be symmetric, then ${}^{A_i}I_{c_i}$ is diagonal

$${}^{A_i}I_{c_i} = \begin{bmatrix} I_{xx_i} & 0 & 0 \\ 0 & I_{xx_i} & 0 \\ 0 & 0 & I_{zz_i} \end{bmatrix}. \tag{5.136}$$

To simplify and eliminate internal forces in Equations 5.128 through 5.133, we may start using 5.128 to derive $\boldsymbol{f}_{a_i}$ as a function of $\boldsymbol{f}_{c_i}$

$$\boldsymbol{f}_{a_i} = m_{i_1}\boldsymbol{a}_{i_1} + \boldsymbol{f}_{c_i} - m_{i_1}\boldsymbol{g}. \tag{5.137}$$

Substitute $\boldsymbol{f}_{a_i}$ in Equation 5.133. Through some manipulations $\boldsymbol{M}_{c_i}$ can be obtained as

$$\begin{aligned}\boldsymbol{M}_{c_i} &= -c_{i_1}\hat{\boldsymbol{s}}_i \times \left(m_{i_1}\boldsymbol{a}_{i_1} + \boldsymbol{f}_{c_i} - m_{i_1}\boldsymbol{g}\right) + d_{i_1}\left(\hat{\boldsymbol{s}}_i \times -\boldsymbol{f}_{c_i}\right)\\ &\quad - {}^{A}\boldsymbol{I}_{c_{i_1}}\dot{\boldsymbol{\omega}}_i - \boldsymbol{\omega}_i \times {}^{A}\boldsymbol{I}_{c_{i_1}}\boldsymbol{\omega}_i.\end{aligned} \tag{5.138}$$

Moreover, to separate the actuator force $\tau_i\hat{\boldsymbol{s}}_i$ from internal force between cylinder and piston, $\boldsymbol{f}_{c_i}$ may be rewritten as

$$\boldsymbol{f}_{c_i} = \boldsymbol{f}^n_{c_i} + \boldsymbol{f}^s_{c_i} = \boldsymbol{f}^n_{c_i} + \tau_i\hat{\boldsymbol{s}}_i, \tag{5.139}$$

in which $\boldsymbol{f}^n_{c_i}$ denotes the component of the force in a direction normal to $\hat{\boldsymbol{s}}_i$. Substitute $\boldsymbol{f}_{c_i}$, $\boldsymbol{a}_{c_{i_1}}$, and d_{i_1} into Equation 5.138. This immediately leads into an equation for $\boldsymbol{M}_{c_i}$ as a function of $\boldsymbol{f}^n_{c_i}$ and the actuator forces τ_i. By using vector triple product rules and considering the fact that $\hat{\boldsymbol{s}}_i \times \boldsymbol{f}_{c_i} = \boldsymbol{f}^n_{c_i}$, the final formulation for $\boldsymbol{M}_{c_i}$ may be obtained as

$$\begin{aligned}\boldsymbol{M}_{c_i} &= -m_{i_1}c^2_{i_1}\dot{\boldsymbol{\omega}}_i - {}^{A}\boldsymbol{I}_{c_{i_1}}\dot{\boldsymbol{\omega}}_i - \boldsymbol{\omega}_i \times {}^{A}\boldsymbol{I}_{c_{i_1}}\boldsymbol{\omega}_i - \left(\ell_i - 2c_{i_2}\right)\left(\hat{\boldsymbol{s}}_i \times \boldsymbol{f}^n_{c_i}\right)\\ &\quad + m_{i_1}c_{i_1}\hat{\boldsymbol{s}}_i \times \boldsymbol{g}.\end{aligned} \tag{5.140}$$

The simplified form of $\hat{\boldsymbol{s}}_i \times \boldsymbol{M}_{c_i}$ is also given for later use as

$$\begin{aligned}\hat{\boldsymbol{s}}_i \times \boldsymbol{M}_{c_i} &= -m_{i_1}c^2_{i_1}\left(\hat{\boldsymbol{s}}_i \times \dot{\boldsymbol{\omega}}_i\right) - \hat{\boldsymbol{s}} \times \left({}^{A}\boldsymbol{I}_{c_{i_1}}\dot{\boldsymbol{\omega}}_i\right)\\ &\quad - \left(\hat{\boldsymbol{s}}_i . \left({}^{A}\boldsymbol{I}_{c_{i_1}}\boldsymbol{\omega}_i\right)\right)\boldsymbol{\omega}_i + \left(\ell_i - 2c_{i_2}\right)\boldsymbol{f}^n_{c_i} + m_{i_1}c_{i_1}\hat{\boldsymbol{s}}_i \times \left(\hat{\boldsymbol{s}}_i \times \boldsymbol{g}\right).\end{aligned} \tag{5.141}$$

To derive $\boldsymbol{f}_{b_i}$ in a vector form, it is separated into two components

$$\boldsymbol{f}_{b_i} = \boldsymbol{f}^n_{b_i} + \boldsymbol{f}^s_{b_i}, \tag{5.142}$$

in which $\boldsymbol{f}^s_{b_i}$ and $\boldsymbol{f}^n_{b_i}$ are the contributions of $\boldsymbol{f}_{b_i}$ along $\hat{\boldsymbol{s}}_i$ and normal to $\hat{\boldsymbol{s}}_i$, respectively. This separation results into a better elimination of internal forces and thus simplification of dynamic equations for each limb. Note that while $\boldsymbol{f}^n_{b_i}$ and $\boldsymbol{f}^n_{c_i}$ are normal to $\hat{\boldsymbol{s}}_i$, they are not generally parallel to each other. Substituting $\boldsymbol{f}_{c_i}$, $\boldsymbol{f}_{b_i}$, and $\boldsymbol{a}_{c_{i_2}}$ into Equation 5.132, with some simplifications the following equation for the piston is derived:

$$\begin{aligned}\boldsymbol{f}^n_{b_i} - \boldsymbol{f}^n_{c_i} + \boldsymbol{f}^s_{b_i} &= \tau_i\hat{\boldsymbol{s}}_i + m_{i_2}\boldsymbol{g} - m_{i_2}\left(\left(\ell_i - c_{i_2}\right)\right.\\ &\quad \left.\times \left(\dot{\boldsymbol{\omega}}_i \times \hat{\boldsymbol{s}}_i - (\boldsymbol{\omega}_i \cdot \boldsymbol{\omega}_i)\hat{\boldsymbol{s}}_i\right) + 2\dot{l}_i\left(\boldsymbol{\omega}_i \times \hat{\boldsymbol{s}}_i\right) + \ddot{l}_i\hat{\boldsymbol{s}}_i\right).\end{aligned} \tag{5.143}$$

Equation 5.143 reveals the fact that separating $\boldsymbol{f}_{b_i}$ into its two n and s components helps greatly to simplify the equations. Note that by dot multiplying Equation 5.143 by $\hat{\boldsymbol{s}}_i$, the component $f^s_{b_i}$ can be derived as a function of only the actuator force and kinematic parameters of the limb

$$f^s_{b_i} = \tau_i \hat{s}_i + m_{i_2}\left(\ell_i - c_{i_2}\right)(\omega_i \cdot \omega_i)\hat{s}_i - m_{i_2}\ddot{\ell}_i\hat{s}_i + m_{i_2}\left(g \cdot \hat{s}_i\right)\hat{s}_i. \tag{5.144}$$

Cross multiply $\hat{s}_i$ by Equations 5.128 and 5.132, and reorder them to find the following equations:

$$\hat{s}_i \times f_{c_i} = \hat{s}_i \times f_{b_i} + m_{i_2}\hat{s}_i \times (a_{i_2} - g) \tag{5.145}$$

$$\begin{aligned}\hat{s}_i \times f_{a_i} &= \hat{s}_i \times f_{c_i} + m_{i_1}\hat{s}_i \times (a_{i_1} - g)\\ &= \hat{s}_i \times f_{b_i} + m_{i_2}\hat{s}_i \times (a_{i_2} - g) + m_{i_1}\hat{s}_i \times (a_{i_1} - g).\end{aligned} \tag{5.146}$$

Now add Equations 5.129 and 5.133, and substitute $\hat{s}_i \times f_{c_i}$ and $\hat{s}_i \times f_{a_i}$

$$\begin{aligned}&\left({}^A I_{c_{i_1}} + {}^A I_{c_{i_2}}\right)\dot{\omega}_i + \omega_i \times \left({}^A I_{c_{i_1}} + {}^A I_{c_{i_2}}\right)\omega_i\\ &= -c_{i_1}(\hat{s}_i \times f_{a_i}) - (\ell_i - c_{i_1} - c_{i_2})(\hat{s}_i \times f_{c_i}) - c_{i_2}(\hat{s}_i \times f_{b_i})\\ &= -c_{i_1} m_{i_1}\hat{s}_i \times (a_{i_1} - g) - (\ell_i - c_{i_2}) m_{i_2}\hat{s}_i \times (a_{i_2} - g) - \ell_i(\hat{s}_i \times f_{b_i}).\end{aligned} \tag{5.147}$$

This equation can be simplified to

$$\begin{aligned}{}^A I_{\text{eq}}\dot{\omega}_i + \omega_i \times {}^A I_{\text{eq}}\omega_i = &-c_{i_1} m_{i_1}\hat{s}_i \times a_{i_1} - (\ell_i - c_{i_2}) m_{i_2}\hat{s}_i\\ &\times a_{i_2} - \ell_i(\hat{s}_i \times f_{b_i}) + \ell_i m_{g_e}\hat{s}_i \times g,\end{aligned} \tag{5.148}$$

in which

$${}^A I_{\text{eq}} = \left({}^A I_{c_{i_1}} + {}^A I_{c_{i_2}}\right), \quad m_{g_e} = \frac{1}{\ell_i}\left(m_{i_1} c_{i_1} + m_{i_2}\left(\ell_i - c_{i_2}\right)\right). \tag{5.149}$$

The dynamic equation of the limbs can be further simplified considering the symmetricity assumption in limb segments. Through the following manipulations it can be seen that the left-hand side of Equation 5.148 may be significantly simplified. Note that if both the cylinder and piston are symmetric, their inertia matrix with respect to the frame $\{A_i\}$, which is attached to the limb, is diagonal. Hence, the equivalent moment of inertia ${}^{A_i}I_{\text{eq}}$, defined as the summation of the inertia matrices of these two segments is also diagonal and has the following form:

$${}^{A_i}I_{\text{eq}} = {}^{A_i}I_{c_{i_1}} + {}^{A_i}I_{c_{i_2}} = \begin{bmatrix} I_{xx_i} & 0 & 0\\ 0 & I_{xx_i} & 0\\ 0 & 0 & I_{zz_i}\end{bmatrix}, \quad \begin{aligned} I_{xx_i} &= I_{xx_{i_1}} + I_{xx_{i_2}}\\ I_{zz_i} &= I_{zz_{i_1}} + I_{zz_{i_2}}.\end{aligned} \tag{5.150}$$

Furthermore, note that ω_i and $\dot{\omega}_i$ have no component along the $\hat{z}_i$ direction, therefore

$$\begin{aligned}{}^A I_{\text{eq}}\dot{\omega}_i &= \left({}^A R_{A_i}\,{}^{A_i} I_{\text{eq}}\,{}^A R^T_{A_i}\right){}^A\dot{\omega}_i = {}^A R_{A_i}\,{}^{A_i} I_{\text{eq}}\,{}^{A_i}\dot{\omega}_i\\ &= {}^A R_{A_i} I_{xx_i}\,{}^{A_i}\dot{\omega}_i = I_{xx_i}\,{}^A\dot{\omega}_i.\end{aligned} \tag{5.151}$$

Similarly, it can be shown that

$$\omega_i \times {}^A I_{\text{eq}}\dot{\omega}_i = 0. \tag{5.152}$$

These relations significantly simplify Equation 5.148 to

$$I_{xx_i}\dot{\boldsymbol{\omega}}_i = -\ell_i(\hat{\mathbf{s}}_i \times \boldsymbol{f}_{b_i}) - c_{i_1}m_{i_1}\hat{\mathbf{s}}_i \times \boldsymbol{a}_{i_1} - (\ell_i - c_{i_2})m_{i_2}\hat{\mathbf{s}}_i \times \boldsymbol{a}_{i_2} + \ell_i m_{g_e}\hat{\mathbf{s}}_i \times \boldsymbol{g}. \tag{5.153}$$

To find $f^n_{b_i}$, cross multiply Equation 5.153 by $\hat{\mathbf{s}}_i$, reorder and simplify.

$$\boldsymbol{f}_{b_i} = \frac{1}{\ell_i}(I_{xx_i} + \ell_i^2 m_{c_e})\hat{\mathbf{s}}_i \times \dot{\boldsymbol{\omega}}_i + \frac{2}{\ell_i} m_{i_2}c_{i_2}\dot{\ell}_i\hat{\mathbf{s}}_i \times \boldsymbol{\omega}_i - m_{g_e}\hat{\mathbf{s}}_i \times (\hat{\mathbf{s}}_i \times \boldsymbol{g}) \tag{5.154}$$

in which m_{c_e} is defined as

$$m_{c_e} = \frac{1}{\ell_i^2}\left(m_{i_1}c_{i_1}^2 + m_{i_2}c_{i_2}^2\right). \tag{5.155}$$

Equations 5.144 and 5.154 describe two components of the interaction force $\boldsymbol{f}_{b_i}$ as a function of kinematic parameters of each limb. This interaction force $\boldsymbol{f}_{b_i}$ is used in the Newton–Euler formulation of the moving platform, detailed in the following section. If the other interacting forces and torques like $\boldsymbol{f}_{a_i}, \boldsymbol{f}_{c_i}$, and $\boldsymbol{M}_{c_i}$ need to be evaluated for design purposes, it can be readily done, similar to that carried out for $\boldsymbol{f}_{b_i}$.

5.3.3.3 Dynamic Formulation of the Moving Platform

Assume that the moving platform center of mass is located at the center point P and it has a mass of m and moment of inertia ${}^A\boldsymbol{I}_P$. Furthermore, consider that gravitational force and external disturbance wrench are applied on the moving platform, $\boldsymbol{\mathcal{F}}_d = [\boldsymbol{F}_d, \boldsymbol{n}_d]^T$ as depicted in the free-body diagram (Figure 5.19). The Newton–Euler formulation of the moving platform is as follows:

$$\sum \boldsymbol{F}_{\text{ext}} = \sum_{i=1}^{6} \boldsymbol{f}_{b_i} + m\boldsymbol{g} + \boldsymbol{F}_d = m\,\boldsymbol{a}_p \tag{5.156}$$

$$\sum {}^p\boldsymbol{n}_{\text{ext}} = \boldsymbol{n}_d + \sum_{i=1}^{6} \boldsymbol{b}_i \times \boldsymbol{f}_{b_i} = {}^A\boldsymbol{I}_P\,\dot{\boldsymbol{\omega}} + \boldsymbol{\omega} \times {}^A\boldsymbol{I}_P\boldsymbol{\omega}, \tag{5.157}$$

in which ${}^A\boldsymbol{I}_P$ is considered in the fixed frame $\{\boldsymbol{A}\}$ and can be calculated by

$${}^A\boldsymbol{I}_P = {}^A\boldsymbol{R}_B\,{}^B\boldsymbol{I}_P\,{}^A\boldsymbol{R}_B{}^T. \tag{5.158}$$

These equations can be rewritten in an implicit form as

$$m\,(\boldsymbol{a}_p - \boldsymbol{g}) - \boldsymbol{F}_d - \sum_{i=1}^{6} \boldsymbol{f}_{b_i} = \mathbf{0} \tag{5.159}$$

$${}^A\boldsymbol{I}_P\,\dot{\boldsymbol{\omega}} + \boldsymbol{\omega} \times {}^A\boldsymbol{I}_P\boldsymbol{\omega} - \boldsymbol{n}_d - \sum_{i=1}^{6} \boldsymbol{b}_i \times \boldsymbol{f}_{b_i} = \mathbf{0}. \tag{5.160}$$

The two vector Equations 5.159 and 5.160 are the governing dynamic formulation of the Stewart–Gough platform, in which $\boldsymbol{\mathcal{F}}_d = [\boldsymbol{F}_d, \boldsymbol{n}_d]^T$ denotes the disturbance wrench exerted on the moving platform, while the interaction forces between the limbs and the moving platform f_{b_i} are given in Equation 5.154. Equations 5.159 and 5.160 can be viewed in an implicit vector form of

$$f(\mathcal{X}, \dot{\mathcal{X}}, \ddot{\mathcal{X}}, \mathcal{F}_d, \boldsymbol{\tau}) = \mathbf{0} \tag{5.161}$$

in which

$$\mathcal{X} = \begin{bmatrix} x_p \\ \boldsymbol{\theta} \end{bmatrix} \tag{5.162}$$

is the motion variable of the moving platform consisting of the linear position of point P denoted by $\boldsymbol{x}_p = [x_p \;\; y_p \;\; z_p]^T$, and the moving platform orientation represented by screw coordinates

$$\boldsymbol{\theta} = \theta \begin{bmatrix} s_x & s_y & s_z \end{bmatrix}^T = \begin{bmatrix} \theta_x & \theta_y & \theta_z \end{bmatrix}^T. \tag{5.163}$$

Note that of which contrary to the Euler angles, screw coordinates $\boldsymbol{\theta}$ is a vector, the derivative is also a vector, representing angular velocity of the moving platform, $\dot{\boldsymbol{\theta}} = \boldsymbol{\omega}$, and therefore, $\dot{\mathcal{X}} = [\boldsymbol{v}_p, \boldsymbol{\omega}]^T$ represents the moving platform twist. Furthermore, $\mathcal{F}_d = [\boldsymbol{F}_d, \boldsymbol{n}_d]^T$ is the external disturbance wrench applied to the moving platform, and $\boldsymbol{\tau} = [\tau_1 \;\; \tau_2 \;\; \ldots \;\; \tau_6]^T$ is the vector of the actuator forces. These equations can be implemented in forward or inverse dynamics simulations, similar to that reported in Section 5.3.1.

5.3.4 Closed-Form Dynamics

While dynamic formulation in the form of Equation 5.161 can be used to simulate inverse dynamics of the Stewart–Gough platform, its implicit nature makes it unpleasant for the dynamic analysis and control. Hence, in this section we introduce a method to reformulate dynamic formulation into a closed form comparable to that usually obtained for serial manipulators [145].

5.3.4.1 Closed-Form Dynamics of the Limbs

To derive a closed-form dynamic formulation for the Stewart–Gough platform as

$$M(\mathcal{X})\ddot{\mathcal{X}} + C(\mathcal{X}, \dot{\mathcal{X}})\dot{\mathcal{X}} + G(\mathcal{X}) = \mathcal{F}, \tag{5.164}$$

first consider an intermediate generalized coordinate $\boldsymbol{x}_i$, which is in fact the position of point b_i. This generalized coordinate is used to harmonize the limb and the moving platform dynamic formulation and to derive a closed-form structure for the whole manipulator. Now, manipulate each limb dynamic equations to convert them into the closed form. Let us first introduce some relations to substitute kinematic parameters like $\dot{\omega}_i, \ddot{\omega}_i, \ddot{\ell}_i$ with the intermediate generalized coordinate $\boldsymbol{x}_i$ and its time derivatives. To do that, vector multiplications such as dot and cross products should be transformed into their corresponding matrix multiplication. Note that for any three arbitrary vectors $\boldsymbol{a}, \boldsymbol{b}$, and $\boldsymbol{c}$ the following equalities hold

$$(\boldsymbol{a} \cdot \boldsymbol{b})\, \boldsymbol{c} = (\boldsymbol{a}^T \boldsymbol{b})\boldsymbol{c} = \boldsymbol{c}(\boldsymbol{a}^T \boldsymbol{b}) = (\boldsymbol{c}\boldsymbol{a}^T)\boldsymbol{b} = \boldsymbol{c}\boldsymbol{a}^T \boldsymbol{b}. \tag{5.165}$$

Similarly, for cross multiplication of $\boldsymbol{a} \times \boldsymbol{b}$ we can use matrix multiplication

$$\boldsymbol{a} \times \boldsymbol{b} = \boldsymbol{a}_{\times} \boldsymbol{b} = -\boldsymbol{b}_{\times} \boldsymbol{a} \tag{5.166}$$

in which $\boldsymbol{a}_{\times}$ and $\boldsymbol{b}_{\times}$ denote skew-symmetric matrices derived from elements of vectors $\boldsymbol{a}$ and $\boldsymbol{b}$, denoted by $\boldsymbol{a} = \begin{bmatrix} a_x & a_y & a_z \end{bmatrix}^T$ and $\boldsymbol{b} = \begin{bmatrix} b_x & b_y & b_z \end{bmatrix}^T$, respectively by

$$\boldsymbol{a}^{\times} = \boldsymbol{a}_{\times} \doteq \begin{bmatrix} 0 & -a_z & a_y \\ a_z & 0 & -a_x \\ -a_y & a_x & 0 \end{bmatrix}. \tag{5.167}$$

As defined in Equation 4.6, this matrix representation can be used to replace vector cross product by matrix multiplication. To relate $\boldsymbol{x}_i$ and its derivatives to the limb kinematic variables, the following relations are useful:

$$\begin{aligned} \dot{l}_i &= \hat{\mathbf{s}}_i^T \dot{\boldsymbol{x}}_i \Rightarrow \dot{l}_i \hat{\mathbf{s}}_i = \hat{\mathbf{s}}_i \hat{\mathbf{s}}_i^T \dot{\boldsymbol{x}}_i \\ \boldsymbol{\omega}_i &= \frac{1}{\ell_i} \hat{\mathbf{s}}_{i\times} \dot{\boldsymbol{x}}_i \Rightarrow |\boldsymbol{\omega}_i|_2^2 \, \hat{\mathbf{s}}_i = \frac{-1}{{\ell_i}^2} \hat{\mathbf{s}}_i \dot{\boldsymbol{x}}_i^T \hat{\mathbf{s}}_{i\times}^2 \dot{\boldsymbol{x}}_i. \end{aligned} \tag{5.168}$$

Furthermore, for linear and angular accelerations, we can derive

$$\begin{aligned} \ddot{l}_i - \ell_i \boldsymbol{\omega}_i \cdot \boldsymbol{\omega}_i &= \hat{\mathbf{s}}_i^T \ddot{\boldsymbol{x}}_i \Rightarrow \ddot{l}_i \hat{\mathbf{s}}_i - \ell_i \left(\boldsymbol{\omega}_i \cdot \boldsymbol{\omega}_i\right) \hat{\mathbf{s}}_i = \hat{\mathbf{s}}_i \hat{\mathbf{s}}_i^T \ddot{\boldsymbol{x}}_i \\ l_i \dot{\boldsymbol{\omega}}_i + 2\dot{l}_i \boldsymbol{\omega}_i &= \hat{\mathbf{s}}_{i\times} \ddot{\boldsymbol{x}}_i \Rightarrow \dot{\boldsymbol{\omega}}_i = \frac{1}{\ell_i} \left(\hat{\mathbf{s}}_{i\times} \ddot{\boldsymbol{x}}_i - 2\dot{l}_i \boldsymbol{\omega}_i\right). \end{aligned} \tag{5.169}$$

Transformations 5.168 and 5.169 are the main simplifying tools for deriving closed-form dynamic formulations of the limbs from their implicit formulations. Substituting kinematic parameters into dynamic equations 5.144 and 5.154 through some simplifications, $\boldsymbol{f}_{b_i}^s$ and $\boldsymbol{f}_{b_i}^n$ can be converted into the following form:

$$\boldsymbol{f}_{b_i}^s - \tau_i \hat{\mathbf{s}}_i = -m_{i_2} \hat{\mathbf{s}}_i \hat{\mathbf{s}}_i^T \ddot{\boldsymbol{x}}_i + \frac{m_{i_2} c_{i_2}}{\ell_i^2} \hat{\mathbf{s}}_i \dot{\boldsymbol{x}}_i^T \hat{\mathbf{s}}_{i\times}^2 \dot{\boldsymbol{x}}_i + m_{i_2} \hat{\mathbf{s}}_i \hat{\mathbf{s}}_i^T \boldsymbol{g} \tag{5.170}$$

$$\boldsymbol{f}_{b_i}^n = \frac{1}{\ell_i^2} (I_{xx_i} + \ell_i^2 m_{c_e}) \hat{\mathbf{s}}_{i\times}^2 \ddot{\boldsymbol{x}}_i + \frac{2}{\ell_i} m_{c_o} \dot{\ell}_i \mathbf{s}_{i\times}^2 \dot{\boldsymbol{x}}_i - m_{g_e} \mathbf{s}_{i\times}^2 \boldsymbol{g}, \tag{5.171}$$

in which

$$m_{c_o} = \frac{1}{\ell_i} m_{i_2} c_{i_2} - \frac{1}{\ell_i^2} (I_{xx_i} + \ell_i^2 m_{c_e}), \quad m_{g_e} = \frac{1}{\ell_i} \left(m_{i_1} c_{i_1} + m_{i_2} \left(\ell_i - c_{i_2}\right)\right). \tag{5.172}$$

Note that as shown in the free-body diagram of the piston in Figure 5.19, $-\boldsymbol{f}_{b_i}$ is considered to be acting on the piston. Therefore, add Equations 5.170 and 5.171, and negate the resulting equation to derive $-\boldsymbol{f}_{b_i}$. Then factor $\ddot{\boldsymbol{x}}_i$ and $\dot{\boldsymbol{x}}_i$ from the resulting equation to yield to a single equation for the dynamics of the limb i based on the intermediate generalized coordinate $\boldsymbol{x}_i$. This equation may be written in a closed form as follows:

$$\boldsymbol{M}_i \ddot{\boldsymbol{x}}_i + \boldsymbol{C}_i \dot{\boldsymbol{x}}_i + \boldsymbol{G}_i = \boldsymbol{F}_i, \tag{5.173}$$

the corresponding mass matrix $\boldsymbol{M}_i$, the Coriolis and centrifugal matrix $\boldsymbol{C}_i$, and the gravity vector G_i can be simplified into the following form:

$$\begin{aligned}
\boldsymbol{M}_i &= m_{i_2}\hat{\boldsymbol{s}}_i\hat{\boldsymbol{s}}_i^T - \frac{1}{\ell_i^2}I_{xx_i}\hat{\boldsymbol{s}}_{i\times}^2 - m_{c_e}\hat{\boldsymbol{s}}_{i\times}^2 \\
\boldsymbol{C}_i &= -\frac{2}{\ell_i}m_{c_o}\dot{\ell}_i\hat{\boldsymbol{s}}_{i\times}^2 - \frac{1}{\ell_i^2}m_{i_2}c_{i_2}\hat{\boldsymbol{s}}_i\dot{\boldsymbol{x}}_i^T\hat{\boldsymbol{s}}_{i\times}^2 \\
\boldsymbol{G}_i &= \left(m_{g_e}\hat{\boldsymbol{s}}_{i\times}^2 - m_{i_2}\hat{\boldsymbol{s}}_i\hat{\boldsymbol{s}}_i^T\right)\boldsymbol{g} \\
\boldsymbol{F}_i &= -\boldsymbol{f}_{b_i} + \tau_i\hat{\boldsymbol{s}}_i,
\end{aligned} \tag{5.174}$$

in which all the intermediate parameters given in Equations 5.149, 5.155, and 5.172 are given here for convenience.

$$\begin{aligned}
\boldsymbol{I}_{\text{eq}} &= \left({}^A\boldsymbol{I}_{c_{i_1}} + {}^A\boldsymbol{I}_{c_{i_2}}\right), \quad m_{c_e} = \frac{1}{\ell_i^2}\left(m_{i_1}c_{i_1}^2 + m_{i_2}c_{i_2}^2\right), \\
m_{c_o} &= \frac{1}{\ell_i}m_{i_2}c_{i_2} - \frac{1}{\ell_i^2}(I_{xx_i} + \ell_i^2 m_{c_e}), \quad m_{g_e} = \frac{1}{\ell_i}\left(m_{i_1}c_{i_1} + m_{i_2}\left(\ell_i - c_{i_2}\right)\right).
\end{aligned} \tag{5.175}$$

5.3.4.2 *Closed-Form Dynamics of the Moving Platform*

In this section, the moving platform equations of motion given in Equations 5.159 and 5.160 are transformed into the following closed-form formulation

$$\boldsymbol{M}_p\ddot{\boldsymbol{\mathcal{X}}} + \boldsymbol{C}_p\dot{\boldsymbol{\mathcal{X}}} + \boldsymbol{G}_p = \boldsymbol{\mathcal{F}}_p, \tag{5.176}$$

in which $\boldsymbol{\mathcal{X}}$ denotes a set of generalized coordinates for the position and orientation of the moving platform. As mentioned hereinbefore, the motion variable of the moving platform consists of six coordinates, the first three coordinates $\boldsymbol{x}_p$ represent linear motion of the moving platform, and the last three coordinates $\boldsymbol{\theta}$ its angular motion. It is preferable to use screw coordinates defined in Equation 5.163 for angular representation. Since, contrary to any Euler angles the screw coordinates $\boldsymbol{\theta}$ is a vector, the derivative of which is also a vector representing angular velocity of the moving platform, $\dot{\boldsymbol{\theta}} = \boldsymbol{\omega}$, and furthermore, its second time derivative represents angular acceleration of the moving platform $\ddot{\boldsymbol{\theta}} = \dot{\boldsymbol{\omega}}$. Therefore

$$\boldsymbol{\mathcal{X}} = \begin{bmatrix} \boldsymbol{x}_p \\ \boldsymbol{\theta} \end{bmatrix}; \quad \dot{\boldsymbol{\mathcal{X}}} = \begin{bmatrix} \boldsymbol{v}_p \\ \boldsymbol{\omega} \end{bmatrix}; \quad \ddot{\boldsymbol{\mathcal{X}}} = \begin{bmatrix} \boldsymbol{a}_p \\ \dot{\boldsymbol{\omega}} \end{bmatrix}. \tag{5.177}$$

Equations 5.159 and 5.160 can be simply converted into a closed form of Equation 5.176 with the following terms:

$$\begin{aligned}
\boldsymbol{M}_p &= \begin{bmatrix} m\boldsymbol{I}_{3\times3} & \boldsymbol{0}_{3\times3} \\ \boldsymbol{0}_{3\times3} & {}^A\boldsymbol{I}_p \end{bmatrix}_{6\times6}; \quad \boldsymbol{C}_p = \begin{bmatrix} \boldsymbol{0}_{3\times3} & \boldsymbol{0}_{3\times3} \\ \boldsymbol{0}_{3\times3} & \boldsymbol{\omega}_\times{}^A\boldsymbol{I}_p \end{bmatrix}_{6\times6} \\
\boldsymbol{G}_p &= \begin{bmatrix} -m\boldsymbol{g} \\ \boldsymbol{0}_{3\times1} \end{bmatrix}_{6\times1}; \quad \boldsymbol{\mathcal{F}}_p = \begin{bmatrix} \boldsymbol{F}_d + \sum\boldsymbol{f}_{b_i} \\ \boldsymbol{n}_d + \sum\boldsymbol{b}_{i\times}\boldsymbol{f}_{b_i} \end{bmatrix}_{6\times1},
\end{aligned} \tag{5.178}$$

in which $\boldsymbol{I}_{3\times3}$ denotes 3×3 identity matrix, $\boldsymbol{0}_{3\times3}$ denotes 3×3 zero matrix, and $\boldsymbol{0}_{3\times1}$ denotes 3×1 zero vector.

5.3.4.3 Closed-Form Dynamics of the Stewart–Gough Manipulator

To derive the closed-form dynamic formulation for the whole manipulator, a transformation is required to map the intermediate generalized coordinates x_i into the principal generalized coordinates $\mathcal{X}$ [117]. Using such a transformation, and by adding the resulting equations of the limbs and the moving platform, the internal forces f_{b_i} can be eliminated, and closed-form dynamic formulation for the whole manipulator can be derived. To generate such a transformation define a Jacobian matrix J_i relating the intermediate coordinates to that of the principal generalized coordinate. Equation 5.108 can be written in the form of

$$\dot{x}_i = J_i \dot{\mathcal{X}}, \tag{5.179}$$

in which

$$J_i = \begin{bmatrix} I_{3\times3} & -b_{i\times} \end{bmatrix}. \tag{5.180}$$

Furthermore, the time derivative of Equation 5.179 results in

$$\ddot{x}_i = J_i \ddot{\mathcal{X}} + \dot{J}_i \dot{\mathcal{X}}, \tag{5.181}$$

in which referring to Equation 5.118, the time derivative of J_i can be derived as

$$\dot{J}_i = \begin{bmatrix} 0_{3\times3} & -(\omega_\times b_i)_\times \end{bmatrix} = \begin{bmatrix} 0_{3\times3} & -\omega_\times b_{i\times} + b_{i\times}\omega_\times \end{bmatrix}. \tag{5.182}$$

Now substitute $\dot{x}_i$ from Equation 5.179 into 5.173. Furthermore, multiply J_i^T from left to both sides of the resulting equation, this yields

$$\begin{aligned} J_i^T M_i \left(J_i \ddot{\mathcal{X}} + \dot{J}_i \dot{\mathcal{X}}\right) + J_i^T C_i J_i \dot{\mathcal{X}} + J_i^T G_i &= J_i^T F_i \\ (J_i^T M_i J_i)\ddot{\mathcal{X}} + (J_i^T M_i \dot{J}_i + J_i^T C_i J_i)\dot{\mathcal{X}} + J_i^T G_i &= J_i^T F_i. \end{aligned} \tag{5.183}$$

Equation 5.183 may be written in closed form

$$M_{li}\ddot{x}_i + C_{li}\dot{x}_i + G_{li} = \mathcal{F}_{li}, \tag{5.184}$$

in which

$$\begin{aligned} M_{li} &= J_i^T M_i J_i; \quad C_{li} = J_i^T M_i \dot{J}_i + J_i^T C_i J_i \\ G_{li} &= J_i^T G_i; \quad \mathcal{F}_{li} = J_i^T F_i. \end{aligned} \tag{5.185}$$

A closer look into the last term of Equation 5.184 reveals the fact why J_i^T is multiplied to this equation. Note that due to the special form of intermediate Jacobian J_i given in Equation 5.180, the resulting F_{li} has the form

$$\mathcal{F}_{li} = \begin{bmatrix} I_{3\times3} \\ -b_{i\times} \end{bmatrix} \begin{bmatrix} -f_{b_i} + \tau_i \hat{s}_i \end{bmatrix} = -\begin{bmatrix} f_{b_i} \\ b_{i\times} f_{b_i} \end{bmatrix} + \begin{bmatrix} \tau_i \hat{s}_i \\ \tau_i b_{i\times} \hat{s}_i \end{bmatrix}. \tag{5.186}$$

Denote the last term of Equation 5.186 as $\mathcal{F}_{\tau_i}$

$$\mathcal{F}_{\tau_i} = \tau_i \begin{bmatrix} \hat{s}_i \\ b_{i\times} \hat{s}_i \end{bmatrix}. \tag{5.187}$$

Adding the closed-form Equation 5.184 for all limbs together with Equation 5.176 of the moving platform results in complete elimination of the internal force f_{b_i} from these equations. By this means, the closed-form dynamic formulation for the Stewart–Gough manipulator is finalized in the form given in Equation 5.164 with the following components:

$$\begin{aligned} M(\mathcal{X}) &= M_p + \sum_{i=1}^{i=6} M_{li}; \quad C(\mathcal{X}, \dot{\mathcal{X}}) = C_p + \sum_{i=1}^{i=6} C_{li} \\ G(\mathcal{X}) &= G_p + \sum_{i-1}^{i=6} G_{li}; \quad \mathcal{F}(\mathcal{X}) = \mathcal{F}_d + \sum_{i=1}^{i=6} \mathcal{F}_{\tau_i}, \end{aligned} \tag{5.188}$$

in which $\mathcal{F}_d$ denotes the external disturbance wrench applied to the moving platform.

5.3.4.4 Forward Dynamics Simulations

As explained hereinbefore, dynamic formulations may be used for forward dynamics simulations. As shown in Figure 5.6, in forward dynamics it is assumed that actuator forces and external disturbance wrench applied to the manipulator are given and the resulting trajectory of the moving platform is to be determined. As shown in the block diagram of forward dynamics simulation in Figure 5.6, the closed-form dynamic formulation of the Stewart–Gough platform, namely the set of equations given in Equation 5.164 whose terms are given in Equation 5.188 are integrated using the Runge–Kutta routine. The manipulator examined in these simulations are considered to be a 3–6 Stewart–Gough platform shown in Figure 4.20, with six identical cylinder–piston limbs. The geometrical and inertial parameters of the manipulator are given in Table 5.2.

A number of simulations have been examined to verify the accuracy and integrity of dynamic formulations. In the first simulation, it is considered that the manipulator is at rest in its central position $\mathcal{X}_0 = [0, 0, 1, 0, 0, 0]^T$. Furthermore, it is assumed that the actuator forces are considered to be all zero $\boldsymbol{\tau} = \mathbf{0}$, and it is considered that the only external force

TABLE 5.2

Geometric and Inertial Parameters of a 3–6 Stewart–Gough Platform

Description		Quantity
R_A:	Fixed base radius	2 m
R_B:	Moving platform radius	1 m
θ_{A_i}:	Angle of fixed base points A_i's	[0°, 60°, 120°, 180°, 240°, 300°]
θ_{B_i}:	Angle of moving platform points B_i's	[30°, 30°, 150°, 150°, 270°, 270°]
m:	Moving platform mass	1150 kg
m_{i_1}:	Cylinder mass	85 kg
m_{i_2}:	Piston mass	22 kg
c_{i_1}:	Cylinder center of mass	0.75 m
c_{i_2}:	Piston center of mass	0.75 m
I_p:	Moving platform moment of inertia	diag(570, 285, 285) kg · m^2
$I_{c_{i_1}}$:	Cylinder moment of inertia	diag(16, 16, 0) kg · m^2
$I_{c_{i_2}}$:	Piston moment of inertia	diag(4.1, 4.1, 0) kg · m^2

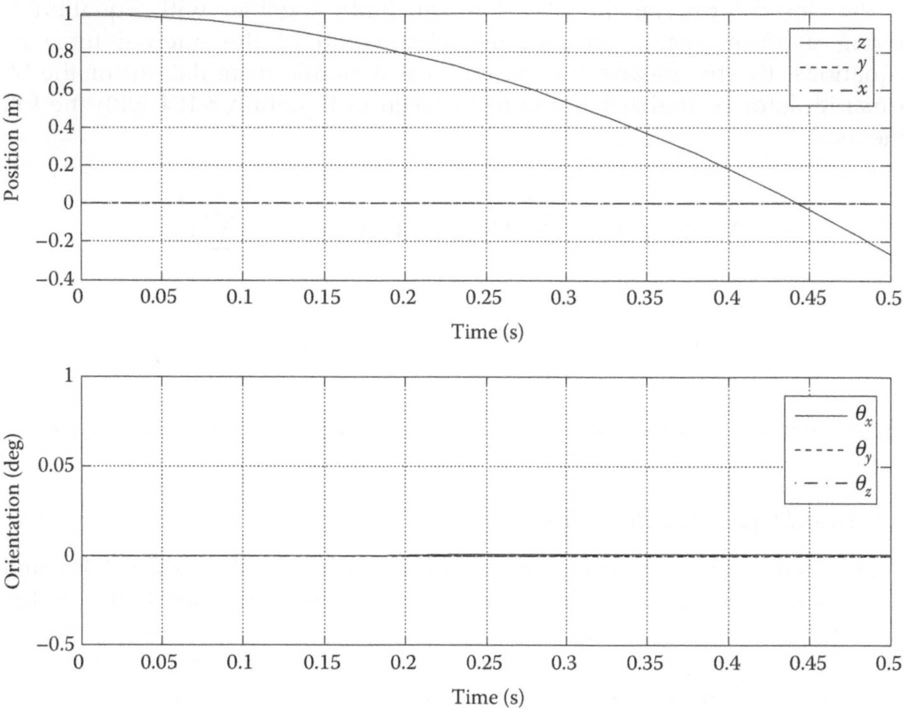

FIGURE 5.20
First simulation scenario for the Stewart–Gough platform with $g = [0, 0, -9.8]^T$ and $\boldsymbol{\tau} = \mathbf{0}$.

acting on the manipulator is the gravitation force $g = [0, 0, -9.8]^T$. For this situation, we expect that the resulting motion of the manipulator be only a translation along the z-axis, and other translations and rotations are all equal to zero. The simulation result given in Figure 5.20 verifies this expectation.

In the second simulation, as illustrated in Figure 5.21, actuator forces are considered to be all equal to $-10kN$, while no gravitational force is considered $g = [0, 0, 0]^T$, to have the resulting force in $-z$ direction. Since no physical constraints are considered in the simulations, the resulting motion is expected to be along $-z$ direction and continue even after the moving platform reaches the fixed base and becomes oscillatory with the same amplitude. The simulation is considered for 1.5 s, and the output motion of the moving platform is illustrated in Figure 5.21. As seen in this figure, positions $x(t)$ and $y(t)$ and orientation angles $\boldsymbol{\theta}$ are all equal to zero, while $z(t)$ has an oscillatory motion. The amplitude of the oscillation is such that the moving platform center of mass passes through the fixed base as expected, and furthermore, it returns to its initial position in about 1.2 s. The output trajectories are in complete agreement to the expected behavior of the system.

In the third simulation, as illustrated in Figure 5.22, the forward dynamics is implemented in a closed-loop block diagram. Suppose it is intended to have a desired trajectory for the motion of the manipulator moving platform as illustrated in Figure 5.23. To produce the actuator forces required for such motion, a closed-loop control system can be designed for the system. In this structure, the actual trajectory of the moving platform $\mathcal{X}(t)$ is measured and compared to the desired trajectory $\mathcal{X}_d(t)$. Furthermore, the tracking error, $e(t) = \mathcal{X}_d(t) - \mathcal{X}(t)$ is fed to a controller to determine the required wrench needed

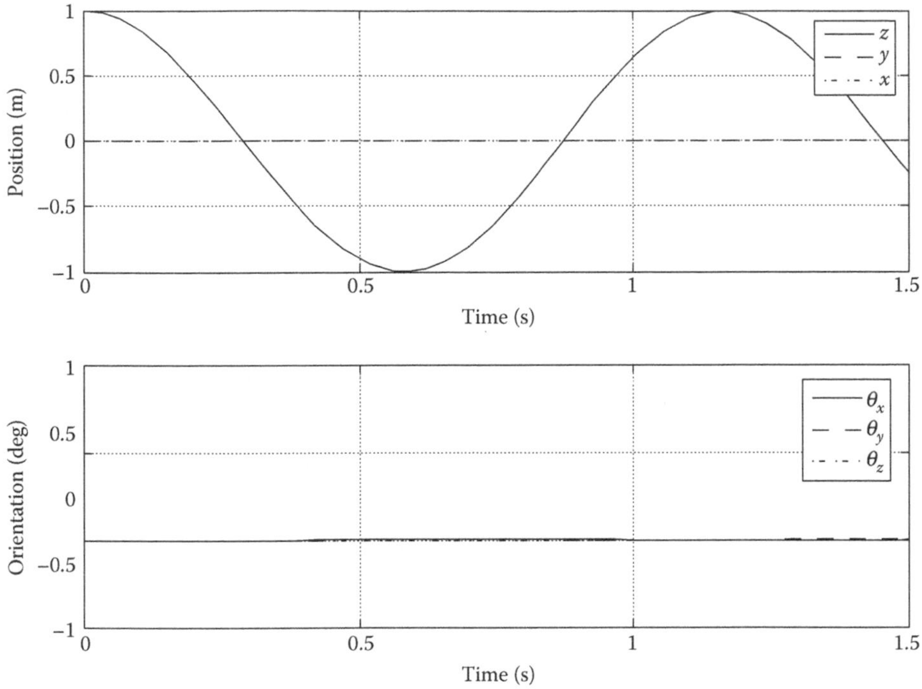

FIGURE 5.21
Second simulation scenario for the Stewart–Gough platform with $g = [0, 0, 0]^T$ and $\tau = [-10, -10, -10, -10, -10, -10]^T$ kN.

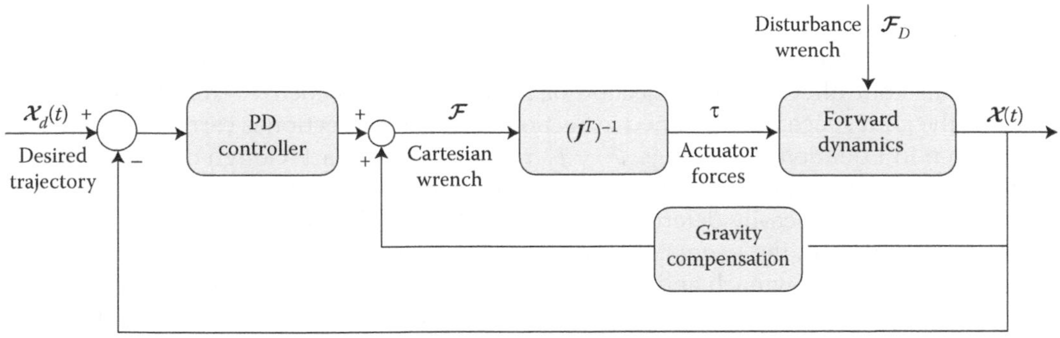

FIGURE 5.22
Block diagram of the forward dynamics implementation for fully parallel manipulators in closed loop.

to be applied to the moving platform. Different control topologies can be used to define such a wrench, which is elaborated in detail in Chapter 6.

In this simulation, for the sake of simplicity, a simple PD controller for each component of the tracking error vector $e(t)$, in addition to a gravity compensation term, is applied to generate the required wrench $\mathcal{F}$ on the manipulator moving platform. To attenuate the effect of gravity on the tracking performance, a gravity compensation term is added to this wrench. This term is indeed the $G(\mathcal{X})$ term derived in the dynamic formulation of the

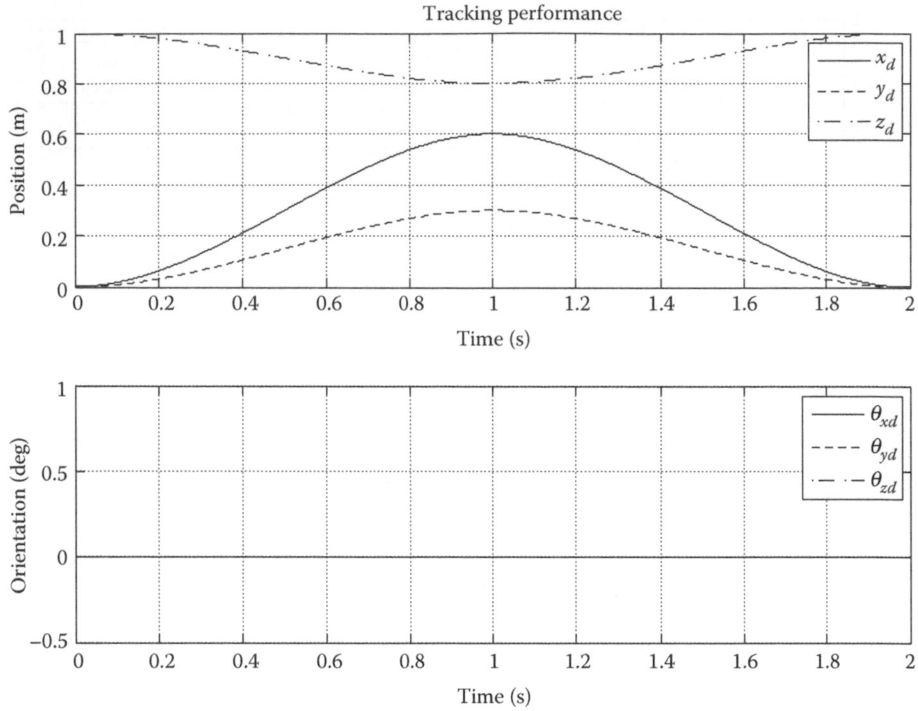

FIGURE 5.23
The desired and output trajectories of the Stewart–Gough manipulator in closed loop.

Stewart–Gough platform given in Equation 5.188. The actuator forces required to accommodate this controller are the projection of the computed wrench $\mathcal{F}$ from the Cartesian space to the joint space. As explained in Section 4.9.2, this projection is carried out through $\boldsymbol{J}^T$ as given in Equation 4.92, that is, $\mathcal{F} = \boldsymbol{J}^T \boldsymbol{\tau}$. Since the Stewart–Gough manipulator is a fully parallel manipulator with no redundancy in actuation, $\boldsymbol{J}^T$ is a squared matrix, and the actuator forces can be easily determined from $\boldsymbol{\tau} = \boldsymbol{J}^{-T} \mathcal{F}$ at nonsingular configurations.

For this simulation, the geometric and inertial parameters given in Table 5.2 are used while the disturbance wrench and the initial conditions are set at zero. The desired trajectory of the moving platform is considered to be smooth and cubic polynomial functions for x_d, y_d, and z_d, as shown in Figure 5.12, while the desired orientation angles are set at zero. The time of simulation is considered to be 2 s, and the PD gains used in the simulations are $\boldsymbol{K}_P = 10^4 \cdot \boldsymbol{I}_6$, and $\boldsymbol{K}_D = 10^6 \cdot \boldsymbol{I}_6$, in which, $\boldsymbol{I}_6$ denotes a 6×6 identity matrix.

The closed-loop tracking performance of the Stewart–Gough manipulator is illustrated in Figures 5.23 and 5.24. As seen in Figure 5.23, the desired and the actual closed-loop motion are identical. As seen in greater detail in Figure 5.13, the PD controller with gravity compensation is capable of reducing the tracking errors to less than 1.5 mm in position and less than 0.001° in orientation. Note that the controller used in this simulation is in its simplest form, and the intention of this example is to see how forward dynamics can be implemented in closed-loop simulations. However, the gravity compensation feedback is the key point to reach such a performance with a simple PD structure in control.

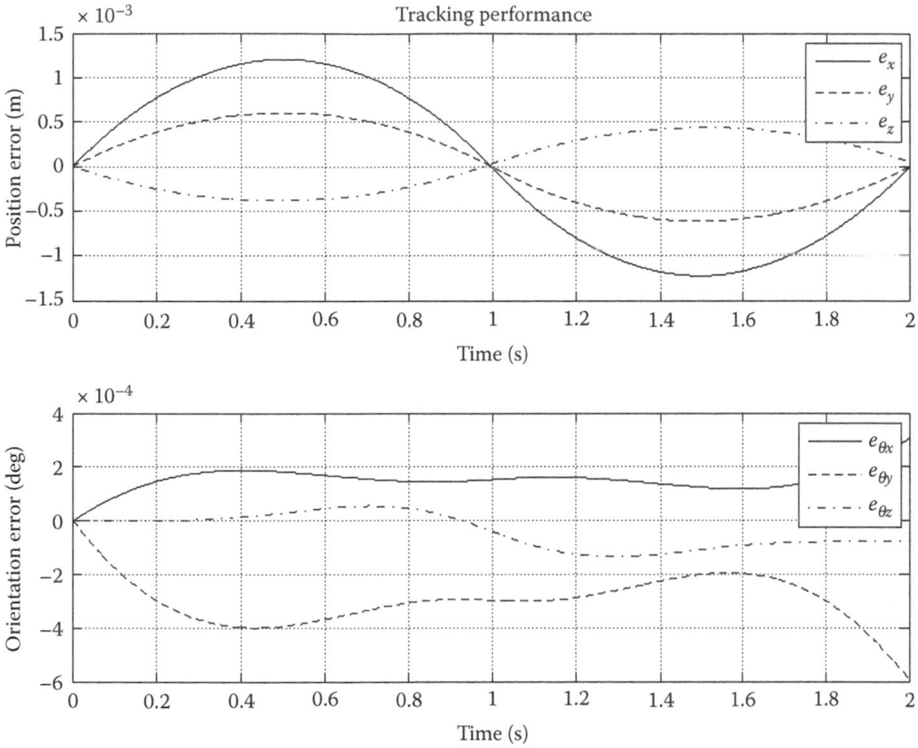

FIGURE 5.24
The closed-loop tracking performance of the Stewart–Gough manipulator.

5.3.4.5 Inverse Dynamics Simulation

In inverse dynamics simulations, it is assumed that the trajectory of the manipulator is given, and the actuator forces required to generate such trajectories are to be determined. As illustrated in Figure 5.25, inverse dynamic formulation is implemented in the following sequence. The first step is trajectory generation for the manipulator moving platform. Many different algorithms are developed for a smooth trajectory generation,* of which a cubic polynomial trajectory is used in these simulations as shown in Figure 5.23. For such a trajectory, $\mathcal{X}_d(t)$ and the time derivatives $\dot{\mathcal{X}}_d(t), \ddot{\mathcal{X}}_d(t)$ are known. The next step is to solve the inverse kinematics of the manipulator and to find the limbs' linear and angular positions, velocity and acceleration as a function of the manipulator trajectory. The manipulator Jacobian matrix J is also calculated in this step. Next, the dynamic matrices given in the closed-form formulations of the limbs and the moving platform are calculated using Equations 5.174 and 5.178, respectively.

To combine the corresponding matrices, and to generate the whole manipulator dynamics, it is necessary to find intermediate Jacobian matrices J_i, given in Equation 5.180, and then compute compatible matrices for the limbs given in Equation 5.185. Now that all the terms required to compute the actuator forces required to generate such a trajectory is computed, let us define $\mathcal{F}$ as the resulting Cartesian wrench applied to the moving

* See Appendix B.

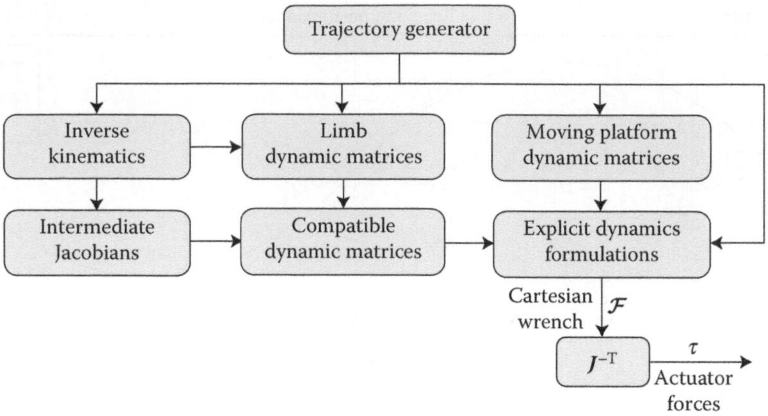

FIGURE 5.25
Flowchart of inverse dynamics implementation sequence.

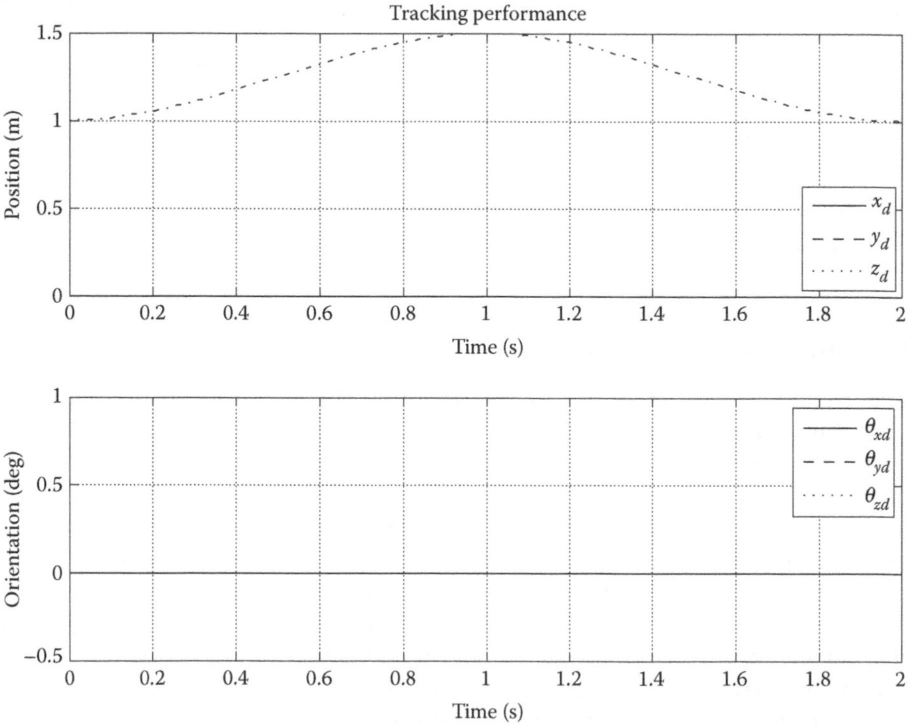

FIGURE 5.26
A typical trajectory that provides motion only in the z direction for the Stewart–Gough platform.

platform. This wrench can be calculated from the summation of all inertial and external forces *excluding the actuator torques* $\boldsymbol{\tau}$ in the closed-form dynamic formulation 5.164. By this definition, $\mathcal{F}$ can be viewed as the projection of the actuator forces acting on the manipulator, mapped to the Cartesian space. Since there is no redundancy in actuation in the

Stewart–Gough manipulator, the Jacobian matrix, J, squared and actuator forces can be uniquely determined from this wrench, by $\tau = J^{-T}\mathcal{F}$, provided J is nonsingular. Therefore, actuator forces τ are computed in the simulations from

$$\tau = J^{-T}\left(M(\mathcal{X})\ddot{\mathcal{X}} + C(\mathcal{X}, \dot{\mathcal{X}})\dot{\mathcal{X}} + G(\mathcal{X}) - \mathcal{F}_d\right). \tag{5.189}$$

Several simulations are given in this section to analyze the dynamic behavior of the Stewart–Gough platform. In the first simulation, a typical trajectory for the moving platform is considered, in which the linear motion is only in the z direction, while the gravity is considered in $-z$ direction. In this trajectory, shown in Figure 5.26, no motion in x and y directions and no orientation in any direction is considered for the moving platform. The inverse dynamics of the manipulator is simulated in this case in the absence of any disturbance forces $\mathcal{F}_D = \mathbf{0}$, and the simulation results are illustrated in Figures 5.27 and 5.28. The Cartesian wrench $\mathcal{F} = [F_x, F_y, F_y, \tau_x, \tau_y, \tau_z]^T$ required to generate such a trajectory is illustrated in Figure 5.27. As seen in this figure, all the components of the Cartesian wrench are equal to zero except for F_z. This component has a similar pattern to the desired trajectory accelerations, which is linear for the cubic trajectories. The effect of gravity can also be seen by the bias of F_z from zero. The required actuator forces to generate the desired

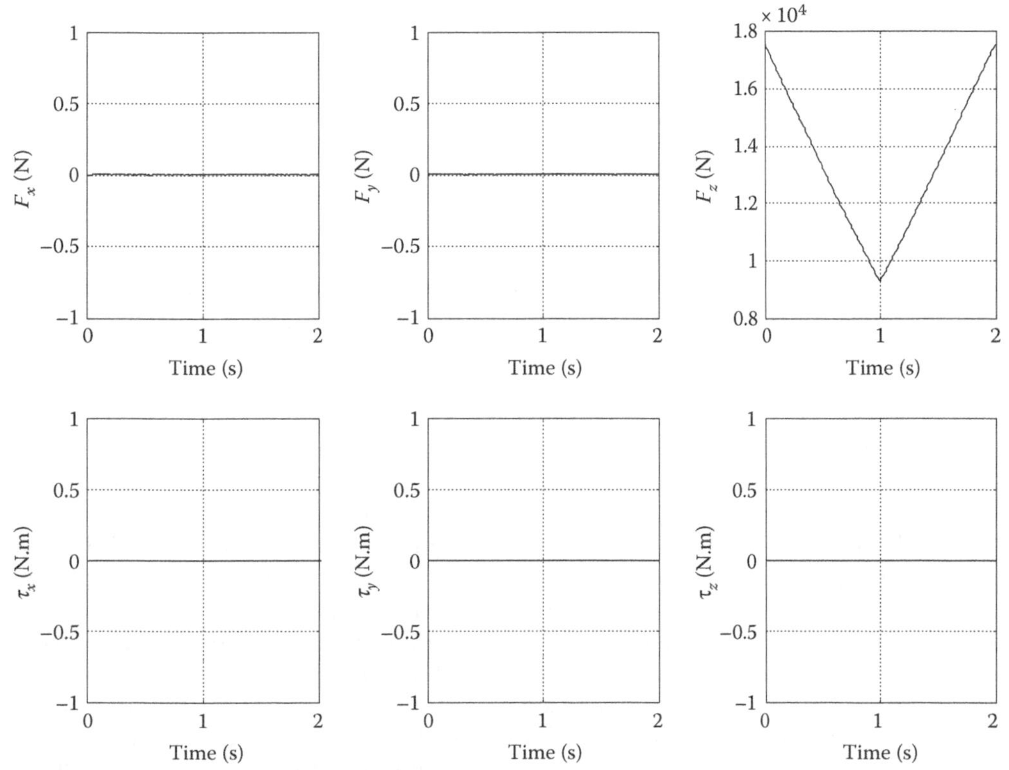

FIGURE 5.27
The Cartesian wrench $\mathcal{F}$ of the Stewart–Gough manipulator required to produce the desired motion depicted in Figure 5.26.

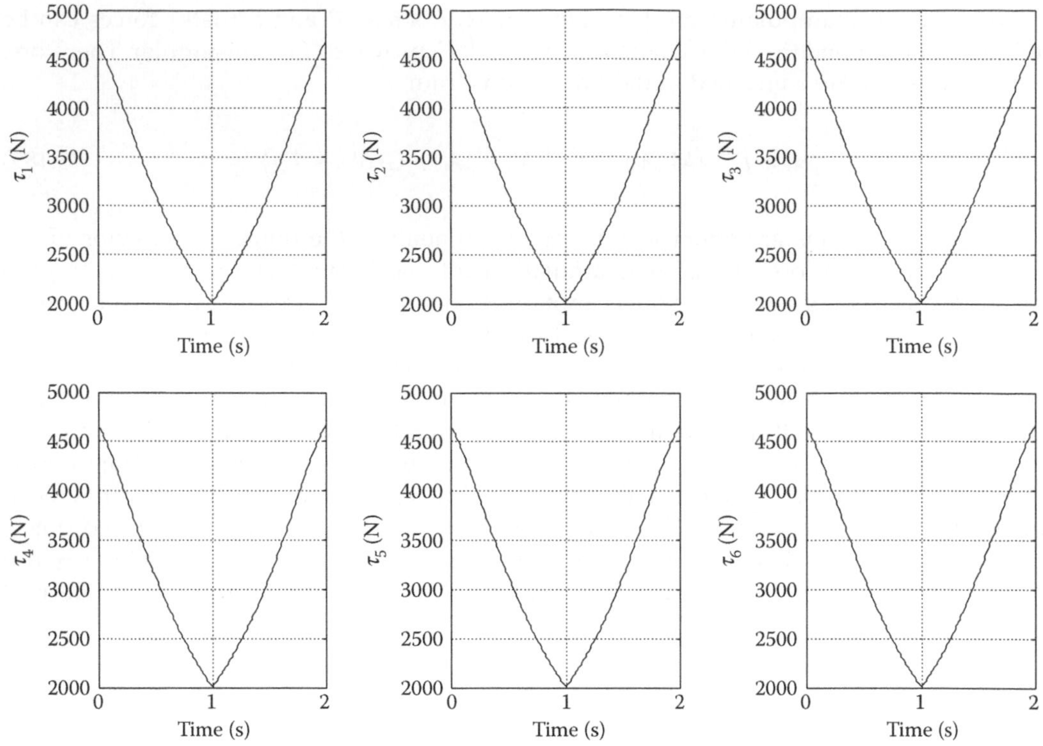

FIGURE 5.28
The actuator forces of the Stewart–Gough manipulator required to produce the desired motion depicted in Figure 5.26.

trajectory is shown in Figure 5.28. As seen in this figure, all the actuator forces are identical due to the symmetrical structure of the Stewart–Gough platform. It can be verified by inspection that equal forces in all actuators are required to provide a pure motion in the z direction even in presence of gravity terms. The simulation results completely comply to this expectation.

Similar pure motion in x and y directions or pure rotations about x-, y-, and z-axes can be examined for the Stewart–Gough platform. However, since in those motions the complete symmetry is not preserved during the motion, the results cannot be easily verified as in the previous simulation. It should be mentioned though in this motion the dominance of the corresponding forces or torques in the direction of the motion compared to the others is clearly observed. As a representative of such simulations Figures 5.29 and 5.30 illustrate the Cartesian wrench and actuator forces required to generate a motion only in the x direction. In these simulations the inverse dynamics is simulated in the absence of gravity and disturbance wrench. As seen in Figure 5.29 the Cartesian force in the x direction F_x, required to produce such motion is in the order of a couple of kN, while the other components of the Cartesian wrench are limited to a couple of N. The corresponding actuator forces shown in Figure 5.30 are though relatively distributed among six actuators.

In the next simulation, inverse dynamics of the manipulator is simulated in the absence of any disturbance forces, for a general and typical motion of the moving platform, already studied in Figure 5.23. This trajectory is considered in a series of simulations to analyze the effect of gravity terms, as well as the limbs' masses in their contribution to the required

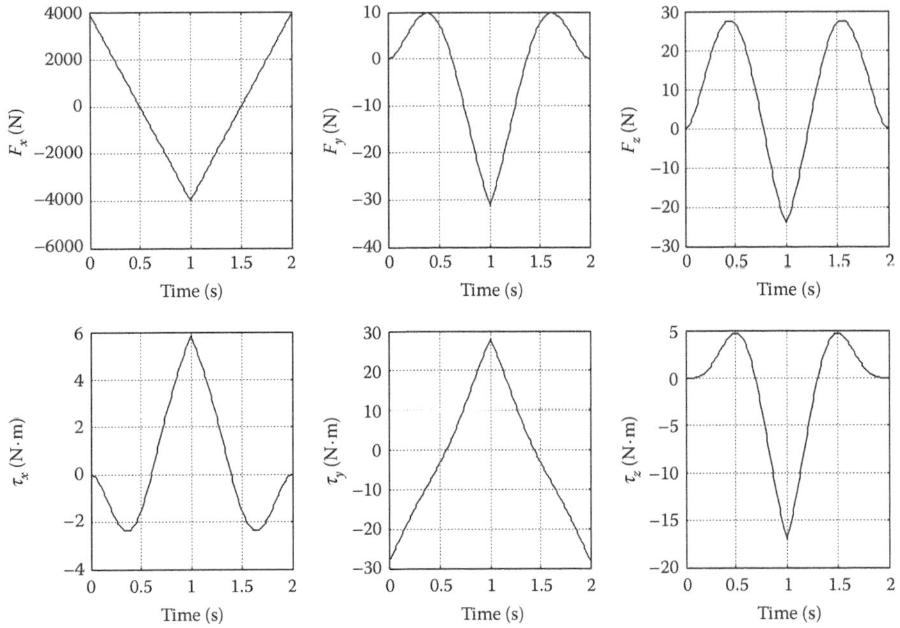

FIGURE 5.29
The Cartesian wrench $\mathcal{F}$ of the Stewart–Gough manipulator required to produce the desired motion in only the x direction.

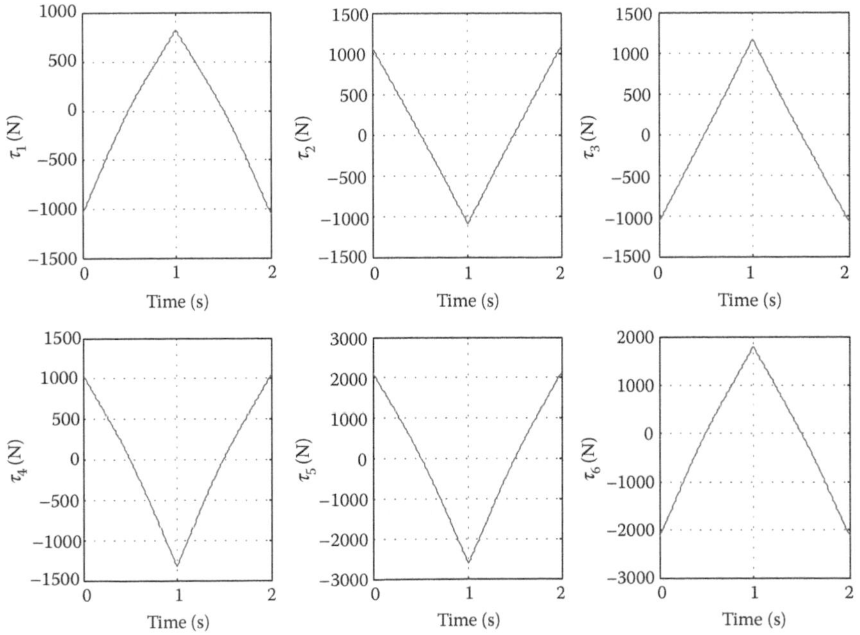

FIGURE 5.30
The actuator forces of the Stewart–Gough manipulator required to produce the desired motion in only the x direction.

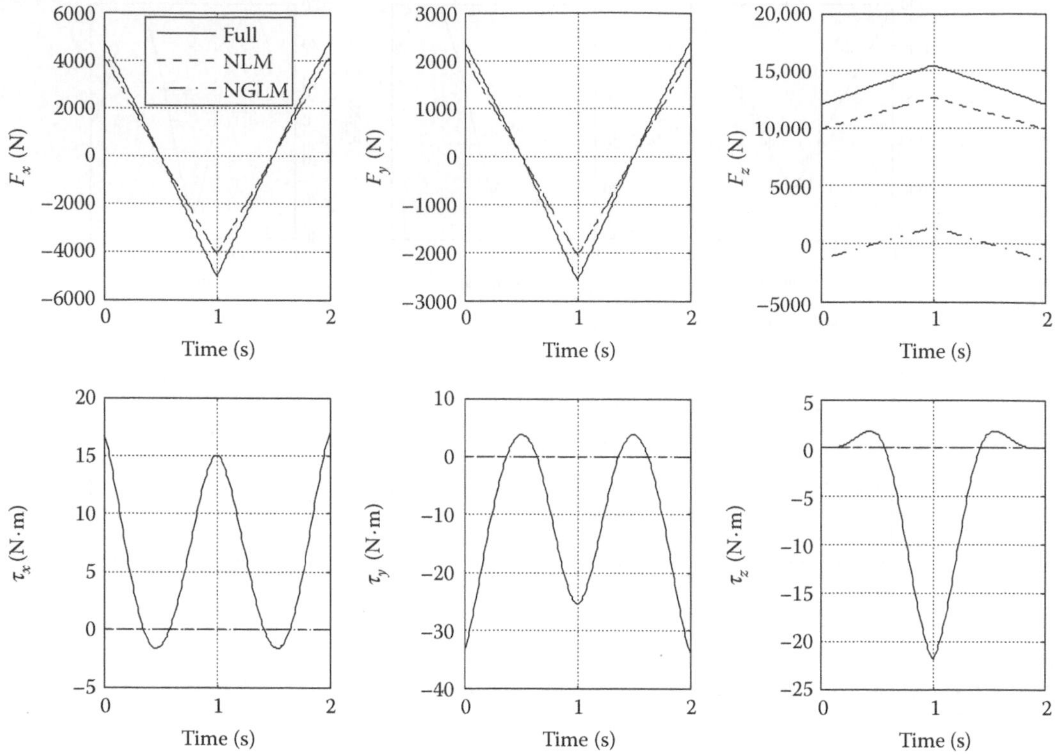

FIGURE 5.31
The Cartesian wrench $\mathcal{F}$ of the Stewart–Gough manipulator required to produce a desired motion depicted in Figure 5.23; The total force (solid), neglecting limb masses (dashed), and neglecting gravity and limb masses (dash-dotted).

actuator forces. The simulation results are illustrated in Figures 5.31 and 5.32. Note that in these figures the total forces are drawn in solid line, while the results of simulations where no limb masses (NLM) are considered are drawn in dashed line. Furthermore, in the last scenario both the gravity terms, and the limb masses are neglected (NGLM) and the results are plotted in dash-dotted line. Note that in this typical trajectory no rotation is considered for the Stewart–Gough platform, therefore, as seen in Figure 5.31, in case the limb masses are neglected no Cartesian torques are needed to be applied to the manipulator (i.e., $\tau_x = \tau_y = \tau_z = 0$ for the dashed lines).

Furthermore, the effect of gravity terms are fully addressed in the illustrations of Cartesian force F_z, in which by neglecting the limb masses 20% of its value drops. This simulation reveals how bulky cylinder and piston actuators contribute in the demand of a pure vertical force, produced by actuator forces. By neglecting gravity terms as well, the value of this force drops dramatically by one order of magnitude. This analysis shows that most of the power consumed by the manipulator is to compensate for its own weight. Comparing the same values for F_x and F_y shows that the contribution of the limb masses in the required Cartesian force for such a trajectory is about 15% of the total force. The actuator forces required for different scenarios are shown in Figure 5.32, in which the force distribution in the actuators are illustrated to perform the desired maneuver with and without considering the limb masses and the gravity terms.

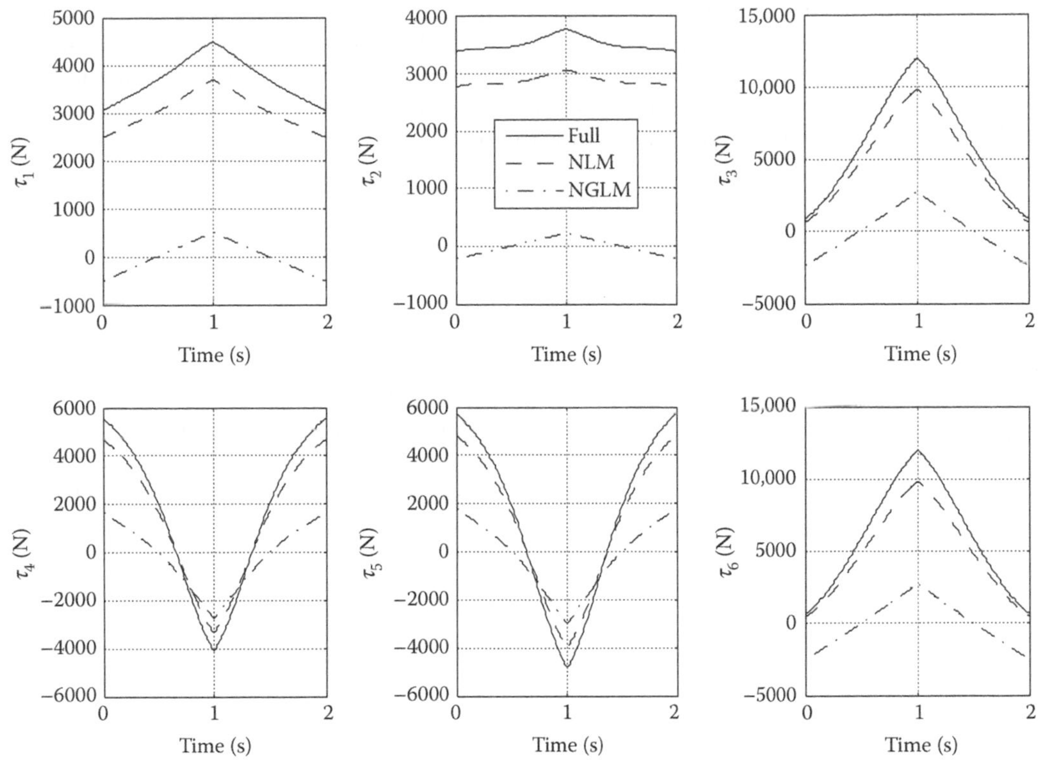

FIGURE 5.32
The actuator forces of the Stewart–Gough manipulator required to produce the desired motion depicted in Figure 5.23; The total force (solid), neglecting limb masses (dashed), and neglecting gravity and limb masses (dash-dotted).

5.4 Virtual Work Formulation

One of the most convenient approaches to derive dynamic formulation of a parallel manipulator is based on the principle of virtual work. In this method, the *inertial* forces and moments are computed using the linear and angular accelerations of each rigid body. Then, the whole manipulator is considered to be in static equilibrium and the principle of virtual work is applied to derive the actuator forces [186]. Since constraint forces and moments do not need to be computed, this approach yields faster computational algorithms compared to that of the Newton–Euler approach.

5.4.1 D'Alembert's Principle

The Newton–Euler laws are based on the fact to balance the summation of external forces and torques applied to a rigid body with the inertial terms resulting from linear or angular accelerations of the rigid body. As explained hereinbefore, it is necessary to derive accelerations with respect to a fixed frame, to apply force and torque balance correctly. When motion of the rigid body is observed from a moving frame attached to the center of mass of the rigid body, the rigid body appears to be at rest or in equilibrium. Thus, the observer who is accelerating with the moving frame may conclude that a *fictitious* inertial force and moment acts on the rigid body to balance the *real* external forces and moments. This point of view allows treatment of dynamic problem by method of statics, and is an extension of the work D'Alembert contained in his book *Traité de Dynamique* published in Paris in 1743. This approach merely focused on rewriting the Newton and Euler Laws as the summation of all real external forces and moments plus the corresponding fictitious inertial forces and moments equal to zero.

$$\sum \boldsymbol{f}_{\text{ext}} - m\,\boldsymbol{a}_c = \boldsymbol{0} \tag{5.190}$$

$$\sum {}^c\boldsymbol{n}_{\text{ext}} - ({}^c\boldsymbol{I}\,\dot{\boldsymbol{\omega}} + \boldsymbol{\omega} \times ({}^c\boldsymbol{I}\,\boldsymbol{\omega})) = \boldsymbol{0}. \tag{5.191}$$

In these equations, $\boldsymbol{a}_c$ is the linear acceleration of the center of mass, $\boldsymbol{\omega}$ and $\dot{\boldsymbol{\omega}}$ are the rigid body angular velocity and acceleration, respectively, and ${}^c\boldsymbol{I}$ is moment of inertia of the rigid body with respect to the moving frame attached to the center of mass. In this framework, inertia term $-m\boldsymbol{a}_c$ may be regarded as a fictitious force $\boldsymbol{f}^* = -m\boldsymbol{a}_c$, and the other inertial terms, namely ${}^c\boldsymbol{n}^* = -({}^c\boldsymbol{I}\,\dot{\boldsymbol{\omega}} + \boldsymbol{\omega} \times ({}^c\boldsymbol{I}\,\boldsymbol{\omega}))$, may be considered as a fictitious torque. Hence, the static balance of forces and moments can be concluded as

$$\sum \hat{\boldsymbol{f}}_{\text{ext}} = \boldsymbol{0} \tag{5.192}$$

$$\sum {}^c\hat{\boldsymbol{n}}_{\text{ext}} = \boldsymbol{0}, \tag{5.193}$$

in which

$$\hat{\boldsymbol{f}}_{\text{ext}} = \boldsymbol{f}_{\text{ext}} + \boldsymbol{f}^* \tag{5.194}$$

$${}^c\hat{\boldsymbol{n}}_{\text{ext}} = {}^c\boldsymbol{n}_{\text{ext}} + {}^c\boldsymbol{n}^*. \tag{5.195}$$

The apparent transformation of a problem in dynamics to one in statics has been known as the *d'Alembert principle*. The basic definitions given above seem to be a very simple view of a dynamic equation to convert it into a static version, and it may be that this formulation does not contribute that much in simplifying the equation of motion. However, in practice, considering the inertial terms as fictitious forces and moments significantly simplifies the derivation of dynamic formulation in occasions where principle of virtual work is being implemented on the problem.

5.4.2 Principle of Virtual Work

In this section, we show how principle of virtual work can be applied to derive the dynamic equation of motion of a parallel robot. In this treatment d'Alembert principle is implemented to represent the dynamic system as a statically balanced system. Following the d'Alembert's principle, the inertial force and moment applied on a rigid body are defined as the force and moment exerted at the center of mass, given by

$$\boldsymbol{f}^* = -m\,\boldsymbol{a}_c \tag{5.196}$$

$${}^c\boldsymbol{n}^* = -({}^c\boldsymbol{I}\,\dot{\boldsymbol{\omega}} + \boldsymbol{\omega} \times ({}^c\boldsymbol{I}\,\boldsymbol{\omega})). \tag{5.197}$$

This force and moment are applied in a direction opposite to the direction of the linear and angular accelerations. Hence, by introducing these fictitious forces and moments into the system, we may consider the rigid body in static equilibrium. If at static equilibrium, a virtual displacement $\delta(\cdot)$ is considered for the system, by application of the principle of virtual work, one can obtain the input forces of the manipulator. Principle of virtual work has its roots on the work–energy relation, that is, for a system with negligible energy loss due to friction, the relation between work and energy of the system can be stated as

$$\mathrm{d}W = \mathrm{d}K + \mathrm{d}P, \tag{5.198}$$

in which $\mathrm{d}W$ represents the total work done by all *active* forces acting on the system during the infinitesimal change in the displacement, and $\mathrm{d}K$ and $\mathrm{d}P$ denote the change in the kinetic and potential energies of the system, respectively. It can be further shown that infinitesimal change in the kinetic and potential energies for a general multibody interconnecting system can be represented by

$$\mathrm{d}K = \sum_i m_i \boldsymbol{a}_{c_i} \cdot \mathrm{d}\boldsymbol{x}_i + \sum_i {}^{c_i}\dot{\boldsymbol{H}}_i \cdot \mathrm{d}\boldsymbol{\theta}_i \tag{5.199}$$

$$\mathrm{d}P = \sum_i W_i \, \mathrm{d}h_i + \sum_i k_i \, x_i \, \mathrm{d}x_i \tag{5.200}$$

where the system consists of a number of bodies with mass m_i, and angular momentum ${}^{c_i}\boldsymbol{H}_i$, while $\mathrm{d}\boldsymbol{x}_i$ and $\mathrm{d}\boldsymbol{\theta}_i$ represent infinitesimal linear and angular displacement of each body center of mass, respectively. Furthermore, h_i represents the vertical motion of the center of mass of representative bodies of weight W_i, and x_i represents the elastic deformation of a representative elastic member of the system (spring) whose stiffness is k_i.

When applying Equation 5.198 for mechanical systems the configuration of which is an unknown function of a prescribed motion, and for systems where there are more than one-degree-of-freedom, it is often convenient to introduce the concept of *virtual work*. The concept of virtual displacement and the resulting virtual work have been introduced for the solution of equilibrium configurations for static systems of interconnected bodies, and then further extended to dynamical systems through the use of the d'Alembert principle.

A virtual displacement $\delta(\cdot)$ is any assumed and arbitrary displacement, linear or angular, away from the natural or actual position of the system. For a system of connected bodies the virtual displacements must be consistent with their kinematic constraints. If a set of virtual displacements satisfying the kinematic constraints is applied to a mechanical system, the proper relationship between coordinates which specify the configuration of the system is established by applying the work–energy relation 5.198. It is customary to use the differential symbol d to refer to differential changes in the *real* displacements, whereas the symbol δ is used to signify differential changes which are assumed or called *virtual* changes. When applying the work–energy principles to an interconnected system of rigid bodies, an active-force diagram of the entire system should be drawn to isolate the system and disclose all forces which do work on the system.

First, consider a static system of interconnected bodies. Since the system is at static equilibrium and the resulting forces and moments applied to the system is equal to zero, the resulting virtual work for the total system is also equal to zero.

$$\delta W = \sum_i \boldsymbol{f}_i \cdot \delta\boldsymbol{x}_i + \sum_i \boldsymbol{n}_i \cdot \delta\boldsymbol{\theta}_i = 0. \tag{5.201}$$

The forces and moments acting on all the bodies consist of the external forces and moments, and the constraint forces and moments acting on each body. However, note that since the virtual displacement should be consistent with the kinematic constraints of the system, the constraint forces and moments do not contribute in any work, since they are orthogonal to the directions of the virtual displacement. Hence, in Equation 5.201, $\boldsymbol{f}_i$ denotes all the external forces acting on the system, including gravity, spring, damping, or any disturbance forces, while $\boldsymbol{n}_i$ denotes all the external moments applied to the system. For shorthand notation we use the definitions of wrench to combine the generalized forces and moments into a 6×1 vector $\mathcal{F}_i = \begin{bmatrix} \boldsymbol{f}_i & \boldsymbol{n}_i \end{bmatrix}^T$, and linear and angular displacements into a screw coordinate $\mathcal{X}_i = \begin{bmatrix} \boldsymbol{x}_i & \boldsymbol{\theta}_i \end{bmatrix}^T$, and rewrite Equation 5.201 as

$$\delta W = \sum_i \mathcal{F}_i \cdot \delta \mathcal{X}_i = 0. \tag{5.202}$$

Similarly, for a dynamical system we use the d'Alembert principle considering the inertial wrench as an external wrench applied to the system

$$\mathcal{F}^* = \begin{bmatrix} \boldsymbol{f}^* \\ {}^c\boldsymbol{n}^* \end{bmatrix} = \begin{bmatrix} -m\,\boldsymbol{a}_c \\ -({}^c\boldsymbol{I}\,\dot{\boldsymbol{\omega}} + \boldsymbol{\omega} \times ({}^c\boldsymbol{I}\,\boldsymbol{\omega})) \end{bmatrix}, \tag{5.203}$$

and apply the principle of virtual work to the system

$$\delta W = \sum_i (\mathcal{F}_i + \mathcal{F}_i^*) \cdot \delta \mathcal{X}_i = 0. \tag{5.204}$$

To derive actuator forces directly from this equation, isolate the actuator forces denoted by $\boldsymbol{\tau}$ from other external and inertial forces and rewrite the principle of virtual work

$$\delta W = \boldsymbol{\tau} \cdot \delta \boldsymbol{q} + \sum_i (\mathcal{F}_i + \mathcal{F}_i^*) \cdot \delta \mathcal{X}_i = 0, \tag{5.205}$$

in which $\mathcal{F}_i$ denotes all the external forces acting on the interconnected bodies *excluding the actuator forces*. Now apply the resulting formulation of the principle of virtual work given in Equation 5.205 for a parallel robot consisting of a number of limbs and a moving platform, and isolate the limbs and moving platform external and inertial wrenches, which results in the following relation:

$$\delta W = \boldsymbol{\tau} \cdot \delta \boldsymbol{q} + (\mathcal{F}_p + \mathcal{F}_p^*) \cdot \delta \mathcal{X}_p + \sum_i (\mathcal{F}_i + \mathcal{F}_i^*) \cdot \delta \mathcal{X}_i = 0, \tag{5.206}$$

in which $\boldsymbol{q} = [q_1, q_2, \ldots, q_m]^T$ denotes the vector of joint variables while the actuator forces are denoted by $\boldsymbol{\tau} = [\tau_1, \tau_2, \ldots, \tau_m]^T$. Furthermore, $\mathcal{F}_p$ and $\mathcal{F}_p^*$ denote the external and inertial wrenches applied to the moving platform center of mass, respectively, $\mathcal{F}_i$ and $\mathcal{F}_i^*$ represent the external and inertial wrenches applied to each limb center of mass, respectively, and $\mathcal{X}_p$ and $\mathcal{X}_i$ denote the virtual displacement of the moving platform and limb center of masses, respectively. Use the d'Alembert shorthand notation given in Equation 5.194 for the moving platform wrench $\hat{\mathcal{F}}_p = \mathcal{F}_p + \mathcal{F}_p^*$, as well as the limb wrenches $\hat{\mathcal{F}}_i = \mathcal{F}_i + \mathcal{F}_i^*$, and convert the dot products in Equation 5.206 to matrix product:

$$\delta \boldsymbol{q}^T \boldsymbol{\tau} + \delta \mathcal{X}_p^T \hat{\mathcal{F}}_p + \sum_i \delta \mathcal{X}_i^T \hat{\mathcal{F}}_i = 0. \tag{5.207}$$

Note that, as explained hereinbefore, the virtual displacements in Equation 5.206 must satisfy the kinematic constraints imposed by the closed-loop chains. Therefore, it is necessary to relate these virtual displacements to a set of independent generalized virtual displacement, where in parallel manipulators the coordinate of the moving platform $\delta\boldsymbol{\mathcal{X}}_p$ can conveniently be chosen as the generalized coordinate. This is because generally virtual displacement of limbs $\delta\boldsymbol{q}$ is related to virtual displacement of the moving platform $\delta\boldsymbol{\mathcal{X}}_p$ by manipulator Jacobian $\boldsymbol{J}$:

$$\delta\boldsymbol{q} = \boldsymbol{J}\delta\boldsymbol{\mathcal{X}}_p. \tag{5.208}$$

Furthermore, virtual displacement of the center of mass of each limb, $\delta\boldsymbol{\mathcal{X}}_i$, can be related to $\delta\boldsymbol{\mathcal{X}}_p$ by a similar Jacobian matrix defined for each limb and denoted by $\boldsymbol{J}_i$:

$$\delta\boldsymbol{\mathcal{X}}_i = \boldsymbol{J}_i\delta\boldsymbol{\mathcal{X}}_p. \tag{5.209}$$

Substituting Equations 5.208 and 5.209 into Equation 5.206 results in

$$\delta\boldsymbol{\mathcal{X}}_p^T\left(\boldsymbol{J}^T\boldsymbol{\tau} + \hat{\boldsymbol{\mathcal{F}}}_p + \sum_i \boldsymbol{J}_i^T\hat{\boldsymbol{\mathcal{F}}}_i\right) = 0. \tag{5.210}$$

Since this is valid for any arbitrary virtual displacement $\delta\boldsymbol{\mathcal{X}}_p$, it follows that

$$\boldsymbol{J}^T\boldsymbol{\tau} + \hat{\boldsymbol{\mathcal{F}}}_p + \sum_i \boldsymbol{J}_i^T\hat{\boldsymbol{\mathcal{F}}}_i = \boldsymbol{0}. \tag{5.211}$$

Equation 5.211 represents the dynamic formulation of a parallel manipulator by the principle of virtual work. Note that all the wrenches in this formulation are considered about the center of mass of each link and the moving platform. Therefore, if an external wrench is applied at a point other than the center of mass, it should be transformed into the center of mass coordinate frame before it is substituted into this equation. Furthermore, this equation is written in a task space coordinate, in which the actuator forces appear in these equations by Jacobian transpose mapping. In general, $\boldsymbol{\mathcal{F}} = \boldsymbol{J}^T\boldsymbol{\tau}$ is the projection of the actuator forces on the moving platform, and can be uniquely determined from Equation 5.211. Furthermore, if the manipulator is fully parallel and has no redundancy in actuation, Jacobian matrix, $\boldsymbol{J}$, is squared, and the actuator forces can be uniquely determined by $\boldsymbol{\tau} = \boldsymbol{J}^{-T}\boldsymbol{\mathcal{F}}$, provided $\boldsymbol{J}$ is nonsingular. For redundant manipulators, however, there are infinitely many solutions for $\boldsymbol{\tau}$ to be projected into $\boldsymbol{\mathcal{F}}$. The simplest solution would be a minimum norm solution, found from the pseudo-inverse of $\boldsymbol{J}^T$, by $\boldsymbol{\tau} = \boldsymbol{J}^{T^\dagger}\boldsymbol{\mathcal{F}}$. Other optimization techniques can be used to find the actuator forces projected from $\boldsymbol{\mathcal{F}}$ which can minimize a user-defined cost function. The details of such redundancy resolution techniques are given in Section 6.7.

Note that although virtual work formulation given in Equation 5.211 is very efficient in computation, a critical step in this formulation is derivation of Jacobian matrices for the moving platform and the limbs. In what follows, the details of the formulation are given for three case studies.

5.4.3 Dynamic Formulation of a Planar Manipulator: Constant Mass Treatment

In this section, dynamic formulation of the planar $4R\underline{P}R$ manipulator the architecture of which is described in Section 3.3.1 and shown in Figure 1.7 is performed using the principle of virtual work. As explained earlier, in this manipulator the moving platform is supported by four limbs of identical $R\underline{P}R$ kinematic structure. A complete Jacobian and singularity analysis of this manipulator is presented in Section 4.6. Furthermore, the dynamic analysis of this manipulator with and without variable mass treatments have been described in Sections 5.3.1 and 5.3.2, respectively. In this section, the constant mass assumptions are considered for the analysis.

To derive the equation of motion for the manipulator by using the principle of virtual work, the external and inertial generalized forces applied to the moving platform $\hat{\mathcal{F}}_p$ and on each limb $\hat{\mathcal{F}}_i$ is first obtained, and then the Jacobian matrices of the limbs are determined. The external and inertial wrenches applied to the moving platform is given by

$$\hat{\mathcal{F}}_p = \begin{bmatrix} \hat{f} \\ {}^C\hat{n} \end{bmatrix} = \begin{bmatrix} f_{d_x} + M(g \cdot \hat{x}) - M\,\ddot{x}_G \\ f_{d_y} + M(g \cdot \hat{y}) - M\,\ddot{y}_G \\ \tau_d - I_m\,\ddot{\phi} \end{bmatrix}. \tag{5.212}$$

The external and inertial wrenches applied to the limb i is determined by

$$\hat{\mathcal{F}}_i = \begin{bmatrix} \hat{f}_i \\ \hat{n}_i \end{bmatrix} = \begin{bmatrix} m_i(g - a_{c_i}) \\ -I_{c_i}\ddot{\alpha}_i \end{bmatrix}, \tag{5.213}$$

in which m_i and I_{c_i} denote the cable mass and moment of inertia about the center of mass, given in Equation 5.53. Writing Equation 5.213 componentwise, in the cable directions $[\hat{S}_i, \hat{N}_i, \hat{K}]^T$, and using the acceleration of the center of mass of the limbs found in Equation 5.52, the following equation is obtained by some manipulation:

$$\hat{\mathcal{F}}_i = \rho_m \begin{bmatrix} L_i g^S - L_i \ddot{L}_i + \frac{1}{2}(L_i\dot{\alpha}_i)^2 \\ L_i g^N - \frac{1}{2}L_i^2\ddot{\alpha}_i - \frac{3}{2}L_i\dot{L}_i\dot{\alpha}_i \\ \frac{1}{12}L_i^3\ddot{\alpha}_i \end{bmatrix}, \tag{5.214}$$

in which g^S and g^N are the components of the gravity vector g along $\hat{S}$ and $\hat{N}$ directions. The next step is to derive the Jacobians. The manipulator Jacobian matrix J has been derived earlier and is given in Equation 4.54. The limb Jacobians J_i's are derived from the linear and angular velocities of the center of mass of the limbs, in the framework where $\hat{\mathcal{F}}_i$ is derived in Equation 5.214:

$$\dot{\mathcal{X}}_i = \begin{bmatrix} v_{c_i} \\ \dot{\alpha}_i\hat{K} \end{bmatrix} = \begin{bmatrix} \dot{L}_i\hat{S}_i \\ \frac{1}{2}L_i\dot{\alpha}_i\hat{N}_i \\ \dot{\alpha}_i\hat{K} \end{bmatrix}. \tag{5.215}$$

Substituting Equation 4.53 and 4.56 into Equation 5.215, the limb Jacobian J_i can be derived in the same $[\hat{S}_i, \hat{N}_i, \hat{K}]^T$ coordinate frame to satisfy the relation $\dot{\mathcal{X}}_i = J_i\dot{\mathcal{X}}_p$ as

$$J_i = \begin{bmatrix} S_{i_x} & S_{i_y} & (E_{i_x}S_{i_y} - E_{i_y}S_{i_x}) \\ -\frac{1}{2}S_{i_y} & \frac{1}{2}S_{i_x} & \frac{1}{2}(E_{i_x}S_{i_x} + E_{i_y}S_{i_y}) \\ -\frac{1}{L_i}S_{i_y} & \frac{1}{L_i}S_{i_x} & \frac{1}{L_i}(E_{i_x}S_{i_x} + E_{i_y}S_{i_y}) \end{bmatrix}. \tag{5.216}$$

Finally, consider the actuator forces as $\boldsymbol{\tau} = \begin{bmatrix} -\tau_1 & -\tau_2 & -\tau_3 & -\tau_4 \end{bmatrix}$ to consider positive actuator forces for tensions. Then, by substitution of Equations 4.54, 5.212, 5.214, and 5.216 into Equation 5.211, the governing equation of motion of the manipulator is derived by some manipulation as follows:

$$f_{d_x} + M(\boldsymbol{g} \cdot \hat{x}) - M\,\ddot{x}_G - \sum_{i=1}^{4} \left\{ (\tau_i - P_i) S_{i_x} + Q_i S_{i_y} \right\} = 0 \tag{5.217}$$

$$f_{d_y} + M(\boldsymbol{g} \cdot \hat{y}) - M\,\ddot{y}_G - \sum_{i=1}^{4} \left\{ (\tau_i - P_i) S_{i_y} - Q_i S_{i_x} \right\} = 0 \tag{5.218}$$

$$\tau_d - I_m\,\ddot{\phi} - \sum_{i=1}^{4} \left\{ (\tau_i - P_i)(E_{i_x} S_{i_y} - E_{i_y} S_{i_x}) - Q_i (\boldsymbol{E}_i \cdot \hat{\boldsymbol{S}}_i) \right\} = 0, \tag{5.219}$$

in which

$$P_i = \rho_m \left(L_i g^S - L_i \ddot{L}_i + \frac{1}{2}(L_i \dot{\alpha}_i)^2 \right), \tag{5.220}$$

$$Q_i = \rho_m \left(\frac{1}{2} L_i g^N - \frac{3}{4} L_i \dot{L}_i \dot{\alpha}_i - \frac{1}{3} L_i^2 \ddot{\alpha}_i \right). \tag{5.221}$$

5.4.4 Formulation Verification

To verify the integrity and correctness of the dynamic formulation obtained for this manipulator, in this section, two formulations derived for the planar manipulator using the Newton–Euler and virtual work approaches are compared. Compare the dynamic formulations for this manipulator obtained by the Newton–Euler approach given in Equations 5.67 through 5.69 to those obtained by the principle of virtual work given in Equations 5.217 through 5.219. It can be easily verified that although the description of these two methods are very different the last result is term by term equivalent.

To show that let us examine and compare just the x-component of both formulations, namely Equations 5.67 and 5.217 here. Substitute $F_{B_i}^N$ and $F_{B_i}^S$ from Equations 5.63 and 5.64, respectively, into Equation 5.67:

$$M\ddot{x}_G - M(\boldsymbol{g} \cdot \hat{x}) - f_{d_x} + \sum_{i=1}^{4} \left(F_{B_i}^S S_{i_x} - F_{B_i}^N S_{i_y} \right) = 0$$

$$\Rightarrow M\ddot{x}_G - M(\boldsymbol{g} \cdot \hat{x}) - f_{d_x} + \sum_{i=1}^{4} \left(\tau_i + \rho_m \left(-L_i g^S + L_i \ddot{L}_i - \frac{1}{2}(L_i \dot{\alpha}_i)^2 \right) \right) S_{i_x}$$

$$- \sum_{i=1}^{4} \rho_m \left(-\frac{1}{2} L_i g^N + \frac{3}{4} L_i \dot{L}_i \dot{\alpha}_i + \frac{1}{3} L_i^2 \ddot{\alpha}_i \right) S_{i_y} = 0.$$

This can be rewritten in the form

$$M\ddot{x}_G - M(\boldsymbol{g}\cdot\hat{\boldsymbol{x}}) - f_{d_x} + \sum_{i=1}^{4}\left(\tau_i - \rho_m\left(L_i g^S - L_i\ddot{L}_i + \frac{1}{2}(L_i\dot{\alpha}_i)^2\right)\right)S_{i_x}$$
$$+\sum_{i=1}^{4}\rho_m\left(\frac{1}{2}L_i g^N - \frac{3}{4}L_i\dot{L}_i\dot{\alpha}_i - \frac{1}{3}L_i^2\ddot{\alpha}_i\right)S_{i_y} = 0. \quad (5.222)$$

On the other side, first negate Equation 5.217

$$M\ddot{x}_G - M(\boldsymbol{g}\cdot\hat{\boldsymbol{x}}) - f_{d_x} + \sum_{i=1}^{4}\left((\tau_i - P_i)S_{i_x} + Q_i S_{i_y}\right) = 0$$

and substitute P_i and Q_i from Equations 5.220 and 5.221 into it. By some simplifications the following equation may be derived:

$$M\ddot{x}_G - M(\boldsymbol{g}\cdot\hat{\boldsymbol{x}}) - f_{d_x} + \sum_{i=1}^{4}\left(\tau_i - \rho_m\left(L_i g^S - L_i\ddot{L}_i + \frac{1}{2}(L_i\dot{\alpha}_i)^2\right)\right)S_{i_x}$$
$$+\sum_{i=1}^{4}\rho_m\left(\frac{1}{2}L_i g^N - \frac{3}{4}L_i\dot{L}_i\dot{\alpha}_i - \frac{1}{3}L_i^2\ddot{\alpha}_i\right)S_{i_y} = 0. \quad (5.223)$$

It is seen that Equations 5.222 and 5.223 are term-by-term identical. It can be shown in a similar manner that the equations of motion in y and z directions are identical as well.

5.4.5 Dynamic Formulation of a Planar Manipulator: Variable Mass Treatment

In this section, dynamic analysis of the planar $4R\underline{P}R$ manipulator considering the variable mass treatment is given in detail. Note that the equation of motion for the manipulator in this case is very similar to that given in Section 5.4.3, while just the inertial forces in the limbs vary. Therefore, referring to Equations 5.89 and 5.90, for variable mass system, the external and inertial wrenches applied to the limb i are determined by

$$\hat{\mathcal{F}}_i = \begin{bmatrix}\hat{\boldsymbol{f}}_i\\ \hat{\boldsymbol{n}}_i\end{bmatrix} = \begin{bmatrix} m_i\boldsymbol{g} - (m_i\ddot{\boldsymbol{r}}_e - \dot{m}_i\dot{\boldsymbol{u}} - \frac{\mathrm{d}^2}{\mathrm{d}t^2}(m_i\boldsymbol{\rho}_e))\\ -\left({}^{c_i}\dot{\boldsymbol{H}} - \boldsymbol{\rho}_e\times\dot{m}_i\dot{\boldsymbol{\rho}}_0\right)\end{bmatrix}, \quad (5.224)$$

in which the components of the inertial terms in Equation 5.224 in the cable directions $[\hat{\boldsymbol{S}}_i, \hat{\boldsymbol{N}}_i, \hat{\boldsymbol{K}}]^T$ are given in Equations 5.101 through 5.103. Hence

$$\hat{\mathcal{F}}_i = \rho_m\begin{bmatrix} L_i g^S - L_i\ddot{L}_i + (L_i\dot{\alpha}_i)^2\\ L_i g^N - \frac{1}{2}L_i^2\ddot{\alpha}_i - 2L_i\dot{L}_i\dot{\alpha}_i\\ \frac{1}{12}L_i^3\ddot{\alpha}_i\end{bmatrix}. \quad (5.225)$$

Finally, by considering the actuator forces as $\boldsymbol{\tau} = \begin{bmatrix} -\tau_1 & -\tau_2 & -\tau_3 & -\tau_4 \end{bmatrix}$, the governing equation of motion of the manipulator considering variable mass treatment results in

$$f_{d_x} + M(\boldsymbol{g} \cdot \hat{\boldsymbol{x}}) - M\,\ddot{x}_G - \sum_{i=1}^{4} \left\{ (\tau_i - P_i) S_{i_x} + Q_i S_{i_y} \right\} = 0 \tag{5.226}$$

$$f_{d_y} + M(\boldsymbol{g} \cdot \hat{\boldsymbol{y}}) - M\,\ddot{y}_G - \sum_{i=1}^{4} \left\{ (\tau_i - P_i) S_{i_y} - Q_i S_{i_x} \right\} = 0 \tag{5.227}$$

$$\tau_d - I_m\,\ddot{\phi} - \sum_{i=1}^{4} \left\{ (\tau_i - P_i)(E_{i_x} S_{i_y} - E_{i_y} S_{i_x}) - Q_i (\boldsymbol{E}_i \cdot \hat{\boldsymbol{S}}_i) \right\} = 0, \tag{5.228}$$

in which

$$P_i = \rho_m \left(L_i g^S - L_i \ddot{L}_i + (L_i \dot{\alpha}_i)^2 \right), \tag{5.229}$$

$$Q_i = \rho_m \left(\frac{1}{2} L_i g^N - L_i \dot{L}_i \dot{\alpha}_i - \frac{1}{3} L_i^2 \ddot{\alpha}_i \right). \tag{5.230}$$

Comparing P_i and Q_i in these equations to those considering constant mass treatment given in Equations 5.220 and 5.221, it is seen that there is only a minor difference between these formulations. Furthermore, as shown in the previous section, it can be shown that the dynamic formulations obtained here using virtual work approach is term-by-term equivalent to those obtained by the Newton–Euler approach.

5.4.6 Dynamic Formulation of the Stewart–Gough Platform

In this section, dynamic analysis of the Stewart–Gough platform the architecture of which is described in Section 3.5.1 and shown in Figure 3.16 is performed by the principle of virtual work. As explained earlier, in this manipulator the spatial motion of the moving platform is generated by six identical piston–cylinder actuators. A complete kinematic analysis of this manipulator is given in Section 3.5 and its Jacobian analysis is presented in Section 4.8. Furthermore, dynamic analysis of this manipulator is described in Section 5.3.3.

To derive the equation of motion for the manipulator by the principle of virtual work, the external and inertial wrenches applied to the moving platform, $\hat{\boldsymbol{\mathcal{F}}}_p$, and on each limbs $\hat{\boldsymbol{\mathcal{F}}}_i$ are first obtained, and then the Jacobian matrices of the limbs are determined. Since, the limb consists of a cylinder and piston, the external and inertial wrenches applied to the limbs are divided into two parts, $\hat{\boldsymbol{\mathcal{F}}}_i = \hat{\boldsymbol{\mathcal{F}}}_{i_1} + \hat{\boldsymbol{\mathcal{F}}}_{i_2}$, in which, as used earlier, subscript i_1 denotes the variable related to the cylinder, while subscript i_2 denotes that related to the piston. Therefore, the principle of virtual work applied to Stewart–Gough platform can be formulated as

$$\boldsymbol{J}^T \boldsymbol{\tau} + \hat{\boldsymbol{\mathcal{F}}}_p + \sum_i \left(\boldsymbol{J}_{i_1}^T \hat{\boldsymbol{\mathcal{F}}}_{i_1} + \boldsymbol{J}_{i_2}^T \hat{\boldsymbol{\mathcal{F}}}_{i_2} \right) = \boldsymbol{0}. \tag{5.231}$$

The Jacobian of the manipulator is derived in Equation 4.84, therefore, the Cartesian wrench projecting the actuator force in the Cartesian space $\boldsymbol{\mathcal{F}} = \boldsymbol{J}^T \boldsymbol{\tau}$ is derived as

$$\boldsymbol{\mathcal{F}} = \boldsymbol{J}^T \boldsymbol{\tau} = \sum_i \begin{bmatrix} \hat{\boldsymbol{s}}_i \tau_i \\ \boldsymbol{b}_{i_\times} \hat{\boldsymbol{s}}_i \tau_i \end{bmatrix}, \tag{5.232}$$

in which $\boldsymbol{b}_{i_\times}$ denotes skew-symmetric matrix derived from elements of vectors $\boldsymbol{b}_i$ as defined in Equation 5.167. Next, the external and inertial wrenches applied to the moving platform is found:

$$\hat{\mathcal{F}}_p = \begin{bmatrix} \hat{f} \\ {}^c\hat{n} \end{bmatrix} = \begin{bmatrix} \boldsymbol{F}_d + m(\boldsymbol{g} - \boldsymbol{a}_p) \\ \boldsymbol{n}_d - {}^A\boldsymbol{I}_P\dot{\boldsymbol{\omega}} - \boldsymbol{\omega} \times {}^A\boldsymbol{I}_P\boldsymbol{\omega} \end{bmatrix}, \tag{5.233}$$

in which $\boldsymbol{a}_p$ denotes the acceleration of point P on the moving platform which is given in Equation 5.117, m denotes the moving platform mass, and ${}^A\boldsymbol{I}_P$ denotes the moving platform moment of inertia about its center of mass. Furthermore, $\boldsymbol{\omega}$ denotes the moving platform angular velocity, $m\boldsymbol{g}$ denotes the gravity force and the external disturbance wrench applied on the moving platform is denoted by $\mathcal{F}_d = [\boldsymbol{F}_d, \boldsymbol{n}_d]^T$. The external and inertial wrenches applied to each segment of the limb is determined by the following two equations:

$$\hat{\mathcal{F}}_{i_1} = \begin{bmatrix} \hat{f}_{i_1} \\ \hat{n}_{i_1} \end{bmatrix} = \begin{bmatrix} m_{i_1}(\boldsymbol{g} - \boldsymbol{a}_{i_1}) \\ -{}^A\boldsymbol{I}_{c_{i_1}}\dot{\boldsymbol{\omega}}_i - \boldsymbol{\omega}_i \times {}^A\boldsymbol{I}_{c_{i_1}}\boldsymbol{\omega}_i \end{bmatrix}, \tag{5.234}$$

and

$$\hat{\mathcal{F}}_{i_2} = \begin{bmatrix} \hat{f}_{i_2} \\ \hat{n}_{i_2} \end{bmatrix} = \begin{bmatrix} m_{i_2}(\boldsymbol{g} - \boldsymbol{a}_{i_2}) \\ -{}^A\boldsymbol{I}_{c_{i_2}}\dot{\boldsymbol{\omega}}_i - \boldsymbol{\omega}_i \times {}^A\boldsymbol{I}_{c_{i_2}}\boldsymbol{\omega}_i \end{bmatrix}. \tag{5.235}$$

Note that in this formulation, ${}^A\boldsymbol{I}_{c_{i_1}}$ and ${}^A\boldsymbol{I}_{c_{i_2}}$ denote the inertia matrix of the cylinder and piston, respectively, evaluated at a frame located at the center points c_{i_1} and c_{i_2}. Furthermore, $\boldsymbol{a}_{i_1}$ and $\boldsymbol{a}_{i_2}$ denote the acceleration of the cylinder and piston centers of masses, respectively, which are given in Equations 5.124 and 5.125. Substituting $\boldsymbol{a}_{i_1}$ and $\boldsymbol{a}_{i_2}$, the equations are simplified to the following equations:

$$\hat{\mathcal{F}}_{i_1} = \begin{bmatrix} m_{i_1}\left(\boldsymbol{g} - c_{i_1}(\dot{\boldsymbol{\omega}}_{i_\times}\hat{\boldsymbol{s}}_i) - c_{i_1}|\boldsymbol{\omega}_i|_2^2\hat{\boldsymbol{s}}_i\right) \\ -{}^A\boldsymbol{I}_{c_{i_1}}\dot{\boldsymbol{\omega}}_i - \boldsymbol{\omega}_{i_\times}{}^A\boldsymbol{I}_{c_{i_1}}\boldsymbol{\omega}_i \end{bmatrix}, \tag{5.236}$$

and

$$\hat{\mathcal{F}}_{i_2} = \begin{bmatrix} m_{i_2}\boldsymbol{g} - m_{i_2}(\ell_i - c_{i_2})\left(\dot{\boldsymbol{\omega}}_{i_\times}\hat{\boldsymbol{s}}_i - |\boldsymbol{\omega}_i|_2^2\hat{\boldsymbol{s}}_i + 2\dot{\ell}_i\boldsymbol{\omega}_{i_\times}\hat{\boldsymbol{s}}_i + \ddot{\ell}_i\hat{\boldsymbol{s}}_i\right) \\ -{}^A\boldsymbol{I}_{c_{i_2}}\dot{\boldsymbol{\omega}}_i - \boldsymbol{\omega}_{i_\times}{}^A\boldsymbol{I}_{c_{i_2}}\boldsymbol{\omega}_i \end{bmatrix}. \tag{5.237}$$

The next step is to derive the limb Jacobians. Limb Jacobians $\boldsymbol{J}_{i_1}$ and $\boldsymbol{J}_{i_2}$ are derived from linear and angular velocities of the cylinder and piston centers of masses as follows:

$$\begin{aligned} \dot{\mathcal{X}}_{i_1} &= \boldsymbol{J}_{i_1}\dot{\mathcal{X}} \\ \begin{bmatrix} \boldsymbol{v}_{i_1} \\ \boldsymbol{\omega}_i \end{bmatrix} &= \boldsymbol{J}_{i_1}\begin{bmatrix} \boldsymbol{v}_p \\ \boldsymbol{\omega} \end{bmatrix} \end{aligned} \tag{5.238}$$

$$\begin{aligned} \dot{\mathcal{X}}_{i_2} &= \boldsymbol{J}_{i_2}\dot{\mathcal{X}} \\ \begin{bmatrix} \boldsymbol{v}_{i_2} \\ \boldsymbol{\omega}_i \end{bmatrix} &= \boldsymbol{J}_{i_2}\begin{bmatrix} \boldsymbol{v}_p \\ \boldsymbol{\omega} \end{bmatrix}. \end{aligned} \tag{5.239}$$

Linear velocity of the cylinder center of mass v_{i_1} can be written in the following form, substituting ω_i from Equation 5.168 and $\dot{x}_i$ from Equation 5.179

$$\begin{aligned} v_{i_1} &= c_{i_1}(\omega_i \times \hat{s}_i) = -c_{i_1}(\hat{s}_i \times \omega_i) \\ &= -c_{i_1}\hat{s}_{i_\times}\omega_i \\ &= -\frac{c_{i_1}}{\ell_i}\hat{s}_{i_\times}^2\dot{x}_i \\ &= -\frac{c_{i_1}}{\ell_i}\hat{s}_{i_\times}^2\begin{bmatrix} I_{3\times 3} & -b_{i_\times}\end{bmatrix}\dot{\mathcal{X}} \\ &= -\frac{c_{i_1}}{\ell_i}\begin{bmatrix} \hat{s}_{i_\times}^2 & -\hat{s}_{i_\times}^2 b_{i_\times}\end{bmatrix}\dot{\mathcal{X}}. \end{aligned} \tag{5.240}$$

Similarly, linear velocity of the piston center of mass v_{i_2} can be written as the following equation, by substitution of ω_i and $\dot{l}\hat{s}_i$ from Equation 5.168, and $\dot{x}_i$ from Equation 5.179.

$$\begin{aligned} v_{i_2} = c_{i_2}(\omega_i \times \hat{s}_i) + \dot{l}\hat{s}_i &= -c_{i_2}\hat{s}_{i_\times}\omega_i + \dot{l}\hat{s}_i \\ &= -\frac{c_{i_2}}{\ell_i}\hat{s}_{i_\times}^2\dot{x}_i + \hat{s}_i\hat{s}_i^T\dot{x}_i \\ &= \left(-\frac{c_{i_2}}{\ell_i}\hat{s}_{i_\times}^2 + \hat{s}_i\hat{s}_i^T\right)\begin{bmatrix} I_{3\times 3} & -b_{i_\times}\end{bmatrix}\dot{\mathcal{X}} \\ &= \begin{bmatrix} -\frac{c_{i_2}}{\ell_i}\hat{s}_{i_\times}^2 + \hat{s}_i\hat{s}_i^T & \frac{c_{i_2}}{\ell_i}\hat{s}_{i_\times}^2 b_{i_\times} - \hat{s}_i\hat{s}_i^T b_{i_\times}\end{bmatrix}\dot{\mathcal{X}}. \end{aligned} \tag{5.241}$$

Furthermore, angular velocity of the limb ω_i can be related to the moving platform generalized coordinate $\dot{\mathcal{X}}$ by Equation 5.112

$$\begin{aligned} \omega_i &= \frac{1}{\ell_i}\left(\hat{s}_i \times v_p + \hat{s}_i \times (\omega \times b_i)\right) \\ &= \frac{1}{\ell_i}\hat{s}_{i_\times}v_p - \frac{1}{\ell_i}\hat{s}_{i_\times}b_{i_\times}\omega \\ &= \frac{1}{\ell_i}\begin{bmatrix}\hat{s}_{i_\times} & -\hat{s}_{i_\times}b_{i_\times}\end{bmatrix}\begin{bmatrix} v_p \\ \omega \end{bmatrix}. \end{aligned} \tag{5.242}$$

Therefore, the final limb Jacobians are derived as follows:

$$J_{i_1} = \frac{1}{\ell_i}\begin{bmatrix} -c_{i_1}\hat{s}_{i_\times}^2 & c_{i_1}\hat{s}_{i_\times}^2 b_{i_\times} \\ \hat{s}_{i_\times} & -\hat{s}_{i_\times}b_{i_\times}\end{bmatrix}, \tag{5.243}$$

and

$$J_{i_2} = \frac{1}{\ell_i}\begin{bmatrix} -c_{i_2}\hat{s}_{i_\times}^2 + \ell_i\hat{s}_i\hat{s}_i^T & c_{i_2}\hat{s}_{i_\times}^2 b_{i_\times} - \ell_i\hat{s}_i\hat{s}_i^T b_{i_\times} \\ \hat{s}_{i_\times} & -\hat{s}_{i_\times}b_{i_\times}\end{bmatrix}. \tag{5.244}$$

The final dynamic formulation of the Stewart–Gough platform can be written in an implicit form by substitution of the corresponding terms in Equation 5.231. In the resulting formulation the actuator wrench $\mathcal{F} = J^T\tau$ is substituted by Equation 5.232, the moving platform wrench $\hat{\mathcal{F}}_p$ is substituted by Equation 5.233, the cylinder and piston wrenches $\hat{\mathcal{F}}_{i_1}$ and $\hat{\mathcal{F}}_{i_2}$ are substituted by Equations 5.236 and 5.237, respectively, and their corresponding Jacobian matrices J_{i_1} and J_{i_2} are substituted by Equations 5.243 and 5.244, respectively. The resulting formulation can be further simplified, however, term-by-term derivation and comparison of the simplified formulation to that derived by the Newton–Euler approach is a prohibitive task. To verify the obtained formulations, one may perform inverse and forward dynamic simulations detailed in Section 5.3.3 and obtain identical results.

5.5 Lagrange Formulation

The Lagrangian mechanics is a reformulation of classical mechanics that combines conservation of momentum with conservation of energy. This formulation is extremely effective in the dynamic analysis of multibody systems, in which many reaction forces and moments exist between the bodies. The Lagrangian method formulates the equations of motion by using a set of *generalized coordinates*, and it eliminates all or some of the unwanted reaction forces and moments at the outset. The use of generalized coordinates may considerably simplify the dynamic analysis of a system. For example, consider a small frictionless bead traveling in a groove. If one is tracking the bead as a particle, calculation of the motion of the bead using the Newtonian mechanics would require solving for the time-varying constraint force required to keep the bead in the groove. Using the Lagrangian mechanics for the same problem, look at the path of the groove and choose a set of independent generalized coordinates that completely characterize the possible motion of the bead. This choice eliminates the need for the constraint forces to enter the resultant system of equations. There are fewer equations since one is not directly calculating the influence of the groove on the bead at a given moment.

In the Lagrangian mechanics, the trajectory of a multi-body system is derived by solving the Lagrange equations in one of two forms, either the Lagrange equations of the first kind [69], which treat constraints explicitly as extra equations, often using Lagrange multipliers [42]; or the Lagrange equations of the second kind, which incorporate the constraints directly by judicious choice of generalized coordinates [149]. In what follows, a general description of both kinds is given.

5.5.1 Generalized Coordinates

A kinematic constraint imposes some conditions on the relative motion between two pair of bodies in a multibody system. In robotic manipulators, the constraints are introduced into the system at the joints, and since there are different joints used in robotic manipulators, different types of constraints exist in the system. A common classification of constraints is holonomic and nonholonomic constraints. A kinematic constraint is called *holonomic* if the conditions of constraints can be expressed as an algebraic equation of their coordinates, and possibly the time, that is

$$f(x_1, x_2, \ldots, t) = 0, \tag{5.245}$$

in which x_i denotes the coordinates of a rigid body. A constraint that cannot be expressed in the foregoing form and requires the inclusion of $\dot{x}_i$ is called *nonholonomic*. The equation of constraint can be derived by the equation of the joint. For example, the constrained motion of a pendulum with length l hinged upon a revolute joint can be formulated as $x^2 + y^2 - l^2 = 0$; and hence, it is holonomic. Similarly the constraints provided by prismatic, universal, spherical, and other common joints in robotic manipulators are all holonomic.

The configuration of a robotic manipulator is completely known, if positions and orientations of all the bodies in the system are known with respect to a reference frame. For a typical parallel manipulator shown in Figure 3.2, each limb may consist of several bodies, being connected to each other and to the moving platform in closed chains. Therefore, many intermediate coordinates and constraints exist to represent the motion of each segment of the manipulator, which are usually dependent to each other. For holonomic constraints, the dependent coordinates can be eliminated through the constraints equations, or by the introduction of m new independent variables denoted by $q_1, q_2, \ldots, q_m$ such that all the intermediate coordinates can be expressed in terms of the m new independent variables. These independent variables produce a set of independent *generalized coordinates*. Thus, the number of independent generalized coordinates is equal to the number of degrees-of-freedom of a parallel robot.

It can be seen from this description that for a parallel robot definition of the generalized coordinates are not unique and they can be defined in several different ways. However, one general approach to define such coordinates is to consider sufficient number of position and orientation variables of the moving platform corresponding to the degrees-of-freedom of the system. Therefore, in general, a generalized coordinate q_j does not have the dimension of length and can be defined as an orientation angle. As we discussed earlier, although the Euler angles can be used to define such coordinates, it is most convenient to assign screw coordinates as the representatives of such coordinates, to directly relate their derivatives to angular velocity of the moving platform.

In the same way that a generalized coordinate q_j is introduced, a generalized force Q_j, can be defined, which plays a conjugate role to the corresponding generalized coordinate. To define generalized forces, apply the principle of virtual work upon the external forces applied to the parallel robot. Consider n forces $\boldsymbol{F}_i$ are applied to the system at different locations, and note that usually the nonconservative forces are treated to define the generalized forces of the system. Depending on the location the forces are applied to the system their virtual work δW_a may be determined by the summation of the work done for each, which is calculated from the dot product of the force and the virtual displacement at the location of the applied force $\delta \boldsymbol{x}_i$

$$\delta W_a = \sum_{i=1}^{n} \boldsymbol{F}_i \cdot \delta \boldsymbol{x}_i. \tag{5.246}$$

The projection of this work on the generalized coordinate can be defined as the work done by each generalized force Q_j along the corresponding generalized coordinate q_j. Therefore, the resulting virtual work can be written in the form of

$$\delta W_a = \sum_{i=1}^{m} Q_j \delta q_j, \tag{5.247}$$

in which the generalized force is equivalent to

$$Q_j = \sum_{i=1}^{n} \boldsymbol{F}_i \cdot \frac{\partial \boldsymbol{x}_i}{\partial q_j}. \tag{5.248}$$

Note that in derivation of the generalized forces, the equivalence of Equation 5.246 to Equation 5.247 is applied for each nonconservative external force applied to the system, and Equation 5.248 is the mathematical definition being used in the Lagrangian formulation derivation that follows. Hence, for a generalized coordinate q_j that is defined by an orientation angle, the corresponding generalized force has the dimension of a torque.

5.5.2 Lagrange Equations of the Second Kind

The equations of motion derived by the Lagrangian mechanics are called the Lagrange equations or the Euler–Lagrange equations. In this section, the derivation of Lagrange equations of the second kind is described briefly. Let us examine the principle of virtual work applied to a system of particles (or in a similar formulation applied to a system of multibody components). Assume there exist n particles with mass m_i, experiencing an acceleration $\boldsymbol{a}_i$, while the external forces applied to the system is denoted by $\boldsymbol{F}_i$. By using the d'Alembert principle and including the inertial forces, the system is considered to be statically balanced, and the principle of virtual work may be written for such a system as

$$\delta W = \sum_{i=1}^{n} (\boldsymbol{F}_i - m_i \boldsymbol{a}_i) \cdot \delta \boldsymbol{r}_i = 0, \tag{5.249}$$

in which δW denotes the total virtual work of the system, and $\delta \boldsymbol{r}_i$ denotes the virtual displacement of the system, consistent with the constraints. Furthermore, consider there exist m independent generalized coordinates q_i for the system, by which the displacement of each particle can be fully defined:

$$\begin{aligned}
\boldsymbol{r}_1 &= \boldsymbol{r}_1(q_1, q_2, \ldots, q_m, t) \\
\boldsymbol{r}_2 &= \boldsymbol{r}_2(q_1, q_2, \ldots, q_m, t) \\
&\vdots \\
\boldsymbol{r}_n &= \boldsymbol{r}_n(q_1, q_2, \ldots, q_m, t).
\end{aligned}$$

Therefore, the expression for the virtual displacement or the differential motion of each particle can be obtained using partial derivatives of $\boldsymbol{r}_i$ with respect to the generalized coordinates:

$$\delta \boldsymbol{r}_i = \sum_{j=1}^{m} \frac{\partial \boldsymbol{r}_i}{\partial q_j} \delta q_j. \tag{5.250}$$

Substitute Equation 5.250 into Equation 5.249, and factor δq_j

$$\delta W = \sum_{j=1}^{m} \sum_{i=1}^{n} \left(\boldsymbol{F}_i \frac{\partial \boldsymbol{r}_i}{\partial q_j} - m_i \boldsymbol{a}_i \frac{\partial \boldsymbol{r}_i}{\partial q_j} \right) \cdot \delta q_j = 0. \tag{5.251}$$

The external forces applied to the system of particles can be divided into two components $\boldsymbol{F}_i = \boldsymbol{F}_i^c + \boldsymbol{F}_i^u$, in which $\boldsymbol{F}_i^c$ denotes the conservative component of the external force, while $\boldsymbol{F}_i^u$ denotes the nonconservative part. Note that a conservative force can be thought of as a force that conserves mechanical energy. The gravitational and spring forces are examples of conservative forces, while other external forces such as friction and air drag are examples of nonconservative forces. For conservative forces, they may be represented by a scalar potential field, P, as $\boldsymbol{F}_i^c = -\nabla P$; therefore, for these forces we can write

$$\sum_{i=1}^{n} \boldsymbol{F}_i^c \cdot \frac{\partial \boldsymbol{r}_i}{\partial q_j} = -\sum_{i=1}^{n} \nabla P \cdot \frac{\partial \boldsymbol{r}_i}{\partial q_j} = -\frac{\partial P}{\partial q_j}. \tag{5.252}$$

Furthermore, the nonconservative components of the external forces may be expressed by generalized forces Q_j, defined in Equation 5.248

$$Q_j = \sum_{i=1}^{n} \boldsymbol{F}_i^u \cdot \frac{\partial \boldsymbol{r}_i}{\partial q_j}. \tag{5.253}$$

The second term in Equation 5.251 may be expressed by the following function of kinetic energy of system K [65]:

$$\sum_{i=1}^{n} m_i \boldsymbol{a}_i \cdot \frac{\partial \boldsymbol{r}_i}{\partial q_j} = \frac{\mathrm{d}}{\mathrm{d}t}\left(\frac{\partial K}{\partial \dot{q}_j}\right) - \frac{\partial K}{\partial q_j}. \tag{5.254}$$

Substitute Equations 5.252 through 5.254 into Equation 5.251, and simplify as

$$\delta W = \sum_{j=1}^{m} \left(-\frac{\partial P}{\partial q_j} + Q_j - \frac{\mathrm{d}}{\mathrm{d}t}\left(\frac{\partial K}{\partial \dot{q}_j}\right) + \frac{\partial K}{\partial q_j}\right) \cdot \delta q_j = 0. \tag{5.255}$$

Since this relation holds for any arbitrary virtual displacement, it can be concluded that, for $j = 1, 2, \ldots, m$

$$\frac{\mathrm{d}}{\mathrm{d}t}\left(\frac{\partial K}{\partial \dot{q}_j}\right) - \frac{\partial K}{\partial q_j} + \frac{\partial P}{\partial q_j} = Q_j. \tag{5.256}$$

The *Lagrangian function* is defined as the difference between the kinetic and potential energies of a mechanical system

$$\mathcal{L} = K - P. \tag{5.257}$$

Since kinetic energy depends on both the generalized coordinates and their time derivatives, and whereas potential energy depends only on the generalized coordinates itself, Equation 5.256 may be formulated in terms of the Lagrangian function as

$$\frac{\mathrm{d}}{\mathrm{d}t}\left(\frac{\partial \mathcal{L}}{\partial \dot{q}_j}\right) - \frac{\partial \mathcal{L}}{\partial q_j} = Q_j, \quad j = 1, 2, \ldots, m, \tag{5.258}$$

in which q_j denotes the generalized coordinate, Q_j denotes the generalized force, and m denotes the number of independent generalized coordinates of the system. The Lagrange Equation 5.258 can be used m times to generate the set of m dynamic equations governing the motion of the system.

5.5.3 Lagrange Equations of the First Kind

Contrary to the Lagrangian formulation of the second kind, the Lagrange equations of the first kind are written in terms of a set of redundant coordinates. Therefore, the formulation requires a set of constraint equations derived from the kinematics of the robot. These constraint equations and their derivatives are adjointed to the main equations of motion using Lagrange multipliers. Therefore, the number of equations derived through this formulation is equal to the number of degrees-of-freedom plus the redundant coordinates.

The Lagrangian formulations of the first type can be written as

$$\frac{\mathrm{d}}{\mathrm{d}t}\left(\frac{\partial \mathcal{L}}{\partial \dot{q}_j}\right) - \frac{\partial \mathcal{L}}{\partial q_j} = Q_j + \sum_{i=1}^{k} \lambda_i \frac{\partial h_i}{\partial q_j}, \quad j = 1, 2, \ldots, n, \tag{5.259}$$

in which h_i denotes the constraint function, k is the number of constraints, λ_i is the Lagrange multiplier, and n is the number of redundant coordinates, which exceeds the number of degrees-of-freedom m by k, that is, $n = m + k$. To derive the dynamic equation, it is more convenient to arrange the Lagrange equations into two sets. The first set contains the Lagrange multipliers as the only unknowns and the other set contains the generalized forces as the unknowns. Through this division, the first k equations are associated with the redundant coordinates, and the remaining $n - k$ equations are associated with the actuated joint variables. By this means, the first set of equations may be written in the following form:

$$\sum_{i=1}^{k} \lambda_i \frac{\partial h_i}{\partial q_j} = \frac{\mathrm{d}}{\mathrm{d}t}\left(\frac{\partial \mathcal{L}}{\partial \dot{q}_j}\right) - \frac{\partial \mathcal{L}}{\partial q_j} - \hat{Q}_j, \quad j = 1, 2, \ldots, k. \tag{5.260}$$

In this set of equations, if any $\hat{Q}_j$ exists, it represents the generalized force contributed by any external force, excluding the gravitational and actuator forces, while λ_i are the unknowns. Given $\hat{Q}_j$, the right-hand side of Equation 5.260 is known, and by writing this equation for each redundant coordinate, a set of k linear equations are obtained that can be solved for Lagrange multipliers.

Once the Lagrange multipliers are found, the dynamic formulation written in terms of actuator-contributed generalized forces can be derived from the remaining equations. These equations can be written in the form of

$$Q_j = \frac{\mathrm{d}}{\mathrm{d}t}\left(\frac{\partial \mathcal{L}}{\partial \dot{q}_j}\right) - \frac{\partial \mathcal{L}}{\partial q_j} - \sum_{i=1}^{k} \lambda_i \frac{\partial h_i}{\partial q_j}, \quad j = k + 1, \ldots, n. \tag{5.261}$$

Although the Lagrange formulations given here look very different from what is given for the Lagrange equations of the second kind, the resulting equation of motion derived by both the methods result in the same conclusion.

Note that in the formulation of the first kind, there is no need to consider a set of independent generalized coordinates equal to the number of the degrees-of-freedom to derive the dynamic equation, and hence, at first glance this formulation seems to be more convenient. However, for any redundant equation, one Lagrange multiplier is to be determined from Equation 5.260, which needs an excess of manipulation in the derivation of the final dynamic equation. On the other hand, the number of Lagrange equations of the second type is directly equal to the number of degrees-of-freedom, but to derive the kinetic and potential energies of the robot, the intermediate limb variables are needed, and a nonlinear inverse kinematics solution is needed to eliminate those variables. Thus, the manipulation required to derive the equations of motion by either means are almost the same. In what follows, we use the Lagrange formulation of the second type to derive the governing equations of motion of some typical parallel manipulators.

5.5.4 Dynamic Formulation Properties

Before deriving the governing equations of motion of some typical parallel manipulators, it is worth reconsidering the dynamic equations obtained for a robot manipulator in terms of its structure and properties. Further development and implementation of dynamic equation characteristics is elaborated in the next chapter, where their properties would be directly used in different control schemes.

Consider an n-degrees-of-freedom fully parallel manipulator, whose generalized coordinate is denoted by the vector $\mathcal{X} = \left[q_1, q_2, \ldots, q_n\right]^T$. If the set of dynamic equation derived from Equation 5.258 is written for all the generalized coordinates in a vector form, this yield

$$\frac{\mathrm{d}}{\mathrm{d}t}\left(\frac{\partial \mathcal{L}}{\partial \dot{\mathcal{X}}}\right) - \frac{\partial \mathcal{L}}{\partial \mathcal{X}} = \mathcal{F}, \tag{5.262}$$

in which $\mathcal{F} = \left[Q_1, Q_2, \ldots, Q_n\right]^T$ denotes the vector of generalized forces of the manipulator, and $\mathcal{L}$ denotes the Lagrangian that is defined as the difference between kinetic and potential energies, $\mathcal{L} = K - P$. Note that for a six-degrees-of-freedom parallel manipulator, it is suggested to use the position of the moving platform center of mass and its orientation variables represented in screw coordinate, as the generalized coordinates. Furthermore, if there is no friction and disturbance wrench applied to the system, the generalized force can be determined from the actuator forces by the Jacobian relation given in Equation 4.92, which reads as $\mathcal{F} = J^T \boldsymbol{\tau}$.

First examine the Lagrangian function. Let us define the *mass matrix* $\boldsymbol{M}(\mathcal{X})$ directly from the kinetic energy of the manipulator

$$K(\mathcal{X}, \dot{\mathcal{X}}) = \frac{1}{2}\dot{\mathcal{X}}^T \boldsymbol{M}(\mathcal{X})\dot{\mathcal{X}}. \tag{5.263}$$

Hence, the Lagrangian function can be written as

$$\mathcal{L}(\mathcal{X}, \dot{\mathcal{X}}) = K(\mathcal{X}, \dot{\mathcal{X}}) - P(\mathcal{X}) = \frac{1}{2}\dot{\mathcal{X}}^T \boldsymbol{M}(\mathcal{X})\dot{\mathcal{X}} - P(\mathcal{X}). \tag{5.264}$$

It is a fundamental property that kinetic energy is a quadratic function of the velocity vector $\dot{\mathcal{X}}$, while potential energy is independent of velocity $\dot{\mathcal{X}}$. Now examine Lagrange equation terms given in Equation 5.262

$$\begin{aligned}\frac{\mathrm{d}}{\mathrm{d}t}\left(\frac{\partial \mathcal{L}}{\partial \dot{\mathcal{X}}}\right) &= \frac{\mathrm{d}}{\mathrm{d}t}\left(\frac{\partial K}{\partial \dot{\mathcal{X}}}\right),\\ &= \frac{\mathrm{d}}{\mathrm{d}t}\left(\boldsymbol{M}(\mathcal{X})\dot{\mathcal{X}}\right),\end{aligned}$$

$$= M(\mathcal{X})\ddot{\mathcal{X}} + \dot{M}(\mathcal{X})\dot{\mathcal{X}}, \tag{5.265}$$

$$\frac{\partial \mathcal{L}}{\partial \mathcal{X}} = \frac{1}{2}\frac{\partial}{\partial \mathcal{X}}\left(\dot{\mathcal{X}}^T M(\mathcal{X})\dot{\mathcal{X}}\right) - \frac{\partial P(\mathcal{X})}{\partial \mathcal{X}}. \tag{5.266}$$

Therefore, Lagrange equation may be written as

$$M(\mathcal{X})\ddot{\mathcal{X}} + \left(\dot{M}(\mathcal{X})\dot{\mathcal{X}} - \frac{1}{2}\frac{\partial}{\partial \mathcal{X}}\left(\dot{\mathcal{X}}^T M(\mathcal{X})\dot{\mathcal{X}}\right)\right) + \frac{\partial P(\mathcal{X})}{\partial \mathcal{X}} = \mathcal{F}. \tag{5.267}$$

Let us define *gravity vector* $G(\mathcal{X})$ as

$$G(\mathcal{X}) = \frac{\partial P(\mathcal{X})}{\partial \mathcal{X}}, \tag{5.268}$$

and the *Coriolis and centrifugal vector* $V(\mathcal{X}, \dot{\mathcal{X}})$ as

$$\begin{aligned} V(\mathcal{X}, \dot{\mathcal{X}}) &= \dot{M}(\mathcal{X})\dot{\mathcal{X}} - \frac{\partial K}{\partial \mathcal{X}} \\ &= \dot{M}(\mathcal{X})\dot{\mathcal{X}} - \frac{1}{2}\frac{\partial}{\partial \mathcal{X}}\left(\dot{\mathcal{X}}^T M(\mathcal{X})\dot{\mathcal{X}}\right). \end{aligned} \tag{5.269}$$

Thus, the Lagrange equations may be written in a general form of

$$M(\mathcal{X})\ddot{\mathcal{X}} + V(\mathcal{X}, \dot{\mathcal{X}}) + G(\mathcal{X}) = \mathcal{F}. \tag{5.270}$$

It is worth noting that through the Lagrange method the general form of dynamic equations of parallel manipulators can be written in a closed form, similar to what is usually seen in serial manipulators.

5.5.4.1 Mass Matrix Properties

Mass matrix is defined from kinetic energy of the system by Equation 5.263. Although this $n \times n$ matrix is configuration dependent, since the kinetic energy of the robot is always positive, the mass matrix is always a symmetric and positive definite matrix, and hence, it is invertible for all configurations. Furthermore, it can be shown that

$$\underline{\lambda}\, I_{n\times n} \leq M(\mathcal{X}) \leq \overline{\lambda}\, I_{n\times n}, \tag{5.271}$$

in which $I_{n\times n}$ denotes $n \times n$ identity matrix, and $\underline{\lambda}$ and $\overline{\lambda}$ are positive constants denoting the lower and upper bounds of the mass matrix eigenvalues for all configurations. Similarly, the inverse of the mass matrix is also bounded:

$$\frac{1}{\underline{\lambda}} I_{n\times n} \geq M^{-1}(\mathcal{X}) \geq \frac{1}{\overline{\lambda}} I_{n\times n}. \tag{5.272}$$

The bounds on the mass matrix may also be represented by matrix norm as follows:

$$\underline{M} \leq \|M(\mathcal{X})\| \leq \overline{M}, \tag{5.273}$$

in which any induced norm can be used to define the positive scalars of $\underline{M}$ and $\overline{M}$.

5.5.4.2 Linearity in Parameters

The dynamic formulation of a parallel manipulator may be given in a general form represented by Equation 5.270. To implement this dynamic formulation in dynamics simulation or use it in model-based control routines, it is necessary to identify the kinematic and dynamic parameters of the manipulator. This is usually done for the manipulators in a formal calibration process. To determine the parameters with a certain accuracy, many calibration techniques are developed for robotic manipulators [8,119,126]. However, the complexity of manipulator dynamics makes the calibration process more challenging for parallel manipulators.

Fortunately, it can be shown that although the dynamic formulation of a parallel manipulator is nonlinear and coupled with respect to the motion variable $\boldsymbol{\mathcal{X}}$ and its derivatives, it can be rewritten in a linear form with respect to the kinematics and dynamic parameters. Define an ℓ-dimensional vector from the calibration parameters denoted by $\boldsymbol{\theta} \in \mathbb{R}^{\ell}$. Note that the dimension of the parameter space ℓ is not unique. In general, a given rigid body is described by 10 parameters, namely the mass, the three coordinates of center of mass, and the six independent parameters of inertia matrix. However, since the limb motions are constrained and coupled by closed kinematic chains, the independent parameters are much fewer. However, finding a minimal set of parameters that can parameterize the dynamic formulation in general is a challenging task. If such a parameter set is found, the linear model with respect to the parameters may be written as follows:

$$\boldsymbol{M}(\boldsymbol{\mathcal{X}})\ddot{\boldsymbol{\mathcal{X}}} + \boldsymbol{V}(\boldsymbol{\mathcal{X}}, \dot{\boldsymbol{\mathcal{X}}}) + \boldsymbol{G}(\boldsymbol{\mathcal{X}}) = \boldsymbol{Y}(\boldsymbol{\mathcal{X}}, \dot{\boldsymbol{\mathcal{X}}}, \ddot{\boldsymbol{\mathcal{X}}})\boldsymbol{\theta}. \tag{5.274}$$

This formulation is written in a linear regression form, in which $\boldsymbol{Y}(\boldsymbol{\mathcal{X}}, \dot{\boldsymbol{\mathcal{X}}}, \ddot{\boldsymbol{\mathcal{X}}})$ is called the regressor, and $\boldsymbol{\theta}$ is the parameter vector to be identified by experiments. If such a representation is derived for a parallel manipulator, calibration techniques based on least-square solution may be used to find the manipulator parameters $\boldsymbol{\theta}$, from calibration experiments. Furthermore, this linear regression representation may be used in advanced adaptive control techniques developed in the literature, and described in Section 6.4.2.

5.5.4.3 Coriolis and Centrifugal Vector Properties

Coriolis and centrifugal vectors in the Lagrange equations, defined in Equation 5.269, is the most complex part to be determined, and it includes all the inertial forces caused by Coriolis and centrifugal accelerations. To determine the components of this term it is required to differentiate the mass matrix components with respect to the motion variable $\boldsymbol{\mathcal{X}}$, which has n components. Such derivatives result in third-order tensors, which makes the manipulations prohibitive. To get around this problem, the Kronecker product of matrices are introduced [17], and well implemented on the dynamic equation of robotic manipulators* [143].

The Kronecker product of two matrices is denoted by $\otimes$, and is defined by term-by-term multiplication of the components of those matrices. For two matrices $A \in \mathbb{R}^{n \times m}$, and $B \in \mathbb{R}^{p \times r}$, the Kronecker product is defined as

$$\boldsymbol{A} \otimes \boldsymbol{B} = [a_{ij}\boldsymbol{B}] \in \mathbb{R}^{np \times mr}, \tag{5.275}$$

in which a_{ij} denotes the elements of matrix $\boldsymbol{A}$, and the resulting matrix has a dimension $np \times mr$, composed of $p \times r$ blocks, each block determined by term-by-term multiplication

* See Appendix A for more details.

of $a_{ij}\boldsymbol{B}$. The Kronecker product is bilinear and associative, but it is not commutative, that is, in general, $\boldsymbol{A} \otimes \boldsymbol{B}$ is different from $\boldsymbol{B} \otimes \boldsymbol{A}$. Moreover, the inverse and transpose of the product follows the following rules. It follows that $\boldsymbol{A} \otimes \boldsymbol{B}$ is invertible, if and only if $\boldsymbol{A}$ and $\boldsymbol{B}$ are invertible, in which case the inverse is given by

$$(\boldsymbol{A} \otimes \boldsymbol{B})^{-1} = \boldsymbol{A}^{-1} \otimes \boldsymbol{B}^{-1}. \tag{5.276}$$

The operation of transposition is also distributive over the Kronecker product:

$$(\boldsymbol{A} \otimes \boldsymbol{B})^{T} = \boldsymbol{A}^{T} \otimes \boldsymbol{B}^{T}. \tag{5.277}$$

There is another property of Kronecker product, used in the simplification of dynamic equations. Consider an n-dimensional vector $\boldsymbol{x}$, since the Kronecker product is not commutative, in general $\boldsymbol{I}_{n\times n} \otimes \boldsymbol{x} \neq \boldsymbol{x} \otimes \boldsymbol{I}_{n\times n}$, in which $\boldsymbol{I}_{n\times n}$ denotes the $n \times n$ identity matrix. However, it can be shown that the following equality holds for any arbitrary vector $\boldsymbol{x} \in \mathbb{R}^n$

$$(\boldsymbol{I}_{n\times n} \otimes \boldsymbol{x})\,\boldsymbol{x} = (\boldsymbol{x} \otimes \boldsymbol{I}_{n\times n})\,\boldsymbol{x}. \tag{5.278}$$

Finally, define partial derivative of a matrix $\boldsymbol{A}(\boldsymbol{\mathcal{X}})$, with respect to the vector $\boldsymbol{\mathcal{X}}$ as a block-column matrix:

$$\frac{\partial \boldsymbol{A}}{\partial \boldsymbol{\mathcal{X}}} = \begin{bmatrix} \frac{\partial A}{\partial \mathcal{X}_1} \\ \vdots \\ \frac{\partial A}{\partial \mathcal{X}_n} \end{bmatrix}. \tag{5.279}$$

It can be proved that the partial derivative of matrix product can be determined by the following Kronecker product:

$$\frac{\partial}{\partial \boldsymbol{\mathcal{X}}}(\boldsymbol{A}(\boldsymbol{\mathcal{X}})\boldsymbol{B}(\boldsymbol{\mathcal{X}})) = (\boldsymbol{I}_{n\times n} \otimes \boldsymbol{A})\frac{\partial \boldsymbol{B}}{\partial \boldsymbol{\mathcal{X}}} + \frac{\partial \boldsymbol{A}}{\partial \boldsymbol{\mathcal{X}}}\boldsymbol{B}. \tag{5.280}$$

To simplify the Coriolis and centrifugal vectors, use the Kronecker product to calculate the partial derivatives appear in the second term of $\boldsymbol{V}$ given in Equation 5.269, and note that $\partial \dot{\boldsymbol{\mathcal{X}}}^T / \partial \boldsymbol{\mathcal{X}} = \boldsymbol{0}$

$$\begin{aligned} \boldsymbol{V}(\boldsymbol{\mathcal{X}}, \dot{\boldsymbol{\mathcal{X}}}) &= \dot{\boldsymbol{M}}(\boldsymbol{\mathcal{X}})\dot{\boldsymbol{\mathcal{X}}} - \frac{1}{2}\left[\frac{\partial}{\partial \boldsymbol{\mathcal{X}}}\left(\dot{\boldsymbol{\mathcal{X}}}^T \boldsymbol{M}(\boldsymbol{\mathcal{X}})\right)\right]\dot{\boldsymbol{\mathcal{X}}} \\ &= \left[\dot{\boldsymbol{M}}(\boldsymbol{\mathcal{X}}) - \frac{1}{2}\left(\boldsymbol{I}_{n\times n} \otimes \dot{\boldsymbol{\mathcal{X}}}^T\right)\frac{\partial \boldsymbol{M}}{\partial \boldsymbol{\mathcal{X}}}\right]\dot{\boldsymbol{\mathcal{X}}}. \end{aligned} \tag{5.281}$$

Hence, this term can be written in terms of the multiplication of an $n \times n$ matrix $\boldsymbol{C}_1$ to the velocity vector $\dot{\boldsymbol{\mathcal{X}}}$ as

$$\boldsymbol{V}(\boldsymbol{\mathcal{X}}, \dot{\boldsymbol{\mathcal{X}}}) = \boldsymbol{C}_1(\boldsymbol{\mathcal{X}}, \dot{\boldsymbol{\mathcal{X}}})\dot{\boldsymbol{\mathcal{X}}}, \tag{5.282}$$

where

$$\boldsymbol{C}_1(\boldsymbol{\mathcal{X}}, \dot{\boldsymbol{\mathcal{X}}}) = \dot{\boldsymbol{M}}(\boldsymbol{\mathcal{X}}) - \frac{1}{2}\boldsymbol{U}(\boldsymbol{\mathcal{X}}, \dot{\boldsymbol{\mathcal{X}}}), \tag{5.283}$$

and

$$\mathcal{U}(\mathcal{X},\dot{\mathcal{X}}) = \left(\boldsymbol{I}_{n\times n}\otimes\dot{\mathcal{X}}^{T}\right)\frac{\partial \boldsymbol{M}}{\partial \mathcal{X}}. \tag{5.284}$$

Note that in Equation 5.282, although the left-hand side is uniquely determined for any manipulator, it may be decomposed into a matrix multiplication where matrix $\boldsymbol{C}_1$ is not unique and can have several variations. In what follows, two more forms of such a matrix are derived, where the last variation is the most suitable form in control law implementations. To find an equivalent expression, simplify the derivative of mass matrix. It can be shown that

$$\dot{\boldsymbol{M}}(\mathcal{X}) = \sum_{i=1}^{n}\frac{\partial \boldsymbol{M}}{\partial \mathcal{X}_i}\dot{\mathcal{X}} \tag{5.285}$$

$$= (\dot{\mathcal{X}}^{T}\otimes \boldsymbol{I}_{n\times n})\frac{\partial \boldsymbol{M}}{\partial \mathcal{X}} \tag{5.286}$$

$$= \left(\frac{\partial \boldsymbol{M}}{\partial \mathcal{X}}\right)^{T}(\dot{\mathcal{X}}\otimes \boldsymbol{I}_{n\times n}). \tag{5.287}$$

Compare Equation 5.287 to the transpose of the matrix $\mathcal{U}$ using Equation 5.277

$$\mathcal{U}^{T} = \frac{\partial \boldsymbol{M}}{\partial \mathcal{X}}^{T}\left(\boldsymbol{I}_{n\times n}\otimes\dot{\mathcal{X}}\right). \tag{5.288}$$

Hence, in general $\dot{\boldsymbol{M}} \neq \mathcal{U}^{T}$. However, considering the property of the Kronecker product given in Equation 5.278, it may be stated that

$$\left(\boldsymbol{I}_{n\times n}\otimes\dot{\mathcal{X}}\right)\dot{\mathcal{X}} = \left(\dot{\mathcal{X}}\otimes\boldsymbol{I}_{n\times n}\right)\dot{\mathcal{X}}, \tag{5.289}$$

and therefore

$$\mathcal{U}^{T}\dot{\mathcal{X}} = \dot{\boldsymbol{M}}\dot{\mathcal{X}}. \tag{5.290}$$

Using this identity, the Coriolis and centrifugal vectors $\boldsymbol{V}$ may be written as

$$\boldsymbol{V}(\mathcal{X},\dot{\mathcal{X}}) = \dot{\boldsymbol{M}}(\mathcal{X})\dot{\mathcal{X}} - \frac{1}{2}\left[\frac{\partial}{\partial \mathcal{X}}\left(\dot{\mathcal{X}}^{T}\boldsymbol{M}(\mathcal{X})\right)\right]\dot{\mathcal{X}}$$

$$= \left(\mathcal{U}^{T} - \frac{1}{2}\mathcal{U}\right)\dot{\mathcal{X}}. \tag{5.291}$$

Hence, this term can be factored to an $n\times n$ matrix $\boldsymbol{C}_2$ multiplied to the velocity vector $\dot{\mathcal{X}}$ as

$$\boldsymbol{V}(\mathcal{X},\dot{\mathcal{X}}) = \boldsymbol{C}_2(\mathcal{X},\dot{\mathcal{X}})\dot{\mathcal{X}}, \tag{5.292}$$

where

$$\boldsymbol{C}_2(\mathcal{X},\dot{\mathcal{X}}) = \mathcal{U}^{T} - \frac{1}{2}\mathcal{U}. \tag{5.293}$$

Note that in general $C_1 \neq C_2$. At this point we may state an extremely useful identity in constructing advanced control schemes, based on inverse dynamic formulations. This property is called the *skew-symmetric property* [113], and it reveals the fact that the derivative of $M(\mathcal{X})$ and the Coriolis and centrifugal vector V are related in the following particular way:

$$\dot{\mathcal{X}}^T \dot{M} \dot{\mathcal{X}} - 2\dot{\mathcal{X}}^T V \dot{\mathcal{X}} = \dot{\mathcal{X}}^T \left(\dot{M} - 2C_i\right) \dot{\mathcal{X}} = 0. \tag{5.294}$$

In fact, this relation holds for any representation C_i of the Coriolis and centrifugal vectors, since V is multiplied by $\dot{\mathcal{X}}$ from the right. Let us prove this property for C_1 and C_2 as follows:

$$\begin{aligned}
\dot{\mathcal{X}}^T \left(\dot{M} - 2C_1\right) \dot{\mathcal{X}} &= \dot{\mathcal{X}}^T \left(\dot{M} - 2(\dot{M} - \frac{1}{2}U)\right) \dot{\mathcal{X}} \\
&= \dot{\mathcal{X}}^T (U - \dot{M}) \dot{\mathcal{X}} \\
&= \left(\dot{\mathcal{X}}^T U - \dot{\mathcal{X}}^T \dot{M}\right) \dot{\mathcal{X}} \\
&= 0,
\end{aligned}$$

in which the last identity results from the transpose of Equation 5.290, that is, $\dot{\mathcal{X}}^T U = \dot{\mathcal{X}}^T \dot{M}$. Similarly, for $V = C_2 \dot{\mathcal{X}}$, it may be shown that

$$\begin{aligned}
\dot{\mathcal{X}}^T \left(\dot{M} - 2C_2\right) \dot{\mathcal{X}} &= \dot{\mathcal{X}}^T \left(\dot{M} - 2(U^T - \frac{1}{2}U)\right) \dot{\mathcal{X}} \\
&= \dot{\mathcal{X}}^T \left(\dot{M}\dot{\mathcal{X}} + \left(2U^T - U\right) \dot{\mathcal{X}}\right) \\
&= \dot{\mathcal{X}}^T \left(U^T - 2U^T + U\right) \dot{\mathcal{X}} \\
&= \dot{\mathcal{X}}^T \left(U - U^T\right) \dot{\mathcal{X}} \\
&= 0,
\end{aligned}$$

in which the last identity results from the fact that a matrix minus its transpose is always skew symmetric. As it is seen in these proofs, although Equation 5.294 holds for both C_i's and any other C_i induced from Coriolis and centrifugal vectors, the matrix $\dot{M} - 2C_i$ itself is not generally skew symmetric. However, there exists a specific skew–symmetric matrix C, which satisfies $V = C\dot{\mathcal{X}}$, as well as the general property stated in Equation 5.294. This skew-symmetric matrix may be obtained from

$$V(\mathcal{X}, \dot{\mathcal{X}}) = C(\mathcal{X}, \dot{\mathcal{X}})\dot{\mathcal{X}} \tag{5.295}$$

where

$$C(\mathcal{X}, \dot{\mathcal{X}}) = \frac{1}{2}(\dot{M} + U^T - U). \tag{5.296}$$

Although all the C_i's can be used to represent the Coriolis and centrifugal vectors, and for all those variations the property stated in Equation 5.294 holds, it is recommended to use the final form given in Equation 5.296 to have the skew–symmetric property of the matrix $\dot{M} - 2C$ as well. Using this representation, the dynamic equation for any parallel robot may be written in the general closed form of

$$M(\mathcal{X})\ddot{\mathcal{X}} + C(\mathcal{X}, \dot{\mathcal{X}})\dot{\mathcal{X}} + G(\mathcal{X}) = \mathcal{F}. \tag{5.297}$$

The structure given in Equation 5.297 for a general parallel manipulator is called the closed–form dynamic formulation of a manipulator. This structure in representing the dynamic formulations for parallel manipulators is very convenient for applying several modern, adaptive, and robust control schemes on parallel robots.

Note that the closed–form formulation is directly derived from the Lagrange formulations, whereas in other methods described hereinbefore, usually an implicit form of dynamic formulation is obtained. However, with some extra manipulations, those formulations may be rewritten in a closed form. Such manipulations are elaborated in the derivation of the Stewart–Gough platform dynamics by the Newton–Euler approach in Section 5.3.3.

To keep harmony between two formulations and to be able to compare the final results in the derivation of Lagrange equations, it is more convenient to derive the kinetic and potential energies for the limbs, using a suitable generalized coordinate. One such coordinate may be selected as the positions of the moving attachment points, the points where the limbs are connected to the moving platform. Furthermore, the kinetic and potential energies of the moving platform are derived by using their own generalized coordinate. By defining the transformation from the limb coordinates to that of the moving platform, all the terms can be written in terms of one set of generalized coordinate, and the equation of motion of the whole manipulator is derived with respect to those coordinates. By this means, the derived equations are comparable to what is found from the Newton–Euler approach. This comparison is done for the Stewart–Gough platform in Section 5.5.6. In what follows, the dynamic formulation of the planar parallel manipulator is given first, and then the dynamic equations of the Stewart–Gough manipulator is derived by the Lagrange formulation.

5.5.5 Dynamic Formulation of a Planar Manipulator

In this section, dynamic analysis of a planar $4R\underline{P}R$ manipulator using the Lagrange approach is studied in detail. The architecture of this manipulator is described in Section 3.3.1 and the manipulator itself is shown in Figure 1.7. As explained earlier, in this manipulator, the moving platform is supported by four limbs of an identical $R\underline{P}R$ kinematic structure. A complete Jacobian and singularity analysis of this manipulator is presented in Section 4.6. The dynamic analysis of this manipulator based on the Newton–Euler approach considering constant and variable mass treatments are described in Sections 5.3.1 and 5.3.2, respectively. Furthermore, similar analysis are given for both the cases using the principle of virtual work in Sections 5.4.3 and 5.4.5. In this section, the constant mass assumptions are considered for the analysis. For variable mass formulation, refer to Reference 12.

To derive the equation of motion for the manipulator using the Lagrange formulation of the second kind, it is necessary to determine the Lagrangian function for the whole manipulator in terms of an appropriate generalized coordinate. The generalized coordinate of the manipulator may be considered as the position and orientation of the moving platform center of mass, denoted by $\mathcal{X} = [x_G, y_G, \phi]^T$, in which x_G and y_G denote the position of the moving platform center of mass, while ϕ denotes the orientation of the moving platform with respect to a fixed frame.

Note that since the manipulator is planar, only three coordinates are sufficient to represent the motion of the manipulator. Note that limb motions are fully determined by the limb length L_i, and limb directions are fully determined by the vector $\hat{S}_i$, which may be simply represented by the limb angle α_i, as shown in Figure 1.7. Therefore, the kinetic and potential energies of the limbs are a function of their own variable, but can be expressed in terms of the generalized coordinate of the system, namely $\boldsymbol{\mathcal{X}}$, through inverse kinematics and Jacobian relations of the manipulator, which are completely discussed in Sections 3.3, and 4.6, respectively.

To show that deriving Lagrangian function of the manipulator should be done carefully to avoid complications in further Lagrange formulations, let us first derive the kinetic energy of the limbs naively, and illustrate the complication it may cause in terms of the required manipulations. The kinetic energy of the limbs can be written as follows:

$$K_i = \frac{1}{2} m_i v_{c_i}^2 + \frac{1}{2} I_{c_i} \dot{\alpha}_i^2, \tag{5.298}$$

in which v_{c_i} is linear velocity of the limb center of mass, while $\dot{\alpha}_i$ is angular velocity of the limb. Substitute inertial parameters, $m_i = \rho_m L_i$, $I_{c_i} = \frac{1}{12}\rho_m L_i^3$ from Equation 5.53, and linear velocity $v_{c_i} = \dot{L}_i \hat{S}_i + \frac{1}{2} L_i \dot{\alpha}_i (\hat{K} \times \hat{S}_i)$ from Equation 5.51 and simplify Equation 5.298 to derive the kinetic energy of the limb as

$$K_i = \frac{1}{2}\rho_m L_i \left(\dot{L}_i^2 + \frac{1}{3}(L_i \dot{\alpha}_i)^2 \right). \tag{5.299}$$

Note that to apply the Lagrange equation of the second kind (Equation 5.258), partial derivatives of the kinetic energy with respect to generalized coordinate $\boldsymbol{\mathcal{X}}$ are needed to be determined. Hence, it is required to substitute the limb coordinates L_i and α_i, and the time derivatives $\dot{L}_i$ and $\dot{\alpha}_i$ as a function of the generalized coordinate $\boldsymbol{\mathcal{X}}$, and its time derivative $\dot{\boldsymbol{\mathcal{X}}}$. Substitution of the velocities are rather simple, and Jacobian matrices $\dot{\boldsymbol{L}} = \boldsymbol{J}\dot{\boldsymbol{\mathcal{X}}}$ and $\dot{\boldsymbol{\alpha}} = \boldsymbol{J}_\alpha \dot{\boldsymbol{\mathcal{X}}}$ given in Equations 4.54 and 4.58, respectively, may be used for these transformations. However, to transform the limb coordinates L_i and α_i to the generalized coordinate $\boldsymbol{\mathcal{X}} = [x_G, y_G, \phi]^T$, inverse kinematics relations given in Equations 3.10 and 3.11 would be used. The nonlinear nature of inverse kinematics relations causes the resulting equation for kinetic energy of the limbs to become very complicated, and therefore, performing partial derivatives a prohibitive task.

As a general remedy to this problem, many researchers proposed to consider each limb separately, select a set of generalized coordinates that completely defines the limb motion, and attain the dynamic model of the limb by the Lagrange formulation. Next, by means of a coordinates transformation, the selected generalized coordinates are transformed into the generalized coordinates considered for the whole manipulator, and the transformed dynamic model of the limbs and the moving platform are combined to achieve the dynamic model for the whole manipulator. As stated in Reference 1, sensible selection of local generalized coordinates of the limbs significantly affects the computational efficiency and simplicity of the eventuated dynamic model. Although this method of using the Lagrange formulation avoids the necessity of Lagrange multipliers evaluation for closed-chain constrains, usually componentwise derivation is followed for deriving dynamic equations of the limbs [1], and hence, it results into bulky formulations for the parallel manipulator.

In what follows, some tools in matrix derivation of the formulation would be given to reduce the complexity of the equations considerably and keep them manageable without the use of any symbolic software packages. The basis of such a manipulation is given in the closed-form Newton–Euler formulation given earlier in Section 5.3.3. Therefore, to derive the dynamic equation of the planar manipulator in hand, decompose the manipulator into the moving platform and its four limbs. Furthermore, assume that external forces acting on the limbs are the actuator and gravity forces while no disturbance wrench is applied to the limbs.

5.5.5.1 Dynamic Formulation of the Limbs

Consider the generalized coordinate for each limb to be the position of the moving attachment point B_i denoted by $\boldsymbol{x}_i$. Note that although only two x and y components of the position vector $\boldsymbol{x}_i$ are sufficient to fully determine the location of each limb in a planar motion, in what follows, the dimension of $\boldsymbol{x}_i$ is considered to be 3×1, to transform the vector manipulation into usual matrix manipulations. Therefore, note that although a 3×1 vector is used to represent the motion of the limb, its third component is always equal to zero, and does not contribute to the motion of the system. Considering this local coordinate for each limb, the governing equation of motion of each limb may be written as

$$\boldsymbol{M}_i\ddot{\boldsymbol{x}}_i + \boldsymbol{C}_i\dot{\boldsymbol{x}}_i + \boldsymbol{G}_i = \boldsymbol{F}_i, \tag{5.300}$$

in which the mass matrix $\boldsymbol{M}_i(\boldsymbol{x}_i)$ may be derived directly from the kinetic energy of each limb

$$K_i = \frac{1}{2}\dot{\boldsymbol{x}}_i^T\boldsymbol{M}_i\dot{\boldsymbol{x}}_i. \tag{5.301}$$

The kinetic energy of each limb K_i is found from the summation of linear and angular kinetic energies of the limb

$$K_i = \frac{1}{2}\boldsymbol{v}_{C_i}^T m_i \boldsymbol{v}_{C_i} + \frac{1}{2}\boldsymbol{\omega}_i^T\, \boldsymbol{I}_{C_i}\, \boldsymbol{\omega}_i. \tag{5.302}$$

Now represent the velocity vector of the limb center of mass as a function of the local generalized coordinate $\dot{\boldsymbol{x}}_i$. Recall Equation 5.51 for linear velocity and write it in terms of angular velocity vector of the limb $\boldsymbol{\omega}_i$ by

$$\boldsymbol{v}_{C_i} = \dot{L}_i\hat{\boldsymbol{S}}_i + \frac{1}{2}L_i\boldsymbol{\omega}_i \times \hat{\boldsymbol{S}}_i. \tag{5.303}$$

Furthermore, both L_i and $\boldsymbol{\omega}_i$ may be represented as functions of $\dot{\boldsymbol{x}}_i$ through the following derivations:

$$\dot{L}_i = \hat{\boldsymbol{S}}_i^T\,\dot{\boldsymbol{x}}_i \Rightarrow \dot{L}_i\hat{\boldsymbol{S}}_i = \hat{\boldsymbol{S}}_i\hat{\boldsymbol{S}}_i^T\,\dot{\boldsymbol{x}}_i \tag{5.304}$$

$$L_i\boldsymbol{\omega}_i = \hat{\boldsymbol{S}}_i \times \dot{\boldsymbol{x}}_i \Rightarrow \boldsymbol{\omega}_i = \frac{1}{L_i}\hat{\boldsymbol{S}}_{i_\times}\dot{\boldsymbol{x}}_i \tag{5.305}$$

$$L_i\boldsymbol{\omega}_i \times \hat{\boldsymbol{S}}_i = (\hat{\boldsymbol{S}}_i \times \dot{\boldsymbol{x}}_i) \times \hat{\boldsymbol{S}}_i = -\hat{\boldsymbol{S}}_i \times (\hat{\boldsymbol{S}}_i \times \dot{\boldsymbol{x}}_i) = -\boldsymbol{S}_{i_\times}^2\dot{\boldsymbol{x}}_i. \tag{5.306}$$

Hence, the linear velocity of the limb may be written as

$$\boldsymbol{v}_{C_i} = \left(\hat{\boldsymbol{S}}_i\hat{\boldsymbol{S}}_i^T - \frac{1}{2}\hat{\boldsymbol{S}}_{i_\times}^2\right)\dot{\boldsymbol{x}}_i. \tag{5.307}$$

Since the motion of the system is in a plane, all rotations are along the $\hat{z}$ direction, and therefore, the only component in inertia matrix contributing in the angular momentum is I_{zz_i}, which is simply denoted by I_{c_i} as in Section 5.3.1. Therefore, similar to the derivation of Equation 5.305, it can be shown that

$$\begin{aligned} \boldsymbol{\omega}_i^T I_{c_i} \boldsymbol{\omega}_i &= I_{c_i} \boldsymbol{\omega}_i^T \boldsymbol{\omega}_i \\ &= \frac{1}{L_i^2} I_{c_i} \dot{\boldsymbol{x}}_i^T \hat{\boldsymbol{S}}_{i_\times}^T \hat{\boldsymbol{S}}_{i_\times} \dot{\boldsymbol{x}}_i \\ &= -\frac{1}{L_i^2} I_{c_i} \dot{\boldsymbol{x}}_i^T \hat{\boldsymbol{S}}_{i_\times}^2 \dot{\boldsymbol{x}}_i. \end{aligned} \tag{5.308}$$

Substitute these equations into Equation 5.302 and simplify

$$K_i = \frac{1}{2} \boldsymbol{x}_i^T \left[m_i \left(\hat{\boldsymbol{S}}_i \hat{\boldsymbol{S}}_i^T - \frac{1}{2} \hat{\boldsymbol{S}}_{i_\times}^2 \right) - \frac{1}{L_i^2} I_{c_i} \hat{\boldsymbol{S}}_{i_\times}^2 \right] \boldsymbol{x}_i. \tag{5.309}$$

Substitute $m_i = \rho_m L_i$ and $I_{c_i} = \frac{\rho_m}{12} L_i^3$ and simplify Equation 5.309 to identify the limb mass matrix as

$$\boldsymbol{M}_i = \rho_m L_i \left(\hat{\boldsymbol{S}}_i \hat{\boldsymbol{S}}_i^T - \frac{7}{12} \hat{\boldsymbol{S}}_{i_\times}^2 \right). \tag{5.310}$$

As another variation to this representation, use the following relation:

$$\hat{\boldsymbol{S}}_i \hat{\boldsymbol{S}}_i^T = \boldsymbol{I}_{3\times 3} + \hat{\boldsymbol{S}}_{i_\times}^2, \tag{5.311}$$

and derive the mass matrix as

$$\boldsymbol{M}_i = \rho_m L_i \left(\boldsymbol{I}_{3\times 3} + \frac{5}{12} \hat{\boldsymbol{S}}_{i_\times}^2 \right). \tag{5.312}$$

The gravity vector $\boldsymbol{G}_i$ is the next term being evaluated. This term may be determined from partial derivative of the potential energy with respect to generalized coordinate $\boldsymbol{x}_i$, or by applying the principle of virtual work on the gravitational forces. Here, the first method is applied. The potential energy of the limbs can be determined by the following equation:

$$P_i = -\boldsymbol{g}^T m_i c_i \hat{\boldsymbol{S}}_i = \frac{1}{2} \boldsymbol{g}^T \rho_m L_i^2 \hat{\boldsymbol{S}}_i. \tag{5.313}$$

Note that the gravitational vector is considered to be in any arbitrary direction to the plane of motion, and the fixed attachment point $\boldsymbol{A}_i$ is considered as the reference for zero potential energy. Since the partial derivative of the potential energy with respect to generalized coordinate $\boldsymbol{x}_i$ is needed to be determined, the differential of potential energy $\mathrm{d}P$ is taken and all the differential variables are written as a function of the generalized coordinate.

$$\mathrm{d}P_i = -\frac{\rho_m}{2} \boldsymbol{g}^T \left(2 L_i \mathrm{d}L_i \hat{\boldsymbol{S}}_i + L_i^2 \mathrm{d}\hat{\boldsymbol{S}}_i \right). \tag{5.314}$$

To convert $\mathrm{d}P$ into a function of generalized coordinates, it is required to determine $\mathrm{d}\hat{\mathbf{S}}_i$, and $\mathrm{d}L_i\hat{\mathbf{S}}_i$ as a function of $\mathrm{d}\boldsymbol{x}_i$. These terms are derived in the following equations:

$$\begin{aligned} \boldsymbol{x}_i &= \boldsymbol{a}_i + L_i\hat{\mathbf{S}}_i \\ \mathrm{d}\boldsymbol{x}_i &= L_i\mathrm{d}\hat{\mathbf{S}}_i + \mathrm{d}L_i\hat{\mathbf{S}}_i. \end{aligned} \tag{5.315}$$

Therefore

$$\mathrm{d}\hat{\mathbf{S}}_i = \frac{1}{L_i}(\mathrm{d}\boldsymbol{x}_i - \mathrm{d}L_i\hat{\mathbf{S}}_i). \tag{5.316}$$

Now, dot multiply Equation 5.315 by $\hat{\mathbf{S}}_i$ to eliminate $\mathrm{d}\hat{\mathbf{S}}_i$, and note that $\hat{\mathbf{S}}_i \cdot \mathrm{d}\hat{\mathbf{S}}_i = 0$.

$$\hat{\mathbf{S}}_i \cdot \mathrm{d}\boldsymbol{x}_i = \mathrm{d}L_i\hat{\mathbf{S}}_i \cdot \hat{\mathbf{S}}_i = \mathrm{d}L_i. \tag{5.317}$$

Hence

$$\mathrm{d}L_i\hat{\mathbf{S}}_i = (\hat{\mathbf{S}}_i \cdot \mathrm{d}\boldsymbol{x}_i)\hat{\mathbf{S}}_i = \hat{\mathbf{S}}_i\hat{\mathbf{S}}_i^T\,\mathrm{d}\boldsymbol{x}_i. \tag{5.318}$$

Next, cross multiply Equation 5.315 by $\hat{\mathbf{S}}_i$ to eliminate $\mathrm{d}L_i$:

$$\hat{\mathbf{S}}_i \times \mathrm{d}\boldsymbol{x}_i = L_i(\hat{\mathbf{S}}_i \times \mathrm{d}\hat{\mathbf{S}}_i). \tag{5.319}$$

Once again, cross multiply both the sides by $\hat{\mathbf{S}}_i$

$$\begin{aligned} \hat{\mathbf{S}}_i \times (\hat{\mathbf{S}}_i \times \mathrm{d}\boldsymbol{x}_i) &= L_i\hat{\mathbf{S}}_i \times (\hat{\mathbf{S}}_i \times \mathrm{d}\hat{\mathbf{S}}_i) \\ &= L_i\left(\hat{\mathbf{S}}_i(\hat{\mathbf{S}}_i \cdot \mathrm{d}\hat{\mathbf{S}}_i) - \mathrm{d}\hat{\mathbf{S}}_i(\hat{\mathbf{S}}_i \cdot \hat{\mathbf{S}}_i)\right) \\ &= -L_i\mathrm{d}\hat{\mathbf{S}}_i. \end{aligned} \tag{5.320}$$

Hence

$$\mathrm{d}\hat{\mathbf{S}}_i = -\frac{1}{L_i}\hat{\mathbf{S}}_{i_\times}^2\,\mathrm{d}\boldsymbol{x}_i. \tag{5.321}$$

Substitute Equations 5.321 and 5.318 into the differential of potential energy given in Equation 5.314, and simplify as

$$\mathrm{d}P_i = \rho_m\,L_i\boldsymbol{g}^T\left(\frac{1}{2}\hat{\mathbf{S}}_{i_\times}^2 - \hat{\mathbf{S}}_i\hat{\mathbf{S}}_i^T\right)\mathrm{d}\boldsymbol{x}_i. \tag{5.322}$$

Partial derivative of the potential energy P with respect to generalized coordinate $\boldsymbol{x}_i$ may be determined by the following equation. Note that the potential energy itself is scalar, but by definition $\partial P_i/\partial \boldsymbol{x}_i$, which is equal to the gravity vector $\boldsymbol{G}_i$, it becomes an $n \times 1$

column vector. Therefore, the transpose of Equation 5.322 must be used to result in the right dimensions:

$$
\begin{aligned}
\frac{\partial P_i}{\partial \boldsymbol{x}_i} = \boldsymbol{G}_i &= \rho_m L_i \left[g^T \left(\frac{1}{2} \hat{S}^2_{i_\times} - \hat{S}_i \hat{S}_i^T \right) \right]^T \\
&= \rho_m L_i \left(\frac{1}{2} \hat{S}^2_{i_\times} - \hat{S}_i \hat{S}_i^T \right)^T g \\
&= \rho_m L_i \left(\frac{1}{2} \left(\hat{S}^2_{i_\times} \right)^T - (\hat{S}_i^T)^T \hat{S}_i \right) g \\
&= \rho_m L_i \left(\frac{1}{2} \hat{S}^2_{i_\times} - \hat{S}_i \hat{S}_i^T \right) g.
\end{aligned} \tag{5.323}
$$

The last term to be determined is the Coriolis and centrifugal vectors $\boldsymbol{V}_i = \boldsymbol{C}_i \dot{\boldsymbol{x}}_i$. This vector may be determined directly from the time derivative of the Lagrangian function based on the Lagrange Equation 5.259. It may also be equivalently determined from relations based on time derivative of the mass matrix and the skew symmetric matrices introduced in the previous section. Use time derivative of the Lagrangian and determine the Coriolis and centrifugal vector $\boldsymbol{V}_i$ from the general formulation given in Equation 5.269. Rewrite Equation 5.269 for the limb-generalized coordinate as follows:

$$
\boldsymbol{V}_i(\boldsymbol{x}_i, \dot{\boldsymbol{x}}_i) = \dot{\boldsymbol{M}}_i(\boldsymbol{x}_i)\dot{\boldsymbol{x}}_i - \frac{1}{2}\frac{\partial}{\partial \boldsymbol{x}_i}\left(\dot{\boldsymbol{x}}_i^T \boldsymbol{M}_i(\boldsymbol{x}_i)\dot{\boldsymbol{x}}_i \right). \tag{5.324}
$$

As seen in this equation, to compute the Coriolis and centrifugal vector, time derivative of the mass matrix, as well as the derivative of kinetic energy with respect to the generalized coordinate, is required. Calculation of these terms requires to convert the derivatives of many intermediate variables being determined as a function of the generalized coordinate $\boldsymbol{x}_i$. Let us first derive the required formulations, and then simplify Equation 5.324. First, find the following relations which would be used to simplify the second term of Equation 5.324

$$
\frac{\partial L_i}{\partial \boldsymbol{x}_i} = \hat{S}_i; \quad \frac{\partial}{\partial \boldsymbol{x}_i}\left(\frac{1}{L_i^2} \right) = -\frac{2}{L_i^3}\hat{S}_i; \quad \frac{\partial \hat{S}_i}{\partial \boldsymbol{x}_i} = -\frac{1}{L_i}\hat{S}^2_{i_\times}. \tag{5.325}
$$

Furthermore, by some manipulations, the following equations are derived

$$
\begin{aligned}
\frac{\mathrm{d}}{\mathrm{d}t}\left(\hat{S}^2_{i_\times} \right) &= -\frac{1}{L_i}\hat{S}_i \dot{\boldsymbol{x}}_i \hat{S}^2_{i_\times} - \frac{1}{L_i}\hat{S}^2_{i_\times} \dot{\boldsymbol{x}}_i \hat{S}_i^T, \\
\frac{\partial}{\partial \boldsymbol{x}_i}\left(\dot{\boldsymbol{x}}_i^T \hat{S}_i \hat{S}_i^T \dot{\boldsymbol{x}}_i \right) &= -\frac{2}{L_i}\hat{S}^2_{i_\times} \dot{\boldsymbol{x}}_i \hat{S}_i^T \dot{\boldsymbol{x}}_i, \\
\frac{\partial}{\partial \boldsymbol{x}_i}\left(\dot{\boldsymbol{x}}_i^T \hat{S}^2_{i_\times} \dot{\boldsymbol{x}}_i \right) &= -\frac{2}{L_i}\dot{L}_i \hat{S}^2_{i_\times} \dot{\boldsymbol{x}}_i.
\end{aligned} \tag{5.326}
$$

Simplify the first term of Coriolis and centrifugal vector in Equation 5.324 as follows:

$$\begin{aligned}\dot{M}_i &= \frac{\mathrm{d}}{\mathrm{d}t}\left(\rho_m L_i \left(I_{3\times 3} + \frac{5}{12}\hat{S}_{i_\times}^2\right)\right) \\ &= \rho_m \dot{L}_i \left(I_{3\times 3} + \frac{5}{12}\hat{S}_{i_\times}^2\right) - \frac{5}{12}\rho_m L_i \left(\frac{1}{L_i}\hat{S}_i \dot{x}_i \hat{S}_{i_\times}^2 - \frac{1}{L_i}\hat{S}_{i_\times}^2 \dot{x}_i \hat{S}_i^T\right).\end{aligned} \tag{5.327}$$

Furthermore, note that

$$\dot{L}_i = \frac{\partial L_i}{\partial x_i} \cdot \dot{x}_i = \hat{S}_i \cdot \dot{x}_i = \hat{S}_i^T \dot{x}_i. \tag{5.328}$$

Hence, the term $\hat{S}_{i_\times}^2 \dot{x}_i \hat{S}_i^T \dot{x}_i$ is simplified to $\dot{L}_i \hat{S}_{i_\times}^2 \dot{x}_i$. Substitute this equality into Equation 5.327, and simplify as

$$\begin{aligned}\dot{M}_i \dot{x}_i &= \left(\rho_m \dot{L}_i + \frac{5}{12}\rho_m \dot{L}_i \hat{S}_{i_\times}^2 - \frac{5}{12}\rho_m \hat{S}_i \dot{x}_i^T \hat{S}_{i_\times}^2 - \frac{5}{12}\rho_m \dot{L}_i \hat{S}_{i_\times}^2\right)\dot{x}_i \\ &= \rho_m \dot{L}_i \dot{x}_i - \frac{5}{12}\rho_m \hat{S}_i \dot{x}_i^T \hat{S}_{i_\times}^2 \dot{x}_i.\end{aligned} \tag{5.329}$$

The second term of the Coriolis and centrifugal vector given in Equation 5.324 can be simplified as follows:

$$\begin{aligned}-\frac{1}{2}\frac{\partial}{\partial x_i}\left(\dot{x}_i^T M_i(x_i)\dot{x}_i\right) &= -\frac{1}{2}\frac{\partial}{\partial x_i}\left[\rho_m L_i \dot{x}_i^T \left(I_{3\times 3} + \frac{5}{12}\hat{S}_{i_\times}^2\right)\dot{x}_i\right] \\ &= -\frac{\rho_m}{2}\left[\hat{S}_i \dot{x}_i^T \left(I_{3\times 3} + \frac{5}{12}\hat{S}_{i_\times}^2\right)\dot{x}_i + \frac{5}{12}L_i\left(-\frac{2}{L_i}\dot{L}_i \hat{S}_{i_\times}^2 \dot{x}_i\right)\right] \\ &= -\frac{\rho_m}{2}\left(\hat{S}_i \dot{x}_i^T + \frac{5}{12}\hat{S}_i \dot{x}_i^T \hat{S}_{i_\times}^2 - \frac{5}{6}\dot{L}_i \hat{S}_{i_\times}^2\right)\dot{x}_i.\end{aligned} \tag{5.330}$$

Therefore, by simply adding Equations 5.329 and 5.330 and some simplification, the Coriolis and centrifugal vector is obtained as follows:

$$V_i(x_i, \dot{x}_i) = \rho_m \left(\dot{L}_i - \frac{1}{2}\hat{S}_i \dot{x}_i^T - \frac{5}{8}\hat{S}_i \dot{x}_i^T \hat{S}_{i_\times}^2 + \frac{5}{12}\dot{L}_i S_{i_\times}^2\right)\dot{x}_i. \tag{5.331}$$

Hence, the matrix C_i satisfying the relation $V_i = C_i \dot{x}_i$ is given by

$$C_i(x_i, \dot{x}_i) = \rho_m \left(\dot{L}_i - \frac{1}{2}\hat{S}_i \dot{x}_i^T - \frac{5}{8}\hat{S}_i \dot{x}_i^T \hat{S}_{i_\times}^2 + \frac{5}{12}\dot{L}_i S_{i_\times}^2\right). \tag{5.332}$$

5.5.5.2 Dynamic Formulation of the Moving Platform

In this section, the Lagrangian of the moving platform is derived, and the corresponding mass matrix M_p, Coriolis and centrifugal matrix C_p, and gravity vector G_p are derived.

Start with the mass matrix, and derive kinetic energy of the moving platform. Kinetic energy of the moving platform may be written as

$$K_p = \frac{1}{2} M v_G^2 + \frac{1}{2} I_m \dot{\phi}^2, \tag{5.333}$$

in which v_G denotes linear velocity of the moving platform center of mass $\dot{\phi}$ denotes the moving platform angular velocity, M and I_m are the moving platform mass and moment of inertia, respectively. Rewrite this equation into matrix form and use the derivative of the manipulator-generalized coordinate $\boldsymbol{\mathcal{X}} = [x_G, y_G, \phi]^T$. This immediately results in

$$K_p = \frac{1}{2} \dot{\boldsymbol{\mathcal{X}}}^T \boldsymbol{M}_p \dot{\boldsymbol{\mathcal{X}}}, \tag{5.334}$$

in which the moving platform mass matrix is given by

$$\boldsymbol{M}_p = \begin{bmatrix} M & 0 & 0 \\ 0 & M & 0 \\ 0 & 0 & I_m \end{bmatrix}. \tag{5.335}$$

Note that the moving platform mass matrix is a constant diagonal matrix; therefore

$$\dot{\boldsymbol{M}}_p \dot{\boldsymbol{\mathcal{X}}} = \boldsymbol{0}; \quad -\frac{1}{2} \frac{\partial}{\partial \boldsymbol{\mathcal{X}}} \left(\dot{\boldsymbol{\mathcal{X}}}^T \boldsymbol{M}_p \dot{\boldsymbol{\mathcal{X}}} \right) = \boldsymbol{0}. \tag{5.336}$$

Hence, the Coriolis and centrifugal vector $\boldsymbol{V}_p(\boldsymbol{\mathcal{X}}, \dot{\boldsymbol{\mathcal{X}}})$ is equal to zero. The gravity term of the moving platform may be determined from the partial derivative of the moving platform potential energy with respect to the generalized coordinate $\boldsymbol{\mathcal{X}}$, or by applying the principle of virtual work. The potential energy approach is used here for this derivation. The potential energy of the moving platform may be determined by

$$P_p = -M\boldsymbol{g} \cdot \begin{bmatrix} x_G \\ y_G \\ 0 \end{bmatrix}. \tag{5.337}$$

Since the potential energy of the moving platform is not a function of orientation ϕ, its partial derivative with respect to this coordinate is zero. Hence

$$\boldsymbol{G}_p = \frac{\partial P_p}{\partial \boldsymbol{\mathcal{X}}} = -M\boldsymbol{g} \begin{bmatrix} 1 \\ 1 \\ 0 \end{bmatrix} = -M \begin{bmatrix} \boldsymbol{g} \cdot \hat{x} \\ \boldsymbol{g} \cdot \hat{y} \\ 0 \end{bmatrix}. \tag{5.338}$$

5.5.5.3 Dynamic Formulation of the Whole Manipulator

To derive the closed-form dynamic formulation for the whole manipulator, a transformation is required to map the intermediate generalized coordinates $\boldsymbol{x}_i$ into the principal generalized coordinates $\boldsymbol{\mathcal{X}}$ [117]. By using such a transformation, and by adding the resulting equations of the limbs and the moving platform together, the closed-form dynamic formulation for the whole manipulator can be derived. To generate such a transformation, define a Jacobian matrix $\boldsymbol{J}_i$, relating the intermediate coordinates to that of the principal generalized coordinate. Equation 5.108 can be written in the form of

$$\dot{x}_i = \boldsymbol{J}_i \dot{\boldsymbol{\mathcal{X}}}, \tag{5.339}$$

in which $\dot{\boldsymbol{\mathcal{X}}} = [\dot{x}_G, \dot{y}_g, \dot{\phi}]^T$. The velocity of the moving attachment point $\boldsymbol{x}_i$ can be written as a function of the generalized coordinate velocity $\dot{\boldsymbol{\mathcal{X}}}$ as follows:

$$\dot{x}_i = v_G + (\hat{\boldsymbol{K}} \times \boldsymbol{E}_i), \tag{5.340}$$

in which $\hat{\boldsymbol{K}}$ is the unit vector in the z direction, which is perpendicular to the plane of motion. Rewrite this equation into a matrix form

$$\dot{x}_i = \begin{bmatrix} 1 & 0 & \\ 0 & 1 & \hat{\boldsymbol{K}} \times \boldsymbol{E}_i \\ 0 & 0 & \end{bmatrix} \begin{bmatrix} \dot{x}_G \\ \dot{y}_G \\ \dot{\phi} \end{bmatrix}. \tag{5.341}$$

Therefore

$$\boldsymbol{J}_i = \begin{bmatrix} 1 & 0 & \\ 0 & 1 & \hat{\boldsymbol{K}} \times \boldsymbol{E}_i \\ 0 & 0 & \end{bmatrix}. \tag{5.342}$$

Furthermore, time derivative of $\boldsymbol{J}_i$ can be derived as

$$\dot{\boldsymbol{J}}_i = \left[\mathbf{0}_{3\times 2} \quad \hat{\boldsymbol{K}} \times \left(\hat{\boldsymbol{K}} \times \boldsymbol{E}_i\right)\dot{\phi}\right] = \left[\mathbf{0}_{3\times 2} \quad -\boldsymbol{E}_i\dot{\phi}\right] \tag{5.343}$$

Now substitute $\dot{x}_i$ from Equation 5.341 into Equation 5.300, and multiply $\boldsymbol{J}_i^T$ from left to both the sides of the resulting equation. This yields

$$\begin{aligned} &\boldsymbol{J}_i{}^T \boldsymbol{M}_i \left(\boldsymbol{J}_i \ddot{\boldsymbol{\mathcal{X}}} + \dot{\boldsymbol{J}}_i \dot{\boldsymbol{\mathcal{X}}}\right) + \boldsymbol{J}_i{}^T \boldsymbol{C}_i \boldsymbol{J}_i \dot{\boldsymbol{\mathcal{X}}} + \boldsymbol{J}_i{}^T \boldsymbol{G}_i = \boldsymbol{J}_i{}^T \boldsymbol{F}_i \\ &(\boldsymbol{J}_i{}^T \boldsymbol{M}_i \boldsymbol{J}_i)\ddot{\boldsymbol{\mathcal{X}}} + (\boldsymbol{J}_i{}^T \boldsymbol{M}_i \dot{\boldsymbol{J}}_i + \boldsymbol{J}_i{}^T \boldsymbol{C}_i \boldsymbol{J}_i)\dot{\boldsymbol{\mathcal{X}}} + \boldsymbol{J}_i{}^T \boldsymbol{G}_i = \boldsymbol{J}_i{}^T \boldsymbol{F}_i. \end{aligned} \tag{5.344}$$

Equation 5.183 can be written as a closed form

$$\boldsymbol{M}_{li} \ddot{\boldsymbol{\mathcal{X}}} + \boldsymbol{C}_{li} \dot{\boldsymbol{\mathcal{X}}} + \boldsymbol{G}_{li} = \boldsymbol{F}_{li}, \tag{5.345}$$

in which

$$\begin{aligned} \boldsymbol{M}_{li} &= \boldsymbol{J}_i^T \boldsymbol{M}_i \boldsymbol{J}_i \\ \boldsymbol{C}_{li} &= \boldsymbol{J}_i^T \boldsymbol{M}_i \dot{\boldsymbol{J}}_i + \boldsymbol{J}_i^T \boldsymbol{C}_i \boldsymbol{J}_i \\ \boldsymbol{G}_{li} &= \boldsymbol{J}_i^T \boldsymbol{G}_i. \end{aligned} \tag{5.346}$$

By this means, the closed-form dynamic formulation for the whole manipulator may be written as

$$\boldsymbol{M}(\boldsymbol{\mathcal{X}})\ddot{\boldsymbol{\mathcal{X}}} + \boldsymbol{C}(\boldsymbol{\mathcal{X}}, \dot{\boldsymbol{\mathcal{X}}})\dot{\boldsymbol{\mathcal{X}}} + \boldsymbol{G}(\boldsymbol{\mathcal{X}}) = \boldsymbol{Q}, \tag{5.347}$$

in which

$$
\begin{aligned}
M(\mathcal{X}) &= M_p + \sum_{i=1}^{i=4} M_{li} \\
C(\mathcal{X}, \dot{\mathcal{X}}) &= C_p + \sum_{i=1}^{i=4} C_{li} \\
G(\mathcal{X}) &= G_p + \sum_{i=1}^{i=4} G_{li}.
\end{aligned}
\tag{5.348}
$$

The only term to be evaluated in Equation 5.347 is the generalized forces acting on the manipulator Q. In the Lagrange approach, there is no need to incorporate any internal forces such as f_{b_i} in the formulations. The only external forces that remain to be accounted for are the actuator forces $\boldsymbol{\tau}$, and the disturbance wrench $\mathcal{F}_d$:

$$
\mathcal{F}_d = \begin{bmatrix} f_{d_x} \\ f_{d_y} \\ \tau_D \end{bmatrix}. \tag{5.349}
$$

The principle of virtual work can be used to determine the resulting generalized forces of the manipulator directly contributed from these two terms. By this means, the projection of the actuator forces acting on the moving platform center of mass can be given by $J^T\boldsymbol{\tau}$, and the disturbance wrench would be seen directly contributing as the generalized forces. Therefore, the generalized force vector Q is determined by the addition of these two terms as follows:

$$
Q = J^T\boldsymbol{\tau} + \mathcal{F}_d. \tag{5.350}
$$

The final equations of motion of planar manipulator is therefore completely derived in a closed form given in Equation 5.347. Although the closed-form formulation given in this section is much more desirable than the componentwise derivation of this formulation derived through the Newton–Euler and virtual work schemes, the resulting equations can be further manipulated to reach identical formulations. These manipulations are exhaustive and are not followed here. However, to verify the formulations, one may perform inverse and forward dynamic simulations detailed in Section 5.3.1 and obtain identical results.

5.5.6 Dynamic Analysis of the Stewart–Gough Platform

In this section, dynamic analysis of the Stewart–Gough platform using the Lagrange formulation has been studied in detail. The architecture of this manipulator is described in detail in Section 3.5.1 and is shown in Figure 3.16. As explained earlier, in this manipulator, spatial motion of the moving platform is generated by six piston–cylinder actuators. A complete kinematic analysis of this manipulator is given in Section 3.5 and its Jacobian analysis is presented in Section 4.8. Dynamic formulation of this manipulator using the Newton–Euler and virtual work approaches are given in Sections 5.3.3 and 5.4.6, respectively.

5.5.6.1 Dynamic Formulation of the Limbs

To derive a dynamic formulation for the Stewart–Gough platform, decompose manipulator into the moving platform and its six limbs. Furthermore, assume that each limb consists of two parts, the cylinder and the piston. The cylinder and piston center of masses are considered to be located at c_{i_1} and c_{i_2} away from their end points, and their masses are denoted by m_{i_1} and m_{i_2}, respectively. The external forces acting on the limbs are the actuator and gravitational forces, while no disturbance wrench is considered to be applied on the limbs.

Consider the generalized coordinate for each limb to be the position of the moving attachment point B_i, denoted by x_i, and the aim of this section is to find the dynamic equation of each limb in a closed form given in Equation 5.173, which is restated here for convenience.

$$M_i\ddot{x}_i + C_i\dot{x}_i + G_i = F_i. \tag{5.351}$$

Mass matrix $M_i(x_i)$ may be derived directly from the kinetic energy of each limb

$$K_i = \frac{1}{2}\dot{x}_i^T M_i \dot{x}_i, \tag{5.352}$$

in which the kinetic energy of each limb K_i is found from the summation of kinetic energy of the cylinder and the piston

$$K_i = \frac{1}{2}v_{i_1}^T m_{i_1} v_{i_1} + \frac{1}{2}\omega_i^T\, {}^A I_{c_{i_1}} \omega_i + \frac{1}{2}v_{i_2}^T m_{i_2} v_{i_2} + \frac{1}{2}\omega_i^T\, {}^A I_{c_{i_2}} \omega_i. \tag{5.353}$$

Substitute I_{eq} defined in Equation 5.175 as the summation of ${}^A I_{c_{i_1}} + {}^A I_{c_{i_2}}$ to simplify the kinetic energy of each limb into the following equation:

$$K_i = \frac{1}{2}v_{i_1}^T m_{i_1} v_{i_1} + \frac{1}{2}v_{i_2}^T m_{i_2} v_{i_2} + \frac{1}{2}\omega_i^T I_{\text{eq}} \omega_i. \tag{5.354}$$

Now represent the velocity vector of the cylinder and piston center of mass as a function of generalized coordinate $\dot{x}_i$. Recall these quantities from Equations 5.240 and 5.241 as

$$v_{i_1} = -\frac{c_{i_1}}{\ell_i}\hat{s}_{i_\times}^2 \dot{x}_i, \tag{5.355}$$

$$v_{i_2} = -\frac{c_{i_2}}{\ell_i}\hat{s}_{i_\times}^2 \dot{x}_i + \hat{s}_i\hat{s}_i^T \dot{x}_i. \tag{5.356}$$

Furthermore, ω_i may be represented as a function of $\dot{x}_i$ as already stated in Equation 5.112

$$\omega_i = \frac{1}{\ell_i}(\hat{s}_i \times \dot{x}_i) = \frac{1}{\ell_i}\hat{s}_{i_\times}\dot{x}_i \tag{5.357}$$

Since it is assumed that the limbs are symmetric with respect to their longitudinal axes, similar to the derivation of Equation 5.151, it can be shown that

$${}^A I_{\text{eq}} \omega_i = I_{xx_i} \omega_i. \tag{5.358}$$

Therefore, substitute Equation 5.357 into the following term and simplify

$$\begin{aligned}\boldsymbol{\omega}_i^T {}^A\boldsymbol{I}_{\text{eq}}\boldsymbol{\omega}_i &= I_{xx_i}\boldsymbol{\omega}_i^T\boldsymbol{\omega}_i \\ &= \frac{1}{\ell_i^2} I_{xx_i}\dot{\boldsymbol{x}}_i^T \hat{\boldsymbol{s}}_{i_\times}^T \hat{\boldsymbol{s}}_{i_\times}\dot{\boldsymbol{x}}_i \\ &= -\frac{1}{\ell_i^2} I_{xx_i}\dot{\boldsymbol{x}}_i^T \hat{\boldsymbol{s}}_{i_\times}^2 \dot{\boldsymbol{x}}_i. \end{aligned} \tag{5.359}$$

Substitute these equations into Equation 5.353 and simplify the resulting equation for kinetic energy:

$$K_i = \frac{1}{2}\boldsymbol{x}_i^T\left(\frac{1}{\ell_i^2} m_{i_1} c_{i_1}^2 \hat{\boldsymbol{s}}_{i_\times}^4 + m_{i_2}\left(-\frac{1}{\ell_i} c_{i_2}\hat{\boldsymbol{s}}_{i_\times}^2 + \hat{\boldsymbol{s}}_i\hat{\boldsymbol{s}}_i^T\right)^2 - \frac{1}{\ell_i^2} I_{xx_i}\hat{\boldsymbol{s}}_{i_\times}^2\right)\boldsymbol{x}_i. \tag{5.360}$$

This equation can be further simplified, using the following properties of $\hat{\boldsymbol{s}}_{i_\times}$. It can be simply verified that

$$\hat{\boldsymbol{s}}_{i_\times}^4 = -\hat{\boldsymbol{s}}_{i_\times}^2 \tag{5.361}$$

$$\hat{\boldsymbol{s}}_i\hat{\boldsymbol{s}}_i^T = \boldsymbol{I}_{3\times 3} + \hat{\boldsymbol{s}}_{i_\times}^2 \tag{5.362}$$

$$\hat{\boldsymbol{s}}_{i_\times}^2\hat{\boldsymbol{s}}_i\hat{\boldsymbol{s}}_i^T = \hat{\boldsymbol{s}}_i\hat{\boldsymbol{s}}_i^T\hat{\boldsymbol{s}}_{i_\times}^2 = \boldsymbol{0}. \tag{5.363}$$

Hence, the mass matrix of the limbs may be derived from further simplification of Equation 5.360

$$\begin{aligned}\boldsymbol{M}_i &= -\frac{1}{\ell_i^2} m_{i_1} c_{i_1}^2 \hat{\boldsymbol{s}}_{i_\times}^2 - \frac{1}{\ell_i^2} m_{i_2} c_{i_2}^2 \hat{\boldsymbol{s}}_{i_\times}^2 + m_{i_2}\hat{\boldsymbol{s}}_i\hat{\boldsymbol{s}}_i^T - \frac{1}{\ell_i^2} I_{xx_i}\hat{\boldsymbol{s}}_{i_\times}^2 \\ &= m_{i_2}\hat{\boldsymbol{s}}_i\hat{\boldsymbol{s}}_i^T - m_{c_e}\hat{\boldsymbol{s}}_{i_\times}^2 - \frac{1}{\ell_i^2} I_{xx_i}\hat{\boldsymbol{s}}_{i_\times}^2, \end{aligned} \tag{5.364}$$

in which m_{c_e} is given in Equation 5.175. The mass matrix obtained here is identical to that evaluated by the Newton–Euler formulation, given in Equation 5.174.

The gravity vector $\boldsymbol{G}_i$ is the next term being evaluated. This term may be determined from partial derivative of potential energy with respect to generalized coordinate $\boldsymbol{x}_i$, or by applying the principle of virtual work on gravitational forces. Here, the first approach is followed. Potential energy of the limbs may be determined by the following equation:

$$P_i = -\boldsymbol{g}^T\left[m_{i_1} c_{i_1}\hat{\boldsymbol{s}}_i + m_{i_2}(\ell_i - c_{i_2})\hat{\boldsymbol{s}}_i\right]. \tag{5.365}$$

Note that fixed attachment point A_i is considered as the reference for zero potential energy. Since partial derivative of potential energy with respect to generalized coordinate $\boldsymbol{x}_i$ is to be determined, the differential of potential energy $\mathrm{d}P$ is taken and all the differential variables are written as a function of the generalized coordinate

$$\mathrm{d}P_i = -\boldsymbol{g}^T\left(m_{i_1} c_{i_1} + m_{i_2}(\ell_i - c_{i_2})\right)\mathrm{d}\hat{\boldsymbol{s}}_i - m_{i_2}\boldsymbol{g}^T\mathrm{d}\ell_i\hat{\boldsymbol{s}}_i. \tag{5.366}$$

To convert $\mathrm{d}P$ into a function of generalized coordinates, it is required to determine $\mathrm{d}\hat{\mathbf{s}}_i$ and $\mathrm{d}\ell_i\hat{\mathbf{s}}_i$ as a function of $\mathrm{d}\boldsymbol{x}_i$. These terms are derived in the following equations:

$$\begin{aligned} \boldsymbol{x}_i &= \boldsymbol{a}_i + \ell_i\hat{\mathbf{s}}_i \\ \mathrm{d}\boldsymbol{x}_i &= \ell_i\mathrm{d}\hat{\mathbf{s}}_i + \mathrm{d}\ell_i\hat{\mathbf{s}}_i. \end{aligned} \tag{5.367}$$

Therefore

$$\mathrm{d}\hat{\mathbf{s}}_i = \frac{1}{\ell_i}(\mathrm{d}\boldsymbol{x}_i - \mathrm{d}\ell_i\hat{\mathbf{s}}_i). \tag{5.368}$$

Now, dot multiply Equation 5.367 by $\hat{\mathbf{s}}_i$ to eliminate $\mathrm{d}\hat{\mathbf{s}}_i$, and note that $\hat{\mathbf{s}}_i \cdot \mathrm{d}\hat{\mathbf{s}}_i = 0$:

$$\hat{\mathbf{s}}_i \cdot \mathrm{d}\boldsymbol{x}_i = \mathrm{d}\ell_i\hat{\mathbf{s}}_i \cdot \hat{\mathbf{s}}_i = \mathrm{d}\ell_i. \tag{5.369}$$

Hence

$$\mathrm{d}\ell_i\hat{\mathbf{s}}_i = (\hat{\mathbf{s}}_i \cdot \mathrm{d}\boldsymbol{x}_i)\hat{\mathbf{s}}_i = \hat{\mathbf{s}}_i\hat{\mathbf{s}}_i^T\mathrm{d}\boldsymbol{x}_i. \tag{5.370}$$

Next, cross multiply Equation 5.367 by $\hat{\mathbf{s}}_i$ to eliminate $\mathrm{d}\ell_i$:

$$\hat{\mathbf{s}}_i \times \mathrm{d}\boldsymbol{x}_i = \ell_i(\hat{\mathbf{s}}_i \times \mathrm{d}\hat{\mathbf{s}}_i). \tag{5.371}$$

Once again, cross multiply both the sides by $\hat{\mathbf{s}}_i$:

$$\begin{aligned} \hat{\mathbf{s}}_i \times (\hat{\mathbf{s}}_i \times \mathrm{d}\boldsymbol{x}_i) &= \ell_i\hat{\mathbf{s}}_i \times (\hat{\mathbf{s}}_i \times d\hat{\mathbf{s}}_i) \\ &= \ell_i\left(\hat{\mathbf{s}}_i(\hat{\mathbf{s}}_i \cdot \mathrm{d}\hat{\mathbf{s}}_i) - \mathrm{d}\hat{\mathbf{s}}_i(\hat{\mathbf{s}}_i \cdot \hat{\mathbf{s}}_i)\right) = -\ell_i\mathrm{d}\hat{\mathbf{s}}_i. \end{aligned} \tag{5.372}$$

Hence

$$\mathrm{d}\hat{\mathbf{s}}_i = -\frac{1}{\ell_i}\hat{\mathbf{s}}_{i_\times}^2\mathrm{d}\boldsymbol{x}_i. \tag{5.373}$$

Substitute Equations 5.373 and 5.370 into Equation 5.366, and simplify as

$$\begin{aligned} \mathrm{d}P_i &= \frac{1}{\ell_i}\left(m_{i_1}c_{i_1} + m_{i_2}(\ell_i - c_{i_2})\right)\boldsymbol{g}^T\hat{\mathbf{s}}_{i_\times}^2\mathrm{d}\boldsymbol{x}_i - m_{i_2}\boldsymbol{g}^T\hat{\mathbf{s}}_i\hat{\mathbf{s}}_i^T\mathrm{d}\boldsymbol{x}_i \\ &= \boldsymbol{g}^T\left(m_{g_e}\hat{\mathbf{s}}_{i_\times}^2 - m_{i_2}\hat{\mathbf{s}}_i\hat{\mathbf{s}}_i^T\right)\mathrm{d}\boldsymbol{x}_i \end{aligned} \tag{5.374}$$

in which m_{g_e} is defined in Equation 5.175. The partial derivative of the potential energy P with respect to generalized coordinate $\boldsymbol{x}_i$ may be determined by the following equation. Note that the potential energy itself is scalar, but by definition $\partial P_i/\partial \boldsymbol{x}_i$, which is equal to the gravity vector $\boldsymbol{G}_i$, is an $n \times 1$ column vector. Therefore, transpose of Equation 5.374 must be used to result in the right dimensions

$$\begin{aligned} \frac{\partial P_i}{\partial \boldsymbol{x}_i} = \boldsymbol{G}_i &= \left[\boldsymbol{g}^T\left(m_{g_e}\hat{\mathbf{s}}_{i_\times}^2 - m_{i_2}\hat{\mathbf{s}}_i\hat{\mathbf{s}}_i^T\right)\right]^T \\ &= \left(m_{g_e}\hat{\mathbf{s}}_{i_\times}^2 - m_{i_2}\hat{\mathbf{s}}_i\hat{\mathbf{s}}_i^T\right)^T\boldsymbol{g} \\ &= \left(m_{g_e}\left(\hat{\mathbf{s}}_{i_\times}^2\right)^T - m_{i_2}(\hat{\mathbf{s}}_i^T)^T\hat{\mathbf{s}}_i\right)\boldsymbol{g} \\ &= \left(m_{g_e}\hat{\mathbf{s}}_{i_\times}^2 - m_{i_2}\hat{\mathbf{s}}_i\hat{\mathbf{s}}_i^T\right)\boldsymbol{g}. \end{aligned} \tag{5.375}$$

The resulting relation for the gravity vector is also identical to what is obtained by the Newton–Euler approach, given in Equation 5.174.

The last term to be determined is the Coriolis and centrifugal vector $\boldsymbol{V}_i = \boldsymbol{C}_i\dot{\boldsymbol{x}}_i$. This vector can be determined directly from the time derivative of the Lagrangian function based on the Lagrange equation 5.259. It may also be equivalently determined from the relations based on the time derivative of the mass matrix and the symmetric matrices introduced in the previous section. Use time derivative of the Lagrangian to determine the Coriolis and centrifugal vector $\boldsymbol{V}_i$ from the general formulation given in Equation 5.269. Rewrite Equation 5.269 for a limb-generalized coordinate as follows:

$$\boldsymbol{V}_i(\boldsymbol{x}_i, \dot{\boldsymbol{x}}_i) = \dot{\boldsymbol{M}}_i(\boldsymbol{x}_i)\dot{\boldsymbol{x}}_i - \frac{1}{2}\frac{\partial}{\partial \boldsymbol{x}_i}\left(\dot{\boldsymbol{x}}_i^T\boldsymbol{M}_i(\boldsymbol{x}_i)\dot{\boldsymbol{x}}_i\right). \tag{5.376}$$

As seen in this equation, to compute the Coriolis and centrifugal vector, time derivative of the mass matrix, as well as the derivative of kinetic energy with respect to the generalized coordinate is required. Calculation of these terms requires to convert the derivatives of many intermediate variables being determined as a function of the generalized coordinate $\boldsymbol{x}_i$. Let us first derive the required formulations, and then simplify Equation 5.376. First rewrite the mass matrix derived in Equation 5.364, in the following form:

$$\begin{aligned}\boldsymbol{M}_i &= m_{i_2}\hat{\mathbf{s}}_i\hat{\mathbf{s}}_i^T - m_{c_e}\hat{\mathbf{s}}_{i_\times}^2 - \frac{1}{\ell_i^2}I_{xx_i}\hat{\mathbf{s}}_{i_\times}^2 \\ &= m_{i_2}(\boldsymbol{I}_{3\times3} + \hat{\mathbf{s}}_{i_\times}^2) - m_{c_e}\hat{\mathbf{s}}_{i_\times}^2 - \frac{1}{\ell_i^2}I_{xx_i}\hat{\mathbf{s}}_{i_\times}^2 \\ &= \left(m_{i_2} - m_{c_e} - \frac{1}{\ell_i^2}I_{xx_i}\right)\hat{\mathbf{s}}_{i_\times}^2 + m_{i_2}\boldsymbol{I}_{3\times3}.\end{aligned} \tag{5.377}$$

Next, find the following relations that would be used to simplify the second term of Equation 5.376

$$\frac{\partial \ell_i}{\partial \boldsymbol{x}_i} = \hat{\mathbf{s}}_i; \quad \frac{\partial}{\partial \boldsymbol{x}_i}\left(\frac{1}{\ell_i^2}\right) = -\frac{2}{\ell_i^3}\hat{\mathbf{s}}_i; \quad \frac{\partial \hat{\mathbf{s}}_i}{\partial \boldsymbol{x}_i} = -\frac{1}{\ell_i}\hat{\mathbf{s}}_{i_\times}^2. \tag{5.378}$$

Furthermore

$$\begin{aligned}\dot{m}_{c_e} &= \frac{\partial}{\partial \boldsymbol{x}_i}\left(\frac{1}{\ell_i^2}(m_{i_1}c_{i_1}^2 + m_{i_2}c_{i_2}^2)\right)\dot{\boldsymbol{x}}_i \\ &= \frac{1}{\ell_i^4}\left(-2\ell_i(m_{i_1}c_{i_1}^2 + m_{i_2}c_{i_2}^2)\hat{\mathbf{s}}_i + 2m_{i_2}c_{i_2}\ell_i^2\hat{\mathbf{s}}_i\right)\dot{\boldsymbol{x}}_i \\ &= \left(-\frac{2}{\ell_i}m_{c_e}\hat{\mathbf{s}}_i + \frac{2}{\ell_i^2}m_{i_2}c_{i_2}\hat{\mathbf{s}}_i\right)\dot{\boldsymbol{x}}_i,\end{aligned} \tag{5.379}$$

and by some manipulations, the following equations are derived:

$$\frac{\mathrm{d}}{\mathrm{d}t}\left(\hat{\mathbf{s}}_{i_\times}\right) = -\frac{1}{\ell_i}\hat{\mathbf{s}}_i\dot{\boldsymbol{x}}_i^T\hat{\mathbf{s}}_{i_\times} - \frac{1}{\ell_i}\hat{\mathbf{s}}_{i_\times}\dot{\boldsymbol{x}}_i^T\hat{\mathbf{s}}_i^T,$$

$$\frac{\mathrm{d}}{\mathrm{d}t}\left(\hat{\mathbf{s}}_{i_\times}^2\right) = -\frac{1}{\ell_i}\hat{\mathbf{s}}_i\dot{\boldsymbol{x}}_i^T\hat{\mathbf{s}}_{i_\times}^2 - \frac{1}{\ell_i}\hat{\mathbf{s}}_{i_\times}^2\dot{\boldsymbol{x}}_i^T\hat{\mathbf{s}}_i^T,$$
$$\frac{\partial}{\partial \boldsymbol{x}_i}\left(\dot{\boldsymbol{x}}_i^T\hat{\mathbf{s}}_i\hat{\mathbf{s}}_i^T\dot{\boldsymbol{x}}_i\right) = -\frac{2}{\ell_i}\hat{\mathbf{s}}_{i_\times}^2\dot{\boldsymbol{x}}_i\hat{\mathbf{s}}_i^T\dot{\boldsymbol{x}}_i,$$
$$\frac{\partial}{\partial \boldsymbol{x}_i}\left(\dot{\boldsymbol{x}}_i^T\hat{\mathbf{s}}_{i_\times}^2\dot{\boldsymbol{x}}_i\right) = -\frac{2}{\ell_i}\dot{\ell}_i\hat{\mathbf{s}}_{i_\times}^2\dot{\boldsymbol{x}}_i. \tag{5.380}$$

Simplify the first term of the Coriolis and centrifugal vector in Equation 5.376 as follows:

$$\begin{aligned}
\dot{\boldsymbol{M}}_i\dot{\boldsymbol{x}}_i &= \frac{\mathrm{d}}{\mathrm{d}t}\left((m_{i_2} - m_{c_e} - \frac{1}{\ell_i^2}I_{xx_i})\hat{\mathbf{s}}_{i_\times}^2 + m_{i_2}\boldsymbol{I}_{3\times3}\right)\dot{\boldsymbol{x}}_i \\
&= -\frac{1}{\ell_i}(m_{i_2} - m_{c_e} - \frac{1}{\ell_i^2}I_{xx_i})(\hat{\mathbf{s}}_i\dot{\boldsymbol{x}}_i^T\hat{\mathbf{s}}_{i_\times}^2 + \hat{\mathbf{s}}_{i_\times}^2\dot{\boldsymbol{x}}_i\hat{\mathbf{s}}_i^T)\dot{\boldsymbol{x}}_i \\
&\quad + \left(\left(\frac{2}{\ell_i}m_{c_e}\hat{\mathbf{s}}_i - \frac{2}{\ell_i^2}m_{i_2}c_{i_2}\hat{\mathbf{s}}_i\right)\dot{\ell}_i + \frac{2}{\ell_i^3}\dot{\ell}_i I_{xx_i}\right)\hat{\mathbf{s}}_{i_\times}^2\dot{\boldsymbol{x}}_i \\
&= -\frac{1}{\ell_i}(m_{i_2} - m_{c_e} - \frac{1}{\ell_i^2}I_{xx_i})(\hat{\mathbf{s}}_i\dot{\boldsymbol{x}}_i^T\hat{\mathbf{s}}_{i_\times}^2\dot{\boldsymbol{x}}_i + \hat{\mathbf{s}}_{i_\times}^2\dot{\boldsymbol{x}}_i\hat{\mathbf{s}}_i^T\dot{\boldsymbol{x}}_i) \\
&\quad + \frac{2}{\ell_i}\left(m_{i_2}c_{i_2} - \frac{1}{\ell_i^2}(I_{xx_i} + \ell_i^2 m_{c_e})\right)\dot{\ell}_i\hat{\mathbf{s}}_{i_\times}^2\dot{\boldsymbol{x}}_i.
\end{aligned} \tag{5.381}$$

To further simplify Equation 5.381, note that

$$\dot{\ell}_i = \frac{\partial \ell_i}{\partial \boldsymbol{x}_i}\cdot\dot{\boldsymbol{x}}_i = \hat{\mathbf{s}}_i\cdot\dot{\boldsymbol{x}}_i = \hat{\mathbf{s}}_i^T\dot{\boldsymbol{x}}_i. \tag{5.382}$$

Hence, the term $\hat{\mathbf{s}}_{i_\times}^2\dot{\boldsymbol{x}}_i\hat{\mathbf{s}}_i^T\dot{\boldsymbol{x}}_i$ is simplified to $\dot{\ell}_i\hat{\mathbf{s}}_{i_\times}^2\dot{\boldsymbol{x}}_i$. Substitute this equality into Equation 5.381, and simplify as

$$\begin{aligned}
\dot{\boldsymbol{M}}_i\dot{\boldsymbol{x}}_i &= \frac{1}{\ell_i}(m_{c_e} - m_{i_2} + \frac{1}{\ell_i^2}I_{xx_i})\hat{\mathbf{s}}_i\dot{\boldsymbol{x}}_i^T\hat{\mathbf{s}}_{i_\times}^2\dot{\boldsymbol{x}}_i \\
&\quad + \frac{1}{\ell_i}\left(3m_{c_e} - m_{i_2} - \frac{2}{\ell_i}m_{i_2}c_{i_2} + \frac{3}{\ell_i^2}I_{xx_i}\right)\dot{\ell}_i\hat{\mathbf{s}}_{i_\times}^2\dot{\boldsymbol{x}}_i.
\end{aligned} \tag{5.383}$$

The second term of the Coriolis and centrifugal vector given in Equation 5.376 may be simplified as follows:

$$\begin{aligned}
-\frac{1}{2}\frac{\partial}{\partial \boldsymbol{x}_i}\left(\dot{\boldsymbol{x}}_i^T\boldsymbol{M}_i(\boldsymbol{x}_i)\dot{\boldsymbol{x}}_i\right) &= -\frac{1}{2}\frac{\partial}{\partial \boldsymbol{x}_i}\left[\dot{\boldsymbol{x}}_i^T\left(m_{i_2}\hat{\mathbf{s}}_i\hat{\mathbf{s}}_i^T - m_{c_e}\hat{\mathbf{s}}_{i_\times}^2 - \frac{1}{\ell_i^2}I_{xx_i}\hat{\mathbf{s}}_{i_\times}^2\right)\dot{\boldsymbol{x}}_i\right] \\
&= \frac{1}{\ell_i}\left(\frac{1}{\ell_i}m_{i_2}(\ell_i - c_{i_2}) - \frac{1}{\ell_i}m_{c_e} - \frac{1}{\ell_i^2}I_{xx_i}\right)\hat{\mathbf{s}}_i\dot{\boldsymbol{x}}_i^T\hat{\mathbf{s}}_{i_\times}^2\dot{\boldsymbol{x}}_i \\
&\quad + \frac{1}{\ell_i}\left(m_{i_2} - m_{c_e} - \frac{1}{\ell_i^2}I_{xx_i}\right)\dot{\ell}_i\hat{\mathbf{s}}_{i_\times}^2\dot{\boldsymbol{x}}_i.
\end{aligned} \tag{5.384}$$

Therefore, by simply adding Equations 5.383 and 5.384, the Coriolis and centrifugal vector is obtained as follows:

$$V_i(x_i, \dot{x}_i) = \frac{1}{\ell_i}\left(\frac{1}{\ell_i}(\ell_i - c_{i_2})m_{i_2} - m_{i_2}\right)\hat{s}_i\dot{x}_i^T\hat{s}_{i_\times}^2\dot{x}_i$$
$$+ \frac{2}{\ell_i}\left(m_{c_e} - \frac{1}{\ell_i}m_{i_2}c_{i_2} + \frac{1}{\ell_i^2}I_{xx_i}\right)\dot{\ell}_i\hat{s}_{i_\times}^2\dot{x}_i$$
$$= \left(-\frac{1}{\ell_i^2}m_{i_2}c_{i_2}\hat{s}_i\dot{x}_i^T\hat{s}_{i_\times}^2\dot{x}_i - \frac{2}{\ell_i}m_{c_o}\dot{\ell}_i\hat{s}_{i_\times}^2\right)\dot{x}_i, \quad (5.385)$$

in which m_{c_o} is given in Equation 5.175. Therefore, the matrix C_i satisfying the relation $V_i = C_i\dot{x}_i$ may be given by

$$C_i(x_i, \dot{x}_i) = -\frac{1}{\ell_i^2}m_{i_2}c_{i_2}\hat{s}_i\dot{x}_i^T\hat{s}_{i_\times}^2\dot{x}_i - \frac{2}{\ell_i}m_{c_o}\dot{\ell}_i\hat{s}_{i_\times}^2, \quad (5.386)$$

and this result is also identical to what is found by the Newton–Euler method, presented in Equation 5.174.

5.5.6.2 Dynamic Formulation of the Moving Platform

In this section, the Lagrangian for the moving platform is derived, and the corresponding mass matrix M_p, Coriolis and centrifugal vector C_p, and gravity vector G_p have been derived. Start with the mass matrix, and derive kinetic energy of the moving platform. The kinetic energy for the moving platform can be written as

$$K_p = \frac{1}{2}v_p^T m\, v_p + \frac{1}{2}\omega^T I_p\omega, \quad (5.387)$$

in which v_p denotes the linear velocity of the moving platform center of mass, ω denotes the moving platform angular velocity, m and I_p are the moving platform mass and inertia matrix, respectively. Rewrite this equation in a matrix form and use the derivative of the manipulator-generalized coordinate $\dot{\mathcal{X}} = [v_p, \omega]^T$. This immediately results in

$$K_p = \frac{1}{2}\dot{\mathcal{X}}^T\begin{bmatrix} m I_{3\times3} & 0_{3\times3} \\ 0_{3\times3} & I_p \end{bmatrix}\dot{\mathcal{X}}, \quad (5.388)$$

in which $I_{3\times3}$ denotes 3×3 identity matrix and $0_{3\times3}$ denotes 3×3 zero matrix. Hence, the moving platform mass matrix is determined by

$$\begin{bmatrix} m I_{3\times3} & 0_{3\times3} \\ 0_{3\times3} & I_p \end{bmatrix}. \quad (5.389)$$

Comparing this equation to Equation 5.178, it is clear that the formulation determined here is identical to the one derived based on the Newton–Euler approach.

Deriving the Coriolis and centrifugal vector $\boldsymbol{V}_p(\boldsymbol{\mathcal{X}}, \dot{\boldsymbol{\mathcal{X}}})$ is indeed more complex than deriving the mass matrix. This is due to the fact that for this derivation we require to differentiate mass matrix with respect to time and kinetic energy with respect to the generalized coordinate

$$\boldsymbol{V}_e(\boldsymbol{\mathcal{X}}, \dot{\boldsymbol{\mathcal{X}}}) = \dot{\boldsymbol{M}}_e \dot{\boldsymbol{\mathcal{X}}} - \frac{\partial}{\partial \boldsymbol{\mathcal{X}}} K_p. \tag{5.390}$$

The first term in this equation may be derived as follows:

$$\dot{\boldsymbol{M}}_p \dot{\boldsymbol{\mathcal{X}}} = \begin{bmatrix} \boldsymbol{0}_{3\times3} & \boldsymbol{0}_{3\times3} \\ \boldsymbol{0}_{3\times3} & \frac{\mathrm{d}}{\mathrm{d}t}\boldsymbol{I}_p \end{bmatrix} \dot{\boldsymbol{\mathcal{X}}}. \tag{5.391}$$

Note that the time derivative of the moving platform inertia matrix $\dot{\boldsymbol{I}}_p$ must be evaluated in the reference frame $\{A\}$. Hence, using Equation 5.158, and the fact that ${}^A\dot{\boldsymbol{R}}_B = \boldsymbol{\omega}_\times {}^A\boldsymbol{R}_B$, the time derivative is obtained as follows:

$$\begin{aligned}
\dot{\boldsymbol{I}}_p &= \frac{\mathrm{d}}{\mathrm{d}t}({}^A\boldsymbol{R}_B\, {}^B\boldsymbol{I}_p\, {}^A\boldsymbol{R}_B^T) \\
&= {}^A\dot{\boldsymbol{R}}_B\, {}^B\boldsymbol{I}_p\, {}^A\boldsymbol{R}_B^T + {}^A\boldsymbol{R}_B\, {}^B\boldsymbol{I}_p\, {}^A\dot{\boldsymbol{R}}_B{}^T \\
&= \boldsymbol{\omega}_\times {}^A\boldsymbol{R}_B\, {}^B\boldsymbol{I}_p\, {}^A\boldsymbol{R}_B^T + {}^A\boldsymbol{R}_B\, {}^B\boldsymbol{I}_p\, {}^A\boldsymbol{R}_B^T \boldsymbol{\omega}_\times^T \\
&= \boldsymbol{\omega}_\times \boldsymbol{I}_p + \boldsymbol{I}_p \boldsymbol{\omega}_\times^T \\
&= \boldsymbol{\omega}_\times \boldsymbol{I}_p - \boldsymbol{I}_p \boldsymbol{\omega}_\times.
\end{aligned} \tag{5.392}$$

Therefore

$$\dot{\boldsymbol{M}}_p \dot{\boldsymbol{\mathcal{X}}} = \begin{bmatrix} \boldsymbol{0}_{3\times3} & \boldsymbol{0}_{3\times3} \\ \boldsymbol{0}_{3\times3} & \boldsymbol{\omega}_\times \boldsymbol{I}_p - \boldsymbol{I}_p \boldsymbol{\omega}_\times \end{bmatrix} \dot{\boldsymbol{\mathcal{X}}}. \tag{5.393}$$

Similar to the above derivation, the second term of Equation 5.390 may be derived as

$$\frac{\partial}{\partial \boldsymbol{\mathcal{X}}} K_p = \begin{bmatrix} \boldsymbol{0}_{3\times3} & \boldsymbol{0}_{3\times3} \\ \boldsymbol{0}_{3\times3} & -\boldsymbol{I}_p \boldsymbol{\omega}_\times \end{bmatrix} \dot{\boldsymbol{\mathcal{X}}}. \tag{5.394}$$

Hence

$$\boldsymbol{V}_p(\boldsymbol{\mathcal{X}}, \dot{\boldsymbol{\mathcal{X}}}) = \boldsymbol{C}_p \dot{\boldsymbol{\mathcal{X}}} = \begin{bmatrix} \boldsymbol{0}_{3\times3} & \boldsymbol{0}_{3\times3} \\ \boldsymbol{0}_{3\times3} & \boldsymbol{\omega}_\times \boldsymbol{I}_p \end{bmatrix} \dot{\boldsymbol{\mathcal{X}}}. \tag{5.395}$$

Comparing this result to Equation 5.178 verifies the identical derivation of $\boldsymbol{C}_p(\boldsymbol{\mathcal{X}}, \dot{\boldsymbol{\mathcal{X}}})$ to the one derived by the Newton–Euler method.

The gravity term of the moving platform may be determined from the partial derivative of the moving platform potential energy with respect to the generalized coordinate $\boldsymbol{\mathcal{X}}$, or by applying the principle of virtual work. The potential energy approach is used here for this derivation. The potential energy of the moving platform can be determined by

$$P_p = -m\boldsymbol{g} \cdot \boldsymbol{x}, \tag{5.396}$$

in which $\boldsymbol{x}$ denotes the position vector of the moving platform. Since the potential energy of the moving platform is not a function of the orientation part of the generalized coordinate $\boldsymbol{\mathcal{X}}$, its partial derivative with respect to $\boldsymbol{\mathcal{X}}$ would be

$$G_p = \frac{\partial P_p}{\partial \boldsymbol{\mathcal{X}}} = \begin{bmatrix} -mg \\ \mathbf{0}_{3\times 1} \end{bmatrix}. \tag{5.397}$$

Comparing this result to Equation 5.178 verifies identical derivation of $\boldsymbol{G}_p$ to that derived by the Newton–Euler method.

5.5.6.3 Dynamic Formulation of the Whole Manipulator

To derive the final dynamic formulation for the whole manipulator, the intermediate generalized coordinates $\boldsymbol{x}_i$ are transformed into the principal generalized coordinates $\boldsymbol{\mathcal{X}}$. This transformation is performed through the Jacobian matrix $\boldsymbol{J}_i$ are defined in Equation 5.180. A similar approach to what is explained in Section 5.3.4.3 should be followed to find the combination of the moving platform and the limb structural matrices, and to obtain the manipulator mass matrix $\boldsymbol{M}$, Coriolis and centrifugal matrix $\boldsymbol{C}$, and gravity vector $\boldsymbol{G}$, as given in Equation 5.188.

The only term that remains to be determined is the generalized forces acting on the manipulator. In the Lagrange approach, there is no need to incorporate the internal forces such as $\boldsymbol{f}_{b_i}$ in the formulations. The only external forces that remain to be accounted for are the actuator forces $\boldsymbol{\tau}$ and the disturbance wrench $\boldsymbol{\mathcal{F}}_d$. The principle of virtual work may be used to determine the resulting generalized forces of the manipulator directly, contributed from these two terms. By this means, the projection of actuator forces acting on the moving platform center of mass can be given by $\boldsymbol{J}^T\boldsymbol{\tau}$, and the disturbance wrench would be seen directly as the generalized forces. Therefore, the generalized force vector $\boldsymbol{Q}$ is determined by the addition of these two terms as follows:

$$\boldsymbol{Q} = \boldsymbol{J}^T\boldsymbol{\tau} + \boldsymbol{\mathcal{F}}_d. \tag{5.398}$$

By careful examination of the components of $\boldsymbol{\mathcal{F}}$ given in Equation 5.188, it is found that the generalized forces obtained in Equation 5.398 is identical to $\boldsymbol{\mathcal{F}}$ derived by the Newton–Euler method. This term was the only remaining component of dynamic formulation, and through this verification, it can be stated that although two different approaches have been used to derive the dynamic formulation of the Stewart–Gough platform, the resulting formulations are completely identical.

PROBLEMS

1. Consider the five-bar mechanism shown in Figure 3.28 and described in Problems 3.1 and 4.1. To simplify the analysis, consider the link i as a slender bar with mass m_i, center of mass located at the center point of the link, and the moment of inertia about center of mass as I_i. Furthermore, consider the gravity vector in the $-y$ direction, and the actuator torques τ_1, and τ_2 applied on joints B, and A, respectively.
   a. For this manipulator, perform a complete acceleration analysis.
   b. Derive a dynamic formulation by the Newton–Euler method.

   c. Give some numerical values to the geometry and inertial parameters of the manipulator, generate a typical trajectory for the tip, and develop a program in MATLAB to solve forward dynamics to determine the actuator forces required to perform such a maneuver.
   d. Plot and analyze trajectory and forces versus time, and justify the results by intuition.

2. Consider the five-bar mechanism shown in Figure 3.29 and described in Problems 3.2 and 4.2. Assume that link i is a slender bar with mass m_i, center of mass located at the center point of the link, and its moment of inertia about center of mass as I_i. Furthermore, consider the gravity vector in $-y$ direction, and the actuator force f_1 applied on the slider pair, and actuator torque τ_2 applied on the joint A.
   a. For this manipulator, perform a complete acceleration analysis.
   b. Derive a dynamic formulation by the Newton–Euler method.
   c. Give some numerical values to the geometry and inertial parameters of the manipulator, generate a typical trajectory for the tip, and develop a program in MATLAB to solve forward dynamics to determine the actuator forces required to perform such a maneuver.
   d. Plot and analyze trajectory and forces versus time, and justify the results by intuition.

3. Consider a planar pantograph mechanism shown in Figure 3.30 and described in Problems 3.3 and 4.3. To simplify the analysis, consider the link i as a slender bar with mass m_i, center of mass located at the center point of the link, and moment of inertia about center of mass as I_i. Furthermore, consider the piston masses denoted by m_{p_1} and m_{p_2}, the gravity vector in the $-y$ direction, and actuator forces f_1 and f_2 applied on the sliders A and B, respectively.
   a. For this manipulator, perform a complete acceleration analysis.
   b. Derive dynamic formulation by the Newton–Euler method.
   c. Give some numerical values to the geometry and inertial parameters of the manipulator, generate a typical trajectory for the tip, and develop a program in MATLAB to solve forward dynamics to determine the actuator forces required to perform such a maneuver.
   d. Plot and analyze trajectory and forces versus time, and justify the results by intuition.

4. Consider the planar three-degrees-of-freedom $3\underline{R}RR$ parallel manipulator shown in Figure 3.31 and described in Problems 3.4 and 4.4. For this manipulator, consider the links as slender bars with mass m_{a_i} and m_{b_i}, center of mass located at the center point of the link, and moment of inertia about center of mass as I_{a_i} and I_{b_i}. Furthermore, consider the moving platform mass denoted by m_e, its center of mass located at the geometrical center of the triangle $\triangle B_1B_2B_3$, and its moments of inertia about the center of mass denoted by I_e. Moreover, consider the manipulator motion is in horizontal plane, while gravity vector is in the $-z$ direction, and actuator torques τ_i's are applied on the joint A_i's.
   a. For this manipulator, perform a complete acceleration analysis and determine the relation for angular accelerations of the actuated limbs $\ddot{\alpha}_i$'s and passive limbs $\ddot{\beta}_i$'s as a function of the moving platform acceleration vector

$\ddot{\mathcal{X}} = [a_x, a_y, \ddot{\phi}]^T$. Furthermore, derive the linear acceleration of the limb center of masses.

b. Apply the Newton–Euler laws for the limbs and moving platform, eliminate the unwanted internal forces and derive the dynamic formulation of the manipulator by the Newton–Euler method.

c. Develop a program in MATLAB with the following numerical values for the manipulator parameters. Generate the following typical trajectory for the moving platform, and solve forward dynamics to determine the required Cartesian and actuator forces to perform such a maneuver.

$$a_i = b_i = 1\,(\text{m}); \quad L = 5\,(\text{m}); \quad \ell = 1.5\,(\text{m})$$
$$m_{a_i} = m_{b_i} = 1\,(\text{kg}); \quad I_{a_i} = I_{b_i} = 0.1\,(\text{kg}\,\text{m}^2)$$
$$m_e = 5\,(\text{kg}); \quad I_e = 1\,(\text{kg}\,\text{m}^2).$$

Consider the manipulator trajectory being a cubic polynomial with the following initial and final positions and time

$$x_o = [0, 0, 0]^T; \quad x_f = [0.2, -0.1, \pi/3]^T; \quad t_o = 0; \quad t_f = 2.$$

d. Plot and analyze the trajectory and the forces versus time, and justify the results by intuition.

5. Consider the planar three-degrees-of-freedom 4$\underline{R}RR$ parallel manipulator shown in Figure 3.32 and described in Problems 3.5 and 4.5. For this manipulator, consider the links as slender bars with mass m_{a_i} and m_{b_i}, center of mass located at the center point of the link, and the moment of inertia about center of mass as I_{a_i} and I_{b_i}. Furthermore, consider the moving platform mass denoted by m_e, its center of mass located at the center of the moving attachment point B_i, and its moments of inertia about the center of mass denoted by I_e. Moreover, consider the manipulator motion in horizontal plane, while the gravity vector in the $-z$ direction, and actuator torques τ_i's applied on the joints A_i's.

   a. For this manipulator, perform a complete acceleration analysis and determine the relation for the angular accelerations of the actuated limbs $\ddot{\alpha}_i$'s and passive limbs $\ddot{\beta}_i$'s as a function of the moving platform acceleration vector $\ddot{\mathcal{X}} = [a_x, a_y, \ddot{\phi}]^T$. Furthermore, derive the linear acceleration of the limb center of masses.

   b. Apply the Newton–Euler laws for the limbs and moving platform, eliminate the unwanted internal forces, and derive the dynamic formulation of the manipulator by the Newton–Euler method.

   c. Develop a program in MATLAB with the following numerical values for the manipulator parameters. Generate the following typical trajectory for the moving platform, and solve forward dynamics to determine the required Cartesian and actuator forces to perform such a maneuver.

$$\theta_{A_i} = \theta_{B_i} = [-3\pi/4, -\pi/4, \pi/4, 3\pi/4]^T;$$
$$a_i = b_i = 1.5\,(\text{m})$$

$$m_{a_i} = m_{b_i} = 1\,(\text{kg}); \quad I_{a_i} = I_{b_i} = 0.1\,(\text{kg}\,\text{m}^2)$$

$$m_e = 5\,(\text{kg}); \quad I_e = 1\,(\text{kg}\,\text{m}^2).$$

Consider the manipulator trajectory being a cubic time trajectory with the following initial and final positions and time

$$x_o = [0,0,0]^T; \quad x_f = [0.2,-0.1,\pi/3]^T; \quad t_o = 0; \quad t_f = 2.$$

d. Plot and analyze the trajectory and the forces versus time, and justify the results by intuition.

6. Consider the spatial parallel manipulator shown in Figure 3.33 and described in Problems 3.6 and 4.6.
   a. For this manipulator, perform a complete acceleration analysis.
   b. Apply the Newton–Euler laws for the limbs and moving platform, eliminate the unwanted internal forces, and derive the dynamic formulation of the manipulator by the Newton–Euler method.
   c. Give some numerical values to the geometry and inertial parameters of the manipulator, generate a typical trajectory for the moving platform, and develop a program in MATLAB to solve forward dynamics to determine the required Cartesian and actuator forces to perform such a maneuver.
   d. Plot and analyze the trajectory and the forces versus time, and justify the results by intuition.
7. Consider the $3U\underline{P}U$ parallel manipulator shown in Figure 3.34 and described in Problems 3.7 and 4.7.
   a. For this manipulator, perform a complete acceleration analysis.
   b. Apply the Newton–Euler laws for the limbs and moving platform, eliminate the unwanted internal forces and derive the dynamic formulation of the manipulator by the Newton–Euler method.
   c. Give some numerical values to the geometry and inertial parameters of the manipulator, generate a typical trajectory for the moving platform, and develop a program in MATLAB to solve forward dynamics to determine the required Cartesian and actuator forces to perform such a maneuver.
   d. Plot and analyze the trajectory and the forces versus time, and justify the results by intuition.
8. Consider the three-degrees-of-freedom $3U\underline{P}S$ manipulator introduced in Reference 47, shown in Figure 3.35, and described in Problems 3.8 and 4.8.
   a. For this manipulator, perform a complete acceleration analysis.
   b. Apply the Newton–Euler laws for the limbs and moving platform, eliminate the unwanted internal forces and derive the dynamic formulation of the manipulator by the Newton–Euler method.
   c. Give some numerical values to the geometry and inertial parameters of the manipulator, generate a typical trajectory for the moving platform, and develop a program in MATLAB to solve forward dynamics to determine the required Cartesian and actuator forces to perform such a maneuver.

d. Plot and analyze the trajectory and the forces versus time, and justify the results by intuition.

9. Consider the three-degrees-of-freedom $3R\underline{P}S$ manipulator represented in Figure 3.36 and described in Problems 3.9 and 4.9.
   a. For this manipulator, perform a complete acceleration analysis.
   b. Apply the Newton–Euler laws for the limbs and moving platform, eliminate the unwanted internal forces, and derive the dynamic formulation of the manipulator by the Newton–Euler method.
   c. Give some numerical values to the geometry and inertial parameters of the manipulator, generate a typical trajectory for the moving platform, and develop a program in MATLAB to solve forward dynamics to determine the required Cartesian and actuator forces to perform such a maneuver.
   d. Plot and analyze the trajectory and the forces versus time, and justify the results by intuition.
10. Consider the three-degrees-of-freedom $3\underline{P}RC$ translational parallel manipulator represented in Figure 3.36 and described in Problems 3.10 and 4.10.
   a. For this manipulator, perform a complete acceleration analysis.
   b. Apply the Newton–Euler laws for the limbs and the moving platform, eliminate the unwanted internal forces and derive the dynamic formulation of the manipulator by the Newton–Euler method.
   c. Give some numerical values to the geometry and inertial parameters of the manipulator, generate a typical trajectory for the moving platform, and develop a program in MATLAB to solve forward dynamics to determine the required Cartesian and actuator forces to perform such a maneuver.
   d. Plot and analyze the trajectory and the forces versus time, and justify the results by intuition.
11. Consider the Delta robot represented in Figure 3.39 and described in Problems 3.11 and 4.11.
   a. For this manipulator, perform a complete acceleration analysis.
   b. Apply the Newton–Euler laws for the limbs and the moving platform, eliminate the unwanted internal forces and derive the dynamic formulation of the manipulator by the Newton–Euler method.
   c. Give some numerical values to the geometry and inertial parameters of the manipulator, generate a typical trajectory for the moving platform, and develop a program in MATLAB to solve forward dynamics to determine the required Cartesian and actuator forces to perform such a maneuver.
   d. Plot and analyze the trajectory and the forces versus time, and justify the results by intuition.
12. Consider the Quadrupteron robot represented in Figure 3.40 and described in Problems 3.12 and 4.12 [157].
   a. For this manipulator, perform a complete acceleration analysis.
   b. Apply the Newton–Euler laws for the limbs and the moving platform, eliminate the unwanted internal forces and derive the dynamic formulation of the manipulator by the Newton–Euler method.

c. Give some numerical values to the geometry and inertial parameters of the manipulator, generate a typical trajectory for the moving platform, and develop a program in MATLAB to solve forward dynamics to determine the required Cartesian and actuator forces to perform such a maneuver.

d. Plot and analyze the trajectory and the forces versus time, and justify the results by intuition.

13. Reconsider the five-bar mechanism shown in Figure 3.28 and described in Problems 3.1 and 4.1. Repeat Problem 5.1 for this mechanism by using the principle of virtual work. Furthermore, compare the dynamic formulation obtained here to that found by the Newton–Euler formulation.
14. Reconsider the five-bar mechanism shown in Figure 3.29 and described in Problems 3.2 and 4.2. Repeat Problem 5.2 for this mechanism by using the principle of virtual work. Furthermore, compare the dynamic formulation obtained here to that found by the Newton–Euler formulation.
15. Reconsider the planar pantograph mechanism shown in Figure 3.30 and described in Problems 3.3 and 4.3. Repeat Problem 5.3 for this mechanism by using the principle of virtual work. Furthermore, compare the dynamic formulation obtained here to that found by the Newton–Euler formulation.
16. Reconsider the planar three-degrees-of-freedom $3\underline{R}RR$ parallel manipulator shown in Figure 3.31 and described in Problems 3.4, and 4.4. Repeat Problem 5.4 for this manipulator by using the principle of virtual work. Furthermore, compare the dynamic formulation obtained here to that found by the Newton–Euler formulation.
17. Reconsider the planar redundant $4\underline{R}RR$ parallel manipulator shown in Figure 3.32 and described in Problems 3.5 and 4.5. Repeat Problem 5.5 for this manipulator by using the principle of virtual work. Furthermore, compare the dynamic formulation obtained here to that found by the Newton–Euler formulation.
18. Reconsider the spatial parallel manipulator shown in Figure 3.33 and described in Problems 3.6 and 4.6. Repeat Problem 5.6 for this manipulator by using the principle of virtual work. Furthermore, compare the dynamic formulation obtained here to that found by the Newton–Euler formulation.
19. Reconsider the $3U\underline{P}U$ parallel manipulator shown in Figure 3.34 and described in Problems 3.7 and 4.7. Repeat Problem 5.7 for this manipulator by using the principle of virtual work. Furthermore, compare the dynamic formulation obtained here to that found by the Newton–Euler formulation.
20. Reconsider the three-degrees-of-freedom $3U\underline{P}S$ manipulator introduced in Reference 47, shown in Figure 3.35, and described in Problems 3.8 and 4.8. Repeat Problem 5.8 for this manipulator by using the principle of virtual work. Furthermore, compare the dynamic formulation obtained here to that found by the Newton–Euler formulation.
21. Reconsider the three-degrees-of-freedom $3R\underline{P}S$ manipulator represented in Figure 3.36 and described in Problems 3.9 and 4.9. Repeat Problem 5.9 for this manipulator by using the principle of virtual work. Furthermore, compare the dynamic formulation obtained here to that found by the Newton–Euler formulation.
22. Reconsider the three-degrees-of-freedom $3\underline{P}RC$ translational parallel manipulator represented in Figure 3.36 and described in Problems 3.10 and 4.10. Repeat

Problem 5.10 for this manipulator by using the principle of virtual work. Furthermore, compare the dynamic formulation obtained here to that found by the Newton–Euler formulation.

23. Reconsider the Delta robot represented in Figure 3.39 and described in Problems 3.11 and 4.11. Repeat Problem 5.11 for this manipulator by using the principle of virtual work. Furthermore, compare the dynamic formulation obtained here to that found by the Newton–Euler formulation.
24. Consider the Quadrupteron robot represented in Figure 3.40 and described in Problems 3.12 and 4.12. Repeat Problem 5.12 for this manipulator by using the principle of virtual work. Furthermore, compare the dynamic formulation obtained here to that found by the Newton–Euler formulation.
25. Reconsider the five bar mechanism shown in Figure 3.28 and described in Problems 3.1 and 4.1. Repeat Problem 5.1 for this mechanism by the Lagrange approach. Furthermore, compare the dynamic formulation obtained here to that found by the Newton–Euler formulation.
26. Reconsider the five-bar mechanism shown in Figure 3.29 and described in Problems 3.2 and 4.2. Repeat Problem 5.2 for this mechanism by the Lagrange approach. Furthermore, compare the dynamic formulation obtained here to that found by the Newton–Euler formulation.
27. Reconsider the planar pantograph mechanism shown in Figure 3.30 and described in Problems 3.3 and 4.3. Repeat Problem 5.3 for this mechanism by the Lagrange approach. Furthermore, compare the dynamic formulation obtained here to that found by the Newton–Euler formulation.
28. Reconsider the planar three-degrees-of-freedom $3\underline{R}RR$ parallel manipulator shown in Figure 3.31 and described in Problems 3.4 and 4.4. Repeat Problem 5.4 for this manipulator by the Lagrange approach. Furthermore, compare the dynamic formulation obtained here to that found by the Newton–Euler formulation.
29. Reconsider the planar redundant $4\underline{R}RR$ parallel manipulator shown in Figure 3.32 and described in Problems 3.5 and 4.5. Repeat Problem 5.5 for this manipulator by the Lagrange approach. Furthermore, compare the dynamic formulation obtained here to that found by the Newton–Euler formulation.
30. Reconsider the spatial parallel manipulator shown in Figure 3.33 and described in Problems 3.6 and 4.6. Repeat Problem 5.6 for this manipulator by the Lagrange approach. Furthermore, compare the dynamic formulation obtained here to that found by the Newton–Euler formulation.
31. Reconsider the $3U\underline{P}U$ parallel manipulator shown in Figure 3.34 and described in Problems 3.7 and 4.7. Repeat Problem 5.7 for this manipulator by the Lagrange approach. Furthermore, compare the dynamic formulation obtained here to that found by the Newton–Euler formulation.
32. Reconsider the three-degrees-of-freedom $3U\underline{P}S$ manipulator introduced in [47], shown in Figure 3.35, and described in Problems 3.8 and 4.8. Repeat Problem 5.8 for this manipulator by the Lagrange approach. Furthermore, compare the dynamic formulation obtained here to that found by the Newton–Euler formulation.

33. Reconsider the three-degrees-of-freedom $3R\underline{P}S$ manipulator represented in Figure 3.36 and described in Problems 3.9 and 4.9. Repeat Problem 5.9 for this manipulator by the Lagrange approach. Furthermore, compare the dynamic formulation obtained here to that found by the Newton–Euler formulation.
34. Reconsider the three-degrees-of-freedom $3\underline{P}RC$ translational parallel manipulator represented in Figure 3.36 and described in Problems 3.10 and 4.10. Repeat Problem 5.10 for this manipulator by the Lagrange approach. Furthermore, compare the dynamic formulation obtained here to that found by the Newton–Euler formulation.
35. Reconsider the Delta robot represented in Figure 3.39 and described in Problems 3.11 and 4.11. Repeat Problem 5.11 for this manipulator by the Lagrange approach. Furthermore, compare the dynamic formulation obtained here to that found by the Newton–Euler formulation.
36. Consider the Quadrupteron robot represented in Figure 3.40 and described in Problems 3.12 and 4.12. Repeat Problem 5.12 for this manipulator by the Lagrange approach. Furthermore, compare the dynamic formulation obtained here to that found by the Newton–Euler formulation.

6

Motion Control

6.1 Introduction

Parallel robots are designed for two different types of applications. In the first type, the moving platform of the robot accurately follows a desired position and orientation path in a specific time frame, while no interacting forces need to be applied to the environment. A typical application of such a case can be seen in the motion of a flight training simulator. In this application the desired position and orientation of the manipulator-moving platform is determined through examination of simulated flight conditions and pilot commands, and the robot controller has to determine the time history of actuator inputs required to cause such motions.

The second type of application include situations where the robot moving platform is in contact with a stiff environment. Using parallel robots for precision machining, accurate grinding, or microassembly may be considered as representatives of such applications. In such applications, the contact force describes the state of interaction more effectively than the position and orientation of the moving platform. The problem of force control can be described as to derive the actuator forces for such a manipulator required to generate a prescribed desired wrench (force/torque) at the manipulator moving platform, while the manipulator is performing its motion.

At two extents of these cases, one may consider execution of motion of a manipulator, while the moving platform moves in a totally free motion subject to no constrained external forces. This problem is treated as motion control problem in this chapter. On the other hand, another situation is when the robot is interacting with a stiff environment, and considerable interacting forces are applied to the robot moving platform. This problem and its extents are treated in the force control schemes given in the succeeding chapter. These problems are treated separately, because of their different objectives, and owing to their different complexities.

There are many control topologies and correspondingly many control techniques developed for robotic manipulators, especially for serial robots. Although a multiple degrees-of-freedom robotic manipulator can usually be represented by a multi-input/multi-output (MIMO) and nonlinear model, many industrial controllers for such robots consist of a number of linear controller designed to control individual joint motions. One of the reasons why such decentralization can perform well in practice is the use of large gear reductions in robot actuators, which significantly reduces the coupling and nonlinear behavior of robot dynamics [113,168]. However, in parallel manipulators, even if such gear reduction is used in the design of the manipulator, the closed-kinematic chains in the structure of the parallel robots significantly limit the performance of simple decentralized controller for such manipulators. Therefore, using advanced techniques in nonlinear and MIMO control are usually developed for such manipulators [99]. In what follows, a number of such topologies and techniques are presented.

6.2 Controller Topology

In motion control of a parallel manipulator it is assumed that the controller computes the required actuator forces or torques to cause the robot motion to follow a desired position and orientation trajectory. Let us use the motion variable as the generalized coordinate of the moving platform defined earlier by $\boldsymbol{\mathcal{X}} = \begin{bmatrix} \boldsymbol{x}_p & \boldsymbol{\theta} \end{bmatrix}^T$, in which the linear motion is represented by $\boldsymbol{x}_p = \begin{bmatrix} x_p & y_p & z_p \end{bmatrix}^T$, while the moving platform orientation is represented by screw coordinates, $\boldsymbol{\theta} = \theta \begin{bmatrix} s_x & s_y & s_z \end{bmatrix}^T = \begin{bmatrix} \theta_x & \theta_y & \theta_z \end{bmatrix}^T$. Note that contrary to the Euler angles, the screw coordinates $\boldsymbol{\theta}$ is a vector, whose derivative is also a vector representing the angular velocity of the moving platform, $\dot{\boldsymbol{\theta}} = \boldsymbol{\omega}$, and therefore, $\dot{\boldsymbol{\mathcal{X}}} = [\boldsymbol{v}_p, \boldsymbol{\omega}]^T$ represents the moving platform twist.

Furthermore, consider the general closed-form dynamic formulation of a parallel robot represented by Equation 5.297, given here for convenience:

$$\boldsymbol{M}(\boldsymbol{\mathcal{X}})\ddot{\boldsymbol{\mathcal{X}}} + \boldsymbol{C}(\boldsymbol{\mathcal{X}}, \dot{\boldsymbol{\mathcal{X}}})\dot{\boldsymbol{\mathcal{X}}} + \boldsymbol{G}(\boldsymbol{\mathcal{X}}) = \boldsymbol{\mathcal{F}}. \tag{6.1}$$

In this formulation, $\boldsymbol{M}(\boldsymbol{\mathcal{X}})$ denotes the mass matrix that can be derived from the kinetic energy of the manipulator by Equation 5.263, $\boldsymbol{C}(\boldsymbol{\mathcal{X}}, \dot{\boldsymbol{\mathcal{X}}})$ denotes the Coriolis and centrifugal matrix given in Equation 5.296, $\boldsymbol{G}(\boldsymbol{\mathcal{X}})$ denotes the gravity vector given in Equation 5.268, and finally, $\boldsymbol{\mathcal{F}}$ denotes the generalized forces applied to the moving platform center of mass. Furthermore, note that if there are no external forces applied to the moving platform, and the robot is considered to have a free motion, generalized forces can be computed by the projection of actuator forces on the moving platform through the Jacobian transpose mapping, $\boldsymbol{\mathcal{F}} = \boldsymbol{J}^T\boldsymbol{\tau}$. However, if there exist any external wrenches applied to the moving platform, such as the disturbance wrench $\boldsymbol{\mathcal{F}}_d$, the generalized forces is computed from the resulting wrenches in the task space, $\boldsymbol{\mathcal{F}} = \boldsymbol{J}^T\boldsymbol{\tau} + \boldsymbol{\mathcal{F}}_d$.

To design a motion controller for parallel manipulator, it is necessary to first introduce possible controller topologies developed for this task, and then introduce different techniques for controller design according to these topologies. In what follows, first the controller topologies are introduced, and then a number of controller designs are presented. Control topology is referred to the structure of the control system used to compute the actuator forces/torques from the measurements, and the required pre- and postprocessing.

For motion control of a manipulator, the controller has to compute the actuator force/torques required to cause the motion of the moving platform according to the desired trajectory. In general, the desired motion of the moving platform may be represented by the desired generalized coordinate of the manipulator, denoted by $\boldsymbol{\mathcal{X}}_d$. This variable has the same dimension and structure of the motion variable of the manipulator $\boldsymbol{\mathcal{X}}$. To perform such motion in a closed-loop structure, it is necessary to measure the output motion of the manipulator by an instrumentation system.

Since, in the motion control of parallel manipulators the final motion of the moving platform is of interest, it is desirable to use instrumentation systems that can directly measure the motion variable $\boldsymbol{\mathcal{X}}$ with the required accuracy. In such instrumentation systems the final position and orientation of the moving platform would be measured. Such instrumentation systems usually consists of two subsystems: the first subsystem may use accurate accelerometers, or global positioning systems to calculate the position of a point on the moving platform; and the second subsystem may use inertial or laser gyros to determine orientation of the moving platform.

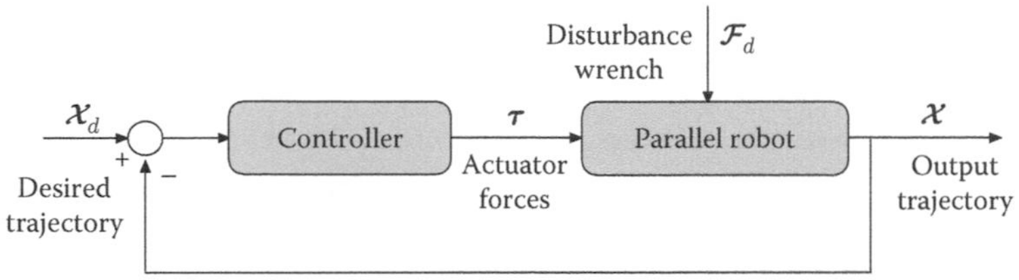

FIGURE 6.1
The general topology of motion feedback control: motion variable $\boldsymbol{\mathcal{X}}$ is measured.

Although there are many other techniques developed to measure the position and orientation of a moving object in space, the technical advancement in this area is still under development, and commercial instrumentation systems to provide such measurements accurately are limited in number, and they are usually expensive. Such devices may be used in applications where the accuracy and integrity of the output measurement is of utmost importance.

Figure 6.1 shows the general topology of a motion controller using direct measurement of motion variable $\boldsymbol{\mathcal{X}}$, as feedback in the closed-loop system. In such a structure, the measured position and orientation of the manipulator is compared to its desired value to generate the motion error vector $\boldsymbol{e}_x$. The controller uses this error information to generate suitable commands for the actuators to minimize the tracking error. Although such controller structure is very typical in terms of feedback control, it should be noted that the feedback measurement is a vector–valued quantity, and the controller has MIMO structure.

Other alternatives for motion control topology may be suggested based on other techniques developed to derive the position and orientation of the moving platform. Consider a general parallel manipulator consisting of a number of limbs with a number of passive and active joints depicted in Figure 3.2. It is usually much easier to measure the active joint variable in such a structure, rather than measuring the final position and orientation of the moving platform.

As defined in Equation 5.206, denote $\boldsymbol{q} = \begin{bmatrix} q_1 & q_2 & \cdots & q_n \end{bmatrix}$ as the joint variables, consisting of the motion variable of active joints. The relation between the joint variable $\boldsymbol{q}$ and motion variable of the moving platform $\boldsymbol{\mathcal{X}}$ has been studied in detail in Chapter 3 through forward and inverse kinematics. Furthermore, the relation between the differential motion variables $\dot{\boldsymbol{q}}$ and $\dot{\boldsymbol{\mathcal{X}}}$ is well studied in Chapter 4 through Jacobian analysis. Note that if $\boldsymbol{q}$ is given, forward kinematic analysis would be carried out to find $\boldsymbol{\mathcal{X}}$. Furthermore, the Jacobian matrix of the parallel manipulator defined in a general form in Equation 4.34 may be used to give relations between $\dot{\boldsymbol{q}}$ and $\dot{\boldsymbol{\mathcal{X}}}$.

If such an analysis is performed for the parallel manipulator in hand, it is possible to use the forward kinematic analysis to calculate $\boldsymbol{\mathcal{X}}$ from the measured joint variables $\boldsymbol{q}$, and one may use the control topology depicted in Figure 6.2 to implement such a controller. As seen in this figure, the joint variable measurement $\boldsymbol{q}$ is used in the feedback, while it is used as the input to forward kinematics to generate the moving platform motion variable $\boldsymbol{\mathcal{X}}$. Similar to what is used in the previous control topology depicted in Figure 6.1, the motion error in task space $\boldsymbol{e}_x$ is generated and is used in an MIMO controller designed suitably to generate the actuator forces/torques $\boldsymbol{\tau}$.

Note that in this topology, the forward kinematic analysis of the manipulator has to be performed to implement the feedback loop. As described earlier, the forward kinematic analysis of parallel manipulators is usually a complex task, requiring to solve a set of non-linear equations. Carrying out this task becomes much difficult if the forward kinematic

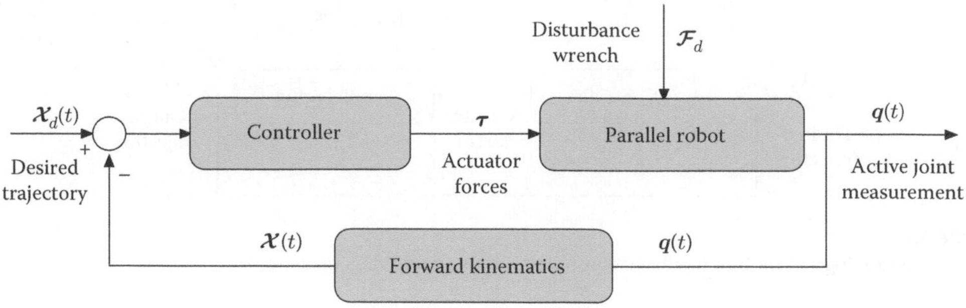

FIGURE 6.2
The general topology of motion feedback control: the active joint variable q is measured.

solution is found in real time. On the other hand, as shown hereinbefore, the inverse kinematic analysis of parallel manipulators is much easier to carry out, and usually a unique and analytic solution to this problem can be found. Therefore, to overcome the implementation problem of the control topology depicted in Figure 6.2, another control topology is usually implemented for parallel manipulators.

In this topology, depicted in Figure 6.3, the desired motion trajectory of the robot $\mathcal{X}_d$ is used in an inverse kinematic analysis to find the corresponding desired values for joint variable q_d. Hence, the controller is designed based on the joint space error e_q, and is totally implemented in the joint space. Therefore, the structure and the characteristics of the controller in this topology is totally different from that given in the first two topologies. Note that in this control topology, the input to the controller is the motion error in the joint space e_q, and the output of the controller is also the actuator force/torque $\boldsymbol{\tau}$, which is also represented in the joint space. However, in the previous topologies, the input to the controller is the motion error in task space e_x, while its output is in the joint space. Therefore, independent controllers for each joint may be suitable for the latter topology, while this may not be suitable for the topologies given in Figures 6.1 and 6.2.

To generate a direct input to output relation in the task space, consider the topology depicted in Figure 6.4. The essentials of this topology is similar to that in the first topology depicted in Figure 6.1; however, the controller output is a wrench calculated in the task space, and is denoted by $\mathcal{F}$. Furthermore, a block is added to this topology denoted by force distribution, which maps the generated wrench in the task space $\mathcal{F}$, to its corresponding actuator forces/torques $\boldsymbol{\tau}$. Note that for a fully parallel manipulator such as the Stewart–Gough platform, in which the number of actuators are equal to the number of degrees-of-freedom, this mapping can be constructed from the Jacobian transpose of the

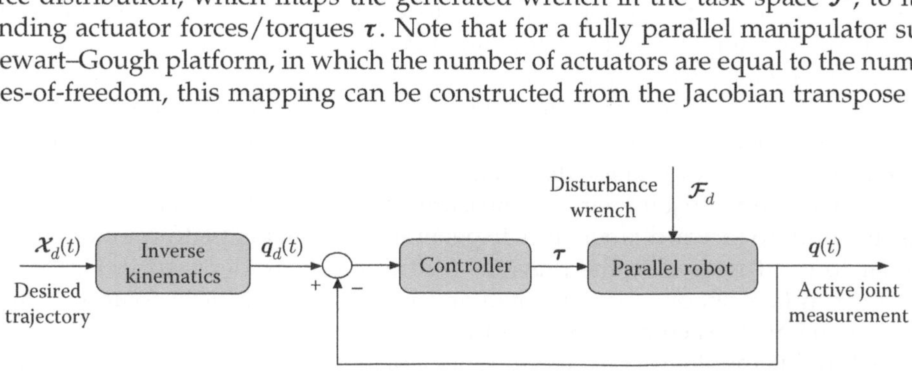

FIGURE 6.3
The general topology of motion feedback control: the active joint variable q is measured, and the inverse kinematic analysis is used.

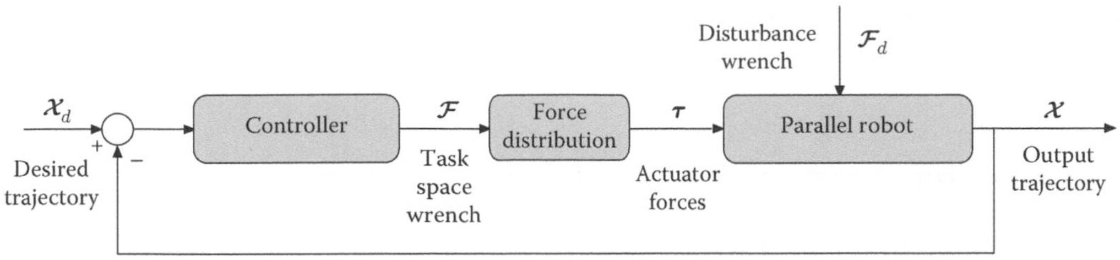

FIGURE 6.4
The general topology of motion feedback control in task space: the motion variable $\mathcal{X}$ is measured, and the controller output generates wrench in task space.

manipulator. It is shown in Equation 4.92 that this mapping is $\mathcal{F} = J^T \tau$. Hence, for a fully parallel manipulator, the actuator forces can be generated by $\tau = J^{-T}\mathcal{F}$ at nonsingular configurations.

The force distributor block may need more computation in case the manipulator is not fully parallel. Consider a redundant parallel manipulator in which the number of actuators denoted by m is greater than the manipulator degrees-of-freedom denoted by n. Redundancy in parallel robots is usually considered to enlarge its singularity-free workspace. However, cable-driven parallel manipulators adhere to redundancy in actuation to keep all the cable forces in tension. In such cases, the Jacobian matrix of the manipulator is not squared, and therefore, the inverse of Jacobian transpose cannot be used for force distribution. In fact, the Jacobian transpose is an $m \times n$, matrix in which $m > n$, and therefore, in such a case infinite solutions exist for actuator efforts τ to generate the required wrench $\mathcal{F}$ in the task space.

There exist many different algorithms developed to use the extra degree-of-actuation optimally, some of them are elaborated in detail in Section 6.7 under the title of *Redundancy Resolution*. The simplest way to solve this problem though is to use the Moore–Penrose pseudo-inverse of the Jacobian transpose denoted by $J^{T^\dagger}$, that is, $\tau = J^{T^\dagger}\mathcal{F}$. In this method, among many possible solutions for the actuator forces that satisfy $\mathcal{F} = J^T\tau$, the minimum norm force distribution denoted by τ_{min} is found to satisfy the wrench projection $\mathcal{F} = J^T\,\tau_{min}$. Although this method is very popular for many applications, it is not recommended for cable-driven manipulators, because it does not guarantee positive tension forces in all the actuators. A complete description of redundancy resolution techniques that guarantee such requirements is given in Section 6.7. Anyways, using any technique for redundancy resolution generates a certain force distribution in actuators, by which the Cartesian wrench relation $\mathcal{F} = J^T\,\tau$ is generated. That is the reason why the block added to the control topology, depicted in Figure 6.4, is called force distribution.

As mentioned earlier, many advanced control techniques are developed for robotic manipulators. In what follows, a number of these techniques are described in a sequence from simple and basic ideas to complete and more complex structures. To categorize the developed controllers corresponding to different implementable topologies introduced in this section, the two latter control topologies are considered mainly. First, the control structures being able to be implemented in the task space topology depicted in Figure 6.4 is introduced. In this topology, the motion variable in task space $\mathcal{X}$ is measured and used in the feedback, and therefore, these techniques are categorized as motion control techniques in the task space. Then, the control structures implementable in the joint space topology, depicted in Figure 6.3 are introduced. In this topology, the active joint variable in the joint space q is measured and used in the feedback, and therefore, these techniques are categorized as motion control techniques in the joint space.

6.3 Motion Control in Task Space

In this section, a number of control techniques for parallel manipulators implementable in task space are described in detail. These control strategies are given in a sequence from simple and basic ideas to more complete and effective structures.

6.3.1 Decentralized PD Control

The first control strategy introduced for parallel robots consists of the simplest form of feedback control in such manipulators. In this control structure, depicted in Figure 6.5, a number of disjoint linear (PD) controllers are used in a feedback structure on each error component. Let us denote the tracking error vector in the task space as $\boldsymbol{e}_x$. In a general six-degrees-of-freedom robot such as the Stewart–Gough platform, this error vector consist of six components as $\boldsymbol{e}_x = \begin{bmatrix} e_x & e_y & e_z & e_{\theta_x} & e_{\theta_y} & e_{\theta_z} \end{bmatrix}$. The decentralized controller therefore consists of six disjoint linear controllers, such as proportional derivative (PD) controller, acting on each error component. The PD controller is denoted by $\boldsymbol{K}_d s + \boldsymbol{K}_p$ block in Figure 6.5, in which $\boldsymbol{K}_d$ and $\boldsymbol{K}_p$ are 6×6 diagonal matrices denoting the derivative and proportional controller gains, respectively. The diagonal elements of these matrices compose the disjoint controller gains for each error term. Furthermore, s denotes the Laplace variable and corresponds to the derivative action in time domain.

Hence, by this structure each tracking error component is treated separately through a disjoint PD controller. The output of the controller is denoted by $\mathcal{F}$, which is defined by a full-scale wrench in the task space. Therefore, for the general case of six-degrees-of-freedom robot such as the Stewart–Gough platform, this wrench consists of the following six components $\mathcal{F} = \begin{bmatrix} F_x & F_y & F_z & \tau_x & \tau_y & \tau_z \end{bmatrix}$. Note that since the output of the controller is defined in the task space, each wrench component directly manipulates the corresponding tracking error component, and therefore, overall tracking performance of the manipulator is suitable if high controller gains are selected.

In practice, the calculated output wrench is transformed into actuator forces through the force distribution block. In this mapping the actuator forces required to generate such a wrench is computed and applied to the manipulator. As explained hereinbefore for

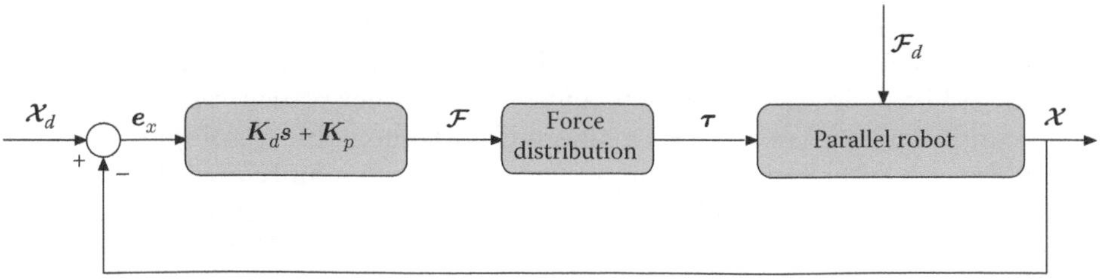

FIGURE 6.5
Decentralized PD controller implemented in task space.

a fully parallel manipulator with no redundancy in actuation, this mapping is implemented through inverse of the manipulator Jacobian transpose by $\boldsymbol{\tau} = J^{-T}\boldsymbol{\mathcal{F}}$. Different alternatives of linear controllers can be used instead of the PD controller used in this structure; however, as seen in the following sections, PD controller is the simplest form which can preserve the manipulator stability, while providing suitable tracking performance.

The proposed decentralized PD controller is very simple in structure and therefore easily implementable. The design of such a controller needs no detailed information on the manipulator dynamic formulation, nor dynamic parameters. The computation complexity of its implementation is also very low, and is just limited to carrying out suitable force distribution calculations in practice. However, the tracking performance of such a controller is relatively poor, and static tracking errors might be unavoidable. On the other hand, owing to the high gain controller design, the required actuation energy is relatively high, while the performance of the closed-loop system is configuration dependent.

The controller gains must be tuned experimentally based on physical realization of the controller in practice by trial and error. Therefore, the final controller gains are obtained as a tradeoff between transient response and steady-state errors at different configurations. As seen in Chapter 5, the dynamic formulation of the parallel manipulator exhibit nonlinear behavior, which is configuration dependent. Therefore, finding suitable controller gains to result in the required performance in all configurations is a difficult task. Usually the designer chooses the controller gains such that the required tracking error of the manipulator at its usual working condition is suitable. This results in varying performance error in other configurations where the tuning process is not applied. The performance of the controller to attenuate the measurement noise, and the external disturbance wrench applied to the manipulator may become poor in practice.

To remedy these shortcomings, some modifications are proposed to be applied to this structure. Two most important improvements are given in the following two sections. However, note that the controller described in this section may be recommended as the first trial in closed-loop experiments.

6.3.2 Feed Forward Control

As seen in Section 6.3.1, the tracking performance of simple PD controller implemented in the task space is not uniform at different configurations, and in the presence of measurement noise and external disturbances applied to the manipulator. To compensate for such effects, a feed forward wrench denoted by $\mathcal{F}_{ff}$ may be added to the structure of the controller as depicted in Figure 6.6. This term is generated from the dynamic model of the

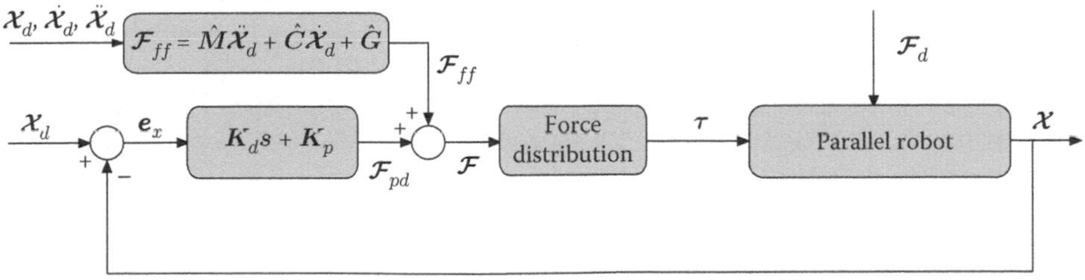

FIGURE 6.6
Feed forward wrench added to the decentralized PD controller in task space.

manipulator in the task space, represented in a closed form by the following equation:

$$\mathcal{F}_{ff} = \hat{M}(\mathcal{X}_d)\ddot{\mathcal{X}}_d + \hat{C}(\mathcal{X}_d, \dot{\mathcal{X}}_d)\dot{\mathcal{X}}_d + \hat{G}(\mathcal{X}_d). \tag{6.2}$$

As shown in Figure 6.6, the desired trajectory in task space $\mathcal{X}_d$ and its derivatives $\dot{\mathcal{X}}_d, \ddot{\mathcal{X}}_d$ are the required inputs for the feed forward block. This term is called feed forward since no online information of the output motion trajectory $\mathcal{X}$ is needed to generate this term. Furthermore, note that to generate this term dynamic formulation of the robot and its kinematic and dynamic parameters are needed. In practice, exact knowledge of dynamic matrices are not available, and therefore, estimates of these matrices are used in this derivation. $\hat{M}, \hat{C}$, and $\hat{G}$ denote approximate estimates of the manipulator mass matrix, Coriolis and centrifugal matrix, and gravity vector, respectively.

The information required to generate the feed forward wrench $\mathcal{F}_{ff}$ is usually available beforehand, and in such a case, the feed forward term corresponding to a given desired trajectory can be derived offline, while the computation of the decentralized feedback term would be executed online. Write the closed-loop dynamic formulation for the manipulator as follows:

$$\begin{aligned} M(\mathcal{X})\ddot{\mathcal{X}} + C(\mathcal{X}, \dot{\mathcal{X}})\dot{\mathcal{X}} + G(\mathcal{X}) &= \mathcal{F} + \mathcal{F}_d \\ &= \mathcal{F}_{pd} + \mathcal{F}_{ff} + \mathcal{F}_d \\ &= K_d\dot{e}_x + K_p e_x + \mathcal{F}_d + \hat{M}\ddot{\mathcal{X}}_d + \hat{C}\dot{\mathcal{X}}_d + \hat{G}. \end{aligned} \tag{6.3}$$

If the knowledge of dynamic matrices is complete, then we may assume that $\hat{M} = M, \hat{C} = C$, and $\hat{G} = G$. Furthermore, consider that the controller of the system performs well and the moving platform track the desired trajectory well; therefore, assume that $\mathcal{X}(t) \simeq \mathcal{X}_d(t), \dot{\mathcal{X}}(t) \simeq \dot{\mathcal{X}}_d(t)$. In such a case, simplify the closed-loop dynamic formulation to

$$\begin{aligned} M(\ddot{\mathcal{X}}_d - \ddot{\mathcal{X}}) + K_d\dot{e}_x + K_p e_x + \mathcal{F}_d &= 0, \\ M\ddot{e}_x + K_d\dot{e}_x + K_p e_x + \mathcal{F}_d &= 0. \end{aligned} \tag{6.4}$$

This equation implies that, if the above-mentioned assumptions hold, the error dynamics satisfies a set of second-order systems in the presence of disturbance. Therefore, by choosing appropriate gains for PD controller, the transient and steady-state performance of tracking error can be designed so as to satisfy the application requirements. Note that if the disturbance wrench applied to the manipulator is zero or decays to zero as time tends to infinity, the steady-state error of the closed-loop system tends to zero for all the tracking error components. Furthermore, note that except for the mass matrix, the error dynamic terms are all configuration independent, and therefore, it is much easier to tune the PD controller gains to work well within the whole workspace of the robot.

However, this method faces a number of limitations in practice. The most important limitation of this control technique is the stringent assumption of a complete knowledge requirement of the dynamic matrices. In practice, derivation of these matrices is a prohibitive task, while accessing to their full information for all configurations is practically infeasible. Therefore, in practice, Equation 6.3 represents the error dynamics better.

In such a case, the difference between the true values of dynamic matrices and their estimated values may be considered as an extra error wrench applied to the error dynamics, and therefore, static errors would be definitely seen in the steady-state behavior of the closed-loop system. Furthermore, even in case of full knowledge of dynamic matrices, true motion of the moving platform is different in practice to that of the desired trajectory, that is, $\mathcal{X}(t) \neq \mathcal{X}_d(t)$, and $\dot{\mathcal{X}}(t) \neq \dot{\mathcal{X}}_d(t)$. This may be seen as a more significant error wrench applied to the error dynamics, which leads to more tracking error at the outset.

Finally, if all the assumptions for Equation 6.4 hold, because of the dependency of the mass matrix to the configuration of the robot, still the error dynamics are not completely decoupled. This means that correction in one error component may be considered as a disturbance effect to the other components. To overcome these limitations inverse dynamic approach is given in the following section.

6.3.3 Inverse Dynamics Control

As seen in Section 6.3.1, the tracking performance of a decentralized PD controller implemented in the task space is not uniform at different configurations, and in the presence of measurement noise and external disturbances applied to the manipulator. To compensate for such effects and as explained in Section 6.3.2, a feed forward wrench is added to the structure of the controller, by which the shortcomings of the decentralized controller is partially remedied. However, the closed-loop performance still suffers from a number of limitations, which cannot be overcome by the feed forward structure of the controller. To overcome these limitations, in this section a control technique is presented based on the inverse dynamics feedback of the manipulator.

In inverse dynamics control (IDC), nonlinear dynamics of the model is used to add a corrective term to the decentralized PD controller introduced in Section 6.3.1. By this means, nonlinear and coupling behavior of the robotic manipulator is significantly attenuated, and therefore, the performance of linear controller is greately improved. This method is known by different names in different research communities. In the Control theory, IDC is referred to as feedback linearization technique, by which the linearization effect of nonlinear dynamics is emphasized in this representation. In the serial robotics research, this method is called the computed torque method, by which the corrective term added to the structure of joint space controller is emphasized. In the parallel robotics research, although other terms are also seen in the literature, this method is usually known as inverse dynamic control. Note that IDC is a general and effective idea developed for the control of robotic manipulators, and therefore, is considered as the basis of many developments of advanced controllers on this topic.

General structure of IDC applied to a parallel manipulator is depicted in Figure 6.7. As seen in this figure a corrective wrench $\mathcal{F}_{fl}$ is added in a feedback structure to the closed-loop system, which is calculated from the Coriolis and centrifugal matrix and gravity vector of the manipulator dynamic formulation. Furthermore, mass matrix is added in the forward path in addition to the desired trajectory acceleration $\ddot{\mathcal{X}}_d$. Note that to generate inverse dynamics control terms, the dynamic formulation of the robot, and its kinematic and dynamic parameters are needed. In practice, exact knowledge of dynamic matrices are not available, and therefore, estimates of these matrices are used in the derivation. $\hat{\boldsymbol{M}}$, $\hat{\boldsymbol{C}}$, and $\hat{\boldsymbol{G}}$ denote an approximate estimate of the manipulator mass matrix, Coriolis and centrifugal matrix, and gravity vector, respectively.

Hence, the controller output wrench applied to the manipulator may be derived as follows:

$$\mathcal{F} = \hat{\boldsymbol{M}}(\mathcal{X})\boldsymbol{a} + \mathcal{F}_{fl},$$

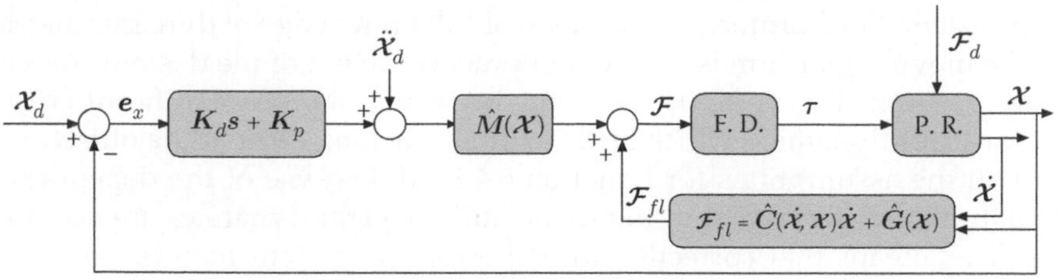

FIGURE 6.7
General configuration of inverse dynamics control implemented in task space.

which in detail reads

$$\mathcal{F} = \hat{M}(\mathcal{X})a + \hat{C}(\mathcal{X}, \dot{\mathcal{X}})\dot{\mathcal{X}} + \hat{G}(\mathcal{X}), \tag{6.5}$$

$$a = \ddot{\mathcal{X}}_d + K_d \dot{e}_x + K_p e_x. \tag{6.6}$$

Note that in a general case of six-degrees-of-freedom manipulator, a is a 6×1 vector, having the dimension of motion variable acceleration. Let us now write the closed-loop dynamic formulation for the manipulator as follows:

$$\begin{aligned} M(\mathcal{X})\ddot{\mathcal{X}} + C(\mathcal{X}, \dot{\mathcal{X}})\dot{\mathcal{X}} + G(\mathcal{X}) &= \mathcal{F} + \mathcal{F}_d \\ &= \hat{M}(\mathcal{X})a + \hat{C}(\mathcal{X}, \dot{\mathcal{X}})\dot{\mathcal{X}} + \hat{G}(\mathcal{X}) + \mathcal{F}_d \\ &= \hat{M}(\mathcal{X})\left(\ddot{\mathcal{X}}_d + K_p e_x + K_d \dot{e}_x\right) \\ &\quad + \hat{C}(\mathcal{X}, \dot{\mathcal{X}})\dot{\mathcal{X}} + \hat{G}(\mathcal{X}) + \mathcal{F}_d. \end{aligned} \tag{6.7}$$

If the knowledge of the dynamic matrices is complete, then we may assume that $\hat{M} = M, \hat{C} = C$, and $\hat{G} = G$. In such a case, simplify the closed-loop dynamic formulation to

$$\hat{M}(\mathcal{X})\left(\ddot{e}_x + K_d \dot{e}_x + K_p e_x\right) + \mathcal{F}_d = \mathbf{0}. \tag{6.8}$$

This equation implies that if there exists full knowledge of dynamic matrices, then error dynamics satisfies a set of second-order systems in the presence of disturbance. Consider the case that the manipulator is moving freely in space and there exists no disturbance wrench applied to the manipulator. Furthermore, notice that the manipulator mass matrix is positive definite for all configurations. Therefore, in such a case error dynamics is simplified to

$$\ddot{e}_x + K_d \dot{e}_x + K_p e_x = \mathbf{0}. \tag{6.9}$$

Hence, by choosing appropriate gains for PD controllers, the tracking error can be designed so as to have a fast transient, while steady-state error is small for the desired trajectory.

This control technique is very popular in practice because of the fact that this technique can significantly linearize and decouple dynamic formulation of the closed-loop error dynamics. Furthermore, the error dynamic terms are all configuration independent, and therefore, it is much easier to tune the PD controller gains for suitable performance in the whole workspace of the robot. However, note that for a good performance, an accurate model of the system is required, and the overall procedure is not robust to modeling uncertainty. Furthermore, this technique is computationally intensive in terms of the online computations needed to carry out the closed-loop control structure. Further modification of this technique is given to reduce the computational cost, while preserving the important characteristics of this approach. This modification is described in Section 6.3.4 under partial linearization IDC.

Now consider the general case, where the knowledge of dynamic matrices is not complete, that is, $\hat{M} \neq M, \hat{C} \neq C, \hat{G} \neq G$, and also an arbitrary disturbance wrench is applied to the moving platform $\mathcal{F}_d \neq \mathbf{0}$. In such a case, the error dynamics may be derived from Equation 6.7 as

$$\ddot{e}_x + K_d \dot{e}_x + K_p e_x = \hat{M}^{-1} \left[(M - \hat{M})\ddot{\mathcal{X}} + (C - \hat{C})\dot{\mathcal{X}} + (G - \hat{G}) - \mathcal{F}_d \right]. \tag{6.10}$$

The right-hand side of Equation 6.10 is not zero in general, but may be treated as an external wrench applied to the error dynamics. The effect of this external wrench can be well attenuated using the high-gain PD controller, or by applying advanced control methods such as robust or adaptive schemes.

In the robust control approach, through the quantitative analysis of modeling uncertainty at different configurations, an upper bound for the right-hand side of Equation 6.10 is derived. Then Lyapunov-based robust control techniques are used to add a corrective term to the PD controller output. By this means, asymptotic stability for the tracking error is guaranteed in the presence of worst-case modeling uncertainty and disturbance wrench. On the other hand, if the estimates of the dynamic formulation matrices can be updated such that the difference between the true values of these matrices to that of their estimates becomes small, the error dynamics of the manipulator will converge to the second-order system described by Equation 6.8. Such convergence becomes feasible through adaptive control schemes. These techniques are presented in Section 6.4.

6.3.4 Partial Linearization IDC

Inverse dynamics control has several features making it very attractive in practice. However, as explained in Section 6.3.3, to apply this method complete knowledge of the dynamic formulation matrices is required. This requirement has two main drawbacks which may practically limit the widespread use of this technique in industrial robots. First, notice that deriving dynamic formulation of a parallel manipulator is a time-consuming task, and furthermore, rewriting the formulation in a closed form, by which the dynamic matrices are analytically derived, is another hard step to be carried out.

To implement all the terms in IDC structure, not only the structure and components of such matrices must be carefully derived, but also all the kinematic and inertial parameters of the robot are needed to be identified and calibrated. This step by itself requires designing careful experiments and use of high-precision calibration equipment, which are not usually accessible in all applications. Finally, if all the terms and parameters are well known, implementation of full inverse dynamic linearization is computationally intensive. These are the reasons why, in practice, IDC control is extended to many different forms to reduce the above-mentioned stringent requirements. In this section, the simplest form of IDC control to reduce the computational cost while preserving the important characteristics of this approach is described, and in the next section, certain methods are developed to use simpler dynamic formulations of the robot manipulator, while compensating for modeling uncertainty through robust and adaptive schemes.

To develop the simplest possible implementable IDC, let us recall dynamic formulation complexities. As explained in detail in Section 5.5.4, the manipulator mass matrix $\boldsymbol{M}(\boldsymbol{\mathcal{X}})$ is derived from kinetic energy of the manipulator by Equation 5.263, while the gravity vector $\boldsymbol{G}(\boldsymbol{\mathcal{X}})$ is derived from potential energy by Equation 5.268. In comparison, calculation of gravity vector is accomplished much easily than that of the mass matrix. Furthermore, computation of the Coriolis and centrifugal matrix $\boldsymbol{C}(\boldsymbol{\mathcal{X}}, \dot{\boldsymbol{\mathcal{X}}})$ defined in Equation 5.296 is significantly more intensive than that of gravity vector and mass matrix. On the other hand, it is shown that certain properties hold for mass matrix $\boldsymbol{M}$, gravity vector $\boldsymbol{G}$, and Coriolis and centrifugal matrix $\boldsymbol{C}$, which might be directly used in the control techniques developed for parallel manipulators. One of the most important properties of dynamic matrices is the skew-symmetric property of the matrix $\dot{\boldsymbol{M}} - 2\boldsymbol{C}$ defined in Equation 5.294. This property may be effectively used to develop the simplest form of IDC as follows.

Consider dynamic formulation of parallel robot given in Equation 6.1, in which the skew–symmetric property of dynamic matrices is satisfied. The simplest form of IDC control effort $\boldsymbol{\mathcal{F}}$ consists of the following two terms:

$$\boldsymbol{\mathcal{F}} = \boldsymbol{\mathcal{F}}_{pd} + \boldsymbol{\mathcal{F}}_{fl}, \tag{6.11}$$

in which the first term $\boldsymbol{\mathcal{F}}_{pd}$ is generated by the simplified PD form on the motion error $\boldsymbol{e}_x$

$$\boldsymbol{\mathcal{F}}_{pd} = \boldsymbol{K}_d \dot{\boldsymbol{e}}_x + \boldsymbol{K}_p \boldsymbol{e}_x, \tag{6.12}$$

where $\boldsymbol{e}_x = \boldsymbol{\mathcal{X}}_d - \boldsymbol{\mathcal{X}}$ denotes the motion tracking error, while $\boldsymbol{K}_p$ and $\boldsymbol{K}_d$ are the proportional and derivative control gain matrices, respectively. The second term in Equation 6.11 is also considered to be only the gravity vector of the manipulator $\boldsymbol{G}(\boldsymbol{\mathcal{X}})$, at any configuration, and the computationally intensive Coriolis and centrifugal term is not used in the feedback:

$$\boldsymbol{\mathcal{F}}_{fl} = \boldsymbol{G}(\boldsymbol{\mathcal{X}}). \tag{6.13}$$

Note that for an appreciable tracking performance with no static error at steady state, it is required to have complete knowledge of only the gravity term, and calculation of the mass matrix and the Coriolis centrifugal matrix is not required in the feedback. By this means, computations required in this control technique are significantly less than that of the general IDC.

Despite the simple structure of such a controller, the resulting control technique is very well performed especially at steady state. Let us examine the characteristics of such a controller by the Lyapunov analysis method.* In what follows, it is shown that this control topology achieves asymptotic tracking for a constant desired trajectory motion, that is, $\dot{\boldsymbol{\mathcal{X}}}_d = \boldsymbol{0}$.

To show that the control technique given in Equation 6.11 achieves asymptotic tracking, consider the overall energy of the closed-loop system as a Lyapunov function candidate:

$$V = \frac{1}{2}\dot{\boldsymbol{\mathcal{X}}}^T \boldsymbol{M} \dot{\boldsymbol{\mathcal{X}}} + \frac{1}{2}\boldsymbol{e}_x^T \boldsymbol{K}_p \boldsymbol{e}_x. \tag{6.14}$$

The Lyapunov function candidate consists of the kinetic energy of the manipulator and the potential energy accounted for the proportional feedback effort $\boldsymbol{K}_p \boldsymbol{e}_x$. Thus, $V(\boldsymbol{\mathcal{X}})$ is a

* The reader is advised to review the discussion given in Appendix C on the Lyapunov stability theorems.

positive scalar function for the whole workspace of the manipulator, except at the configurations where the robot has reached its desired position, that is, $\mathcal{X} = \mathcal{X}_d, \dot{\mathcal{X}} = \mathbf{0}$, at which V is zero. The Lyapunov stability method is based on the concept to show that along any motion trajectory of the robot, the time derivative of the Lyapunov function is a negative definite. If such a condition holds for any motion trajectory, it implies that the total energy of the robot is approaching zero. Furthermore, since zero energy occurs only in a state where the output trajectory of the robot converges to the desired one, it implies achieving asymptotic tracking at the outset.

For the particular constant desired trajectory considered in this analysis, the time derivative of Lyapunov function can be derived as

$$\dot{V} = \dot{\mathcal{X}}^T \boldsymbol{M} \ddot{\mathcal{X}} + \frac{1}{2}\dot{\mathcal{X}}^T \dot{\boldsymbol{M}} \dot{\mathcal{X}} + \dot{e}_x^T \boldsymbol{K}_p e_x. \tag{6.15}$$

Substitute $\boldsymbol{M}\ddot{\mathcal{X}}$ from the closed-loop dynamic formulation 6.1 in Equation 6.15, and note that for constant desired trajectory motion, $\dot{\mathcal{X}}_d = \mathbf{0}$. Hence, the time derivative of the Lyapunov function is simplified to

$$\begin{aligned}\dot{V} &= \dot{\mathcal{X}}^T \left(\mathcal{F}_{pd} + \mathcal{F}_{fl} - C\dot{\mathcal{X}} - G\right) + \frac{1}{2}\dot{\mathcal{X}}^T \dot{\boldsymbol{M}} \dot{\mathcal{X}} + \dot{e}_x^T \boldsymbol{K}_p e_x \\ &= \dot{\mathcal{X}}^T \left(\boldsymbol{K}_p e_x + \boldsymbol{K}_d \dot{e}_x - C\dot{\mathcal{X}}\right) + \frac{1}{2}\dot{\mathcal{X}}^T \dot{\boldsymbol{M}} \dot{\mathcal{X}} + \dot{e}_x^T \boldsymbol{K}_p e_x.\end{aligned}$$

To simplify this equation, note that for a constant desired trajectory $\dot{\mathcal{X}}_d = \mathbf{0}$, and therefore, $\dot{e}_x = (\dot{\mathcal{X}}_d - \dot{\mathcal{X}}) = -\dot{\mathcal{X}}$. Hence,

$$\begin{aligned}\dot{V} &= \dot{\mathcal{X}}^T \boldsymbol{K}_p e_x - \dot{\mathcal{X}}^T \boldsymbol{K}_d \dot{\mathcal{X}} + \frac{1}{2}\dot{\mathcal{X}}^T \left(\dot{M} - 2C\right)\dot{\mathcal{X}} - \dot{\mathcal{X}}^T \boldsymbol{K}_p e_x \\ &= -\dot{\mathcal{X}}^T \boldsymbol{K}_d \dot{\mathcal{X}} + \frac{1}{2}\dot{\mathcal{X}}^T \left(\dot{M} - 2C\right)\dot{\mathcal{X}} \\ &= -\dot{\mathcal{X}}^T \boldsymbol{K}_d \dot{\mathcal{X}} \leq 0,\end{aligned} \tag{6.16}$$

where in the last equality the skew-symmetric property of dynamic matrices is used to simplify the equation. This result indicates that the derivative of the Lyapunov function is negative semi-definite, and this does not guarantee achieving asymptotic tracking.

To prove asymptotic convergence use the Lasalle's Theorem*. Note that V is always decreasing, provided $\dot{\mathcal{X}}$ is not zero. Hence, it is possible that the manipulator reaches a configuration where $\dot{\mathcal{X}} = 0$, but yet $\mathcal{X} \neq \mathcal{X}_d$. To show that this can never occur, note that at such a configuration $\dot{V} \equiv 0$. Therefore, Equation 7.18 implies that $\dot{\mathcal{X}} \equiv \mathbf{0}$, and hence $\ddot{\mathcal{X}} \equiv \mathbf{0}$. In such a case, the closed-loop dynamic formulation

$$\boldsymbol{M}(\mathcal{X})\ddot{\mathcal{X}} + C(\mathcal{X}, \dot{\mathcal{X}})\dot{\mathcal{X}} = \boldsymbol{K}_d \dot{\mathcal{X}} + \boldsymbol{K}_p e_x \tag{6.17}$$

simplifies to

$$\boldsymbol{K}_p e_x = \mathbf{0}, \tag{6.18}$$

* Described in Appendix C.

which implies that for any positive proportional controller gain, $e_x = \mathbf{0}$. Therefore, by the Lasalle's theorem it is guaranteed that $\boldsymbol{\mathcal{X}} = \boldsymbol{\mathcal{X}}_d$ at steady state, and the closed-loop system achieves asymptotic tracking.

This analysis reveals the fact that even if the mass matrix and Coriolis and centrifugal matrix are not used in the feedback, and the closed-loop dynamics is not completely linearized, the PD control structure with gravity compensation can still lead to asymptotic tracking. This analysis is performed for a special case of constant desired trajectory, and is not valid for any arbitrary trajectory. Moreover, the only characteristic being analyzed here is asymptotic stability. Hence no general claim on asymptotic tracking to a more general desired trajectory or transient performance of the closed-loop system can be drawn by such analysis. Therefore, one may conclude that for suitable transient performance, more information of the system dynamics must be used in the linearization technique given in IDC, and the error dynamics must be completely linearized and decoupled. However, this analysis reveals an important fact that the skew-symmetric property of the dynamic formulation matrices provides sufficient assurance for the designer to implement partial linearization IDC, and not to worry about the tracking error divergence.

6.4 Robust and Adaptive Control

Inverse dynamics control faces the stringent requirement that for a good performance an accurate model of the system is required, and the overall procedure is not robust to modeling uncertainty. Furthermore, this technique is computationally intensive in terms of online computation needed to carry out the closed-loop control structure. In Section 6.3.4, a modified inverse dynamics control formulation is given, wherein by partial linearization of dynamic formulation, computational cost is significantly reduced while asymptotic convergence of the tracking error is preserved for a class of constant desired trajectories. Although this modification is beneficial in terms of computational cost, it is not suitable in terms of a closed-loop transient performance, especially for time-varying trajectories.

Another important approach to modify inverse dynamics control is to consider a complete linearization, but assume that complete knowledge of dynamic formulation matrices is not available. To compensate for the lack of knowledge, two advanced control methods, namely robust and adaptive control, are proposed in the literature. In the robust control approach, by quantitative analysis of the modeling uncertainties at different configurations, an upper bound for the right-hand side of Equation 6.10 is found. Then the Lyapunov-based robust control schemes are used to add a corrective term to the PD controller output to guarantee asymptotic stability of error dynamics in the presence of worst-case modeling uncertainty and disturbance wrench. In the adaptive approach, the estimates of dynamic formulation matrices are updated such that the difference between the true values of these matrices to their estimates converges to zero. By this means, the error dynamics of the manipulator converges to the second-order system described by Equation 6.8. These techniques are detailed in Sections 6.4.1 and 6.4.2, respectively.

Note that the goal of both robust and adaptive control techniques is to preserve the closed-loop performance of the manipulator, despite modeling uncertainties and external disturbances. However, the way this goal is accomplished is different in robust and adaptive techniques. In a robust scheme, a fixed controller is designed to satisfy the control objectives for the worst possible case of modeling uncertainties and disturbance wrenches.

Whereas in an adaptive controller, online parameter identification is carried out to tune the controller parameters. Therefore, in robust controllers, a general form of modeling uncertainties, including parametric (structured) and unmodeled dynamics (unstructured uncertainties), may be considered at the design stage. Furthermore, usually a high-gain controller is designed to compensate for such uncertainties in the worst case. Whereas, in the adaptive control scheme the structure of the model is considered to be known, and only the parameters are subject to online identification. As a result, the controller gains are varying, and therefore, the control effort decreases as the estimated parameters converge to their true values. However, the adaptive controller may not perform well in the presence of unmodeled dynamics or applied disturbance wrenches. A global understanding of the trade-offs involved in each method is needed to employ either of them in practice. Adaptive robust controllers (ARC) have been developed recently by combining the two methods to incorporate the advantages of both the methods, at the expense of higher computational cost [176].

6.4.1 Robust Inverse Dynamics Control

Inverse dynamics control performance relies on complete linearization of nonlinear dynamic terms by feedback. This requirement may never be satisfied in practice, and practically limits the widespread use of this technique in industrial robots. First, note that deriving dynamic formulation of a parallel manipulator is a difficult task, and furthermore, rewriting the formulation in a closed form by which the dynamic matrices are analytically derived is another challenging requirement. For a complete feedback linearization and to implement all the dynamic terms in IDC structure, not only the structure and components of such matrices should be carefully derived, but also all the kinematic and inertial parameters of the robot manipulator need to be identified and calibrated. This step, by itself, requires careful design of experiments, and high-precision calibration equipment, which are not usually accessible in all applications. Therefore, various sources of uncertainties such as unmodeled dynamics, unknown parameters, calibration errors, unknown disturbance wrenches, and varying payloads may exist, that are not seen in dynamic model of the manipulator.

To consider these modeling uncertainties in the closed-loop performance of the manipulator, recall the general closed-form dynamic formulation of the manipulator given in Equation 6.1, and modify the inverse dynamics control input $\mathcal{F}$ as

$$\mathcal{F} = \hat{M}(\mathcal{X})a_r + \hat{C}(\mathcal{X},\dot{\mathcal{X}})\dot{\mathcal{X}} + \hat{G}(\mathcal{X}), \tag{6.19}$$

$$a_r = \ddot{\mathcal{X}}_d + K_d\dot{e}_x + K_p e_x + \delta_a, \tag{6.20}$$

in which a_r is the robustified control input. Comparing this equation to the usual IDC given in Equation 6.6, a robustifying term δ_a is added in Equation 6.20 to compensate for modeling uncertainties. Note that, as defined earlier, the notation $\hat{(\cdot)}$ represents the estimated value of $(\cdot)$ terms and $\tilde{(\cdot)}$ is defined as the error or mismatch between the estimated value and the true value of the corresponding term as $\tilde{(\cdot)} = \hat{(\cdot)} - (\cdot)$. Hence,

$$\tilde{M} = \hat{M} - M; \quad \tilde{C} = \hat{C} - C; \quad \tilde{G} = \hat{G} - G.$$

In a similar manner $\tilde{(\cdot)}$ notation may be applied to the motion variables as

$$\tilde{\mathcal{X}} = \mathcal{X} - \mathcal{X}_d = -e_x, \quad \dot{\tilde{\mathcal{X}}} = \dot{\mathcal{X}} - \dot{\mathcal{X}}_d = -\dot{e}_x.$$

Let us examine the closed-loop dynamic formulation of the manipulator given in Equation 6.1, and substitute the robust IDC control effort $\mathcal{F}$ given in 6.19. By some manipulation it yields

$$\ddot{\mathcal{X}} = a_r + \eta\left(\mathcal{X}, \dot{\mathcal{X}}, a_r\right) \tag{6.21}$$

in which

$$\eta = M^{-1}\left(\tilde{M}a_r + \tilde{C}\dot{\mathcal{X}} + \tilde{G}\right) \tag{6.22}$$

is a measure of modeling uncertainty. The closed-loop manipulator dynamics represented by the uncertain system Equation 6.21 is still nonlinear and coupled owing to the modeling uncertainty represented by η. The robust controller would be designed such that the closed-loop system satisfies the desired performance in the presence of the modeling uncertainty η. In what follows, the analysis of the closed-loop tracking performance in the presence of modeling uncertainty is presented based on the Lyapunov theory.

There are many approaches developed for robust control of an uncertain system. In what follows, a nonlinear robust control approach based on the Lyapunov stability criteria is developed. Rewrite the robust control effort defined in Equation 6.20 using the $\tilde{(\cdot)}$ notion, to focus on robustifying corrective term $\boldsymbol{\delta}_a$:

$$a_r = \ddot{\mathcal{X}}_d - K_d\dot{\tilde{\mathcal{X}}} - K_p\tilde{\mathcal{X}} + \boldsymbol{\delta}_a. \tag{6.23}$$

Substitute robust control effort given in Equation 6.23 into the closed-loop dynamic formulation 6.21 and simplify as

$$\ddot{\tilde{\mathcal{X}}} = -K_d\dot{\tilde{\mathcal{X}}} - K_p\tilde{\mathcal{X}} + \boldsymbol{\delta}_a + \eta. \tag{6.24}$$

The closed-loop dynamics may be written in a state-space form by defining the following state vector

$$\boldsymbol{\varepsilon} = \begin{bmatrix} \tilde{\mathcal{X}} \\ \dot{\tilde{\mathcal{X}}} \end{bmatrix} = \begin{bmatrix} \mathcal{X} - \mathcal{X}_d \\ \dot{\mathcal{X}} - \dot{\mathcal{X}}_d \end{bmatrix}. \tag{6.25}$$

Therefore, the closed-form dynamic formulation given in Equation 6.24 may be written in state-space form as follows

$$\dot{\boldsymbol{\varepsilon}} = A\boldsymbol{\varepsilon} + B(\boldsymbol{\delta}_a + \eta), \tag{6.26}$$

in which

$$A = \begin{bmatrix} 0 & I \\ -K_p & -K_d \end{bmatrix}, \quad B = \begin{bmatrix} 0 \\ I \end{bmatrix}. \tag{6.27}$$

The state-space representation of the closed-loop system clarifies the robustifying action of the corrective term $\boldsymbol{\delta}_a$. To have a stable error dynamics, the PD controller gains must be chosen such that the linear part of the state-space representation is stable. Moreover, the nonlinear part of control action, namely $\boldsymbol{\delta}_a$ would be designed so as to overcome the destabilizing effect of uncertainty η in the worst case. Therefore, the stabilizing controller consists of two linear and nonlinear parts.

To stabilize the linear part, it is sufficient to choose the controller gain matrices $\boldsymbol{K}_p$ and $\boldsymbol{K}_d$ such that the matrix $\boldsymbol{A}$ becomes Hurwitz, that is, all its eigenvalues lie in the open left-half of the complex plane. If such suitable gains are selected, there exists a symmetric positive-definite matrix $\boldsymbol{P}$ to satisfy the following matrix Lyapunov equation for any arbitrary symmetric positive-definite matrix $\boldsymbol{Q}$:

$$\boldsymbol{A}^T\boldsymbol{P} + \boldsymbol{P}\boldsymbol{A} = -\boldsymbol{Q}. \tag{6.28}$$

To stabilize the nonlinear part, the corrective term $\boldsymbol{\delta}_a$ would be designed to satisfy the Lyapunov theory for the error dynamics. Consider the following positive definite Lyapunov function for this analysis:

$$V = \boldsymbol{\varepsilon}^T\boldsymbol{P}\boldsymbol{\varepsilon}, \tag{6.29}$$

in which matrix $\boldsymbol{P}$ is the symmetric positive-definite function derived from matrix Lyapunov equation given in Equation 6.28 for the linear part. To stabilize the error dynamics, $\boldsymbol{\delta}_a$ would be designed such that the derivative of the Lyapunov function along any trajectory becomes negative definite. Compute this time derivative as

$$\begin{aligned}\dot{V} &= \dot{\boldsymbol{\varepsilon}}^T\boldsymbol{P}\boldsymbol{\varepsilon} + \boldsymbol{\varepsilon}^T\boldsymbol{P}\dot{\boldsymbol{\varepsilon}} \\ &= \boldsymbol{\varepsilon}^T\left(\boldsymbol{A}^T\boldsymbol{P} + \boldsymbol{P}\boldsymbol{A}\right)\boldsymbol{\varepsilon} + 2\boldsymbol{\varepsilon}^T\boldsymbol{P}\boldsymbol{B}(\boldsymbol{\delta}_a + \boldsymbol{\eta}) \\ &= -\boldsymbol{\varepsilon}^T\boldsymbol{Q}\boldsymbol{\varepsilon} + 2\boldsymbol{\varepsilon}^T\boldsymbol{P}\boldsymbol{B}(\boldsymbol{\delta}_a + \boldsymbol{\eta}).\end{aligned} \tag{6.30}$$

Note that in the first equation the closed-loop dynamic equation 6.26 is substituted, and in the last equation the matrix Lyapunov function 6.28 is used to simplify the first term, which is clearly negative definite. However, the second term in the Equation 6.30 may become positive and destabilize the error dynamics owing to the modeling uncertainty $\boldsymbol{\eta}$. To design the corrective term $\boldsymbol{\delta}_a$, assume that an upper bound on the norm of uncertainty term η is computable. This upper bound is generally a function of the error variables and time. Assume that this upper bound is given as follows:

$$\|\boldsymbol{\eta}\| \leq \rho(\boldsymbol{\varepsilon}, t). \tag{6.31}$$

Furthermore, for the sake of simplicity, denote $\boldsymbol{B}^T\boldsymbol{P}\boldsymbol{\varepsilon}$ by a vector $\boldsymbol{v}$:

$$\boldsymbol{v} = \boldsymbol{B}^T\boldsymbol{P}\boldsymbol{\varepsilon}. \tag{6.32}$$

Using this variable, the second term in the derivative of Lyapunov function 6.30, may be written as $\boldsymbol{v}^T(\boldsymbol{\delta}_a + \boldsymbol{\eta})$. This term vanishes if $\boldsymbol{v} = \boldsymbol{0}$, and may become negative definite for $\boldsymbol{v} \neq \boldsymbol{0}$, if the corrective term is chosen as follows:

$$\boldsymbol{\delta}_a = -\rho\frac{\boldsymbol{v}}{\|\boldsymbol{v}\|}. \tag{6.33}$$

This claim can be proved using the Schwartz inequality as follows:

$$\begin{aligned} v^T(\delta_a + \eta) &= v^T(-\rho\frac{v}{\|v\|} + \eta) \\ &\leq -\rho\|v\| + \|v\|\|\eta\| \\ &= \|v\|(-\rho + \|\eta\|) \\ &\leq 0, \end{aligned} \tag{6.34}$$

since as given in Equation 6.31, $\|\eta\| \leq \rho$. Therefore,

$$\dot{V} \leq -\varepsilon^T Q \varepsilon < 0. \tag{6.35}$$

This calculation confirms that the corrective term given in Equation 6.33 is capable of overcoming the destabilizing effect of the worst-case uncertainty η, represented by its upper bound ρ in Equation 6.31.

To summarize the robust IDC approach, consider the parallel robot model given in Equation 6.1, in which the control effort is calculated by Equations 6.19 and 6.20. This control effort consists of the usual inverse dynamics control law in addition to a corrective term δ_a given by

$$\delta_a = \begin{cases} -\rho\dfrac{v}{\|v\|} & \text{if } \|v\| \neq 0 \\ 0 & \text{if } \|v\| = 0, \end{cases} \tag{6.36}$$

in which as defined in Equation 6.32, $v = B^T P\varepsilon$, P is the symmetric positive-definite matrix found from the solution of matrix Lyapunov Equation 6.28, ε denotes the augmented motion error defined in Equation 6.25, and ρ denotes the least upper bound of the modeling uncertainty, $\|\eta\| \leq \rho$. Figure 6.8 illustrates the robust IDC implementation block diagram on a parallel robot. As seen in this figure, the corrective term is added as a wrench in task space, and the actuator torques are found by force distribution block.

Note that the corrective term is a discontinuous term, since it is found from a multivalued function represented by Equation 6.36. Different variations of this type of robust control scheme are applied in the literature, which are usually referred to as *variable structure control*. This term indicates the switching structure of the control effort, which depends on the sign of input variable $\|v\|$.

Note that in fact the corrective term is a high-gain valued wrench with a negative sign if v is positive, or with a positive sign if v is negative, and zero if the tracking error is

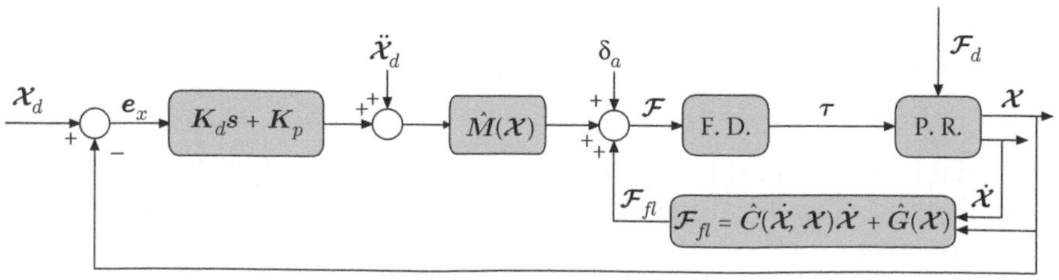

FIGURE 6.8
General configuration of robust inverse dynamics control implemented in the task space.

zero, $v = 0$. The high-gain value is determined according to the norm of uncertainty $\|\eta\|$. If accurate value of the uncertainty measure at different configurations is plausible to be determined, the value of this high gain can be adjusted correspondingly. Otherwise, a high gain larger than the least upper bound of the uncertainty measure would be used in practice. In any case, a high-gain switching term acts on the closed-loop system to guarantee convergence of the tracking error toward zero for all desired configurations, and for the worst-case modeling uncertainty of the system. Even if a norm-bounded disturbance wrench is applied to the manipulator, as depicted in Figure 6.8, its upper bound can be used to adjust for the high-gain value of the corrective term, and to compensate for its effect on the tracking error. By this means, and by the Lyapunov stability analysis performed on the closed system it is ensured that asymptotic tracking is achieved in theory.

Although, the high-gain switching term proves to be very efficient to overcome the modeling uncertainty and disturbance wrench on the tracking performance, it usually results in a phenomenon called *chattering*. Chattering is a high-oscillation behavior observed at the outputs of the system, as a result of nonideal implementation of any discontinuous control function. Note that to implement a discontinuous control function, the sign of input argument v must be determined at each instance. Sign detection of any variable, when its amplitude is relatively low, becomes inaccurate in the presence of measurement noises. In fact, the measurement noise highly contaminates low-amplitude inputs and induces many false zero crossings. This fact is illustrated in Figure 6.9, in which a monotonically decreasing variable contaminated with noise induces several false sign changes. These false sign changes are not critical when they are not fed back through a high-gain controller; however, in a variable structure controller with a high gain, it introduces unpleasant oscillation at the output.

To remove chattering, one may implement a continuous approximation to the discontinuous control as

$$\delta_a = \begin{cases} -\rho \dfrac{v}{\|v\|} & \text{if } \|v\| > \epsilon \\ -\rho \dfrac{v}{\epsilon} & \text{if } \|v\| \leq \epsilon \end{cases} \tag{6.37}$$

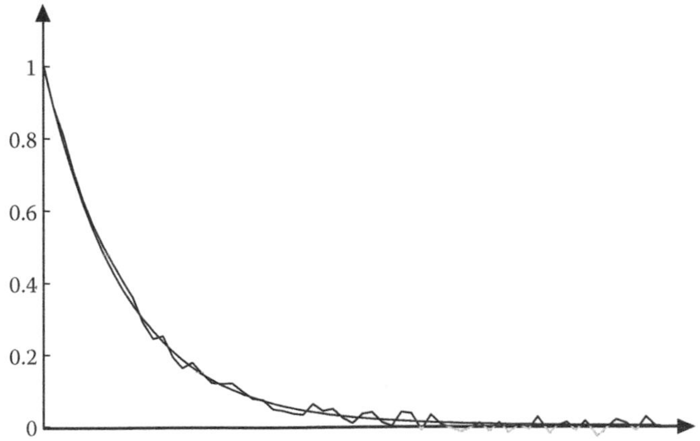

FIGURE 6.9
A monotonically decreasing function contaminated with measurement noise may introduce chattering by high-gain switching controller.

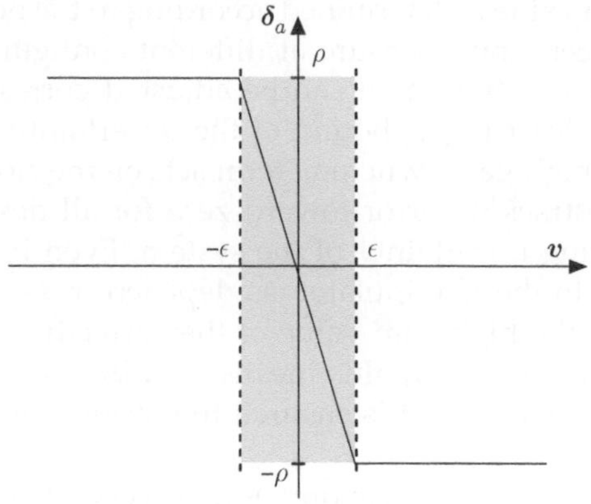

FIGURE 6.10
A threshold ϵ is used to continuously approximate the switching control.

in which ϵ is a threshold width on the v variable. As shown in Figure 6.10, while norm of v lies in the threshold of ϵ, that is, $\|v\| \leq \epsilon$, the switching controller is replaced by a linear control law. If ϵ is suitably chosen to be larger than the measurement noise amplitude, this approximation significantly reduces the output chattering. Although such modification is very effective in practice, the stability analysis performed on the closed-loop system is not valid any more, and this slight change in the control effort may change asymptotic tracking performance.

To analyze stability of the closed-loop system using the modified control law 6.37, consider the same Lyapunov function given in 6.29, and take the derivative along an arbitrary motion trajectory. For $\|v\| \geq \epsilon$, the same argument holds for $\dot{V}$ as before, and we may conclude $\dot{V} < 0$. However, for $\|v\| \leq \epsilon$, the Lyapunov function derivative along any motion trajectory yields

$$\begin{aligned} \dot{V} &\leq -\boldsymbol{\varepsilon}^T Q \boldsymbol{\varepsilon} + 2v^T(\boldsymbol{\delta}_a + \boldsymbol{\eta}) \\ &\leq -\boldsymbol{\varepsilon}^T Q \boldsymbol{\varepsilon} + 2v^T\left(-\rho\frac{v}{\epsilon} + \rho\frac{v}{\|v\|}\right) \\ &\leq -\boldsymbol{\varepsilon}^T Q \boldsymbol{\varepsilon} - 2\frac{\rho}{\epsilon}\|v\|^2 + 2\rho\|v\|. \end{aligned} \tag{6.38}$$

The last two terms in the expression 6.38 attain its maximum value by a positive constant $\frac{1}{2}\epsilon\rho$, when $\|v\| = \epsilon/2$. Thus, Equation 6.38 simplifies to

$$\dot{V} \leq -\boldsymbol{\varepsilon}^T Q \boldsymbol{\varepsilon} + \frac{1}{2}\epsilon\rho < 0 \tag{6.39}$$

provided

$$\boldsymbol{\varepsilon}^T Q \boldsymbol{\varepsilon} > \frac{1}{2}\epsilon\rho. \tag{6.40}$$

Note that, $\boldsymbol{\varepsilon}^T Q \boldsymbol{\varepsilon}$ has the following lower and upper bounds:

$$\underline{\lambda}(Q)\|\boldsymbol{\varepsilon}\| \leq \frac{1}{2}\epsilon\rho \leq \bar{\lambda}(Q)\|\boldsymbol{\varepsilon}\| \tag{6.41}$$

in which $\underline{\lambda}(Q)$ and $\bar{\lambda}(Q)$ denote the minimum and maximum eigenvalues of matrix Q, respectively. Therefore, negative definiteness of $\dot{V}$ is guaranteed with a level of conservatism, provided

$$\underline{\lambda}(Q)\|\boldsymbol{\varepsilon}\| > \frac{1}{2}\epsilon\rho \tag{6.42}$$

or, equivalently

$$\|\boldsymbol{\varepsilon}\| > \left(\frac{\epsilon\rho}{2\underline{\lambda}(Q)}\right)^{\frac{1}{2}} \doteq \delta_e. \tag{6.43}$$

This analysis implies that if the norm of the tracking error is larger than B_δ, which denotes the ball with radius δ_e, the derivative of the Lyapunov function is a negative definite. This guarantees that the tracking error decays toward the origin, and therefore all trajectories eventually enter the ball B_r, which is usually a ball greater than B_δ. However, asymptotic convergence is not guaranteed, since the Lyapunov function is not globally negative definite. This means that the tracking error remains *uniformly ultimately bounded* (UUB).* Note that to determine the radius of ultimate steady-state tracking error it is necessary to find the level sets in which $c_1 < V < c_2$, and to determine the smallest ball containing the level set c_1. However, as seen in Equation 6.43, the radius of the ultimate steady-state tracking error is proportional to the product of the uncertainty bound ρ, and the threshold ϵ. This implies that to decrease the output chattering, one may increase the threshold ϵ. However, this directly results in larger values of the steady-state tracking error.

6.4.2 Adaptive Inverse Dynamics Control

Another important extension of inverse dynamics control is to consider full feedback linearization; however, first use rough estimates of the dynamic formulation matrices, and further update these estimates in an adaptive scheme. Contrary to the robust control approach represented in Section 6.4.1, in the adaptive approach, the estimates of the dynamic formulation matrices have been updated such that the difference between the true values of these matrices to that of their estimates converge to zero. By this means, the error dynamics of the manipulator converges to the second-order system described by Equation 6.8.

Note that although the goal of both robust and adaptive control techniques is the same, the way this goal is accomplished is different. In a robust controller, a fixed controller is designed to satisfy the control objectives for the worst possible case of modeling uncertainties and disturbance wrenches, whereas in an adaptive controller, online parameter identification is carried out to tune the controller parameters. Therefore, in the robust approach, usually a high-gain controller is designed to compensate for such uncertainties in the worst case, whereas in the adaptive control the structure of the model is considered to be known, and only the parameters are subject to online identification. Fortunately, as shown in Section 5.5.4.2 by finding a minimal set of parameters $\boldsymbol{\theta}$, the dynamic formulation of parallel manipulators may be written in a linear regression model represented by Equation 5.274.

* See Appendix C for formal definition of UUB.

To develop the adaptive IDC, recall the general closed-form dynamic formulation of the manipulator given in Equation 6.1, and the inverse dynamics control input $\mathcal{F}$ as

$$\mathcal{F} = \hat{M}(\mathcal{X})a + \hat{C}(\mathcal{X}, \dot{\mathcal{X}})\dot{\mathcal{X}} + \hat{G}(\mathcal{X}), \tag{6.44}$$

$$a = \ddot{\mathcal{X}}_d - K_d\dot{\tilde{\mathcal{X}}} - K_p\tilde{\mathcal{X}}, \tag{6.45}$$

in which a is the usual inverse dynamics control input, while the model estimates would be adapted by a suitable adaption law the details of which are given as follows. Note that as defined hereinbefore the notion $\hat{(\cdot)}$ represents the estimated value of $(\cdot)$ and the notion $\tilde{(\cdot)}$ represents the error or mismatch between the estimated value and the true value of the corresponding term as $\tilde{(\cdot)} = \hat{(\cdot)} - (\cdot)$. Hence

$$\tilde{M} = \hat{M} - M; \quad \tilde{C} = \hat{C} - C; \quad \tilde{G} = \hat{G} - G,$$

while in a similar manner, $\tilde{(\cdot)}$ notation may be applied to the motion variables as

$$\tilde{\mathcal{X}} = \mathcal{X} - \mathcal{X}_d, \quad \dot{\tilde{\mathcal{X}}} = \dot{\mathcal{X}} - \dot{\mathcal{X}}_d.$$

Let us examine the closed-loop dynamic formulation of the manipulator given in Equation 6.1, and substitute the adaptive IDC control effort $\mathcal{F}$ given in Equation 6.44.

$$\begin{aligned} M\ddot{\mathcal{X}} + C\dot{\mathcal{X}} + G &= \mathcal{F} \\ &= \hat{M}\left(\ddot{\mathcal{X}}_d - K_d\dot{\tilde{\mathcal{X}}} - K_p\tilde{\mathcal{X}}\right) + \hat{C}\dot{\mathcal{X}} + \hat{G}. \end{aligned}$$

Add and subtract $\ddot{\mathcal{X}}$ into the parenthesis, and simplify to

$$\tilde{M}\ddot{\mathcal{X}} + \tilde{C}\dot{\mathcal{X}} + \tilde{G} = \hat{M}\left(\ddot{\tilde{\mathcal{X}}} - K_d\dot{\tilde{\mathcal{X}}} - K_p\tilde{\mathcal{X}}\right).$$

Therefore,

$$\begin{aligned} \ddot{\tilde{\mathcal{X}}} - K_d\dot{\tilde{\mathcal{X}}} - K_p\tilde{\mathcal{X}} &= \hat{M}^{-1}\left(\tilde{M}\ddot{\mathcal{X}} + \tilde{C}\dot{\mathcal{X}} + \tilde{G}\right) \\ &= \hat{M}^{-1}Y(\mathcal{X}, \dot{\mathcal{X}}, \ddot{\mathcal{X}})\tilde{\theta}. \end{aligned} \tag{6.46}$$

Note that the linear regression model $Y(\mathcal{X}, \dot{\mathcal{X}}, \ddot{\mathcal{X}})$ represented in Equation 5.274 is used to simplify the last equation. The closed-loop dynamics may be written in a state space form by the previously defined state vector:

$$\varepsilon = \begin{bmatrix} \tilde{\mathcal{X}} \\ \dot{\tilde{\mathcal{X}}} \end{bmatrix} = \begin{bmatrix} \mathcal{X} - \mathcal{X}_d \\ \dot{\mathcal{X}} - \dot{\mathcal{X}}_d \end{bmatrix}. \tag{6.47}$$

Therefore, the closed-form dynamic formulation given in Equation 6.46 may be written in a state space form as follows:

$$\dot{\varepsilon} = A\varepsilon + B\Phi\tilde{\theta}, \tag{6.48}$$

in which

$$A = \begin{bmatrix} 0 & I \\ -K_p & -K_d \end{bmatrix}, \quad B = \begin{bmatrix} 0 \\ I \end{bmatrix}, \quad \Phi = \hat{M}^{-1} Y(\mathcal{X}, \dot{\mathcal{X}}, \ddot{\mathcal{X}}). \tag{6.49}$$

To have a stable error dynamics, the PD controller gains must be chosen such that the linear part of the state space representation remains stable. Moreover, the adaptation law in the nonlinear part of control action, namely $\boldsymbol{B\Phi\tilde{\theta}}$, would be designed so as the total tracking error remains asymptotically stable.

To stabilize the linear part, it is sufficient to choose the controller gain matrices $\boldsymbol{K}_p$ and K_d such that the matrix $\boldsymbol{A}$ becomes Hurwitz, that is, all its eigenvalues lie in open left-half of the complex plane. If such suitable gains are selected, there exists a symmetric positive definite matrix $\boldsymbol{P}$ to satisfy the following matrix Lyapunov equation for any arbitrary symmetric positive definite matrix $\boldsymbol{Q}$:

$$A^T P + PA = -Q. \tag{6.50}$$

To stabilize the tracking error, the adaptation law would be designed to satisfy the Lyapunov theorem for the error dynamics. Consider the following positive definite Lyapunov function for this analysis:

$$V = \boldsymbol{\varepsilon}^T \boldsymbol{P}\boldsymbol{\varepsilon} + \tilde{\boldsymbol{\theta}}^T \boldsymbol{\Gamma} \tilde{\boldsymbol{\theta}}, \tag{6.51}$$

in which matrix $\boldsymbol{P}$ is the symmetric positive definite function derived from matrix Lyapunov equation given in Equation 6.50 for the linear part, and matrix $\boldsymbol{\Gamma}$ is another constant, symmetric, positive definite matrix defined for the adaptation law. To stabilize the error dynamics, the adaptation law would be designed such that the derivative of the Lyapunov function along any arbitrary trajectory becomes negative definite. Compute this derivative as

$$\begin{aligned} \dot{V} &= \boldsymbol{\varepsilon}^T \left(A^T P + PA\right) \boldsymbol{\varepsilon} + 2\tilde{\boldsymbol{\theta}}^T \boldsymbol{\Phi}^T \boldsymbol{B}^T \boldsymbol{P}\boldsymbol{\varepsilon} + 2\tilde{\boldsymbol{\theta}}^T \boldsymbol{\Gamma}\dot{\tilde{\boldsymbol{\theta}}} \\ &= -\boldsymbol{\varepsilon}^T \boldsymbol{Q}\boldsymbol{\varepsilon} + 2\tilde{\boldsymbol{\theta}}^T \left(\boldsymbol{\Phi}^T \boldsymbol{B}^T \boldsymbol{P}\boldsymbol{\varepsilon} + \boldsymbol{\Gamma}\dot{\tilde{\boldsymbol{\theta}}}\right). \end{aligned} \tag{6.52}$$

Note that in the first equation, the closed-loop dynamic Equation 6.48 is substituted, and in the last equation, matrix Lyapunov function 6.50 is used to simplify the first term, which is clearly negative definite. However, if the adaptation law is not suitably designed, the second term in the last equation may become positive and destabilize the error dynamics. A suitable adaptation law may be designed such that the second term vanishes. This can be accomplished by the following adaptation law, noting that $\dot{\tilde{\boldsymbol{\theta}}} = \dot{\hat{\boldsymbol{\theta}}}$:

$$\dot{\hat{\boldsymbol{\theta}}} = -\boldsymbol{\Gamma}^{-1} \boldsymbol{\Phi}^T \boldsymbol{B}^T \boldsymbol{P}\boldsymbol{\varepsilon}. \tag{6.53}$$

Using this adaptation law, the second term in the derivative of the Lyapunov function 6.52 vanishes and results in

$$\dot{V} = -\boldsymbol{\varepsilon}^T \boldsymbol{Q}\boldsymbol{\varepsilon} < 0. \tag{6.54}$$

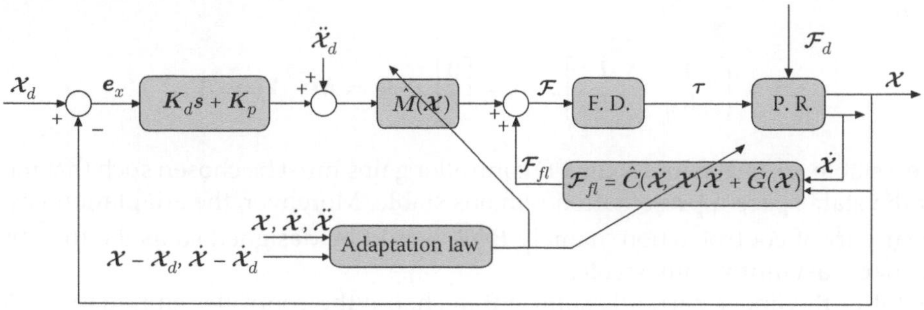

FIGURE 6.11
General configuration of adaptive inverse dynamics control implemented in task space.

This calculation confirms that the closed-loop system achieves asymptotic tracking performance, while the parameter estimation errors remain bounded.

Figure 6.11 illustrates general implementation of adaptive inverse dynamics control in task space. As illustrated in this figure, the implementation is very similar to a regular inverse dynamics control, while the dynamic formulation matrices are adapted through the adaptation law given in Equation 6.53. To implement this adaptive law, however, note that $\ddot{\mathcal{X}}$ must be given and $\hat{M}$ must be invertible for the calculation of the matrix ϕ in the adaptation law. The need for motion acceleration in the adaptation law is a severe shortcoming for the implementation. One may consider using $\ddot{\mathcal{X}}_d$ instead of $\ddot{\mathcal{X}}$ in the implementations, since if the controller performance is good enough, there is not much difference between these two variables in practice. However, even small mismatch between these two variables needs to be examined in the Lyapunov stability analysis performed hereinbefore, to ensure that the parameter estimation errors are bounded. There are several approaches proposed in the literature to remedy this shortcoming, while the passivity-based approaches [168] are a suitable alternative to remove both these implementation barriers.

6.5 Motion Control in Joint Space

Although the motion control schemes developed in Section 6.3 are very effective in terms of achieving asymptotic tracking performance, they suffer from an implementation constraint that the motion variable $\mathcal{X}$ must be measured in practice. If this measurement is available without any doubt, such topologies are among the best routines to be implemented in practice. However, as explained in Section 6.2, in many practical situations measurement of the motion variable $\mathcal{X}$ is difficult or expensive, and usually just the active joint variables q are measured. In such cases, the controllers developed in the joint space may be recommended for practical implementation.

To generate a direct input to output relation in the joint space, consider the topology depicted in Figure 6.3. In this topology, the controller input is the joint variable error vector $e_q = q_d - q$, and the controller output is directly the actuator force vector τ, and hence there exists a one-to-one correspondence between the controller input to its output. However, the general form of dynamic formulation of parallel robot is usually given in the task space as in Equation 6.1. While what is suitable for the tracking error analysis in the joint space is a dynamic formulation which is given in the joint space, by which the actuator forces τ are directly related to the active joint variables q. In what follows, first the dynamic formulation of parallel robots is transformed into the joint space, and then a number of control schemes in the joint space would be developed.

6.5.1 Dynamic Formulation in the Joint Space

The relation between the task space variables to their counterparts in the joint space can be derived by forward and inverse kinematic relations. In forward kinematic analysis, it is assumed that the motion variable $\mathcal{X}$ is given and the joint variable q needs to be determined, while in the inverse kinematic analysis, the motion variable $\mathcal{X}$ is determined based on the knowledge of variable q. Although both analyses involve solution to a set of nonlinear equations, for parallel manipulators, inverse kinematic solution proves to be much easier to obtain than that of forward kinematic solution. This relation in differential kinematics is much simpler and can be completely determined by the Jacobian matrix. The Jacobian matrix relates $\dot{\mathcal{X}}$ to $\dot{q}$ in a direct matrix relation as defined in Equation 4.34, which is given in the following equation to be more accessible:

$$\dot{q} = J\dot{\mathcal{X}}. \tag{6.55}$$

Furthermore, the relation between the actuator force vector τ to the corresponding task space wrench is given in Equation 4.92, which is also given here for convenience:

$$\mathcal{F} = J^T \tau. \tag{6.56}$$

These equations show that Jacobian matrix not only reveals the relation between the joint variable velocities $\dot{q}$ and the moving platform velocities $\dot{\mathcal{X}}$, it constructs the transformation needed to find the actuator forces τ from the wrench acting on the moving platform $\mathcal{F}$. By using these relations, the dynamic formulation of parallel robot given in the task space by Equation 6.1 can be transformed into joint space formulations.

This transformation is valid for a completely parallel manipulator in which the number of actuators are equal to the number of manipulator degrees-of-freedom. In such manipulators, the Jacobian matrix is squared and in nonsingular configurations, it is invertible. Considering these assumptions, the velocity and acceleration variables in the task space can be determined by the following equations:

$$\dot{q} = J\dot{\mathcal{X}} \rightarrow \dot{\mathcal{X}} = J^{-1}\dot{q} \tag{6.57}$$

$$\ddot{q} = \dot{J}\dot{\mathcal{X}} + J\ddot{\mathcal{X}} \rightarrow \ddot{\mathcal{X}} = J^{-1}\ddot{q} - J^{-1}\dot{J}\dot{\mathcal{X}}. \tag{6.58}$$

Substituting $\dot{\mathcal{X}}$ and $\ddot{\mathcal{X}}$ from the above equation into the dynamic formulation of the robot 6.1, while considering that an external disturbance wrench is applied to the moving platform. This results in

$$M\left(J^{-1}\ddot{q} - J^{-1}\dot{J}\dot{\mathcal{X}}\right) + CJ^{-1}\dot{q} + G + \mathcal{F}_d = J^T\tau.$$

Multiply J^{-T} from left to both the sides and simplify as

$$\left(J^{-T}MJ^{-1}\right)\ddot{q} + J^{-T}\left(C - MJ^{-1}\dot{J}\right)J^{-1}\dot{q} + J^{-T}G + J^{-T}\mathcal{F}_d = \tau.$$

Therefore, the dynamic formulation of a parallel robot in the joint space can be represented by the following equation:

$$M_q\ddot{q} + C_q\dot{q} + G_q + \tau_d = \tau, \quad (6.59)$$

in which τ denotes the actuator force vector and

$$M_q = J^{-T}MJ^{-1}, \quad (6.60)$$

$$C_q = J^{-T}\left(C - MJ^{-1}\dot{J}\right)J^{-1}, \quad (6.61)$$

$$G_q = J^{-T}G, \quad (6.62)$$

$$\tau_d = J^{-T}\mathcal{F}_d. \quad (6.63)$$

Equation 6.59 represents the closed form of dynamic formulation of a general parallel robot in the joint space. Note that although in this formulation the dynamic matrices are found in a closed form, they are not represented explicitly in terms of the joint variable vector q. In fact, to fully derive these matrices, the Jacobian matrix J, its inverse J^{-1}, its transpose inverse J^{-T}, and its time derivative $\dot{J}$ are required. Whereas the Jacobian matrix is usually derived as a function of the motion variable $\mathcal{X}$. Furthermore, the main dynamic formulation matrices are all functions of the motion variable $\mathcal{X}$. Hence, in practice, to find the dynamic matrices represented in the joint space, forward kinematics should be solved to find the motion variable $\mathcal{X}$ for any given joint motion vector q.

Since in parallel robots the forward kinematic analysis is a computationally intensive algorithm, there exist inherent difficulties to find the dynamic matrices in the joint space as an explicit function of q. This is the main reason why the control schemes developed for parallel robots are mainly in task space. In fact, if it is possible to solve forward kinematics of parallel manipulators in an online manner, it is recommended to use the control topology depicted in Figure 6.2, and implement control laws designed in the task space.

However, one implementable alternative to calculate the dynamic matrices represented in the joint space is though to use the desired motion trajectory $\mathcal{X}_d$ instead of the true values of motion vector $\mathcal{X}$ in the calculations. This approximation significantly reduces the computational cost, with the penalty of having mismatch between the estimated values of these matrices to their true values. Robust and adaptive approaches can then be used to treat the modeling uncertainty of the robot in a systematic way to compensate for these modeling uncertainties, and to achieve asymptotic tracking. In what follows, a number of control schemes based on the dynamic formulations of the robot represented in the joint space by Equation 6.59 are developed.

6.5.2 Decentralized PD Control

The first control strategy introduced in the joint space consists of the simplest form of feedback control in such manipulators. In this control structure, depicted in Figure 6.12, a number of disjoint linear (PD) controllers are used in a feedback structure on each error component. Let us denote the tracking error vector in the joint space as e_q. The decentralized controller, therefore, consists of n disjoint linear controllers such as proportional derivative (PD) acting on each error component, in which n denotes the number of actuators in a fully parallel manipulator. The PD controller is denoted by $K_d s + K_p$ block in

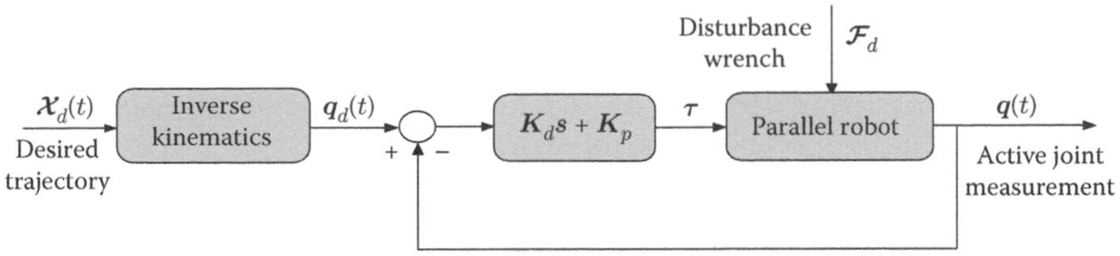

FIGURE 6.12
Decentralized PD controller implemented in joint space.

Figure 6.12, in which K_d and K_p are $n \times n$ diagonal matrices denoting the derivative and proportional controller gains, respectively. The diagonal elements of these matrices are the disjoint controller gains for each error component. Furthermore, s denotes the Laplace variable indicating the differentiation action in the time domain.

Hence, by this structure, each tracking error component is treated separately by its disjoint PD controller. The output of the controller is denoted by $\boldsymbol{\tau}$, which is the control effort needed to be applied by the actuators in the joint space. Note that since the output of the controller is defined in the joint space, each actuator force component directly manipulates the corresponding tracking error component, and therefore, the overall tracking performance of the manipulator is suitable if high controller gains are considered.

The proposed decentralized PD controller is very simple in structure, and therefore very easy to be implemented on the manipulator. The design of such a controller needs no detailed information on the manipulator dynamic formulation, nor manipulator dynamic parameters. The computation complexity of its implementation is also very low. However, the tracking performance of such a controller is relatively poor, and static tracking errors might be unavoidable. On the other hand, owing to high-gain controller design, the required actuation energy is high, while the performance of the closed-loop system is configuration dependent.

The controller gains must be tuned experimentally based on physical realization of the controller in practice by trial and error. Therefore, the final controller gains are obtained as a trade-off between the transient response and the steady-state errors at different configurations. As seen in Section 6.5.1, the dynamic formulation of parallel manipulator verifies nonlinear behavior of the system, which is configuration dependent. Therefore, finding suitable controller gains to result in required performance in all configurations is a difficult task. Usually the designer chooses the controller gains such that the required tracking error of the manipulator at its usual working condition is satisfied. This results in varying performance error in other configurations wherein the tuning process is not applied to. The performance of the controller to attenuate the measurement noise and external disturbance wrenches applied to the manipulator are also poor in practice.

To remedy these shortcomings, some modifications have been proposed to this structure. Two most important improvements have been given in the following two sections. However, notice that the controller described in this section may be recommended for the first trial in terms of closed-loop experiments where the implementation is in the joint space.

6.5.3 Feed Forward Control

As seen in Section 6.5.2, the tracking performance of the simple PD controller implemented in the joint space is not sufficient at different configurations, and in the presence of measurement noise and external disturbances applied to the manipulator. To compensate for such effects, a feed forward actuator force denoted by $\boldsymbol{\tau}_{ff}$ may be added to the structure of the controller as depicted in Figure 6.13. This term is generated through the mapping of a feed forward wrench calculated from the dynamic formulation of the manipulator in the task space 6.1, into the joint space by the following equation:

$$\begin{aligned}\boldsymbol{\tau}_{ff} &= \boldsymbol{J}^{-T}\boldsymbol{\mathcal{F}}_{ff} \\ &= \boldsymbol{J}^{-T}\left(\hat{\boldsymbol{M}}(\boldsymbol{\mathcal{X}}_d)\ddot{\boldsymbol{\mathcal{X}}}_d + \hat{\boldsymbol{C}}(\boldsymbol{\mathcal{X}}_d, \dot{\boldsymbol{\mathcal{X}}}_d)\dot{\boldsymbol{\mathcal{X}}}_d + \hat{\boldsymbol{G}}(\boldsymbol{\mathcal{X}}_d)\right). \end{aligned} \tag{6.64}$$

As shown in Figure 6.13, the desired trajectory in the task space $\boldsymbol{\mathcal{X}}_d$ and its derivatives $\dot{\boldsymbol{\mathcal{X}}}_d, \ddot{\boldsymbol{\mathcal{X}}}_d$ are the required inputs to generate the feed forward wrench. Furthermore, the feed forward actuator forces are generated by the required mapping of the wrench from the task space into the actuator force in the joint space through $\boldsymbol{J}^{-T}(\boldsymbol{\mathcal{X}}_d)$. This mapping is valid for fully parallel manipulators at nonsingular configurations.

In such a case, the number of actuators is equal to the number of degrees-of-freedom; hence, the Jacobian matrix is squared and it is invertible at nonsingular configurations. The corrective term in this control scheme is calculated in feed forward, since no online information of the output motion of the robot $\boldsymbol{q}$ is needed to generate $\boldsymbol{\tau}_{ff}$. Furthermore, note that to generate this term, the dynamic formulation of the robot, and its kinematic and dynamic parameters are needed. In practice, exact knowledge of dynamic matrices are not available, and therefore, estimates of these matrices are used in this derivation, in which $\hat{\boldsymbol{M}}, \hat{\boldsymbol{C}}$, and $\hat{\boldsymbol{G}}$ denote, respectively, an approximate estimate of the manipulator mass matrix, Coriolis and centrifugal matrix, and the gravity vector represented in the task space.

The information required to generate the feed forward actuator force $\boldsymbol{\tau}_{ff}$ is usually available beforehand, and in such a case, the feed forward term corresponding to a given desired trajectory can be determined off-line, while the computation of the decentralized feedback term would be executed online. Write the closed-loop dynamic formulation for the manipulator in the joint space as follows:

$$\begin{aligned}\boldsymbol{M}_q\ddot{\boldsymbol{q}} + \boldsymbol{C}_q\dot{\boldsymbol{q}} + \boldsymbol{G}_q + \boldsymbol{\tau}_d &= \boldsymbol{\tau}_{pd} + \boldsymbol{\tau}_{ff} \\ &= \boldsymbol{K}_d\dot{\boldsymbol{e}}_q + \boldsymbol{K}_p\boldsymbol{e}_q + \boldsymbol{J}^{-T}\left(\hat{\boldsymbol{M}}\ddot{\boldsymbol{\mathcal{X}}}_d + \hat{\boldsymbol{C}}\dot{\boldsymbol{\mathcal{X}}}_d + \hat{\boldsymbol{G}}\right). \end{aligned} \tag{6.65}$$

If complete information of the dynamic matrices is available, then we may assume that $\hat{\boldsymbol{M}} = \boldsymbol{M}, \hat{\boldsymbol{C}} = \boldsymbol{C}$, and $\hat{\boldsymbol{G}} = \boldsymbol{G}$. Furthermore, consider the controller of the system is performing well and the moving platform is suitably tracking the desired trajectory; therefore, assume $\boldsymbol{\mathcal{X}}(t) \simeq \boldsymbol{\mathcal{X}}_d(t)$, $\dot{\boldsymbol{\mathcal{X}}}(t) \simeq \dot{\boldsymbol{\mathcal{X}}}_d(t)$. In such a case, the closed-loop dynamic formulation

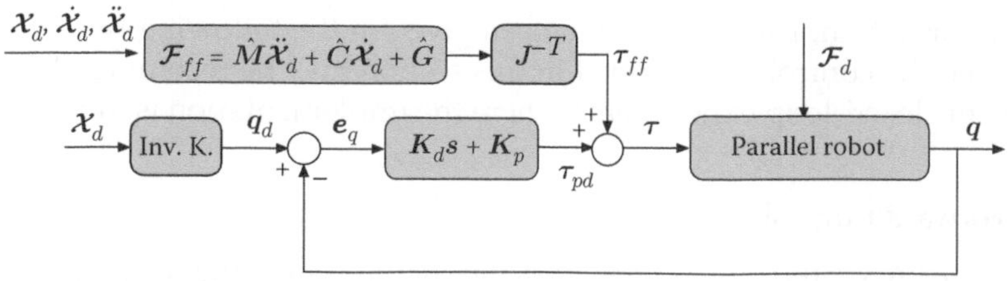

FIGURE 6.13
Feed forward actuator force added to the decentralized PD controller in joint space.

may be simplified to the following equation, considering Equation 6.59:

$$\begin{aligned} M_q(\ddot{q}_d - \ddot{q}) + K_d\dot{e}_q + K_p e_q - \tau_d &= 0 \\ M_q\ddot{e}_q + K_d\dot{e}_q + K_p e_q &= \tau_d. \end{aligned} \tag{6.66}$$

This equation implies that if the above-mentioned restricting assumptions hold, the error dynamics satisfy a set of second-order differential equations in the presence of disturbance. Therefore, by choosing appropriate gains for the PD controller, the transient and steady-state performance of the tracking error can be suitably designed. Note that if the disturbance wrench applied to the manipulator is zero, or it is converging to zero, the steady-state error of the closed-loop system tends to zero for all the tracking error components. Furthermore, note that except for the mass matrix, the error dynamics terms are all configuration independent, and therefore, it is much easier to tune the PD controller gains to work well in the whole workspace of the robot in such a structure.

However, this method suffers from a number of limitations in practice. The most important limitation of this control technique is the stringent assumption of the complete information requirement of dynamic matrices. In practice, derivation of these matrices is a prohibitive task, while accessing to their complete information is practically infeasible. Therefore, in practice, Equation 6.65 is representing the error dynamics in a better way. In such a case, the difference between the true values of dynamic matrices and their estimated values may be considered as an extra error torque vector applied to the error dynamics, and therefore, static errors would be seen in the steady-state behavior of the closed-loop system. Furthermore, even in case of complete knowledge of dynamic matrices, the true motion of the moving platform is in practice different to that of the desired trajectory, that is, $\mathcal{X}(t) \neq \mathcal{X}_d(t)$ and $\dot{\mathcal{X}}(t) \neq \dot{\mathcal{X}}_d(t)$. This may be seen as a more significant error term applied to the error dynamics, which influences the tracking error at the outset. Finally, if all the assumptions needed to generate the error dynamics Equation 6.66 hold, because of the configuration dependence of the mass matrix in this equation, the error dynamics is still not completely decoupled for all the components. This means that correction in one component may be considered as a disturbance effect to the other components. To overcome these limitations, the inverse dynamic approach has been developed and is given in the following section.

6.5.4 Inverse Dynamics Control

As seen in Section 6.5.2, the tracking performance of a decentralized PD controller implemented in the joint space is not uniform at different configurations, and in the presence of measurement noise, and external disturbances applied to the manipulator. To compensate for such effects, as explained in Section 6.5.3, a feed forward torque is added to the structure of the controller, by which the shortcomings of the decentralized controller is partially remedied. However, the closed-loop performance still faces a number of limitations, which cannot be completely remedied because of the inherent conditions on feed forward structure of that proposed controller. To overcome these limitations, in this section, a control technique based on inverse dynamic feedback of the manipulator in the joint space is presented.

In the inverse dynamics control (IDC) strategy, the nonlinear dynamics of the model is used to add a corrective term to the decentralized PD controller introduced in Section 6.5.2. By this means, the nonlinear and coupling characteristics of robotic manipulator

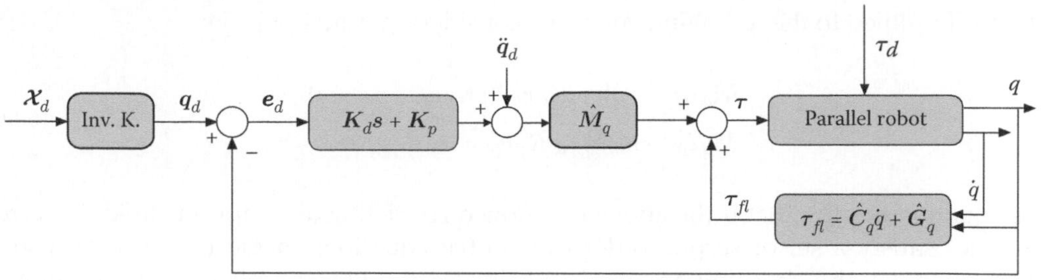

FIGURE 6.14
General configuration of inverse dynamics control implemented in joint space.

is significantly attenuated, and therefore, the performance of linear controller is significantly improved. Note that IDC is a general and effective idea developed for the control of robotic manipulators, and therefore, it is considered as the basis for many other advanced controllers developed in the joint space.

The general structure of inverse dynamics control applied to a parallel manipulator in the joint space is depicted in Figure 6.14. As seen in this figure, a corrective torque $\boldsymbol{\tau}_{fl}$ is added in a feedback structure to the closed-loop system, which is calculated from the Coriolis and centrifugal matrix, and the gravity vector of the manipulator dynamic formulation in the joint space. Furthermore, the mass matrix is acting in the forward path, in addition to the desired trajectory acceleration $\ddot{\boldsymbol{q}}_d$. Note that to generate this term, the dynamic formulation of the robot, and its kinematic and dynamic parameters are needed. In practice, exact knowledge of dynamic matrices are not available, and therefore, estimates of these matrices are used in the derivation, in which $\hat{\boldsymbol{M}}_q, \hat{\boldsymbol{C}}_q$, and $\hat{\boldsymbol{G}}_q$ denote, respectively, an approximate estimate of the manipulator mass matrix, Coriolis and centrifugal matrix, and gravity vector in the joint space.

Hence, the controller output torque applied to the manipulator may be calculated by

$$\boldsymbol{\tau} = \hat{\boldsymbol{M}}_q \boldsymbol{a}_q + \boldsymbol{\tau}_{fl},$$

which in detail reads

$$\boldsymbol{\tau} = \hat{\boldsymbol{M}}_q \boldsymbol{a}_q + \hat{\boldsymbol{C}}_q \dot{\boldsymbol{q}} + \hat{\boldsymbol{G}}_q \tag{6.67}$$

$$\boldsymbol{a}_q = \ddot{\boldsymbol{q}}_d + \boldsymbol{K}_d \dot{\boldsymbol{e}}_q + \boldsymbol{K}_p \boldsymbol{e}_q. \tag{6.68}$$

Note that in the general case of six-degrees-of-freedom manipulator, $\boldsymbol{a}_q$ is a 6×1 vector corresponding to the actuator acceleration components. Let us now write the closed-loop dynamic formulation for the manipulator as follows:

$$\begin{aligned} \boldsymbol{M}_q \ddot{\boldsymbol{q}} + \boldsymbol{C}_q \dot{\boldsymbol{q}} + \boldsymbol{G}_q &= \boldsymbol{\tau} + \boldsymbol{\tau}_d \\ &= \hat{\boldsymbol{M}}_q \boldsymbol{a}_q + \hat{\boldsymbol{C}}_q \dot{\boldsymbol{q}} + \hat{\boldsymbol{G}}_q + \boldsymbol{\tau}_d \\ &= \hat{\boldsymbol{M}}_q \left(\ddot{\boldsymbol{q}}_d + \boldsymbol{K}_p \boldsymbol{e}_q + \boldsymbol{K}_d \dot{\boldsymbol{e}}_q \right) + \hat{\boldsymbol{C}}_q \dot{\boldsymbol{q}} + \hat{\boldsymbol{G}}_q + \boldsymbol{\tau}_d. \end{aligned} \tag{6.69}$$

If the knowledge of dynamic matrices is complete, then we may assume that $\hat{\boldsymbol{M}}_q = \boldsymbol{M}_q$, $\hat{\boldsymbol{C}}_q = \boldsymbol{C}_q$, and $\hat{\boldsymbol{G}}_q = \boldsymbol{G}_q$. In such a case, simplify the closed-loop dynamic formulation to

$$\hat{\boldsymbol{M}}_q \left(\ddot{\boldsymbol{e}}_q + \boldsymbol{K}_d \dot{\boldsymbol{e}}_q + \boldsymbol{K}_p \boldsymbol{e}_q \right) + \boldsymbol{\tau}_d = \boldsymbol{0}. \tag{6.70}$$

This equation implies that if there exists complete knowledge of the dynamic matrices, the tracking error dynamic equation satisfies a set of second-order systems in the presence of disturbance. Consider the case that the manipulator is moving freely in the space and there exist no disturbance wrench applied to the manipulator. Furthermore, note that the manipulator mass matrix $\boldsymbol{M}_q$ is positive definite at all nonsingular configurations. Therefore, in such a case, the error dynamics is simplified to

$$\ddot{\boldsymbol{e}}_q + \boldsymbol{K}_d\dot{\boldsymbol{e}}_q + \boldsymbol{K}_p\boldsymbol{e}_q = \boldsymbol{0}. \tag{6.71}$$

Hence, by choosing appropriate gains for the PD controllers, the transient performance of the tracking error can be designed to have a fast transient convergence toward zero with a suitable steady-state performance. This control technique is very popular in practice because of the fact that it can significantly linearize and decouple the dynamic formulation of the closed-loop system for error dynamics components. Furthermore, the error dynamic terms are all configuration independent, and therefore, it is much easier to tune the PD controller gains to perform well in the whole workspace of the robot. However, note that for a good performance, an accurate model of the system is required, and the overall procedure is not robust to model uncertainties. Furthermore, this technique is computationally intensive in terms of the online computation needed to perform the control law.

Now consider the general case where the knowledge of dynamic matrices is not complete, that is $\hat{\boldsymbol{M}}_q \neq \boldsymbol{M}_q$, $\hat{\boldsymbol{C}}_q \neq \boldsymbol{C}_q$, $\hat{\boldsymbol{G}}_q \neq \boldsymbol{G}_q$, and also a disturbance force is applied to the moving platform $\boldsymbol{\tau}_d \neq \boldsymbol{0}$. In such a case, the error dynamics may be derived from Equation 6.69 as

$$\ddot{\boldsymbol{e}}_q + \boldsymbol{K}_d\dot{\boldsymbol{e}}_q + \boldsymbol{K}_p\boldsymbol{e}_q = \hat{\boldsymbol{M}}_q^{-1}\left[(\boldsymbol{M}_q - \hat{\boldsymbol{M}}_q)\ddot{\boldsymbol{q}} + (\boldsymbol{C}_q - \hat{\boldsymbol{C}}_q)\dot{\boldsymbol{q}} + (\boldsymbol{G}_q - \hat{\boldsymbol{G}}_q) - \boldsymbol{\tau}_d\right]. \tag{6.72}$$

The right-hand side of Equation 6.72 is not zero in general but may be treated as an external disturbance torque applied to the error dynamics. The effect of this disturbance vector can be well attenuated, using a high-gain PD controller, or by applying advanced control methods, such as the robust or adaptive control. A detailed treatment of such techniques is developed for task space implementation, and similar treatment may be considered for joint space.

6.6 Summary of Motion Control Techniques

In this chapter, a number of control techniques have been developed for parallel robots. Based on the dynamic formulation given for the parallel robot in Chapter 5, many model-based control techniques have been developed for implementation in the task space as well as in the joint space. These control techniques are presented from the simplest form of decentralized PD control to more advanced robust and adaptive inverse dynamics control. To have an overall view of the techniques developed so far, a summary of these techniques is given below.

6.6.1 Dynamic Formulations

The dynamic formulation of a parallel robot may be directly represented as a function of motion variable $\mathcal{X}$ in the task space as follows:

$$M(\mathcal{X})\ddot{\mathcal{X}} + C(\mathcal{X},\dot{\mathcal{X}})\dot{\mathcal{X}} + G(\mathcal{X}) = \mathcal{F} + \mathcal{F}_d.$$

The dynamic formulation may be represented as a function of actuator motion variable q as

$$M_q\ddot{q} + C_q\dot{q} + G_q = \tau + \tau_d,$$

in which these two formulations are closely related to each other by the following relations:

$$\begin{aligned} M_q &= J^{-T}MJ^{-1}, \\ C_q &= J^{-T}\left(C - MJ^{-1}\dot{J}\right)J^{-1}, \\ G_q &= J^{-T}G, \\ \tau_d &= J^{-T}\mathcal{F}_d. \end{aligned}$$

6.6.2 Decentralized PD Control

The simplest controller for a parallel robot can be considered as a decentralized PD controller being implemented individually on each error component. If such a structure is implemented in the task space, the control effort is calculated by

$$\mathcal{F} = K_d\dot{e}_x + K_p e_x,$$

and the actuator effort can be generally determined through a force distribution scheme. A general treatment of such a force distribution is elaborated in Section 6.7. However, for a completely parallel manipulator, the actuator forces can be generated by $\tau = J^{-T}\mathcal{F}$ at nonsingular configurations. Decentralized PD control can be directly implemented in the joint space by the following equation:

$$\tau = K_d\dot{e}_q + K_p e_q.$$

6.6.3 Feed Forward Control

To reduce the performance limitations of simple PD control, the control effort may be enforced with a feed forward wrench given by

$$\mathcal{F} = \mathcal{F}_{pd} + \mathcal{F}_{ff},$$

in which

$$\mathcal{F}_{ff} = K_d\dot{e}_x + K_p e_x + \hat{M}(\mathcal{X}_d)\ddot{\mathcal{X}}_d + \hat{C}(\mathcal{X}_d,\dot{\mathcal{X}}_d)\dot{\mathcal{X}}_d + \hat{G}(\mathcal{X}_d).$$

This controller can be implemented in joint space as follows:

$$\begin{aligned} \tau &= \tau_{pd} + \tau_{ff} \\ &= K_d\dot{e}_q + K_p e_q + J^{-T}\mathcal{F}_{ff}. \end{aligned}$$

6.6.4 Inverse Dynamics Control

In the inverse dynamics control, the nonlinear dynamics of the model is used to add a corrective term to the decentralized PD controller. If such a structure is implemented in the task space, the control effort is calculated by

$$\begin{aligned}\mathcal{F} &= \hat{M}(\mathcal{X})a + \hat{C}(\mathcal{X},\dot{\mathcal{X}})\dot{\mathcal{X}} + \hat{G}(\mathcal{X})\\ a &= \ddot{\mathcal{X}}_d + K_d\dot{e}_x + K_p e_x.\end{aligned}$$

In general, the tracking error dynamics can be represented by

$$\ddot{e}_x + K_d\dot{e}_x + K_p e_x + \hat{M}^{-1}\left[\tilde{M}\ddot{\mathcal{X}} + \tilde{C}\dot{\mathcal{X}} + \tilde{G} + \mathcal{F}_d\right] = 0.$$

This controller can be implemented in the joint space as follows:

$$\begin{aligned}\tau &= \hat{M}_q a_q + \hat{C}_q\dot{q} + \hat{G}_q\\ a_q &= \ddot{q}_d + K_d\dot{e}_q + K_p e_q,\end{aligned}$$

by which the tracking error dynamics is summarized as

$$\ddot{e}_q + K_d\dot{e}_q + K_p e_q + \hat{M}_q^{-1}\left[\tilde{M}_q\ddot{q} + \tilde{C}_q\dot{q} + \tilde{G}_q + \tau_d\right] = 0.$$

6.6.5 Partial Linearization IDC

To reduce the computational cost of the inverse dynamic control, it is possible to use partial linearization of dynamic formulation, just by gravity compensation, while keeping asymptotic tracking stability of the closed-loop system. In such a case, the control input wrench in the task space is simplified to

$$\mathcal{F} = K_d\dot{e}_x + K_p e_x + \hat{G}(\mathcal{X}).$$

The following Lyapunov function may be used to analyze the stability of tracking dynamics of the closed-loop system:

$$\dot{V} = \dot{\mathcal{X}}^T M\ddot{\mathcal{X}} + \frac{1}{2}\dot{\mathcal{X}}^T\dot{M}\dot{\mathcal{X}} + e_x^T K_p e_x.$$

Stability analysis of the closed-loop system in this case reveals the fact that this simplified version of inverse dynamics control can lead to asymptotic tracking for constant desired trajectories.

6.6.6 Robust Inverse Dynamics Control

To accommodate modeling uncertainties in inverse dynamic control, the following robust control scheme in the task space is developed:

$$\begin{aligned}\mathcal{F} &= \hat{M}(\mathcal{X})a_r + \hat{C}(\mathcal{X},\dot{\mathcal{X}})\dot{\mathcal{X}} + \hat{G}(\mathcal{X})\\ a_r &= \ddot{\mathcal{X}}_d + K_d\dot{e}_x + K_p e_x + \delta_a,\end{aligned}$$

in which the robustifying corrective term $\boldsymbol{\delta}_a$ is found through a Lyapunov stability analysis of tracking error dynamics. The tracking error dynamics can be represented by the following linear and nonlinear components:

$$\dot{\boldsymbol{\varepsilon}} = \boldsymbol{A}\boldsymbol{\varepsilon} + \boldsymbol{B}(\boldsymbol{\delta}_a + \boldsymbol{\eta}),$$

where $\boldsymbol{\eta}$ encapsulates the norm-bounded modeling uncertainties and

$$A = \begin{bmatrix} 0 & I \\ -K_p & -K_d \end{bmatrix}, \quad B = \begin{bmatrix} 0 \\ I \end{bmatrix}.$$

The corrective term $\boldsymbol{\delta}_a$ can be found as

$$\boldsymbol{\delta}_a = \begin{cases} -\rho \frac{\boldsymbol{v}}{\|\boldsymbol{v}\|} & \text{if } \|\boldsymbol{v}\| > \epsilon \\ -\rho \frac{\boldsymbol{v}}{\epsilon} & \text{if } \|\boldsymbol{v}\| \leq \epsilon \end{cases},$$

in which $\boldsymbol{v}$ is defined by $\boldsymbol{v} = \boldsymbol{B}^T\boldsymbol{P}\boldsymbol{\varepsilon}$, where $\boldsymbol{P}$ is the solution to the matrix Lyapunov equation 6.28 and ϵ is a smoothing threshold. It is shown that by adding this corrective term to the regular inverse dynamics control, the closed-loop system achieves uniform ultimate bounded tracking errors.

6.6.7 Adaptive Inverse Dynamics Control

In the adaptive version of the inverse dynamics control, full feedback linearization is considered through adaptive update of dynamic formulation matrices. The error dynamics in this case is

$$\dot{\boldsymbol{\varepsilon}} = \boldsymbol{A}\boldsymbol{\varepsilon} + \boldsymbol{B}\boldsymbol{\Phi}\tilde{\boldsymbol{\theta}},$$

in which

$$A = \begin{bmatrix} 0 & I \\ -K_p & -K_d \end{bmatrix}, \quad B = \begin{bmatrix} 0 \\ I \end{bmatrix}, \quad \boldsymbol{\Phi} = \hat{M}^{-1} Y(\boldsymbol{\mathcal{X}}, \dot{\boldsymbol{\mathcal{X}}}, \ddot{\boldsymbol{\mathcal{X}}}).$$

Based on the Lyapunov stability analysis, by using the following Lyapunov function

$$V = \boldsymbol{\varepsilon}^T\boldsymbol{P}\boldsymbol{\varepsilon} + \tilde{\boldsymbol{\theta}}^T\boldsymbol{\Gamma}\boldsymbol{\theta},$$

the following parameter adaptation law is derived for updates

$$\dot{\hat{\boldsymbol{\theta}}} = -\boldsymbol{\Gamma}^{-1}\boldsymbol{\Phi}^T\boldsymbol{B}^T\boldsymbol{P}\boldsymbol{\varepsilon}.$$

By this means, the closed-loop system achieves asymptotic tracking performance, while the parameter estimation errors remain bounded.

6.7 Redundancy Resolution

6.7.1 Introduction

Redundancy is a major characteristic of robot manipulators in performing tasks that require dexterity. Most robot manipulators have the same number of actuators as the number of required degrees-of-freedom to perform their task. Although this nonredundant structure of the manipulator is cost-efficient, it limits its manipulability to perform the required task. Many biological mechanisms, such as human shoulder, arm, or hand benefit from redundancy in actuation. In general, redundant manipulators hire more number of actuators than the required number of degrees-of-freedom to perform the required task with some extra desirable user-defined characteristic(s). In other words, the extra actuators are efficiently used to fulfill some desirable kinematic characteristics, such as singularity avoidance [139], poster control [104], collision avoidance [57], or motion under joint limits [165]. Although in many redundant manipulators redundancy is introduced to increase the performance of the robot, in cable–driven manipulators redundancy is a stringent requirement without which the manipulator cannot perform its required task. The reason for such a requirement is that the cables can only pull and they are unable to push, and therefore, redundancy is necessary to perform the required maneuver for the moving platform, while keeping all the cable forces under tension. The methods developed to optimally manage the extra degree(s)-of-freedom for a redundant manipulator are technically called *redundancy resolution* schemes.

Redundancy resolution techniques have been extensively worked out during the past three decades. Despite this long history, earlier investigations were often focused on the Jacobian pseudo-inverse approach proposed originally by Whitney [191] in 1969 and improved subsequently by Liegeois [116] in 1977. This approach resolves the redundancy at the velocity level by optimization routines applied on some objective functions. Owing to the structure of serial manipulators, this method works well to produce minimum norm forces in the joint space, and therefore, the required energy from the actuators is minimized. By using the Jacobian pseudo-inverse approach, Hollerbach and Suh [77] have suggested methods for minimizations of instantaneous joint torques. Khatib has proposed a scheme to reduce joint torques through inertia-weighted Jacobian pseudo-inverse [98]. Dubey and Luh [40] and Chiu [23] used the pseudo-inverse approach to optimize the manipulator mechanical advantage and velocity ratio using the force and velocity manipulability hyperellipsoids. Furthermore, Seraji proposed a configuration control approach for redundancy resolution of serial manipulators [164,165]. In other researches, optimization techniques such as the Pontryagin's maximum principle has been used to find the solution [140]. Some of these methods would automatically generate trajectories that avoid kinematic singularities [121], while other approaches maximize some function of the joint angles, such as the manipulability measure [196].

Redundancy resolution of parallel manipulators presents an inherent complexity owing to their dynamic constraints, particularly when parallel manipulator is cable-driven and the tension force requirement in cables are needed. Very few works in redundancy resolution of cable-driven parallel robots are reported based on wrench-feasibility of cable robots [16,146]. Barrette and Gosselin also showed that it is possible to study redundancy resolution of cable robots as an optimization problem with equality and inequality constraints [11]. Hassan and Khajepour in Reference 70 studied actuators force distribution in cable robot as a projection problem. In this work, they have presented two numerical solutions. A minimum-norm solution is presented first by minimizing the Euclidean norm of all forces in the cables and redundant limbs, and another solution is given to minimize the Euclidean norm of just the cable forces. The optimization problem is expressed as a projection on an intersection of convex sets, and the Dykstra's projection scheme is used to obtain the solutions. In some recent works, the redundancy resolution of planar cable robot is studied at kinematic and dynamic levels [2], in which the redundancy resolution problem is studied as an optimization problem to minimize the Euclidean norm of actuator forces and in addition to minimize the Euclidean norm of the mobile platform velocity, subject to positive tension in the cables.

6.7.2 Problem Formulation

As mentioned in Section 6.2, some controllers generate wrench commands in the task space, and to apply them to the manipulator it is required to project them into the joint space. This projection, in its general form, is depicted in Figure 6.4 as the force distribution block. The force distribution block projects the required wrench in the task space $\mathcal{F}$ to its corresponding actuator forces/torques $\boldsymbol{\tau}$. Note that for a fully parallel manipulator such as the Stewart–Gough platform, in which the number of actuators are equal to the number of degrees-of-freedom, this mapping can be constructed from the Jacobian transpose of the manipulator $\boldsymbol{J}^T$ by $\boldsymbol{\tau} = \boldsymbol{J}^{-T}\mathcal{F}$ at nonsingular configurations.

The force distributor block may need more computation if the parallel manipulator is not fully parallel. Redundancy resolution techniques would be used to implement force distribution block physically for redundant manipulator. Let us define a redundant manipulator as a robot the joint space dimension of which, denoted by m, is greater than its degrees-of-freedom, denoted by n. Furthermore, define $r = m - n$ $(r \geq 1)$ as the degree-of-redundancy. In such a case, the Jacobian matrix of the manipulator is nonsquare, and therefore, the inverse of its transpose cannot be used for force distribution. Indeed, the Jacobian transpose is an $m \times n$ matrix, in which $m > n$, and therefore, for redundant manipulators, infinite solutions exist for actuator efforts $\boldsymbol{\tau}$ to generate the required wrench $\mathcal{F}$ in the task space. Redundancy resolution techniques seek for a set of solutions among the many possible solutions, which optimize a user-defined cost function in the presence of kinematic constraints.

From the Jacobian analysis of parallel manipulators, the general projection map between the task space and joint space velocity variables are given in Equation 6.73. Furthermore, the Jacobian transpose relates the generalized force variable in the task space to the actuator forces in joint space by

$$\dot{q} = \boldsymbol{J}\dot{\mathcal{X}}, \quad \mathcal{F} = \boldsymbol{J}^T\boldsymbol{\tau} \tag{6.73}$$

in which q and $\boldsymbol{\tau}$ denote the motion and force variables in the joint space, respectively, while $\mathcal{X}$ and $\mathcal{F}$ denote the variables in the task space. Note that for redundant manipulators, $q \in \mathbb{R}^m$ and $\mathcal{X} \in \mathbb{R}^n$, in which $m > n$, and therefore, the Jacobian matrix is a nonsquare matrix, which belongs to $\mathbb{R}^{m\times n}$. For the sake of simplicity, let us define the structure matrix of the redundant parallel manipulator [46], denoted by $\boldsymbol{A}$, as follows:

$$\boldsymbol{A} = \boldsymbol{J}^T \rightarrow \mathcal{F} = \boldsymbol{A}\boldsymbol{\tau}, \tag{6.74}$$

where $\boldsymbol{A}$ is also a nonsquare matrix belonging to $\mathbb{R}^{n\times m}, m > n$. This matrix generates the required map to transform joint space actuator forces $\boldsymbol{\tau}$ into the task space wrench $\mathcal{F}$. If the manipulator is fully parallel, then the mapping that projects $\mathcal{F}$ to $\boldsymbol{\tau}$ is easily obtained by the structure matrix $\boldsymbol{A}$ at nonsingular configurations:

$$\boldsymbol{\tau} = \boldsymbol{A}^{-1}\mathcal{F}, \quad \text{if } m = n. \tag{6.75}$$

To generalize the solution to redundant manipulators, one may consider the pseudo-inverse of the structure matrix, denoted by $\boldsymbol{A}^{\dagger}$, as a suitable mapping:

$$\boldsymbol{\tau}_o = \boldsymbol{A}^{\dagger}\mathcal{F}, \quad \text{if } m > n. \tag{6.76}$$

Note that to solve $\boldsymbol{\tau}$ for a given $\mathcal{F}$ in this case, the set of equations represented by the vector equation $\mathcal{F} = \boldsymbol{A}\boldsymbol{\tau}$ is an underdetermined set of linear equations, in which, since $m > n$, the number of unknowns are greater than the number of equations. Therefore, the structure matrix $\boldsymbol{A}$ is a rectangular matrix, in which the number of columns is greater than the number of rows. In such a case, there exist infinitely many solutions $\boldsymbol{\tau}$ to satisfy Equation 6.74, of them the pseudo-inverse solution has a particular characteristic that minimizes the Euclidean norm of the output vector $\|\boldsymbol{\tau}\|_2$. The pseudo-inverse solution for a rectangular matrix $\boldsymbol{A}$, the number of columns m of which is greater than the number of rows n, is determined by*

$$\boldsymbol{A}^{\dagger} \doteq \boldsymbol{A}^T(\boldsymbol{A}\boldsymbol{A}^T)^{-1}, \quad \text{if } m > n. \tag{6.77}$$

This solution may be rigorously found by solving a constrained optimization problem in which the cost function to be minimized is the Euclidean norm of the output $\boldsymbol{\tau}$. Furthermore, the optimization problem is subject to the constraint that the kinematic relation between the joint space variable $\boldsymbol{\tau}$, and its corresponding task space projection $\mathcal{F}$ is definitely set to Equation 6.74. This optimization problem is the simplest form of redundancy resolution formulation for a redundant manipulator, in which the extra degree(s)-of-freedom in choosing $\boldsymbol{\tau}$ among the infinite solutions is used to minimize the needed energy of the actuator effort $\|\boldsymbol{\tau}\|_2$. Therefore, in its simplest form, the redundancy resolution problem for a redundant manipulator can be formulated into the following constrained optimization problem:

Find $\boldsymbol{\tau}$ such that

$$\begin{cases} \min \|\boldsymbol{\tau}\|_2, \\ \text{subject to } \mathcal{F} = \boldsymbol{A}\boldsymbol{\tau}. \end{cases} \tag{6.78}$$

Although this problem is the simplest form of redundancy resolution, it provides a suitable solution for the actuators, in which the force distribution minimizes the required actuator energy to perform the required task. In fact, if no other user–defined objective is enforced, and no other limitations are needed to be satisfied, this optimization formulation is very suitable, and the Moore–Penrose pseudo-inverse solution to the problem $\boldsymbol{\tau}_o$ may be considered as the base solution for redundancy resolution scheme. Furthermore, all other solutions of the above mapping can be generated using this base solution $\boldsymbol{\tau}_o$ given in Equation 6.76.

To find the set of all solutions $\boldsymbol{\tau}$ that satisfies Equation 6.74, let us add some practical limitations to the problem. As mentioned hereinbefore, in cable-driven parallel manipulators, keeping the cable forces in tension is a stringent requirement. Let us formulate

* For more details read Section A.4 in Appendix A.

the redundancy resolution scheme for such a case as an optimization problem and seek its solution from the base solution $\boldsymbol{\tau}_o$. The positive force requirement may be written as another inequality constraint to the optimization problem as follows:

Find $\boldsymbol{\tau}$ such that

$$\min \|\boldsymbol{\tau}\|_2, \qquad \text{subject to} \begin{cases} \boldsymbol{\mathcal{F}} &= \boldsymbol{A}\boldsymbol{\tau} \\ \boldsymbol{\tau} &\geq \boldsymbol{\tau}_{\min} \end{cases}, \tag{6.79}$$

in which $\boldsymbol{\tau}_{\min}$ is a user-defined lower bound for the actuator force, where for a cable-driven manipulator it satisfies $\boldsymbol{\tau}_{\min} > \mathbf{0}$. Note that the base solution $\boldsymbol{\tau}_o$ may not satisfy the positive tension requirement, represented by the last inequality constraint in 6.79, and therefore it might not be the right solution for this optimization problem.

The set of all solutions to satisfy Equation 6.74 may be given by the following equation*

$$\boldsymbol{\tau} = \boldsymbol{\tau}_o + \left(\boldsymbol{I} - \boldsymbol{A}^{\dagger}\boldsymbol{A}\right)\boldsymbol{y}, \quad \boldsymbol{y} \in \mathbb{R}^m, \tag{6.80}$$

in which $\boldsymbol{\tau}_o$ denotes the base pseudo-inverse solution given in Equation 6.76, $\boldsymbol{I}$ denotes an $m \times m$ identity matrix, and $\boldsymbol{y}$ is any arbitrary vector in $\mathbb{R}^m$. This general solution consists of two parts, the first part is the base solution $\boldsymbol{\tau}_o$ and the second part spans the null space of matrix $\boldsymbol{A}$. To see this, let us prove that $\boldsymbol{\tau}$ given in Equation 6.80 satisfies the projection requirement given in Equation 6.74. Multiply $\boldsymbol{A}$ to both sides of Equation 6.80 and simplify as follows:

$$\begin{aligned} \boldsymbol{A}\boldsymbol{\tau} &= \boldsymbol{A}\boldsymbol{\tau}_o + \boldsymbol{A}(\boldsymbol{I} - \boldsymbol{A}^{\dagger}\boldsymbol{A})\boldsymbol{y} \\ &= \boldsymbol{\mathcal{F}} + (\boldsymbol{A} - \boldsymbol{A}\boldsymbol{A}^{\dagger}\boldsymbol{A})\boldsymbol{y} \\ &= \boldsymbol{\mathcal{F}} + (\boldsymbol{A} - \boldsymbol{A}\boldsymbol{A}^T(\boldsymbol{A}\boldsymbol{A}^T)^{-1}\boldsymbol{A})\boldsymbol{y} \\ &= \boldsymbol{\mathcal{F}} + (\boldsymbol{A} - \boldsymbol{A})\boldsymbol{y} \\ &= \boldsymbol{\mathcal{F}}. \end{aligned}$$

This completes the proof and shows that the second part of the solution contributes to zero task space wrench. Hence, the subspace generated by $\boldsymbol{M}\boldsymbol{y} = (\boldsymbol{I} - \boldsymbol{A}^{\dagger}\boldsymbol{A})\boldsymbol{y}$ spans the null space of matrix $\boldsymbol{A}$ for any arbitrary vector $\boldsymbol{y}$.

Among all possible solutions generated by the arbitrary selection of $\boldsymbol{y}$, you may search for the one that can satisfy the inequality constraint introduced in Equation 6.79, while still minimizing the required actuator energy. Many effective search routines have been developed to solve such an optimization problem with equality and inequality constraints. These techniques may be developed on gradient-based iterative schemes like active set and interior point methods [154], or soft computing-based techniques such as genetic algorithm [49]. An iterative-analytic scheme has also been developed recently to solve this problem [173]. The details of some of these techniques are given in the next section.

The variety and effectiveness of numerical techniques developed to solve the optimization problem is such that more challenging optimization problems, with multiple cost functions, and multiple equality and inequality constraints are solvable. In what follows,

* Read Appendix A.4 for more details and proofs.

two other redundancy resolution schemes have been represented by the optimization problem to broaden the scope of such techniques for parallel manipulators. First, consider the case where the actuator forces are bound to some actuator saturation limits, denoted by $\boldsymbol{\tau}_{max}$, in addition to previously detailed constraints. This additional requirement may be encapsulated by another inequality constraint to the optimization problem as follows:

Find $\boldsymbol{\tau}$ such that

$$\min \|\boldsymbol{\tau}\|_2,$$

$$\text{subject to} \begin{cases} \mathcal{F} = A\boldsymbol{\tau} \\ \boldsymbol{\tau}_{\min} \leq \boldsymbol{\tau} \leq \boldsymbol{\tau}_{\max} \end{cases} \tag{6.81}$$

In this redundancy resolution scheme, more constraints are forced on the actuator forces, and to fulfill these requirements, more degrees-of-redundancy are usually required. Note that if the redundancy degree is not sufficient, it might not be possible to find any feasible solution to the optimization problem, while satisfying all the constraints.

As a final case, consider that a user-defined cost function is given to resolve for the redundancy, and denote $f(\boldsymbol{\tau})$ as a functional representing the user-defined cost function. This case can also be represented in the following optimization problem and found feasible for numerical solution:

Find $\boldsymbol{\tau}$ such that

$$\min f(\boldsymbol{\tau}),$$

$$\text{subject to} \begin{cases} \mathcal{F} = A\boldsymbol{\tau} \\ \boldsymbol{\tau}_{\min} \leq \boldsymbol{\tau} \leq \boldsymbol{\tau}_{\max} \\ \vdots \end{cases} \tag{6.82}$$

6.7.3 Lagrange and Karush–Kuhn–Tucker Multipliers

As shown in the previous section, redundancy resolution schemes lead to optimization problems with equality and inequality constraints. To find the optimal solution, which satisfies the required constraints, it is customary to use some multipliers to augment the constraints into the main objective function, and use analytical or numerical approach to search for the local minimum of the augmented cost function. Furthermore, necessary conditions to have a feasible solution for the augmented problem can be set on the introduced multipliers. To show the idea, let us first consider the simplest optimization formulation introduced in Equation 6.78, which requires finding the minimum of a cost function subject to a number of equality constraints. Rewrite this problem in the following general form:

Find $\boldsymbol{x}$ such that

$$\begin{cases} \min f(\boldsymbol{x}), \\ \text{subject to } \; g(\boldsymbol{x}) = \mathbf{0} \end{cases} \tag{6.83}$$

in which the cost function is considered any arbitrary functional $f : \mathbb{R}^m \to \mathbb{R}$ of the optimization variable $\boldsymbol{x} \in \mathbb{R}^m$, while the equality constraints are represented by a vector valued function $g(\boldsymbol{x}) : \mathbb{R}^m \to \mathbb{R}^n$. Note that in our redundancy resolution scheme, the functional f is the Euclidean norm of the optimization vector $\boldsymbol{x} = \boldsymbol{\tau}$, while the equality constraints are represented as a linear function of τ in terms of the structure matrix A. However, the representation given in Equation 6.83 is a general form of such optimization problem subject to equality constraints. To study the optimization solution, let us define *regular point* and *stationary value*.

Definition 6.1

A point x_o that satisfies the equality constraints given in Equation 6.83 is said to be a *regular point*, if the following vectors are linearly independent at $x = x_o$.

$$\left(\frac{\partial g_i}{\partial x}\right)^T, \quad i = 1, 2, \ldots, n. \tag{6.84}$$

To have x_o satisfying Equation 6.84, one may calculate the rank of the Jacobian matrix of g, with respect to x evaluated at $x = x_o$. If the rank of this matrix $\left.\frac{\partial g}{\partial x}\right|_{x=x_o}$ is equal to n, then x_o is a regular point.

Definition 6.2

A functional f given in a constraint optimization problem 6.83 is said to have a *stationary value* at $x = x_o$ if its partial derivative with respect to x becomes equal to zero at $x = x_o$, that is

$$\left.\frac{\partial f}{\partial x}\right|_{x=x_o} = 0. \tag{6.85}$$

A stationary point of a functional describes a local extremum of it, and the above condition is a necessary one to seek for a local extremum.

Theorem 6.1 (Lagrange Multiplier)

Assume that a point x_o is an extremum of the optimization problem subject to equality constraints, represented by Equation 6.83. If x_o is a regular point of the constraints, then there exist scalars $\lambda_i, i = 1, 2, \ldots, n$ that provide the stationary value for the augmented functional f_a at x_o:

$$f_a(x) = f(x) + \lambda^T g(x), \quad \lambda = \begin{bmatrix} \lambda_1 \\ \lambda_2 \\ \vdots \\ \lambda_n \end{bmatrix} \in \mathbb{R}^n. \tag{6.86}$$

The scalars λ_i are called the Lagrange multipliers.

The importance of this theorem is the way it treats the equality constraints. The notion of stationary values for multivariable functionals provides local extremum for the functional by computing partial derivatives of the functional with respect to the optimization variable x, and equating them to zero. This is a well-known principle that at local extremum, the slope of the functional with respect to all its variables is zero. By introducing the Lagrange multiplier as a new set of unknown variables, and considering the augmented variable $x_a = \left[x, \quad \lambda\right]^T$, and the augmented functional f_a, seeking for the extremum becomes similar to the one hereinbefore. The extremum point must satisfy the zero slope for the functional at all coordinates x_i:

$$\left.\frac{\partial f_a}{\partial x}\right|_{x=x_o} = 0$$

and furthermore, it should provide zero slope with respect to coordinates λ_i:

$$\left.\frac{\partial f_a}{\partial \lambda}\right|_{x=x_o} = g(x_o) = 0.$$

The latter condition assures that the equality constraints are all satisfied at the extremum point. Note that the first condition provides m equation in which both the functional f_a and the constraint function g contribute.

The Lagrange multiplier theorem reduces an optimization problem subject to equality constraints to a stationary value problem with no constraints. This result has been extended to optimization problems subject to inequality constraints. This extension was originally named after Harold W. Kuhn and Albert W. Tucker, who published the conditions in Reference 107. Later scholars discovered that the necessary conditions for this problem had been stated earlier by William Karush in his master's thesis [94]. Let us consider the most general optimization formulation introduced in Equation 6.82, which requires finding the minimum of a cost function subject to a number of equality and inequality constraints. Rewrite this problem in the following general form:

Find x such that

$$\min f(x),$$

$$\text{subject to} \begin{cases} g(x) = 0, & g : \mathbb{R}^m \rightarrow \mathbb{R}^n \\ h(x) \leq 0, & h : \mathbb{R}^m \rightarrow \mathbb{R}^\ell \end{cases}, \tag{6.87}$$

in which the cost function is considered as any arbitrary functional $f : \mathbb{R}^m \rightarrow \mathbb{R}$ on the optimization variable $x \in \mathbb{R}^m$, while equality constraints are represented by a vector-valued function $g(x) : \mathbb{R}^m \rightarrow \mathbb{R}^n$, and inequality constraints are represented by a vector-valued function $h(x) : \mathbb{R}^m \rightarrow \mathbb{R}^\ell$. To study the optimization solution, let us extend the definition of a *regular point* for this general problem.

Definition 6.3

Let J be the set of indices of active inequality constraints in which $h_j(x) = 0$, for $j \in J$. A point x_o that satisfies the equality constraints given in Equation 6.87 is said to be a *regular point*, if the following vectors are linearly independent at $x = x_o$:

$$\left(\frac{\partial g_i}{\partial x}\right)^T, \quad i = 1, 2, \ldots, n; \quad \left(\frac{\partial h_j}{\partial x}\right)^T, \quad j \in J. \tag{6.88}$$

Using this definition, the sufficient condition for a stationary point x_o is given by the following theorem.

Theorem 6.2 (Karush–Kuhn–Tucker (KKT) Theorem)

Assume that a point x_o is an extremum of the optimization problem subject to equality and inequality constraints represented by Equation 6.87. If x_o is a regular point of the constraints, then there exist scalars $\lambda_i, i = 1, 2, \ldots, n$ and nonnegative scalars $\mu_j \geq 0, j = 1, 2, \ldots, \ell$ that provide the stationary value for the following augmented functional f_a at x_o:

$$f_a(\boldsymbol{x}) = f(\boldsymbol{x}) + \boldsymbol{\lambda}^T \boldsymbol{g}(\boldsymbol{x}) + \boldsymbol{\mu}^T \boldsymbol{h}(\boldsymbol{x}), \tag{6.89}$$

in which

$$\boldsymbol{\mu}^T \boldsymbol{h}(\boldsymbol{x}_o) = 0, \tag{6.90}$$

and

$$\boldsymbol{\lambda} = \begin{bmatrix} \lambda_1 \\ \lambda_2 \\ \vdots \\ \lambda_n \end{bmatrix} \in \mathbb{R}^n, \quad \boldsymbol{\mu} = \begin{bmatrix} \mu_1 \\ \mu_2 \\ \vdots \\ \mu_\ell \end{bmatrix} \in \mathbb{R}^\ell. \tag{6.91}$$

The scalars λ_i and μ_i are called the KKT multipliers.

Note that since $\mu_i \geq 0$ and $h_i \leq 0$, Equation 6.90 implies that $\mu_i = 0$ for $h_i(\boldsymbol{x}) < 0$ and $\mu_i \geq 0$ for $h_i(\boldsymbol{x}) = 0$. The importance of the KKT theorem is that it reduces an optimization problem subject to equality and inequality constraints to a stationary value problem with no constraints. Furthermore, it provides the necessary conditions for a feasible solution to exist. In the next section, a review on the iterative techniques to solve this problem is given.

6.7.4 Iterative Solutions

To solve a constrained optimization problem, many different iterative routines have been developed. The general aim of these techniques is to transform the constrained optimization problem into an easier subproblem that can be used as the basis of an iterative process to search for the optimal solution. One general approach then is the translation of the constrained problem to a basic unconstrained problem by using a penalty function for constraints that are near or beyond the constraint boundary. Many state-of-the-art techniques focus on the solution of the Karush–Kuhn–Tucker (KKT) conditions.

6.7.4.1 Numerical Methods

As stated in the previous subsection, the KKT equations are necessary conditions for optimality of a constrained optimization problem. For a convex programming problem in which the cost function $f(\boldsymbol{x})$ and the constraint functions $g_i(\boldsymbol{x}), i = 1, 2, \ldots, n$ and $h_i(\boldsymbol{x}), i = 1, 2, \ldots, \ell$ are convex functions, it can be shown that the conditions are both necessary and sufficient for a global solution point $\boldsymbol{x}^\star$. Referring to the KKT theorem, for the optimization problem represented by Equation 6.87, the following KKT equations may be stated for the optimal solution $\boldsymbol{x}^\star$, in addition to the original constraints given in Equation 6.87:

$$\nabla f(\boldsymbol{x}^\star) + \sum_{i=1}^{n} \lambda_i \nabla g_i(\boldsymbol{x}^\star) + \sum_{j=1}^{\ell} \mu_j \nabla h_j(\boldsymbol{x}^\star) = 0 \tag{6.92}$$

and

$$\lambda_i g_i(x^\star) = 0, \quad i = 1, 2, \ldots, n \tag{6.93}$$

$$\mu_j \geq 0, \qquad j = 1, 2, \ldots, \ell. \tag{6.94}$$

To have an optimal solution $x^\star$, Equation 6.92 provides the required conditions for the gradients between the objective function and the active constraints to be canceled out at the solution point. To satisfy this condition, it is necessary for the KKT multipliers to balance the deviations in the magnitude of the objective function and constrained gradients. Because only active constraints are included from this canceling operation, constraints that are not active must be excluded, and therefore, the corresponding KKT multipliers must be equal to zero. This is stated implicitly in the last two equations.

The solution of KKT equations forms the basis of many nonlinear programming algorithms, where these algorithms attempt to compute KKT multipliers directly. As a representative to these algorithms, the active-set method can be introduced here. The active-set algorithm closely mimics the Newton's method for constrained optimization problem just as is done for unconstrained optimization. At each major iteration, an approximation of the Hessian of the Lagrangian function represented in Equation 6.92 is made using a quasi–Newton updating method. This is then used to generate a quadratic programming subproblem, whose solution is used to form a search direction for a line search procedure. The details of such numerical techniques are beyond the scope of this book and an overview of this method may be found in References 49,55,154.

Another simple yet powerful concept used in many optimization routines is based on the trust-region approach. The basic idea is to approximate the minimization function f represented in 6.87 with a simpler function q, which reasonably reflects the behavior of function f in a neighborhood $\mathcal{N}$ around the point x. This neighborhood is called the trust region. A trial step s is then computed by minimizing or approximately minimizing over the region $\mathcal{N}$, or

$$\min\{q(s), s \in \mathcal{N}\}. \tag{6.95}$$

The current point is updated to $x + s$, if $f(x + s) < f(x)$; otherwise, the current point remains unchanged and $\mathcal{N}$, the region of trust, is shrunk and the trial step computation is repeated. The key questions in defining a specific trust-region approach to minimizing $f(x)$ are how to choose and compute the approximation q, defined at the current point x, how to choose and modify the trust region $\mathcal{N}$, and how to solve the trust-region subproblem accurately. In the standard trust-region method [137], the quadratic approximation q is defined by the first two terms of the Taylor approximation to f at x; the neighborhood $\mathcal{N}$ is usually spherical or ellipsoidal in shape. Mathematically, the trust-region subproblem is stated as

$$\begin{cases} \min \frac{1}{2} s^T H s + s^T g, \\ \text{such that } \|Ds\| \leq \Delta. \end{cases} \tag{6.96}$$

where g is the gradient of f at current point x, H is the Hessian matrix of second derivatives, D is a diagonal scaling matrix, and Δ is a positive scalar. Tractable algorithms exist for solving Equation 6.96 quite accurately [137]. However, the elapsed time to compute several factorizations of H is relatively high.

One of the most efficient methods to solve the constrained optimization problem represented in Equation 6.87 is the interior–point approach. In this method, the constrained optimization problem is solved in a sequence of approximate minimization problems. The original optimization problem 6.87 is approximated by the following problem:

Find x, s such that

$$\min f_\nu(x) = \min f(x) - \nu \sum_{i=1}^{\ell} \ln s_i, \tag{6.97}$$

$$\text{subject to} \begin{cases} g(x) = 0, \\ h(x) + s = 0. \end{cases} \tag{6.98}$$

in which s has the same dimension as the inequality constraints: $s \in \mathbb{R}^\ell$ and is called the slack variable. By adding slack variables into the cost function, s_i's are restricted to be positive to keep $\ln s_i$ bounded, and therefore this logarithmic term is called a barrier function. As ν decreases to zero, the minimum of f_ν approaches the minimum of f. The approximate problem represented in Equation 6.97 is a sequence of equality constrained problems, which are easier to solve than the original inequality constrained problem represented in Equation 6.87.

To solve the approximate problem, the algorithm performs one of the two main types of steps at each iteration, either a Newton step in (x, s) or a conjugate gradient step using a trust region. The Newton step attempts to solve the KKT equations 6.92 and 6.93 for the approximate problem via a linear approximation. The conjugate gradient approach adjusts both x and s, while keeping the slack variable s positive. The approach is to minimize a quadratic approximation of the approximate problem in a trust region, subject to linearized constraints. By default, the algorithm first attempts to take a Newton step, and if it cannot converge, it attempts a conjugate gradient step.

The numerical methods described briefly here are implemented in many commercial software, including the optimization toolbox of MATLAB. Several functions are available to the user to perform different approaches and to find the solution of an optimization technique in this toolbox. One of the suitable functions to solve a general constrained optimization problem represented by Equation 6.82 is `fmincon`. In this function, different numerical routines such as the trust-region, active-set, and interior-point methods are implemented to find the solution of an inequality constrained optimization problem. Readers are recommended to review the details of such routines in the user manual [30].

6.7.4.2 An Iterative-Analytical Method

It is important to note that, in the implementation of all numerical methods, iterative calculations are the only means to find the optimization solution. As seen in Section 6.2 and illustrated in Figure 6.4, redundancy resolution schemes would be used to carry out force distribution in a closed loop. To use such techniques in a closed loop, it is required to solve the problem in real time, and therefore, the optimization routine must converge to a solution in a fixed and small period of time. However, this is in direct contrast to generic numerical algorithms, which take a variable time step and exit only when a certain precision has been achieved.

To overcome this problem, an iterative–analytic method has been developed recently [13]. The main benefit of having an iterative–analytic solution to the redundancy resolution problem is to guarantee that the amount of time required for the overall solution remains within an acceptable and small period of time that can be used in real-time implementation. In this method, the Karush–Kuhn–Tucker theorem is used to analyze the optimization problem and to generate a set of possible analytic solutions. Subsequently, a tractable and iterative algorithm is given to find a suitable solution effectively. To describe how this method works in practice, let us rewrite the constraint optimization problem given in Equation 6.79 as follows:

Find $\boldsymbol{\tau}$ such that

$$\min \|\boldsymbol{\tau}\|_2, \qquad \text{subject to} \begin{cases} \mathcal{F} &= \boldsymbol{A\tau} \\ \boldsymbol{\tau} &\geq \boldsymbol{\tau}_{\min}, \end{cases} \tag{6.99}$$

in which the set of all solutions to this problem that satisfies the equality constraint may be formulated by Equation 6.80, which is rewritten here for more convenience:

$$\boldsymbol{\tau} = \boldsymbol{\tau}_o + \left(\boldsymbol{I} - \boldsymbol{A}^{\dagger}\boldsymbol{A}\right)\boldsymbol{y}, \quad \boldsymbol{y} \in \mathbb{R}^m. \tag{6.100}$$

As detailed earlier, in this equation, $\boldsymbol{A}$ denotes the manipulator Jacobian matrix transpose $\boldsymbol{J}^T$, $\boldsymbol{\tau}_o$ denotes the base pseudo-inverse solution given in Equation 6.76, $\boldsymbol{I}$ denotes an $m \times m$ identity matrix, and $\boldsymbol{y}$ is any arbitrary vector in $\mathbb{R}^m$. This general solution consists of two parts, the first part is the base solution $\boldsymbol{\tau}_o$, and the subspace generated by $\boldsymbol{My} = (\boldsymbol{I} - \boldsymbol{A}^{\dagger}\boldsymbol{A})\boldsymbol{y}$ spans the null space of matrix $\boldsymbol{A}$, for any arbitrary vector $\boldsymbol{y}$. Among all possible solutions generated by arbitrary selection of $\boldsymbol{y}$, one may search for the required solution that can satisfy the inequality constraint introduced in 6.99, while still minimizing the required actuator energy. Let us define matrix $\boldsymbol{B}$, generated by collecting and orthonormalizing the linearly independent column vectors of matrix $\boldsymbol{M} = (\boldsymbol{I} - \boldsymbol{A}^{\dagger}\boldsymbol{A})$. Hence, the set of all solutions to the optimization problem that satisfies the equality constraint may be written as

$$\boldsymbol{\tau} = \boldsymbol{\tau}_o + \boldsymbol{By}, \quad \boldsymbol{y} \in \mathbb{R}^m. \tag{6.101}$$

By using this definition, the optimization problem with equality and inequality constraints is reduced to the following problem with just the inequality constraint:

Find $\boldsymbol{y}$ such that

$$\begin{aligned} \min f(\boldsymbol{y}) &= \|\boldsymbol{\tau}\|_2^2 = \boldsymbol{\tau}^T\boldsymbol{\tau} \\ &= (\boldsymbol{\tau}_o + \boldsymbol{By})^T(\boldsymbol{\tau}_o + \boldsymbol{By}) \end{aligned} \tag{6.102}$$

$$\text{subject to} \qquad \boldsymbol{h}(\boldsymbol{y}) = \boldsymbol{\tau}_{\min} - (\boldsymbol{\tau}_o + \boldsymbol{By}) \leq \boldsymbol{0}.$$

To find any solution $\boldsymbol{y}^{\star}$, use the Karush–Kuhn–Tucker theorem by defining the augmented functional f_a, and the KKT multipliers $\boldsymbol{\mu} = \begin{bmatrix}\mu_1 & \mu_2 & \cdots & \mu_m\end{bmatrix}^T$ as follows:

$$\begin{aligned} f_a(\boldsymbol{y}) &= f(\boldsymbol{y}) + \boldsymbol{\mu}^T\boldsymbol{h}(\boldsymbol{y}) \\ &= \boldsymbol{\tau}_o^T\boldsymbol{\tau} + \boldsymbol{\tau}_o^T\boldsymbol{By} + \boldsymbol{y}^T\boldsymbol{B}^T\boldsymbol{\tau}_o + \boldsymbol{y}^T\boldsymbol{B}^T\boldsymbol{By} + \boldsymbol{\mu}^T(\boldsymbol{\tau}_{min} - \boldsymbol{\tau}_o - \boldsymbol{By}). \end{aligned} \tag{6.103}$$

The optimal point $y = y^\star$ must satisfy the following conditions:

$$\left.\frac{\partial f_a}{\partial y}\right|_{y=y^\star} = \mathbf{0} \tag{6.104}$$

and furthermore, the following second condition shall be held for positive KKT multipliers $\boldsymbol{\mu}$:

$$\boldsymbol{\mu}^T\, \boldsymbol{h}(y)\Big|_{y=y^\star} = 0. \tag{6.105}$$

These two conditions may be simplified to

$$2\boldsymbol{\tau}_o^T \boldsymbol{B} + 2y^{\star T} \boldsymbol{B}^T \boldsymbol{B} - \boldsymbol{\mu}^T \boldsymbol{B} = \mathbf{0} \tag{6.106}$$

$$\boldsymbol{\mu}^T (\boldsymbol{\tau}_{min} - \boldsymbol{\tau}_o - \boldsymbol{B} y^\star) = 0. \tag{6.107}$$

Note that matrix $\boldsymbol{B}$ is generated by orthonormal column vectors of matrix $\boldsymbol{M} = (\boldsymbol{I} - \boldsymbol{A}^\dagger \boldsymbol{A})$; therefore, $\boldsymbol{B}^T\boldsymbol{B} = \boldsymbol{I}$. Use this fact and write the transpose of Equation 6.106 in the following form:

$$2\boldsymbol{B}^T \boldsymbol{\tau}_o + 2y^\star - \boldsymbol{B}^T \boldsymbol{\mu} = \mathbf{0}, \tag{6.108}$$

$$\boldsymbol{\mu}^T (\boldsymbol{\tau}_{min} - \boldsymbol{\tau}_o - \boldsymbol{B} y^\star) = 0. \tag{6.109}$$

Equation 6.109 may be written componentwise as follows, in which $y^\star$ is a feasible solution if all $\mu_i \geq 0$:

$$\sum_{i=1}^{m} \mu_i h_i(y^\star) = 0. \tag{6.110}$$

Since all μ_i's must be positive for the regular point $y^\star$ to be a feasible solution for the optimization problem, it may be concluded that

$$\mu_i = 0 \quad \text{for } h_i(y^\star) < 0, \tag{6.111}$$

$$\mu_i \geq 0 \quad \text{for } h_i(y^\star) = 0. \tag{6.112}$$

This equation implies that $\mu_i > 0$ may only hold at instances where the corresponding inequality constraint reaches its boundary $h_i(y^\star) = 0$, in which the corresponding actuator force reaches its lower limit $\tau_i = \tau_{\min_i}$. Furthermore, for the instances where the inequality constraint hold $\tau_{\min_i} - \tau_i < 0$, the corresponding KKT multiplier must be equal to zero $\mu_i = 0$ to satisfy Equation 6.110. Considering these relations, the solution of the optimization problem can be derived from three different cases:

- Case 1: Assume that all forces are inside the solution set defined by the inequality constraint and $\mu_i = 0,\ \forall i = 1, 2, \ldots, m$. Therefore, in such a case, $h_i(y^\star) > 0,\ \forall i = 1, 2, \ldots, m$, and the solution to the optimization problem can be found by simplifying Equation 6.108 as

$$y^\star = -\boldsymbol{B}^T \boldsymbol{\tau}_o. \tag{6.113}$$

In this case, $\boldsymbol{\mu} = \mathbf{0}$, and therefore the actuator forces may be simply derived by

$$\boldsymbol{\tau} = \boldsymbol{\tau}_o + \boldsymbol{B}\boldsymbol{y}^\star, \tag{6.114}$$

in which $\boldsymbol{y}^\star$ is given in Equation 6.113.

- Case 2: Consider the case in which for the optimal solution $\mu_i > 0$ for some i's within $i = 1, 2, \ldots, m$, and $\mu_j = 0$ for the rest of them, namely $j \neq i$ within $j = 1, 2, \ldots, m$. In such a case, for the components that $\mu_i > 0$, the corresponding actuator forces lie at the boundary $\boldsymbol{\tau}_i = \boldsymbol{\tau}_{\min_i}$, and for the rest of μ_j's the corresponding actuator forces may be calculated from a linear equation deduced from Equation 6.108 by elimination of the rows and columns of matrix $\boldsymbol{B}$ corresponding to the zero μ_i's. In such a case, $\boldsymbol{y}^\star$ and μ_j's may be obtained by solving the following linear equation:

$$\begin{bmatrix} -2I_{m\times m} & \begin{bmatrix} \vec{b}_1^T & \cdots & \vec{b}_j^T \end{bmatrix} \\ \begin{bmatrix} \vec{b}_1 \\ \vdots \\ \vec{b}_j \end{bmatrix} & \mathbf{0}_{j\times j} \end{bmatrix} \cdot \begin{bmatrix} \boldsymbol{y}^\star \\ \mu_1 \\ \vdots \\ \mu_j \end{bmatrix} = \begin{bmatrix} 2\boldsymbol{B}^T \tau_o \\ \tau_{min_1} - \tau_{o_1} \\ \vdots \\ \tau_{min_j} - \tau_{o_j} \end{bmatrix}. \tag{6.115}$$

In this equation, matrix $\boldsymbol{B}$ is decomposed by its rows $\boldsymbol{B} = \begin{bmatrix} \vec{b}_1 \\ \vdots \\ \vec{b}_m \end{bmatrix}$, in which each $\vec{b}_j$ is a row vector, and only the row vectors corresponding to nonzero μ_j's are left in this equation. It is shown in Reference 173 that the left-hand side matrix in Equation 6.115 is always invertible, and therefore, a solution can be found in this case. However, this solution is feasible only if all the KKT multipliers are strictly positive, that is, $\mu_j > 0, \ \forall j$.

- Case 3: In this case, all the forces lie on the boundary of the inequality constraints, and therefore all the KKT multipliers are positive, $\mu_i \geq 0, \ \forall i = 1, 2, \ldots, m$. Therefore, $h_i(\boldsymbol{y}^\star) = 0, \ \forall i = 1, 2, \ldots, m$, and thus, the optimal actuator forces for all joints are simply calculated by

$$\boldsymbol{\tau} = \boldsymbol{\tau}_{\min}. \tag{6.116}$$

In this case, the solution may be simply found by Equation 6.116, and it is not necessary to solve Equation 6.108 for $\boldsymbol{y}^\star$.

To seek the optimal solution $\boldsymbol{y}^\star$, a search algorithm is proposed in Reference 173, in which, in the first loop, it is assumed that all forces lie on the boundary of the solution set defined by the inequality constraints, that is $\mu_i > 0, \ \forall i = 1, 2, \ldots, m$, as categorized in case 3. In such a case, the solution is found by Equation 6.116, and this solution is feasible if it can satisfy the equality constraint $\boldsymbol{\mathcal{F}} = \boldsymbol{A}\boldsymbol{\tau}$. Otherwise, the combinations of actuator forces that may lie on the boundaries of the inequality constraint must be searched through. For this search, there exist $\binom{\boldsymbol{m}}{S}$ combinations, in which $\boldsymbol{m} = \begin{bmatrix} 1 & 2 & \cdots & m \end{bmatrix}^T$ and $S = 1$ to m may be the possible solutions for the problem. These solutions may be searched through by sweeping all possible combinations through changing S in a loop. However, the solution is a feasible solution for the optimization problem only if $\mu_j > 0, \ \forall j$. After searching through all these combinations, if no feasible solution exists, the last step is to check case 1, in which, all the actuator forces lie inside the boundary. This can be checked if the actuator forces derived by Equation 6.114 are all positive.

This search algorithm is effectively implemented in Reference 173 and compared to the three numerical routines described hereinbefore, namely the active set, trust region, and interior point methods. It is shown through simulations that the elapsed time required to implement the analytic–iterative scheme is considerably less than that of other numerical optimization methods for a special case study on a planar cable-driven robot [173].

6.8 Motion Control of a Planar Manipulator

In this section, the motion control of a planar $4R\underline{P}R$ manipulator is studied in detail. Different control topologies detailed in this chapter were implemented on this manipulator, and a thorough comparison is carried out among the closed-loop performances. The architecture of the planar manipulator under study is described in detail in Section 3.3.1 and is shown in Figure 3.3. As explained earlier, in this manipulator, the moving platform is supported by four limbs of identical $R\underline{P}R$ kinematic structure. Complete Jacobian and singularity analyses of this manipulator is presented in Section 4.6, and the dynamic analysis of the manipulator under constant mass and variable mass assumptions is given in Sections 5.3.1 and 5.3.2, respectively. Since dynamic analyses of the system using these two approaches are quite similar, the controllers are being applied only on constant-mass dynamic models. In what follows, different control strategies in the task space are given, and then the control topologies in the joint space are implemented on the manipulator. These control strategies are given in a sequence as in Section 6.3 from simple to more complete and their closed-loop performances have been analyzed and compared, respectively.

6.8.1 Decentralized PD Control

The first control strategy introduced for parallel robots consists of the simplest form of feedback control in such manipulators. In this control structure depicted in Figure 6.5, a number of disjoint linear controllers are used in a feedback structure at each error component. For planar manipulators, the motion variable in task space is a three-dimensional vector consisting of the two position components in addition to one orientation angle, $\mathcal{X} = \begin{bmatrix} x_G & y_G & \phi \end{bmatrix}^T$; therefore, the tracking error is defined as $e_x = \begin{bmatrix} e_x & e_y & e_\phi \end{bmatrix}$. The decentralized controller, therefore, consists of three disjoint proportional derivative (PD) controllers each acting on an error component. The PD controller is denoted by $\boldsymbol{K}_d s + \boldsymbol{K}_p$ block in Figure 6.5, in which $\boldsymbol{K}_d$ and $\boldsymbol{K}_p$ are 3×3 diagonal matrices denoting the derivative and proportional controller gains, respectively.

The output of the controller is denoted by $\mathcal{F}$, which is defined by a three-dimensional wrench in the task space. Therefore, for a planar manipulator of three-degrees-of-freedom, this wrench consists of the following three components: $\mathcal{F} = \begin{bmatrix} F_x & F_y & \tau_\phi \end{bmatrix}^T$. Note that since the output of the controller is defined in the task space, each wrench component directly manipulates the corresponding tracking error component, and therefore, the overall tracking performance of the manipulator is suitable if high controller gains are considered. In practice, the calculated output wrench is transformed into actuator forces

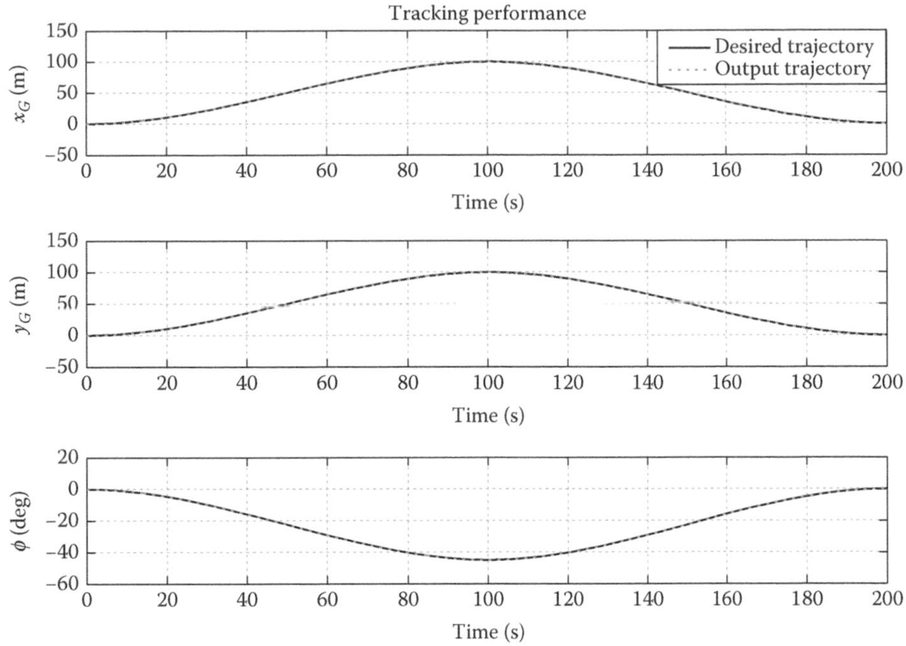

FIGURE 6.15
The desired and output trajectories of the planar $4R\underline{P}R$ manipulator in closed loop for decentralized PD control.

through the force distribution block. In this mapping, the required actuator forces required to generate such a wrench is computed and applied to the manipulator. As explained hereinbefore, for a cable-driven redundant manipulator as the planar manipulator under consideration, redundancy resolution schemes detailed in Section 6.7 may be used for this projection.

For the simulations, the geometric and inertial parameters given in Table 5.1 and the initial conditions for the states are all set at zero. Furthermore, the gravity vector is considered to be perpendicular to the plane of motion as $g = [0,\ 0,\ -9.81]^T$. For the first set of simulations, the disturbance wrench applied to the moving platform is considered to be zero, while, as shown in Figure 6.15, the desired trajectory of the moving platform is generated by cubic polynomial functions for x_d, y_d, and ϕ_d. The time of simulation is considered to be 200 s, and the PD gains used in the simulations are set to $\boldsymbol{K}_p = \boldsymbol{K}_d = 10^4 \cdot \text{diag}[1, 1, 100]$, after a number of trial-and-error nominations.

The tracking error of the moving platform is illustrated in Figure 6.16. As seen in Figure 6.15, the desired and final closed-loop motions of the system are not distinguishable, and as seen in greater detail in Figure 6.16, the decentralized PD controller for the manipulator is capable of reducing the tracking errors to less than 0.02 m in position and less than 0.012° in orientation.

The Cartesian wrench $\mathcal{F} = [F_x, F_y, \tau_\phi]^T$ generated by the PD control action at the moving platform are illustrated in Figure 6.17. As seen in this figure, Cartesian wrenches have similar pattern to the desired trajectory accelerations, which are linear for cubic trajectories.

Similarly, the actuator forces of the manipulator are illustrated in Figure 6.18. In this figure, two sets of actuator forces have been illustrated. The dashed lines illustrate the base solution of the optimization problem introduced in redundancy resolution scheme by Equation 6.76, while the solid line is the final optimization solution of redundancy

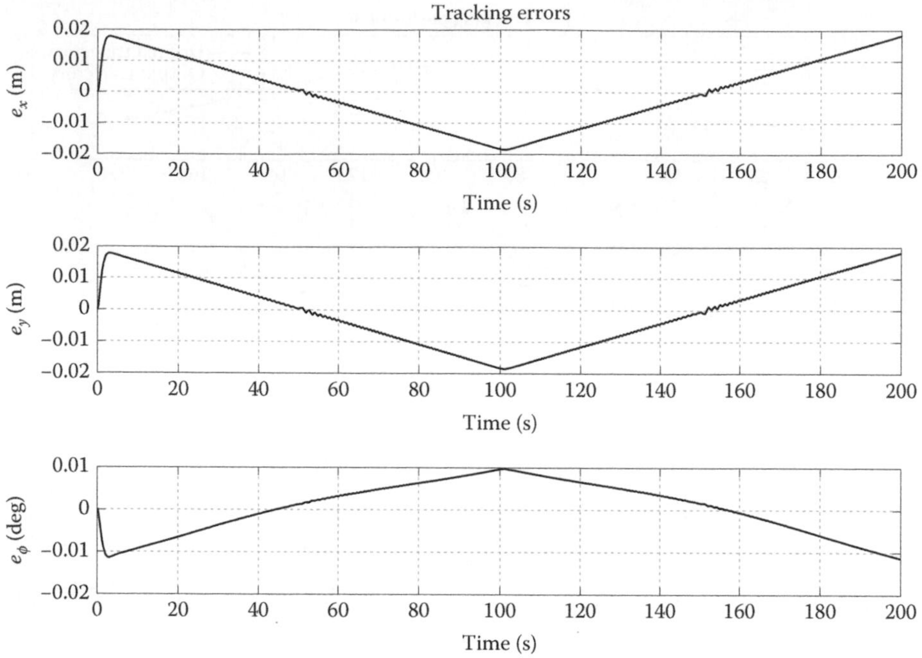

FIGURE 6.16
The closed-loop tracking performance of the planar 4$R\underline{P}R$ manipulator for decentralized PD control.

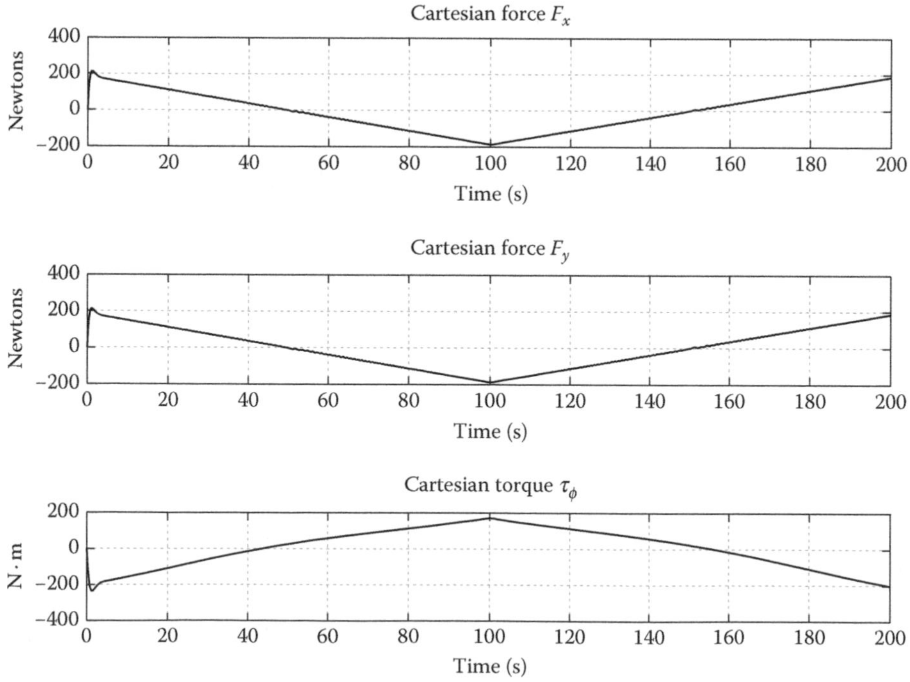

FIGURE 6.17
Cartesian wrench of the planar 4$R\underline{P}R$ manipulator generated by PD controller to generate trajectory shown in Figure 5.12.

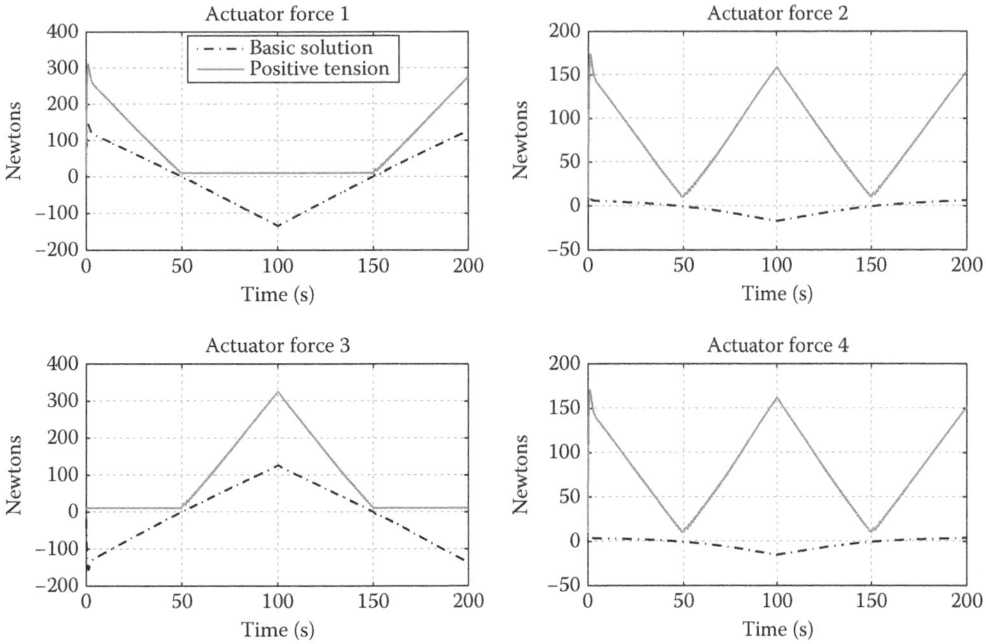

FIGURE 6.18
The actuator forces of the planar 4$R\underline{P}R$ manipulator generated by PD controller to generate trajectories shown in Figure 6.15.

resolution problem detailed in Equation 6.79. The solution is obtained through iterative numerical method. As seen in Figure 6.18, it is observed that since the manipulator moves in positive x and y directions, the actuator forces of the first and third limbs are dominant. Furthermore, the tension forces derived from the redundancy resolution scheme remain positive for all configurations.

To examine the PD controller performance in practice, another desired trajectory for the manipulator is considered, in which the same maneuvers required to be carried out in 200 s in Figure 6.15 are requested to be performed only in 20 s. The tracking error of the PD controller for such a trajectory is illustrated in Figure 6.19 while the actuator forces are given in Figure 6.20. As seen in Figure 6.19, the tracking errors increase 100 times, whereas the actuator forces required are also 100 times more than before.

This is due to the fact that the controller gains must be tuned experimentally based on physical realization of the controller in practice by trial and error. Therefore, the final controller gains are obtained as a trade-off between the transient response and the steady-state errors at different configurations. Since the dynamic behavior of the manipulator is configuration dependent, finding suitable controller gains to result in the required performance in all configurations is a difficult task, and usually the designer chooses the controller gains such that the required tracking error of the manipulator at its usual working condition is satisfied. This yields varying performance in other configurations where the tuning process is not applied to, as seen in Figure 6.19.

To examine the performance of the PD controller to attenuate the effect of external disturbance and measurement noise, two more simulations are executed for the closed-loop system. In the first simulation, a unit step disturbance force is applied to the system under the original trajectory shown in Figure 6.15. To examine the effect of disturbance force and

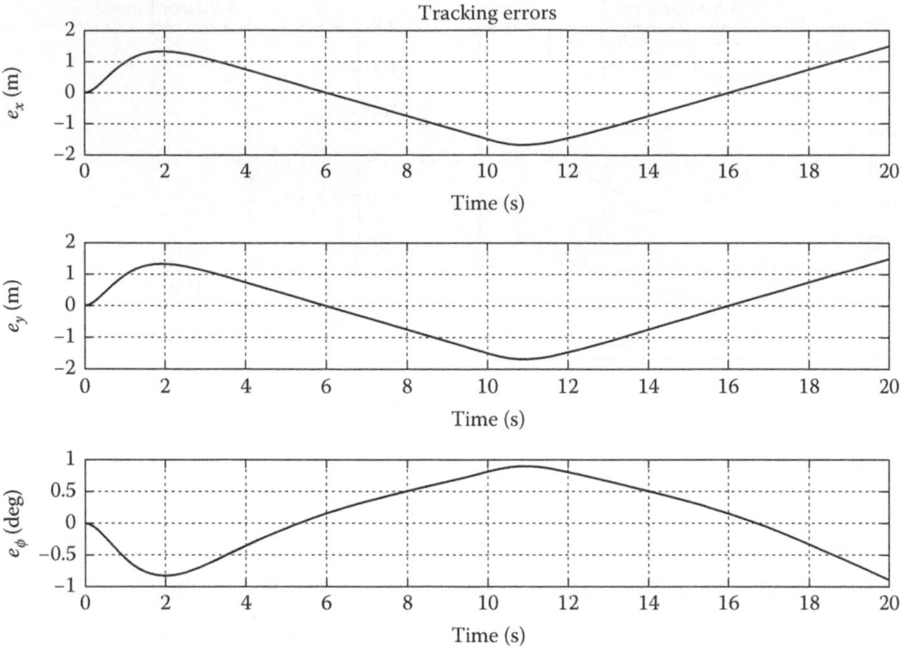

FIGURE 6.19
The closed-loop tracking performance of the planar 4$R\underline{P}R$ manipulator for decentralized PD Control for a more rapid maneuver.

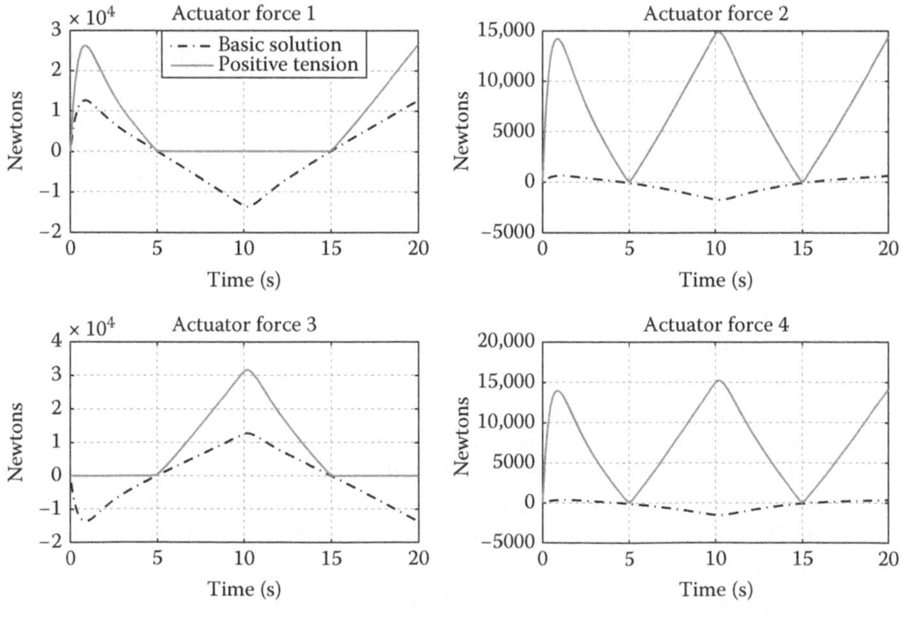

FIGURE 6.20
The actuator forces of the planar 4$R\underline{P}R$ manipulator required to generate a more rapid trajectory with a PD controller.

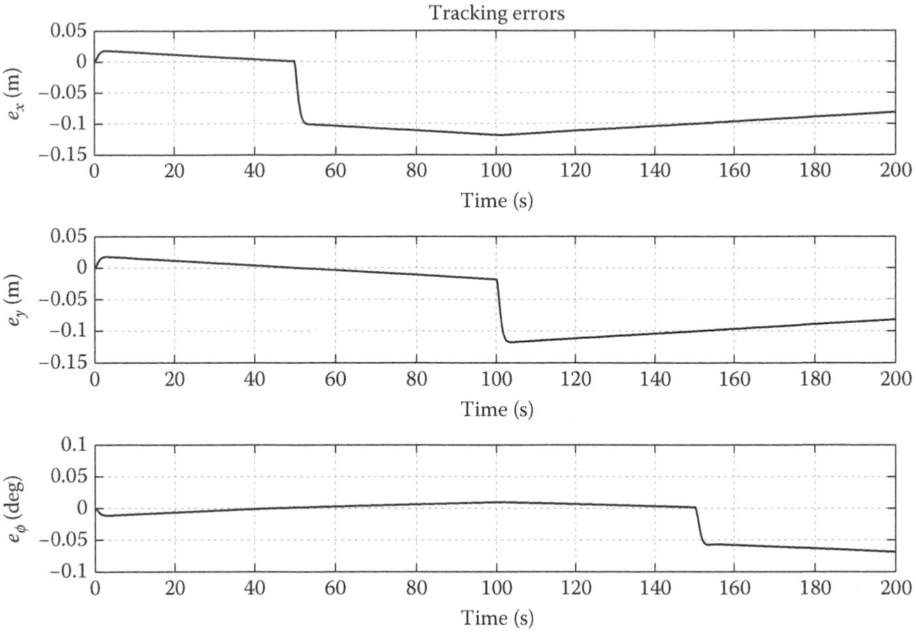

FIGURE 6.21
The closed-loop tracking performance of the planar $4R\underline{P}R$ manipulator for decentralized PD control in the presence of disturbance.

torques separately, a 1 kN step disturbance force in the x direction F_{d_x} is applied at time 50 s, then another 1 kN step disturbance force in the y direction F_{d_y} is applied at time 100 s, and finally a 1 kN.m disturbance torque in the ϕ direction τ_{d_ϕ} is applied at time 150 s. The controller gains are set at $\boldsymbol{K}_p = \boldsymbol{K}_d = 10^4 \cdot \text{diag}[1, 1, 100]$ as hereinbefore.

The closed-loop tracking error of the system is shown in Figure 6.21. By comparing this figure to Figure 6.16, it can be seen that the effect of external disturbance is not completely attenuated by PD controller, and the abrupt change in the tracking error at the instances when the disturbance wrench are applied is clearly seen in the response. Furthermore, as a result of nonvanishing disturbances applied to the system, static errors are accumulated in the closed-loop response of the system. Figure 6.22 illustrates the Cartesian wrench generated by the PD controller to carry out the required maneuver in the presence of disturbance. The effect of step disturbance wrench on the required controller forces is also clearly seen at times 50, 100, and 150 s.

The effect of measurement noise is considered next by considering that all the measured variables $\boldsymbol{\mathcal{X}}$ are contaminated with a Gaussian noise with an amplitude of %0.01 peak values of the original signals. Although the amount of noise is very limited in this simulation, since high-gain controllers have to be used in the decentralized PD controller, the effect of noise on tracking performance is significant. As shown in Figures 6.23 and 6.24, although the tracking error is not much increased compared to that of the system without noise, the required actuator forces to carry out such a maneuver are very oscillatory. This is due to the fact that high PD gains are used to accommodate the required tracking performance, and although the amplitude of noise is about 10^{-4} times the peak values of the measurement signals, since the PD gains are higher than 10^4, the signature of noise is apparently seen in

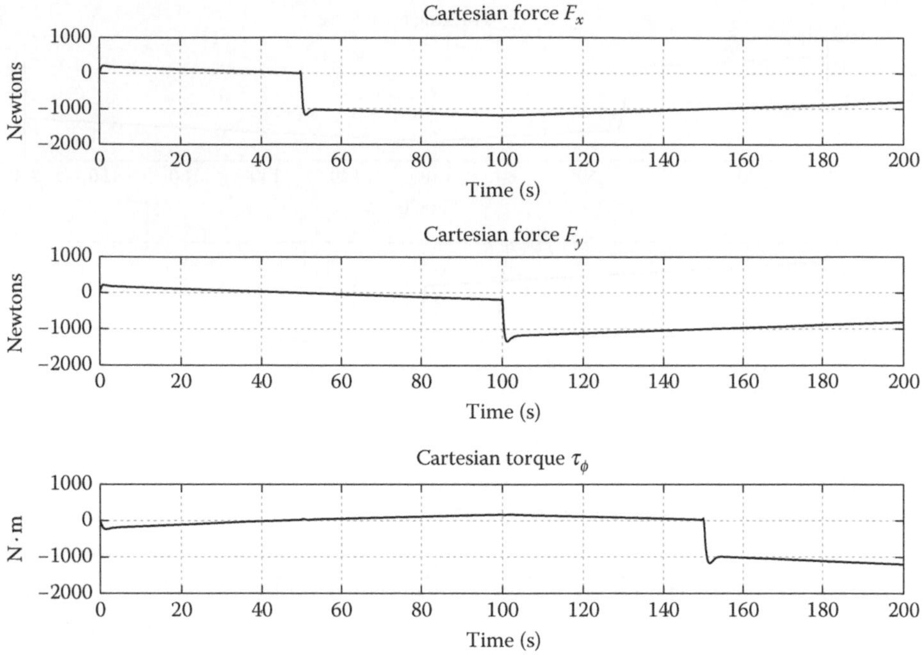

FIGURE 6.22
Cartesian wrench of the planar $4R\underline{P}R$ manipulator generated by PD controller to generate trajectory shown in Figure 6.15 in the presence of disturbance.

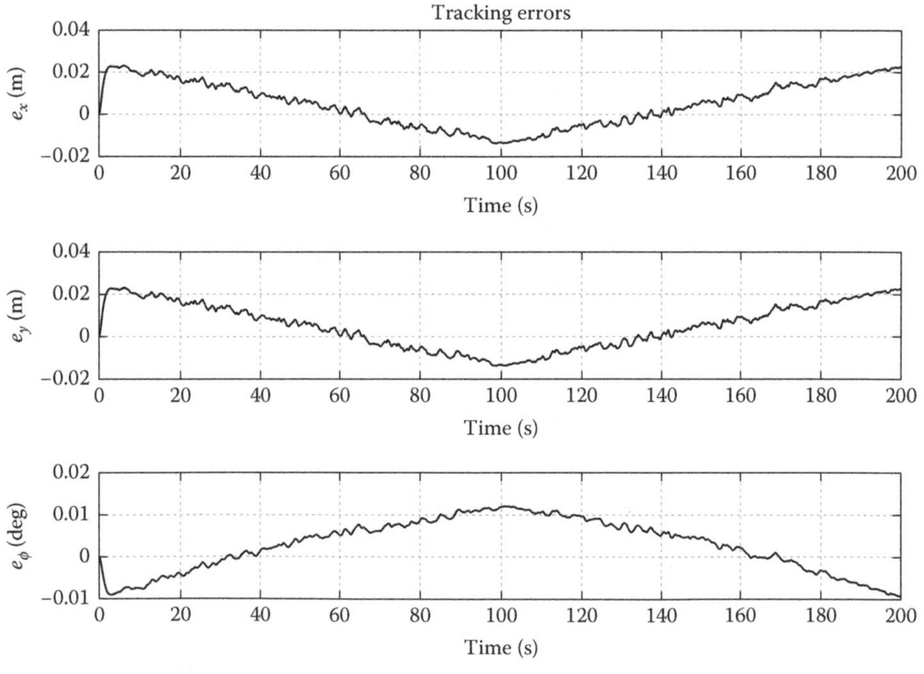

FIGURE 6.23
The closed-loop tracking performance of the planar $4R\underline{P}R$ manipulator for decentralized PD control in the presence of measurement noise.

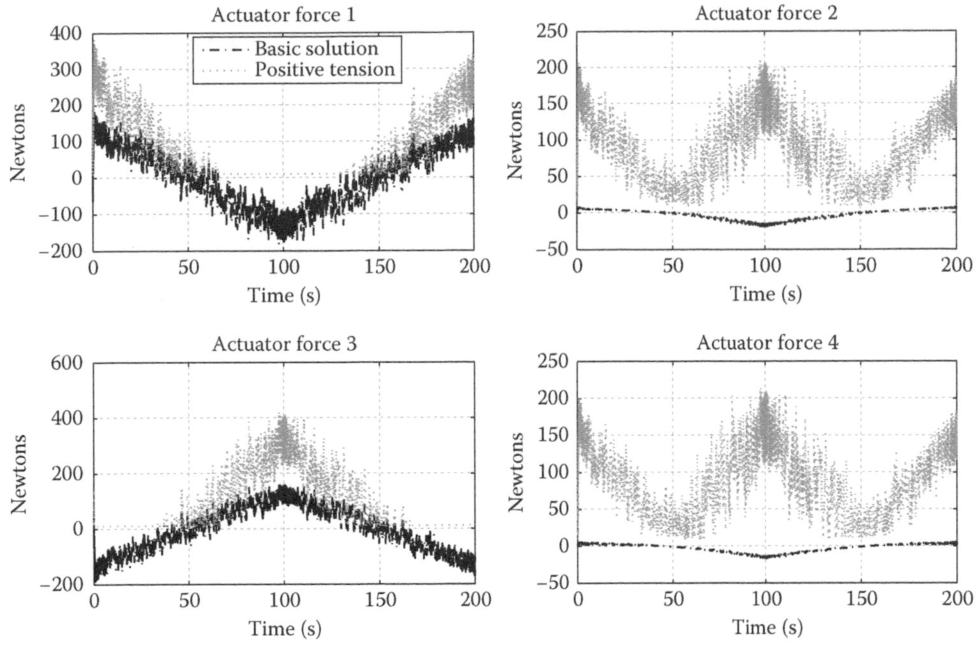

FIGURE 6.24
The actuator forces of the planar $4R\underline{P}R$ manipulator in the presence of measurement noise.

the required actuator forces. In fact, generating such actuator forces are infeasible in practice, and hence, reaching the tracking performance shown in Figure 6.23 is not possible.

6.8.2 Feed Forward Control

As seen in the previous section, the tracking performance of the PD controller implemented in the task space is not sufficient for different configurations, and in the presence of measurement noise and external disturbances applied to the manipulator. To compensate for such effects, a feed forward wrench denoted by $\mathcal{F}_{ff}$ may be added to the structure of the controller as depicted in Figure 6.6. This term is generated from the dynamic model of the manipulator in task space by Equation 6.2. Furthermore, note that to generate this term, the dynamic formulation of the robot and its kinematic and dynamic parameters are needed.

In practice, exact information on dynamic matrices is not available, and therefore, in simulations, estimates of these matrices with %10 perturbation in all kinematic and inertial parameters are used in this derivation. For the sake of comparison, the PD controller gains are set to $\boldsymbol{K}_p = \boldsymbol{K}_d = 10^4 \cdot \text{diag}\,[1, 1, 100]$, as in the case of decentralized PD control. Figure 6.25 illustrates the closed-loop tracking performance of the planar $4R\underline{P}R$ manipulator with feed forward control for the typical trajectory depicted in Figure 6.15. Comparing this figure to that of PD control illustrated in Figure 6.16, it can be seen that the feed forward action is capable of improving the tracking performance 5–10 times of that of pure PD control.

The Cartesian wrench generated by the feed forward controller for such a maneuver is given in Figure 6.26. In this figure, the total Cartesian wrench is shown with a solid line, while its components, namely the PD action and the feed forward action are given in

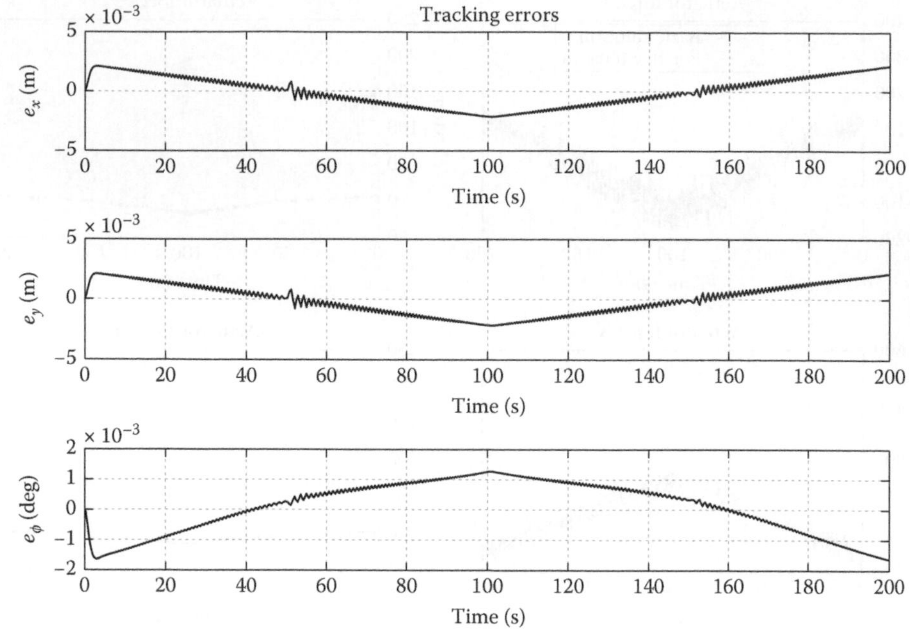

FIGURE 6.25
The closed-loop tracking performance of the planar $4R\underline{P}R$ manipulator with feed forward control.

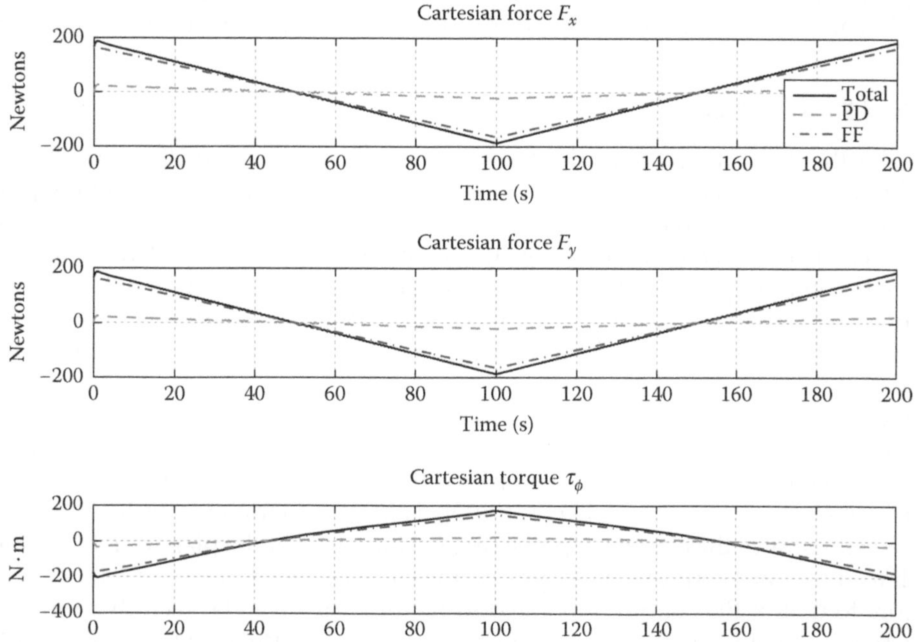

FIGURE 6.26
Cartesian wrench of the planar $4R\underline{P}R$ manipulator generated by feed forward controller to generate trajectories shown in Figure 6.15.

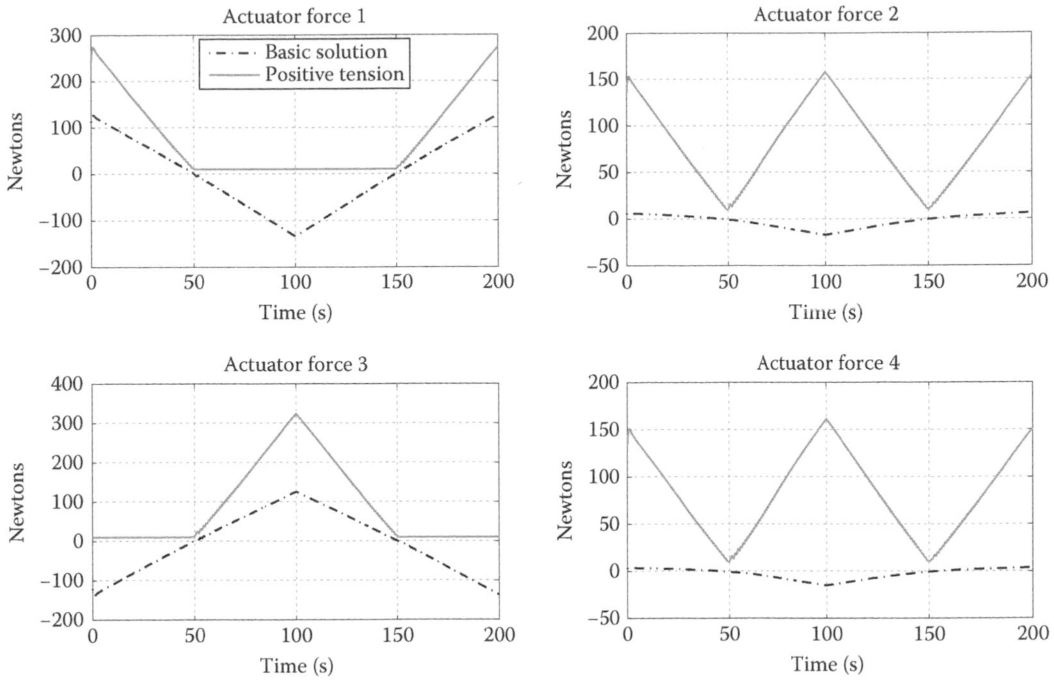

FIGURE 6.27
The actuator forces of the planar 4$R\underline{P}R$ manipulator with feed forward controller required to generate trajectories shown in Figure 6.15.

dashed and dashed-dotted line, respectively. As seen in this figure, since the knowledge of the system model is quite accurate, and only %10 perturbation in kinematic and inertial parameters of the model are considered in this simulation, the main component of the Cartesian wrench contributing is the feed forward term, which generates about %90 of the generated wrench. The PD action in this simulation contributes to only about %10 of the generated force. The total force needed to produce the required motion is almost the same as in pure PD control, while the tracking performance is significantly improved. This is mainly due to the contribution of feed forward term to the accurate motion of the manipulator, as expected. The actuator forces, which are also illustrated in Figure 6.27, are obtained by numerical solution of the redundancy resolution problem in this case. As seen in this figure, the tension forces derived from the redundancy resolution scheme in this simulation remain positive for all configurations.

To understand the significance of the feed forward term and how it improves the tracking performance, in the next simulation, the rapid trajectory is examined, in which the whole maneuver is traversed in only 20 s. The tracking performance of the closed-loop system for this rapid motion is shown in Figure 6.28. Comparing this response to that of pure PD control depicted in Figure 6.19 shows 10 times improvement of tracking error in position and 5 times improvement in orientation error with feed forward control.

In the next simulation, it is assumed that the knowledge of the designer about the system model is very poor, and %50 perturbation in all kinematic and inertial parameters are considered in derivation of the feed forward control term. It is expected that the tracking performance be significantly affected because of this lack of knowledge of system model. Figures 6.29 and 6.30 illustrate the tracking performance and contribution of the PD and feed forward term in the total generated Cartesian wrench. As seen in Figure 6.29, lack

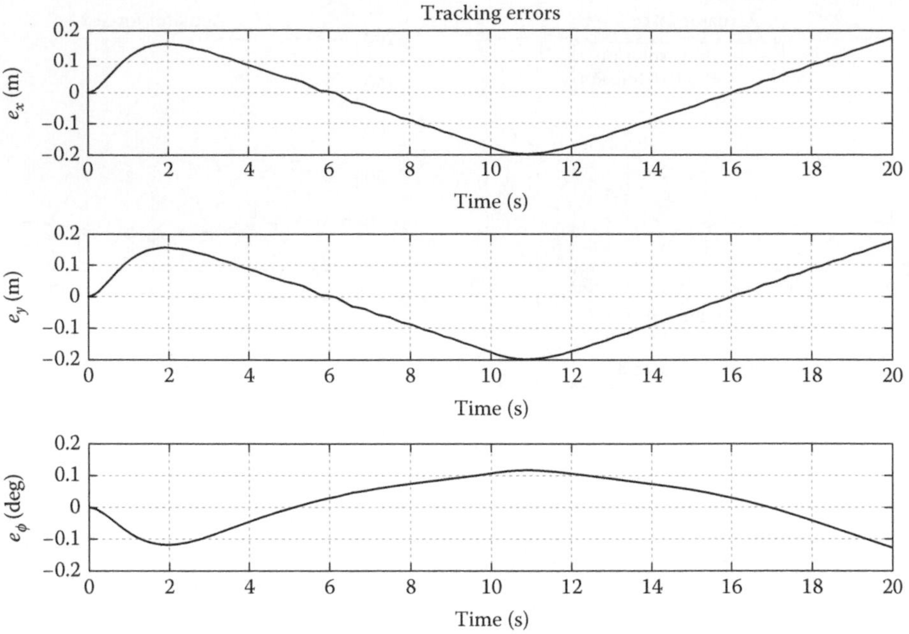

FIGURE 6.28
The closed-loop tracking performance of the planar 4$R\underline{P}R$ manipulator with feed forward control for a rapid trajectory.

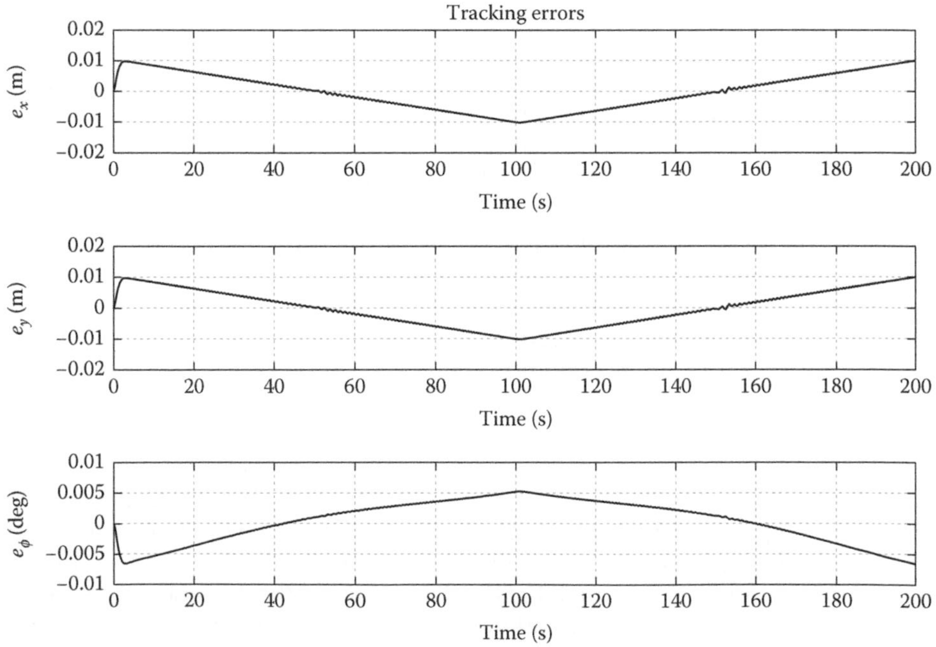

FIGURE 6.29
The closed-loop tracking performance of the planar 4$R\underline{P}R$ manipulator with feed forward control; %50 perturbation in kinematic and inertial parameters.

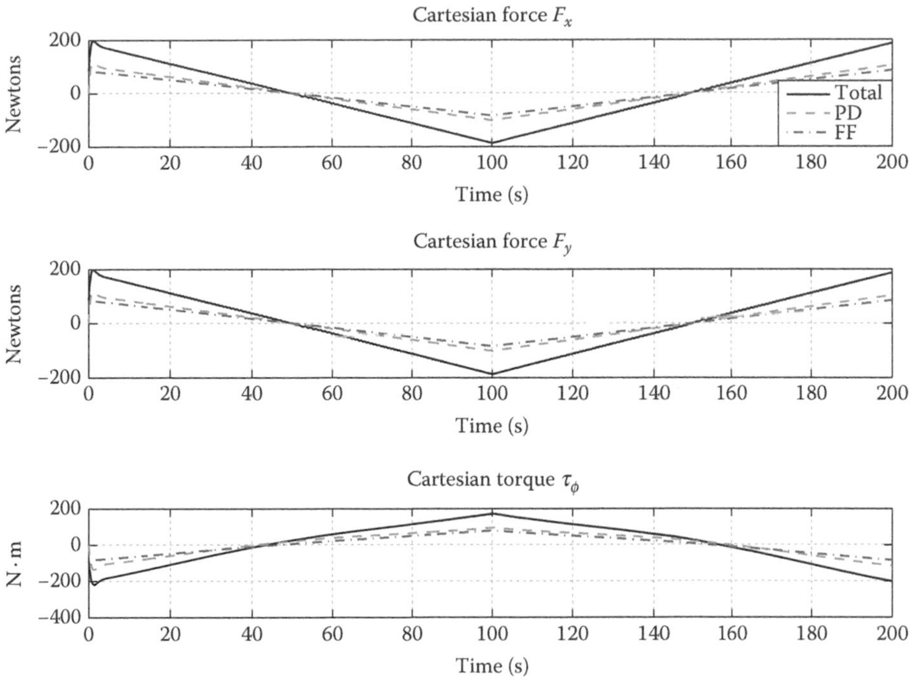

FIGURE 6.30
Cartesian wrench of the planar $4R\underline{P}R$ manipulator generated by feed forward controller; %50 perturbation in kinematic and inertial parameters.

of knowledge of the model parameters significantly affect the tracking errors. Comparing this figure to Figure 6.25, it can be concluded that the tracking errors for the positions are increased by a factor of 5 while this factor is 2.5 for the orientation error.

Figure 6.30 illustrates the total Cartesian wrench generated by this control law, and its PD and feed forward components. As seen in this figure, since the knowledge of model parameters is poor in this simulation, the contribution of the PD control term is increased to almost %50 of the total force. This is in complete agreement to the fact that for a model-based control structure such as the feed forward scheme under consideration, if the knowledge of the model is poor, the contribution of PD term becomes more significant. Finally, the effect of disturbance and noise is also examined on this controller, and similar results to that of pure PD control is observed. Since this controller structure is model-based, and in these simulations high PD gains are considered for the controller, poor performance in the presence of noise and disturbance is expected.

6.8.3 Inverse Dynamics Control

As seen in the previous sections, the tracking performance of a decentralized PD controller implemented in the task space is not uniform at different configurations and in the presence of measurement noise and external disturbances. To compensate for such effects, a feed forward wrench is added to the structure of the controller, by which the shortcomings of the decentralized controller is partially remedied. However, the closed-loop performance still faces a number of limitations that cannot be overcome.

To examine these limitations on performance, inverse dynamics control is implemented on the planar manipulator. The general structure of the inverse dynamics control applied

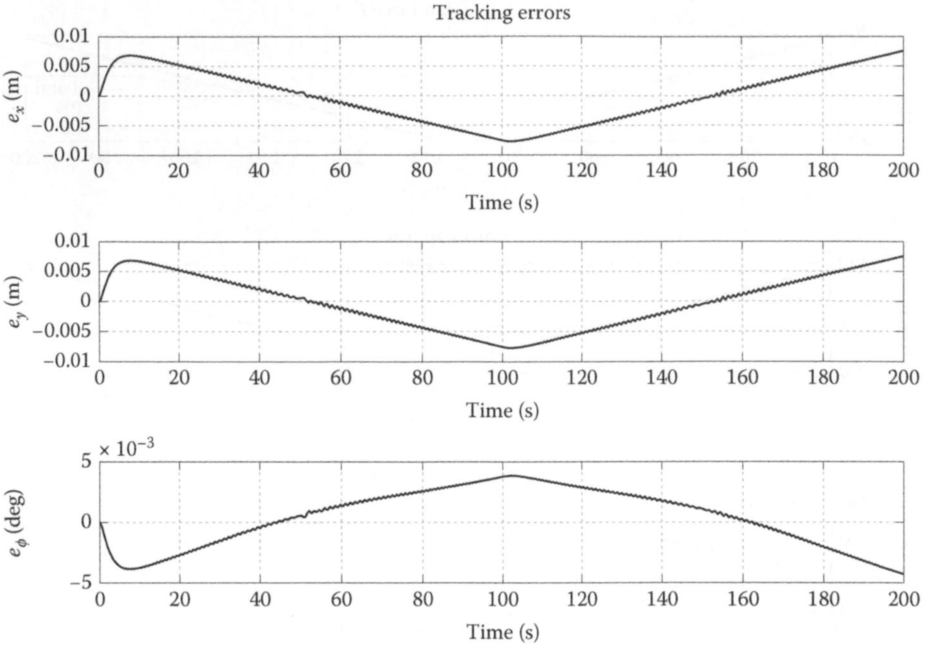

FIGURE 6.31
The closed-loop tracking performance of the planar 4$R\underline{P}R$ manipulator with inverse dynamics control.

to a parallel manipulator is depicted in Figure 6.7. As seen in this figure, a corrective wrench $\mathcal{F}_{fl}$, is added in a feedback structure to the closed-loop system, which is calculated from the Coriolis and centrifugal matrix, and the gravity vector of the manipulator dynamic formulation. Furthermore, the mass matrix acts in the forward path, in addition to the desired trajectory acceleration $\ddot{\mathcal{X}}_d$.

Note that to generate this term, the dynamic formulation of the robot and its kinematic and dynamic parameters are needed. In practice, exact knowledge of dynamic matrices are not available, and therefore a %10 perturbation in all kinematic and inertial parameters are considered in the following simulations. For the sake of comparison to the previously examined controllers, the PD controller gains are adjusted by trial and error to $\boldsymbol{K}_p = \boldsymbol{I}_{3\times3}$ and $\boldsymbol{K}_d = 2\sqrt{2}\,\boldsymbol{I}_{3\times3}$, in which $\boldsymbol{I}_{3\times3}$ denotes a 3×3 identity matrix. Note that as given in Equation 6.5, the PD controller terms are multiplied by the manipulator mass matrix, and therefore the same dimensions used in pure PD or feed forward controllers are not applicable here. However, tuning the PD controller gains in an IDC controller structure is very easy, and with a few trials, appropriate gains may be selected.

Figure 6.31 illustrates the closed-loop tracking performance of the planar 4$R\underline{P}R$ manipulator with the inverse dynamic control for the typical trajectory depicted in Figure 6.15. Comparing this figure to that of the PD control illustrated in Figure 6.16, it can be seen that the PD controller gains are tuned to obtain a similar performance as the pure decentralized PD control with much easier controller gain tuning process.

The Cartesian wrench generated by inverse dynamics controller for such a maneuver is given in Figure 6.32. In this figure, the total Cartesian wrench generated by inverse dynamics controller is shown with a solid line, while its PD component is given in a dashed line. As seen in this figure, since the knowledge of the system model is quite accurate, and only

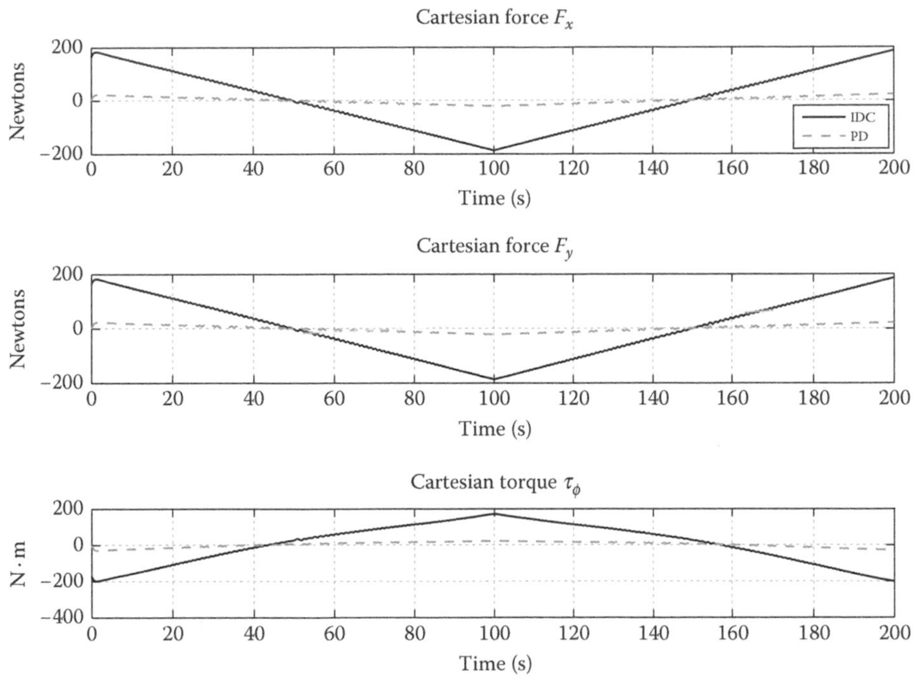

FIGURE 6.32
Cartesian wrench of the planar 4$R\underline{P}R$ manipulator generated by inverse dynamics controller to generate trajectories shown in Figure 6.15.

%10 perturbation in kinematic and inertial parameters of the model is considered in this simulation, the main component of Cartesian wrench is the inverse dynamic term, which generates about %90 of the generated wrench. The PD action in this simulation contributes to only about %10 of the generated wrench. The total force needed to produce the required motion is almost the same as in the pure PD control, while the tracking performance is also almost the same.

Figure 6.33 illustrates the actuator forces obtained from a numerical solution of the redundancy resolution problem in this case. As seen in this figure, the tension forces derived from the redundancy resolution scheme remain positive for all configurations.

To investigate the significance of inverse dynamic control and how it improves the tracking performance, in the next simulation the rapid trajectory is examined, in which the whole maneuver is traversed in only 20 s. The tracking performance of the closed-loop system for this rapid motion is shown in Figure 6.34. Comparing this response to that of pure PD control depicted in Figure 6.19 shows about 2 times improvement. However, the performance improvement compared to that of feed forward controller is not significant.

In the next simulation, it is assumed that the knowledge of the system model is very poor, and %50 perturbation in all kinematic and inertial parameters are considered in derivation of the inverse dynamics control law. It is expected that the tracking performance be significantly affected because of this lack of knowledge of system model. Figures 6.35 and 6.36 illustrate the tracking performance and the contribution of the PD term with respect to the total generated Cartesian wrench.

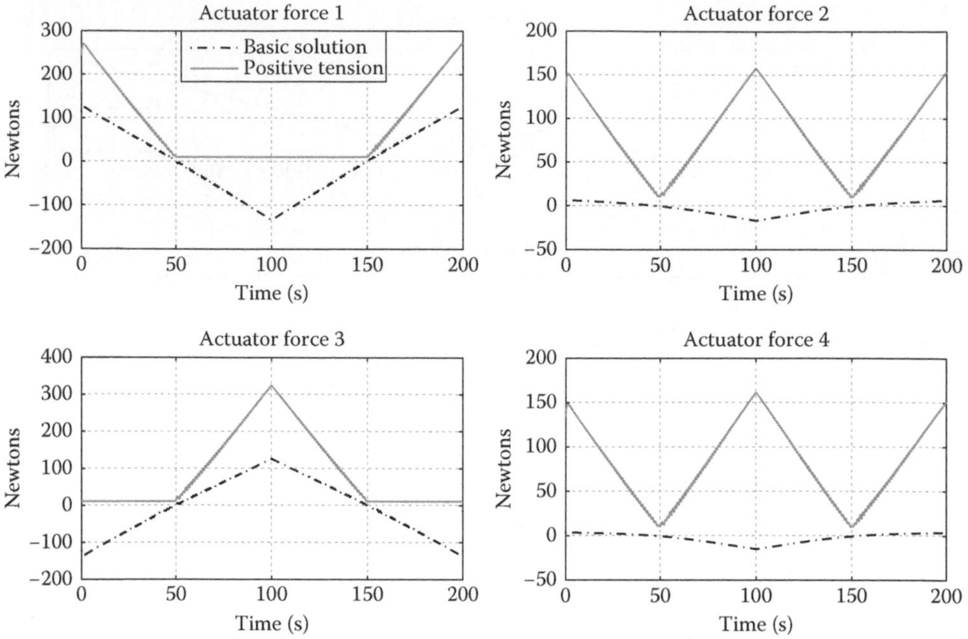

FIGURE 6.33
The actuator forces of the planar 4$R\underline{P}R$ manipulator with inverse dynamic controller required to generate trajectories shown in Figure 6.15.

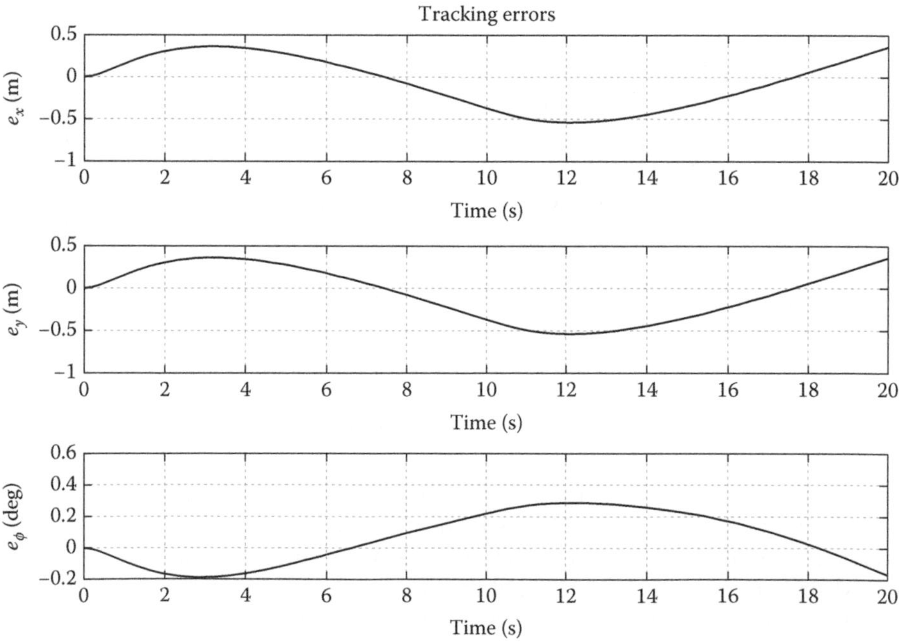

FIGURE 6.34
The closed-loop tracking performance of the planar 4$R\underline{P}R$ manipulator with inverse dynamics control for a rapid trajectory.

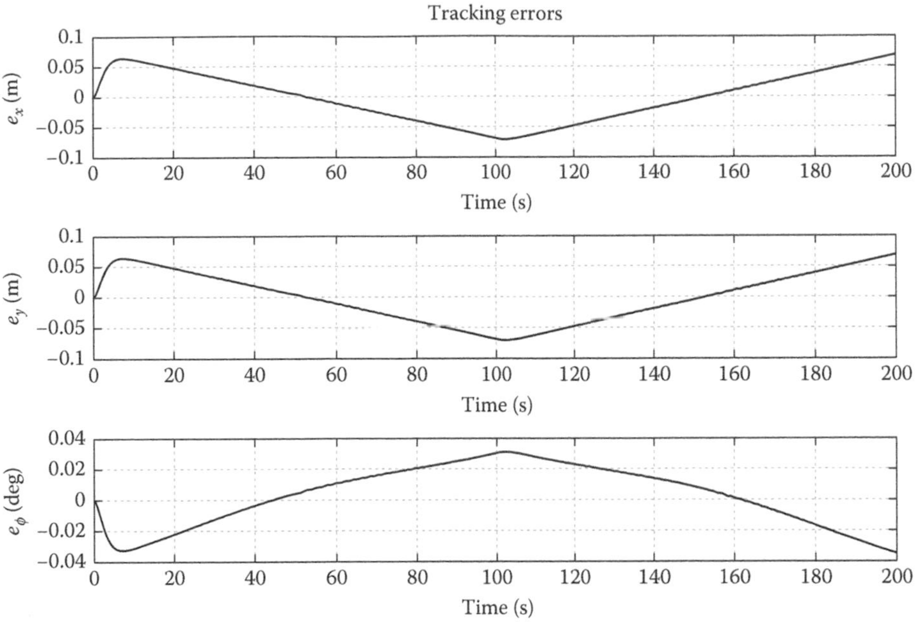

FIGURE 6.35
The closed-loop tracking performance of the planar $4R\underline{P}R$ manipulator with inverse dynamics control; %50 perturbation in kinematic and inertial parameters.

As seen in Figure 6.35, lack of knowledge of the model parameters significantly affects the tracking errors. Comparing this figure to Figure 6.31, it can be concluded that the tracking errors for the positions and orientation increase by a factor of 10. Figure 6.36 illustrates the total Cartesian wrench generated by this control law and its PD component. As seen in this figure, since the knowledge of model parameters is poor in this simulation, the contribution of the PD control term is increased to almost %50 of the total wrench. This is in complete agreement to the fact that for a model-based control structure such as IDC under consideration, if the knowledge of the model is poor, the contribution of the PD term becomes more significant.

Finally, the effect of disturbance and noise is also examined on this controller, and similar results to that of pure PD control is observed. In first simulation, a sequence of unit step disturbance forces at times 50, 100, and 150 s is applied to the system, similar to that applied to the PD controller structure. The closed-loop tracking error of the system is shown in Figure 6.37. By comparing this figure to the PD controller performance in the presence of disturbance, namely Figure 6.21, it can be seen that the effect of external disturbance is even more in IDC than in the pure PD controller. This is because of the fact that in inverse dynamics controller the controller gains are reduced to smaller values, and since the only terms counteracting the effect of disturbance is the PD term, the tracking performance is declined. Furthermore, as a result of nonvanishing disturbances applied to the system, static errors are accumulated in the closed-loop response of the system.

The effect of measurement noise, however, is reduced in the required control actions in inverse dynamics control, since the PD gains are selected much smaller. Nevertheless, as shown in Figure 6.38, the required actuator bandwidth is not affordable.

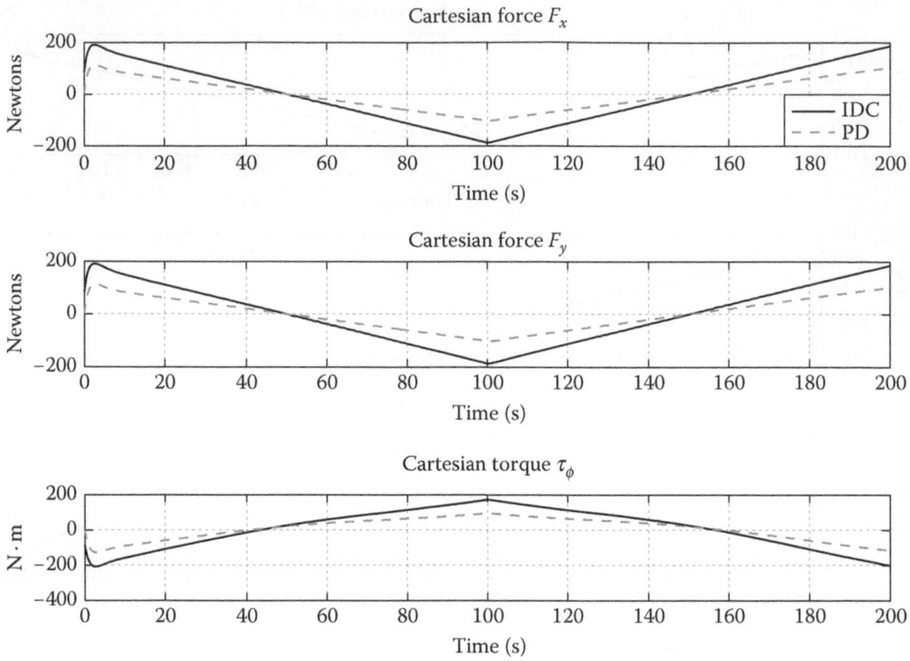

FIGURE 6.36
Cartesian wrench of the planar 4$R\underline{P}R$ manipulator generated by inverse dynamics controller; %50 perturbation in kinematic and inertial parameters.

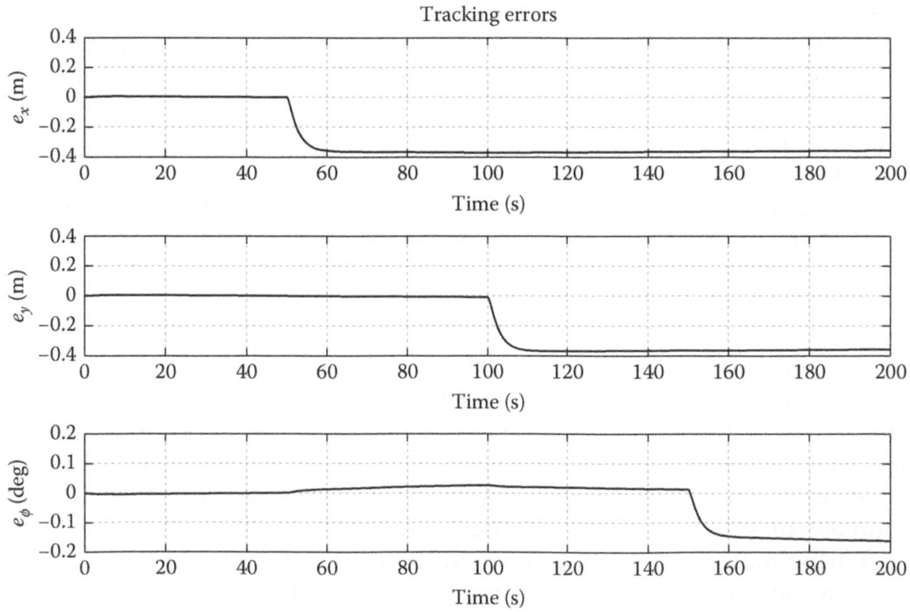

FIGURE 6.37
The closed-loop tracking performance of the planar 4$R\underline{P}R$ manipulator for IDC in the presence of disturbance.

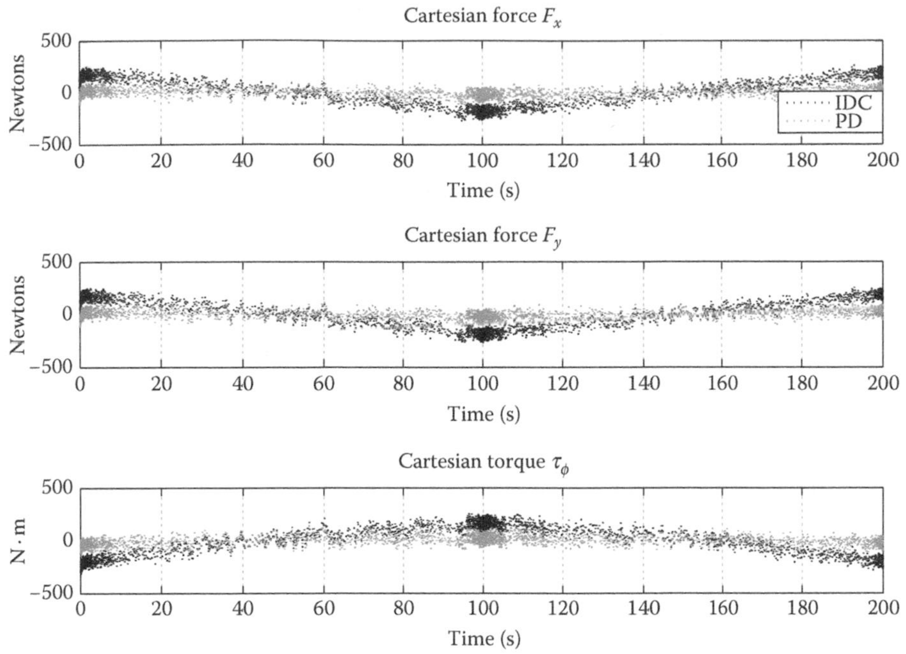

FIGURE 6.38
Cartesian wrench required to perform the desired maneuver in the presence of measurement noise.

6.8.4 Partial Linearization IDC

Inverse dynamics control has several features making it quite attractive in practice. However, as explained in Section 6.3.3, to apply this method, complete knowledge of the dynamic formulation matrices is required. To develop the simplest possible implementable IDC, let us recall the partial linearization IDC developed in Section 6.3.4. In the simplest form of IDC, the control effort $\mathcal{F}$ consists of two terms as given in Equation 6.11. The first term $\mathcal{F}_{pd}$ is generated by the PD controller implemented on the motion error e_x, given in Equation 6.12, while the second term is considered to be only the gravity vector of the manipulator at any configuration $G(\mathcal{X})$, given by Equation 6.13. In this implementation, the computationally intensive Coriolis and centrifugal term is not used in the feedback. For the sake of comparison to previously examined controller structure, in the following simulations, a %10 perturbation is considered in all kinematic and inertial parameters and the PD controller gains are set at $K_p = I_{3\times3}$ and $K_d = 2\sqrt{2}\, I_{3\times3}$ in which $I_{3\times3}$ denotes a 3×3 identity matrix, similar to that of inverse dynamics control.

The tracking performance of the closed-loop system with partially linearized IDC is shown in Figures 6.39 and 6.40, in which in the first simulation the typical trajectory depicted in Figure 6.15 is considered, while in the second simulation the rapid trajectory, in which the whole maneuver is traversed in only 20 s is examined. The tracking errors in both cases are very similar to the case where in a complete linearization is applied, and as it can be seen in the comparison of these two figures to their corresponding cases illustrated in Figures 6.31 and 6.34, only the error in orientation ϕ slightly increases for the partial linearized IDC.

These simulations confirm the analysis given in Section 6.3.4, which shows that even if the Coriolis and centrifugal matrix is not used in the feedback, and the closed-loop

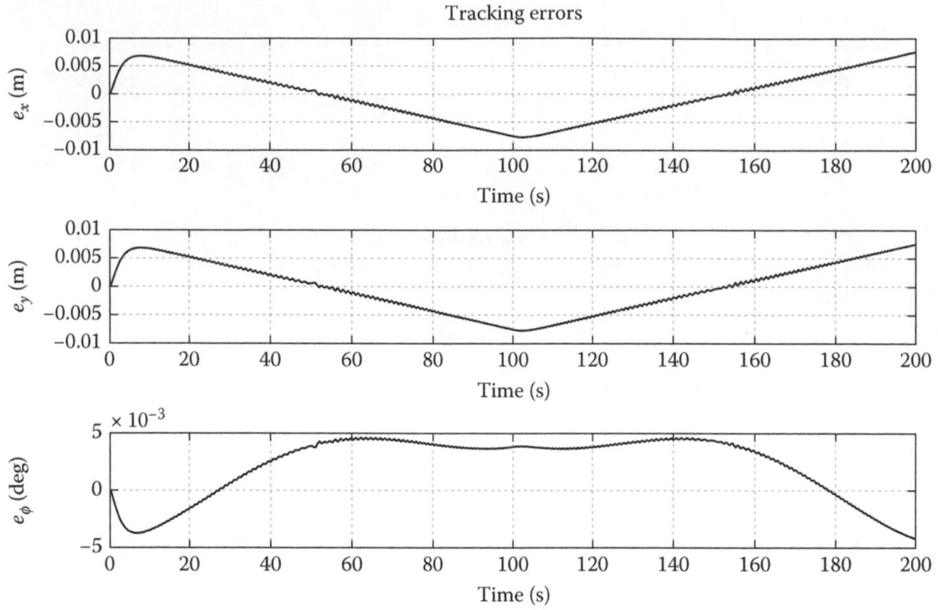

FIGURE 6.39
The closed-loop tracking performance of the planar $4R\underline{P}R$ manipulator with partially linearized IDC.

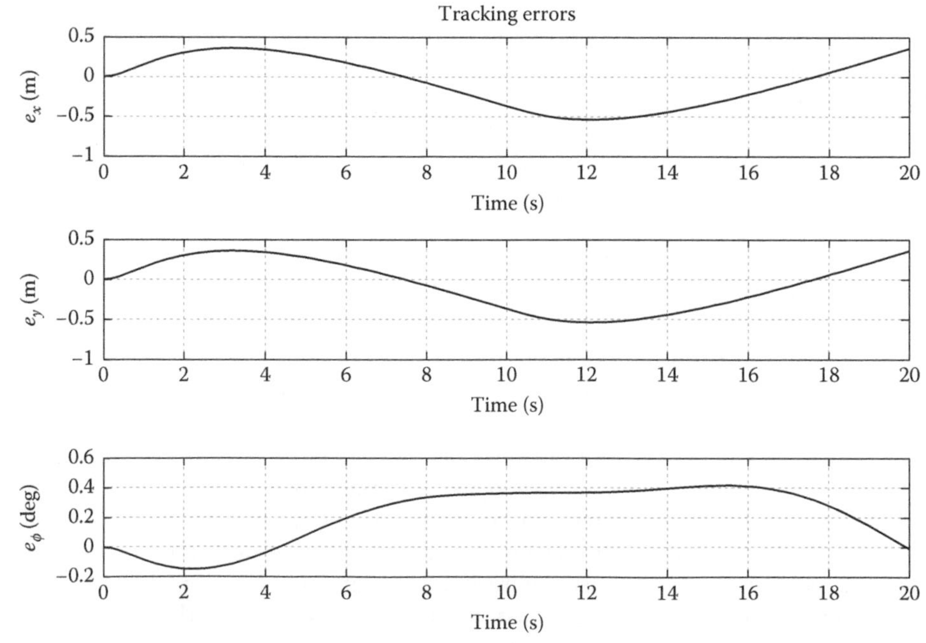

FIGURE 6.40
The closed-loop tracking performance of the planar $4R\underline{P}R$ manipulator with partially linearized IDC for a rapid trajectory.

dynamics is not completely linearized, the PD control structure with gravity compensation can still lead to a suitable transient performance. Furthermore, it verifies that the skew-symmetric property of the dynamic formulation matrices provides sufficient assurance that partial linearization IDC does not result in tracking divergence in practice.

6.8.5 Robust Inverse Dynamics Control

The inverse dynamics control performance relies on a complete linearization of nonlinear dynamic terms by feedback. This requirement may never be satisfied in practice due to modeling uncertainty, and therefore, it may practically limit the performance of IDC control in parallel robots. To compensate for such modeling uncertainties in the closed-loop performance of the manipulator, a robustifying term $\boldsymbol{\delta}_a$ is introduced in Section 6.4.1, which is added to the usual IDC control structure. Figure 6.8 illustrates the robust IDC implementation block diagram on a parallel robot, in which the corrective term $\boldsymbol{\delta}_a$ is added as a wrench in task space. This corrective term is given in Equation 6.37, in which the vector $\boldsymbol{v}$ is defined in Equation 6.32.

In the following simulations, robust inverse dynamics control is applied to the planar manipulator, in which to examine high amount of uncertainties in the model, %50 perturbation in all kinematic and inertial parameters are considered. For the sake of comparison to previously examined controller structure, the PD controller gains are set at $\boldsymbol{K}_p = \boldsymbol{I}_{3\times3}$ and $\boldsymbol{K}_d = 2\sqrt{2}\,\boldsymbol{I}_{3\times3}$ as given in IDC simulations, in which $\boldsymbol{I}_{3\times3}$ denotes a 3×3 identity matrix. Furthermore, the robust controller parameters are set at $\rho = 50$, $\epsilon = 10$ by careful examination of different simulations for typical trajectory maneuvers depicted in Figure 6.15.

Figure 6.41 illustrates the closed-loop tracking performance of the planar $4R\underline{P}R$ manipulator with robust inverse dynamics control in comparison to that for regular IDC for the typical trajectory depicted in Figure 6.15. This comparison clearly indicates that, by using robust control, the tracking errors are reduced by a factor of 2 compared to that of the regular IDC. This significant improvement is obtained through the contribution of corrective term added in the robust control structure, the amount of which is illustrated in Figure 6.42. In this figure, the total robust inverse dynamics control effort is plotted by solid line, while its PD and corrective terms are plotted in dashed and dashed-dotted lines, respectively. As can be seen in this figure, about %20 of the total force is generated by the PD term, while the corrective term is contributing by about the same amount, and the remaining %60 of the control effort is generated by inverse dynamic terms.

The transient performance of robust IDC may be improved by further tuning of PD gains and the robust controller gain ρ. To show that the derivative gains of the PD controller are doubled, that is, $\boldsymbol{K}_d = 4\sqrt{2}\,\boldsymbol{I}_{3\times3}$, to reduce the overshoot in the transient response of the closed-loop system. The tracking performance for this case is shown in Figure 6.43. As it can be seen in this figure, all the responses have a smooth transient with almost the same amount of steady-state errors.

Finally, the effect of disturbance and noise is also examined on this controller, and similar results to that of IDC are observed. In the first simulation, a sequence of unit step disturbance forces at times 50, 100, and 150 s is applied to the system, similar to that applied to IDC structure. The closed-loop tracking error of the system is shown in Figure 6.44. It can be seen that like previous controllers, the effect of nonvanishing external disturbance cannot be well attenuated with this structure, and yet significant steady-state errors are observed in the performance. The effect of measurement noise is also very

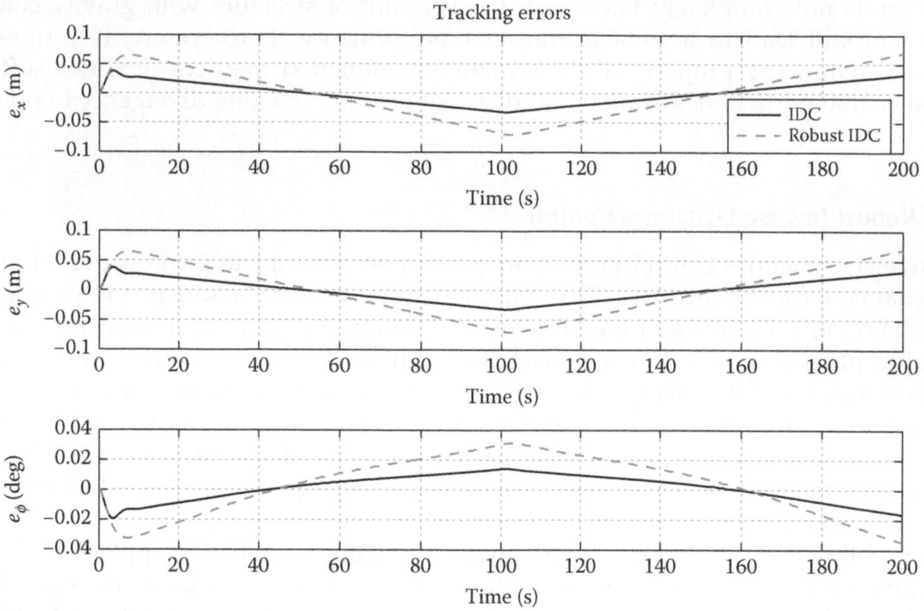

FIGURE 6.41
The closed-loop tracking performance of the planar $4R\underline{P}R$ manipulator with robust IDC compared to that of IDC; %50 perturbation in all kinematic and inertial parameters.

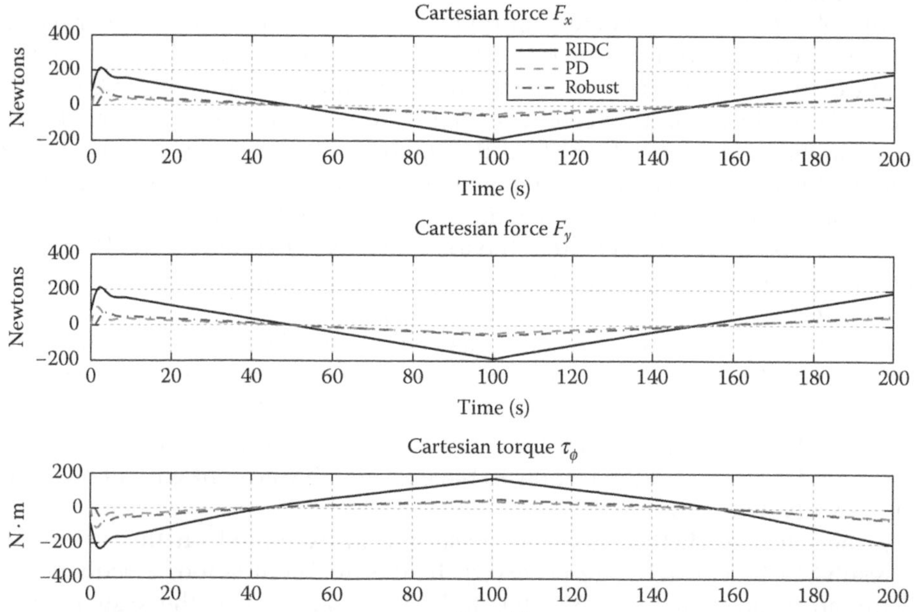

FIGURE 6.42
Cartesian wrench of the planar $4R\underline{P}R$ manipulator generated by robust inverse dynamics controller; solid: total force, dashed: PD term, dashed-dotted: robust corrective term.

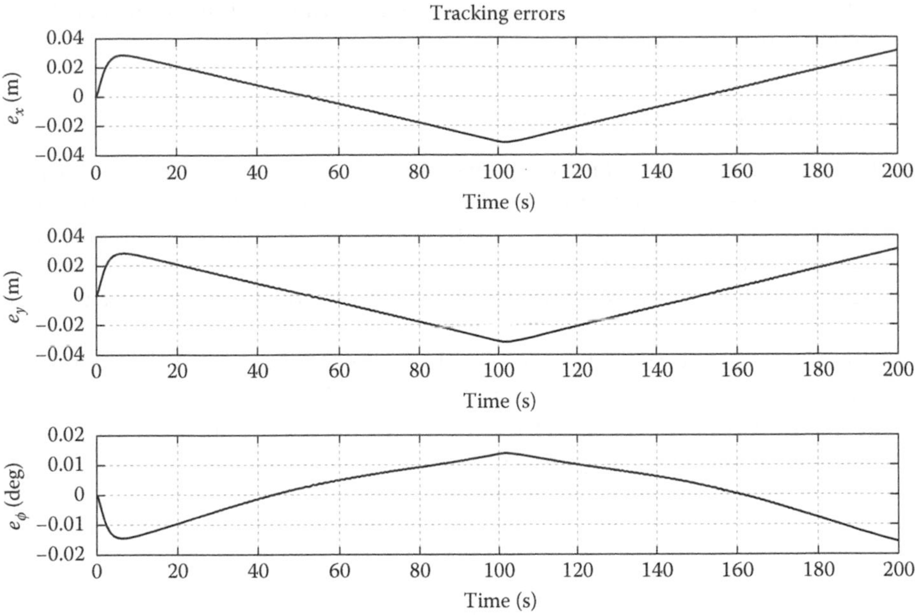

FIGURE 6.43
The closed-loop tracking performance of the planar $4R\underline{P}R$ manipulator with inverse dynamics control for a rapid trajectory.

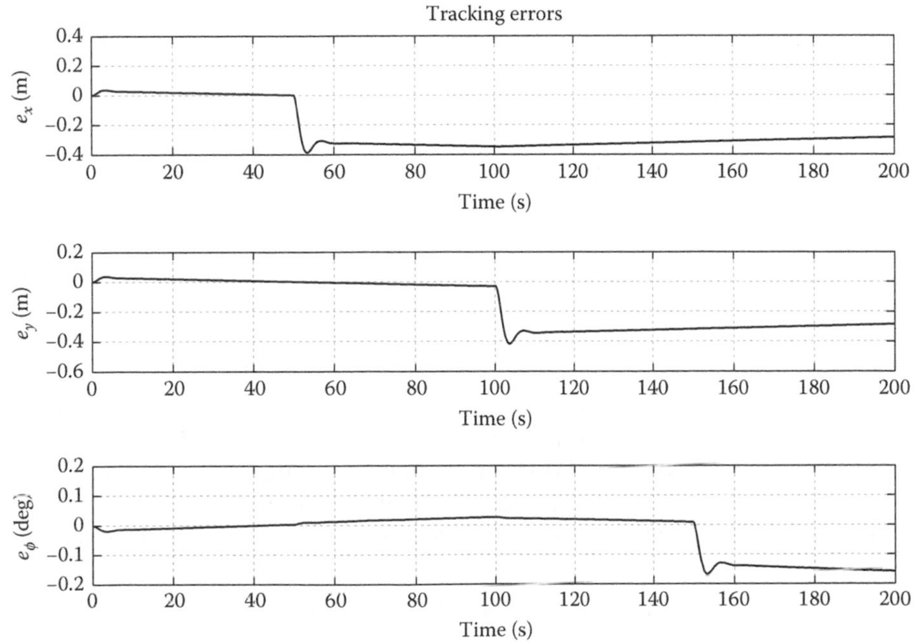

FIGURE 6.44
The closed-loop tracking performance of the planar $4R\underline{P}R$ manipulator for robust IDC in the presence of disturbance.

similar to that of inverse dynamics control, and still actuators with high bandwidth are required to perform such maneuvers.

6.8.6 Adaptive Inverse Dynamics Control

As described hereinbefore, in adaptive inverse dynamics control, first a rough estimate of the dynamic parameters is used, and then the estimate is updated such that the difference between the true values of the dynamic matrices to that of their estimates converges to zero. Hence, in an adaptive controller, the structure of the dynamic model is considered to be known, and only the parameters are subject to online identification to tune the controller parameters.

To develop the adaptive IDC, it is required to write dynamic formulation into a linear regression model represented in Equation 5.274 by $\boldsymbol{Y}(\boldsymbol{X}, \dot{\boldsymbol{X}}, \ddot{\boldsymbol{X}})$ in terms of the adaptive parameters $\boldsymbol{\theta}$. For the planar manipulator in hand, the adaptive parameters are the moving platform mass M, the moving platform moment of inertia I_m, and the cable density ρ_m. Therefore, the adaptive parameter vector is assigned by

$$\boldsymbol{\theta} = \begin{bmatrix} M \\ I_m \\ \rho_m \end{bmatrix}. \tag{6.117}$$

By careful examination of the dynamic formulation of the planar manipulator given in Equation 5.347, it can be observed that the left-hand side of this equation may easily be written in a linear regression form with respect to $\boldsymbol{\theta}$. The linear regression matrix $\boldsymbol{Y}$ has three columns, in which first and second columns correspond to dynamic terms related to the moving platform, while the last column is related to cable dynamics. The first two columns may be derived by equation $\boldsymbol{M}_p\ddot{\boldsymbol{\mathcal{X}}} + \boldsymbol{G}_p$, in which the mass matrix $\boldsymbol{M}_p$ and the gravity vector $\boldsymbol{G}_p$ are given in Equations 5.335 and 5.338, respectively. Hence, the two first columns of the linear regression matrix are given by

$$\boldsymbol{Y} = \begin{bmatrix} Y_1 & Y_2 & Y_3 \end{bmatrix}; \quad Y_1 = \begin{bmatrix} \ddot{x}_G - g \cdot \hat{x} \\ \ddot{y}_G - g \cdot \hat{y} \\ 0 \end{bmatrix}; \quad Y_2 = \begin{bmatrix} 0 \\ 0 \\ \ddot{\phi} \end{bmatrix}. \tag{6.118}$$

The last column of the linear regression matrix is related to cable dynamic terms, and may be written as

$$Y_3 = \boldsymbol{M}_{li_\rho}\ddot{\boldsymbol{\mathcal{X}}} + C_{li_\rho}\dot{\boldsymbol{\mathcal{X}}} + G_{li_\rho}. \tag{6.119}$$

In this derivation, the terms $\boldsymbol{M}_{i_\rho}, \boldsymbol{G}_{i_\rho}$, and $\boldsymbol{C}_{i_\rho}$ are determined from their corresponding terms $\boldsymbol{M}_i, \boldsymbol{G}_i$, and $\boldsymbol{C}_i$ excluding the ρ_m term from Equations 5.312, 5.323, and 5.332, and then these terms are substituted in Equation 5.346 to generate $\boldsymbol{M}_{li_\rho}, \boldsymbol{C}_{li_\rho}$, and $\boldsymbol{G}_{li_\rho}$. By this means, the parameter ρ_m is factorized from the equation and may be written into the linear regression form. Finally, dynamic equation of the planar manipulator may be written in the form of

$$\boldsymbol{M}\ddot{\boldsymbol{\mathcal{X}}} + \boldsymbol{C}\dot{\boldsymbol{\mathcal{X}}} + \boldsymbol{G} = \boldsymbol{Y}(\boldsymbol{\mathcal{X}}, \dot{\boldsymbol{\mathcal{X}}}, \ddot{\boldsymbol{\mathcal{X}}}) \cdot \boldsymbol{\theta}, \tag{6.120}$$

in which $\boldsymbol{\theta}$ is given in Equation 6.117 and $\boldsymbol{Y}$ is given in Equation 6.118. Note that the implementation of this equation in control law requires the knowledge of moving platform acceleration vector $\ddot{\boldsymbol{\mathcal{X}}}$. Acceleration sensors are usually noisy and add additional cost. Furthermore, numerical differentiation of the velocity signals are also not advisable in practice owing to the noise amplification. However, other filtering schemes such as the Kalman filter may be used to estimate the acceleration vector from position or velocity signals effectively. In the simulations performed on the planar manipulator, however, $\ddot{\boldsymbol{\mathcal{X}}}_d$ has been used to replace $\ddot{\boldsymbol{\mathcal{X}}}$, since with a well-tuned controller, there is no significant difference between these two variables in practice.

Implementation of adaptive inverse dynamics control is performed based on the block diagram illustrated in Figure 6.11. As illustrated in this figure, the implementation is very similar to the regular inverse dynamics control; however, the values of the dynamic formulation matrices are updated by the adapted inertial parameters which are found online through the adaptation law given in Equation 6.53. Hence, in the following simulations, the adaptation matrix $\boldsymbol{\Gamma}^{-1}$ is set at a suitable positive definite matrix, and Equation 6.53 is integrated in the main routine by augmentation of $\hat{\boldsymbol{\theta}}$ to the robot state variables. To examine high amount of uncertainties in the model, %50 perturbation in the dynamic parameters $\boldsymbol{\theta}$ is considered.

For the sake of comparison to previously examined controller structures, the PD controller gains are set to $\boldsymbol{K}_p = \boldsymbol{I}_{3\times3}$ and $\boldsymbol{K}_d = 2\sqrt{2}\,\boldsymbol{I}_{3\times3}$ as given in robust IDC simulations, in which $\boldsymbol{I}_{3\times3}$ denotes a 3×3 identity matrix. Furthermore, the adaptive controller parameters are set at $\boldsymbol{\Gamma}^{-1} = \text{diag}[40 \cdot M_o^2, 2 \times 10^5 I_{m_o}^2, 150\rho_{m_o}^2]$ by careful examination of different simulations for typical trajectory maneuvers depicted in Figure 6.15. Note that in this derivation, the square of the nominal values for the moving platform mass $M_o = 2500$ kg, the moving platform moment of inertia $I_{m_o} = 3.5 \times 10^5$, and the cable density $\rho_{m_o} = 0.215$ are first used as the diagonal elements of the adaptation matrix $\boldsymbol{\Gamma}^{-1}$ to normalize the adaptation parameters, and then appropriate gains are added to these diagonal elements for suitable adaptation rate for all the parameters.

Figure 6.45 illustrates the closed-loop tracking performance of the planar $4R\underline{P}R$ manipulator with adaptive inverse dynamics control, for the typical trajectory depicted in Figure 6.15. This comparison clearly indicates that adaptive control enables the system to significantly reduce the steady-state tracking errors after a suitable transient in 20 s. The steady-state error in positions is less than 0.2 mm, while this value is about 2.5×10^{-3} degrees for orientation, which are both extremely suitable.

This significant improvement is obtained by a fast convergence of the dynamic parameters $\hat{\boldsymbol{\theta}}$ to their nominal values as depicted in Figure 6.46. The adaptation gains are tuned such that the inertial parameter estimates, which have the initial values of %50 of their nominal values, converge to their nominal values in about 100 s. The convergence of the parameters M and ρ_m to their true nominal values is perfect, while this convergence for the I_m parameter is with less accuracy. A better performance for this parameter may be obtained by increasing the corresponding gain in $\boldsymbol{\Gamma}^{-1}$ diagonal elements. However, this is not done in this simulation, since it causes adverse effect on other simulations that have been explained later in this section.

The contribution of the PD term added in the adaptive control structure is illustrated in Figure 6.47, in which the total adaptive inverse dynamics control effort is plotted by a solid line, while its PD term is plotted in a dashed line. As seen in this figure, by a better prediction of the inertial parameters after the first 10 s, the contribution of the PD term in total control effort is significantly reduced, and the suitable performance of the tracking error is mainly due to adaptive IDC term.

To appreciate the adaptive inverse dynamic performance, the variation in inertial terms $\boldsymbol{\theta}$ is increased to %80. As seen in Figures 6.48 through 6.50, although more tracking errors are

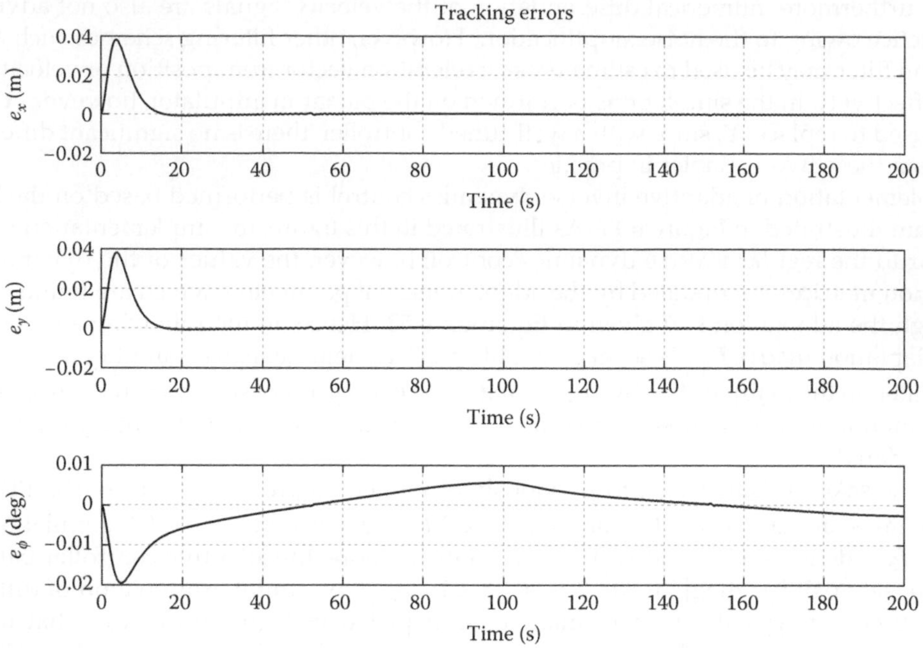

FIGURE 6.45
The closed-loop tracking performance of the planar 4$R\underline{P}R$ manipulator with adaptive IDC; %50 perturbation in all inertial parameters.

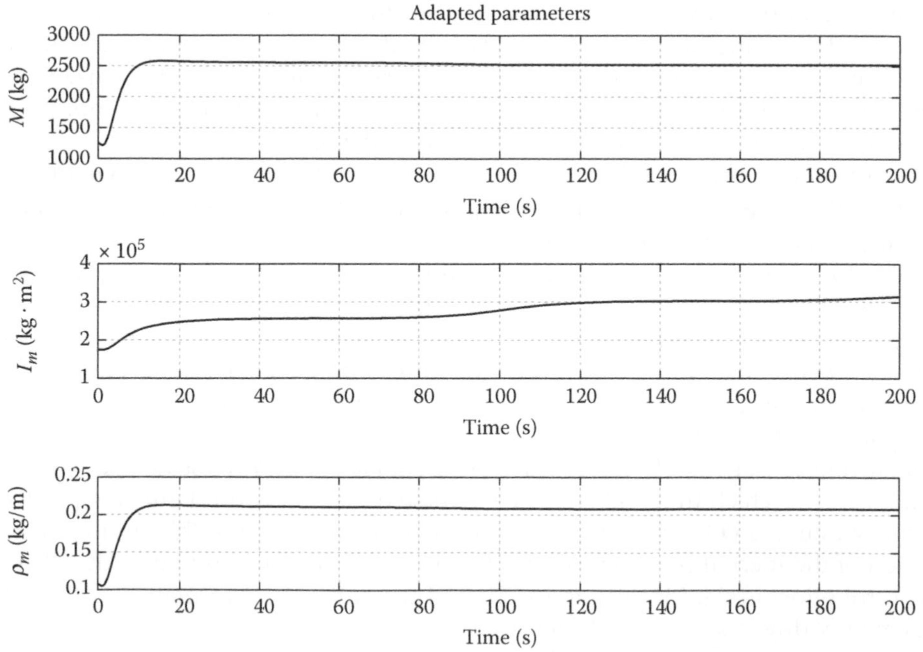

FIGURE 6.46
The adaptation of inertial parameters in adaptive IDC with %50 perturbation in all inertial parameters.

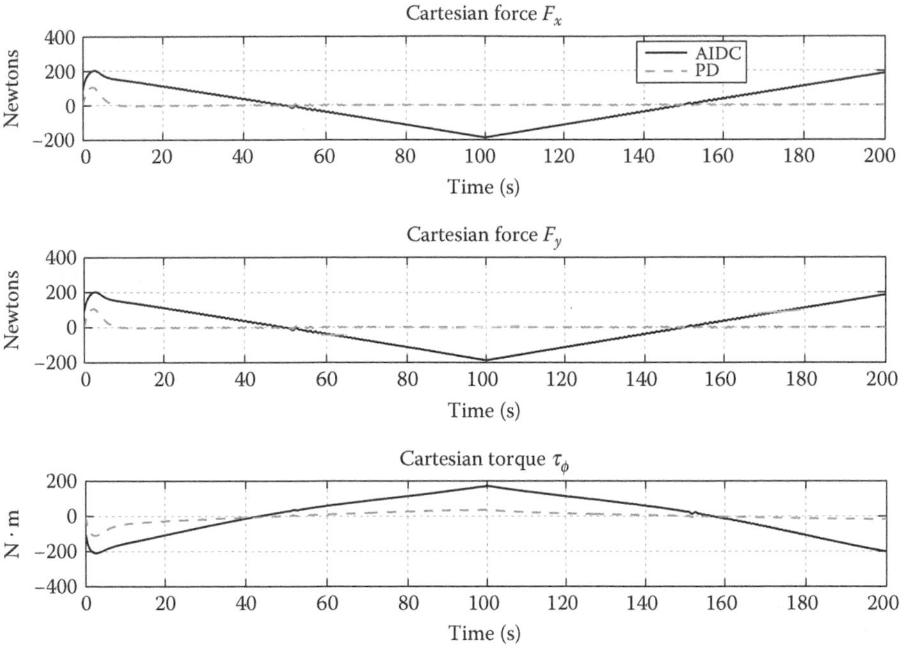

FIGURE 6.47
Cartesian wrench of the planar 4$R\underline{P}R$ manipulator generated by adaptive IDC; solid: total force, dashed: PD term.

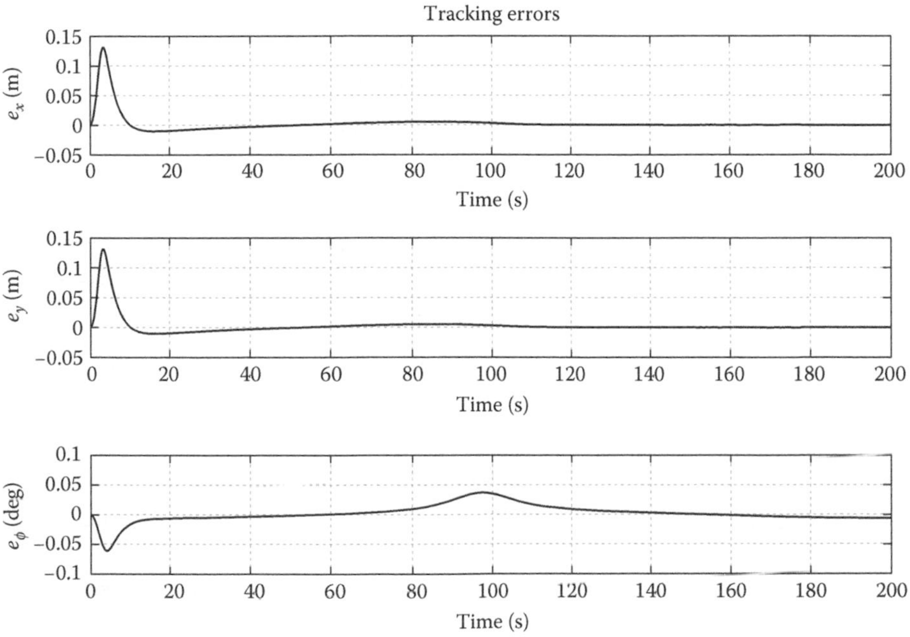

FIGURE 6.48
The closed-loop tracking performance of the planar 4$R\underline{P}R$ manipulator with adaptive IDC; %80 perturbation in all inertial parameters.

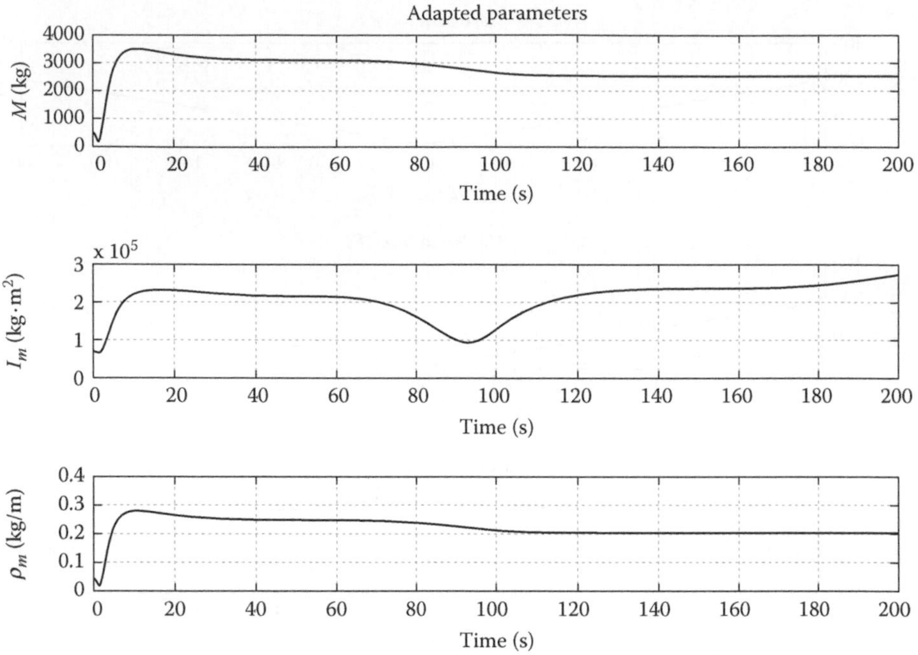

FIGURE 6.49
The adaptation of inertial parameters in adaptive IDC with %80 perturbation in all inertial parameters.

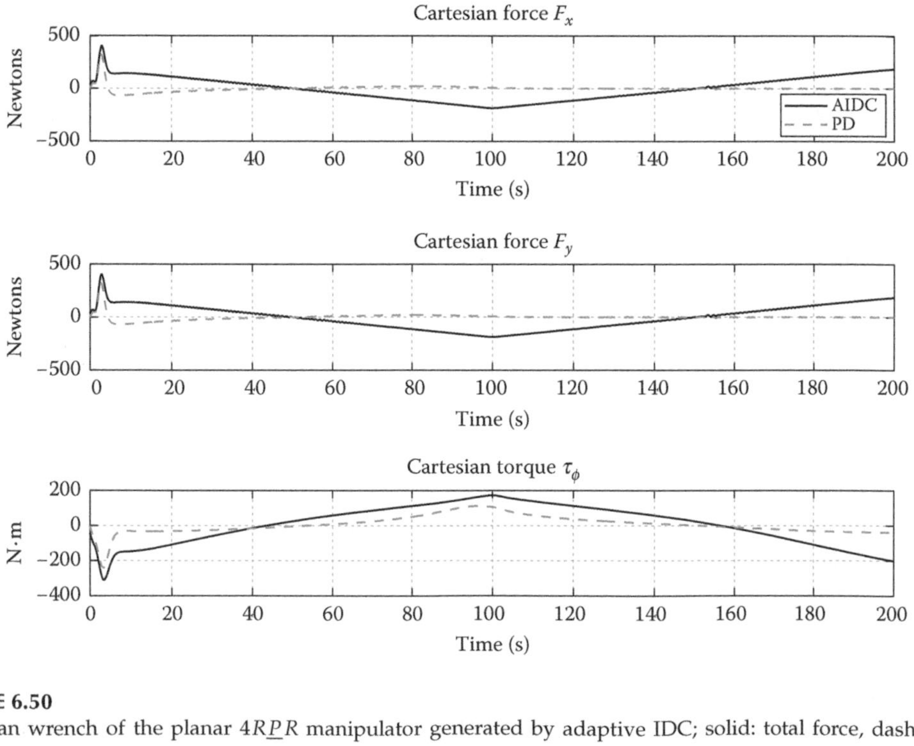

FIGURE 6.50
Cartesian wrench of the planar $4R\underline{P}R$ manipulator generated by adaptive IDC; solid: total force, dashed: PD term.

observed in transient, the same performance in the steady state is observed for this amount of perturbation in inertial parameters. Figure 6.49 illustrates the suitable adaptation of the estimated parameters to their nominal values. Figure 6.50 illustrates that although the initial contribution of the PD term is more than that of other terms, the suitable performance of the tracking error in the steady state is mainly due to adaptive IDC term.

Finally, the effect of disturbance and noise is also examined on this controller, and similar results to that of IDC is observed. It is observed that like previous controllers, the effect of nonvanishing external disturbance cannot be well attenuated with this structure, and due to the convergence of the inertial parameters to some inaccurate values, significant steady-state errors are observed in the performance. The effect of measurement noise is also very similar to that of inverse dynamics control, and still actuators with high bandwidth are required to perform such maneuvers.

6.8.7 Motion Control in Joint Space

As mentioned hereinbefore, although the motion control schemes in the task space are very effective in terms of achieving asymptotic tracking performance, they suffer from an implementation constraint that the motion variable $\mathcal{X}$ must be measured in practice. However, as explained in Section 6.2, in many practical situations, measurement of the motion variable $\mathcal{X}$ is difficult or expensive, and usually just the active joint variables q are measured. In such a case, the controllers developed in the joint space may be used in practice.

To implement such controller topologies on the planar manipulator, it is first required that the dynamic formulation of this manipulator is rewritten in the joint space as given in Equation 6.59. However, to derive the dynamic matrices formulated in the joint space, as seen in Equations 6.60 through 6.63, it is required that the manipulator be fully parallel, and the inverse of the Jacobian matrix is derivable at different configurations. However, the parallel manipulator in hand is a redundant manipulator, and the Jacobian matrix is not square; hence, this inverse cannot be determined. Therefore, this controller topology is not implementable on this manipulator.

An effective solution to the forward kinematic problem of this manipulator is obtained in an online routine [100]. Therefore, even for the case where just the active joint variables q are measured, it is recommended to use the control topology depicted in Figure 6.2, and implement control laws designed in the task space.

6.9 Motion Control of the Stewart–Gough Platform

In this section, motion control of the Stewart–Gough platform is studied in detail, and the control topologies detailed in this chapter would be implemented on this manipulator. The architecture of this manipulator is described in detail in Section 3.5.1 and shown in Figure 3.16. As explained earlier, in this manipulator the spatial motion of the moving platform is generated by six piston–cylinder actuators. A complete kinematic analysis of this manipulator is given in Section 3.5 and its Jacobian analysis is presented in Section 4.8. The dynamic analysis of this manipulator using the Newton–Euler and virtual work approaches is given in Section 5.3.3 and Section 5.4.6, respectively. The closed-form dynamic formulation being used in this section for the simulations is derived in Section 5.5.6. In what follows, different control strategies in task space, and in joint space are implemented on this manipulator. These control strategies are given in a sequence as in Section 6.3 from simple to more complete and their closed-loop performances are analyzed and compared.

6.9.1 Decentralized PD Control

The first control strategy introduced for parallel robots consists of the simplest form of feedback control. In this control structure depicted in Figure 6.5, a number of disjoint PD controllers have been used in a feedback structure at each error component. For a fully parallel manipulator with six-degrees-of-freedom like the Stewart–Gough platform, the motion variable in the task space is a six-dimensional vector

$$\mathcal{X} = \begin{bmatrix} x_p \\ \theta \end{bmatrix} \tag{6.121}$$

consisting of the linear motion represented by the position vector of the moving platform center of mass $x_p = [x_p \;\; y_p \;\; z_p]^T$ and the moving platform orientation represented by screw coordinates

$$\theta = \theta \begin{bmatrix} s_x & s_y & s_z \end{bmatrix}^T = \begin{bmatrix} \theta_x & \theta_y & \theta_z \end{bmatrix}^T. \tag{6.122}$$

Therefore, the tracking error is defined as $e = \begin{bmatrix} e_x & e_y & e_z & e_{\theta_x} & e_{\theta_y} & e_{\theta_z} \end{bmatrix}^T$. The decentralized controller, therefore, consists of six disjoint proportional derivative (PD) controllers acting on each error component. The PD controller is denoted by $K_d s + K_p$ block in Figure 6.5, in which K_d and K_p are 6×6 diagonal matrices denoting the derivative and proportional controller gains, respectively.

The output of the controller is denoted by $\mathcal{F}$, which is defined by a full-scale wrench in the task space, $\mathcal{F} = \begin{bmatrix} F_x & F_y & F_z & \tau_x & \tau_y & \tau_z \end{bmatrix}^T$. Note that since the output of the controller is defined in the task space, each wrench component directly manipulates the corresponding tracking error component, and therefore, the overall tracking performance of the manipulator is suitable if high controller gains are used.

In practice, the calculated output wrench is transformed into actuator forces through the force distribution block. In this mapping, the actuator forces required to generate such a wrench is computed, and is applied to the manipulator. As explained hereinbefore, for fully parallel manipulators such as the Stewart–Gough platform under study, the inverse of Jacobian transpose may be used for this projection at nonsingular configurations.

For the simulations, the geometric and inertial parameters have been given in Table 5.2, and the initial conditions for the states are all set at zero, except for the initial state of the moving platform height, which is set at $z_o = 1\,\text{m}$. For the first set of simulations, the gravity vector is neglected $g = [0, 0, 0]^T$, and the disturbance wrench applied to the moving platform is considered to be zero, while the desired trajectory of the moving platform is generated by cubic polynomials for x_d, y_d, z_d, as well as for $\theta_x, \theta_y, \theta_z$ as shown in Figure 6.51. As seen in this figure, a typical trajectory in which the moving platform is moved 0.25 m in x direction, 0.5 m in y direction, and -0.25 m in z direction. The orientation of the moving platform is also a compound rotation in all x, y, and z directions with the values shown in Figure 6.51. The time of simulation is considered to be 2 s to examine high accelerative motions for the moving platform. The PD gains used in the simulations

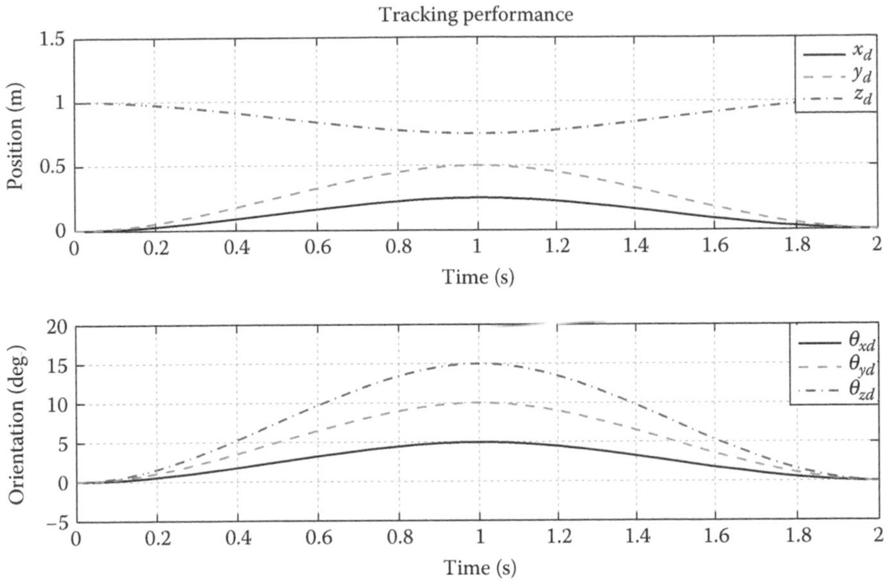

FIGURE 6.51
The desired and output trajectories of the Stewart–Gough platform in closed loop for decentralized PD control with no gravitational force.

are set at $\boldsymbol{K}_p = 10^6 \cdot \text{diag}[1, 1, 1, 10, 10, 10]$ and $\boldsymbol{K}_d = 10^5 \cdot \text{diag}[1, 1, 1, 1, 1, 1]$ after a number of trial-and-error nominations.

The tracking performance of the moving platform is illustrated in Figures 6.51 and 6.52. As seen in Figure 6.51, if the gravity forces are neglected, the desired and final closed-loop motion of the system are not distinguishable, and as it is seen in greater detail in Figure 6.52, a decentralize PD controller for the manipulator is capable of reducing the tracking errors to less than 4 mm in position and less than 0.005° in orientation.

The Cartesian wrench $\mathcal{F} = [F_x, F_y, F_z, \tau_x, \tau_y, \tau_z]^T$ generated by the PD control action at the moving platform is illustrated in Figure 6.53. As seen in this figure, the Cartesian wrench has a similar pattern to the desired trajectory acceleration, which is linear for a cubic polynomial trajectory. The amount of Cartesian wrench are relatively high and it reaches to about 4 kN for the forces, and about 1 kN.m for the torques. High force and torques are generated by the PD controller to accommodate high accelerations required in the desired trajectory. Similarly, the actuator forces of the manipulator are illustrated in Figure 6.54, and it is observed that relatively high actuator forces are also required to carry out the required maneuver.

To examine the effect of the gravitational forces on the performance of the PD controller in closed loop, another simulation is performed identical to the previous simulation; however, the gravity acceleration is considered in the $-z$ direction: $\boldsymbol{g} = [0, 0, -9.81]^T$. The tracking errors for the same controller in the presence of gravitational forces are illustrated in Figure 6.55. As clearly seen in this figure, gravitational forces significantly increase the tracking error in the z direction to about 15 mm, while this error in the steady-state tends to 10 mm. This has caused the required Cartesian force in the z direction for a steady-state offset of about 10 kN as shown in Figure 6.56.

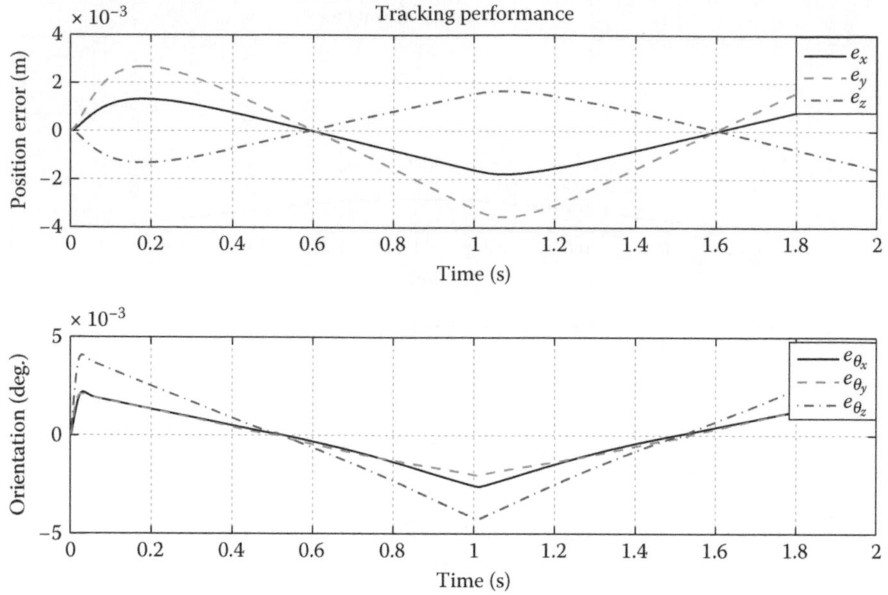

FIGURE 6.52
The closed-loop tracking performance of the Stewart–Gough platform for decentralized PD control with no gravitational force.

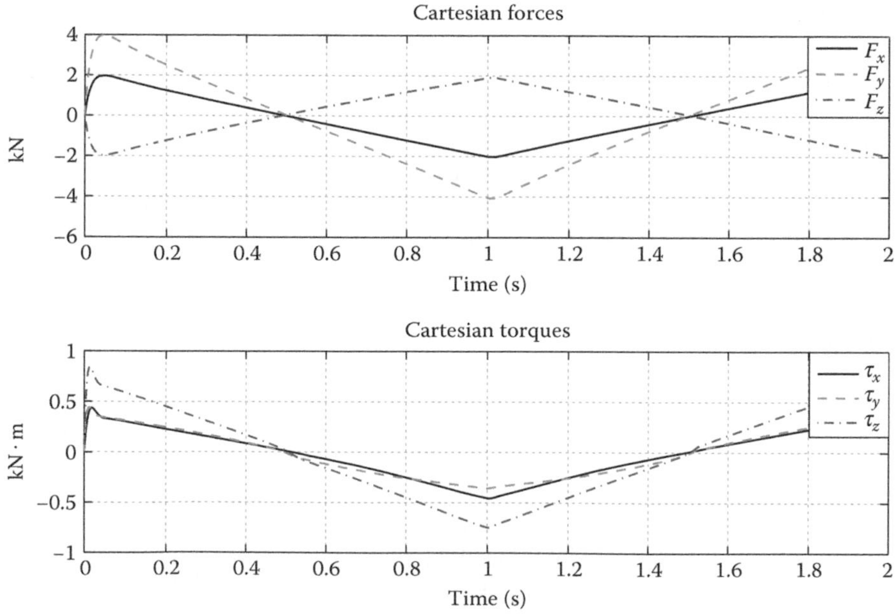

FIGURE 6.53
The Cartesian wrench of the Stewart–Gough platform generated by PD controller with no gravitational force to generate trajectories shown in Figure 6.51.

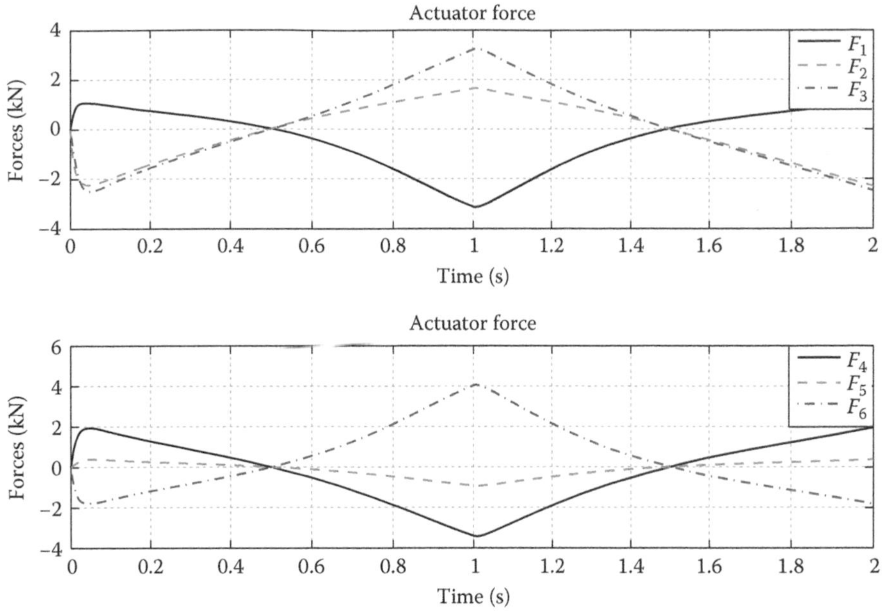

FIGURE 6.54
The actuator forces of the Stewart–Gough platform required to generate trajectory shown in Figure 6.51; no gravitational force.

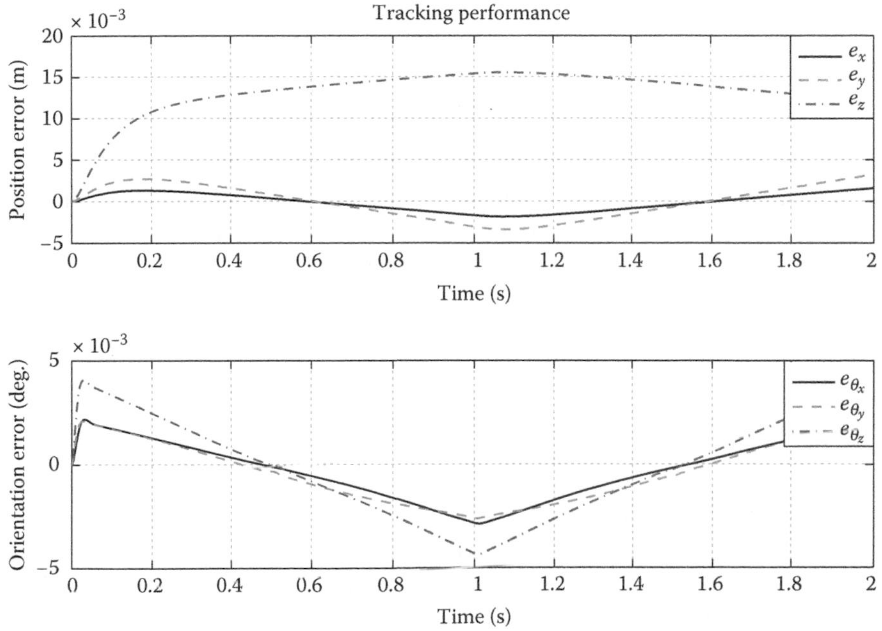

FIGURE 6.55
The closed-loop tracking performance of the Stewart–Gough platform for decentralized PD control with gravity.

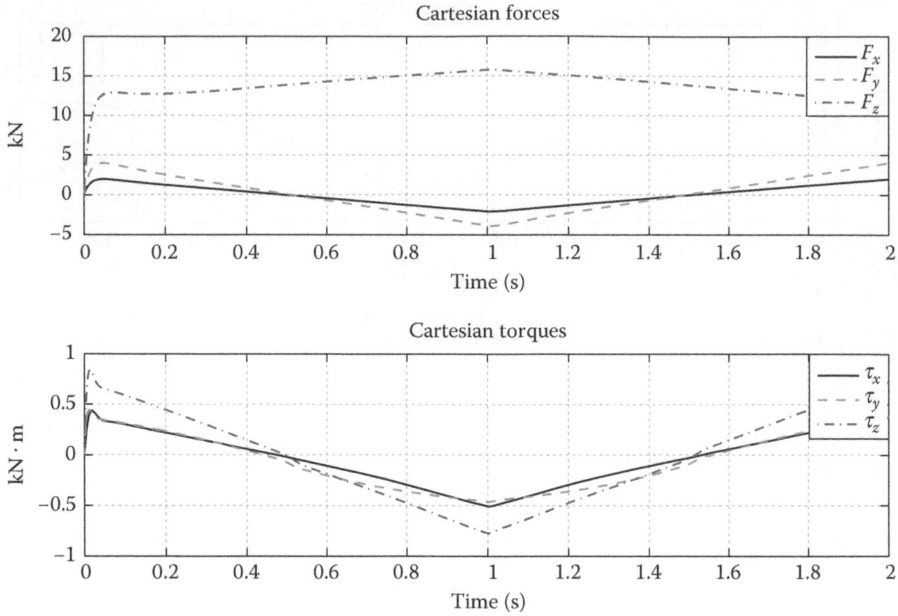

FIGURE 6.56
The Cartesian wrench of the Stewart–Gough platform generated by PD controller with gravity to generate trajectory shown in Figure 6.51.

To examine the performance of the PD controller to attenuate the effect of external disturbance and measurement noise, two more simulations have been considered for the closed-loop system. In the first simulation, a unit step disturbance force is applied to the system under original trajectories shown in Figure 6.51. To examine the effect of disturbance forces and torques separately, a 10 kN step disturbance force in x, y, z directions is applied at time 0.5 s, then another 10 kN.m step disturbance torque in x, y, z directions is applied at time 1 s. The controller gains are set to $\boldsymbol{K}_p = 10^6 \cdot \text{diag}[1, 1, 1, 10, 10, 10]$ and $\boldsymbol{K}_d = 10^5 \cdot \text{diag}[1, 1, 1, 1, 1, 1]$ as before.

The closed-loop tracking error of the system is shown in Figure 6.57. By comparing this figure to Figure 6.55, it is seen that the effect of external disturbance cannot be completely attenuated by PD controller, and the abrupt change in the tracking error at the instances when the disturbance forces/torques are applied is clearly seen in the response. Furthermore, as a result of nonvanishing disturbances applied to the system, static errors are accumulated in the closed-loop response of the system. Figure 6.58 illustrates the Cartesian wrench generated by PD controller to carry out the required maneuver in the presence of disturbance. The effect of step disturbance wrench on the required controller forces is also clearly seen at times 0.5 and 1 s.

The effect of measurement noise is considered next by considering that all the measured variables $\boldsymbol{\mathcal{X}}$ are contaminated with a Gaussian noise with an amplitude of %0.1 peak values of the original signals. Although the amount of noise is very limited in this simulation, since high gains have to be used in decentralized PD controllers, the effect of noise on tracking performance is significant. As shown in Figures 6.59 and 6.60, although the tracking error is not increased much compared with that of the system without noise, the obtained actuator forces to carry out such a maneuver require high bandwidth. This is owing to the fact that high PD gains have been used to accommodate the required tracking

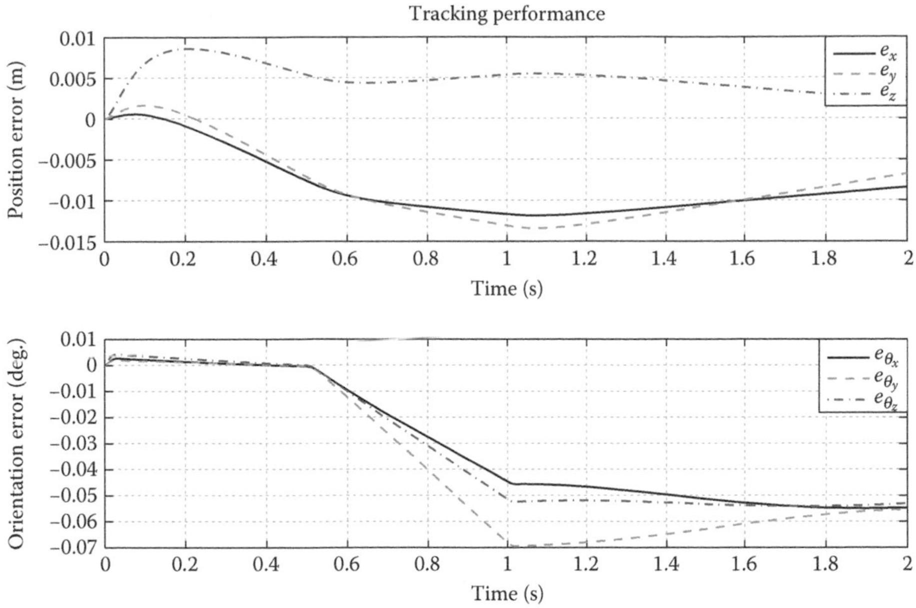

FIGURE 6.57
The closed-loop tracking performance of the Stewart–Gough platform for decentralized PD control in the presence of disturbance.

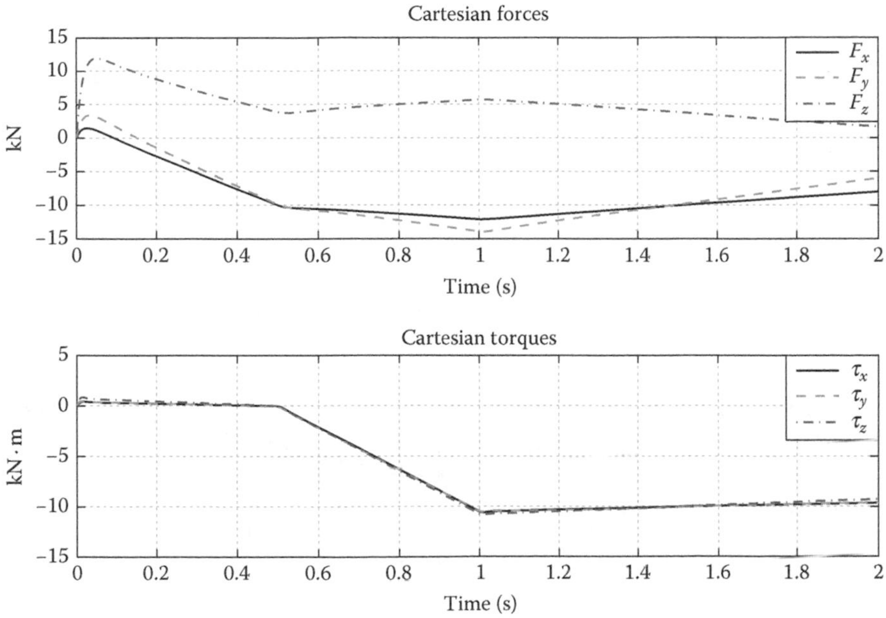

FIGURE 6.58
The Cartesian wrench of the Stewart–Gough platform generated by PD controller for the maneuver shown in Figure 6.51 in the presence of disturbance.

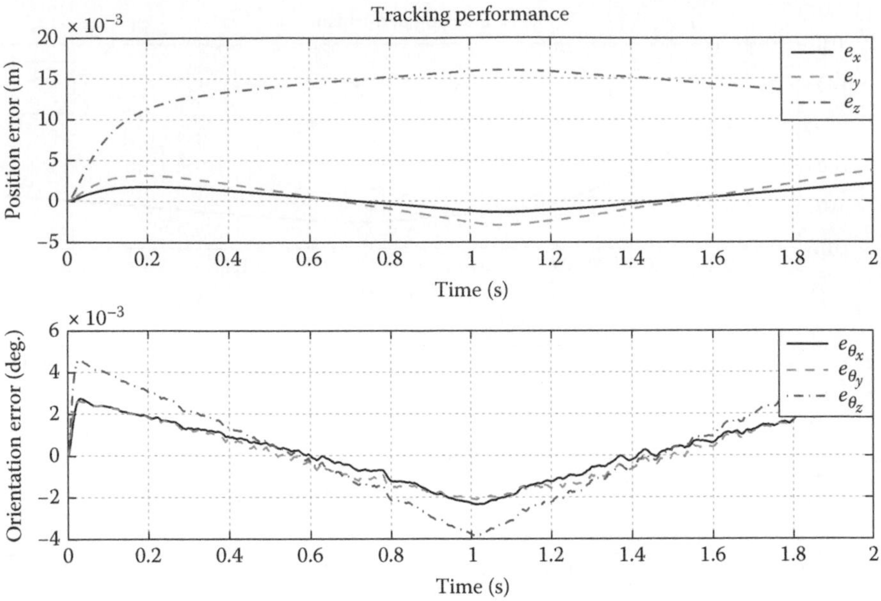

FIGURE 6.59
The closed-loop tracking performance of the Stewart–Gough platform for decentralized PD control in the presence of measurement noise.

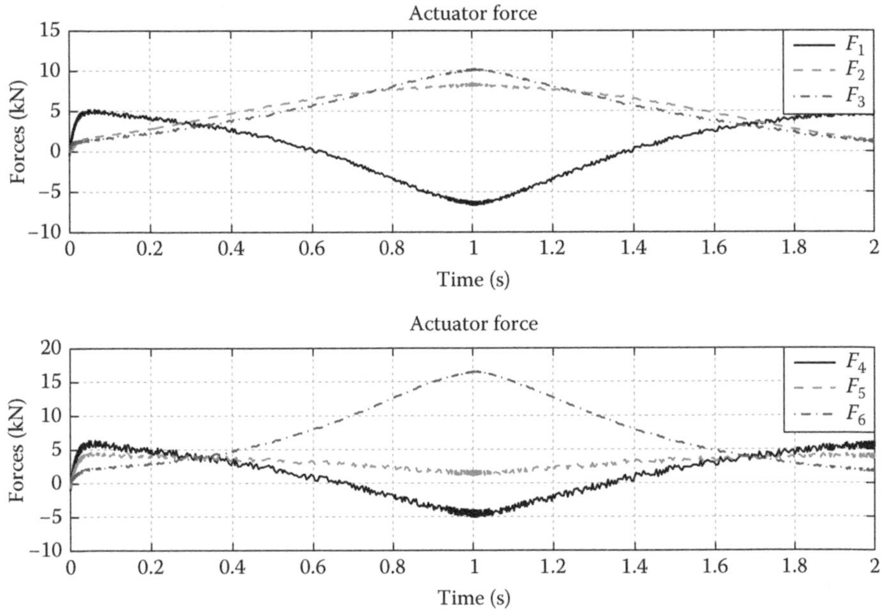

FIGURE 6.60
The actuator forces generated by the PD control of the Stewart–Gough platform in the presence of measurement noise.

performance, and although the amplitude of noise is about 10^{-3} times the peak values of the measurement signals, since the PD gains are higher than 10^5, the signature of the noise is apparently seen in the required actuator force. In fact, generating such actuator forces are very costly in practice, and hence the tracking performance shown in Figure 6.59 is not easily achievable.

6.9.2 Feed Forward Control

As seen in the previous section, the tracking performance of PD controller implemented in the task space is not sufficient in the presence of gravitational forces, measurement noise, and external disturbances applied to the manipulator. To compensate for such effects, a feed forward wrench denoted by $\mathcal{F}_{ff}$ may be added to the structure of the controller as depicted in Figure 6.6. This term is generated from the dynamic model of the manipulator in the task space by Equation 6.2. Furthermore, note that to generate this term, the dynamic formulation of the robot and its kinematic and dynamic parameters are needed.

In practice, exact information on dynamic matrices is not available, and therefore in simulations, estimates of these matrices with %10 perturbation in all kinematic and inertial parameters are used in the derivations. For the sake of comparison, the PD controller gains are set at $\boldsymbol{K}_p = 10^6 \cdot \text{diag}[1, 1, 1, 10, 10, 10]$ and $\boldsymbol{K}_d = 10^5 \cdot \text{diag}[1, 1, 1, 1, 1, 1]$ as in the case of decentralized PD control.

Figure 6.61 illustrates the closed-loop tracking performance of the Stewart–Gough platform with feed forward control for the typical trajectory depicted in Figure 6.51. Comparing this figure to that of the PD control illustrated in Figure 6.55, it is seen that the feed forward action is capable of improving the positional tracking performance 10 times and the rotational tracking performance 5 times that of decentralized PD control in the presence of gravitational forces.

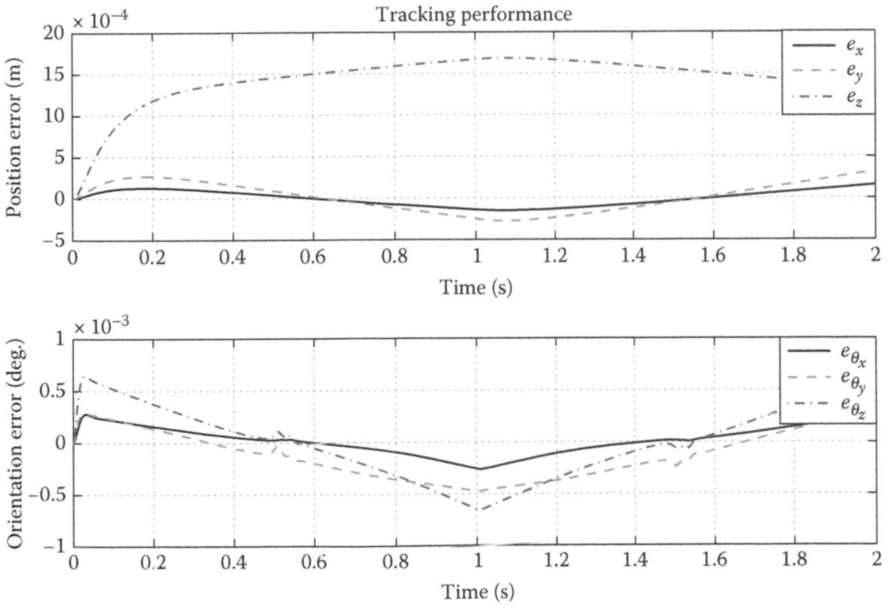

FIGURE 6.61
The closed-loop tracking performance of the Stewart–Gough platform with feed forward control.

The Cartesian wrench generated by feed forward controller for such a maneuver is given in Figure 6.62. In this figure, the total Cartesian wrench is indicated with a solid line, while its PD component in a dashed line. As seen in this figure, since the information on the system model is quite accurate, and only %10 perturbation in kinematic and inertial parameters of the model is considered in this simulation, the main component of the Cartesian force contributing is the feed forward term, which generates about %90 of the generated force, and the PD action in this simulation contributes to only about %10 of that. The total force needed to produce the required motion is almost the same as in pure PD control, while the tracking performance is significantly improved. This is mainly owing to the contribution of feed forward term into accurate motion of the manipulator, as expected. The actuator forces are illustrated in Figure 6.63, in which the amount of required actuator forces is almost the same as that in the pure PD control.

In the next simulation, it is assumed that the information on the system model is very poor, and %50 perturbation in all kinematic and inertial parameters are considered in derivation of the feed forward control term. It is expected that the tracking performance is significantly affected because of this lack of information on the system model. Figures 6.64 and 6.65 illustrate the tracking performance and the contribution of the PD and feed forward term in the total generated Cartesian wrench. As seen in Figure 6.64, lack of knowledge of the model parameters significantly affects the tracking errors. Comparing this figure to Figure 6.61, it can be concluded that the tracking errors are increased by a factor of 5.

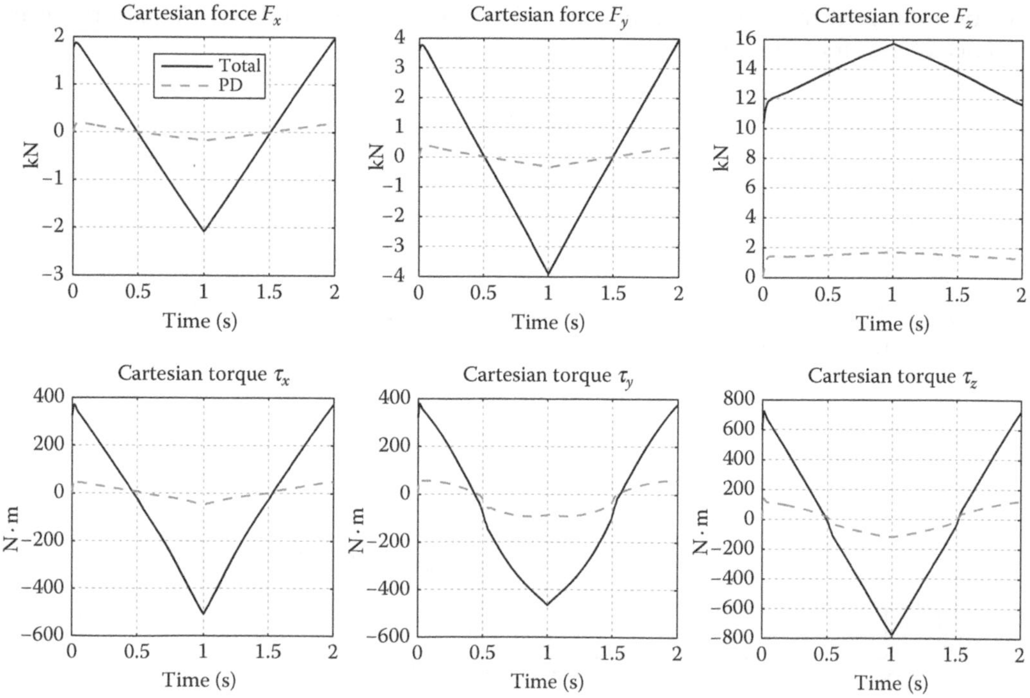

FIGURE 6.62
Cartesian wrench of the Stewart–Gough platform generated by feed forward controller to generate trajectories shown in Figure 6.51.

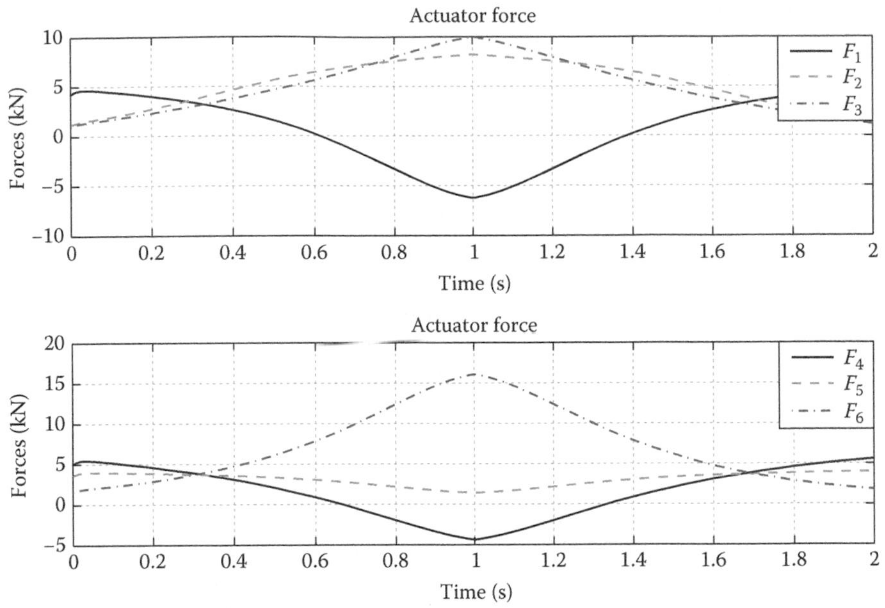

FIGURE 6.63
The actuator forces of the Stewart–Gough platform with feed forward controller required to generate trajectories shown in Figure 6.51.

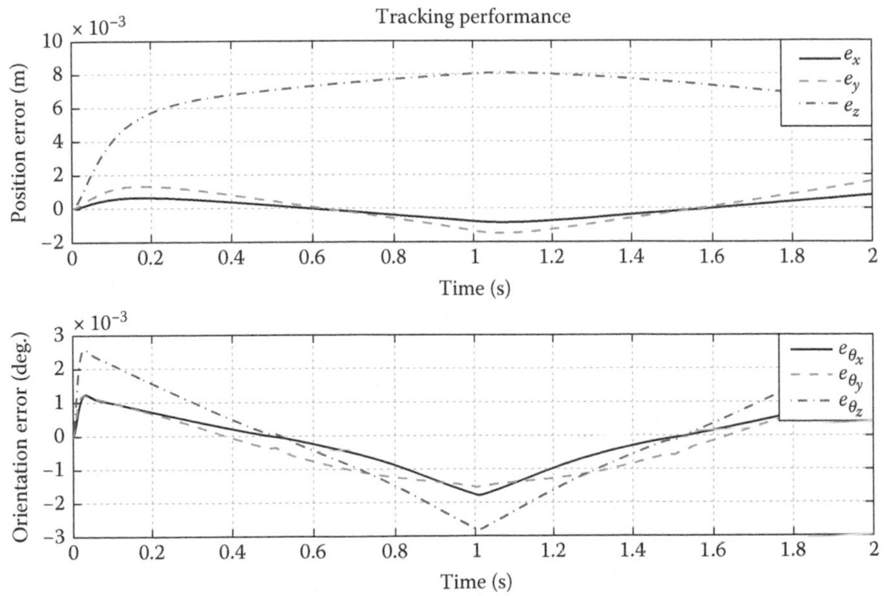

FIGURE 6.64
The closed-loop tracking performance of the Stewart–Gough platform with feed forward control; %50 perturbation in kinematic and inertial parameters.

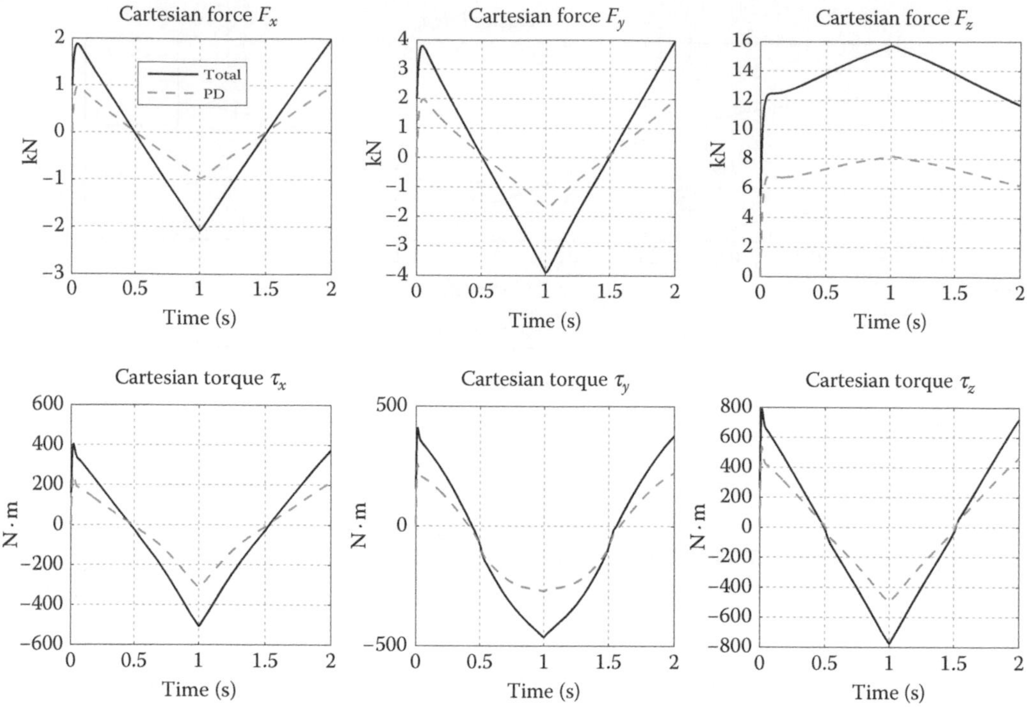

FIGURE 6.65
Cartesian wrench of the Stewart–Gough platform generated by feed forward controller; %50 perturbation in kinematic and inertial parameters.

Figure 6.65 illustrates the total Cartesian wrench generated by this control law. As seen in this figure, since the knowledge of model parameters is poor in this simulation, the contribution of the PD control term increases to almost %50 of the total wrench. This is in complete agreement to the fact that for a model-based control structure such as the feed forward scheme under consideration, if the information on the model is poor, the contribution of the PD term becomes more significant.

Finally, the effect of disturbance and noise is also examined on this controller, and similar results to those of the pure PD control is observed. Since this controller structure is model based, and in these simulations high PD gains are considered for the controller, such a performance in the presence of noise and disturbance is expected.

6.9.3 Inverse Dynamics Control

As seen in the previous sections, the tracking performance of a decentralized PD controller is not uniform in the presence of gravitational forces, measurement noise, and external disturbances applied to the manipulator. To compensate for such effects, a feed forward wrench is added to the structure of the controller, by which the shortcomings of the decentralized controller is partially remedied. However, the closed-loop performance still faces a number of limitations.

To examine these limitations on performance, inverse dynamics control is implemented on the Stewart–Gough platform. The general structure of the inverse dynamics control

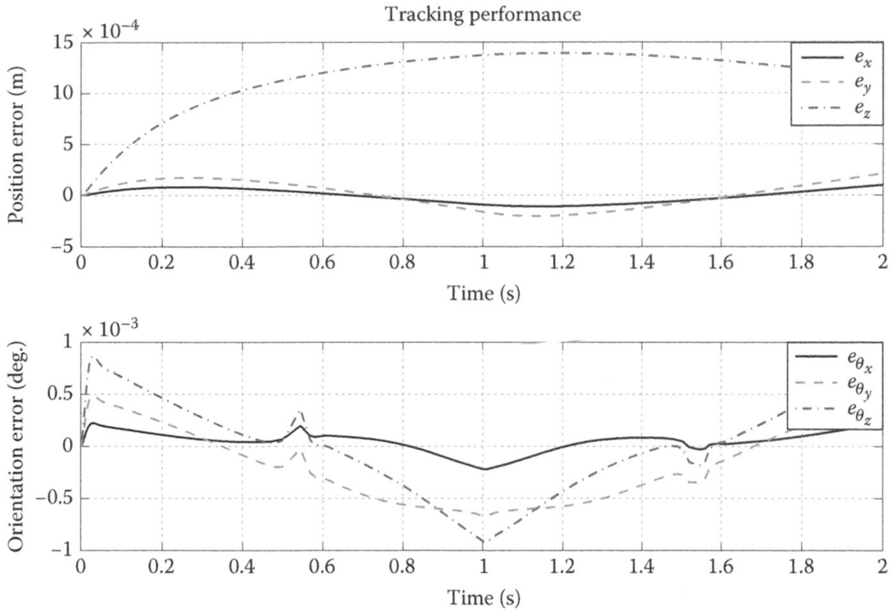

FIGURE 6.66
The closed-loop tracking performance of the Stewart–Gough platform with inverse dynamics control.

applied to a parallel manipulator is depicted in Figure 6.7. As seen in this figure, a corrective wrench $\mathcal{F}_{fl}$ is added in a feedback structure to the closed-loop system, which is calculated from the Coriolis and centrifugal matrix, and gravity vector of the manipulator dynamic formulation. Furthermore, the mass matrix is acting in the forward path, in addition to the desired trajectory acceleration $\ddot{\mathcal{X}}_d$.

Note that to generate this term, the dynamic formulation of the robot and its kinematic and dynamic parameters are needed. In practice, exact knowledge of dynamic matrices are not available, and therefore, a %10 perturbation in all kinematic and inertial parameters are used in the following simulations. For the sake of comparison to previously examined controller structures, the PD controller gains are adjusted by trial and error to $\boldsymbol{K}_p = 10^3 \cdot \text{diag}[1, 1, 1, 20, 20, 20]$ and $\boldsymbol{K}_d = 200 \cdot \text{diag}[1, 1, 1, 1, 1, 1]$ to have the same order of tracking error as to that of the feed forward control. Note that as detailed in Equation 6.5, the PD controller terms are multiplied by the manipulator mass matrix, and therefore, the same dimensions of controller gains used in pure PD or feed forward controller are not applicable here. However, tuning the PD controller gains in an IDC controller structure is very easy, and with a few trials over the desired trajectory, appropriate gains are selected.

Figure 6.66 illustrates the closed-loop tracking performance of the Stewart–Gough platform with inverse dynamics control for the typical trajectory depicted in Figure 6.51. Comparing this figure to that of feed forward control illustrated in Figure 6.61, it is seen that the PD controller gains are tuned so to obtain a similar performance as the feed forward control with much easier controller gain tuning process.

The Cartesian wrench generated by the inverse dynamics controller for such a maneuver is given in Figure 6.67. In this figure, the total Cartesian wrench generated by the inverse dynamic controller is shown with a solid line, while its PD component is shown by a dashed line. As seen in this figure, since the knowledge of the system model is quite accurate, and only %10 perturbation in kinematic and inertial parameters of the model is

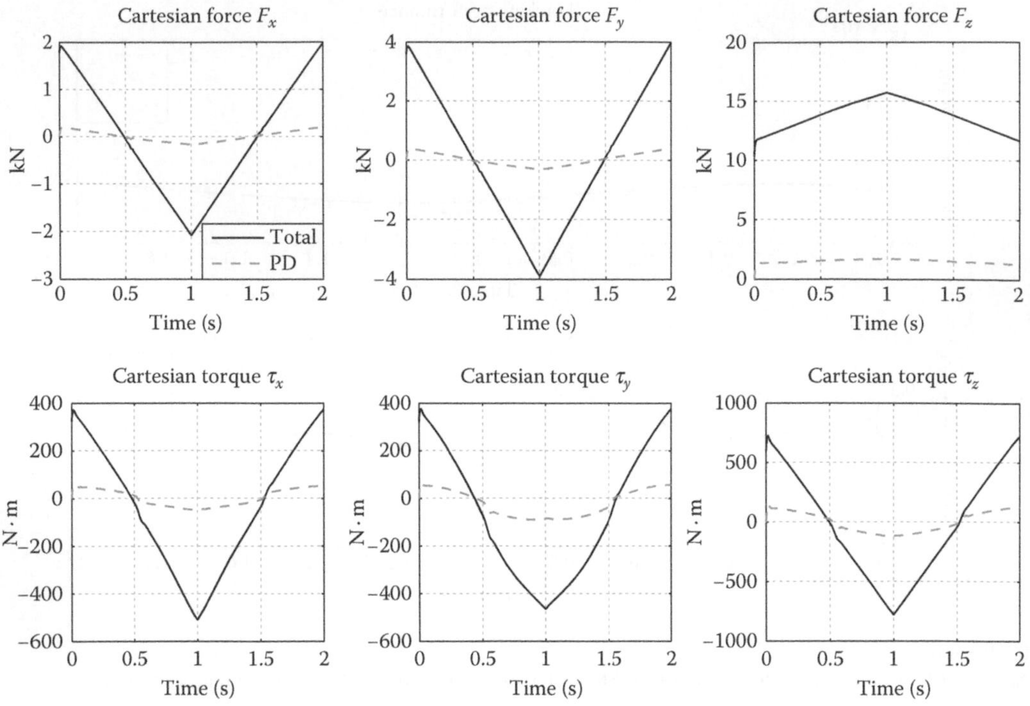

FIGURE 6.67
Cartesian wrench of the Stewart–Gough platform generated by inverse dynamic controller to generate trajectories shown in Figure 6.51.

considered in this simulation, the main component of the Cartesian force is the inverse dynamic term, which generates about %90 of the total force. The PD action in this simulation contributes to only about %10 of the total wrench. The total force needed to produce the required motion is almost the same as in feed forward control, while the tracking performance is also almost the same. Similarly, the actuator forces of the manipulator are illustrated in Figure 6.68, and it is observed that relatively high actuator forces are required to carry out the required maneuver. This is mainly because of high accelerations that are required to carry out the desired trajectory.

In the next simulation, it is assumed that the knowledge of the system model is very poor, and %50 perturbation in all kinematic and inertial parameters are considered in derivation of the inverse dynamics control law. It is expected that the tracking performance is significantly affected because of this lack of information on the system model. Figures 6.69 and 6.70 illustrate the tracking performance and the contribution of the PD term with respect to the total inverse dynamics control effort. As seen in Figure 6.69, lack of information on model parameters significantly affects the tracking errors. Comparing this figure with Figure 6.66, it is observed that the tracking errors for both the positions and the orientations are increased by a factor of 10.

Figure 6.70 illustrates the total Cartesian wrench generated by this control law and its PD component. As seen in this figure, since the information on model parameters is relatively poor in this simulation, the contribution of PD control term increases to almost %50 of the total force. This is in complete agreement to the fact that for a model-based control

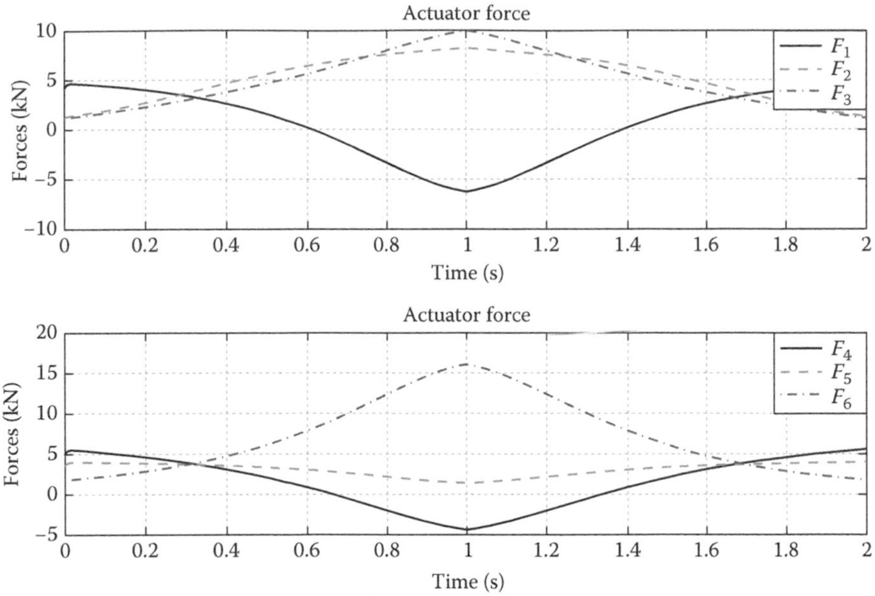

FIGURE 6.68
The actuator forces of the Stewart–Gough platform with inverse dynamics controller required to generate trajectories shown in Figure 6.51.

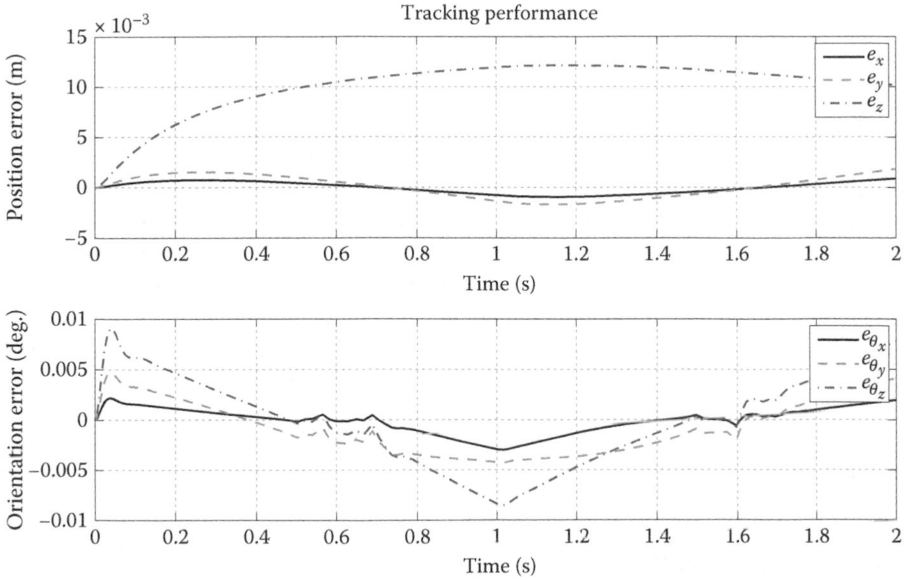

FIGURE 6.69
The closed-loop tracking performance of the Stewart–Gough platform with inverse dynamics control; %50 perturbation in kinematic and inertial parameters.

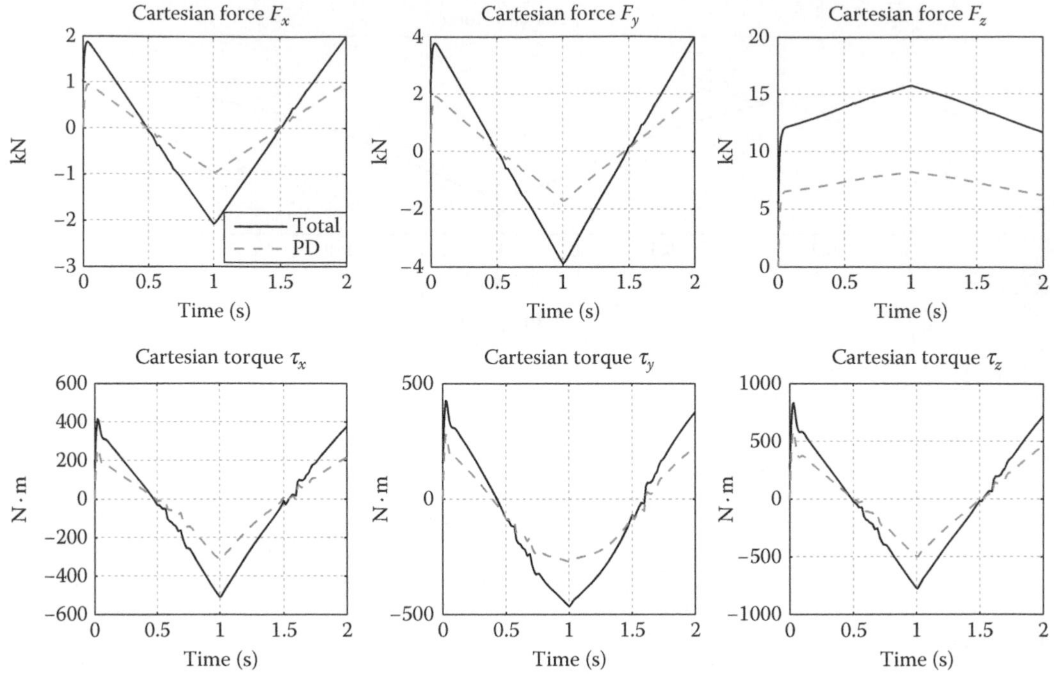

FIGURE 6.70
Cartesian wrench of the Stewart–Gough platform generated by inverse dynamic controller; %50 perturbation in kinematic and inertial parameters.

structure such as the IDC under consideration, if the information on the model is poor, then the contribution of the PD term becomes more significant.

Finally, the effect of disturbance and noise is also examined on this controller performance, and similar results to those of the pure PD control is observed. To examine the effect of disturbance forces and torques separately as before, a 10 kN step disturbance force in x, y, and z directions is applied at time 0.5 s, then another 10 kN.m step disturbance torque in x, y, and z directions is added at time 1.0 s. The controller gains are set at $\boldsymbol{K}_p = 10^3 \cdot \text{diag}[1, 1, 1, 20, 20, 20]$ and $\boldsymbol{K}_d = 200 \cdot \text{diag}[1, 1, 1, 1, 1, 1]$ as hereinbefore.

The closed-loop tracking error of the system is shown in Figure 6.71. By comparing this figure to the PD controller performance in the presence of disturbance, namely Figure 6.57, is seen that the effect of external disturbance on the tracking error is almost the same as that in the pure PD controller. Furthermore, as a result of nonvanishing disturbances applied to the system, static errors are accumulated in the closed-loop response of the system. In the presence of the measurement noise, however, very suitable tracking performance can be achieved. Nevertheless, as shown in Figure 6.72, the required actuator bandwidth is still relatively high.

6.9.4 Partial Linearization IDC

Inverse dynamics control has several features making it very attractive in practice. However, as explained in Section 6.3.3, to apply this method, complete knowledge of the dynamic formulation matrices is required. To develop the simplest possible implementable

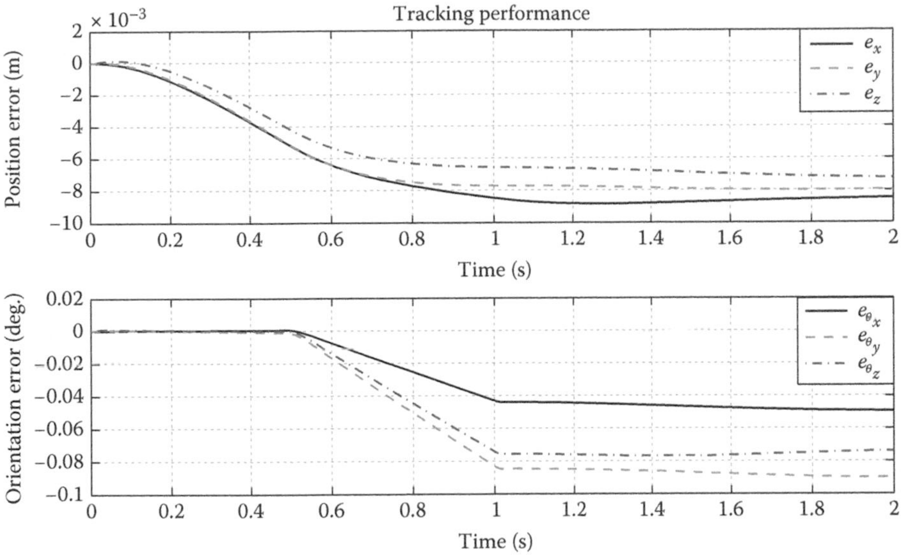

FIGURE 6.71
The closed-loop tracking performance of the Stewart–Gough platform, for IDC in the presence of disturbance.

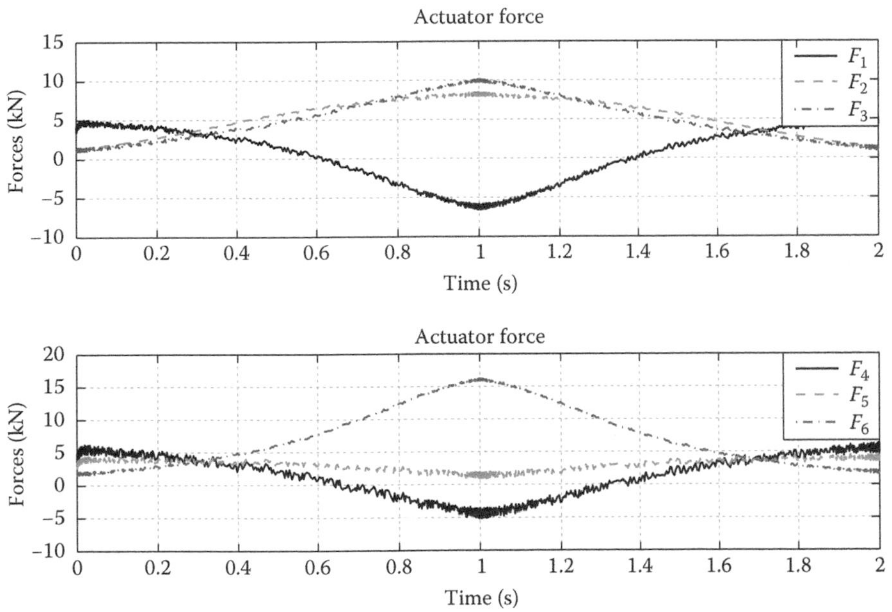

FIGURE 6.72
Actuator forces required to perform the desired maneuver in the presence of measurement noise.

IDC, let us recall the partial linearization IDC developed in Section 6.3.4. This simplest form of IDC control wrench $\mathcal{F}$ consists of two terms as given in Equation 6.11. The first term $\mathcal{F}_{pd}$ is generated by a PD control on the motion error e_x, given in Equation 6.12, while the second term is considered to be only the gravity vector of the manipulator $G(\mathcal{X})$,

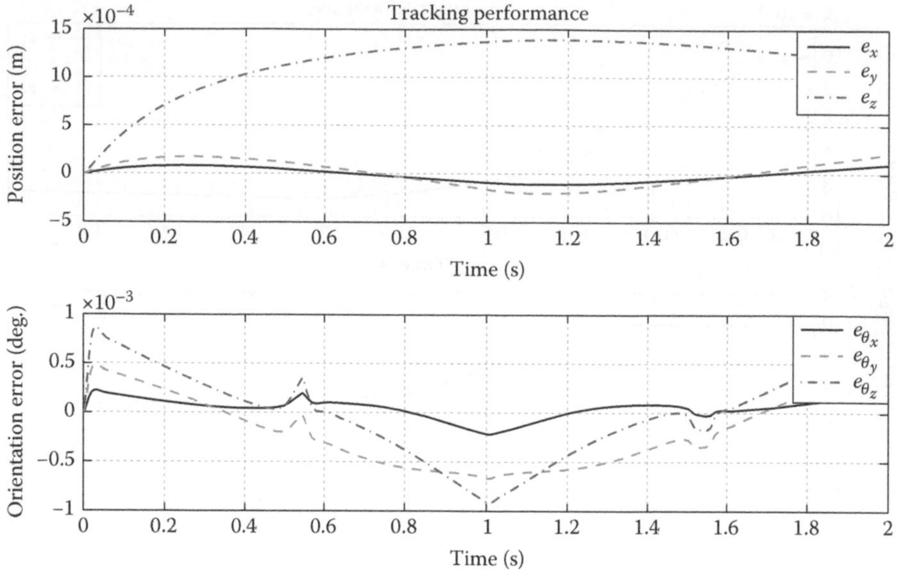

FIGURE 6.73
The closed-loop tracking performance of the Stewart–Gough platform with partially linearized IDC and %10 perturbation in kinematic and inertial parameters.

given by Equation 6.13. In this implementation the computationally intensive Coriolis and centrifugal term is not used in the feedback.

For the sake of comparison to the previously examined controller structures, in the first simulation, %10 perturbation is considered in all kinematic and inertial parameters, while in the second simulation, %50 perturbation is considered. The PD controller gains are set at $\boldsymbol{K}_p = 10^3 \cdot \text{diag}[1, 1, 1, 20, 20, 20]$ and $\boldsymbol{K}_d = 200 \cdot \text{diag}[1, 1, 1, 1, 1, 1]$, as for that in inverse dynamics control. The tracking performances of the closed-loop system with partially linearized IDC for these two cases are shown in Figures 6.73 and 6.74, respectively. The tracking errors in both the cases are very similar to the case wherein complete linearization is applied, and as seen in the comparison of these two figures to their corresponding cases illustrated in Figures 6.66 and 6.69, only a slight increase in positional tracking errors is observed for the partial linearized IDC.

This simulation confirms the analysis given in Section 6.3.4, which shows that even if the Coriolis and centrifugal matrix is not used in the feedback, and the closed-loop dynamics is not fully linearized, the PD control structure with gravity compensation can still lead to a suitable performance. Furthermore, it verifies that the skew-symmetric property of the dynamic formulation matrices provides sufficient assurance that partial linearization IDC does not result in tracking divergence in practice.

6.9.5 Robust Inverse Dynamics Control

The inverse dynamics control performance relies on complete linearization of nonlinear dynamic terms by feedback. This requirement may never be satisfied in practice due to the modeling uncertainty, and therefore, it may practically limit the performance of IDC control in parallel robots. To compensate for such modeling uncertainties in the closed-loop

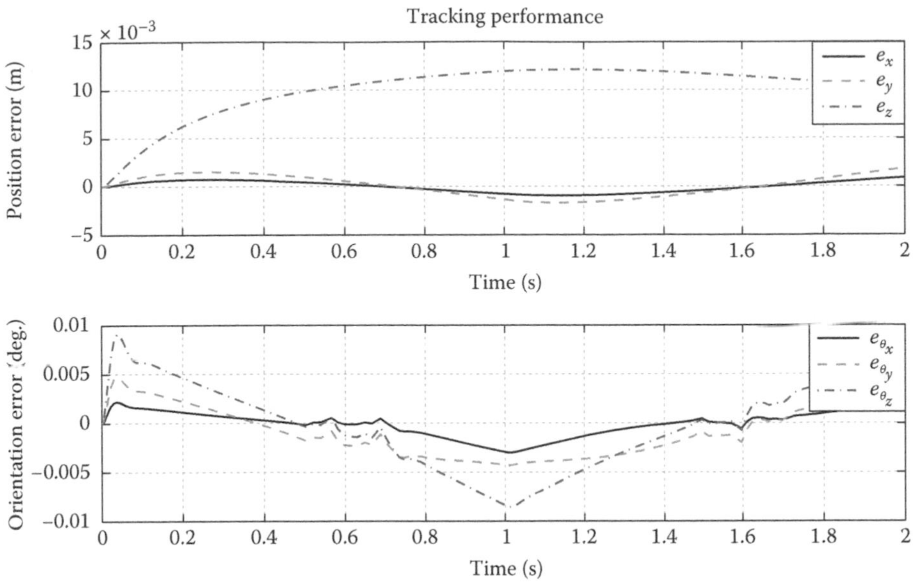

FIGURE 6.74
The closed-loop tracking performance of the Stewart–Gough platform with partially linearized IDC and %50 perturbation in kinematic and inertial parameters.

performance of the manipulator, a robustifying term $\boldsymbol{\delta}_a$ is introduced in Section 6.4.1, which is added to the usual IDC control structure. Figure 6.8 illustrates the robust IDC implementation block diagram on a parallel robot, in which the corrective term $\boldsymbol{\delta}_a$ is added as a wrench in task space. This corrective term is given in Equation 6.37, in which the vector $\boldsymbol{v}$ is defined in Equation 6.32.

In the following simulations, the robust inverse dynamic approach is applied to the Stewart–Gough platform, in which, to examine high amount of uncertainties in the model, %50 perturbation in all kinematic and inertial parameters are considered. For the sake of comparison the previously examined controller structures, the PD controller gains are set at $\boldsymbol{K}_p = 10^3 \cdot \text{diag}[1,1,1,20,20,20]$ and $\boldsymbol{K}_d = 200 \cdot \text{diag}[1,1,1,1,1,1]$, as given in IDC simulations. Furthermore, the robust controller parameters are set at $\rho = 500$, $\epsilon = 80$ by careful examination of different simulations for the typical trajectory depicted in Figure 6.51.

Figure 6.75 illustrates the closed-loop tracking performance of the Stewart–Gough platform with robust inverse dynamic control. To compare the obtained results to those of the regular IDC better, the tracking errors of these two cases have been compared in Figure 6.76 for the typical trajectory depicted in Figure 6.51. This comparison clearly indicates that by using the robust control, the precision of tracking errors are not changed that much; however, the robust IDC is capable of reducing the sharp variations in tracking error that is experienced by regular IDC. The smoothness of the results is obtained through the contribution of the corrective term added in the robust control structure, the amount of which is illustrated in Figure 6.77. In this figure, the total robust inverse dynamics control wrench is plotted by a solid line, while its PD and corrective terms are plotted in dashed and dashed-dotted lines, respectively. As seen in this figure, about %50 of the total force is generated by the PD term, while the corrective term contributes only at the instances where the tracking error changes rapidly, and the remaining control effort is generated by the inverse dynamic terms.

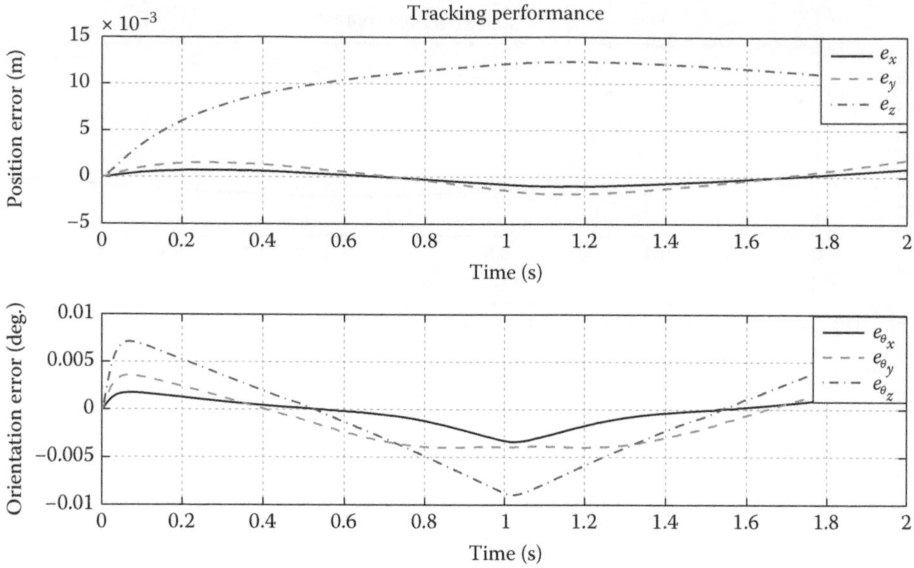

FIGURE 6.75
The closed-loop tracking performance of the Stewart–Gough platform with robust inverse dynamics control; %50 perturbation in kinematic and inertial parameters.

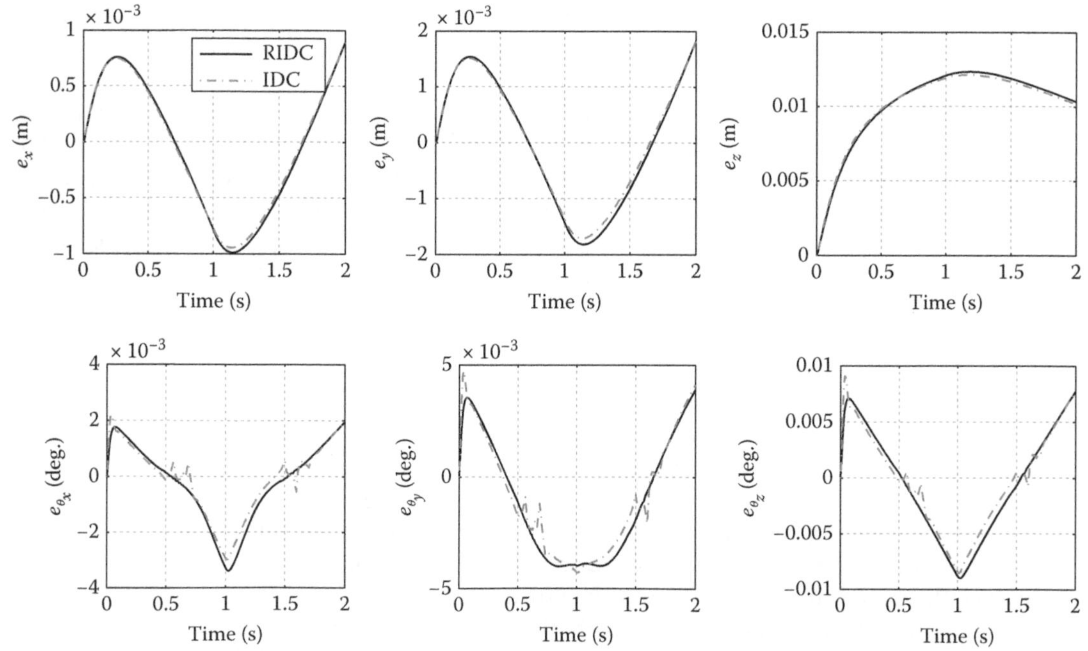

FIGURE 6.76
The closed-loop tracking performance of the Stewart–Gough platform with robust IDC compared to that of regular IDC; %50 perturbation in all kinematic and inertial parameters.

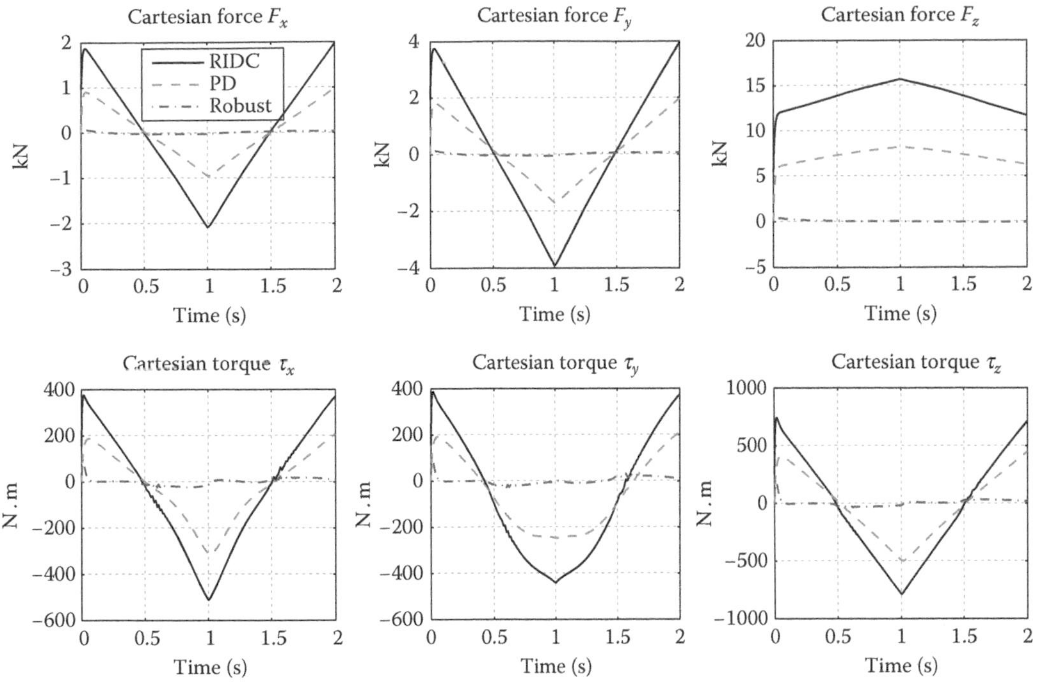

FIGURE 6.77
The Cartesian wrench of the Stewart–Gough platform generated by robust inverse dynamics controller; solid: total force, dashed: PD term, dashed-dotted: robust corrective term.

The transient and steady-state performances of the robust IDC may be improved by further tuning of PD gains and the robust controller gain ρ and ϵ. To show that, while considering %50 perturbation in all kinematic and inertial parameters, the PD controller gains are increased to $\boldsymbol{K}_p = 4 \times 10^3 \cdot \text{diag}[1, 1, 1, 20, 20, 20]$ and $\boldsymbol{K}_d = 400 \cdot \boldsymbol{I}_{6\times 6}$, to reduce the steady-state error of the closed-loop system. Furthermore, the robust controller gains are increased to $\rho = 1000$ and $\epsilon = 200$ by careful examination of different simulations, to obtain a smooth and robust tracking performance for the closed-loop system. The tracking performance for this case is shown in Figure 6.78. As seen in this figure, the tracking performance improve significantly in this tuning example, and the tracking precision increased about five times in both position and orientation.

Finally, the effect of disturbance and noise is also examined on this controller, and similar results to those of the IDC are observed. In the first simulation, a 10 kN step disturbance force in x, y, and z directions is applied at time 0.5 s, then another 10 kN.m step disturbance torque in x, y, and z directions is applied at time 1 s. The closed-loop tracking error of the system is shown in Figure 6.79. It is seen that like previous controllers, the effect of nonvanishing external disturbance cannot be well attenuated with this structure, and still significant steady-state errors are observed in the performance. The closed-loop performance of the robust IDC in the presence of noise is illustrated in Figure 6.80. As seen in this figure, the effect of measurement noise is also very similar to that of the inverse dynamics control, and still actuators with high bandwidth are required to perform such maneuvers.

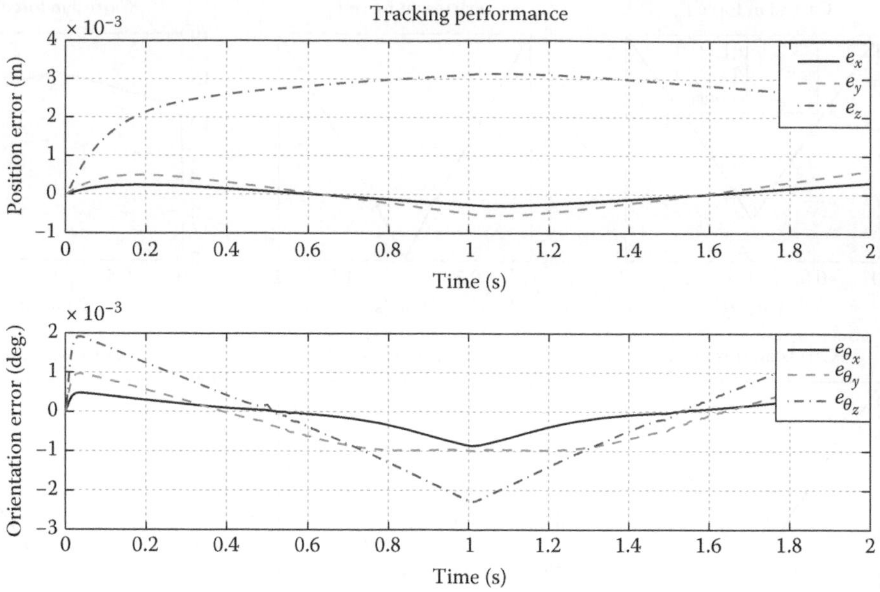

FIGURE 6.78
The closed-loop tracking performance of the Stewart–Gough platform with robust inverse dynamics control; tuning the gains for performance.

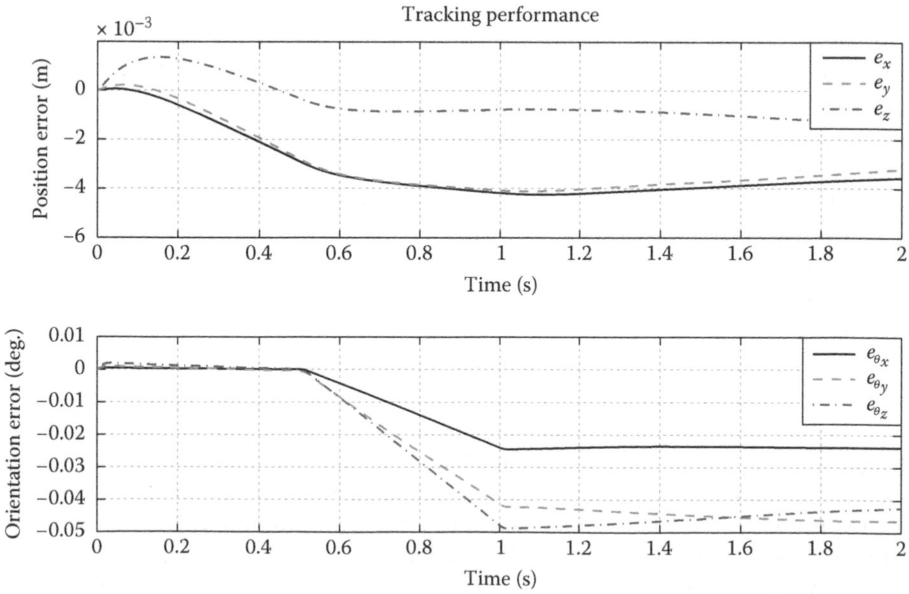

FIGURE 6.79
The closed-loop tracking performance of the Stewart–Gough platform for robust IDC in the presence of disturbance.

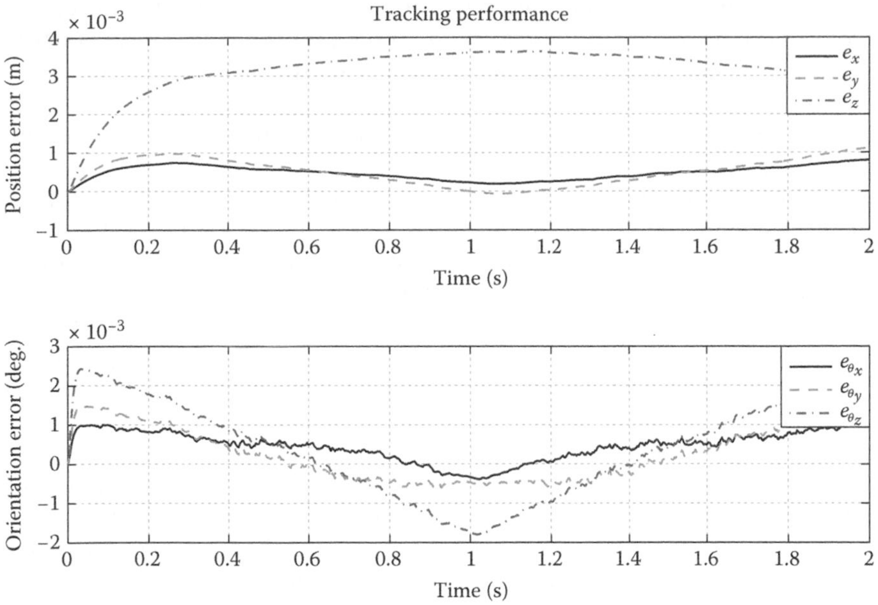

FIGURE 6.80
The closed-loop tracking performance of the Stewart–Gough platform for robust IDC in the presence of measurement noise.

6.9.6 Motion Control in Joint Space

As mentioned hereinbefore, although the motion control schemes in the task space are very effective in terms of achieving asymptotic tracking performance, they suffer from an implementation constraint, that the motion variable $\mathcal{X}$ must be measured in practice. However, as explained in Section 6.2, in many practical situations, measurement of the motion variable $\mathcal{X}$ is difficult or expensive, and usually just the active joint variables q are measured. In such a case, the controllers developed in the joint space may be used in practice. The closed-form dynamic formulation being used in this section is the one given in the joint space and derived in Section 6.5.1. In what follows, different control strategies in the joint space are implemented on the Stewart–Gough platform. These control strategies are given in a sequence as in Section 6.5 from simple to more complete and their closed-loop performances have been analyzed and compared, respectively.

6.9.6.1 Decentralized PD Control

The first control strategy introduced for parallel robots consists of the simplest form of feedback control in such manipulators. In this control structure depicted in Figure 6.12, a number of disjoint linear (PD) controllers have been used in a feedback structure at each error component. For a fully parallel manipulator with six-degrees-of-freedom like the Stewart–Gough platform, the motion variable in task space is a six-dimensional vector

$$\mathcal{X} = \begin{bmatrix} x_p \\ \theta \end{bmatrix}$$

consisting of the linear motion represented by the position vector of the moving platform center of mass, and the moving platform orientation represented by screw coordinates. The joint variable $\boldsymbol{q}(t)$ is also a six-dimensional vector consisting of the limb lengths denoted by $\boldsymbol{q} = [\ell_1, \ell_2, \ldots, \ell_6]^T$. Therefore, the tracking error is defined as $\boldsymbol{e}_q = \boldsymbol{q}_d - \boldsymbol{q}$, in which the desired motion variable in the joint space $\boldsymbol{q}_d$ is determined by the solution of inverse kinematics, and $\boldsymbol{q}$ is given by direct measurement of the limb lengths.

The decentralized controller, therefore, consists of six disjoint PD controllers acting on each error component. The PD controller is denoted by $\boldsymbol{K}_d s + \boldsymbol{K}_p$ block in Figure 6.12, in which $\boldsymbol{K}_d$ and $\boldsymbol{K}_p$ are 6×6 diagonal matrices denoting the derivative and proportional controller gains, respectively. The output of the controller directly generates the actuator torques denoted by, $\boldsymbol{\tau}$. Note that since the output of the controller is defined in the joint space, each actuator directly manipulates the corresponding tracking error in the joint space, and therefore, the overall tracking performance of the manipulator is suitable if high controller gains are considered.

For the simulations, the geometric and inertial parameters given in Table 5.2 and the initial conditions for the states are all set at zero, except for the initial state of the moving platform height, which is set at $z_o = 1$ m. For the first set of simulations, the gravity vector is neglected $\boldsymbol{g} = [0, 0, 0]^T$, and the disturbance wrench applied to the moving platform is considered to be zero, while the desired trajectory of the moving platform is generated by cubic polynomials for x_d, y_d, and z_d, as well as for θ_x, θ_y, and θ_z, as shown in Figure 6.51. This trajectory is considered the same for the controllers designed in the joint space and those designed in the task space for a comparison of the performance in the two cases. As seen in Figure 6.51, a typical trajectory is considered, in which the moving platform moves 0.25 m in x direction, 0.5 m in the y direction, and -0.25 in the z direction. The orientation of the moving platform is also a compound rotation in all x, y, and z directions with the values in Figure 6.51. The time of simulation is considered to be 2 s, given to examine high accelerations for the moving platform.

The PD gains used in the simulations are set at $\boldsymbol{K}_p = 2 \times 10^6 \cdot \boldsymbol{I}_{6\times 6}$ and $\boldsymbol{K}_d = 5 \times 10^4 \cdot \boldsymbol{I}_{6\times 6}$ after a number of trial-and-error nominations. Figure 6.81 illustrates the tracking error in terms of the limb lengths, and as seen in this figure, the tracking performance in joint space is uniform. As seen in Figure 6.82, a decentralize PD controller for the manipulator is capable of reducing the tracking errors in the task space to less than 4 mm in position and less than 0.5° in orientation. Comparing these results to those of the PD controller in the task space, illustrated in Figure 6.52, it is observed that the orientation error significantly increased from 0.005° to 0.5° in the joint space control scheme. The main reason is that the controller gains directly penalize the position error of limb lengths, and not the orientation errors, and therefore, there is no direct controller action to be suitably tuned to reduce the orientation error.

The actuator forces $\boldsymbol{\tau}$ generated by the PD control action have been illustrated in Figure 6.83. Furthermore, the projection of the actuator forces on the Cartesian space is evaluated by $\mathcal{F} = J^T \boldsymbol{\tau}$ and is depicted in Figure 6.84. As seen in these figures, the amount of actuator forces and correspondingly the Cartesian wrench required to perform this task are relatively high and it reaches about 4 kN for the forces, and about 1 kN.m for the torques. High forces and torques are generated by the PD controller to accommodate high accelerations required in the desired trajectory.

To examine the effect of gravitational forces on the performance of PD controller in closed loop, another simulation is performed which is identical to the previous simulation, however, the gravity acceleration is considered in the $-z$ direction: $\boldsymbol{g} = [0, 0, -9.81]^T$. The tracking error for the same controller in the presence of gravitational forces are

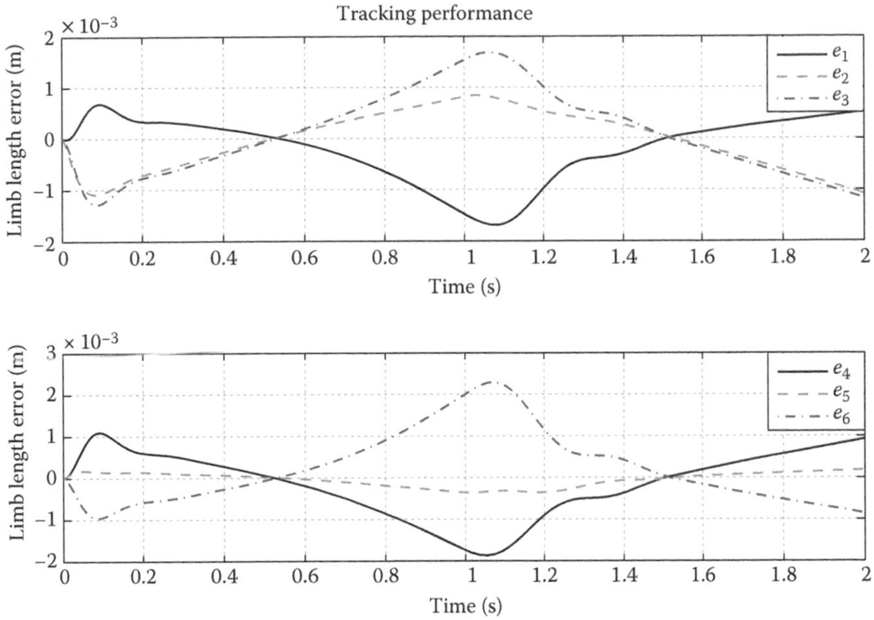

FIGURE 6.81
The closed-loop tracking error of the Stewart–Gough platform in joint space for decentralized PD control designed in joint space with no gravitational force.

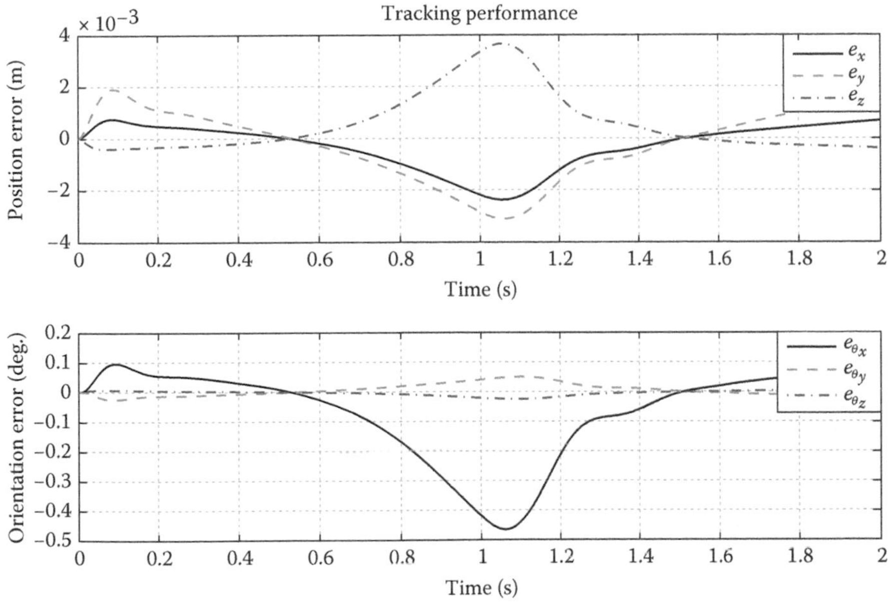

FIGURE 6.82
The closed-loop tracking error of the Stewart–Gough platform in task space for decentralized PD control designed in joint space with no gravitational force.

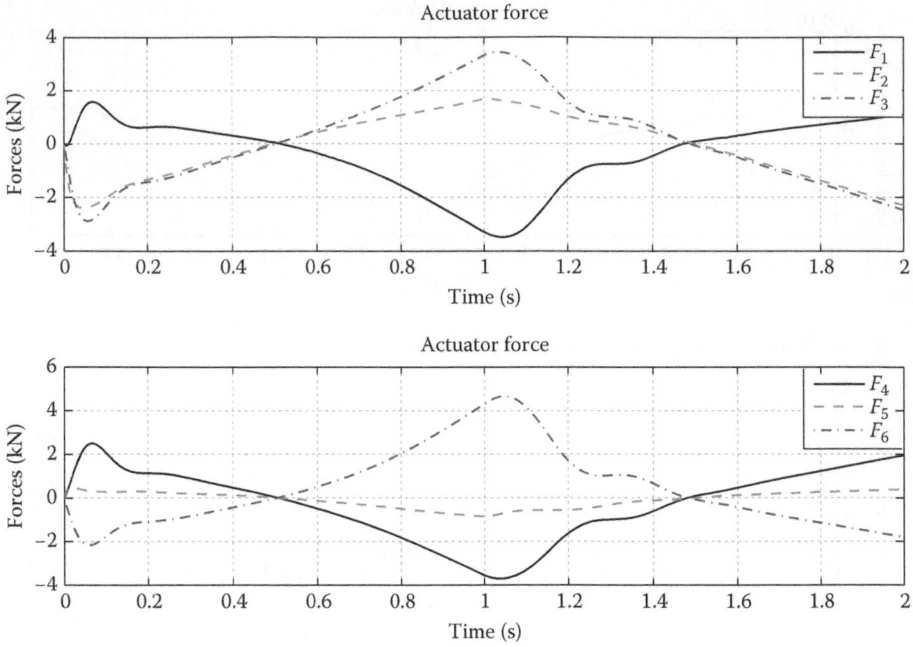

FIGURE 6.83
The actuator forces of the Stewart–Gough platform with no gravitational force generated by the PD controller in the joint space.

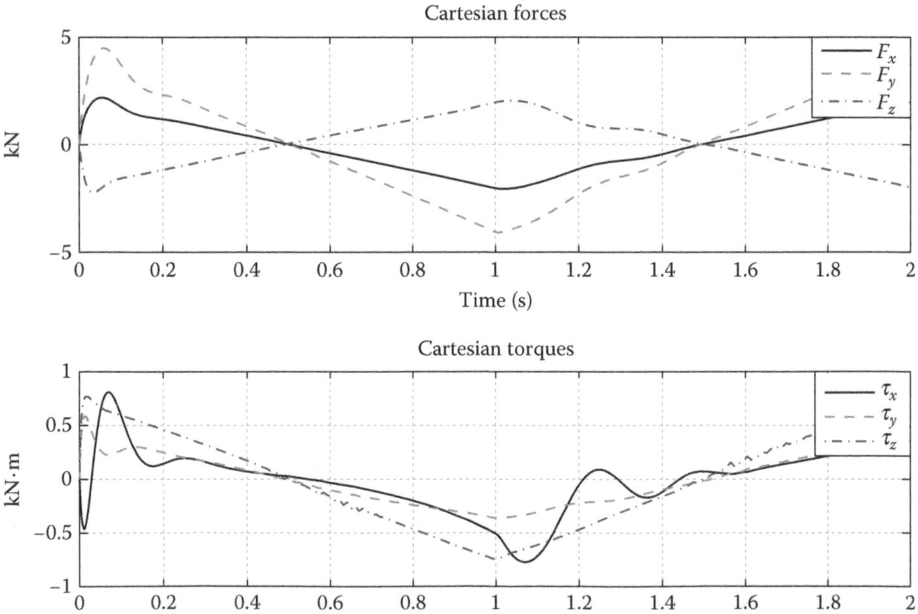

FIGURE 6.84
The Cartesian wrench of the Stewart–Gough platform with no gravitational force generated by PD controller in the joint space.

illustrated in Figure 6.85. As clearly seen in this figure, gravitational forces have significantly increased the tracking error in the z direction to about 15 mm, while this error for orientation it increased to 1.5°. This has caused the required Cartesian force in z direction to have a steady-state offset of about 10 kN as shown in Figure 6.86.

To examine the performance of the PD controller to attenuate the effect of external disturbance and measurement noise, two more simulations are executed for the closed-loop system. In the first simulation, a unit step disturbance force is applied to the system under original trajectories shown in Figure 6.51. To examine the effect of disturbance forces and torques separately, a 10 kN step disturbance force in x, y, and z directions is applied at time 0.5 s, then another 10 kN.m step disturbance torque in x, y, and z directions is applied at time 1 s. The controller gains are set at $\boldsymbol{K}_p = 2 \times 10^6 \cdot \boldsymbol{I}_{6\times6}$, and $\boldsymbol{K}_d = 5 \times 10^4 \cdot \boldsymbol{I}_{6\times6}$ as hereinbefore.

The closed-loop tracking error of the system is shown in Figure 6.87. By comparing this figure with Figure 6.85, it is seen that the effect of external disturbance cannot be completely attenuated by the PD controller, and the abrupt change in the tracking error at the instances when the disturbance forces/torques are applied is clearly seen in the response, especially for the position error in the z direction, that is, e_z, and orientation error in the x direction, that is, e_{θ_x}. Furthermore, as a result of nonvanishing disturbances applied to the system, static errors are accumulated in the closed-loop response of the system. The actuator forces $\boldsymbol{\tau}$ generated by the PD control action have been illustrated in Figure 6.88. Furthermore, the projection of the actuator forces on the Cartesian space is evaluated by $\boldsymbol{F} = \boldsymbol{J}_T\boldsymbol{\tau}$ and is depicted in Figure 6.89. As seen in these figures, the amount of actuator forces and correspondingly the Cartesian wrench required to perform this task are relatively high and it reaches about 4 kN for the forces, and about 1 kN.m for the torques. High

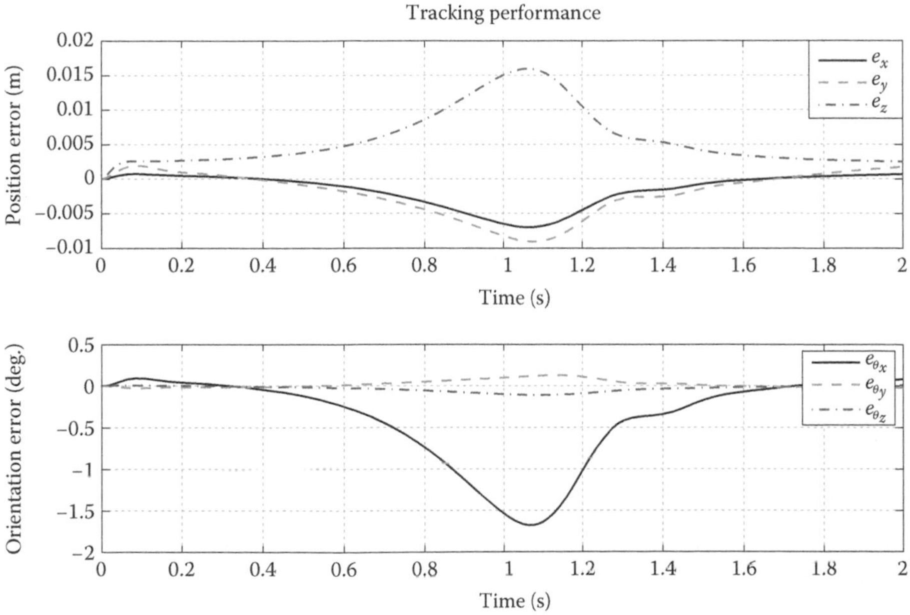

FIGURE 6.85
The closed-loop tracking performance of the Stewart–Gough platform considering the gravitational forces, for decentralized PD control in the joint space.

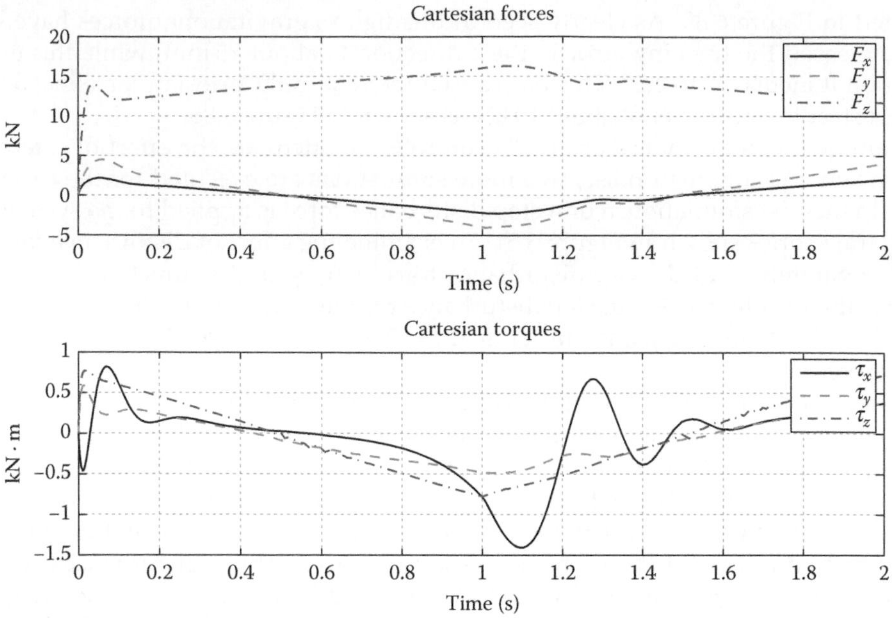

FIGURE 6.86
The Cartesian wrench of the Stewart–Gough platform considering the gravitational forces and generated by the PD controller in the joint space.

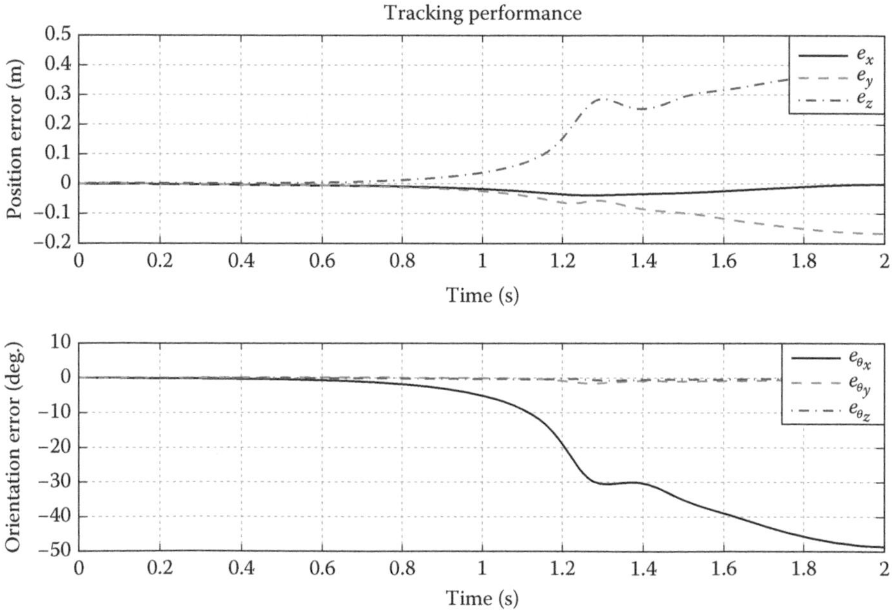

FIGURE 6.87
The closed-loop tracking performance of the Stewart–Gough platform for decentralized PD control designed in the joint space in the presence of disturbance.

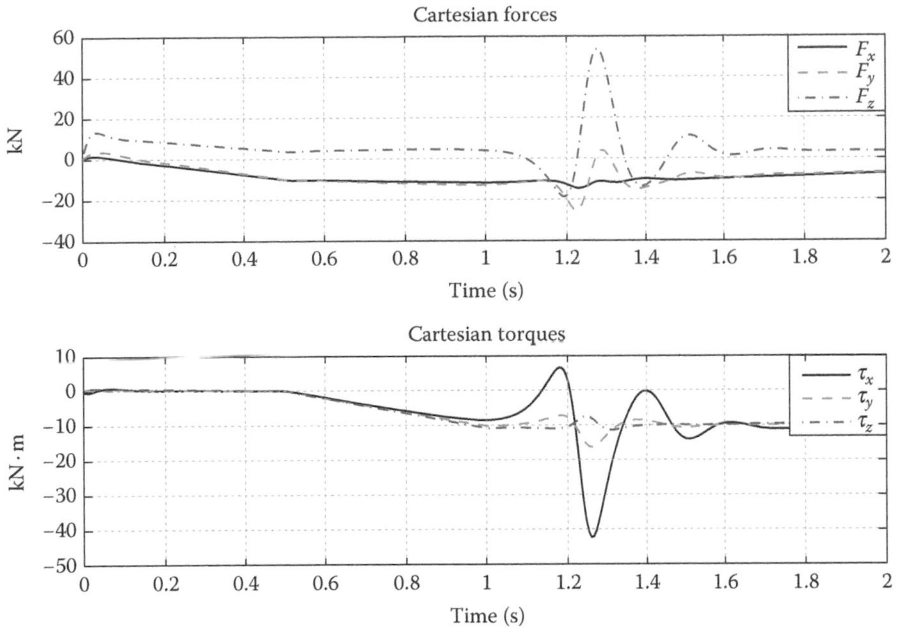

FIGURE 6.88
Cartesian wrench of the Stewart–Gough platform for decentralized PD controller designed in the joint space in the presence of disturbance.

forces and torques are generated by the PD controller to accommodate high accelerations required in the desired trajectory. Figure 6.88 illustrates the Cartesian wrench generated by the projection of PD controller torques to carry out the required maneuver in the presence of disturbance. The effect of the step disturbance wrench on the required controller forces is also clearly seen at times 0.5 and 1 s in Figure 6.88.

The effect of measurement noise is considered next by considering that all the measured limb lengths are contaminated with a Gaussian noise with an amplitude of 5 mm. Although the amount of noise is limited in this simulation, since high-gain controllers have to be used in the decentralized PD controller, the effect of noise on tracking performance is significant. As shown in Figures 6.89 and 6.90, although the tracking error is not increased much compared to that of the system without noise, the actuator forces require high bandwidth. This is owing to the fact that high PD gains are used to accommodate the required tracking performance, and although the amplitude of noise is about 10^{-3} times the peak values of the measurement signals, since the PD gains are about 10^5, the signature of the noise is apparently seen in the required actuator forces. In fact, generating such actuator forces is an expensive affair in practice, and the tracking performance shown in Figure 6.89 is not easily achievable.

Finally, comparing the closed-loop performance of the PD controllers designed in the joint space to those designed in the task space, it can be concluded that tuning of the PD gains for a suitable performance is much easier in task space designs. Furthermore, a very small error signature in the joint space may be accumulated to produce relatively larger tracking errors in the task space. This is clearly observed in orientation tracking errors in the controller designs in the joint space. Hence, it is recommended to design and implement controllers in the task space, if the required motion variables can be directly measured or the forward kinematic solution can be calculated in an online routine.

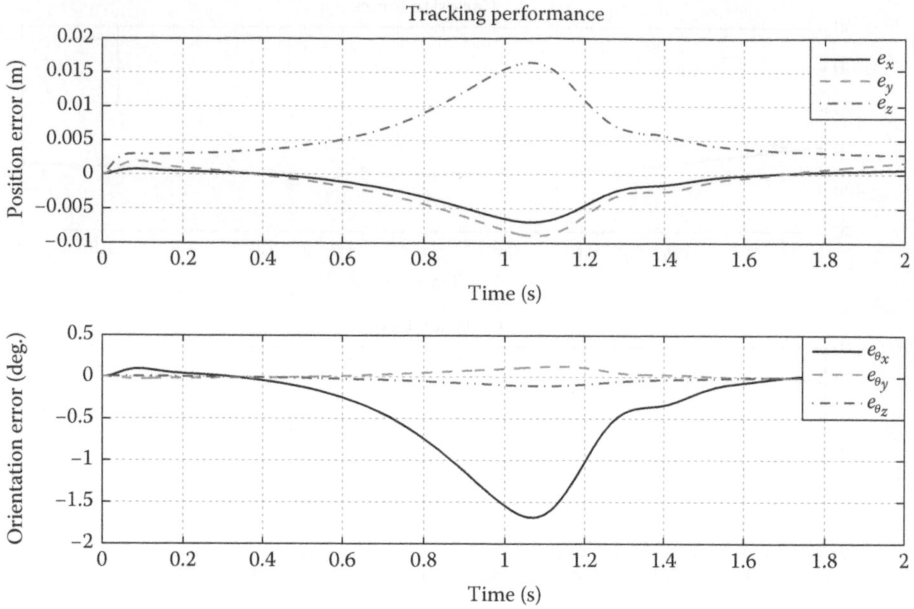

FIGURE 6.89
The closed-loop tracking performance of the Stewart–Gough platform for decentralized PD control in the joint space in the presence of measurement noise.

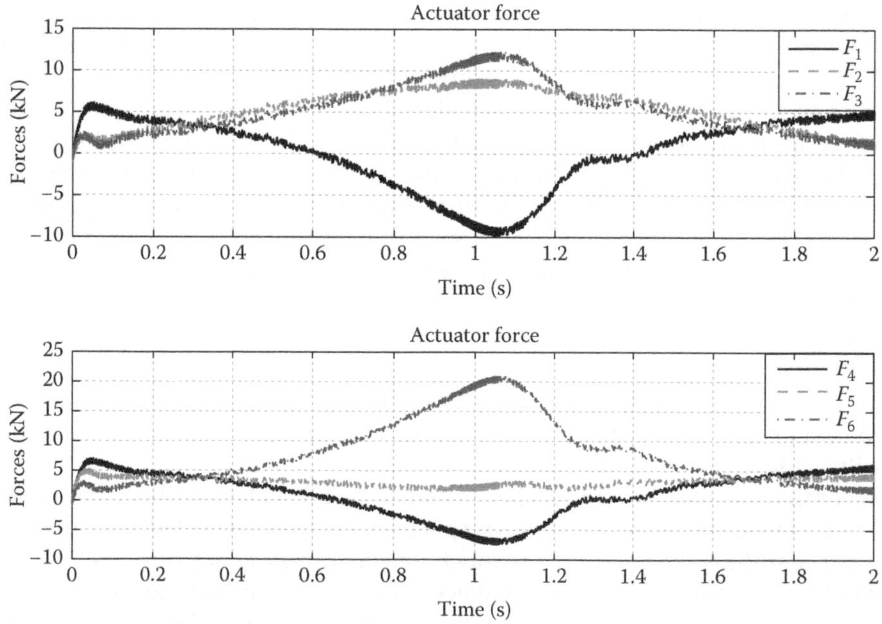

FIGURE 6.90
The actuator forces of the Stewart–Gough platform generated by the PD controller in the joint space in the presence of measurement noise.

6.9.6.2 Feed Forward Control

As seen in the previous section, the tracking performance of the PD controller implemented in the joint space is not sufficient in the presence of gravitational forces, measurement noise, and external disturbances applied to the manipulator. To compensate for such effects, a feed forward term denoted by $\boldsymbol{\tau}_{ff}$ may be added to the structure of the controller as depicted in Figure 6.64. This term is generated from the projection of dynamic model of the manipulator in the task space into the joint space by Equation 6.64. Furthermore, note that to generate this term, the dynamic formulation of the robot and its kinematic and dynamic parameters are needed. In practice, exact information on dynamic matrices is not available, and therefore, estimates of these matrices with %10 perturbation in all kinematic and inertial parameters are used in simulations. For the sake of comparison, PD controller gains are set at $\boldsymbol{K}_p = 2 \times 10^6 \cdot \boldsymbol{I}_{6\times 6}$ and $\boldsymbol{K}_d = 5 \times 10^4 \cdot \boldsymbol{I}_{6\times 6}$ as in the case of decentralized PD control.

Figure 6.91 illustrates the closed-loop tracking performance of the Stewart–Gough platform with feed forward control for typical trajectories depicted in Figure 6.51. Comparing the results with those of the PD control illustrated in Figure 6.85, it is seen that the feed forward action is capable of improving the orientation tracking performance more than 10 times in the presence of the gravitational forces.

The actuator forces generated by the feed forward controller for such a maneuver is given in Figure 6.92. In this figure, the total actuator force is indicated with a solid line, while its components, namely the PD action and the feed forward action are in dashed and dashed-dotted lines, respectively. As seen in this figure, since the knowledge of the system model is quite accurate, and only %10 perturbation in kinematic and inertial parameters is considered, the main contributing component of the actuator force is the feed forward

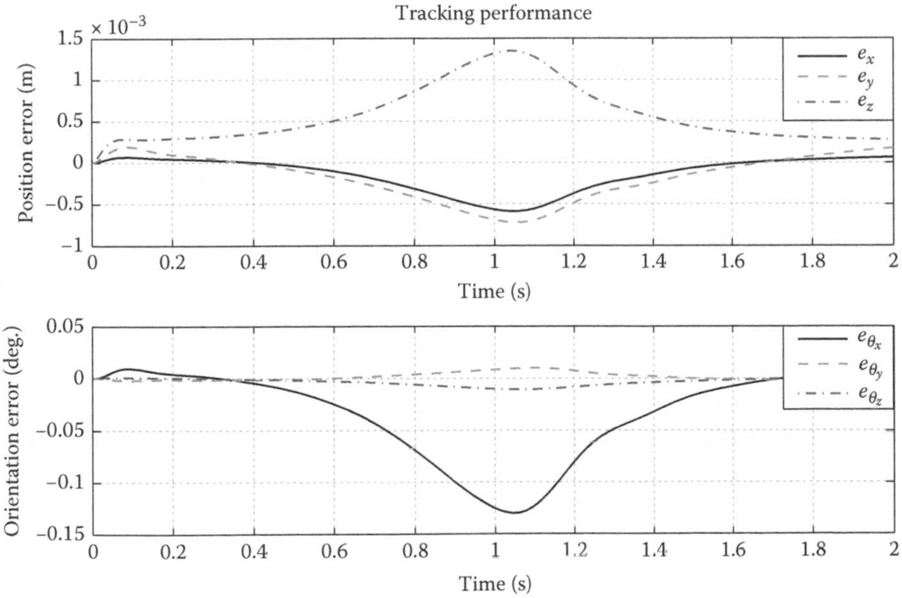

FIGURE 6.91
The closed-loop tracking performance of the Stewart–Gough platform with feed forward control in the joint space considering the gravitational forces.

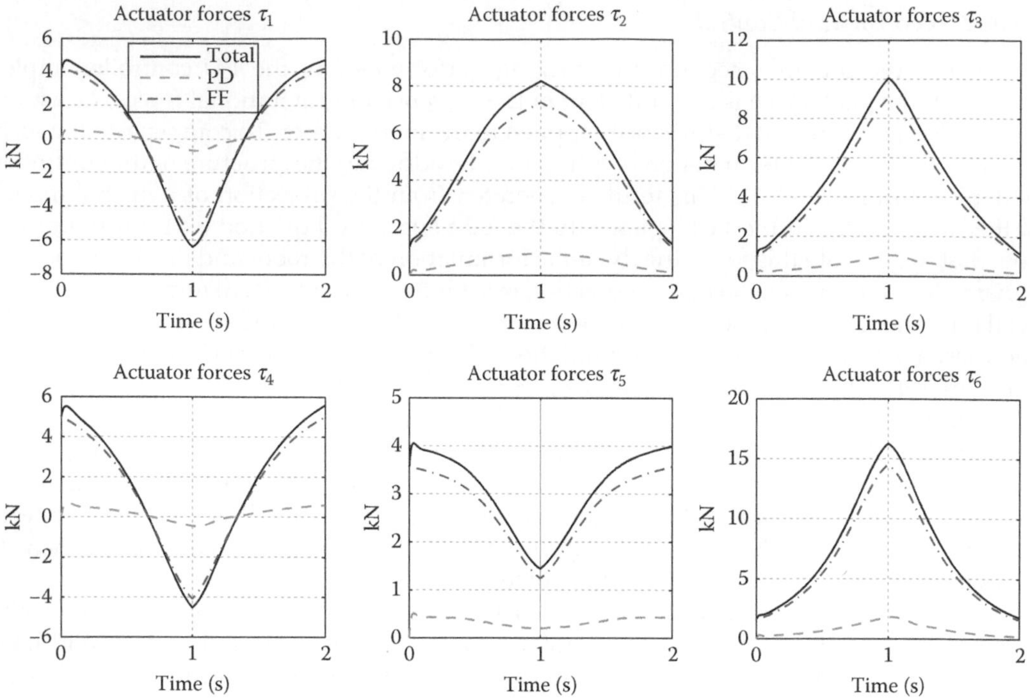

FIGURE 6.92
The actuator forces of the Stewart–Gough platform with feed forward controller in the joint space considering the gravitational forces.

term, which generates about %90 of the generated force. The PD action in this simulation contributes to only about %10 of the generated force. The total force needed to produce the required motion is even smaller than a pure PD control, while the tracking performance is significantly improved. This is mainly due to the contribution of feed forward term into the accurate motion of the manipulator as expected. The Cartesian wrench is illustrated in Figure 6.93, in which the amount of required actuator forces is also smaller than that in pure PD control.

In the next simulation, it is assumed that the knowledge of the system model parameters is very poor, and %50 perturbation in all kinematic and inertial parameters is considered in derivation of the feed forward control term. It is expected that the tracking performance is significantly affected because of this lack of knowledge of the system model. Figures 6.94 and 6.95 illustrate the tracking performance and the contribution of the PD and feed forward term in the total generated Cartesian wrench. As seen in Figure 6.94, lack of knowledge of the model parameters significantly affect the tracking errors. Comparing this figure with Figure 6.91, it can be concluded that the tracking errors are increased by a factor of 5.

Figure 6.95 illustrates the actuator forces generated by this control law and its PD and feed forward components. As seen in this figure, since the information on model parameters is poor in this simulation, the contribution of PD control term increase to almost %50 of the total force. This is in complete agreement to the fact that for a model-based control structure such as the feed forward scheme under consideration, if the information on the model is poor, the contribution of the PD term becomes more significant. Finally, the effects

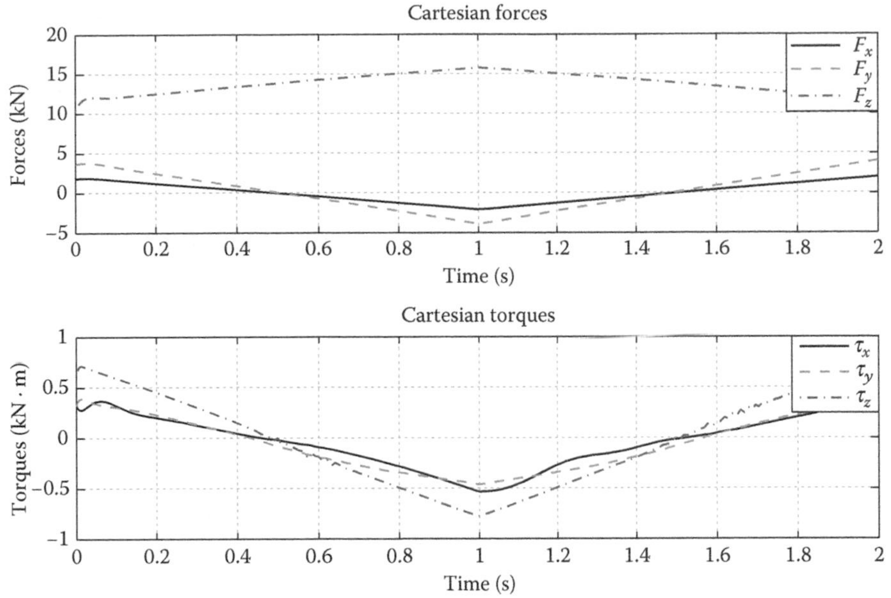

FIGURE 6.93
Cartesian wrench of the Stewart–Gough platform generated by feed forward controller in the joint space, considering the gravitational forces.

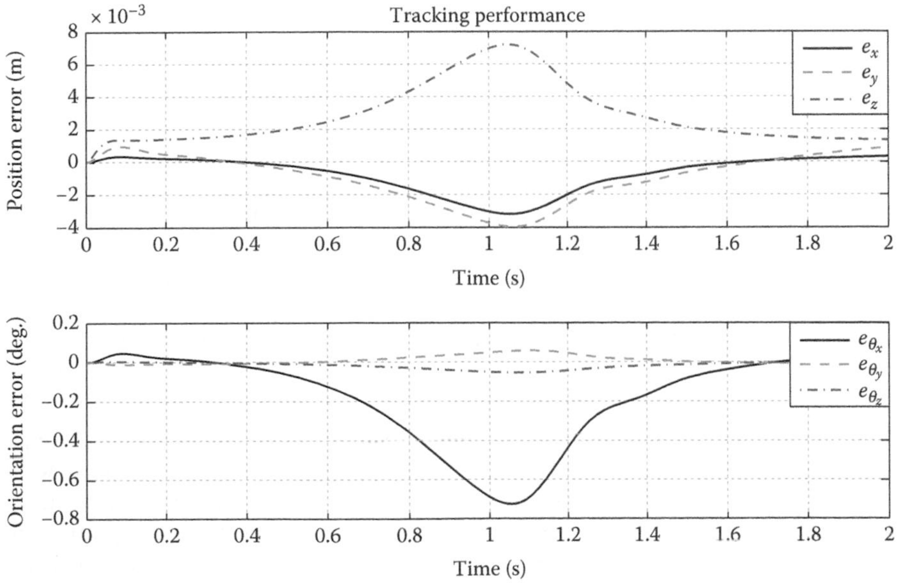

FIGURE 6.94
The closed-loop tracking performance of the Stewart–Gough platform with feed forward control in the joint space; %50 perturbation in kinematic and inertial parameters.

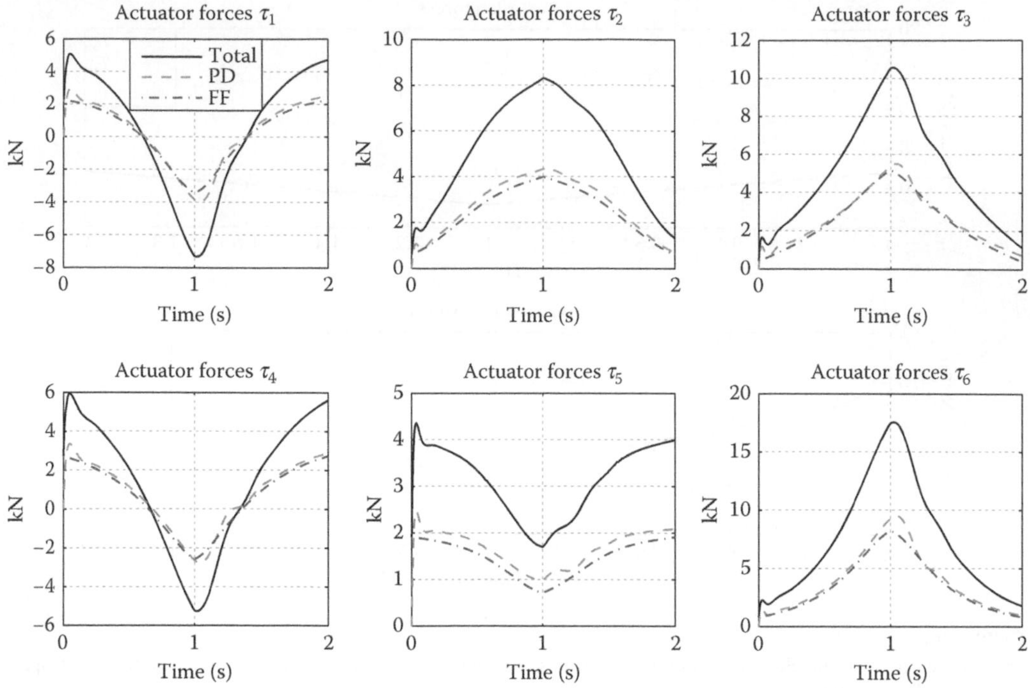

FIGURE 6.95
The actuator forces of the Stewart–Gough platform generated by feed forward controller in the joint space; %50 perturbation in kinematic and inertial parameters.

of disturbance and noise are also examined on this controller, and similar results to those of the pure PD control is observed. Since this controller structure is model based, and in these simulations high PD gains are considered for the controller, such performances in the presence of noise and disturbance are expected.

6.9.6.3 Inverse Dynamics Control

As seen in the previous sections, the tracking performance of a decentralized PD controller is not uniform in the presence of gravitational forces, measurement noise, and external disturbances applied to the manipulator. To compensate for such effects, a feed forward term is added to the structure of the controller, by which the shortcomings of the decentralized controller is partially remedied. However, the closed-loop performance still faces a number of limitations. To examine these limitations on performance, inverse dynamics controller in the joint space is implemented on the Stewart–Gough platform. The general structure of inverse dynamic control applied to a parallel manipulator is depicted in Figure 6.14. As seen in this figure, a corrective force τ_{fl} is added in a feedback structure to the closed-loop system, which is calculated from the Coriolis and centrifugal matrix, and the gravity vector of the manipulator dynamic formulation in the joint space. Furthermore, the mass matrix acts in the forward path manipulating the PD control term, in addition to the desired trajectory acceleration in the joint space $\ddot{q}_d$.

Note that to generate this term, dynamic formulation of the robot in the joint space, given in Equation 6.59, and its kinematic and dynamic parameters are needed. In practice, exact information on dynamic matrices is not available, and therefore, a %10 perturbation in

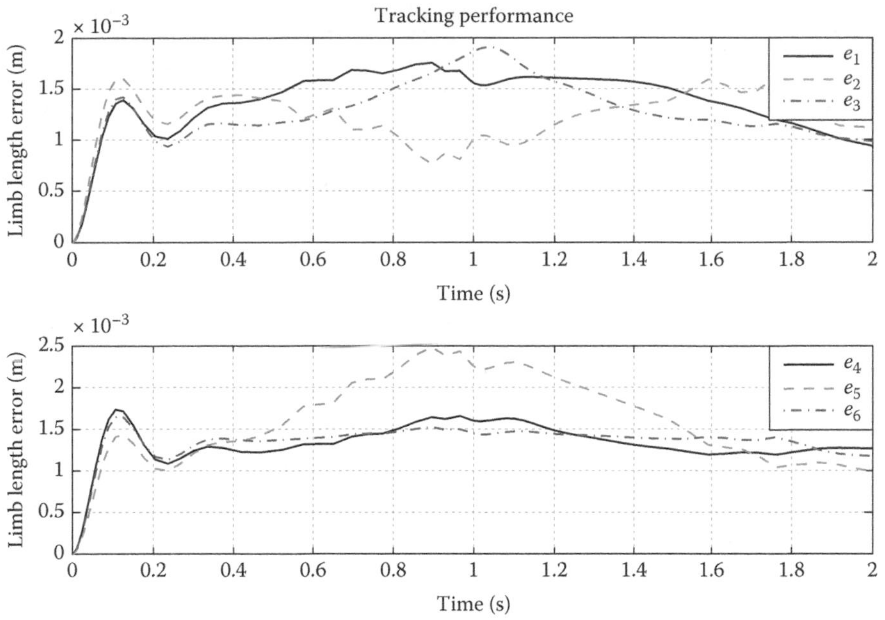

FIGURE 6.96
The closed-loop tracking performance of the Stewart–Gough platform with inverse dynamics control designed in the joint space.

all kinematic and inertial parameters are used in the following simulations. Furthermore, to compute the dynamic matrices in the joint space by Equations 6.60 through 6.62, it is required to calculate $\dot{J}$. This can be derived by differentiation of Equation 4.84 with respect to time as follows:

$$\dot{J}^T = \begin{bmatrix} \dot{\hat{s}}_i \\ \dot{b}_i \times \hat{s}_i + b_i \times \dot{\hat{s}}_i \end{bmatrix}_{i=1}^{6} = \begin{bmatrix} \omega_i \times \hat{s}_i \\ (\omega \times b_i) \times \hat{s}_i + b_i \times (\omega_i \times \hat{s}_i) \end{bmatrix}_{i=1}^{6}.$$

Hence

$$\dot{J}^T = \begin{bmatrix} \omega_{i_\times} \hat{s}_i \\ -\hat{s}_{i_\times} \omega_x b_i + b_{i_\times} \omega_{i_\times} \hat{s}_i \end{bmatrix}_{i=1}^{6}, \tag{6.123}$$

where ω denotes the angular velocity of the moving platform and ω_i denotes the angular velocity of limb i. The calculation of the term $\ddot{q}_d$ may also be carried out by substitution of $\dot{J}$ in Equation 6.58.

For the sake of comparison to previously examined controller structure, PD controller gains are adjusted by trial and error to $K_p = 2 \times 10^3 \cdot I_{6\times 6}$ and $K_d = 50 \cdot I_{6\times 6}$ to have the same order of tracking error as to that of feed forward control. Note that as shown in Figure 6.14, the PD controller terms are multiplied by the manipulator mass matrix, and therefore, the same dimensions of the controller gains used in pure PD or feed forward controller are not applicable here. However, tuning the PD controller gains in an IDC controller structure is very easy, and with a few trials-and-errors over the desired trajectory appropriate gains may be suitably selected.

Figure 6.96 illustrates the closed-loop tracking performance of the Stewart–Gough platform with inverse dynamics control in the joint space for the typical trajectory depicted

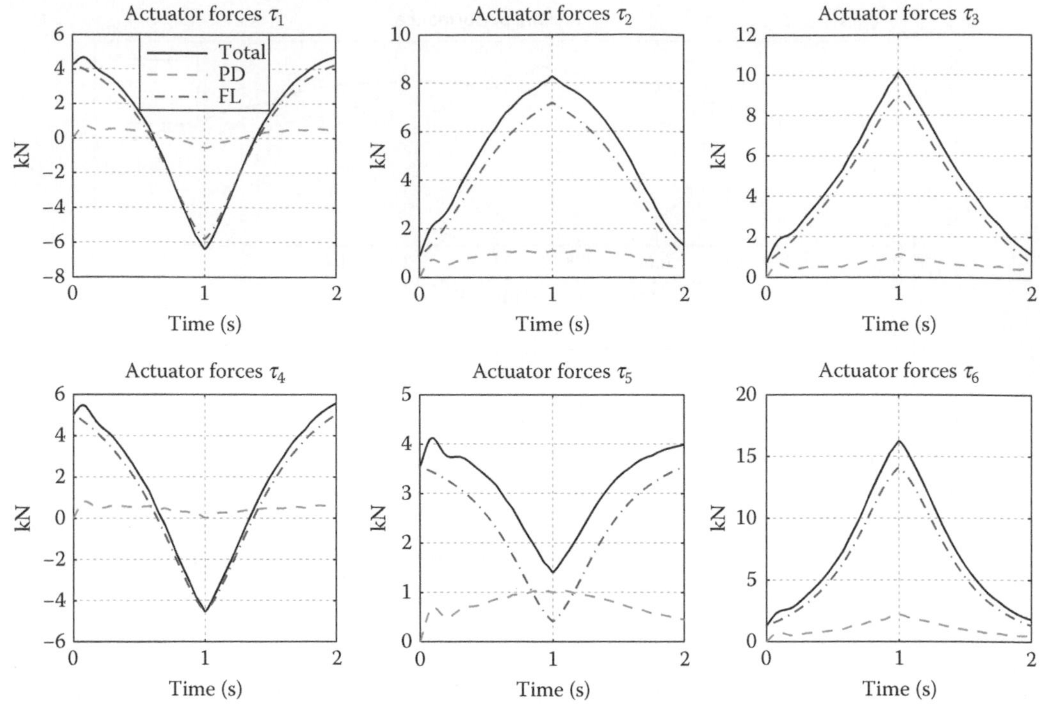

FIGURE 6.97
The actuator forces of the Stewart–Gough platform generated by inverse dynamic controller designed in the joint space.

in Figure 6.51. Comparing this figure with that of feed forward control illustrated in Figure 6.91, it is seen that the PD controller gains are tuned to obtain a similar performance as the feed forward control with a much easier process. However, the orientation errors in the IDC controller are more favorable.

The actuator forces generated by inverse dynamics controller for such a maneuver are given in Figure 6.97. In this figure, the total actuator force generated by inverse dynamic controller is plotted with a solid line, while its PD component is plotted in a dashed line, and the feedback linearizing term is plotted in a dashed–dotted line. As seen in this figure, since the knowledge of system model is quite accurate, and only %10 perturbation in kinematic and inertial parameters of the model is considered in this simulation, the main component of the actuator force is the inverse dynamic term, which generates about %90 of the total force. The PD action in this simulation contributes to only about %10 of the total force. The total force needed to produce the required motion is almost the same as in feed forward control, while the tracking performance is also almost the same for positions while that is more suitable for orientations. Similarly, the Cartesian wrench of the manipulator are illustrated in Figure 6.98, and it is observed that relatively higher wrench is required to carry out the required maneuver. This is mainly because of high accelerations that are required to carry out the desired trajectory.

In the next simulation, it is assumed that the information on system model is very poor, and %50 perturbation in all kinematic and inertial parameters are considered in derivation of inverse dynamic control law. It is expected that the tracking performance is significantly affected because of this lack of knowledge of system model. Figures 6.99 and 6.100 illustrate the tracking performance and the contribution of the PD and feedback linearizing

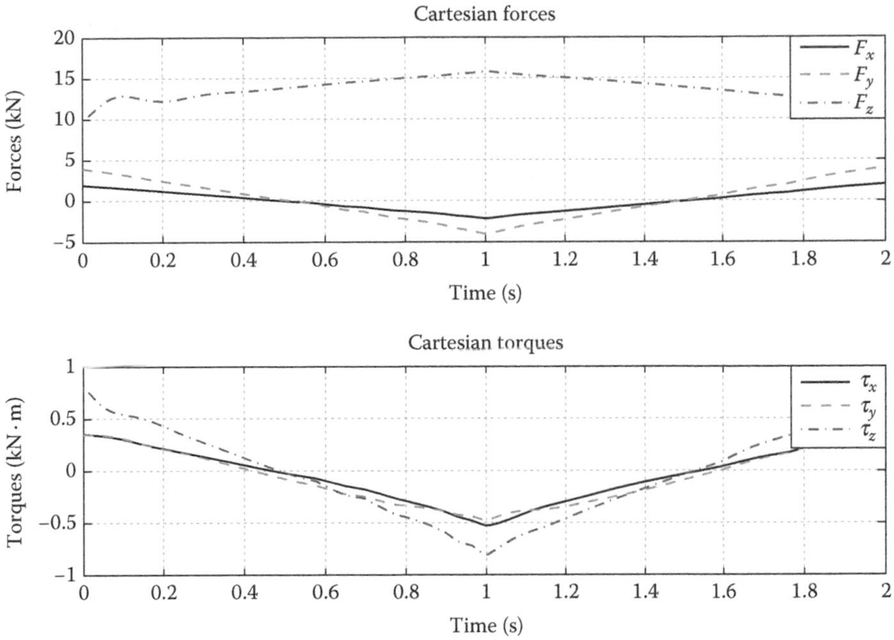

FIGURE 6.98
The Cartesian wrench of the Stewart–Gough platform with inverse dynamic controller designed in the joint space.

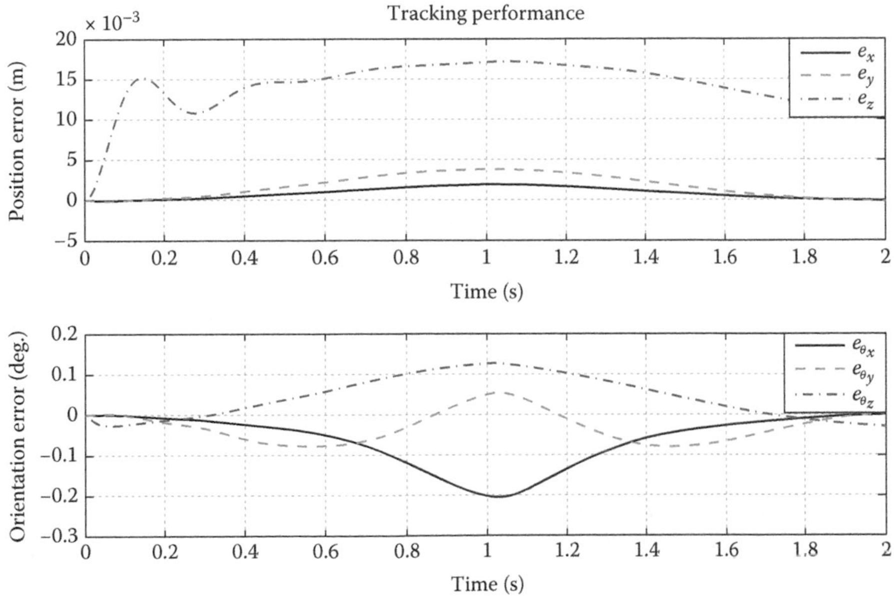

FIGURE 6.99
The closed-loop tracking performance of the Stewart–Gough platform with inverse dynamics control in the joint space; %50 perturbation in kinematic and inertial parameters.

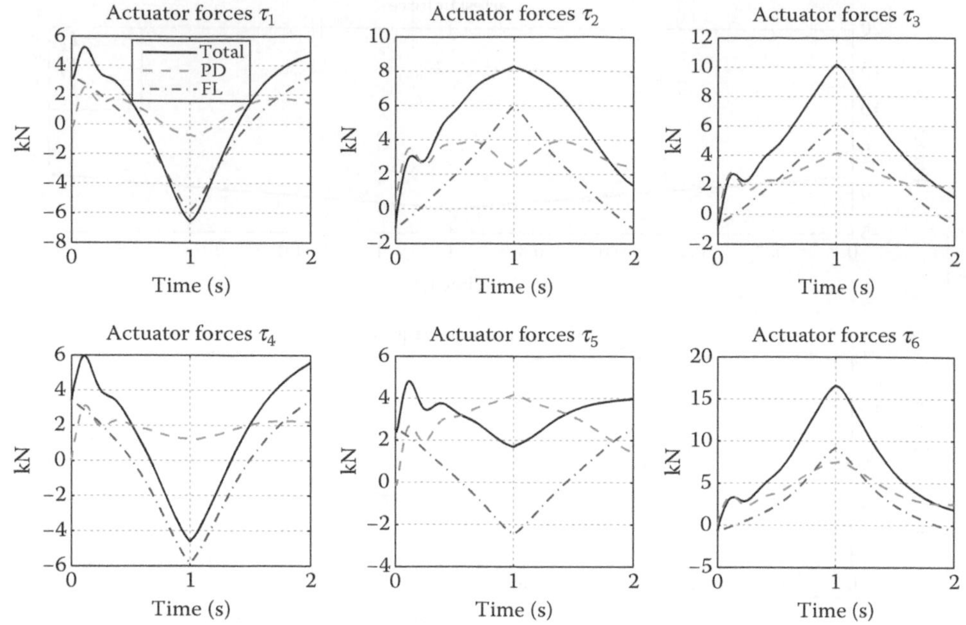

FIGURE 6.100
The Cartesian wrench of the Stewart–Gough platform generated by inverse dynamics controller in the joint space; %50 perturbation in kinematic and inertial parameters.

terms with respect to the total actuator effort. As seen in Figure 6.99, lack of information on the model parameters significantly affects the tracking errors. By comparing this figure with Figure 6.96, it is observed that the tracking errors increase by a factor of 10.

Figure 6.100 illustrates the total actuator forces generated by this control law and its PD component. As seen in this figure, since the information on model parameters is relatively poor in this simulation, the contribution of the PD control term increases to almost %50 of the total force. This is in complete agreement with the fact that for a model-based control structure such as the IDC under consideration, if the knowledge of the model is poor, then the contribution of the PD term becomes more significant.

Finally, the effect of disturbance and noise is also examined on this controller performance, and similar results to those of pure PD control is observed. To examine the effect of disturbance force and torques separately, as hereinbefore a 10 kN step disturbance force in x, y, and z directions is applied at time 0.5 s, then another 10 kN.m step disturbance torque in x, y, and z directions is applied at time 1.0 s. The controller gains are set at $\boldsymbol{K}_p = 2 \times 10^3 \cdot \boldsymbol{I}_{6\times 6}$ and $\boldsymbol{K}_d = 50 \cdot \boldsymbol{I}_{6\times 6}$ as earlier.

The closed-loop tracking error of the system is shown in Figure 6.101. By comparing this figure with that of the PD controller performance in the presence of disturbance depicted in Figure 6.87, it is seen that IDC controller significantly reduces the effect of external disturbance on the tracking error compared with that of pure PD controller. However, as a result of nonvanishing disturbances applied to the system, static errors are still observed in the closed-loop response of the system. In the presence of the measurement noise, however, very suitable tracking performance can be achieved. Nevertheless, as shown in Figure 6.102, the required actuator bandwidth is still relatively high.

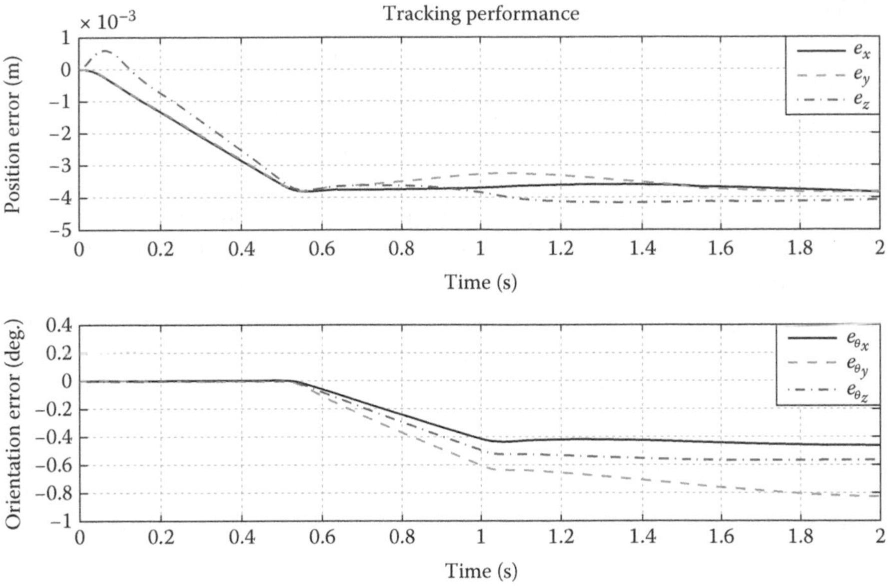

FIGURE 6.101
The closed-loop tracking performance of the Stewart–Gough platform with inverse dynamics control in the joint space in the presence of disturbance.

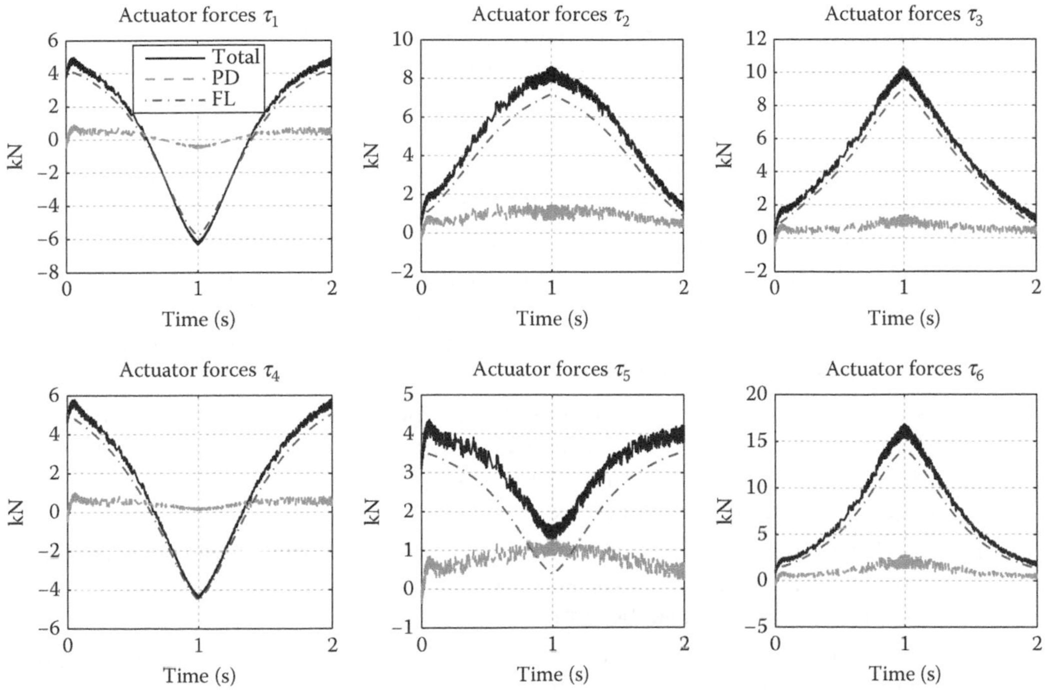

FIGURE 6.102
Actuator forces required to carry out the desired maneuver with inverse dynamics control in the joint space in the presence of measurement noise.

PROBLEMS

1. Consider the five-bar mechanism shown in Figure 3.28 and described in Problems 3.1 and 4.1. Consider the dynamic formulation with the geometrical and inertial parameters and the typical trajectory used in Problem 5.1.
   a. For this mechanism, design a decentralized PD controller in the task space and implement the required force distribution in terms of the mechanism Jacobian matrix. Develop a program in MATLAB to simulate the closed-loop tracking performance of the mechanism. Find proper choices for the controller gains to obtain suitable tracking performance with a limited control effort.
   b. Improve the performance by a feed forward control scheme. Simulate the closed-loop behavior and consider %10 perturbation in all geometrical and inertial parameters. Tune the controller gains to reach a suitable tracking performance with a limited control effort.
   c. Design an inverse dynamics controller for the mechanism and consider %10 perturbation in all geometrical and inertial parameters. Simulate the closed-loop performance and tune the controller gains to reach a suitable tracking performance with a limited control effort.
   d. Consider a partial feedback linearization in the IDC control and consider %10 perturbation in all geometrical and inertial parameters. Simulate the closed-loop performance and use the controller gains considered in full IDC.
   e. Plot and analyze motion trajectories, tracking errors, and the Cartesian and actuator forces versus time for different controllers and compare the results.
2. Consider the five-bar mechanism shown in Figure 3.29 and described in Problems 3.2 and 4.2. Consider the dynamic formulation with the geometrical and inertial parameters, and the typical trajectory used in Problem 5.2.
   a. For this mechanism, design a decentralized PD controller in the task space and implement the required force distribution in terms of the mechanism Jacobian matrix. Develop a program in MATLAB to simulate the closed-loop tracking performance of the mechanism. Find proper choices for the controller gains to obtain suitable tracking performance with a limited control effort.
   b. Improve the performance by a feed forward control scheme. Simulate the closed-loop behavior and consider %10 perturbation in all geometrical and inertial parameters. Tune the controller gains to reach a suitable tracking performance with a limited control effort.
   c. Design an inverse dynamics controller for the mechanism and consider %10 perturbation in all geometrical and inertial parameters. Simulate the closed-loop performance and tune the controller gains to reach a suitable tracking performance with a limited control effort.
   d. Consider a partial feedback linearization in the IDC control and consider %10 perturbation in all geometrical and inertial parameters. Simulate the closed-loop performance and use the controller gains considered in full IDC.
   e. Plot and analyze motion trajectories, tracking errors, and the Cartesian and actuator forces versus time for different controllers and compare the results.
3. Consider a planar pantograph mechanism shown in Figure 3.30 and described in Problems 3.3 and 4.3. Consider the dynamic formulation with the geometrical and inertial parameters, and the typical trajectory used in Problem 5.3.

a. For this mechanism, design a decentralized PD controller in the task space, and implement the required force distribution in terms of the mechanism Jacobian matrix. Develop a program in MATLAB to simulate the closed-loop tracking performance of the mechanism. Find proper choices for the controller gains to obtain suitable tracking performance with a limited control effort.

b. Improve the performance by a feed forward control scheme. Simulate the closed-loop behavior and consider %10 perturbation in all geometrical and inertial parameters. Tune the controller gains to reach a suitable tracking performance with a limited control effort.

c. Design an inverse dynamics controller for the mechanism and consider %10 perturbation in all geometrical and inertial parameters. Simulate the closed-loop performance and tune the controller gains to reach a suitable tracking performance with a limited control effort.

d. Consider a partial feedback linearization in the IDC control and consider %10 perturbation in all geometrical and inertial parameters. Simulate the closed-loop performance and use the controller gains considered in full IDC.

e. Plot and analyze motion trajectories, tracking errors, and the Cartesian and actuator forces versus time for different controllers and compare the results.

4. Consider the planar three-degrees-of-freedom $3\underline{R}RR$ parallel manipulator shown in Figure 3.31 and described in Problems 3.4 and 4.4. Consider the dynamic formulation with the geometrical and inertial parameters and the typical trajectory used in Problem 5.4.

a. For this manipulator, design a decentralized PD controller in the task space, and implement the required force distribution in terms of the mechanism Jacobian matrix. Develop a program in MATLAB to simulate the closed-loop tracking performance of the mechanism. Find proper choices for the controller gains to obtain suitable tracking performance with a limited control effort.

b. Improve the performance by a feed forward control scheme. Simulate the closed-loop behavior and consider %10 perturbation in all geometrical and inertial parameters. Tune the controller gains to reach a suitable tracking performance with a limited control effort.

c. Design an inverse dynamics controller for the mechanism and consider %10 perturbation in all geometrical and inertial parameters. Simulate the closed-loop performance and tune the controller gains to reach a suitable tracking performance with a limited control effort.

d. Consider a partial feedback linearization in the IDC control and consider %10 perturbation in all geometrical and inertial parameters. Simulate the closed-loop performance and use the controller gains considered in full IDC.

e. Plot and analyze motion trajectories, tracking errors, and the Cartesian and actuator forces versus time for different controllers and compare the results.

5. Consider the planar three-degrees-of-freedom $4\underline{R}RR$ parallel manipulator shown in Figure 3.32 and described in Problems 3.5 and 4.4. Consider the dynamic formulation with the geometrical and inertial parameters and the typical trajectory used in Problem 5.5.

a. For this manipulator, design a decentralized PD controller in the task space, and implement the required force distribution in terms of the mechanism Jacobian matrix. Develop a program in MATLAB to simulate the closed-loop tracking performance of the mechanism. Find proper choices for the controller gains to obtain suitable tracking performance with a limited control effort.

b. Improve the performance by a feed forward control scheme. Simulate the closed-loop behavior and consider %10 perturbation in all geometrical and inertial parameters. Tune the controller gains to reach a suitable tracking performance with a limited control effort.

c. Design an inverse dynamics controller for the mechanism and consider %10 perturbation in all geometrical and inertial parameters. Simulate the closed-loop performance and tune the controller gains to reach a suitable tracking performance with a limited control effort.

d. Consider a partial feedback linearization in the IDC control and consider %10 perturbation in all geometrical and inertial parameters. Simulate the closed-loop performance and use the controller gains considered in full IDC.

e. Plot and analyze motion trajectories, tracking errors, and the Cartesian and actuator forces versus time for different controllers and compare the results.

6. Consider the spatial parallel manipulator shown in Figure 3.33 and described in Problems 3.6 and 4.6. Consider the dynamic formulation with the geometrical and inertial parameters and the typical trajectory used in Problem 5.6.

   a. For this manipulator, design a decentralized PD controller in the task space, and implement the required force distribution in terms of the mechanism Jacobian matrix. Develop a program in MATLAB to simulate the closed-loop tracking performance of the mechanism. Find proper choices for the controller gains to obtain suitable tracking performance with a limited control effort.

   b. Improve the performance by a feed forward control scheme. Simulate the closed-loop behavior and consider %10 perturbation in all geometrical and inertial parameters. Tune the controller gains to reach a suitable tracking performance with a limited control effort.

   c. Design an inverse dynamics controller for the mechanism and consider %10 perturbation in all geometrical and inertial parameters. Simulate the closed-loop performance and tune the controller gains to reach a suitable tracking performance with a limited control effort.

   d. Consider a partial feedback linearization in the IDC control and consider %10 perturbation in all geometrical and inertial parameters. Simulate the closed-loop performance and use the controller gains considered in full IDC.

   e. Plot and analyze motion trajectories, tracking errors, and the Cartesian and actuator forces versus time for different controllers and compare the results.

7. Consider the $3U\underline{P}U$ parallel manipulator shown in Figure 3.34 and described in Problems 3.7 and 4.7. Consider the dynamic formulation with the geometrical and inertial parameters and the typical trajectory used in Problem 5.7.

   a. For this manipulator, design a decentralized PD controller in the task space, and implement the required force distribution in terms of the mechanism Jacobian matrix. Develop a program in MATLAB to simulate the closed-loop tracking performance of the mechanism. Find proper choices for the controller gains to obtain suitable tracking performance with a limited control effort.

b. Improve the performance by a feed forward control scheme. Simulate the closed-loop behavior and consider %10 perturbation in all geometrical and inertial parameters. Tune the controller gains to reach a suitable tracking performance with a limited control effort.

c. Design an inverse dynamics controller for the mechanism and consider %10 perturbation in all geometrical and inertial parameters. Simulate the closed-loop performance and tune the controller gains to reach a suitable tracking performance with a limited control effort.

d. Consider partial feedback linearization in the IDC control and consider %10 perturbation in all geometrical and inertial parameters. Simulate the closed-loop performance and use the controller gains considered in full IDC.

e. Plot and analyze motion trajectories, tracking errors, and the Cartesian and actuator forces versus time for different controllers and compare the results.

8. Consider the manipulator introduced in Reference 47, shown in Figure 3.35, and described in Problems 3.8 and 4.8. Consider the dynamic formulation with the geometrical and inertial parameters and the typical trajectory used in Problem 5.8.

a. For this manipulator, design a decentralized PD controller in the task space, and implement the required force distribution in terms of the mechanism Jacobian matrix. Develop a program in MATLAB to simulate the closed-loop tracking performance of the mechanism. Find proper choices for the controller gains to obtain suitable tracking performance with a limited control effort.

b. Improve the performance by a feed forward control scheme. Simulate the closed-loop behavior and consider %10 perturbation in all geometrical and inertial parameters. Tune the controller gains to reach a suitable tracking performance with a limited control effort.

c. Design an inverse dynamics controller for the mechanism and consider %10 perturbation in all geometrical and inertial parameters. Simulate the closed-loop performance and tune the controller gains to reach a suitable tracking performance with a limited control effort.

d. Consider a partial feedback linearization in the IDC control and consider %10 perturbation in all geometrical and inertial parameters. Simulate the closed-loop performance and use the controller gains considered in full IDC.

e. Plot and analyze motion trajectories, tracking errors, and the Cartesian and actuator forces versus time for different controllers and compare the results.

9. Consider the three-degrees-of-freedom $3R\underline{P}S$ manipulator represented in Figure 3.36 and described in Problems 3.9 and 4.9. Consider the dynamic formulation with the geometrical and inertial parameters and the typical trajectory used in Problem 5.9.

a. For this manipulator, design a decentralized PD controller in the task space and implement the required force distribution in terms of the mechanism Jacobian matrix. Develop a program in MATLAB to simulate the closed-loop tracking performance of the mechanism. Find proper choices for the controller gains to obtain suitable tracking performance with a limited control effort.

b. Improve the performance by a feed forward control scheme. Simulate the closed-loop behavior and consider %10 perturbation in all geometrical and inertial parameters. Tune the controller gains to reach a suitable tracking performance with a limited control effort.

c. Design an inverse dynamics controller for the mechanism and consider %10 perturbation in all geometrical and inertial parameters. Simulate the closed-loop performance and tune the controller gains to reach a suitable tracking performance with a limited control effort.

d. Consider a partial feedback linearization in the IDC control and consider %10 perturbation in all geometrical and inertial parameters. Simulate the closed-loop performance and use the controller gains considered in full IDC.

e. Plot and analyze motion trajectories, tracking errors, and the Cartesian and actuator forces versus time for different controllers and compare the results.

10. Consider the three-degrees-of-freedom $3\underline{P}RC$ translational parallel manipulator represented in Figure 3.36 and described in Problems 3.10 and 4.10. Consider the dynamic formulation with the geometrical and inertial parameters and the typical trajectory used in Problem 5.10.

a. For this manipulator, design a decentralized PD controller in the task space and implement the required force distribution in terms of the mechanism Jacobian matrix. Develop a program in MATLAB to simulate the closed-loop tracking performance of the mechanism. Find proper choices for the controller gains to obtain suitable tracking performance with a limited control effort.

b. Improve the performance by a feed forward control scheme. Simulate the closed-loop behavior and consider %10 perturbation in all geometrical and inertial parameters. Tune the controller gains to reach a suitable tracking performance with a limited control effort.

c. Design an inverse dynamics controller for the mechanism and consider %10 perturbation in all geometrical and inertial parameters. Simulate the closed-loop performance and tune the controller gains to reach a suitable tracking performance with a limited control effort.

d. Consider a partial feedback linearization in the IDC control and consider %10 perturbation in all geometrical and inertial parameters. Simulate the closed-loop performance and use the controller gains considered in full IDC.

e. Plot and analyze motion trajectories, tracking errors, and the Cartesian and actuator forces versus time for different controllers and compare the results.

11. Consider the Delta robot represented in Figure 3.39 and described in Problems 3.11 and 4.11. Consider the dynamic formulation with the geometrical and inertial parameters and the typical trajectory used in Problem 5.11.

a. For this manipulator, design a decentralized PD controller in the task space and implement the required force distribution in terms of the mechanism Jacobian matrix. Develop a program in MATLAB to simulate the closed-loop tracking performance of the mechanism. Find proper choices for the controller gains to obtain suitable tracking performance with a limited control effort.

b. Improve the performance by a feed forward control scheme. Simulate the closed-loop behavior and consider %10 perturbation in all geometrical and inertial parameters. Tune the controller gains to reach a suitable tracking performance with a limited control effort.

c. Design an inverse dynamics controller for the mechanism and consider %10 perturbation in all geometrical and inertial parameters. Simulate the closed-loop performance and tune the controller gains to reach a suitable tracking performance with a limited control effort.

d. Consider a partial feedback linearization in the IDC control and consider %10 perturbation in all geometrical and inertial parameters. Simulate the closed-loop performance and use the controller gains considered in full IDC.

e. Plot and analyze motion trajectories, tracking errors, and the Cartesian and actuator forces versus time for different controllers and compare the results.

12. Consider the Quadrupteron robot represented in Figure 3.40 and described in Problems 3.12 and 4.7 [157]. Consider the dynamic formulation with the geometrical and inertial parameters and the typical trajectory used in Problem 5.12.

    a. For this manipulator, design a decentralized PD controller in the task space, and implement the required force distribution in terms of the mechanism Jacobian matrix. Develop a program in MATLAB to simulate the closed-loop tracking performance of the mechanism. Find proper choices for the controller gains to obtain suitable tracking performance with a limited control effort.

    b. Improve the performance by a feed forward control scheme. Simulate the closed-loop behavior and consider %10 perturbation in all geometrical and inertial parameters. Tune the controller gains to reach a suitable tracking performance with a limited control effort.

    c. Design an inverse dynamics controller for the mechanism and consider %10 perturbation in all geometrical and inertial parameters. Simulate the closed-loop performance and tune the controller gains to reach a suitable tracking performance with a limited control effort.

    d. Consider a partial feedback linearization in the IDC control and consider %10 perturbation in all geometrical and inertial parameters. Simulate the closed-loop performance and use the controller gains considered in full IDC.

    e. Plot and analyze motion trajectories, tracking errors, and the Cartesian and actuator forces versus time for different controllers and compare the results.

13. Reconsider the five-bar mechanism shown in Figure 3.28 and described in Problems 3.1 and 4.1. Consider the dynamic formulation with the geometrical and inertial parameters and the typical trajectory used in Problem 5.1.

    a. For this mechanism, design a robust inverse dynamics controller in the task space, and implement the required force distribution in terms of the mechanism Jacobian matrix. Develop a program in MATLAB to simulate the closed-loop tracking performance of the mechanism. Consider %50 perturbation in all geometrical and inertial parameters and tune the controller gains to reach a suitable tracking performance with a limited control effort.

    b. Plot and analyze motion trajectories, tracking errors, and the Cartesian and actuator forces versus time for this controller and compare the results to those of Problem 6.1.

14. Reconsider the five-bar mechanism shown in Figure 3.29 and described in Problems 3.2 and 4.2. Consider the dynamic formulation with the geometrical and inertial parameters and the typical trajectory used in Problem 5.2.

a. For this mechanism, design a robust inverse dynamics controller in task space and implement the required force distribution in terms of the mechanism Jacobian matrix. Develop a program in MATLAB to simulate the closed-loop tracking performance of the mechanism. Consider %50 perturbation in all geometrical and inertial parameters, and tune the controller gains to reach a suitable tracking performance with a limited control effort.

b. Plot and analyze motion trajectories, tracking errors, and the Cartesian and actuator forces versus time for this controller and compare the results to those of Problem 6.2.

15. Reconsider a planar pantograph mechanism shown in Figure 3.30 and described in Problems 3.3 and 4.3. Consider the dynamic formulation with the geometrical and inertial parameters and the typical trajectory used in Problem 5.3.

    a. For this mechanism, design a robust inverse dynamics controller in the task space, and implement the required force distribution in terms of the mechanism Jacobian matrix. Develop a program in MATLAB to simulate the closed-loop tracking performance of the mechanism. Consider %50 perturbation in all geometrical and inertial parameters and tune the controller gains to reach a suitable tracking performance with a limited control effort.

    b. Plot and analyze motion trajectories, tracking errors, and the Cartesian and actuator forces versus time for this controller and compare the results to those of Problem 6.3.

16. Reconsider the planar three-degrees-of-freedom $3\underline{R}RR$ parallel manipulator shown in Figure 3.31 and described in Problems 3.4 and 4.4. Consider the dynamic formulation with the geometrical and inertial parameters and the typical trajectory used in Problem 5.4.

    a. For this manipulator, design a robust inverse dynamics controller in task space and implement the required force distribution in terms of the mechanism Jacobian matrix. Develop a program in MATLAB to simulate the closed-loop tracking performance of the mechanism. Consider %50 perturbation in all geometrical and inertial parameters, and tune the controller gains to reach a suitable tracking performance with a limited control effort.

    b. Plot and analyze motion trajectories, tracking errors, and the Cartesian and actuator forces versus time for this controller and compare the results to those of Problem 6.4.

17. Reconsider the planar three-degrees-of-freedom $4\underline{R}RR$ parallel manipulator shown in Figure 3.32 and described in Problems 3.5 and 4.4. Consider the dynamic formulation with the geometrical and inertial parameters and the typical trajectory used in Problem 5.5.

    a. For this manipulator, design a robust inverse dynamics controller in task space, and implement the required force distribution in terms of the mechanism Jacobian matrix. Develop a program in MATLAB to simulate the closed-loop tracking performance of the mechanism. Consider %50 perturbation in all geometrical and inertial parameters and tune the controller gains to reach a suitable tracking performance with a limited control effort.

   b. Plot and analyze motion trajectories, tracking errors, and the Cartesian and actuator forces versus time for this controller and compare the results to those of Problem 6.5.

18. Reconsider the spatial parallel manipulator shown in Figure 3.33 and described in Problems 3.6 and 4.6. Consider the dynamic formulation with the geometrical and inertial parameters and the typical trajectory used in Problem 5.6.
   a. For this manipulator, design a robust inverse dynamics controller in the task space and implement the required force distribution in terms of the mechanism Jacobian matrix. Develop a program in MATLAB to simulate the closed-loop tracking performance of the mechanism. Consider %50 perturbation in all geometrical and inertial parameters, and tune the controller gains to reach a suitable tracking performance with a limited control effort.
   b. Plot and analyze motion trajectories, tracking errors, and the Cartesian and actuator forces versus time for this controller and compare the results to those of Problem 6.6.

19. Reconsider the $3U\underline{P}U$ parallel manipulator shown in Figure 3.34 and described in Problems 3.7 and 4.7. Consider the dynamic formulation with the geometrical and inertial parameters and the typical trajectory used in Problem 5.7.
   a. For this manipulator, design a robust inverse dynamics controller in the task space and implement the required force distribution in terms of the mechanism Jacobian matrix. Develop a program in MATLAB to simulate the closed-loop tracking performance of the mechanism. Consider %50 perturbation in all geometrical and inertial parameters, and tune the controller gains to reach a suitable tracking performance with a limited control effort.
   b. Plot and analyze motion trajectories, tracking errors, and the Cartesian and actuator forces versus time for this controller and compare the results to those of Problem 6.7.

20. Reconsider the manipulator introduced in Reference 47, shown in Figure 3.35, and described in Problems 3.8 and 4.8. Consider the dynamic formulation with the geometrical and inertial parameters and the typical trajectory used in Problem 5.8.
   a. For this manipulator, design a robust inverse dynamics controller in the task space and implement the required force distribution in terms of the mechanism Jacobian matrix. Develop a program in MATLAB to simulate the closed-loop tracking performance of the mechanism. Consider %50 perturbation in all geometrical and inertial parameters and tune the controller gains to reach a suitable tracking performance with a limited control effort.
   b. Plot and analyze motion trajectories, tracking errors, and the Cartesian and actuator forces versus time for this controller and compare the results to those of Problem 6.8.

21. Reconsider the three-degrees-of-freedom $3R\underline{P}S$ manipulator represented in Figure 3.36 and described in Problems 3.9 and 4.9. Consider the dynamic formulation with the geometrical and inertial parameters and the typical trajectory used in Problem 5.9.

a. For this manipulator, design a robust inverse dynamics controller in the task space and implement the required force distribution in terms of the mechanism Jacobian matrix. Develop a program in MATLAB to simulate the closed-loop tracking performance of the mechanism. Consider %50 perturbation in all geometrical and inertial parameters and tune the controller gains to reach a suitable tracking performance with a limited control effort.

b. Plot and analyze motion trajectories, tracking errors, and the Cartesian and actuator forces versus time for this controller and compare the results to those of Problem 6.9.

22. Reconsider the three-degrees-of-freedom 3$\underline{P}RC$ translational parallel manipulator represented in Figure 3.36 and described in Problems 3.10 and 4.10. Consider the dynamic formulation with the geometrical and inertial parameters and the typical trajectory used in Problem 5.10.

    a. For this manipulator, design a robust inverse dynamics controller in the task space and implement the required force distribution in terms of the mechanism Jacobian matrix. Develop a program in MATLAB to simulate the closed-loop tracking performance of the mechanism. Consider %50 perturbation in all geometrical and inertial parameters and tune the controller gains to reach a suitable tracking performance with a limited control effort.

    b. Plot and analyze motion trajectories, tracking errors, and the Cartesian and actuator forces versus time for this controller and compare the results to those of Problem 6.10.

23. Reconsider the Delta robot represented in Figure 3.39 and described in Problems 3.11 and 4.11. Consider the dynamic formulation with the geometrical and inertial parameters and the typical trajectory used in Problem 5.11.

    a. For this manipulator, design a robust inverse dynamics controller in the task space, and implement the required force distribution in terms of the mechanism Jacobian matrix. Develop a program in MATLAB to simulate the closed-loop tracking performance of the mechanism. Consider %50 perturbation in all geometrical and inertial parameters, and tune the controller gains to reach a suitable tracking performance with a limited control effort.

    b. Plot and analyze motion trajectories, tracking errors, and the Cartesian and actuator forces versus time for this controller and compare the results to those of Problem 6.11.

24. Reconsider the Quadrupteron robot represented in Figure 3.40 and described in Problems 3.12 and 4.7 [157]. Consider the dynamic formulation with the geometrical and inertial parameters, and the typical trajectory used in Problem 5.12.

    a. For this manipulator, design a robust inverse dynamics controller in the task space, and implement the required force distribution in terms of the mechanism Jacobian matrix. Develop a program in MATLAB to simulate the closed-loop tracking performance of the mechanism. Consider %50 perturbation in all geometrical and inertial parameters, and tune the controller gains to reach a suitable tracking performance with a limited control effort.

    b. Plot and analyze motion trajectories, tracking errors, and the Cartesian and actuator forces versus time for this controller and compare the results to those of Problem 6.12.

7

Force Control

7.1 Introduction

In the previous chapter, the motion control of parallel robots has been analyzed through a variety of different control topologies and control schemes. These controllers are suitable when the moving platform of the robot accurately tracks a desired motion trajectory, while no interacting forces are needed to be applied. However, in many applications, it may occur that the robot moving platform is in contact with a stiff environment and specific interacting wrench is required. Parallel robots used in precision machining, grinding, or microassembly may be seen as representatives of such applications. In such applications, the contact wrench describes the state of interaction more effectively than the position and orientation of the moving platform. The problem of force control can be described as to derive the actuator forces required to generate a prescribed desired wrench (force/torque) at the manipulator moving platform, when the manipulator is carrying out its desired motion. In such a situation, the robot is interacting with a stiff environment, and a considerable interacting wrench is applied to the robot moving platform. This problem and its extents are treated in the force control algorithms described in this chapter. A force control strategy is one that modifies position trajectories based on the sensed wrench, or force–motion relations.

When the manipulator moving platform is in contact with a stiff environment, owing to the shape and stiffness of the environment, some constraints are set at the geometric paths that can be traversed by the moving platform. In such cases, using pure motion controllers would either fail or even become dangerous if the robot is interacting with humans. Accurate positioning in motion control schemes are indebted to the stiffness of the closed-loop manipulators, and in practice, high-gain controllers and near-rigid structures are used to obtain high accuracies. These characteristics perform well while the robot is moving freely in space. However, and as soon as the robot touches a stiff environment, its motion is constrained and positioning errors are induced in the control scheme.

In case of having high gains and stiff structure for the robot, like what is usually seen in parallel manipulators, the motion control scheme produced high actuator forces to overcome the tracking errors. This means that if a pure motion control scheme is used for a manipulator, in case it contacts an environment, the robot does not sense the presence of the environment, and its driving forces become harshly high to reduce the tracking errors. In such a case, the robot may break the object it is in contact with or will break its internal structure, if the environment is more rigid than the robot. Both cases are not suitable, especially when the interacting object is a human, which may cause injuries. In such cases, the motion control schemes are not suitable, and moreover, the manipulator should be equipped with more sensors than the regular motion detectors to be able to feel the interaction and to control the interacting forces. Different wrench sensors are developed for such applications, where six-degrees-of-freedom wrench sensors are mainly used for complete information of the interacting forces and torques. It is also possible to use joint torque or link force measurement units to determine the projection of the interacting forces in the joint space.

The use of wrench sensors either in the task space or in the joint space opens horizons to use different force control topologies for the manipulators. Using such sensors does not imply that regular motion sensors used in motion control schemes are not necessary. The use of motion sensors and the usual corresponding control topologies are usually necessary, since the motion of the manipulator is one of the outputs to be controlled. However, these measurements are not usually sufficient if force control schemes need to be implemented. Depending on the type and configuration of the wrench sensors used in the manipulator, different force control topologies are developed, which are elaborated in Section 7.2. In many of such developed topologies, both motion and wrench measurements are used together with a hierarchy that corresponds to the prime control objective.

Although the use of force control schemes are necessary only if the parallel manipulator is in contact with a stiff environment, in cable-driven parallel manipulators this becomes a stringent requirement even for pure motion control scheme. As explained earlier, when cables are used for force transmission in parallel manipulators, an inherent limitation of positive tension in cables is imposed on the manipulator control system. Theoretical techniques to use the redundant degree(s)-of-freedom of the manipulator to guarantee that the cables are all in tension are elaborated in Section 6.7. However, to implement such routines in practice, it is usually required to measure the transmission forces through the cables and use them in the control topologies. Although these control schemes fall solely into motion control topologies, they are treated in this chapter and in Section 7.3 owing to their resemblance to other force control topologies.

7.2 Controller Topology

In force control of a parallel manipulator, it is assumed that the controller computes the required actuator forces and torques to cause the robot motion to follow a desired motion trajectory, when the robot is in contact with a stiff environment. Moreover, the controller modifies the motion trajectories based on the sensed wrench, or force–motion relations. Let us use the motion variable as the generalized coordinate of the moving platform defined earlier as $\boldsymbol{\mathcal{X}} = \begin{bmatrix} \boldsymbol{x}_p & \boldsymbol{\theta} \end{bmatrix}^T$, in which the linear motion is represented by $\boldsymbol{x}_p = [x_p \;\; y_p \;\; z_p]^T$, while the moving platform orientation is represented by screw coordinates $\boldsymbol{\theta} = \theta \begin{bmatrix} s_x & s_y & s_z \end{bmatrix}^T = \begin{bmatrix} \theta_x & \theta_y & \theta_z \end{bmatrix}^T$.

Furthermore, consider the force/torque variable in the task space defined earlier as the wrench $\boldsymbol{\mathcal{F}} = \begin{bmatrix} \boldsymbol{F} & \boldsymbol{n} \end{bmatrix}^T$, in which the force is represented by $\boldsymbol{F} = [f_x \;\; f_y \;\; f_z]^T$, while the torque applied to the moving platform is represented by $\boldsymbol{n} = \begin{bmatrix} n_x & n_y & n_z \end{bmatrix}^T$. The motion variable in the joint space may be represented by $\boldsymbol{q} = \begin{bmatrix} q_1 & q_2 & \dots & q_m \end{bmatrix}^T$, while the actuator forces in the joint space is denoted by $\boldsymbol{\tau} = \begin{bmatrix} \tau_1 & \tau_2 & \dots & \tau_m \end{bmatrix}^T$, in which m represents the number of actuators. Note that the robot Jacobian matrix constructs the required mapping between the motion twist and the joint velocities by $\dot{q} = J\dot{\mathcal{X}}$, and the projection of the moving platform twists to the actuator forces vector by $\mathcal{F} = J^T \boldsymbol{\tau}$.

Now consider the general closed-form dynamic formulation of a parallel robot represented by Equation 6.1, extended to the case where the robot is in contact with the environment, and furthermore, disturbance wrenches are applied to the moving platform:

$$M(\mathcal{X})\ddot{\mathcal{X}} + C(\mathcal{X}, \dot{\mathcal{X}})\dot{\mathcal{X}} + G(\mathcal{X}) + \mathcal{F}_e = \mathcal{F} = J^T \boldsymbol{\tau}. \tag{7.1}$$

In this equation, $M(\mathcal{X})$ denotes the mass matrix that can be derived from the kinetic energy of the manipulator by Equation 5.263, $C(\mathcal{X}, \dot{\mathcal{X}})$ denotes the Coriolis and centrifugal matrix given in Equation 5.296, and $G(\mathcal{X})$ denotes the gravity vector given in Equation 5.268. Furthermore, $\mathcal{F}_e$ denotes the interacting wrench between the moving platform and the environment at center of mass, and finally, $\boldsymbol{\tau}$ represents the actuator force vector, while $\mathcal{F}$ is the projection of the actuator force vector into the task space.

To design a force controller for parallel manipulators, first it is necessary to introduce possible controller topologies developed to accomplish this task, and then introduce different techniques for controller design according to these topologies. In what follows, first the controller topologies are introduced, and then a number of controller design techniques are described. Control topology is referred to the structure of the control system that is used to compute the actuator forces/torques from the measurements, and the required pre- and postprocessing. For force control of parallel manipulators, the controller computes the required actuator forces and torques to track a desired position and orientation trajectory, when the robot is in contact with a stiff environment, and the controller modifies the motion trajectories based on the sensed wrench or force–motion relations.

In a general case, the desired motion of the moving platform may be represented by the desired generalized coordinate of the manipulator, denoted by $\mathcal{X}_d$. This variable has the same dimension and structure of the manipulator motion variable $\mathcal{X}$. To carry out such motion in a closed-loop structure, it is necessary to measure the output motion of the manipulator through an instrumentation system. Since in the motion control of parallel manipulators, the final motion of the moving platform is of interest, it would be desirable to use instrumentation systems that can directly provide the motion variable $\mathcal{X}$ with the required accuracy at the outset. In such instrumentation systems, the final position and orientation of the moving platform would be measured. However, typical sensors used for motion measurements are the linear or angular position sensors embedded on the actuated joints, by which the joint motion variable q is directly measured or can be easily determined from the measured quantity.

For a force control scheme, the desired interacting wrench of the moving platform and the environment may be of interest. This quantity may be denoted by $\mathcal{F}_d$, which has the same dimension and structure of the manipulator wrench $\mathcal{F}$. To carry out such a control task in a closed-loop structure, it is necessary to measure the output wrench of the manipulator through an instrumentation system. Since in the force control of parallel manipulators, the final interacting force of the moving platform may be of interest; it is desirable to use instrumentation systems that can directly provide the interacting wrench $\mathcal{F}$ with the required accuracy at the outset. Commercial measurement units to measure a six-degrees-of-freedom wrench are available, and they are called six-degrees-of-freedom wrench sensors. Such instruments are commercially used in the wrist of serial manipulators for applications where the robot has force interaction with the environment. The same apparatus may be used on the moving platform of a parallel robot where the final end effector of the manipulator is installed.

Although there are many commercial six-degrees-of-freedom wrench sensors available in the market, they are usually more expensive than single joint force or torque measurement units. A six-degrees-of-freedom wrench sensor may be used in applications where the interaction force must be exactly measured and controlled. Another alternative for force measurement is direct measurement of the actuator forces or torques. Many commercial linear actuators are available in the market in which embedded force measurement is considered in their design. Moreover, many servo motors have included torque measurement units in their assembly. Therefore, it might be preferable in some applications to use direct actuator force/torque measurements to carry out the feedback control. In such cases, denote $\boldsymbol{\tau}_m$ as the measured variable. Depending on the measurement units we incorporate in the structure of the robot, different control topologies are advised for force control schemes, which have been discussed in the following sections.

7.2.1 Cascade Control

In a general force control scheme, the prime objective is tracking of the interacting wrench between the moving platform and the environment. However, note that the motion control of the robot when the robot is in interaction with the environment is also another less-important objective, and when the contact of the robot moving platform to the environment is released, motion control becomes the prime objective. To follow two objectives with different priorities in one control system, usually a hierarchy of two feedback loops is used in practice [7]. This kind of control topology is called *cascade control*, which is used when there are several measurements and one prime control variable. This control topology can completely address the prime objectives, while other objectives may be considered through a careful design of the system controller(s). Cascade control is implemented by nesting the control loops, as shown in Figure 7.1. The control system shown in this figure is constructed by two control loops, in which the outer control loop is called the *primary loop*, while the inner loop is called the secondary loop and is used to fulfil a secondary objective in the closed-loop system. It is possible to have a cascade control system with more nested loops.

The performance of the system may be improved by using more measurement variables in the structure, whereas, in the state-feedback structure, the feedback of all state variables by which the dynamics of the system is represented is used. In such a case, the control system may be viewed as a cascade control system with n-nested loop, in which n denotes the number of system states. However, in industrial applications, usually two nested loops are used, as in the force control schemes of parallel manipulators. In the cascade control schemes used for these applications, the measured variables are the motion and interacting wrench. As explained earlier, these quantities may be measured in the task space or in

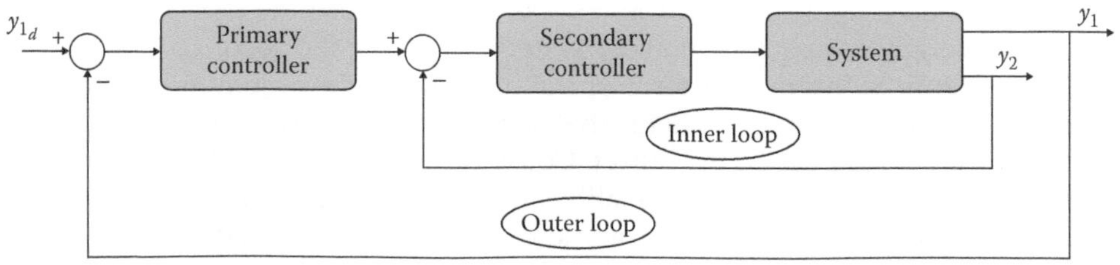

FIGURE 7.1
Block diagram of a closed-loop system with cascade control.

the joint space, and therefore, different control topologies may be advised for each set of measurement variables.

To improve the performance of the control system for a particular objective, it is important to choose the right variables for internal and external feedback loops, and to design suitable controllers for each feedback system. Although these differ in different topologies described in the following sections, some general rules are applied to design a well-performing cascade control system. A general idea in cascade control design is the ideal case, in which the inner loop is designed so tight that the secondary loop behaves as a perfect servo, and responds very quickly to the internal control command. This idea is effectively used in many applications, wherein a nearly-perfect actuator to respond to the requested commands is designed by using an inner control feedback. The design criteria for the inner loop is to have possibly high gain in the inner loop, by which the time response of the secondary variable is at least 5 times more than that of the primary variable, and it can overcome the effect of disturbances and unmodeled dynamics in the internal feedback structure [7]. It is also necessary to have a well-defined relation between the primary and secondary variables, to have harmony in the objectives followed in the primary and secondary loops.

7.2.2 Force Feedback in Outer Loop

Consider the force control schemes, in which force tracking is the prime objective. In such a case it is advised that the outer loop of cascade control structure is constructed by wrench feedback, while the inner loop is based on position feedback. Since different types of measurement units may be used in parallel robots, different control topologies may be constructed to implement such a cascade structure. Consider the case wherein both the wrench, and position variables are measured in the task space configuration. In such cases, the moving platform wrench (applied on the center of mass) $\mathcal{F}$ might be measured through a six-degrees-of-freedom wrench sensor, and the motion variable $\mathcal{X}$ may be measured by a measurement unit. Although these measurements are relatively expensive, in cases that they are both implemented in the task space, the closed-loop system may return accurate force and positioning performance for the manipulator.

Figure 7.2 illustrates the first cascade control topology for the system in which the measured variables are both in the task space. These measured variables are the resulting wrench measured at the moving platform center of mass $\mathcal{F}$, and the motion variable of the moving platform denoted by $\mathcal{X}$. The inner loop is constructed by position feedback while the outer loop is based on force feedback. By this means, the primary controller is

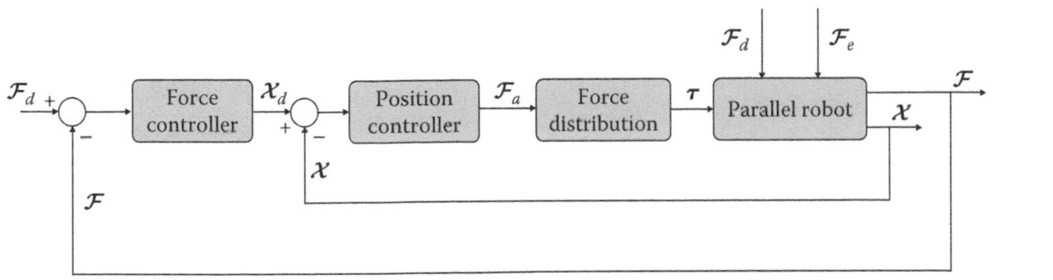

FIGURE 7.2
First cascade topology of force feedback control: position in inner loop and force in outer loop, moving platform wrench $\mathcal{F}$, and motion variable $\mathcal{X}$ are measured in the task space.

a force controller to guarantee the force-tracking objectives, while the secondary objective is a position-tracking requirement. As seen in Figure 7.2, the controller is constructed in the task space, and output of the force controller block is fed to the motion controller. This might be seen as the generated desired motion trajectory for the inner loop.

The output of motion controller is also designed in the task space, and to convert it to implementable actuator force $\boldsymbol{\tau}$, the force distribution block is considered in this topology. As completely explained in the motion control topologies, the force distribution block maps the generated wrench in the task space denoted by $\mathcal{F}_a$ to its corresponding actuator forces/torques $\boldsymbol{\tau}$. For a fully parallel manipulator such as the Stewart–Gough platform, in which the number of actuators are equal to the number of degrees-of-freedom, this mapping can be constructed by the Jacobian transpose of the manipulator $\boldsymbol{J}^T$. Since this mapping for parallel robots is determined by $\mathcal{F}_a = \boldsymbol{J}^T\boldsymbol{\tau}$ for a fully parallel manipulator, the actuator forces may be generated by $\boldsymbol{\tau} = \boldsymbol{J}^{-T}\mathcal{F}_a$ at nonsingular configurations. The force distribution block may need more computations in case the parallel manipulator is not fully parallel. For such cases, redundancy resolution techniques elaborated in Section 6.7 may be used as force distribution routine.

Other alternatives for force control topology may be suggested based on the variations of position and force measurements. Consider a general parallel manipulator consisting of a number of limbs with a number of passive and active joints depicted in Figure 3.2. As explained earlier, it is usually much easier to measure the actuator forces/torques $\boldsymbol{\tau}$ directly, or derive them from joint space measurements. Assume that still the motion variable in the task space $\boldsymbol{\mathcal{X}}$ is measured. Then, the topology suggested in Figure 7.3 is advised for the control system. In this topology, the measured actuator force/torque vector $\boldsymbol{\tau}$ is mapped into its corresponding wrench in the task space by the Jacobian transpose mapping, since it is known that $\mathcal{F} = \boldsymbol{J}^T\boldsymbol{\tau}$. Through this mapping, the controller structure remains in the task space and the force and position controllers are designed by the same techniques developed for the first topology.

Consider the case where the force and motion variables are both measured in the joint space. Figure 7.4 suggests the force control topology in the joint space, in which the inner loop is based on measured motion variable in the joint space $\boldsymbol{q}$, and the outer loop uses the measured actuator force/torque vector. In this topology, it is advised that the force controller is designed in the task space, and the Jacobian transpose mapping is used to project the measured actuator force/torque vector into its corresponding wrench in the task space. However, the inner loop is constructed in the joint space, since mapping of motion variable in the joint space to its corresponding motion variable in the task space requires forward

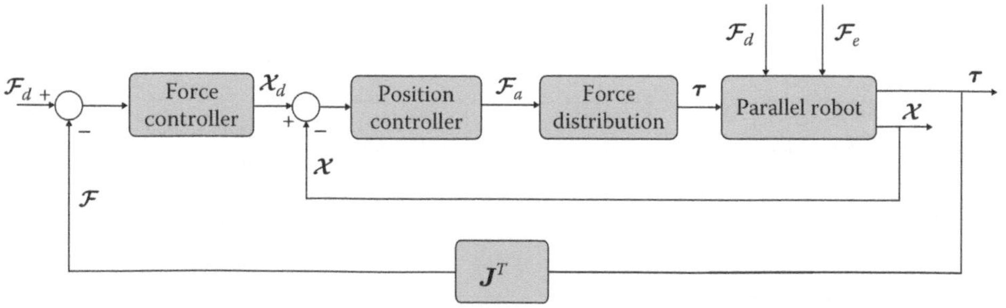

FIGURE 7.3
Second cascade topology of force feedback control: position in inner loop and force in outer loop, actuator forces/torques $\boldsymbol{\tau}$, and motion variable $\boldsymbol{\mathcal{X}}$ are measured.

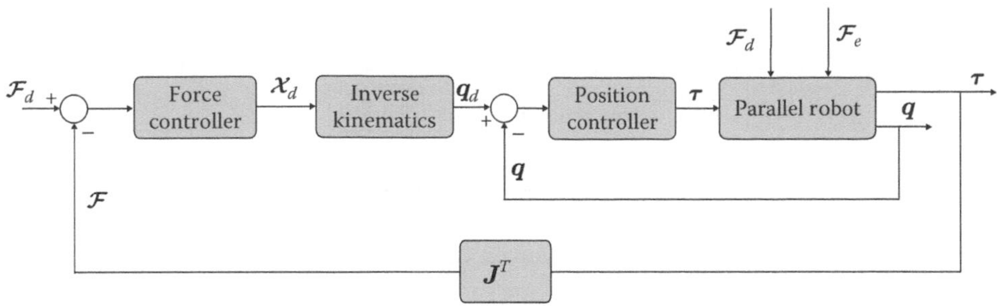

FIGURE 7.4
Third cascade topology of force feedback control: position in inner loop and force in outer loop, actuator force/torques τ, and active joint motion variable q are measured in the joint space.

kinematic solution, which is relatively complex for parallel manipulators. Therefore, the desired motion variable $\mathcal{X}_d$ is mapped into joint space using inverse kinematic solution depicted in Figure 7.4. Hence, the controller is designed based on the joint space error e_q, and is totally implemented in the joint space.

Therefore, the structure and the characteristics of the position controller in this topology is totally different from that given in the first two topologies. Note that in this control topology, the input to the controller is the motion error in the joint space e_q, and the output of the controller is the actuator force/torque τ, which is also represented in the joint space. However, in the previous topologies, the input to the motion controller is the motion error in the task space e_x, while its output is in the joint space. Therefore, independent controllers for each joint may be suitable for the third topology depicted in Figure 7.4. Although many other control topologies may be seen in the literature, all of them are not elaborated in this section to keep the complexity of analysis at a manageable level. Further topologies may be developed based on the ideas given in this section.

The topologies given in this section are mainly used in pure force control schemes, which have been elaborated in Section 7.4. By pure force control, it is meant that the primary objective is force tracking, while the manipulator is in contact with the environment. The secondary objective in these control algorithms is to preserve the stability of motion and to some extents having motion tracking at the times where the robot end-effector is released from the environment. Different control schemes have been developed to provide the required force-tracking requirements in the case where the manipulator has force interaction with the environment. These control strategies are given in a sequence from simple and basic ideas to more complete and effective structures in Section 7.4.

7.2.3 Force Feedback in Inner Loop

Consider the force control schemes for a parallel manipulator in which the motion–force relation is the prime objective. In such a case, force tracking is not the primary objective, and it is advised that the outer loop of cascade control structure consists of a motion control feedback while the inner loop is based on force feedback. Since different types of measurement units may be used in parallel robots, different control topologies may be constructed to implement such cascade controllers. Consider the case where both the force/torque and position variables are measured in the task space. In such cases, the wrench applied to the moving platform center of mass $\mathcal{F}$ might be measured through a six-degrees-of-freedom wrench sensor, and the motion variable $\mathcal{X}$ may be measured by a measurement unit.

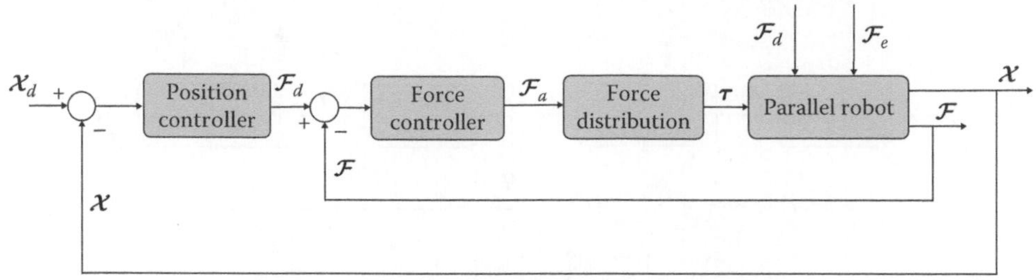

FIGURE 7.5
Fourth cascade topology of force feedback control: force in inner loop and position in outer loop, moving platform wrench $\mathcal{F}$, and motion variable $\mathcal{X}$ are measured in the task space.

Although these measurements are relatively expensive, in cases that they are both implemented in the task space, the closed-loop system may return to accurate motion–force relation for the manipulator.

Figure 7.5 illustrates the cascade control topology for the system in which the measured variables are both in the task space. These measured variables are the resulting wrench at the moving platform center of mass denoted by $\mathcal{F}$, and the motion variable of the moving platform denoted by $\mathcal{X}$. The inner loop is constructed by force feedback while the outer loop is based on position feedback. The primary controller is a motion controller, while the secondary one is a force-tracking controller. By this means, when the manipulator is not in contact with a stiff environment, position tracking is guaranteed through the primary controller. However, when there is interacting wrench $\mathcal{F}_e$ applied to the moving platform, this structure controls the force–motion relation. Furthermore, the controller is constructed in the task space, and the output of the motion controller block is fed to the force controller. This configuration may be seen as if the outer loop generates a desired force trajectory for the inner loop.

The output of the force controller is also designed in the task space, and to convert it into the actuator force $\boldsymbol{\tau}$, the force distribution block is considered in this topology. As completely described earlier, the force distribution block maps the generated wrench in the task space denoted by $\mathcal{F}_a$, to its corresponding actuator forces/torques $\boldsymbol{\tau}$. For a fully parallel manipulator such as the Stewart–Gough platform, in which the number of actuators are equal to the number of degrees-of-freedom, this mapping can be constructed from the Jacobian transpose of the manipulator $\boldsymbol{J}^T$. Since this mapping for parallel robots is determined by $\mathcal{F}_a = \boldsymbol{J}^T \boldsymbol{\tau}$, for a fully parallel manipulator, the actuator forces can be generated by $\boldsymbol{\tau} = \boldsymbol{J}^{-T} \mathcal{F}_a$ at nonsingular configurations. The force distribution block may need more computations in case parallel manipulator is not fully parallel. For such a case, redundancy resolution techniques elaborated in Section 6.7 may be used as a force distribution routine.

Other alternatives for force control topology may be suggested based on the variations of position and force measurements. Consider a general parallel manipulator consisting of a number of limbs with a number of passive and active joints depicted in Figure 3.2. As explained earlier, it is usually much easier to measure the actuator forces/torques $\boldsymbol{\tau}$ directly, or derive them from joint space measurements. If the motion variable in the task space $\mathcal{X}$ is still measured, the topology suggested in Figure 7.6 is advised for the control system. In this topology, the measured actuator force/torque vector $\boldsymbol{\tau}$ is mapped to its corresponding wrench in the task space, through the Jacobian transpose mapping, since it is known that $\mathcal{F} = \boldsymbol{J}^T \boldsymbol{\tau}$. Through this mapping, the controller structure remains in the task space and the force and position controllers are designed in the same way as that for the previous topology.

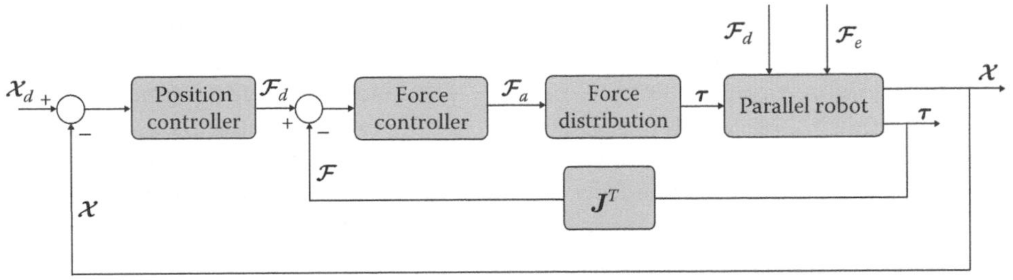

FIGURE 7.6
Fifth cascade topology of force feedback control: force in inner loop and position in outer loop, actuator forces/torques $\boldsymbol{\tau}$, and motion variable $\boldsymbol{\mathcal{X}}$ are measured.

Consider the case where the force and motion variables are both measured in the joint space. Figure 7.7 suggests the force control topology in the joint space, in which the inner loop is based on the measured actuator force/torque vector in the joint space $\boldsymbol{\tau}$, and the outer loop uses the measured actuated joint position vector $\boldsymbol{q}$. In this topology, the desired motion in the task space is mapped into the joint space using inverse kinematic solution depicted in Figure 7.4, and both the position and force feedbacks are designed in the joint space. Hence, the position controller is designed based on the joint space error $\boldsymbol{e}_q$, and is totally implemented in the joint space. Therefore, the structure and the characteristics of the position controller in this topology is totally different from the two topologies suggested earlier.

Note that in this control topology, the input to the controller is the motion error in joint space $\boldsymbol{e}_q$, and the output of the controller is also the actuator force/torque vector $\boldsymbol{\tau}$, which is also represented in the joint space. However, in the previous topologies, the input to the motion controller is the motion error in the task space $\boldsymbol{e}_x$, while its output is also in the task space. Therefore, independent controllers for each joint may be suitable for this topology. If model-based controllers such as inverse dynamics-control need to be implemented in this topology, the computational cost would be much higher, since the dynamic formulation for parallel manipulators in the joint space is needed. Although many other control topologies may be seen in the literature, all of them are not elaborated in this section to keep the complexity of analysis in a manageable level. Other control topologies may be developed based on the ideas given in this section.

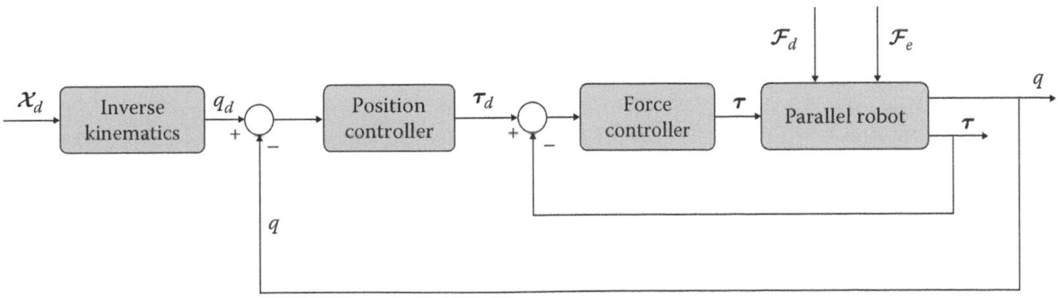

FIGURE 7.7
Sixth cascade topology of force feedback control: force in inner loop and position in outer loop, actuator force/torques $\boldsymbol{\tau}$, and active joint motion variable $\boldsymbol{q}$ are measured in joint space.

The topologies given in this section are mainly used in impedance control schemes, which have been elaborated in Section 7.5. By impedance control, it is meant that neither force nor position requirement, is not solely considered as the prime objective, but the dynamic relation between the interacting force and robot motion is controlled by the structure. Various control schemes have been developed to provide the impedance requirements where the manipulator has force interaction with the environment. These control strategies are given in a sequence from simple and basic ideas to more complete and effective structures in Section 7.5. In what follows, first the idea of force control techniques is developed for simple one-loop control scheme through stiffness control algorithm. Then, pure force control and impedance control techniques follow.

7.3 Stiffness Control

Increasing position accuracy demands of industrial robots have forced mechanical designers to use much rigid robotic manipulators. Parallel robots are more suitable for this purpose, since the closed-kinematic structure inherently increases the stiffness of the robot. When high-gain controllers are applied to the rigid structure of the robots, the resulting manipulator provides high-precision motion at the outset. This is desirable when the robot moving freely in space and experiences no contact to any objects in its entire workspace. To satisfy this condition, the robot is usually operated in an isolated area framed in a safety fence. For such a robot manipulator that uses high-gain motion controller and a very stiff structure, it is very difficult to interact with a stiff environment and it is even dangerous if it experiences interaction with a human operator.

Consider that in the course of motion of such a robot, an obstacle appears in front of its preplanned trajectory. The robot hits the obstacle, and since it is designed to track the preplanned trajectory with high accuracy, it may damage the obstacle, if the obstacle is more compliant than the robot, or otherwise break its own components. This case is very dangerous if the obstacle is a human operator who unintentionally appeares on the motion trajectory of the robot. For this reason, introducing some compliance in the robot structure is suitable for avoiding these drawbacks. The compliance may be considered in the mechanical structure of the robot, or in its control structure.

Using compliance in the structure of the robot is a newly developed approach that designers are considering in industrial applications. Compliance in the robot structure may originate from using light weight and relatively long links in robots such as the robots being used in space [156] and minimal invasive surgery applications [136], or by use of state-of-the-art joint transmission systems [174,175] and tendon-type components in the structure of the robots [54]. Although the induced compliance in the structure of the robot eases its interaction with a stiff environment, design of suitable control algorithms to accomplish high tracking precisions for such robots is a challenging task.

Another way to add compliance to the robot is through its control structure. The compliant performance of the robot may originate from its mechanical structure and/or its control. By using appropriate control topology and designing suitable controller gains, a robot that is quite stiff in its mechanical structure may experience compliant response. Adding compliance in the control structure might be more beneficial for industrial robots, since the compliance characteristics of the robot may be easily adjusted by the software. In such a case, the mechanical design of the robot is the same as the robots used in high-precision motion-tracking applications, while its controller gains can be designed so as to provide suitable compliance for the applications wherein the robot is in interaction with a stiff environment. This kind of control algorithm is called *stiffness control* [109,111] or *compliance control* [25,200], where the former expression is adopted in this book.

The control topology used in stiffness control is very similar to that of motion control topologies elaborated in Section 6.2, since in this control structure, only motion variables are measured and used in the feedback. The interacting force is then estimated using compliant models for the environment and this force is included in the dynamic formulation of the robot and closed-loop characteristics analysis of the system. The prime objective of this control algorithm is to keep the closed-loop stiffness of the manipulator at a desired level, for a suitable interaction with a stiff environment. To develop the idea and carry out the analysis at a manageable level, let us first present the basis of stiffness control for a single-degree-of-freedom robot in the proceeding section, and then elaborate on the general case in Section 7.3.2.

7.3.1 Single-Degree-of-Freedom Stiffness Control

Consider a parallel robot with one-degree-of-freedom motion interacting with a stiff environment. As shown in Figure 7.8, the motion of such a robot is considered in the x direction, while the robot has touched the environment at the x_e position. It is considered that the robot stiffness is higher than that of the environment, and due to this interaction the environment experience a deformation in the x direction with an amount of $x - x_e$. If the compliance of the environment is modeled by a linear relation, the interacting force in x direction may be formulated as

$$f_e = k_e \, \Delta x = k_e(x - x_e), \tag{7.2}$$

in which Δx denotes the compliant deformation of the body, k_e represents the linear stiffness coefficient of the environment in the x direction, and f_e denotes the interacting force between the robot and the environment. The general dynamic formulation of parallel robots, given in Equation 7.1, may be simplified to a scalar equation represented by the following equation in the absence of disturbance forces:

$$m(x)\ddot{x} + c(x, \dot{x})\dot{x} + g(x) + f_e = f, \tag{7.3}$$

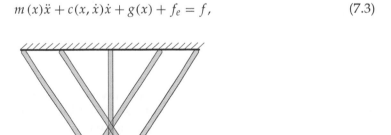

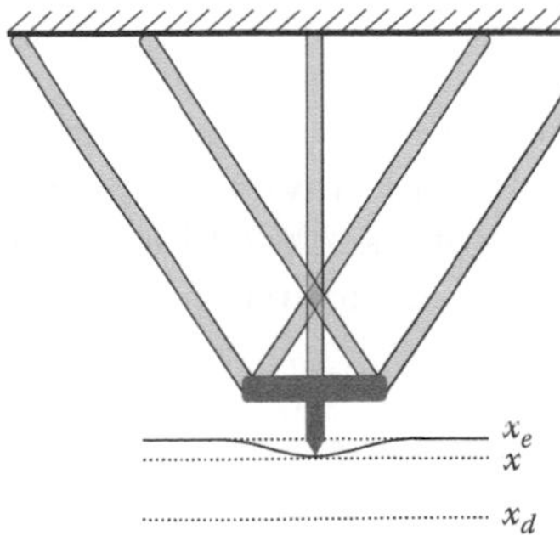

FIGURE 7.8
Compliant environment in stiffness control scheme; one-degree-of-freedom analysis.

in which $m(x)$ denotes the effective mass of the robot in the x direction, $c(x, \dot{x})\dot{x}$ denotes the effective Coriolis and centrifugal terms in the x direction, $g(x)$ denotes the gravity term in the x direction, f_e denotes the interaction force with a stiff environment in the x direction, and f is the projection of the actuator forces in the x direction.

It can be shown that different control algorithms developed for pure motion control of the robot can be used for stiffness control with some adjustments. To develop the concept, consider the general inverse dynamics control strategy developed in Section 6.3.3, for this analysis. This control structure is the base of development of more advanced robust and adaptive control structures that are completely addressed in Section 6.4, and therefore, is used as the base controller for stiffness analysis of parallel robots in this section. To implement such a controller, only motion variable would be measured. This measurement may be done in the task space, or in the joint space as explained earlier. However, for a single-degree-of-freedom case, it is considered that the motion variable x is measured in the task space, and is used in the feedback. The topology of stiffness control scheme used for such a case is, therefore, the one depicted in Figure 6.4.

Now, consider full information feedback linearization and a proportional-derivative (PD) control for inverse dynamics control (IDC) in the task space. Hence, the projection of actuator forces in the task space would be determined by

$$f = c(x, \dot{x})\dot{x} + g(x) + k_p e_x + k_d \dot{e}_x, \tag{7.4}$$

in which k_p and k_d denote the scalar PD gains and the tracking error e_x is determined from $e_x = x_d - x$, where x_d denotes the desired motion trajectory. As explained earlier, to implement such a controller, only the motion variable feedback is required, and furthermore, the control command is generated based on the desired trajectory of the robot. For a suitable interaction with the environment, special care is required for the desired trajectory x_d, which will be elaborated later in this section by the following analysis.

To examine the characteristics of the control law, examine the behavior of the closed-loop system by substitution of the control law, given in Equation 7.4, into the manipulator dynamics 7.3. By assuming full information for centrifugal and gravity terms, those terms are canceled in the closed-loop dynamic formulation as

$$m(x)\ddot{x} + k_e(x - x_e) = k_p(x_d - x) + k_d(\dot{x}_d - \dot{x}). \tag{7.5}$$

The closed-loop dynamics for the robot may be further simplified to the following equation by factorizing the motion variable x:

$$m(x)\ddot{x} + k_d\dot{x} + (k_p + k_e)x = k_e x_e + k_p x_d + k_d \dot{x}_d. \tag{7.6}$$

Note that the inverse dynamics controller is able to linearize the closed-loop dynamics, if complete information on the Coriolis and gravity terms are available. In such a case, the transient and steady-state behaviors of the closed-loop system may be analyzed by the Laplace transform. Apply the Laplace transform to both sides of Equation 7.6, and find the motion variable $x(s)$ in s-domain by

$$x(s) = \frac{(k_p + k_d s)x_d(s) + k_e x_e(s)}{m s^2 + k_d s + (k_p + k_e)}. \tag{7.7}$$

To analyze the steady-state behavior of the motion variable, consider constant input variables applied to the system. For a constant desired trajectory with a steady-state value of

$\bar{x}$, we may formulate the input by a step function in s-domain, $x(s) = \bar{x}/s$. Similarly, consider a smooth and flat obstacle in the environment, represented by a constant x_e with a steady-state value of $\bar{x}_e$ in time domain, and a step input $x_e(s) = \bar{x}_e/s$, in s-domain. The steady-state behavior of the motion variable in time domain can be represented by the final value theorem in s-domain as [38]

$$\bar{x} = \lim_{t\to\infty} x(t) = \lim_{s\to 0} s\, x(s). \tag{7.8}$$

Therefore, the steady-state value for motion variable $\bar{x}$ may be calculated from the following equation:

$$\begin{aligned}\bar{x} &= \lim_{s\to 0} \frac{(k_p + k_d s)\bar{x}_d + k_e\bar{x}_e}{m s^2 + k_d s + (k_p + k_e)} \\ &= \frac{k_p\bar{x}_d + k_e\bar{x}_e}{k_p + k_e}.\end{aligned} \tag{7.9}$$

Let us examine the steady-state behavior of interaction force denoted by $\bar{f}_e$. The steady-state value of the interaction force with the environment is calculated by Equation 7.2 as

$$\begin{aligned}\bar{f}_e &= k_e(\bar{x} - \bar{x}_e) \\ &= \frac{k_e(k_p\bar{x}_d + k_e\bar{x}_e)}{k_p + k_e} - \frac{k_e\bar{x}_e(k_p + k_e)}{k_p + k_e} \\ &= \frac{k_e k_p(\bar{x}_d - \bar{x}_e)}{k_p + k_e}.\end{aligned} \tag{7.10}$$

However, note that the environment stiffness coefficient k_e, is in orders of magnitude higher than the controller gain k_p, namely $k_e \gg k_p$. Hence, the steady-state value of the interaction force may be approximated by the following equation:

$$\begin{aligned}\bar{f}_e &\simeq \frac{k_e k_p}{k_e}(\bar{x}_d - \bar{x}_e) \\ &\simeq k_p(\bar{x}_d - \bar{x}_e).\end{aligned} \tag{7.11}$$

Equation 7.11 reveals an important fact of stiffness control scheme that although the environment stiffness coefficient k_e is usually very high, by using the proposed structure for the controller, the interaction force at the steady-state $\bar{f}_e$ is related to the corresponding environment deflection at the steady-state $\bar{x}_d - \bar{x}_e$ with the controller gain k_p. Hence, the interacting stiffness of the manipulator with respect to even very stiff environments can be well controlled by choosing moderate controller gain k_p. Therefore, the interaction force of the robot with a stiff environment may be tuned such that there is embedded flexibility considered in the structure of the robot for even very rigid robot designs. The main reason why this controller structure is called stiffness control stems out of this fact.

Furthermore, notice the importance and the influence of the desired trajectory definition in such a structure by examination of Equation 7.11. To come up with a moderate stiffness for the manipulator at the outset, it is sufficient to define the desired trajectory a bit inside the environment surface. This idea is illustrated in Figure 7.8, in which the desired trajectory x_d is considered inside the body surface, in cases where the robot is in

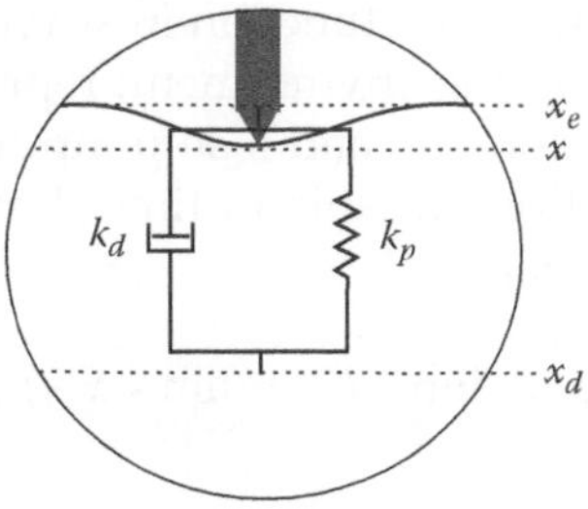

FIGURE 7.9
Compliant environment in stiffness control scheme; virtual spring–damper model.

contact with the stiff environment. To visualize this concept, a graphical representation of Equation 7.11 is given in Figure 7.9. As illustrated in Figure 7.9, a virtual spring–damper with the controller gains, k_p and k_d is attached to the tip of the end effector and is acting between x_e and x_d. When the robot is in contact with the environment, the resulting force induced by the virtual spring–damper keeps the end effector in contact with the surface, and the force–motion relation is forced to satisfy the required stiffness. However, when the robot is moving freely and is not in contact with any surface, then $x = x_e$, and this virtual spring–damper forces the end effector to track the desired trajectory. By this means, the interacting forces can be well tuned to accommodate the required manipulator stiffness by designing appropriate gain for the controller gain k_p, and careful design of the desired motion trajectory of the robot.

Although the structure of stiffness control scheme given in this section is very similar to the position control scheme given in Section 6.3.3, there are clear distinctions in the implementation of these schemes in practice. In a pure motion control, it is considered that the robot moves freely in the environment, and there is no contact with any surface within its whole maneuver. Therefore, high PD gains are recommended to provide the required trajectory-tracking performance for the manipulator. However, in stiffness control routine, it is considered that the robot comes in contact with a stiff environment, and therefore, interaction forces significantly affect the motion. Therefore, when the robot is in contact with the environment, the desired trajectory is adjusted according to the contact surface. Furthermore, during the contact, the prime objective of the controller is to keep the end effector in contact with the environment, while tuning the interaction force to behave through a desired stiffness relation. By this means, the robot carries out a moderate trajectory tracking, when it is not in contact with the environment, since the controller gains are selected not as high as in a pure motion controller. However, when the robot is in contact with the environment, a smooth and safe interaction of the robot and the environment results by controlling the robot stiffness at the desired level.

This control scheme is very favorable in cases in which the robot is in contact with a predetermined object during its motion. The interaction forces are well tuned through stiffness control scheme, without the need of any force measurement. However, the accuracy of motion measurements must be higher than that used in a pure motion control scheme. This is a stringent requirement to detect desired motion trajectories that are intentionally assigned a bit inside the contact surface. The order of magnitude of the desired trajectory with respect to the object surface is designed higher than the elastic deformation of the object during contact; however, it is much smaller than the overall motion of the robot. Therefore, a motion measurement unit with high resolution in a wide range of operation is required to successfully implement stiffness control routines in practice.

7.3.2 General Stiffness Control

The concept of stiffness control is developed for a single-degree-of-freedom manipulator, and the behavior of the control system in the steady-state is analyzed through Laplace transform. To generalize this concept for a multiple-degrees-of-freedom manipulator, consider a parallel robot with multiple-degrees-of-freedom motion interacting with a stiff environment. In such a case, the motion of the robot end-effector is represented by the motion vector $\boldsymbol{\mathcal{X}}$, in which for a general six-degrees-of-freedom manipulator, the motion vector consists of the linear motion vector $\boldsymbol{x}$ as well as the orientation variable $\boldsymbol{\theta}$ preferably represented by screw coordinates, that is, $\boldsymbol{\mathcal{X}} = \begin{bmatrix} x & \theta \end{bmatrix}^T$.

In the course of robot interaction with the environment, an interaction wrench $\boldsymbol{\mathcal{F}}_e$ should be added to the dynamics formulation. For a general six-degrees-of-freedom manipulator, the interaction wrench consists of the contact force $\boldsymbol{F}_e$ and torque $\boldsymbol{n}_e$, $\boldsymbol{\mathcal{F}}_e = \begin{bmatrix} F_e & n_e \end{bmatrix}^T$. It is considered that the robot stiffness is higher than that of the environment, and due to this interaction, the environment experiences a positional and orientational deformation represented by $\boldsymbol{\mathcal{X}} - \boldsymbol{\mathcal{X}}_e$. If the linear and angular compliance of the environment is modeled by linear relation, the interacting force may be formulated as

$$\boldsymbol{\mathcal{F}}_e = \boldsymbol{K}_e \Delta(\boldsymbol{\mathcal{X}}) = \boldsymbol{K}_e(\boldsymbol{\mathcal{X}} - \boldsymbol{\mathcal{X}}_e), \tag{7.12}$$

in which $\Delta\boldsymbol{\mathcal{X}}$ denotes the compliant deformation of the body, $\boldsymbol{K}_e$ represents the stiffness matrix of the environment incorporating the linear stiffness coefficients corresponding to $\boldsymbol{x}$, as well as angular stiffness coefficients corresponding to $\boldsymbol{\theta}$. Furthermore, $\boldsymbol{\mathcal{F}}_e$ denotes the interacting wrench between the robot and the environment. The general dynamic formulation of parallel robots, given in Equation 7.1, may be written by the following equation in the absence of any disturbance wrench:

$$\boldsymbol{M}(\boldsymbol{\mathcal{X}})\ddot{\boldsymbol{\mathcal{X}}} + \boldsymbol{C}(\boldsymbol{\mathcal{X}}, \dot{\boldsymbol{\mathcal{X}}})\dot{\boldsymbol{\mathcal{X}}} + \boldsymbol{G}(\boldsymbol{\mathcal{X}}) + \boldsymbol{K}_e(\boldsymbol{\mathcal{X}} - \boldsymbol{\mathcal{X}}_e) = \boldsymbol{J}^T\boldsymbol{\tau} = \boldsymbol{\mathcal{F}}, \tag{7.13}$$

in which $\boldsymbol{M}(\boldsymbol{\mathcal{X}})$ denotes the robot mass matrix, $\boldsymbol{C}(\boldsymbol{\mathcal{X}}, \dot{\boldsymbol{\mathcal{X}}})$ denotes the Coriolis and centrifugal matrix, $\boldsymbol{G}(\boldsymbol{\mathcal{X}})$ denotes the gravity vector, $\boldsymbol{\tau}$ represents the actuator force vector, $\boldsymbol{\mathcal{F}}$ is the projection of the actuator force vector into the task space, and finally, $\boldsymbol{\mathcal{F}}_e$ denotes the interaction wrench.

As discussed in Section 7.3.1, different motion control schemes may be used for stiffness control with some consideration on the controller gain and adjustment of the desired trajectory. To analyze stiffness control characteristics for multiple-degrees-of-freedom robot, consider partial linearization IDC developed in Section 6.3.4. Note that as for the single-degree-of-freedom case, to implement such a controller only motion variable is measured, and it is considered that this measurement is performed in the task space. The topology of stiffness control scheme used for such a measurement is, therefore, the one depicted in Figure 6.4.

By considering partial feedback linearization and a PD control in the task space, the projection of actuator forces in the task space would be determined by

$$\boldsymbol{\mathcal{F}} = \boldsymbol{G}(\boldsymbol{\mathcal{X}}) + \boldsymbol{K}_p \boldsymbol{e}_x + \boldsymbol{K}_d \dot{\boldsymbol{e}}_x, \tag{7.14}$$

in which $\boldsymbol{K}_p$ and $\boldsymbol{K}_d$ denote the PD gain matrices and the tracking error $\boldsymbol{e}_x$ is determined from $\boldsymbol{e}_x = \boldsymbol{\mathcal{X}}_d - \boldsymbol{\mathcal{X}}$, where $\boldsymbol{\mathcal{X}}_d$ denotes the desired motion trajectory in the task space. As explained earlier, to implement such a controller, only the motion variable feedback is required, and furthermore, the control command is generated based on the desired trajectory of the robot.

To examine the characteristics of the control law, let us examine the behavior of a closed loop system by substitution of the control law given in Equation 7.14 into the manipulator dynamics 7.13. Assuming partial linearization in the control structure, the Coriolis and centrifugal terms remain in the closed-loop system dynamics, while the gravity term is completely canceled out through feedback linearization.

$$M(\mathcal{X})\ddot{\mathcal{X}} + C(\mathcal{X}, \dot{\mathcal{X}})\dot{\mathcal{X}} + K_e(\mathcal{X} - \mathcal{X}_e) = K_p e_x + K_d \dot{e}_x. \tag{7.15}$$

Note that although for partial linearization IDC if complete information on the gravity vector is available, this term vanishes from the closed-loop dynamics; Equation 7.15 still represents a nonlinear and coupled dynamics. In such a case, the steady-state behavior of the closed-loop system may be analyzed through the Lyapunov analysis.*

To analyze the steady-state behavior of the motion variable, consider constant input variables applied to the system. For such an analysis, consider the desired motion trajectory of the robot $\mathcal{X}_d$ is constant, and furthermore, the contact surface in robot interaction $\mathcal{X}_e$ is also constant. For asymptotic stability analysis of the error dynamics, consider the following Lyapunov function candidate for the system:

$$V(\mathcal{X}) = \frac{1}{2}\dot{\mathcal{X}}^T M \dot{\mathcal{X}} + \frac{1}{2} e_x^T K_p e_x + \frac{1}{2}(\mathcal{X} - \mathcal{X}_e)^T K_e (\mathcal{X} - \mathcal{X}_e). \tag{7.16}$$

The Lyapunov function candidate consists of the kinetic energy of the manipulator and the potential energy accounted for the proportional feedback effort $K_p e_x$, as well as the elastic deformation $K_e(\mathcal{X} - \mathcal{X}_e)$. Thus, $V(\mathcal{X})$ is a positive definite scalar function. The Lyapunov stability method is based on the concept to show that along any motion trajectory of the robot, the time derivative of the Lyapunov function is a negative definite. For the particular constant desired trajectory and constant contact surface considered in this analysis, the time derivative of the Lyapunov function may be derived as

$$\dot{V}(\mathcal{X}) = \dot{\mathcal{X}}^T M \ddot{\mathcal{X}} + \frac{1}{2}\dot{\mathcal{X}}^T \dot{M} \dot{\mathcal{X}} + \dot{e}_x^T K_p e_x + \dot{\mathcal{X}}^T K_e (\mathcal{X} - \mathcal{X}_e). \tag{7.17}$$

Substitute $M\ddot{\mathcal{X}}$ from closed-loop dynamic formulation 7.15 in Equation 7.17, and note that for constant desired trajectory motion, $\dot{\mathcal{X}}_d = \mathbf{0}$. Hence, the time derivative of the Lyapunov function is simplified to

$$\begin{aligned} \dot{V} = {} & \dot{\mathcal{X}}^T \left(-C\dot{\mathcal{X}} - K_e(\mathcal{X} - \mathcal{X}_e) + K_p e_x + K_d \dot{e}_x\right) \\ & + \frac{1}{2}\dot{\mathcal{X}}^T \dot{M} \dot{\mathcal{X}} + \dot{e}_x^T K_p e_x + \dot{\mathcal{X}}^T K_e (\mathcal{X} - \mathcal{X}_e). \end{aligned}$$

* The reader is advised to review the discussion given in Appendix C on the Lyapunov stability.

To simplify this equation, note that for a constant-desired trajectory $\dot{\mathcal{X}}_d = \mathbf{0}$, and therefore, $\dot{e}_x = (\dot{\mathcal{X}}_d - \dot{\mathcal{X}}) = -\dot{\mathcal{X}}$. Hence

$$\begin{aligned}\dot{V} &= \frac{1}{2}\dot{\mathcal{X}}^T \left(\dot{M} - 2C\right)\dot{\mathcal{X}} - \dot{\mathcal{X}}^T K_e(\mathcal{X} - \mathcal{X}_e) + \dot{\mathcal{X}}^T K_p e_x \\ &\quad - \dot{\mathcal{X}}^T K_d \dot{\mathcal{X}} - \dot{\mathcal{X}}^T K_p e_x + \dot{\mathcal{X}}^T K_e(\mathcal{X} - \mathcal{X}_e) \\ &= -\dot{\mathcal{X}}^T K_d \dot{\mathcal{X}} + \frac{1}{2}\dot{\mathcal{X}}^T \left(\dot{M} - 2C\right)\dot{\mathcal{X}} \\ &= -\dot{\mathcal{X}}^T K_d \dot{\mathcal{X}} \le 0 \end{aligned} \tag{7.18}$$

where in the last equality, the skew-symmetric property of the dynamic formulation matrices is used to simplify the equation. This result indicates that the derivative of the Lyapunov function is negative semidefinite. Although this result ensures stability in the sense of Lyapunov, it does not guarantee achieving asymptotic tracking.

To analyze asymptotic behavior, use the Lasalle's Theorem.* Note that V is always decreasing, provided $\dot{\mathcal{X}}$ is nonzero. Hence, the steady-state behavior may be viewed by asymptotic analysis of Equation 7.15, where $\dot{\mathcal{X}}$ and $\ddot{\mathcal{X}}$ are zero at the steady state. In such a case, from Equation 7.15 it is concluded that

$$\lim_{t\to\infty}\left[K_e(\mathcal{X} - \mathcal{X}_e) - K_p(\mathcal{X}_d - \mathcal{X})\right] = 0 \tag{7.19}$$

or

$$\lim_{t\to\infty}\left[(K_e + K_p)\mathcal{X} - K_e\mathcal{X}_e - K_p\mathcal{X}_d\right] = 0. \tag{7.20}$$

Let us denote the steady-state value of the motion variable by $\bar{\mathcal{X}}$. Furthermore, considering the constant desired trajectory assumption, the steady-state value of the desired trajectory is denoted by $\bar{\mathcal{X}}_d$, and that for the contact surface is represented by $\bar{\mathcal{X}}_e$. Hence, Equation 7.20 is used to derive the steady-state value of the motion variable as

$$\bar{\mathcal{X}} = [K_e + K_p]^{-1}[K_e\bar{\mathcal{X}}_e + K_p\bar{\mathcal{X}}_d]. \tag{7.21}$$

Let us examine the steady-state behavior of interaction wrench denoted by $\bar{\mathcal{F}}_e$. The steady-state value of the interaction wrench with the environment is calculated by Equation 7.12 as

$$\bar{\mathcal{F}}_e = K_e(\bar{\mathcal{X}} - \bar{\mathcal{X}}_e). \tag{7.22}$$

Substitute Equation 7.21 into Equation 7.22 and simplify

$$\bar{\mathcal{F}}_e = K_e[K_e + K_p]^{-1}K_p(\bar{\mathcal{X}}_d - \bar{\mathcal{X}}_e). \tag{7.23}$$

However, note that the environment stiffness coefficient K_e is in orders of magnitude higher than the controller matrix gain K_p, that is, $K_e \gg K_p$. Therefore, the matrix $[K_e + K_p]^{-1}$ may be approximated by K_e^{-1}, and the steady-state value of the interaction force may be approximated by the following equation:

$$\bar{\mathcal{F}}_e \simeq K_p(\bar{\mathcal{X}}_d - \bar{\mathcal{X}}_e). \tag{7.24}$$

---

* Refer to Appendix C.

Similar to a single-degree-of-freedom case, Equation 7.24 reveals an important fact of the stiffness control scheme, that although the environment stiffness coefficient $\boldsymbol{K}_e$ is usually very large, by using the proposed structure for the controller, the interaction force at the steady-state $\bar{\mathcal{F}}_e$ is related to the corresponding environment deflection at the steady-state $\bar{\mathcal{X}}_d - \bar{\mathcal{X}}_e$, through the controller gain $\boldsymbol{K}_p$. Hence, the interacting stiffness of the manipulator with respect to even very stiff environments can be well controlled by choosing moderate controller matrix gain $\boldsymbol{K}_p$. Therefore, the interaction of the robot with a stiff environment may be tuned such that there is embedded flexibility considered in the structure of the robot for even very rigid mechanical designs.

This analysis reveals another important fact that even if complete information of the mass matrix and Coriolis and centrifugal matrix is not used in the IDC structure, and the closed-loop dynamics is not completely linearized, the PD control structure with gravity compensation can still lead to the required stiffness control strategy at steady state. In this analysis, the skew-symmetric property of dynamic formulation matrices provides sufficient assurance that the Lyapunov function candidate will monotonically converge to zero along all motion trajectories.

Although the structure of stiffness control scheme given in this section is very similar to the position control scheme given in Section 6.3.4, there are clear distinctions in the implementation of these schemes in practice. In a pure motion control, it is considered that the robot moves freely in space, and there is no contact with any surface within its whole maneuver. Therefore, high PD gains are recommended to provide the required trajectory tracking for the manipulator. However, in stiffness control routine, it is considered that the robot experiences contact with a stiff environment, and therefore, interaction forces significantly affect the motion. Therefore, in the segments of motion that the robot is in contact with the environment, the desired trajectory would be adjusted according to the contact surface. Furthermore, in these parts of the motion, the prime objective of the controller is to keep the end effector in touch with the environment, while the interaction force acts in accordance with the desired stiffness. By this means, the robot carries out a moderate trajectory tracking when it is not in contact with the environment, since the controller gains are selected not as high as the pure motion controller. However, when the robot is in contact with the environment, a smooth and safe interaction with the environment results by controlling the robot stiffness at the desired level. This control scheme is very favorable in cases in which the robot comes in contact with a predetermined object during its motion, and the interaction forces are well tuned through stiffness control scheme, without any need of force measurement.

7.3.3 Stiffness Control of a Planar Manipulator

In this section, stiffness control of a planar $4R\underline{P}R$ manipulator is studied in detail. The architecture of a planar manipulator under this study is described in detail in Section 3.3.1 and shown in Figure 3.3. As explained earlier, in this manipulator, the moving platform is supported by four limbs of identical $R\underline{P}R$ kinematic structure. A complete Jacobian and singularity analyses of this manipulator is presented in Section 4.6, and dynamic analysis of the manipulator under constant mass assumption and variable mass assumption is given in Sections 5.3.1 and 5.3.2, respectively. Since dynamic analysis of the system using these two approaches are quite similar, the controllers are being applied only on constant mass dynamics. Different motion control structures are implemented on this manipulator.

The performance of these controllers are examined in detail in Section 6.8. In what follows, the stiffness control strategy is implemented on this manipulator.

The first stiffness control strategy applied on the system has the structure of partial feedback linearization in addition to PD stiffness control. The control effort applied to the system is given in Equation 7.14. Note that the motion variable in the task space is defined as $\boldsymbol{\mathcal{X}} = \begin{bmatrix} x_G & y_G & \phi \end{bmatrix}^T$, and the tracking error is defined as $\boldsymbol{e}_x = \begin{bmatrix} e_x & e_y & e_\phi \end{bmatrix}$. The PD controller is denoted by $(\boldsymbol{K}_d s + \boldsymbol{K}_p)\boldsymbol{e}_x$, in which $\boldsymbol{K}_d$ and $\boldsymbol{K}_p$, are 3×3 diagonal matrices, denoting the derivative and proportional stiffness controller gains, respectively. The output of the controller is denoted by $\boldsymbol{\mathcal{F}}$, which is composed of $\boldsymbol{\mathcal{F}} = \begin{bmatrix} F_x & F_y & \tau_\phi \end{bmatrix}^T$. In practice, the calculated output wrench is transformed into actuator forces through redundancy resolution schemes detailed in Section 6.7.

As schematically shown in Figure 7.10, it is considered that a flat surface is placed in the workspace of the robot such that the center of robot moving platform experiences contact with a stiff environment in a segment of its maneuver. In the previously considered maneuvers for the planar manipulator, the desired trajectory of the moving platform was considered to be a smooth and cubic polynomial function for x_d, y_d, starting from zero initial conditions to a maximum of 100 m, as shown in Figure 6.15. The time of simulation was considered to be 200 s, while the desired position for ϕ_d, is also a smooth and cubic polynomial function starting from $0°$ to $45°$. This base trajectory is changed in the simulations performed for stiffness control to examine collision to the flat object shown in Figure 7.11 within the workspace.

As shown in Figure 7.11, the desired trajectory in the following simulations is considered to start from 0 initial conditions to 100 m final condition in 100 s, while remaining at this position until 120 s. The object is considered to be placed within the workspace of the robot 75 m from zero initial position. As schematically shown in Figure 7.10, the flat object is considered to have a linear contact surface modeled by a line passing through a point denoted by $\boldsymbol{x}_l = (x_l, y_l)$, and a slope s_l in the x–y plane. Therefore, the object surface is represented by the following line equation:

$$y = y_l + s_l(x - x_l). \tag{7.25}$$

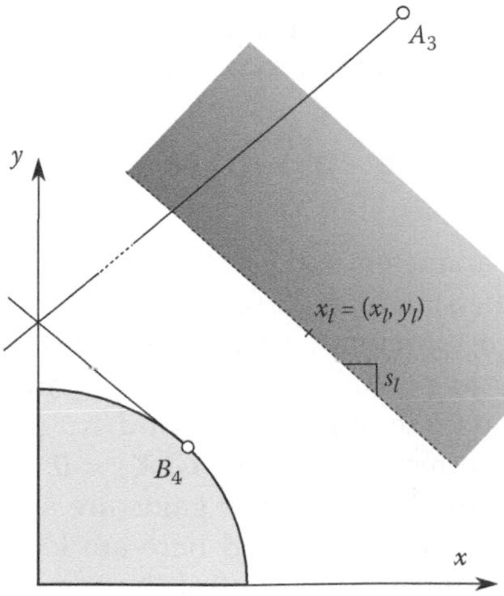

FIGURE 7.10
Schematic of a flat object within the workspace of the planar 4<u>R</u>PR manipulator.

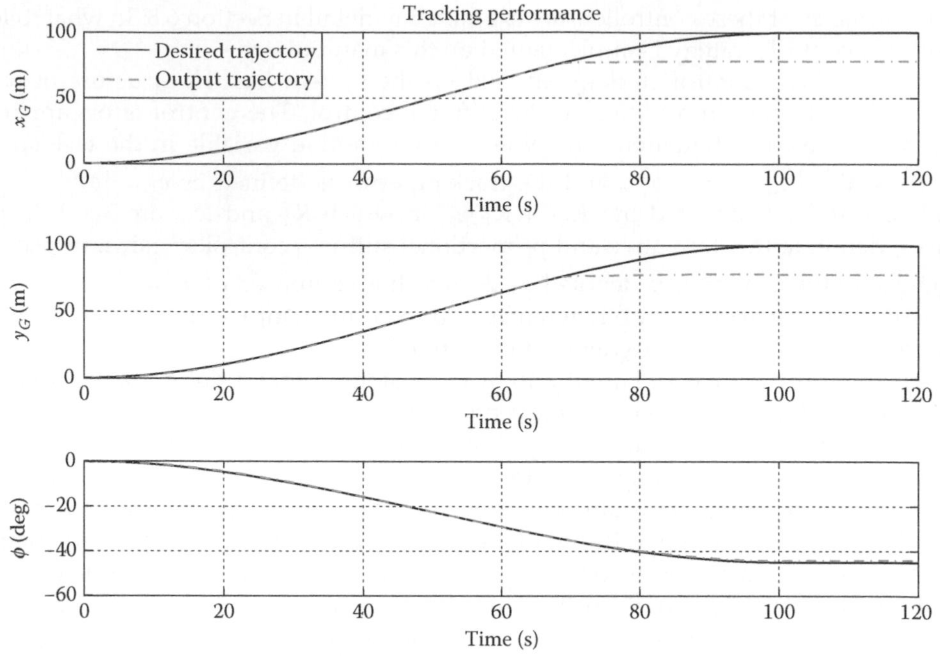

FIGURE 7.11
The desired and output trajectories of the planar $4R\underline{P}R$ manipulator colliding with a flat object; partial IDC stiffness control.

For the simulations, it is considered that $x_l = (75, 75)$ and the slope of the contact surface is $s_l = -1$. By this means, in the prescribed trajectory shown in Figure 7.11, the center of moving platform collides to the flat object at about 67 s, while the contact is held until the end of simulation time. The following time of 20 s is considered at the end of simulation, where all desired trajectories are considered to be constant to examine the steady-state behavior of the closed-loop force and position trajectories. The stiffness of the object is modeled by linear stiffness gain $\boldsymbol{K}_e = 10^3 \cdot \boldsymbol{I}_{3\times3}$ (N/m), with a linear structural damping of $\boldsymbol{C}_e = 10^3 \cdot \boldsymbol{I}_{3\times3}$ (N · m/s). Thus, the contact force is represented by

$$\mathcal{F}_e = \boldsymbol{K}_e(\boldsymbol{\mathcal{X}} - \boldsymbol{\mathcal{X}}_e) + \boldsymbol{C}_e(\dot{\boldsymbol{\mathcal{X}}} - \dot{\boldsymbol{\mathcal{X}}}_e). \tag{7.26}$$

The geometric and inertial parameters used in the simulations are given in Table 5.1, and the initial conditions for the states are all set at zero. For the following simulations, the disturbance wrench applied to the moving platform is considered to be zero, while the gravity vector is considered as $\boldsymbol{g} = [0, 0, -9.81]^T$. The control effort is calculated from Equation 7.14 and since the gravity term has no effect in xy directions, the gravity compensation term is zero in these directions, that is, $\boldsymbol{G}(\boldsymbol{\mathcal{X}}) = \boldsymbol{0}$. To compare the result to a complete IDC implementation, the PD controller gains are set at $\boldsymbol{K}_d = 0.1 \cdot \boldsymbol{M}(\boldsymbol{\mathcal{X}}) \cdot \boldsymbol{I}_{3\times3}$ and $\boldsymbol{K}_p = \boldsymbol{M}(\boldsymbol{\mathcal{X}}) \cdot \boldsymbol{I}_{3\times3}$. The controller gains used here are less than that considered for pure position control, since it is intended to control the force motion relation when the moving platform is in contact with a stiff environment. Therefore, with such controllers, high accuracy in positioning is not the main objective.

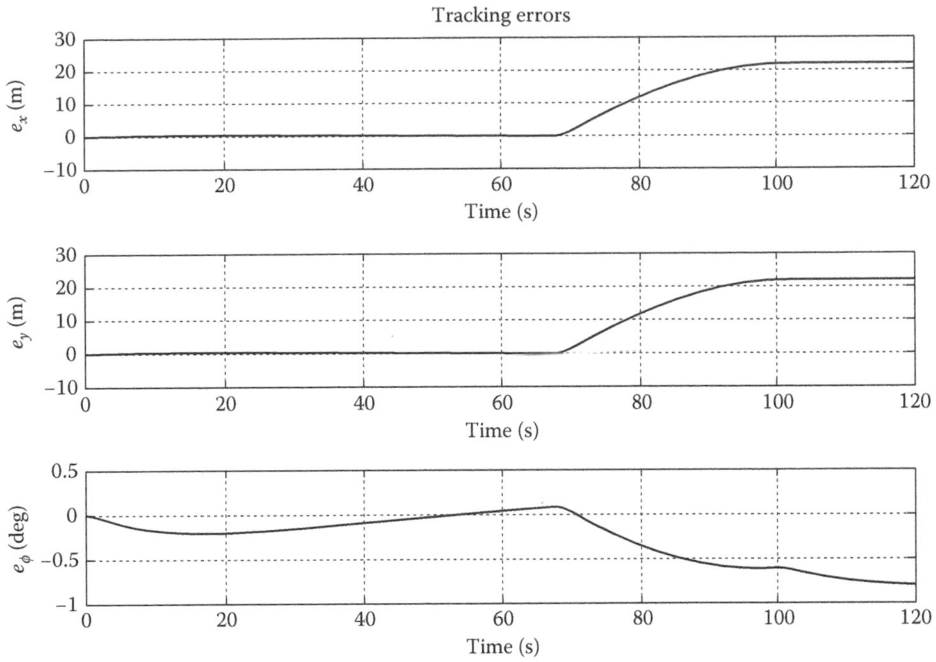

FIGURE 7.12
The closed-loop tracking performance of the planar $4R\underline{P}R$ manipulator colliding with a flat object; partial IDC stiffness control.

The final motion of the moving platform and the tracking errors are illustrated in Figures 7.11 and 7.12. As seen in Figure 7.11, the desired and final closed-loop motion of the system are almost the same before collision of the moving platform to the object. At the time of 67 s, the collision occurs, and the motion of the moving platform remains almost constant on the surface of the object. More details of the moving platform motion at this time is seen in Figure 7.12, in which it is observed that after collision, a smooth and nondestructive motion of the moving platform on the object surface is experienced. The moving platform has caused minor deflection to the object. Furthermore, note that no interaction torque is considered in the simulations, and this is confirmed by relatively uniform tracking error in the orientation before and after the collision. The tracking performance of this controller is lower than that of a pure position controller, since the PD gains are set at relatively lower values to preserve a smooth and nondestructive force–motion interaction during the contact.

The Cartesian wrench $\mathcal{F} = [F_x, F_y, \tau_\phi]^T$ generated by the stiffness controller is illustrated in Figure 7.13. As seen in Figure 7.13, the Cartesian forces in x and y directions rapidly increase after collision at time 67 s, while no interaction torques are observed in ϕ direction. Similarly, the actuator forces of the manipulator are illustrated in Figure 7.14. In this figure, two sets of actuator forces are illustrated. The dash–dotted lines illustrate the base solution of the optimization problem introduced in redundancy resolution scheme by Equation 6.76, while the solid line is the final optimization solution of the redundancy resolution problem detailed in Equation 6.79. The redundancy resolution solution is obtained through the iterative numerical method. As seen in Figure 7.14, the actuator forces rapidly increase after collision, while the tension forces derived from the redundancy resolution scheme remain positive for all time.

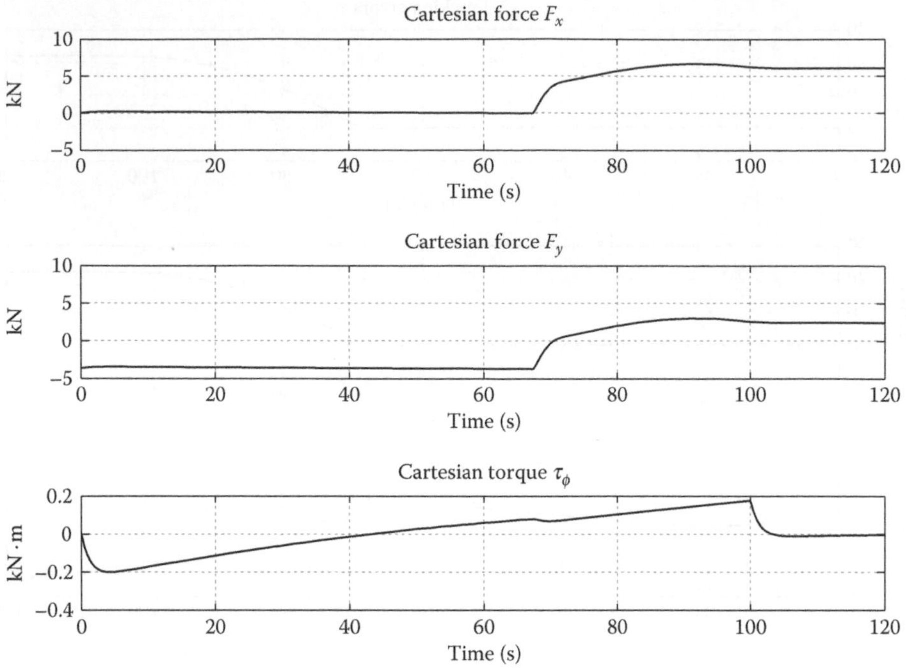

FIGURE 7.13
The Cartesian wrench of the planar $4R\underline{P}R$ manipulator colliding with a flat object; partial IDC stiffness control.

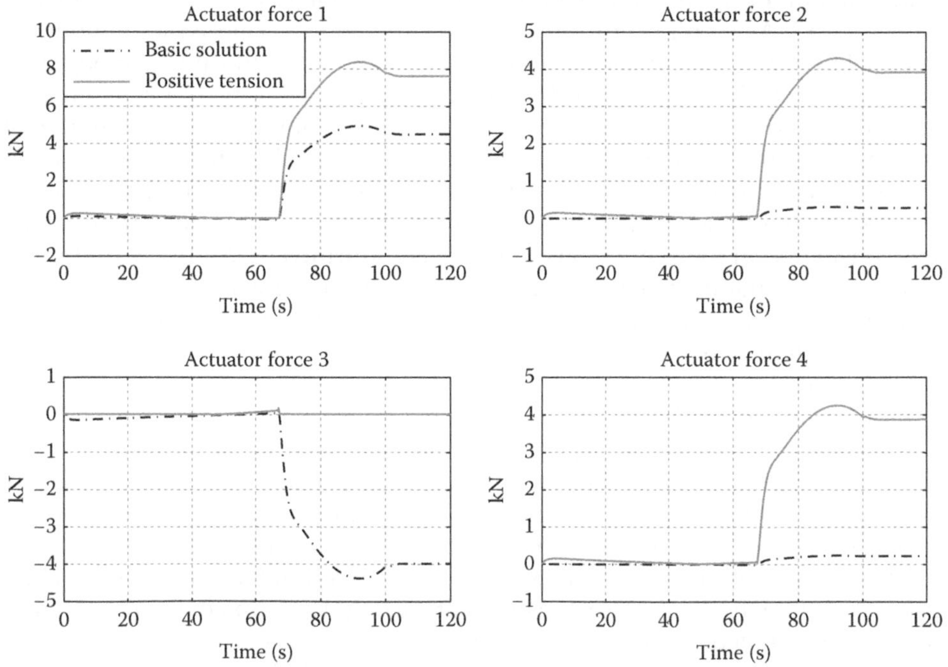

FIGURE 7.14
The actuator forces of the planar $4R\underline{P}R$ manipulator colliding with a flat object; partial IDC stiffness control.

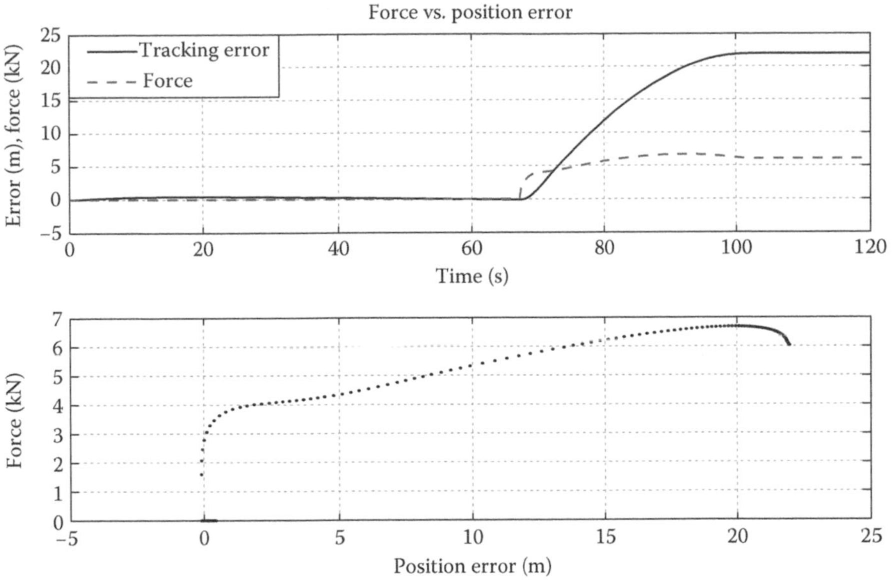

FIGURE 7.15
The tracking error e_x versus interacting force F_x of the planar $4R\underline{P}R$ manipulator colliding with a flat object; partial IDC stiffness control.

To examine the stiffness property of the manipulator in greater detail, the tracking error e_x and interacting force F_x are plotted together in Figure 7.15. In the top plot of Figure 7.15, the tracking error and force trajectories are plotted with respect to time, while in the bottom plot of Figure 7.15, the tracking error and force trajectories are plotted with respect to each other. As seen in the top figure, the tracking error follows the interacting force except at the transient time after collision. The force–motion relation is clearly displayed in the bottom figure, where except at the transient, it shows a quasi-linear behavior. At the steady state, the force position behavior converges to a particular stiffness value for the system, which may be suitably tuned by stiffness controller parameters.

To show how a designer can suitably assign the required stiffness, consider another simulation in which all the parameters are the same as before, while the stiffness coefficients are increased by a factor of 5, that is, $\boldsymbol{K}_d = 0.5 \cdot \boldsymbol{M}(\boldsymbol{\mathcal{X}})$ and $\boldsymbol{K}_p = 5 \cdot \boldsymbol{M}(\boldsymbol{\mathcal{X}})$. The tracking error e_x and interacting force $\boldsymbol{F}_x$ in this case are shown in Figure 7.16. As seen in Figure 7.16, the steady-state tracking error is decreased from about 21.96 m in the previous case to about 14.8 m in this case. This means that by requiring higher stiffness for the manipulator, the closed-loop system becomes more stiff, and as a result, the object deflection is decreased. The interaction force, however, has been increased from 6.05 kN in the previous case to about 20.38 kN in this condition. The steady-state stiffness of the system can be determined from $K = F_x/e_x$ to be 0.275 kN/m in the previous case and 1.38 kN/m in this condition, which is increased by a factor of 5 as required.

For the next simulation, stiffness control strategy is invigorated by complete feedback linearization. In this controller, a corrective wrench $\boldsymbol{\mathcal{F}}_{fl}$ is added in a feedback structure to the closed-loop system, which is calculated from the Coriolis and centrifugal matrix and gravity vector of the manipulator dynamic formulation. Furthermore, the mass matrix acts in the forward path in addition to the desired trajectory acceleration $\ddot{\boldsymbol{\mathcal{X}}}_d$. Thus, the control

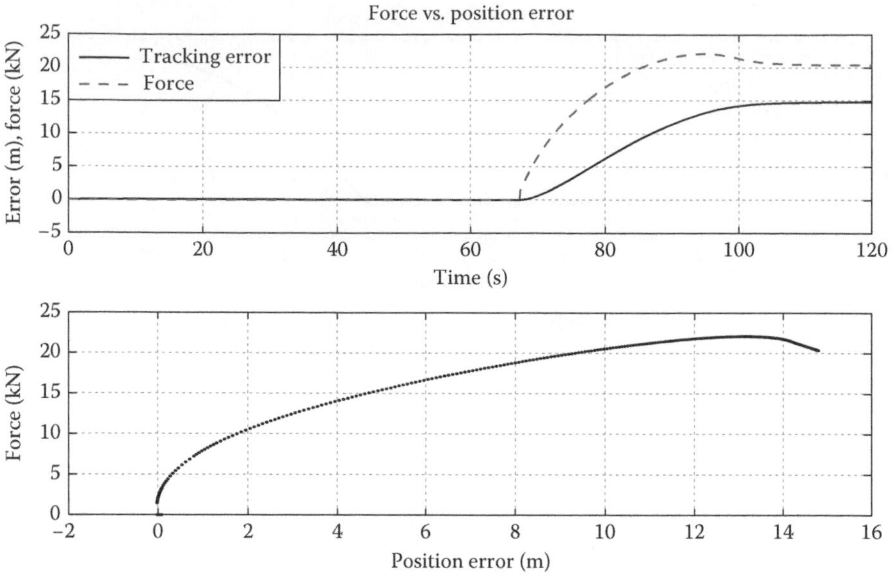

FIGURE 7.16
The tracking error e_x versus interacting force F_x of the planar $4R\underline{P}R$ manipulator colliding with a flat object; partial IDC stiffness control with 5 times more stiffness requirement.

effort is determined from the following equation:

$$\mathcal{F} = \hat{C}(\dot{\mathcal{X}}, \mathcal{X})\dot{\mathcal{X}} + \hat{G}(\mathcal{X}) + \hat{M}(\mathcal{X})(\ddot{\mathcal{X}}_d + K_p e_x + K_d \dot{e}_x). \tag{7.27}$$

Note that to generate this term, the dynamic formulation of the robot and its kinematic and dynamic parameters are needed. In practice, exact information on dynamic matrices is not available, and therefore, a %10 perturbation in all kinematic and inertial parameters is used in the following simulations. For the sake of comparison to the first examined controller, the PD controller gain is adjusted to $K_d = 0.1 \cdot I_{3\times3}$ and $K_p = I_{3\times3}$. The geometric and inertial parameters used in the simulations are given in Table 5.1, and the initial conditions for the states are all set at zero. The disturbance wrench applied to the moving platform is considered to be zero, while the gravity vector is considered as $g = [0, 0, -9.81]^T$. The desired trajectory and the shape and placement of the contact object is considered the same as the previous simulation.

It is observed that the final motion of the moving platform and the tracking errors are very similar to those of the partial IDC controller. The Cartesian wrench are illustrated in Figure 7.17, in which the total control effort is plotted in a solid line while its PD component is plotted in a dashed line. As seen in Figure 7.17, the control effort in x and y directions is merely generated by the PD controller, and the effect of the feedback linearizing term is negligible. This is mainly because of the significant amount of contact force compared to that of the inertial forces. Thus, the feedback linearizing terms are not that important in the performance of the closed-loop system after collision. This is the main reason why no meaningful difference can be seen in the performance of full IDC compared to that of partial IDC.

To examine the stiffness property of the manipulator in greater detail, the tracking error e_x and interacting force F_x are also plotted together in Figure 7.18. This figure is almost

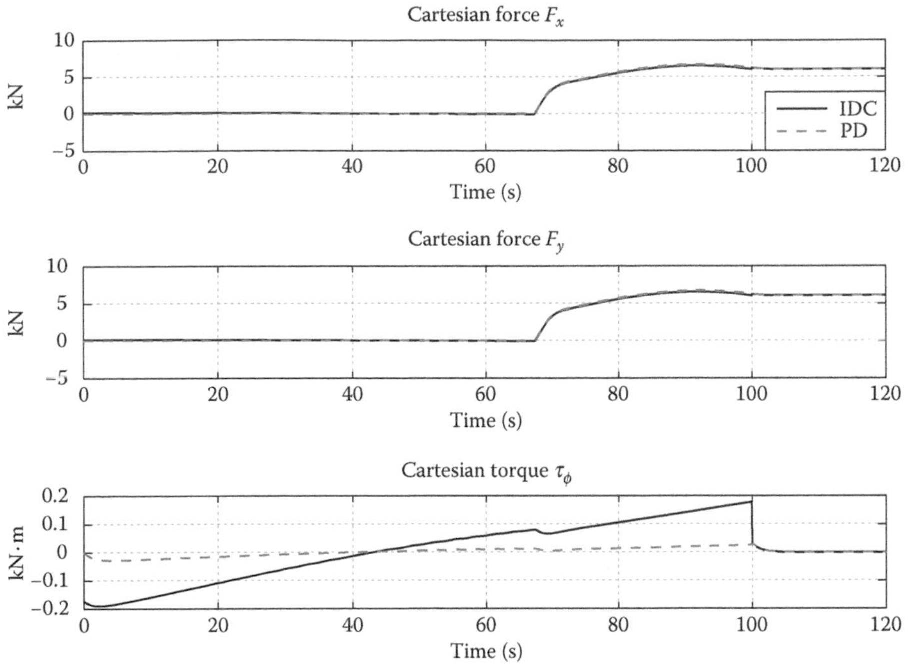

FIGURE 7.17
The Cartesian wrench of the planar 4$R\underline{P}R$ manipulator colliding with a flat object; full IDC stiffness control.

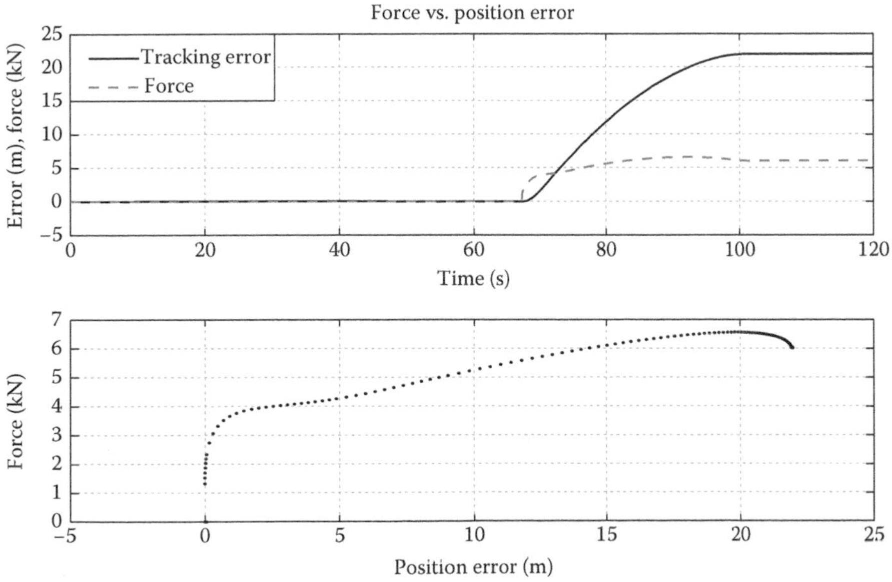

FIGURE 7.18
The tracking error e_x versus interacting force F_x of the planar 4$R\underline{P}R$ manipulator colliding with a flat object; full IDC stiffness control.

identical to Figure 7.16. Therefore, it may be concluded that when the robot is in contact with a stiff environment, there is no need to completely linearize the dynamic equation, and the controller design given in Equation 7.14, is quite favorable.

7.3.4 Stiffness Control of the Stewart–Gough Platform

In this section, stiffness control of the Stewart–Gough platform has been studied in detail. The architecture of this manipulator is described in detail in Section 3.5.1 and shown in Figure 3.16. As explained earlier, in this manipulator, the spatial motion of the moving platform is generated by six piston–cylinder actuators. Complete kinematic analysis of this manipulator is given in Section 3.5 and its Jacobian analysis is presented in Section 4.8. The dynamic analyses of this manipulator using the Newton–Euler and virtual work approaches are given in Sections 5.3.3 and 5.4.5, respectively. The closed-form dynamic formulation being used in this section for the simulations, is derived in Section 5.5.6. Various motion control strategies are implemented in Section 6.9. In what follows, stiffness control strategy is implemented on this manipulator.

The stiffness control strategy applied on the system has the structure of a full feedback linearization in addition to PD stiffness control. The control effort applied to the system is given in Equation 7.14. Note that the motion variable for a completely parallel manipulator with six-degrees-of-freedom, is defined by $\boldsymbol{\mathcal{X}} = \begin{bmatrix} \boldsymbol{x}_p \\ \boldsymbol{\theta} \end{bmatrix}$, which consists of the linear position vector $\boldsymbol{x}_p = [x_p \;\; y_p \;\; z_p]^T$ and the orientation vector preferably represented by screw coordinates $\boldsymbol{\theta} = \theta \begin{bmatrix} s_x & s_y & s_z \end{bmatrix}^T$. The positional tracking error is defined as $\boldsymbol{e} = \begin{bmatrix} e_x & e_y & e_z & e_{\theta_x} & e_{\theta_y} & e_{\theta_z} \end{bmatrix}^T$. The PD controller is denoted by $(\boldsymbol{K}_d s + \boldsymbol{K}_p)\boldsymbol{e}$, in which $\boldsymbol{K}_d$ and $\boldsymbol{K}_p$ are 6×6 diagonal matrices, denoting the derivative and proportional stiffness controller gains, respectively. The output of the controller is denoted by $\boldsymbol{\mathcal{F}}$, which is composed of $\boldsymbol{\mathcal{F}} = \begin{bmatrix} F_x & F_y & F_z & \tau_x & \tau_y & \tau_z \end{bmatrix}^T$. In practice, the calculated output wrench is transformed into actuator forces through the Jacobian transpose inverse projection.

As schematically shown in Figure 7.19, it is considered that a flat surface is placed in the workspace of the robot such that the center of robot-moving platform experiences contact with a stiff environment in a segment of its maneuver. In the previously considered maneuvers for the planar manipulator, the desired trajectory of the moving platform was considered to be a smooth and cubic polynomial function for motion variables x_d, y_d, and z_d, as shown in Figure 6.51, where the simulation time was considered to be 2 s. This base trajectory is changed in the simulations performed for stiffness control to examine collision to the flat object within the workspace. As shown in Figure 7.20, the desired trajectory in the following simulations is considered to start from zero for x_d and y_d, and from 1 m for z_d with a positive 0.25 motion in 1 s, while remaining at this position until 2 s. The object is considered to be placed within the workspace of the robot 0.2 m above the initial position of the moving platform. As schematically shown in Figure 7.19, the flat object is considered to have a planar contact surface, modeled by a plane passing through a point denoted by $\boldsymbol{x}_l = (x_l, y_l, z_l)$ and normal to a vector $\hat{\boldsymbol{n}}_l = [n_x, n_y, n_z]^T$ in space. Therefore, the object surface is represented by the following plane equation:

$$n_x(x - x_l) + n_y(y - y_l) + n_z(z - z_l) = 0. \tag{7.28}$$

For the simulations, it is considered that $\boldsymbol{x}_l = [0.2, 0.2, 1.2]^T$ and the normal vector to the plane is $\boldsymbol{n}_l = [1, 1, 1]^T$. By this means, in the prescribed trajectory shown in Figure 7.20, the center of moving platform collides to the flat object at about 0.7 s, while the contact is

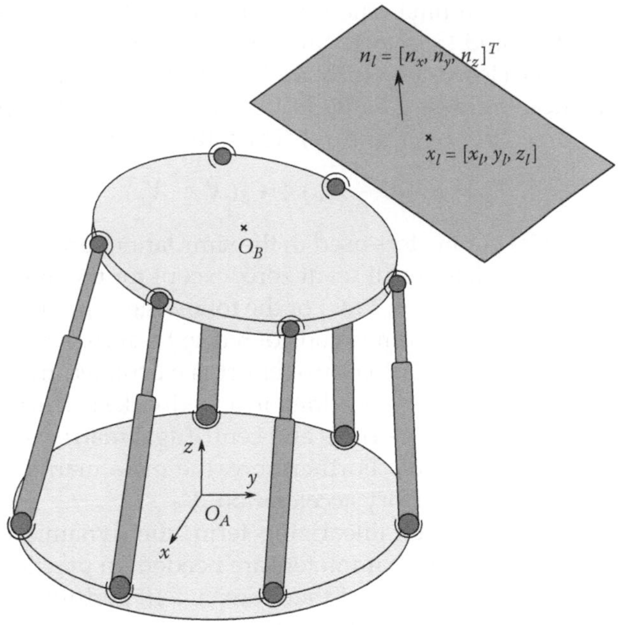

FIGURE 7.19
Schematic of a flat object within the workspace of the Stewart–Gough platform.

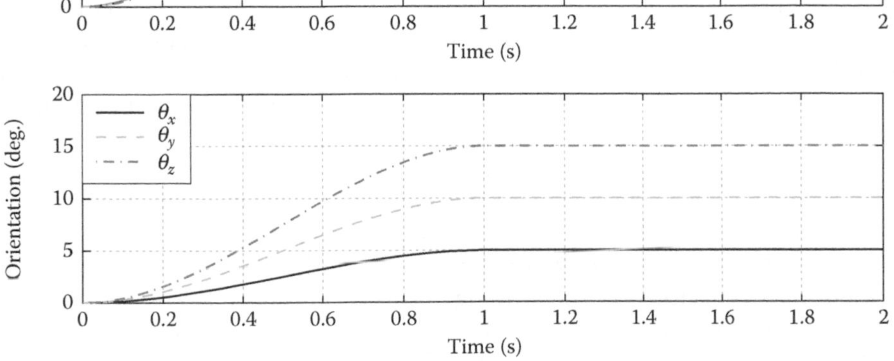

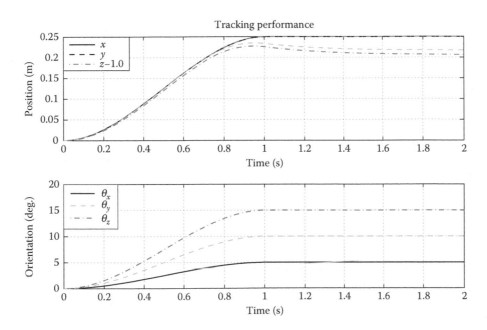

FIGURE 7.20
The desired and output trajectories of the Stewart–Gough platform colliding with a flat object; partial IDC stiffness control.

held until the end of simulation time. The following time of 1 s is considered at the end of simulation, where all desired trajectories are considered to be constant to examine the steady-state behavior of the closed-loop force and position trajectories. The stiffness of the object is modeled by linear stiffness gain $K_e = 10^5 \cdot I_{6\times6}(\mathrm{N/m})$, while possessing a linear structural damping of $C_e = 10^4 \cdot I_{6\times6}(\mathrm{N \cdot m/s})$. Thus, the contact force is represented by

$$\mathcal{F}_e = K_e(\mathcal{X} - \mathcal{X}_e) + C_e(\dot{\mathcal{X}} - \dot{\mathcal{X}}_e). \tag{7.29}$$

The geometric and inertial parameters used in the simulations are given in Table 5.2, and the initial conditions for the states are all set at zero, except for the initial state of the moving platform height, which is set at $z_o = 1$ m. For the following simulations, the disturbance wrench applied to the moving platform is considered to be zero, while the gravity vector is considered as $g = [0, 0, -9.81]^T$. The control effort is calculated from Equation 7.27. In this controller, a corrective wrench $\mathcal{F}_{fl}$, is added in a feedback structure to the closed-loop system, which is calculated from the Coriolis and centrifugal matrix, and gravity vector of the manipulator dynamic formulation. Furthermore, the mass matrix acts in the forward path, in addition to the desired trajectory acceleration $\ddot{\mathcal{X}}_d$.

Note that to generate the feedback linearizing term, the dynamic formulation of the robot, and its kinematic and dynamic parameters are needed. In practice, exact knowledge of dynamic matrices are not available, and therefore, a %10 perturbation in all kinematic and inertial parameters are used in the following simulations. The PD gains used in the simulations are set at $K_p = 100 \cdot \mathrm{diag}[1, 1, 1, 10, 10, 10]$ and $K_d = 100 \cdot \mathrm{diag}[1, 1, 1, 1, 1, 1]$, after a number of trial-and-error nominations. The controller gains used here is significantly lower than that considered for pure position control, since it is intended to control the force–motion relation when the moving platform is in contact with a stiff environment. Therefore, with such controllers, high accuracy in positioning is not the main objective.

The final motion of the moving platform and the tracking errors are illustrated in Figures 7.20 and 7.21. As seen in Figure 7.20, the desired and final closed-loop motion of the system are almost the same before collision of the moving platform to the object. At almost 0.7 s, the collision occurs and the motion of the moving platform remains almost constant on the surface of the object. Greater details of the moving platform motion at this time is seen in Figure 7.21, in which it is observed that after collision, a smooth and nondestructive motion of the moving platform on the object surface is generated. The moving platform causes minor deflection to the object. Furthermore, note that no interaction torques are considered in the simulations, and this is confirmed by relatively uniform tracking error in the orientation before and after the collision. The tracking performance of this controller is lower than that of a pure position controller, since the PD gains are set at relatively lower values to preserve a smooth and nondestructive force–motion interaction during the contact.

The Cartesian wrench $\mathcal{F}$, generated by the stiffness controller is illustrated in Figure 7.22. As seen in Figure 7.22, the Cartesian forces F_x, F_y, and F_z rapidly increases after collision at time 0.7 s, while no interaction torques τ_x, τ_y, and τ_z are observed in this figure. The torques generated in this figure correspond to the linear accelerations requested for the cubic polynomial orientation trajectories of the moving platform. Similarly, the actuator forces of the manipulator are illustrated in Figure 7.23. As seen in Figure 7.23, the actuator forces rapidly increase after collision. The abrupt change of the actuator forces, seen in this figure at 1 s, is due to abrupt change in the requested acceleration in the cubic polynomial trajectories.

To examine the stiffness property of the manipulator in greater detail, tracking error and interacting force are plotted together in Figure 7.24. The left-hand-side figures illustrate the force and tracking errors in the x direction, while the right-hand-side figures represent that

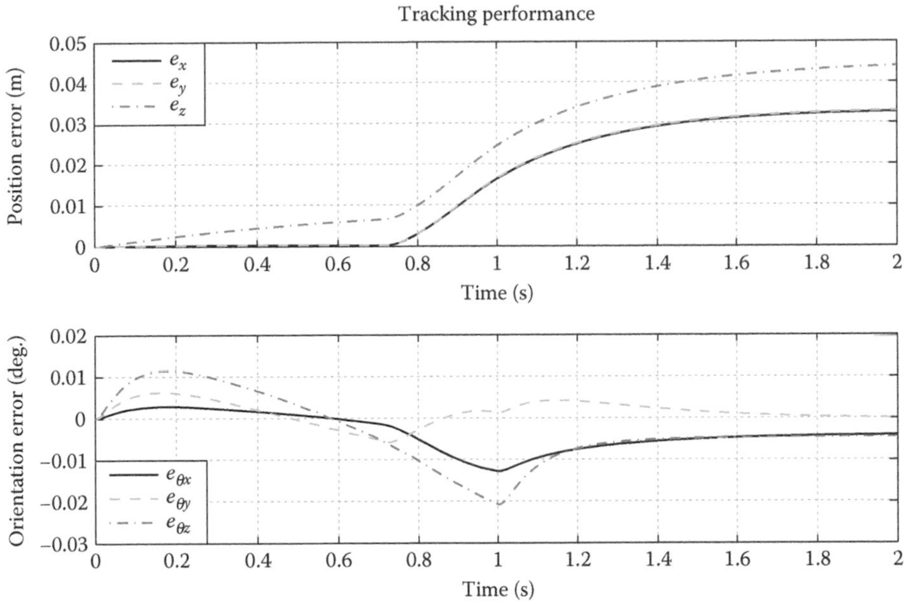

FIGURE 7.21
The closed-loop tracking performance of the Stewart–Gough platform colliding with a flat object; partial IDC stiffness control.

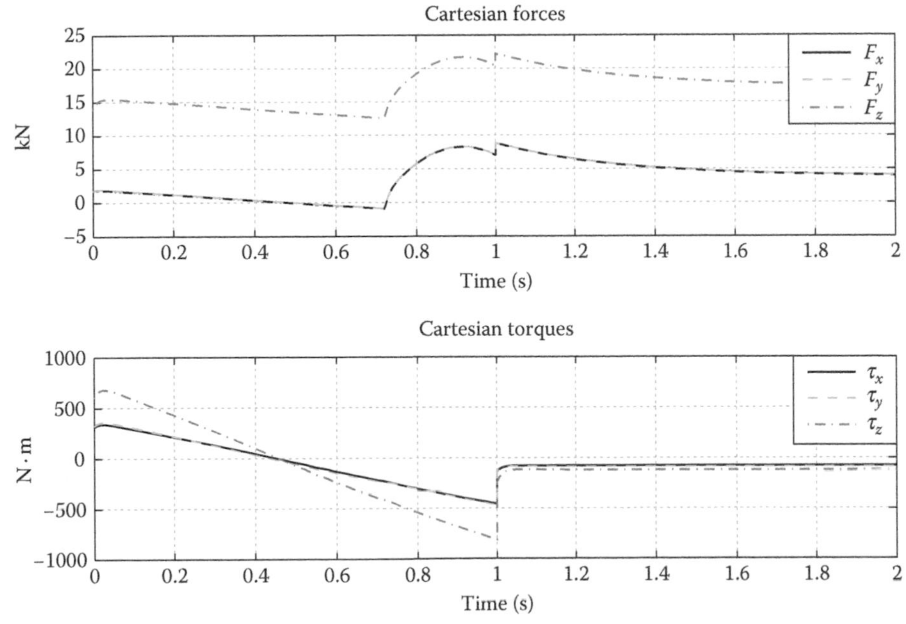

FIGURE 7.22
The Cartesian wrench of the Stewart–Gough platform colliding with a flat object; partial IDC stiffness control.

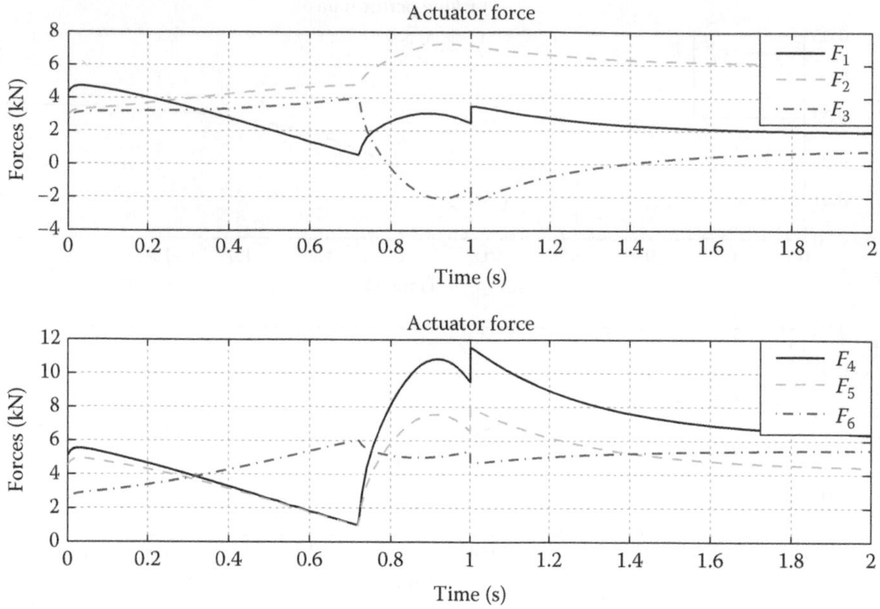

FIGURE 7.23
The actuator forces of the Stewart–Gough platform colliding with a flat object; partial IDC stiffness control.

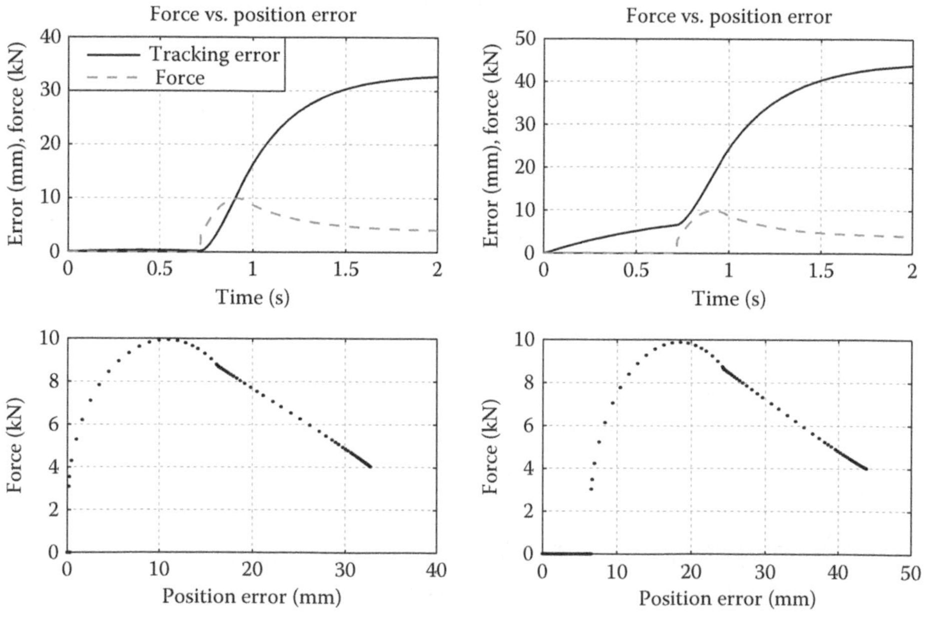

FIGURE 7.24
The tracking errors e_x and e_z versus interacting forces F_x and F_z of the Stewart–Gough platform colliding with a flat object; partial IDC stiffness control.

for the z direction. In the top plots of Figure 7.24, the tracking error and force trajectories are plotted with respect to time, while in the bottom plots, force trajectories are plotted with respect to the tracking errors. As seen in the top figures, the tracking errors monotonically increase, while the interacting forces increase rapidly after collision and then converge to their steady-state values. The force–motion relation is clearly displayed in the bottom figures, where except at the transient, it shows a quasi-linear behavior. At the steady state, the force position behavior converges to a particular stiffness value for the system, which may be suitably tuned by stiffness controller parameters.

To show how a designer can suitably assign the required stiffness, consider another simulation in which all the parameters are the same as earlier, while the stiffness coefficients increase by a factor of 5, that is, $\boldsymbol{K}_p = 500 \cdot \text{diag}[1, 1, 1, 10, 10, 10]$ and $\boldsymbol{K}_d = 500 \cdot \text{diag}[1, 1, 1, 1, 1, 1]$. The tracking errors and interacting forces in x and z directions are shown in Figure 7.25 for this case. As seen in Figure 7.25, the steady-state tracking error in the x direction decrease from about 32.7 mm in previous case to about 14 mm in this case. This means that by requiring higher stiffness for the manipulator, the closed-loop system becomes more stiff, and as a result, the object deflection decreases. The interaction force in the x direction, however, has increased from 4.08 kN in the previous case to about 10.5 kN in this condition. The steady-state stiffness of the system in the previous case can be determined from $K = F_x/e_x$ to be 0.125 kN/mm in the previous case and 0.75 kN/mm in this condition, which is increased by a factor of 5 as required.

In the next simulation, it is assumed that the knowledge of the designer of the system model is very poor, and %50 perturbation in all kinematic and inertial parameters are considered in the derivation of the inverse dynamics control term. It is expected that the tracking performance is significantly affected because of this lack of knowledge of the system model. Figures 7.26 through 7.28 illustrate the tracking performance and the

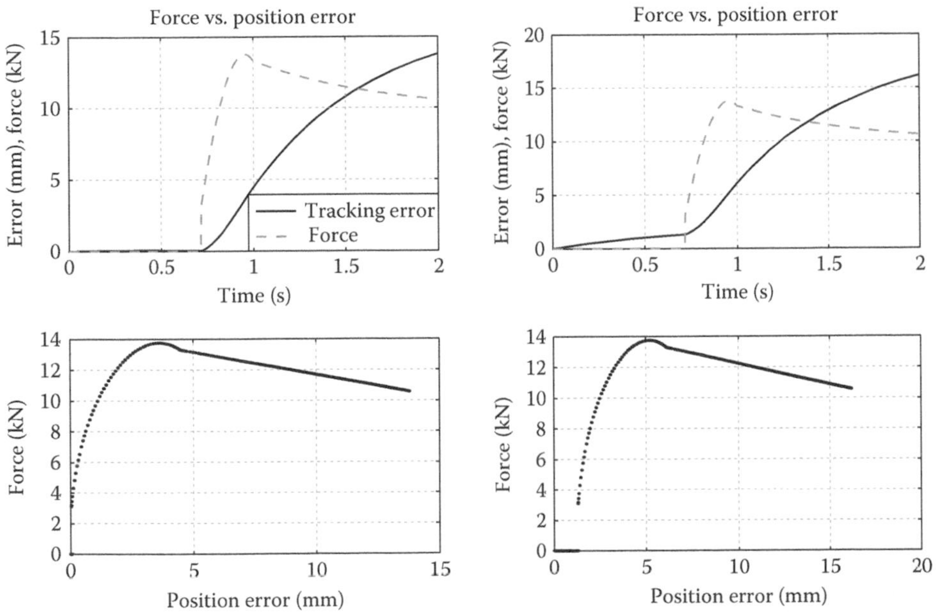

FIGURE 7.25
The tracking errors e_x and e_z versus interacting forces F_x and F_z of the Stewart–Gough platform colliding with a flat object; partial IDC stiffness control with 5 times more stiffness requirement.

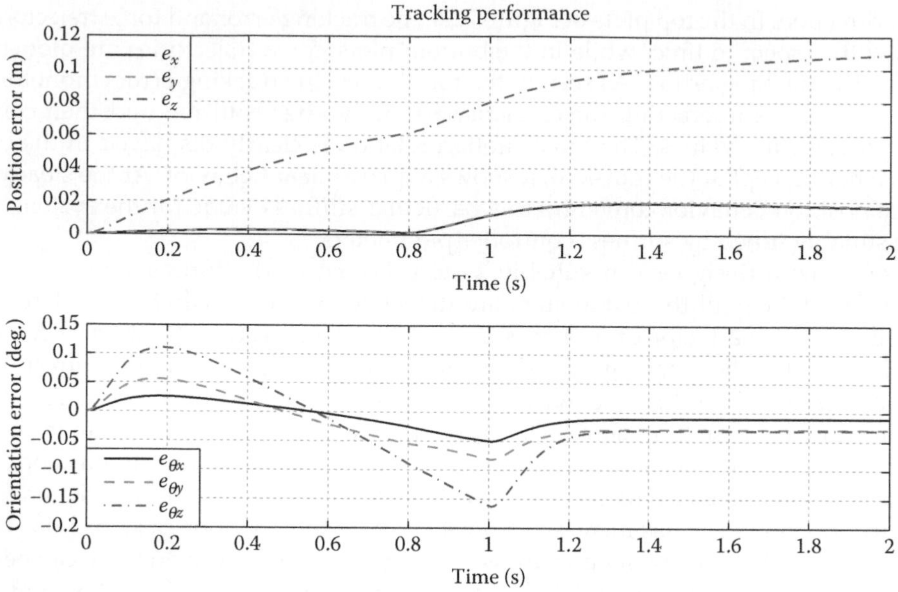

FIGURE 7.26
The tracking errors of the Stewart–Gough platform colliding with a flat object; full IDC stiffness control with %50 perturbation in parameters.

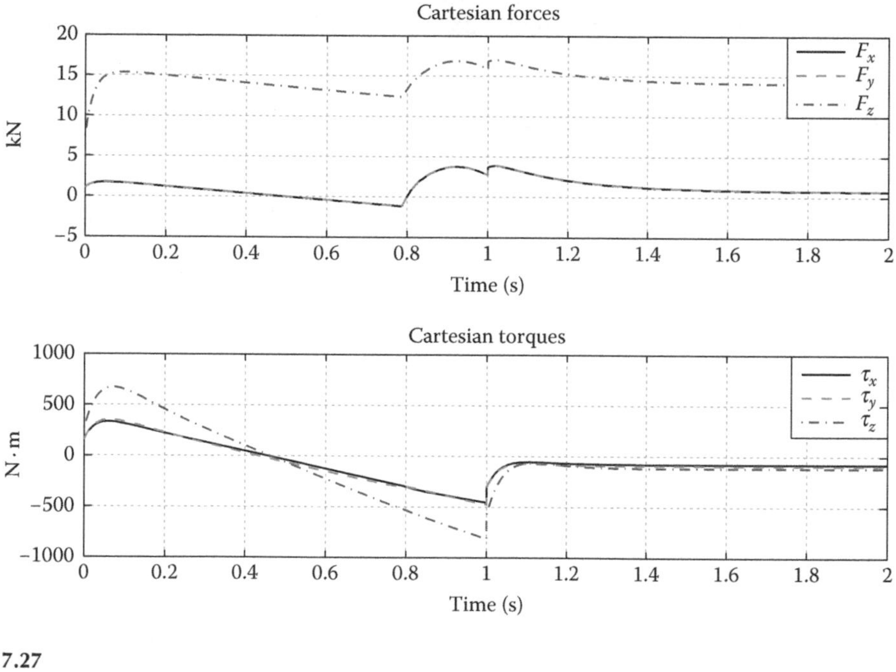

FIGURE 7.27
The Cartesian wrench of the Stewart–Gough platform colliding with a flat object; full IDC stiffness control with %50 perturbation in parameters.

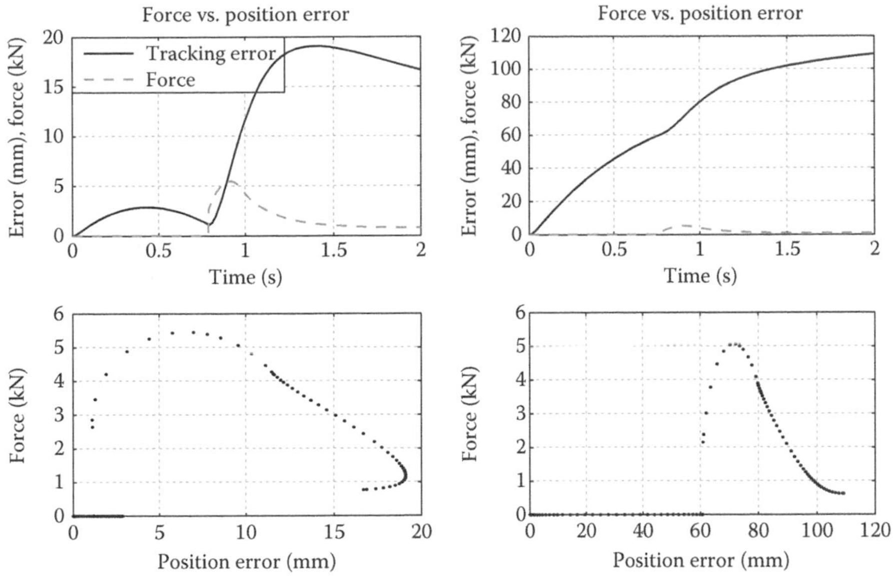

FIGURE 7.28
The tracking errors e_x and e_z versus interacting forces F_x and F_z of the Stewart–Gough platform with a flat object; full IDC stiffness control with %50 perturbation in parameters.

stiffness characteristics of the closed-loop system in this case. As seen in Figure 7.26, lack of knowledge of the model parameters has significantly affected the tracking errors. Comparing Figure 7.26 with Figure 7.21, it can be clearly seen that the tracking error in the z direction increases by a factor of 2, while the orientation error increases by a factor of 10. The effect of uncompensated gravity term in this case can be clearly seen in increased tracking error in the z direction. The Cartesian wrench generated by the stiffness control scheme in this case is shown in Figure 7.27 and the increase of the generated forces in the z direction is also clearly seen in this figure.

To examine the stiffness property of the manipulator in this case, the tracking error and interacting force are plotted together in Figure 7.28. As seen in the top figures, the interacting forces increase rapidly after collision and then converge to their steady-state values. The force–motion relation is clearly displayed in the bottom figures, wherein in contrast to the previous case, the stiffness in x and in z directions performs significantly differently. As seen in the right-side plots of Figure 7.28, the tracking error in the z direction is 5 times more than that in x direction. Hence, the stiffness of the closed-loop system in the z direction is significantly less than that in the x direction, although similar stiffness characteristics are requested in different directions. Therefore, it may be concluded that to obtain the required stiffness of the closed-loop system in a stiffness control scheme, the knowledge of the dynamic matrices, especially the gravity vector, plays an important role.

7.4 Direct Force Control

Stiffness control scheme developed in Section 7.3 is more favorable than pure motion control schemes in cases where the robot is in contact with a stiff environment. In this control scheme, the force–motion relation is well tuned to track a desired stiffness requirement at steady state, without the need of any force measurement. However, stiffness control suffers from a number of drawbacks, which limits its widespread use in applications in which the robot accurately controls the contacts force during its course of motion. Since there is no force measurement incorporated in this control scheme, accurate information on contact force may be only obtained by very accurate position sensors with high resolution. Furthermore, the contact force is estimated by linear compliance models of the environment, and therefore, its accuracy depends directly on the elastic behavior of the contact object and accurate knowledge of its linear and angular compliance coefficients.

Moreover, the steady-state relation between the force and position of the robot moving platform obeys a desired stiffness relation by a set-point control. In other words, the desired end-effector manipulator position and force exerted on the environment must be constant to experience the required stiffness relation. In many robotic applications such as grinding, the moving platform tracks a desired motion trajectory along the object surface, while it is necessary to apply a desired force trajectory upon the object. In this type of applications, stiffness control scheme is not able to satisfy the required objectives, and direct control schemes are frequently used by practitioners.

If it is desired to monitor and control the contact force accurately, it is necessary to measure the interaction force directly, and advise control schemes based on force measurements. Many different force control schemes have been developed for such applications. In this section, a typical direct force control scheme is presented based on inverse dynamics control. As mentioned earlier, inverse dynamics control scheme is the basis for further development of robust and adaptive force control schemes, and therefore, it is chosen as the primary control scheme to be presented in this section.

Note that although the prime objective in direct force control scheme is desired force tracking, however, motion of the moving platform would be considered as the secondary objective. For this reason, consider the cascade control topology shown in Figure 7.2, in which the outer loop is based on force measurement, and the inner loop is based on position feedback. As explained earlier, in this force control topology, it is assumed that both force and position measurements are performed in the task space. In this configuration, the moving platform wrench $\mathcal{F}$ may be measured through a six-degrees-of-freedom wrench sensor, and the motion variable $\mathcal{X}$ may be measured by a position measurement unit.

Consider a parallel robot with multiple-degrees-of-freedom interacting with a stiff environment. In such a case, the motion of the robot end-effector is represented by the motion vector $\mathcal{X}$, in which for a general six-degrees-of-freedom manipulator such as the Stewart–Gough platform, the motion vector consists of the linear motion vector $\boldsymbol{x}$ as well as the orientation variable $\boldsymbol{\theta}$ preferably represented by screw coordinates, that is, $\mathcal{X} = \begin{bmatrix} \boldsymbol{x} & \boldsymbol{\theta} \end{bmatrix}^T$. In the course of robot interaction with the environment, a wrench $\mathcal{F}_e$ is applied to the robot end effector, which consists of the contact force $\boldsymbol{F}_e$ and torque $\boldsymbol{n}_e$, as $\mathcal{F}_e = \begin{bmatrix} \boldsymbol{F}_e & \boldsymbol{n}_e \end{bmatrix}^T$. It is considered that the interacting force $\mathcal{F}_e$ is measured in the task space and is used in an outer force feedback loop of a cascade controller structure. Furthermore, it is considered that the motion variable $\mathcal{X}$ is measured and is used in the inner feedback loop.

The general dynamic formulation of parallel robots given in Equation 7.1 in the absence of any disturbance wrench is rewritten here for reader convenience:

$$\boldsymbol{M}(\mathcal{X})\ddot{\mathcal{X}} + \boldsymbol{C}(\mathcal{X}, \dot{\mathcal{X}})\dot{\mathcal{X}} + \boldsymbol{G}(\mathcal{X}) + \mathcal{F}_e = \boldsymbol{J}^T\boldsymbol{\tau} = \mathcal{F}, \tag{7.30}$$

in which $\boldsymbol{M}(\mathcal{X})$ denotes the robot mass matrix, $\boldsymbol{C}(\mathcal{X}, \dot{\mathcal{X}})$ denotes the Coriolis and centrifugal matrix, $\boldsymbol{G}(\mathcal{X})$ denotes the gravity vector, $\boldsymbol{\tau}$ represents the actuator force vector, $\mathcal{F}$ is

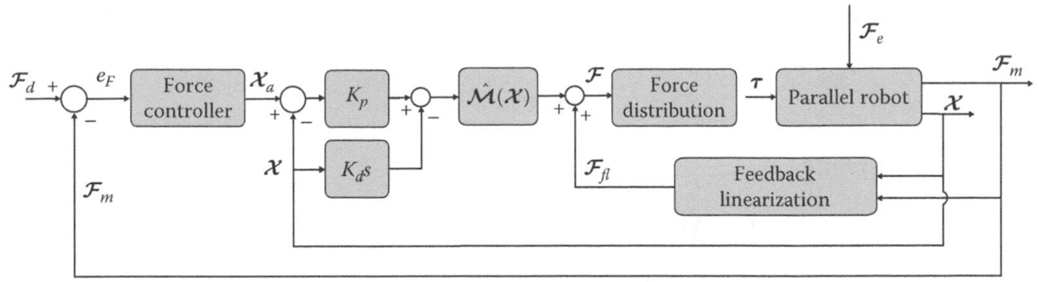

FIGURE 7.29
Direct force control scheme, force feedback in the outer loop and motion feedback in the inner loop.

the projection of the actuator force vector into the task space, and finally, $\mathcal{F}_e$ denotes the interaction wrench with a stiff environment.

As a possible direct force control scheme, consider the closed-loop system depicted in Figure 7.29, in which the force feedback is considered in the outer loop, while position feedback is used in the inner loop. As shown in Figure 7.29, the force-tracking error is directly determined from force measurement by $e_F = \mathcal{F}_d - \mathcal{F}_m$ in the outer loop and the force controller is designed to satisfy the required force tracking performance. However, direct motion-tracking objective is not assigned in this control scheme, since the desired motion trajectory $\mathcal{X}_d$ is absent. Nevertheless, an auxiliary motion trajectory $\mathcal{X}_a$ is generated from the force control law and is used as the reference for the motion tracking. By this means, no prescribed motion trajectory is tracked, while the force control scheme would advise a motion trajectory for the robot to ensure that the desired force is tracked. As explained earlier in stiffness control scheme, when the robot is in contact with a stiff environment, the auxiliary trajectory is automatically designed along the object surface to ensure suitable force interaction with the environment.

As shown in Figure 7.29, the feedback control law generates the required wrench in the task space $\mathcal{F}$ to be applied by the actuators $\boldsymbol{\tau}$. To implement the control effort $\boldsymbol{\tau}$, it is required to use a suitable force distribution scheme in practice. As explained earlier, the force distribution block projects the required wrench in the task space $\mathcal{F}$ to its corresponding actuator force vector $\boldsymbol{\tau}$. Note that for a completely parallel manipulator such as the Stewart–Gough platform, in which the number of actuators are equal to the number of degrees-of-freedom, this mapping can be constructed from the Jacobian transpose of the manipulator J^T by $\boldsymbol{\tau} = J^{-T}\mathcal{F}$, at nonsingular configurations. However, the force distribution scheme may need more details if the manipulator is not completely parallel. Redundancy resolution techniques, which are completely elaborated in Section 6.7, would be used to physically implement force distribution scheme for redundant manipulators.

The required wrench $\mathcal{F}$ in the direct force control scheme depicted in Figure 7.29 is based on inverse dynamics control and consists of three main parts. In the inner loop, the motion control scheme is based on a feedback linearization part in addition to a linear control scheme, while in the outer loop usually a linear force controller is sufficient for suitable force tracking. Although, advanced methods such as robust control schemes may be considered to design the linear parts of the motion controller, and that of the force controller, to develop the idea and analyze the performance, simple PD and proportional-integrator (PI) controllers are considered for the motion and force controllers, respectively.

Hence, the controller output wrench $\mathcal{F}$ applied to the manipulator may be derived as follows:

$$\mathcal{F} = \hat{M}(\mathcal{X})a + \mathcal{F}_{fl}, \tag{7.31}$$

which in detail reads

$$\mathcal{F} = \hat{M}(\mathcal{X})a + \hat{C}(\mathcal{X}, \dot{\mathcal{X}})\dot{\mathcal{X}} + \hat{G}(\mathcal{X}) + \mathcal{F}_m \tag{7.32}$$

$$a = K_p(\mathcal{X}_a - \mathcal{X}) - K_d\dot{\mathcal{X}}, \tag{7.33}$$

in which K_p and K_d denote the motion controller proportional and derivative gain matrices. Note that in general case of six-degrees-of-freedom manipulator, a is a 6×1 vector. Furthermore, the auxiliary motion trajectory $\mathcal{X}_a$ is generated from the force-tracking error $e_F = \mathcal{F}_d - \mathcal{F}_m$, by the force control law. Let us denote the general force controller by the function C_F. Therefore, in general, the auxiliary motion trajectory $\mathcal{X}_a$, is found by

$$\mathcal{X}_a = C_F(\mathcal{F}_d - \mathcal{F}_m) = C_F(e_F). \tag{7.34}$$

Although advanced control scheme, such as a robust controller design may be used to determine the function C_F, to develop the idea and analyze the performance, let us examine simple proportional (P) and proportional-integrator (PI) controllers for the force feedback loop. The simplest possible linear controller to perform in force feedback loop is a pure gain controller. The derivative of force trajectory error is not advised for this loop because of two reasons. First, force measurements are usually contaminated with measurement noise, and derivative action may amplify the noise amplitude at the outset. Second, comparing force with motion variables, force is related to the acceleration of the motion, and hence, force measurement is inherently comparable to acceleration measurement. Therefore, in terms of measurement characteristics, there is no need to include the derivative of force-tracking error in the feedback loop. This would be comparable to motion jerk feedback, which is also not advised in the usual motion control schemes.

However, using the integrator controller is recommended, by which the steady-state performance of the force trajectory is regulated. Adding integrator action to the proportional controller assures that the steady-state error of the force trajectory converges to zero, and since force is comparable to the acceleration, force integral action is comparable to velocity feedback in the motion controller, which is recommended to suitably shape the force trajectory performance. Hence, for such cases, the controller function $C_F(e_F)$ may be derived in time domain from the following equation:

$$C_F(e_F) = K_p^{-1}\left[K_{F_p}e_F(t) + K_{F_i}\int_0^t e_F(\eta)\mathrm{d}\eta\right]. \tag{7.35}$$

Take Laplace transform of Equation 7.35 to represent the controller in s-domain by

$$C_F(e_F) = K_p^{-1}\left(K_{F_p} + \frac{1}{s}K_{F_i}\right)e_F(s), \tag{7.36}$$

in which K_{F_p} and K_{F_i} are the proportional and integrator controller gain matrices, respectively. The controller is multiplied by K_p^{-1}, where K_p is the motion controller proportional gain matrix to compensate for this gain in the forward path of the feedback control system, as illustrated in Figure 7.29. Note that this controller is simplified to a pure gain controller for the force feedback if it is assumed that $K_{F_i} = \mathbf{0}$.

Let us now write the closed-loop dynamic formulation for the manipulator in the absence of any disturbance wrench as follows:

$$\begin{aligned} M(\mathcal{X})\ddot{\mathcal{X}} + C(\mathcal{X},\dot{\mathcal{X}})\dot{\mathcal{X}} + G(\mathcal{X}) + \mathcal{F}_e &= \hat{M}(\mathcal{X})a + \mathcal{F}_{fl} \\ &= \hat{M}(\mathcal{X})\{K_p(\mathcal{X}_a - \mathcal{X}) - K_d\dot{\mathcal{X}}\} + \hat{C}(\mathcal{X},\dot{\mathcal{X}})\dot{\mathcal{X}} + \hat{G}(\mathcal{X}) + \mathcal{F}_m. \end{aligned}$$

If the knowledge of dynamic matrices is complete, then we may assume that $\hat{M} = M, \hat{C} = C$, and $\hat{G} = G$. Furthermore, if force measurements are noise free, then $\mathcal{F}_m = \mathcal{F}_e$. In such cases, simplify the closed-loop dynamic formulation to

$$M(\mathcal{X})\ddot{\mathcal{X}} = M(\mathcal{X})\{-K_d\dot{\mathcal{X}} - K_p\mathcal{X} + K_pC_F(e_F)\}. \tag{7.37}$$

Finally, substitute 7.35 into 7.37 and simplify to derive the closed-loop dynamic equation as follows:

$$\ddot{\mathcal{X}} + K_d\dot{\mathcal{X}} + K_p\mathcal{X} = K_{F_p}e_F(t) + K_{F_i}\int_0^t e_F(\eta)\mathrm{d}\eta. \tag{7.38}$$

This equation implies that if complete information of dynamic matrices is assumed, and force measurements are noise free, the closed-loop motion dynamics satisfy a set of second-order systems in the presence of proportional integrator action on the force error. Therefore, by choosing appropriate positive definite gains for the motion PD controllers, the transient performance of the motion can be shaped for a fast but stable motion. The motion suitably converges to zero, only if the force-tracking error becomes zero, otherwise, the manipulator continues its motion to converge the force trajectory error toward zero.

This can be viewed from a physical interpretation of the control structure depicted in Figure 7.29. As mentioned earlier, the motive for motion control in the inner loop of the control structure is the auxiliary motion variable $\mathcal{X}_a$, which is directly generated from the force error e_F. This motive is dynamically represented by the right-hand side of Equation 7.38. The inverse dynamics control structure used in the inner loop motion control guarantees a stable motion for the manipulator, as long as the motive $\mathcal{X}_a$ is not zero. Furthermore, this motive tends to zero only when the force error converges to zero. By this means, as long as the error force is not zero, the manipulator moves toward the interacting object or away from it to satisfy zero force tracking. Therefore, it can be concluded that the force tracking asymptotically converges to zero. Adding integral action in the force controller scheme satisfies zero steady-state error in the presence of input disturbance or measurement noise.

This control technique is very popular in practice, in cases where force tracking is the prime objective. This is because of the fact that this technique can significantly linearize and decouple the dynamic formulation of the closed-loop system for motion dynamics. Furthermore, the closed-loop motion dynamic terms are all configuration independent, and therefore, it is much easier to tune the controller gains to suitably perform within the whole workspace of the robot. However, note that for a good performance, an accurate model of the system is required, and the overall procedure is not robust to modeling uncertainty. Furthermore, in this technique, motion trajectory tracking is not directly considered as the secondary objective, since no direct desired motion trajectory is given in the closed-loop system. Hence, this topology is not advisable for the cases where the robot may loose its contact with the environment. In such cases, the interacting forces vanish and to reach the desired force trajectory very high acceleration is induced to the robot, which might be very dangerous in practice. Therefore, it is necessary to provide sufficient means to keep the robot in contact with the environment for the entire foreseen maneuvers.

7.4.1 Force Control of a Planar Manipulator

In this section, direct force control schemes of a planar $4R\underline{P}R$ manipulator has been studied in detail. The architecture of planar manipulator under this study is described in detail in Section 3.3.1 and is shown in Figure 3.3. The Jacobian and singularity analyses of this manipulator is presented in Section 4.6, and the dynamic analysis of the manipulator under constant mass assumption and variable mass assumption is given in Sections 5.3.1 and 5.3.2, respectively. Since the dynamic analyses of the system using these two approaches are quite similar, the controllers are being applied only to constant mass dynamic models. Different motion control structures are implemented on this manipulator. The performances of these controllers have been examined in detail in Section 6.8. Stiffness control of this manipulator is studied in Section 7.3.3. In what follows, the force control scheme developed in Section 7.4 is implemented on this manipulator.

The force control scheme applied to the system has the structure shown in Figure 7.29, in which the force feedback is considered in the outer loop, while position feedback is used in the inner loop. As shown in Figure 7.29, the force-tracking error is directly determined from force measurement by $e_F = \mathcal{F}_d - \mathcal{F}_m$ in the outer loop and the force controller is designed to satisfy the required force-tracking performance. However, direct motion-tracking objective is not assigned in this control scheme, and the desired motion trajectory $\mathcal{X}_d$ is absent. On the other hand, an auxiliary motion trajectory $\mathcal{X}_a$ is generated from the force control law and is used as the reference to the motion tracking. By this means, no prescribed motion trajectory is followed, while the force control scheme would advise a motion trajectory for the robot to ensure that the desired force tracking is satisfied.

The required wrench $\mathcal{F}$ in this force control scheme is based on inverse dynamics control and consists of three main parts. In the inner loop, the motion control scheme is based on a feedback linearization part in addition to a PD controller, while in the outer loop, a PI force controller is used to obtain a suitable force tracking. The control effort applied to the system is given in Equation 7.32, in which $\boldsymbol{a}$ is given in Equation 7.33, and $\mathcal{X}_a$ is given in Equation 7.34. Note that to generate these terms, the dynamic formulation of the robot and its kinematic and dynamic parameters are needed. In practice, exact information on dynamic matrices is not available, and therefore, a %50 perturbation in all kinematic and dynamic parameters is used in the following simulations.

As shown in Figure 7.30, the desired force trajectory in x and y directions is considered to start from zero initial conditions to 100 N final condition in 100 s. Furthermore, as schematically shown in Figure 7.10, it is considered that a flat surface is placed in the workspace of the robot such that the center of robot-moving platform experiences collision to it in a segment of its maneuver. The shape and position of the object is considered the same as that in stiffness control scheme. For simulations, it is considered that $\boldsymbol{x}_l = (75, 75)$ and the slope of the contact surface is $s_l = -1$. By this means and as seen in Figure 7.30, the moving platform collides with the object at about 46 s, and the contact is held until the end of the simulation time.

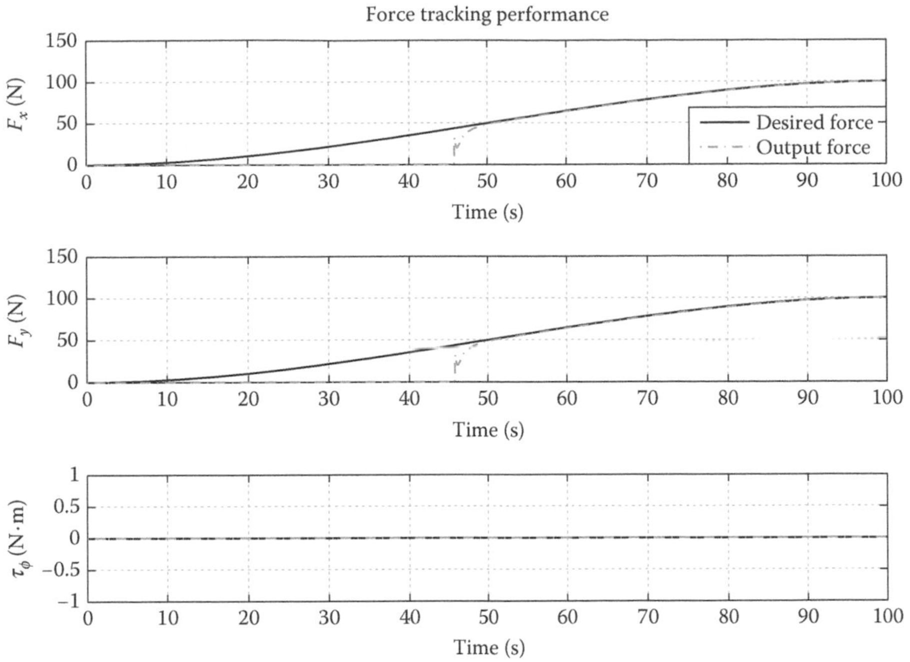

FIGURE 7.30
The desired and output force trajectories of the planar 4$R\underline{P}R$ manipulator colliding with a flat object; direct force control.

The stiffness of the object is modeled by Equation 7.26 in which the linear stiffness gain is considered as $\boldsymbol{K}_e = 10 \cdot \boldsymbol{I}_{3\times3}$(N/m), while the linear structural damping is set at $\boldsymbol{C}_e = 10 \cdot \boldsymbol{I}_{3\times3}$(N · m/s). The geometrical and inertial parameters used in the simulations are given in Table 5.1, and the initial conditions for the states are all set at zero. For the following simulations, the disturbance wrench applied to the moving platform is considered to be zero, while the gravity vector is considered as $\boldsymbol{g} = [0, 0, -9.81]^T$. The control effort applied to the system is given in Equation 7.32, in which, $\boldsymbol{a}$ is given in Equation 7.33, and $\boldsymbol{\mathcal{X}}_a$ is given in Equation 7.34. The force controller gains are set at $\boldsymbol{K}_{F_i} = 0.1 \cdot \boldsymbol{I}_{3\times3}$ and $\boldsymbol{K}_{F_p} = 0.05 \cdot \boldsymbol{I}_{3\times3}$, while the inner loop position controller gains are adjusted to $\boldsymbol{K}_p = 20 \cdot \boldsymbol{I}_{3\times3}$ and $\boldsymbol{K}_d = 5 \cdot \boldsymbol{I}_{3\times3}$.

The final force trajectory of the robot and the force-tracking errors are illustrated in Figures 7.30 and 7.31. As seen in Figure 7.30, although the desired force trajectory increases monotonically before collision, the output forces experienced by the moving platform are very small and not tracking the desired forces. At almost 46 s the collision occurs and the interaction forces rapidly converge to the desired force trajectory after a short transient. Greater details of the force-tracking performance is seen in Figure 7.31, in which it is observed that after collision, very rapid force convergence to the desired values is obtained. Note that the desired force in x and y directions is considered the same, and therefore, the final forces in these directions are almost the same. Furthermore, no interaction torque is considered in the simulations, and this is confirmed by zero torque in ϕ direction before and after the collision.

The motion trajectory of the robot is illustrated in Figure 7.32. In this figure, the actual motion of the moving platform is plotted in a solid line, while the auxiliary motion $\boldsymbol{\mathcal{X}}_a$ generated by force feedback law is plotted in a dash–dotted line. Note that in direct force

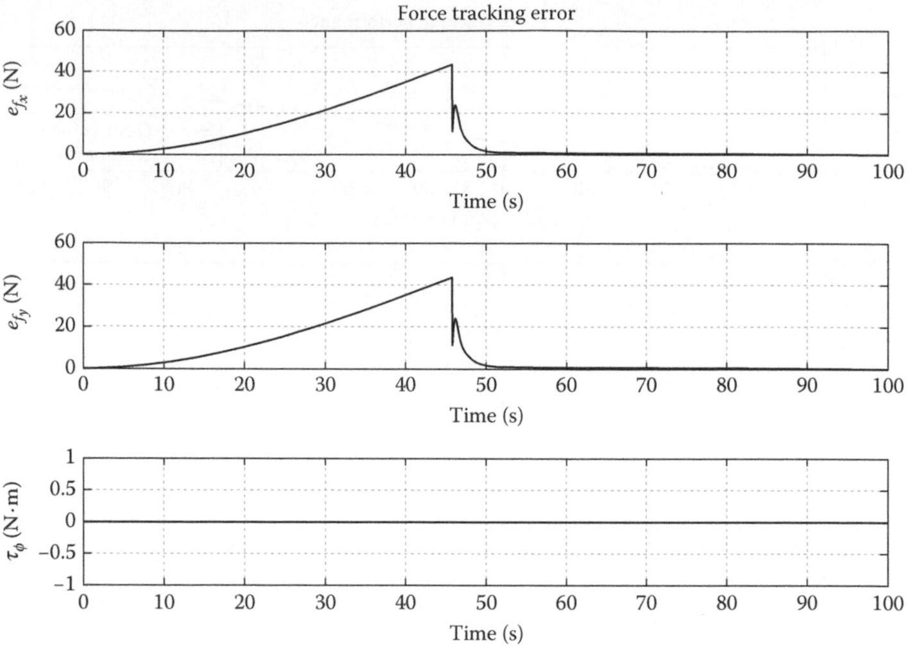

FIGURE 7.31
The closed-loop force tracking performance of the planar $4R\underline{P}R$ manipulator colliding with a flat object; direct force control.

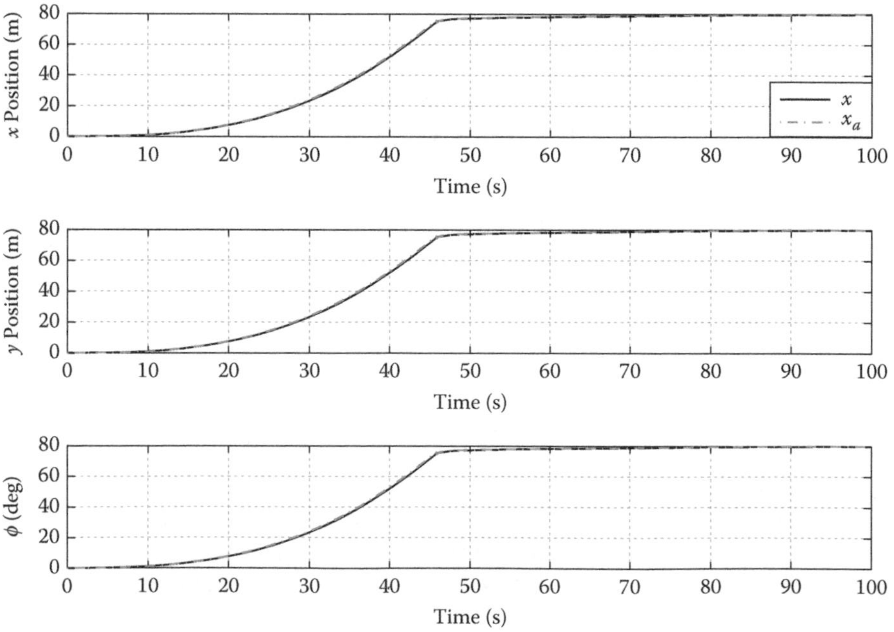

FIGURE 7.32
The motion trajectories of the planar $4R\underline{P}R$ manipulator colliding with a flat object; direct force control.

control technique, motion trajectory tracking is not directly considered as the secondary objective, since no direct desired motion trajectory is given in the closed-loop system. Therefore, in practice, the motion of the moving platform follows the auxiliary motion generated by the force feedback law. As seen in Figure 7.32, the moving platform perfectly follows the auxiliary motion variable $\mathcal{X}_a$, before and after collision. The interacting forces are small before collision, and force tracking error is relatively large. This causes the auxiliary motion variable to increase in x, y, and ϕ directions, before collision, and hence, the moving platform carries out relatively large motions. However, after collision, the force-tracking error converges rapidly to zero, and therefore, the motion of the manipulator is very small. This small motion causes a small deflection to the object such that the elastic forces are tuned to the required force command.

The Cartesian wrench $\mathcal{F} = [F_x, F_y, \tau_\phi]^T$ generated by the direct force controller is illustrated in Figure 7.33, in which the total force is plotted in a solid line while the PD term is plotted in a dash–dotted line. As seen in Figure 7.33, the Cartesian wrench increases monotonically before collision, causing relatively large motion for the manipulator. During the transient time of collision, high forces are experienced by the robot, and the force controller reacts to this transient by producing relatively high forces at the outset. This ensures that the interacting forces are rapidly tuned to the desired value, and then the actuator forces required to keep the interacting forces close to their desired values are relatively small.

Furthermore, it is observed that PD term is the major part of the control effort before collision, but relatively it is a small portion of that after collision. This is owing to the fact that the feedback-linearizing force $\mathcal{F}_{fl}$, given in Equation 7.32, includes the measurement force

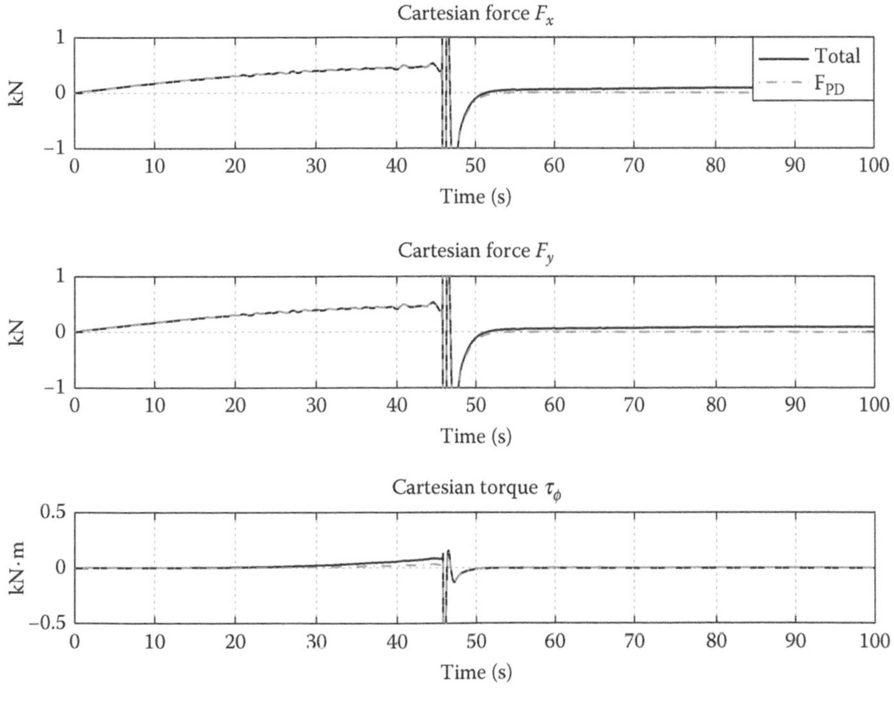

FIGURE 7.33
The Cartesian wrench of the planar $4R\underline{P}R$ manipulator colliding with a flat object; direct force control.

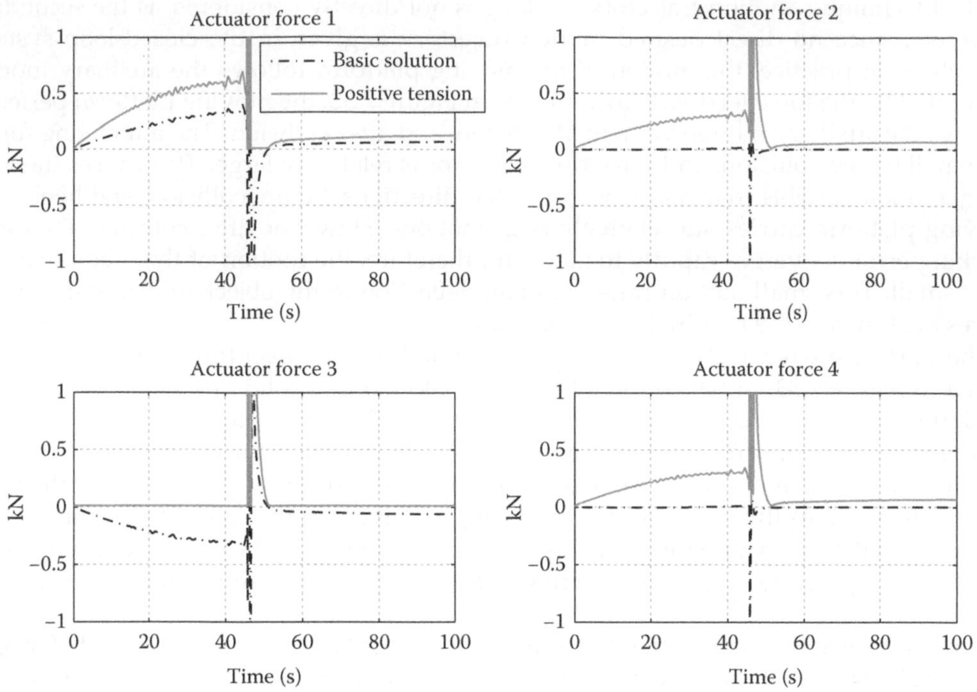

FIGURE 7.34
The actuator forces of the planar 4$R\underline{P}R$ manipulator colliding with a flat object; direct force control.

$\mathcal{F}_m$, where it rapidly dominates the controller force after collision. Similarly, the actuator forces of the manipulator are illustrated in Figure 7.34. In this figure, two sets of actuator forces are illustrated. The dash–dotted lines illustrate the base solution of the optimization problem introduced in redundancy resolution scheme by Equation 6.76, while the solid line is the final optimization solution of the redundancy resolution problem detailed in Equation 6.79. The redundancy resolution is performed through an iterative numerical method. As seen in Figure 7.53, the actuator forces increase significantly during the collision transient, while the tension forces derived from the redundancy resolution scheme remain positive at all times.

To examine the performance of the force controller in the presence of external disturbance and measurement noise, two more simulations are carried out. In the first simulation, a unit step disturbance force is applied to the system at time 75 s. At this time, 1 kN step disturbance force in x and y directions and 1 kN · m disturbance torque in ϕ direction are applied to the moving platform. The force controller gains are set at $\boldsymbol{K}_{F_i} = 0.1 \cdot \boldsymbol{I}_{3\times3}$ and $\boldsymbol{K}_{F_p} = 0.05 \cdot \boldsymbol{I}_{3\times3}$, while the inner loop position controller gains are adjusted to $\boldsymbol{K}_p = 20 \cdot \boldsymbol{I}_{3\times3}$ and $\boldsymbol{K}_d = 5 \cdot \boldsymbol{I}_{3\times3}$, as earlier.

The closed-loop force-tracking error of the system is shown in Figure 7.35. This figure is identical to Figure 7.31 where no disturbance was considered. This implies that the effect of external disturbance is completely rejected by this structure of the force controller. Although the disturbance forces and torques applied to the system are nonvanishing, static errors are completely compensated in the closed-loop response of the system. This is mainly because of the fact that force measurement is used in the feedback. Figure 7.36 illustrates the Cartesian wrench generated by the direct force controller in the presence

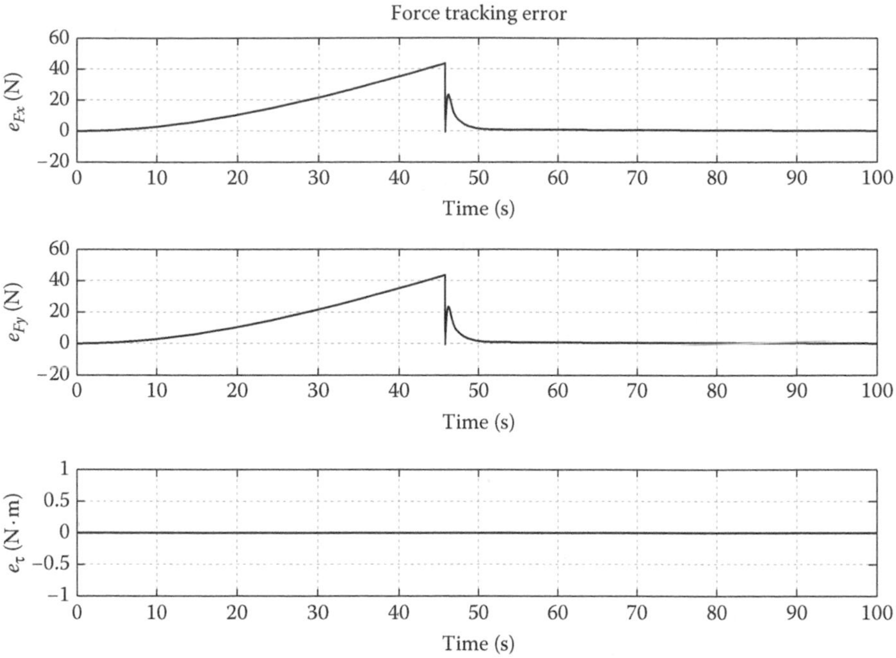

FIGURE 7.35
The closed-loop force-tracking performance of the planar $4R\underline{P}R$ manipulator colliding with a flat object; direct force control in the presence of disturbance.

of disturbance. By comparing Figure 7.36 with Figure 7.33, the effect of step disturbance wrench on the required controller forces is clearly seen at the steady-state values of the required control efforts.

The effect of measurement noise is simulated next by considering that all the force measurements are contaminated by a Gaussian noise with an amplitude of %0.1 peak values of the force signals. Although the amount of noise is very limited in this simulation, since relatively high-gain controller is used, the effect of noise is amplified in the required control effort. As shown in Figures 7.37 and 7.38, although the tracking error is identical to that of the system without noise, the actuator forces to carry out such an activity require high bandwidth. In fact, generating such actuator forces are infeasible in practice, and reaching the tracking performance shown in Figure 7.37 is not practically possible. In practice, appropriate filters have to be used in all force measurements to reduce the requirement of high-bandwidth actuators.

7.4.2 Force Control of the Stewart–Gough Platform

In this section, direct force control of the Stewart–Gough platform has been studied in detail. The architecture of this manipulator is described in detail in Section 3.5.1 and shown in Figure 3.16. As explained earlier, in this manipulator, the spatial motion of the moving platform is generated by six piston–cylinder actuators. Complete kinematic analysis of this manipulator are given in Section 3.5 and its Jacobian analysis is presented in Section 4.8. The dynamic analyses of this manipulator using Newton–Euler and virtual work

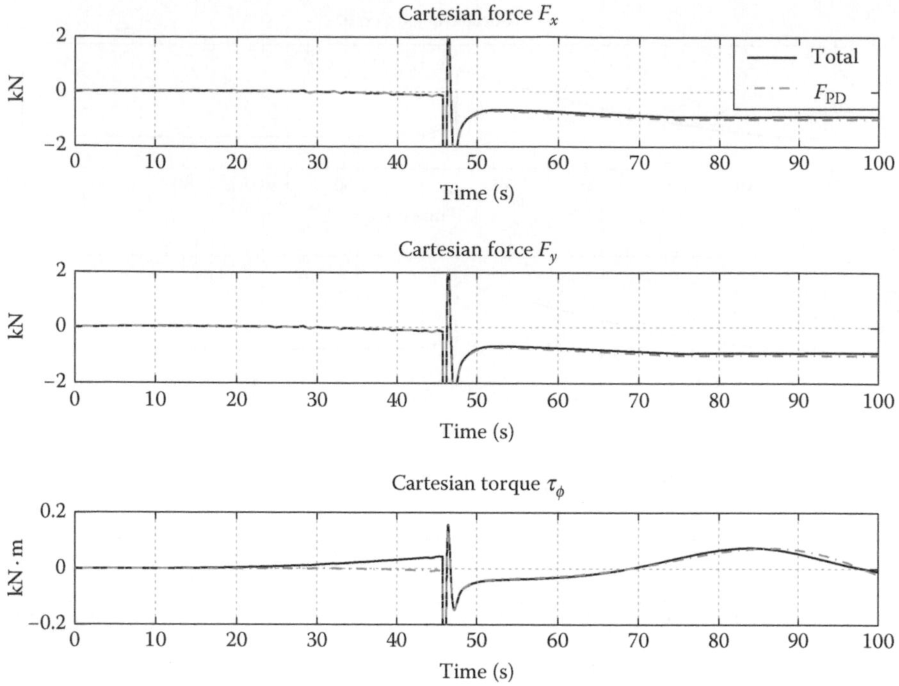

FIGURE 7.36
The Cartesian wrench of the planar $4R\underline{P}R$ manipulator colliding with a flat object; direct force control in the presence of disturbance.

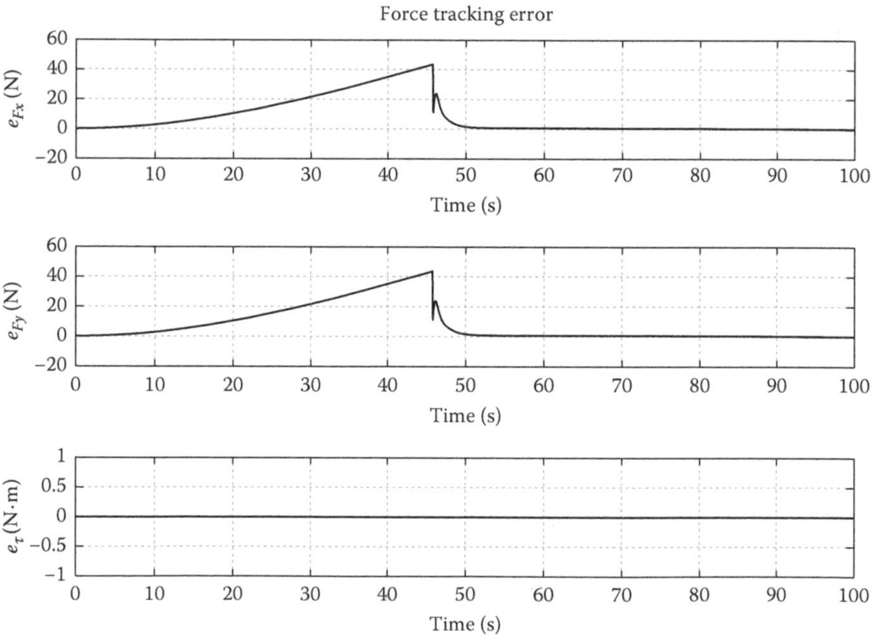

FIGURE 7.37
The closed-loop force-tracking performance of the planar $4R\underline{P}R$ manipulator colliding with a flat object; direct force control in the presence of measurement noise.

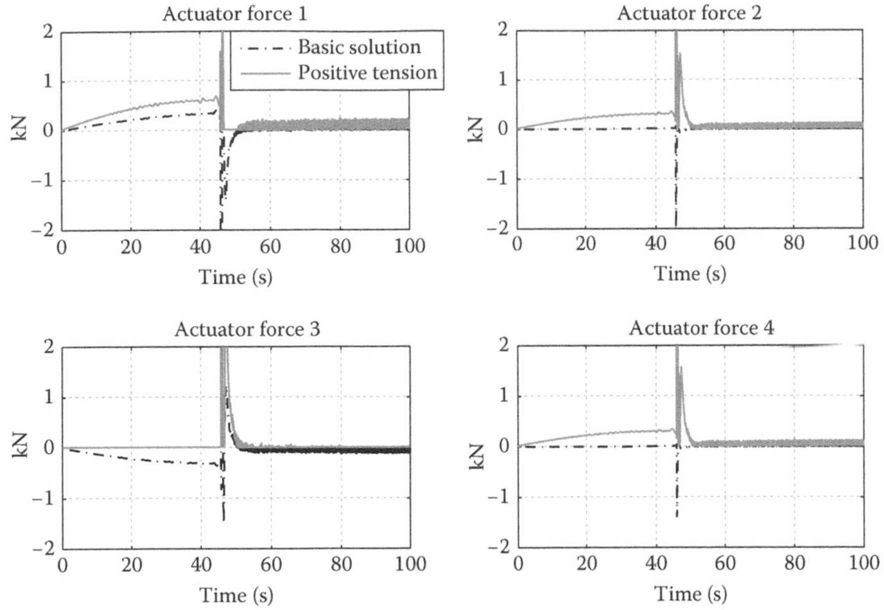

FIGURE 7.38
The actuator forces of the planar 4$R\underline{P}R$ manipulator colliding with a flat object; direct force control in the presence of measurement noise.

approaches are given in Sections 5.3.3 and 5.4.6, respectively. The closed-form dynamic formulation being used in this section for the simulations, is derived in Section 5.5.6. Different motion control strategies are implemented in Section 6.9. Stiffness control strategy applied on this manipulator is studied in details in Section 7.3.4. In what follows, direct force control strategy is implemented on this manipulator.

The force control scheme applied on the system has the structure shown in Figure 7.29, in which the force feedback is considered in the outer loop, while position feedback is used in the inner loop. As shown in Figure 7.29, the force-tracking error is directly determined from force measurement by $\boldsymbol{e}_F = \boldsymbol{\mathcal{F}}_d - \boldsymbol{\mathcal{F}}_m$ in the outer loop and the force controller is designed to satisfy the required force-tracking performance. However, direct motion-tracking objective is not assigned in this control scheme, and the desired motion trajectory $\boldsymbol{\mathcal{X}}_d$ is absent. On the other hand, an auxiliary motion trajectory $\boldsymbol{\mathcal{X}}_a$ is generated from the force control law and is used as the reference for the motion tracking. By this means, no prescribed motion trajectory is followed, while the force control scheme would advise a motion trajectory for the robot to ensure the desired force tracking.

The required wrench $\boldsymbol{\mathcal{F}}$ in this force control scheme is based on inverse dynamics control, and consists of three main parts. In the inner loop, the motion control scheme is based on a feedback linearization part in addition to a PD controller, while in the outer loop, a PI force controller is used to obtain a suitable force tracking. The control effort applied to the system is given in Equation 7.32, in which $\boldsymbol{a}$ is given in Equation 7.33 and $\boldsymbol{\mathcal{X}}_a$ is given in Equation 7.34. Note that to generate these terms, the dynamic formulation of the robot and its kinematic and dynamic parameters are needed. In practice, exact information on dynamic matrices is not available and therefore a %50 perturbation in all kinematic and inertial parameters is used in the following simulations.

As shown in Figure 7.39, the desired force trajectory in x, y, and z directions is considered to start from zero initial conditions to 500 N final condition in 1 s. Furthermore, as shown in Figure 7.19 schematically, it is considered that a flat surface is placed in the workspace of the robot such that the center of robot-moving platform experiences contact with a stiff environment in a segment of its maneuver. For the simulations, it is considered that $\boldsymbol{x}_l = [0.2, 0.2, 1.2]^T$ and the normal vector to the plane is $\boldsymbol{n}_l = [1,\ 1,\ 1]^T$. By this means, in the prescribed force trajectory shown in Figure 7.39, the moving platform collides with the flat object at about 0.85 s, while the contact is held until the end of simulation time. The stiffness of the object is modeled by linear stiffness gain $\boldsymbol{K}_e = 10^5 \cdot \boldsymbol{I}_{6\times6}$ (N/m), while considering a linear structural damping of $\boldsymbol{C}_e = 10^4 \cdot \boldsymbol{I}_{6\times6}$ (N · m/s) for it.

The geometrical and inertial parameters used in the simulations are given in Table 5.2, and the initial conditions for the states are all set at zero, except for the initial state of the moving platform height, which is set at $z_o = 1$ m. For the following simulations, the disturbance wrench applied to the moving platform is considered to be zero, while the gravity vector is considered as $\boldsymbol{g} = [0,\ 0,\ -9.81]^T$. The control effort applied to the system is given in Equation 7.32, in which $\boldsymbol{a}$ is given in Equation 7.33 and $\boldsymbol{\mathcal{X}}_a$ is given in Equation 7.34. The force controller gains are set at $\boldsymbol{K}_{F_i} = \boldsymbol{I}_{6\times6}$ and $\boldsymbol{K}_{F_p} = 0.01 \cdot \boldsymbol{I}_{6\times6}$, while the inner loop position controller gains are adjusted to $\boldsymbol{K}_p = 5000 \cdot \boldsymbol{I}_{6\times6}$ and $\boldsymbol{K}_d = 50 \cdot \boldsymbol{I}_{6\times6}$.

The final force trajectory of the robot and the force-tracking errors are illustrated in Figures 7.39 and 7.40. As seen in Figure 7.39, although the desired force trajectory increases monotonically prior to collision, the output forces experienced by the moving platform are very small and do not track the desired forces. At almost 0.085 s, the collision occurs and the interaction forces rapidly converge to the desired force trajectory after a short transient.

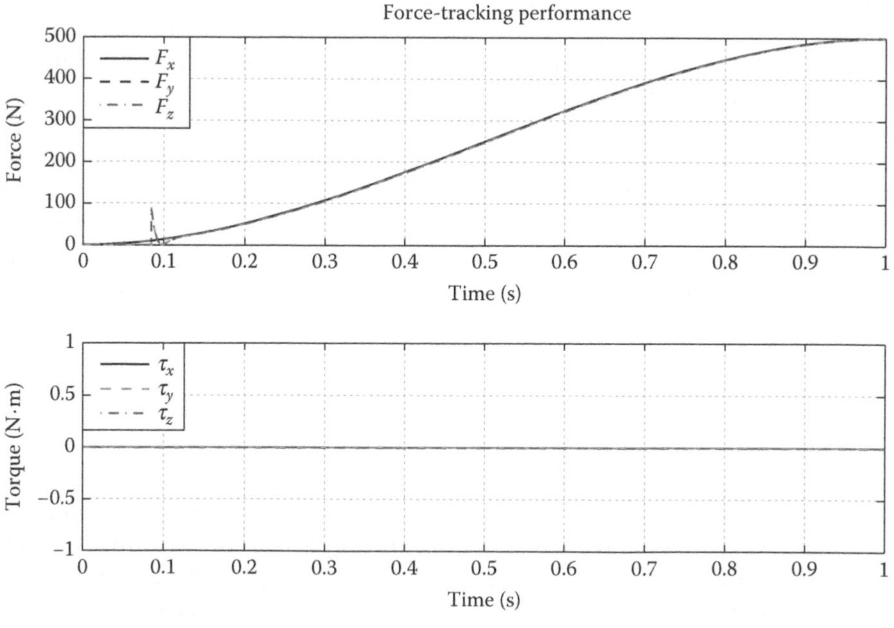

FIGURE 7.39
The desired and output force trajectories of the Stewart–Gough platform colliding with a flat object; direct force control.

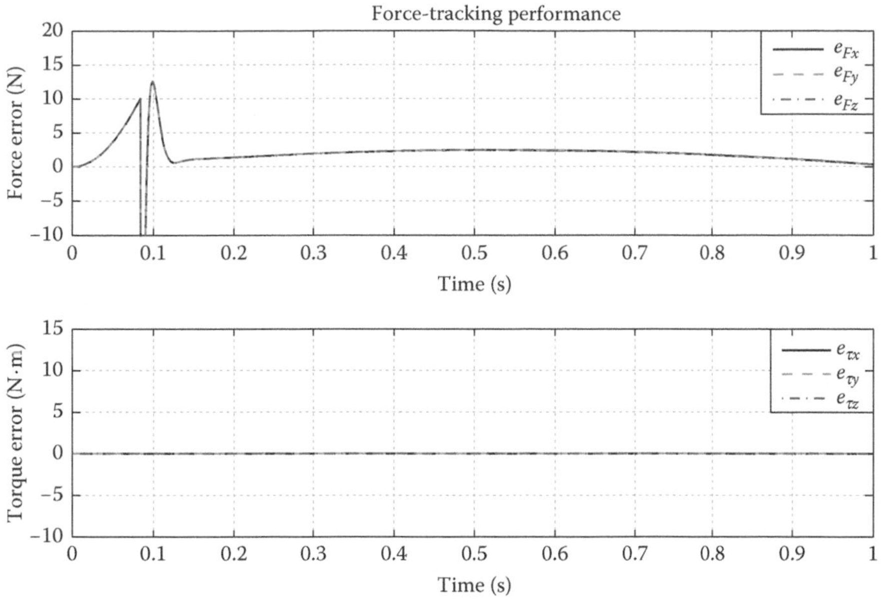

FIGURE 7.40
The closed-loop force-tracking performance of the Stewart–Gough platform colliding with a flat object; direct force control.

Greater details of the force-tracking performance are seen in Figure 7.40, in which, convergence of the actual forces to the desired values after collision is observed with a short transient. Note that the desired forces in x, y, and z directions are considered the same, and therefore, the final interacting forces in these directions are almost the same. Furthermore, no interaction torque is considered in the simulations, and this is confirmed by zero torques in all three directions before and after the collision.

The motion trajectory of the robot is illustrated in Figure 7.41. In this figure, the actual motion of the moving platform is plotted in a solid line, while the auxiliary motion variable $\mathcal{X}_a$ generated by force feedback law is plotted in a dash–dotted line. Note that in direct force control technique, motion trajectory tracking is not directly considered as the secondary objective, since no direct desired motion trajectory is given in the closed-loop system. Therefore, in practice, the motion of the moving platform follows the auxiliary motion generated by the force feedback law. As seen in Figure 7.41, the moving platform suitably tracks the auxiliary motion variable $\mathcal{X}_a$ after collision. The interacting forces are small before collision, and the force-tracking errors are relatively large. This causes the auxiliary motion variable to increase in x, y, and z directions before collision, and hence the moving platform carries out relatively rapid motions. However, after collision, the force-tracking error is relatively small, and therefore, the motion of the manipulator suitably tracks the auxiliary motion variable. This motion causes a small deflection to the object such that the elastic forces are tuned to the required force command.

The Cartesian wrench $\mathcal{F}$ generated by the direct force controller is illustrated in Figure 7.42, in which the total wrench is plotted in a solid line while the PD term is plotted in a dashed line. As seen in Figure 7.42, the Cartesian forces and torques increase monotonically prior to collision, causing a relatively rapid motion for the manipulator. During the transient time of collision, high forces are experienced by the robot, and the

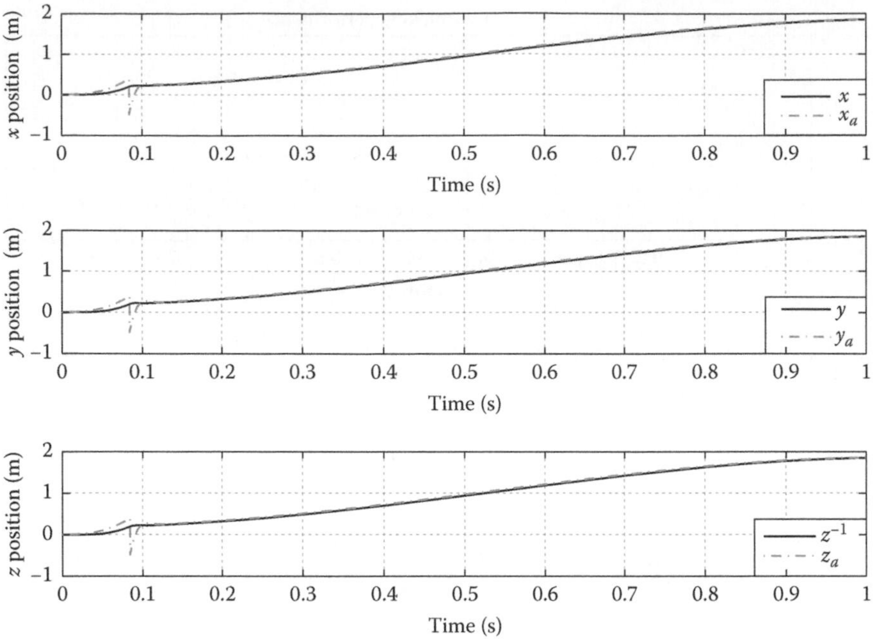

FIGURE 7.41
The motion trajectories of the Stewart–Gough platform colliding with a flat object; direct force control.

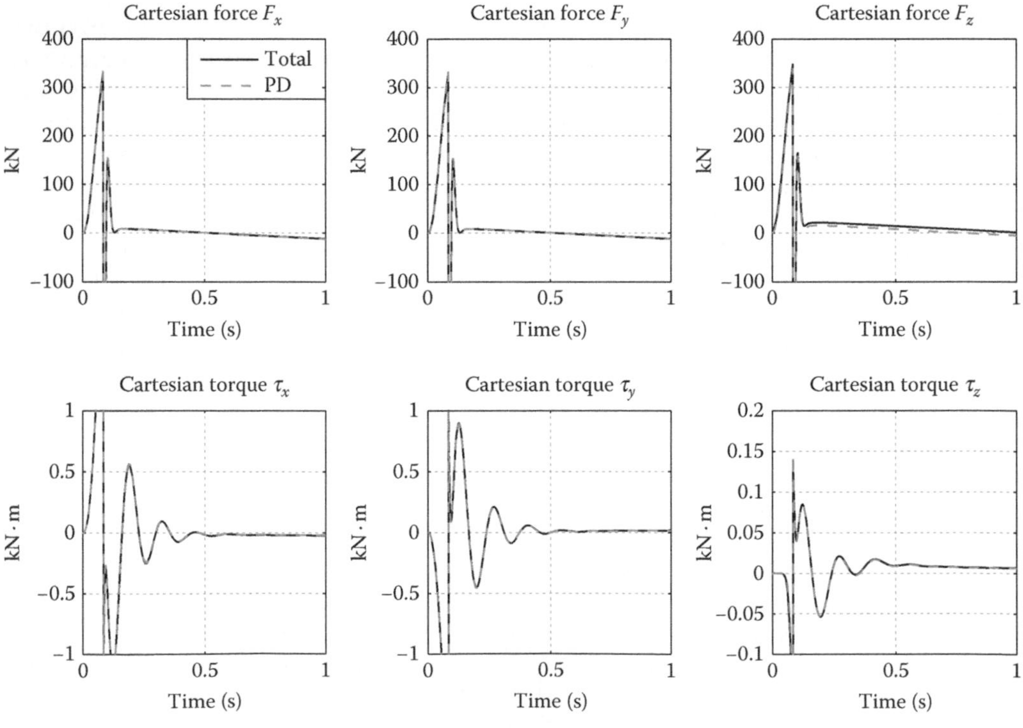

FIGURE 7.42
The Cartesian wrench of the Stewart–Gough platform colliding with a flat object; direct force control.

force controller reacts to this transient by producing relatively high forces at the outset. This ensures that the interacting forces are rapidly tuned to the desired value, and then the actuator forces required to keep the interacting forces close to their desired values are relatively small. Furthermore, it is observed that PD term is the major part of the control effort. Similarly, the actuator forces of the manipulator are illustrated in Figure 7.43. As seen in Figure 7.43, the actuator forces are relatively large before collision, and increase significantly during the collision transient. However, the actuator forces required to keep the interacting forces close to their desired values after collision are relatively small.

To examine the performance of the force controller in the presence of external disturbance and measurement noise, two more simulations are performed. In the first simulation, a step disturbance force is applied to the system at time 0.5 s. At this time, 100 kN step disturbance force in x, y, and z directions is applied to the moving platform. The force controller gains are set at $\boldsymbol{K}_{F_i} = \boldsymbol{I}_{6\times 6}$ and $\boldsymbol{K}_{F_p} = 0.01 \cdot \boldsymbol{I}_{6\times 6}$, while the inner loop position controller gains are adjusted to $\boldsymbol{K}_p = 5000 \cdot \boldsymbol{I}_{6\times 6}$ and $\boldsymbol{K}_d = 50 \cdot \boldsymbol{I}_{6\times 6}$ as before.

The closed-loop force tracking error of the system is shown in Figure 7.44. This figure is very similar to Figure 7.40, where no disturbance was considered, except for a small transient at 0.5 s. This implies that the effect of external disturbance is completely rejected by this structure of the force controller. Although the disturbance forces and torques applied to the system are nonvanishing, static errors are completely compensated in the closed-loop response of the system. This is mainly because of the fact that force measurement is used in the feedback. Figure 7.45 illustrates the Cartesian wrench generated by direct force controller in the presence of disturbance. By comparing Figure 7.45 with Figure 7.42, the effect of step disturbance forces on the required controller forces is clearly seen at the transient time of 0.5 s and at the steady-state values.

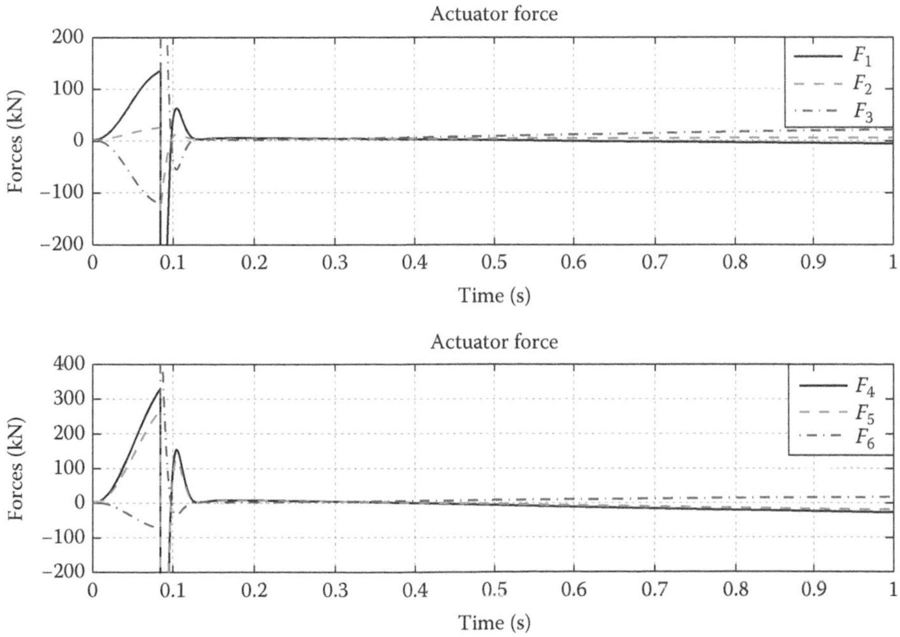

FIGURE 7.43
The actuator forces of the Stewart–Gough platform colliding with a flat object; direct force control.

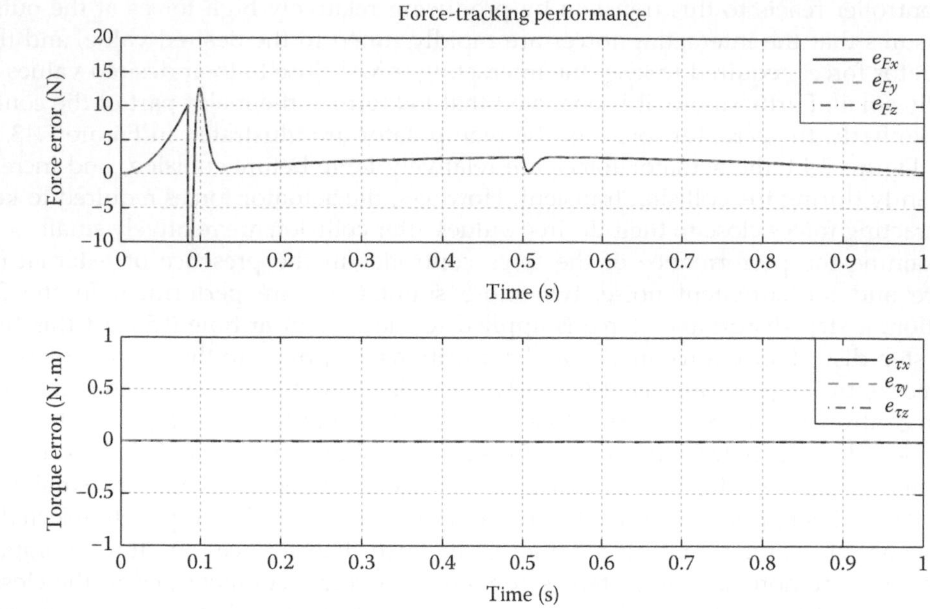

FIGURE 7.44
The closed-loop force-tracking performance of the Stewart–Gough platform colliding with a flat object; direct force control in the presence of disturbance.

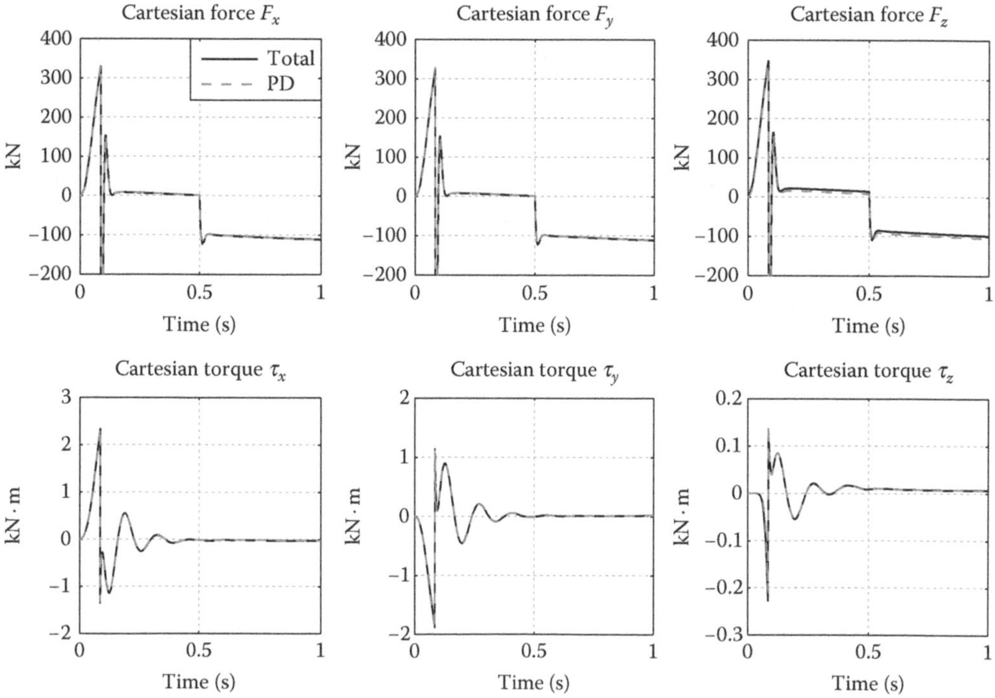

FIGURE 7.45
The Cartesian wrench of the Stewart–Gough platform colliding with a flat object; direct force control in the presence of disturbance.

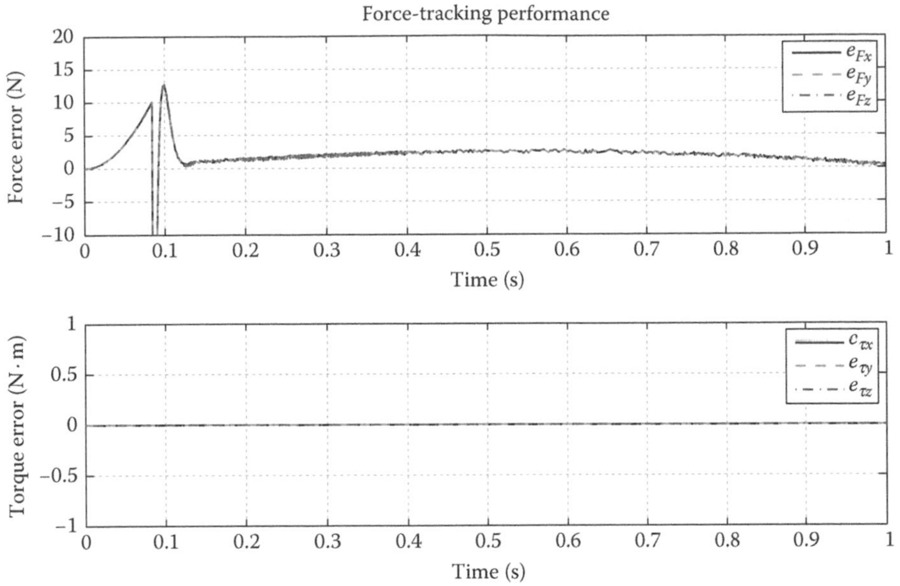

FIGURE 7.46
The closed-loop force-tracking performance of the Stewart–Gough platform colliding with a flat object; direct force control in the presence of measurement noise.

The effect of measurement noise is simulated next by considering that all the force measurements are contaminated by a Gaussian noise with an amplitude of %0.05 peak values of the force signals. Although the amount of noise is very limited in this simulation, since relatively high-gain controller is used, the effect of noise is amplified in the required control effort. As shown in Figures 7.46 and 7.47, although the tracking error is almost identical to that of the system without noise, the actuator forces require high bandwidth to carry out such an activity. In fact, generating such actuator forces are infeasible in practice, and reaching the tracking performance shown in Figure 7.46 is not practically possible. In practice, suitable filters have to be used in all force measurements to reduce the requirement of high-bandwidth actuators.

7.5 Impedance Control

In the previous two sections devoted to the stiffness control and direct force control schemes, it is observed that when the manipulator-moving platform is in contact with a stiff environment, the motion variable $\mathcal{X}$ and the interacting force variable $\mathcal{F}$ are two dynamically dependent quantities. In stiffness control, it is aimed to adjust the static relation between these two quantities. In this scheme, the general topology of the motion control scheme is considered, and no force measurement is required to perform this task; however, careful design on the desired motion trajectory and PD controller gains would be considered to tune the stiffness property of the interaction at steady state. In force control schemes, on the other hand, the force tracking is the prime objective, and force measurement is a stringent requirement to implement such schemes. Although in this

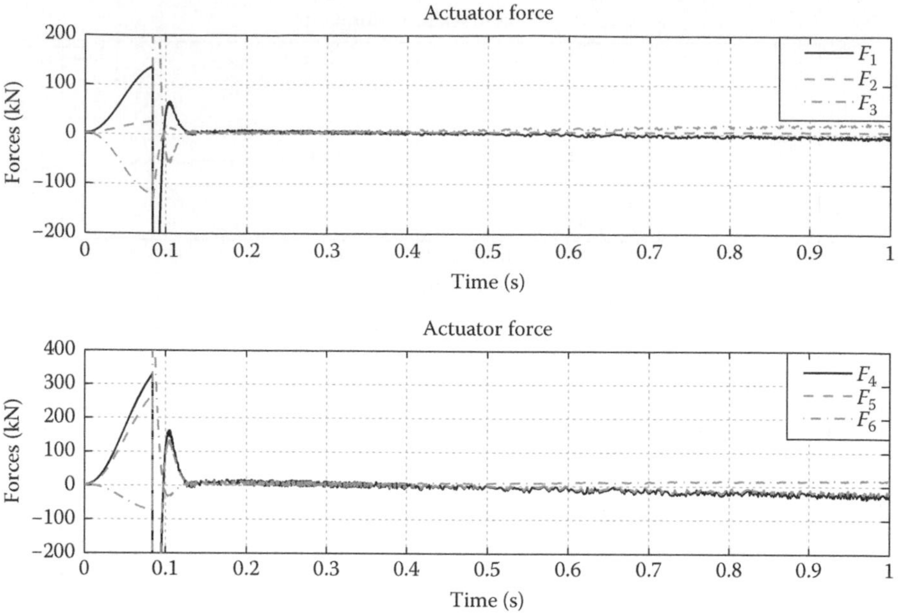

FIGURE 7.47
The actuator forces of the Stewart–Gough platform colliding with a flat object; direct force control in the presence of measurement noise.

control scheme, by including motion feedback in the inner loop of a cascade control topology, the motion control of the manipulator is considered as the secondary objective, careful design of the desired force trajectory and desired motion trajectory should be considered to avoid conflict of these two objectives in practice. The main reason that the motion and force variables are not being able to be controlled independently is that for an n-degrees-of-freedom manipulator, only n-independent control inputs are available, and therefore, only n-independent variables can be controlled, while the force and motion quantities count to $2n$ independent variables. Therefore, independent control of these quantities is not possible in practice.

The key idea behind *impedance control* schemes, developed in this section, is to tune the dynamic relation between the force and the motion variables, and not a hierarchy of tracking objectives in force and in position variables. In this scheme, contrary to stiffness control schemes, both force and position variables are measured and used in the control structure. Furthermore, it is aimed to tune the dynamic relation of the force and position quantities and not just their static relation at steady state.

The impedance control idea has been introduced by Hogan in 1985 [75,76]. The definition of mechanical impedance is given in an analogy of the well-known electrical impedance definition as the relationship between the effort and flow variables. Since this relation can be well determined in the frequency domain, the dynamical relation of force and motion variable may be represented by mechanical impedance. Impedance control schemes provide control topology to tune the mechanical impedance of a system to a desired value. By this means, the force and the motion variables are not controlled independently, or in a hierarchy, but their dynamic relation represented by mechanical impedance is suitably controlled. In what follows, first the definition of mechanical impedance and its characteristics is given. Then, some illustrative examples are given to develop the idea of impedance control, and finally, an impedance control scheme has been developed and analyzed for parallel manipulators.

7.5.1 Impedance

Impedance was first defined in electrical networks as the measure of the opposition that an electrical circuit presents to the passage of a current when a voltage is applied. To generalize the impedance definition in other disciplines, voltage is generalized to the effort in a network and current is generalized to the flow. In quantitative terms, impedance is a complex function defined as the ratio of the Laplace transform of the effort (voltage) to the Laplace transform of the flow (current). In fact, impedance extends the concept of static resistance of the circuits in steady-state analysis to a dynamic relation by a complex function that has both magnitude and phase. Furthermore, this definition reveals the relation between the effort and the flow in all other frequencies, and therefore may lead to transient response analysis unlike the steady-state characteristics of a resistance. Therefore, when an electric circuit is driven by direct current, there is no distinction between impedance and resistance.

Impedance is usually denoted by $\boldsymbol{Z}(s)$ and it may be represented by writing its magnitude and phase in the form of $|\boldsymbol{Z}|\angle\boldsymbol{Z}$. The magnitude of the complex impedance $|\boldsymbol{Z}|$ is the ratio of the effort amplitude to that of the flow, while the phase of the complex impedance $\angle\boldsymbol{Z}$ is the phase shift by which the flow is ahead of the effort. The reciprocal of impedance is called admittance, which represents the flow-to-effort ratio as a complex function of Laplace variable s.

Mechanical impedance is defined in analogy to electrical impedance as follows:

Definition 7.1

Mechanical impedance is defined as the ratio of the Laplace transform of the mechanical effort to the Laplace transform of the mechanical flow

$$\boldsymbol{Z}(s) = \frac{\boldsymbol{F}(s)}{\boldsymbol{v}(s)}, \tag{7.39}$$

in which effort in mechanical systems is represented by force $\boldsymbol{F}$ and flow is represented by velocity $\boldsymbol{v}$.

Note that this definition is given for a single-degree-of-freedom motion for force and motion. For such a case, the motion can be generalized to linear or angular motion, in which in angular motion the effort is represented by torque, while the flow is represented by angular velocity. Furthermore, the impedance may be generalized to multiple-degrees-of-freedom system, in which for a general spatial motion effort is represented by wrench $\boldsymbol{\mathcal{F}}$, while effort is represented by motion twist $\dot{\boldsymbol{\mathcal{X}}}$. Nevertheless, note that Laplace transform is only applicable for linear time invariant systems, and for a parallel manipulator the dynamic formulation of which is nonlinear, the concept of mechanical impedance may be extended to the differential equation relating the mechanical wrench $\boldsymbol{\mathcal{F}}$ to motion twist $\dot{\boldsymbol{\mathcal{X}}}$.

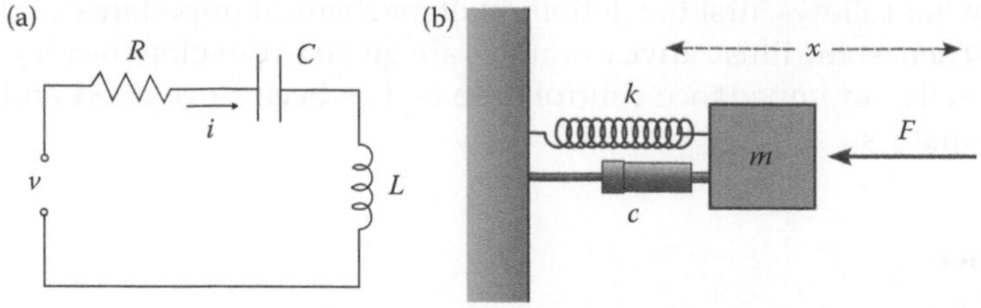

FIGURE 7.48
Analogy of electrical impedance in (a) an electrical RLC circuit to (b) a mechanical body–spring–damper system.

Example 7.1

Consider an RLC circuit depicted in Figure 7.48a. The differential equation relating voltage to the current in this circuit is given by

$$v = L\frac{di}{dt} + Ri + \int_0^t \frac{1}{C} i(\tau)d\tau, \tag{7.40}$$

in which L denotes the inductance, R denotes the resistance, C denotes the capacitance, v denotes the circuit voltage, and i denotes the current passing through the circuit. The impedance of the system may be found from the Laplace transform of Equation 7.40, by

$$Z(s) = \frac{v(s)}{i(s)} = Ls + R + \frac{1}{Cs}. \tag{7.41}$$

Example 7.2

Consider a body–spring–damper system depicted in Figure 7.48b. The governing dynamic formulation for this system is given by

$$m\ddot{x} + c\dot{x} + kx = f, \tag{7.42}$$

in which m denotes the body mass, c denotes the damper viscous coefficient, k denotes the spring stiffness, and f denotes the external force applied to the system. The impedance of the system may be found from the Laplace transform of Equation 7.42, by

$$Z(s) = \frac{f(s)}{v(s)} = ms + c + \frac{k}{s}. \tag{7.43}$$

As inferred from the above two examples, although the physical nature of the system may differ from each other, they may be represented by similar impedances. This observation may help electrical engineers to infer the mechanical impedance of a mechanical system intuitively through the analogy of electrical and mechanical components. From this analogy, a terminology for impedance classification is introduced as follows:

An impedance $\mathbf{Z}(s)$ is called

- *Inductive*, if and only if, its value at $s = 0$ is equal to zero, that is, $|\mathbf{Z}(0)| = 0$
- *Resistive*, if and only if, its value at $s = 0$ is equal to a finite and nonzero constant R, that is, $|\mathbf{Z}(0)| = R$
- *Capacitive*, if and only if, its value tends to infinity, as $s \to 0$, that is, $\lim_{s\to 0} |\mathbf{Z}(s)| = \infty$.

Hence, for the mechanical system represented in Example 7.2, mass represents inductive impedance, viscous friction coefficient represents resistive impedance, and spring stiffness represents capacitive impedance.

The environments that a robot interacts with may be represented by these classified impedance components. A very stiff environment the equivalent stiffness coefficient of which is relatively high may be represented by high-capacitive impedance models. On the other hand, an environment with high structural damping, may be categorized as high-resistive impedance. The force–motion relation of an operator hand holding a haptic device may be modeled by a combination of inductive, resistive, and capacitive impedances. Therefore, the impedance coefficients are different from one operator to the other, or for men-to-women operators. When the operator keeps the haptic device very firmly, the capacitive impedance may dominate the other two components.

A desired force–motion characteristic of a robot interacting with an environment therefore may be adjusted in a closed-loop by seeking a desired impedance model for the system. Two examples are given in the following section to obtain a better insight into impedance characteristics, and the way it may be used to control the force–motion relation.

7.5.2 Impedance Control Concept

As explained earlier, the key idea behind impedance control schemes is to tune the dynamic relation between force and motion variables. Impedance control schemes provide control topology to tune the mechanical impedance of a system toward a desired impedance. To understand the underlying concepts in this control strategy better, in this section, two intuitive examples, have been given to show how impedance control is different from direct position, stiffness or force control.

As the first example, consider using robots for an automated fish-skinning operation. In a fish factory, different fish products are produced, among which fish fillet is one of the most desired products for delicate consumers. Fish fillets are generally obtained by slicing raw fish parallel to the spine. After removing the bones from the fillets, the fish have to be skinned and cut into regular sizes. Fish-skinning operation is the most complex part for an automation system. To visualize the difficulties, consider how this operation is accomplished manually by an expert professional.

While the fish fillet is placed on the cutting board on its skinned side, a human angles a sharp knife very gently down the flesh such that the edge is toward the skin. By this means, the operator ensures that the skin is cut away and only little flesh is removed along with the skin. The operator holds the skin tightly and carefully slides the knife all the way down the fillet, using a gentle back-and-forth sawing motion. This should cut all the flesh from the skin. This sawing motion can easily be performed by an expert, while a nonexpert operator might tear the skin, or leave some flesh on the peeled skin.

The way an expert performs this task may be visualized by carefully sensing the force required on the motion trajectory of the knife. If the force becomes too loose while the motion is easier to carry out, it means that the knife is moved too close to the skin, and it is possible that the skin tears. On the other hand, if the sensed force increases while the motion is slowed down, this means that the knife is moving toward the flesh. An expert operator has been trained to control hand motion such that the ratio of force to velocity of the motion is kept at a desired level, while the knife is moving. By this means, the skin is peeled-off perfectly.

As it can be intuitively seen in this example, replacing an expert professional by a robot requires accurate measurement of both force and motion. A pure motion control scheme is not sufficient, since fish fillets have irregular shape, and the skin is too thin compared with the uncertainty in dimensions. Therefore, it is not possible to assign a predefined desired trajectory for the robot that can work for different fish fillets. Stiffness control or direct force control are more suitable than pure position control. However, a design of the predefined motion trajectory is still required for stiffness control as well as the secondary objective in the cascade control structure in direct force control, and this is not easily implementable. Nevertheless, if it is possible to suitably tune the operational impedance at a desired level, it is possible to carry out this task suitably. The operational impedance is composed from the robot impedance itself, which corresponds to the coupled nonlinear dynamics of the manipulator. However, a desired operational impedance could be adjusted by desired inductive, resistive, and capacitive impedances, which forms a desired linear impedance for the closed-loop system as follows:

$$\boldsymbol{Z}_d(s) = \boldsymbol{M}_d s + \boldsymbol{C}_d + \frac{1}{s}\boldsymbol{K}_d. \tag{7.44}$$

In Equation 7.44, $\boldsymbol{Z}_d$ denotes the desired impedance of the closed-loop system, which is composed of the desired inductive impedance $\boldsymbol{M}_d$, the desired resistive impedance $\boldsymbol{C}_d$, and finally, the desired capacitive impedance $\boldsymbol{K}_d$. Impedance control structures that are developed in the following section may be used to tune the closed-loop impedance of the system suitably to follow such a desired impedance.

As the second example, consider the control system of a flight simulator yoke. In a flight simulator, a number of subsystems are designed to generate an almost real flight feeling for the pilot. Flight simulators are used globally to train pilots for regular and occasional flight conditions. There exist four main subsystems to stimulate real flight conditions for the pilots. A sophisticated visual screen is designed to simulate the scenes that a pilot is watching during the flight. A Stewart–Gough platform is used to generate the accelerations that the pilot experiences during the flight, especially during take-off and landing. Exposing pilots to integrated visual and motion inputs, provide an almost real flight feeling for the pilots. The other two subsystems are the flight computer that simulates the aerodynamics, and the resulting motion, and finally, simulated flight utilities that exactly resemble the gauges and actuators existing in the cockpit. One of the utilities in the cockpit is the yoke which is usually a handle used to control the elevation and attitude of the aeroplane.

For the sake of simplicity, consider just one-degree-of-freedom yoke to control the aeroplane elevation. In practice, this is a vertical handle that is firmly kept by the pilot during the flight. Pulling the handle toward the body demands more elevation, while pushing it away, commands to reduce the flight height. In real aeroplanes, the yoke is connected to the elevons through hydraulic or mechanical components, and aerodynamic forces applied on the elevons are sensed by the pilot.

When the yoke is in its idle position, the elevon surfaces are along the aeroplane wing surface, and therefore, no elevation signals are commanded to the plane. By pushing or pulling the yoke, the elevon surfaces become inclined with respect to the wing surface, and as a result, elevation commands are applied to the aeroplane. On the other hand, the aerodynamic forces applied to the elevon is directly sensed by the pilot who firmly holds the yoke by his/her hand. The more the elevation command, the more the force sensed by the pilot. The actual relation between the yoke angle and the force sensed by the pilot can be determined from the aerodynamic relation applied on the wings and elevons; however, this relation can be simplified to a constant stiffness relation for usual operational elevation commands in a flight. These characteristics may be the first important feature to be considered in the control system design of a flight simulator yoke.

The second requirement for a flight simulator yoke is to resemble the behavior of a real yoke when the pilot releases the yoke during the operation. By releasing the yoke it returns to its idle position naturally, owing to the aerodynamic forces applied to the elevons. In other words, consider that the pilot is keeping the yoke at an angle $\theta \neq 0$ with respect to its idle position. By releasing the yoke, it moves toward the idle position $\theta = 0$ with a smooth and stable motion. No chattering or unwanted oscillation is seen in the motion of a real yoke in practice. Such desired motion may be represented by a second-order system time response whose damping ratio is well tuned to $\xi = 0.707$. For such systems, a very smooth and fast response is obtained with a small overshoot.

A third requirement may be considered in practice by which when the yoke is at its idle position and the pilots hit it accidentally, it remains stable, and no chattering or unwanted oscillation is seen in its motion. Finally, the last requirement is that the force–position relation needed for the flight simulator yoke is independent on how the pilot has held the yoke. In other words, this relation is close to the real yoke for men or women pilots, or for the case in which the pilot holds the yoke firmly or very freely. Technically speaking, the impedance of the operator hand does not change the required impedance characteristics of the flight simulator yoke.

Examine the behavior of different motion and force control schemes for this example. A pure motion controller may work well for the stability and regulation requirements mentioned as the second and third objectives. However, pure motion control scheme is unable to control the force, or the force–motion relation required as the first and main objective of this example. Stiffness control might be considered to be well suited for the first objective. However, since no force measurement is used in this structure, it is not suitable to satisfy the remaining three objectives. Pure force controller seems the most appropriate, since it may consider both force and motion objectives. However, the main objective of force–motion requirement, and its independence to human operator characteristics cannot be well satisfied through this control scheme.

It can be concluded that if an impedance-based controller can be advised by which the operational impedance of the closed-loop system is tuned to a desired value, it is possible to suitably satisfy all the four above-mentioned objectives. A desired operational impedance may be considered by desired resistive and capacitive impedances, which forms a desired linear impedance for the closed-loop system as follows:

$$Z_d(s) = m\,s + c_d + \frac{1}{s}k_d. \tag{7.45}$$

In Equation 7.45, Z_d denotes the desired impedance of the closed-loop system, which is composed of the system-inherent inductive impedance m, the desired resistive impedance c_d, and the desired capacitive impedance k_d, such that the smooth and stable response with $\xi = 0.707$ may be obtained at the outset. Impedance control structures that are developed in the following section may be used to tune the closed-loop impedance of the system suitably to follow such a desired impedance.

7.5.3 Impedance Control Structure

Consider a parallel robot with multiple-degrees-of-freedom interacting with a stiff environment. In such a case, the motion of the robot end effector is represented by the motion vector $\mathcal{X}$, and the interacting wrench applied to the robot end effector is denoted by $\mathcal{F}_e$. It is considered that the interacting force is measured in the task space and is used in the inner force feedback loop. Furthermore, it is considered that the motion variable $\mathcal{X}$ is measured and is used in the outer feedback loop. The general dynamic formulation of parallel robots in the absence of any disturbance wrench is given in Equation 7.30. In the impedance control scheme, regulation of the motion–force dynamic relation is the prime objective, and since force tracking is not the primary objective it is advised to use a cascade control structure with motion control feedback in the outer loop and force feedback in the inner loop. As explained earlier, Figure 7.5 may be considered as a representative of such a cascade control topology for the system in which the measured variables are both in the task space. In this topology, the prime controller is a motion controller and the second controller is a force-tracking controller. Therefore, when the manipulator is not in contact with a stiff environment, position tracking is guaranteed by a primary controller. However, when there is an interacting wrench $\mathcal{F}_e$ applied to the moving platform, this structure may be designed to control the force–motion dynamic relation.

As a possible impedance control scheme, consider the closed-loop system depicted in Figure 7.49, in which the position feedback is considered in the outer loop, while force feedback is used in the inner loop. This structure is advised when a desired impedance relation between the force and motion variables is required that consists of desired inductive, resistive, and capacitive impedances, as given in Equation 7.44. As shown in Figure 7.49, the motion-tracking error is directly determined from motion measurement by $e_x = \mathcal{X}_d - \mathcal{X}$ in the outer loop and the motion controller is designed to satisfy the required impedance structure depicted in Equation 7.44. Moreover, direct force-tracking objective is not assigned in this control scheme, and therefore the desired force trajectory $\mathcal{F}_d$ is absent in this scheme. However, an auxiliary force trajectory $\mathcal{F}_a$ is generated from the motion control law and is used as the reference for the force tracking. By this means, no prescribed

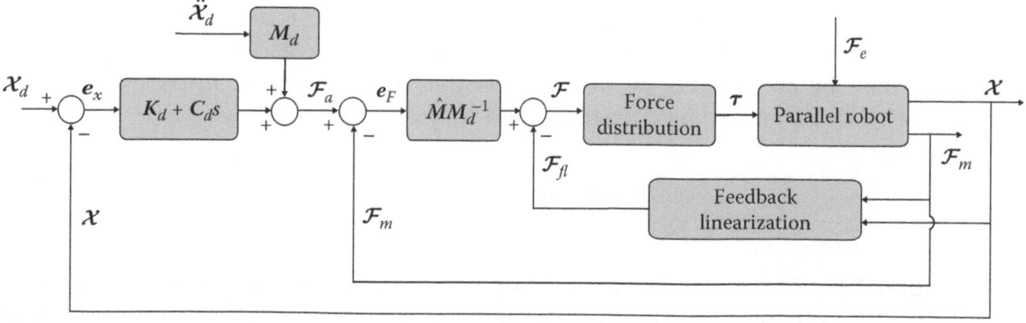

FIGURE 7.49
Impedance control scheme; motion feedback in the outer loop and force feedback in the inner loop.

force trajectory is tracked, while the motion control scheme would advise a force trajectory for the robot to ensure the desired impedance regulation.

The required wrench $\mathcal{F}$ in the impedance control scheme, depicted in Figure 7.49, is based on inverse dynamics control and consists of three main parts. In the inner loop, the force control scheme is based on a feedback linearization part in addition to a mass matrix adjustment, while in the outer loop usually a linear motion controller is considered based on the desired impedance requirement such as in Equation 7.44. Although, many different impedance structures may be considered as the basis of the control law, in Figure 7.49, a linear impedance relation between the force and motion variables is generated that consists of desired inductive $\boldsymbol{M}_d$, resistive $\boldsymbol{C}_d$, and capacitive impedances $\boldsymbol{K}_d$, as given in Equation 7.44.

As shown in Figure 7.49, the feedback control law generates the required wrench $\mathcal{F}$ to be applied by the actuators $\boldsymbol{\tau}$ in the task space. To implement the control effort $\boldsymbol{\tau}$, though, it is required to use a suitable force distribution scheme in practice. As explained earlier, the force distribution block projects the required wrench in the task space $\mathcal{F}$ to its corresponding actuator force vector $\boldsymbol{\tau}$. Note that for a completely parallel manipulator such as the Stewart–Gough platform, in which the number of actuators are equal to the number of degrees-of-freedom, this mapping can be constructed from the Jacobian transpose of the manipulator $\boldsymbol{J}^T$ by $\boldsymbol{\tau} = \boldsymbol{J}^{-T}\mathcal{F}$, at nonsingular configurations. However, the force distribution scheme may need more computations if the manipulator is not completely parallel. Redundancy resolution techniques, which are completely elaborated in Section 6.7, would be used to physically implement force distribution scheme for redundant manipulators.

According to Figure 7.49, the controller output wrench $\mathcal{F}$, applied to the manipulator may be formulated as

$$\mathcal{F} = \hat{\boldsymbol{M}}\boldsymbol{M}_d^{-1} \cdot \boldsymbol{e}_F + \mathcal{F}_{fl}, \tag{7.46}$$

in which

$$\boldsymbol{e}_F = \mathcal{F}_a - \mathcal{F}_m \tag{7.47}$$

$$\mathcal{F}_a = \boldsymbol{M}_d \ddot{\mathcal{X}}_d + \boldsymbol{C}_d \dot{\boldsymbol{e}}_x + \boldsymbol{K}_d \boldsymbol{e}_x, \tag{7.48}$$

and as before, the feedback linearizing term is given by

$$\mathcal{F}_{fl} = \hat{\boldsymbol{C}}(\mathcal{X}, \dot{\mathcal{X}})\dot{\mathcal{X}} + \hat{\boldsymbol{G}}(\mathcal{X}) + \mathcal{F}_m. \tag{7.49}$$

In Equation 7.48, $\boldsymbol{M}_d$ denotes the desired inductive impedance, $\boldsymbol{C}_d$ denotes the desired resistive impedance, and $\boldsymbol{K}_d$ denotes the desired capacitive impedance matrices. Let us now write the closed-loop dynamic formulation for the manipulator in the absence of any disturbance wrench as follows:

$$\begin{aligned}&\boldsymbol{M}(\mathcal{X})\ddot{\mathcal{X}} + \boldsymbol{C}(\mathcal{X}, \dot{\mathcal{X}})\dot{\mathcal{X}} + \boldsymbol{G}(\mathcal{X}) + \mathcal{F}_e \\ &= \hat{\boldsymbol{M}}\boldsymbol{M}_d^{-1}\left\{\boldsymbol{M}_d \ddot{\mathcal{X}}_d + \boldsymbol{C}_d \dot{\boldsymbol{e}}_x + \boldsymbol{K}_d \boldsymbol{e}_x - \mathcal{F}_m\right\} + \hat{\boldsymbol{C}}(\mathcal{X}, \dot{\mathcal{X}})\dot{\mathcal{X}} + \hat{\boldsymbol{G}}(\mathcal{X}) + \mathcal{F}_m.\end{aligned}$$

If the information on the dynamic matrices is complete, then we may assume that $\hat{\boldsymbol{M}} = \boldsymbol{M}, \hat{\boldsymbol{C}} = \boldsymbol{C}$, and $\hat{\boldsymbol{G}} = \boldsymbol{G}$. Furthermore, if force measurements are noise free, then $\mathcal{F}_m = \mathcal{F}_e$. In such a case, simplify the closed-loop dynamic formulation to

$$\boldsymbol{M}\boldsymbol{M}_d^{-1}(\boldsymbol{M}_d \ddot{\mathcal{X}}) = \boldsymbol{M}\boldsymbol{M}_d^{-1}\left\{\boldsymbol{M}_d \ddot{\mathcal{X}}_d + \boldsymbol{C}_d \dot{\boldsymbol{e}}_x + \boldsymbol{K}_d \boldsymbol{e}_x - \mathcal{F}_e\right\}. \tag{7.50}$$

Note that the manipulator mass matrix $M(\mathcal{X})$ is a positive definite matrix, and by choosing positive definite matrix M_d, the concluding matrix MM_d^{-1} can be definitely determined at all configurations, and is nonsingular. Hence, the closed-loop dynamic formulation is simplified to

$$M_d\ddot{e}_x + C_d\dot{e}_x + K_d e_x = \mathcal{F}_e. \tag{7.51}$$

This equation implies that if there exists complete information on the dynamic matrices, and the force measurements are noise free, the closed-loop error dynamic equation satisfies a set of second-order systems with the desired impedance coefficients in relation to the interacting force. In other words, the control structure guarantees that the force and motion relation follows a desired impedance structure given in Equation 7.44. Therefore, by choosing appropriate positive definite gains for the impedance matrices, the transient performance of the force–motion relation can be shaped so as to have a fast but stable interaction. By this means, what is controlled is the dynamic relation between the force and motion variables, and not direct position and motion tracking. However, if the robot is moving freely in space and has no interaction with the environment, $\mathcal{F}_e = \mathbf{0}$, the closed-loop system will provide a suitable motion tracking. This is because of using a motion controller in the outer loop.

Another physical interpretation of impedance control scheme may be constructed by simplifying the impedance requirement as following. Consider that there is no need to adjust the inductive impedance matrix of the closed-loop system, and the inherent mass matrix of the robot is suitable and is not needed to be adjusted. In such a case, set $M_d = M$, and therefore, $MM_d^{-1} = I$, where I denotes the identity matrix. This amendment simplifies the control structure significantly, and resembles it to a stiffness control scheme. Reconsider the control structure given in Figure 7.49, and note that in this case the main force controller block MM_d^{-1} may be removed, since this block is simplified to the identity matrix. Now consider that the measured wrench $\mathcal{F}_m$, is being directly subtracted from the feedback linearizing wrench $\mathcal{F}_{fl}$. The subtraction of these two terms yields

$$\mathcal{F}_{fl} - \mathcal{F}_m = \hat{C}(\mathcal{X}, \dot{\mathcal{X}})\dot{\mathcal{X}} + \hat{G}(\mathcal{X}). \tag{7.52}$$

This implies that to generate the required wrench in this case, no force measurement is required. Therefore, no internal force feedback loop is needed to implement this controller, while the feedback linearizing term is simplified to the left-hand side of Equation 7.52. By these modifications, the impedance control structure depicted in Figure 7.49 is simplified to a pure motion inverse dynamics controller depicted in Figure 6.7, which is the basis of the stiffness control scheme. This observation reveals that the impedance controller is a generalization of the stiffness control scheme, through which higher-order dynamic terms are suitably tuned. On the other hand, if only the resistive and capacitive impedance coefficients are considered in the force–motion relation, the dynamic relation of force–motion is simplified to stiffness and damping terms.

The impedance control scheme is very popular in practice, wherein tuning the force and motion relation in a robot manipulator interacting with a stiff environment is the prime objective. This scheme can be suitably implemented for the examples given in Section 7.5.2. However, note that for a good performance, an accurate model of the system is required, and the obtained force and motion dynamics are not robust to modeling uncertainty.

7.5.4 Impedance Control of a Planar Manipulator

In this section, impedance control of a planar $4R\underline{P}R$ manipulator has been studied in detail. The architecture of the planar manipulator under study is described in detail in Section 3.3.1 and shown in Figure 3.3. The Jacobian and singularity analyses of this manipulator are presented in Section 4.6, and the dynamic analysis of the manipulator under constant mass assumption and variable mass assumption is given in Sections 5.3.1 and 5.3.2, respectively. Since the dynamic analyses of the system using these two approaches are quite similar, the controllers are being applied only on constant mass dynamic models. Different motion control structures are implemented on this manipulator. The performance of these controllers has been examined in detail in Section 6.8. Stiffness control and direct force control of this manipulator is studied in Sections 7.3.3 and 7.4.1, respectively. In what follows, the impedance control strategy is implemented on this manipulator.

The impedance control strategy applied on the system has the structure shown in Figure 7.49, in which position feedback is considered in the outer loop, while force feedback is used in the inner loop. As shown in Figure 7.49, the motion-tracking error is directly determined from motion measurement by $e_x = \mathcal{X}_d - \mathcal{X}$ in the outer loop and the motion controller is designed to satisfy the required impedance structure depicted in Equation 7.44. The required wrench $\mathcal{F}$ in impedance control scheme is based on inverse dynamics control, and consists of three main parts. In the inner loop, the force control scheme is based on a feedback linearization part in addition to a mass matrix adjustment, while in the outer loop usually a linear motion controller is considered based on the desired impedance requirement such as that in Equation 7.44.

A linear impedance relation between the force and motion variables is generated through this structure that consists of a desired inductive $\boldsymbol{M}_d$, resistive C_d, and capacitive $\boldsymbol{K}_d$ impedance matrices as given in Equation 7.44. The control effort applied to the system is given in Equation 7.46, in which $e_F, \mathcal{F}_a$, and $\mathcal{F}_{fl}$ are given in Equations 7.47 through 7.49, respectively. Note that to generate these terms, the dynamic formulation of the robot and its kinematic and dynamic parameters are needed. In practice, exact information on dynamic matrices is not available, and therefore, an exaggerated %50 perturbation in all kinematic and inertial parameters is used in the following simulations.

As schematically shown in Figure 7.10, it is considered that a flat surface is placed in the workspace of the robot such that the robot moving platform experiences contact with a stiff environment in a segment of its maneuver. As shown in Figure 7.50, the desired trajectory in the following simulations is considered the same as in the stiffness control scheme depicted in Figure 7.11, while the position of the object is also considered the same as that in the stiffness control scheme. Thus, for the simulations, it is considered that $\boldsymbol{x}_l = (75, 75)$ and the slope of the contact surface is $s_l = -1$. By this means, in the prescribed trajectory shown in Figure 7.50, the moving platform collides with the flat object at about 67 s, and the contact is held until the end of the simulation time.

The stiffness of the object is modeled by Equation 7.29, in which the linear stiffness gain is considered as $\boldsymbol{K}_e = 10^3 \cdot \boldsymbol{I}_{3\times3}$ (N/m), while the linear structural damping is set at $C_e = 10^3 \cdot \boldsymbol{I}_{3\times3}$ (N · m/s). The geometric and inertial parameters used in the simulations are given in Table 5.1, and the initial conditions for the states are all set at zero. For the following simulations, the disturbance wrench applied to the moving platform is considered to be zero, while the gravity vector is considered as $g = [0, 0, -9.81]^T$. The control effort is calculated from Equation 7.46, in which $e_F, \mathcal{F}_a$, and $\mathcal{F}_{fl}$ are given

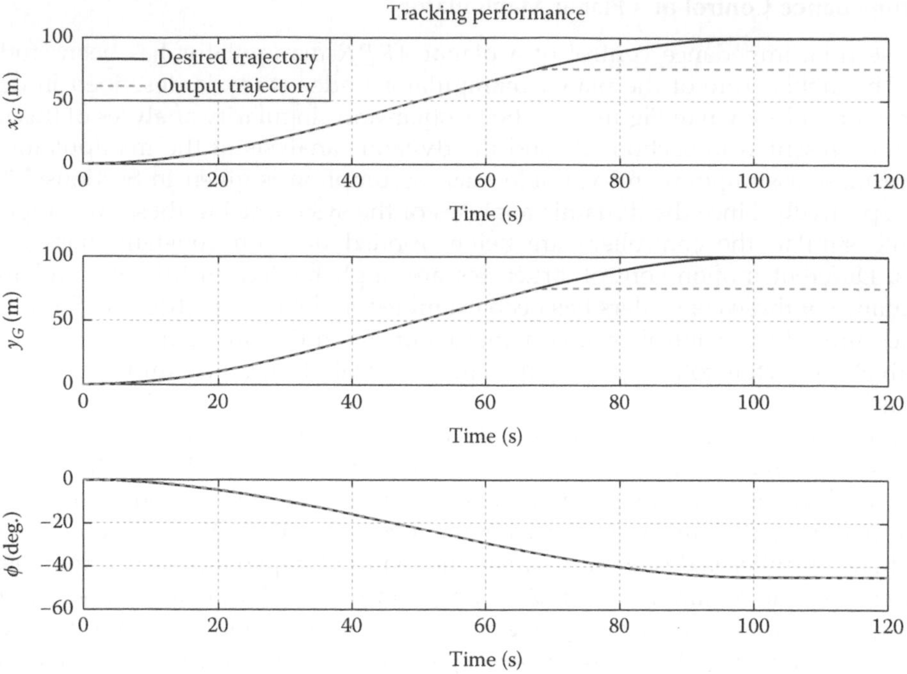

FIGURE 7.50
The desired and output trajectories of the planar $4R\underline{P}R$ manipulator colliding with a flat object; impedance control.

in Equations 7.47 through 7.49, respectively. The desired impedance matrices are set at $\boldsymbol{M}_d = 100 \cdot \boldsymbol{I}_{3\times3}, \boldsymbol{C}_d = 20 \cdot \boldsymbol{I}_{3\times3}$, and $\boldsymbol{K}_d = 100 \cdot \boldsymbol{I}_{3\times3}$.

The final motion of the moving platform and the tracking errors are illustrated in Figures 7.50 and 7.51. As seen in Figure 7.50, the desired and final closed-loop motion of the system are almost identical prior to collision. At almost 67 s, the collision occurs, and the motion of the moving platform remains almost constant on the surface of the object. Greater details of the moving platform motion at this time are given in Figure 7.51, in which it is observed that after collision a smooth motion of the moving platform on the object surface is produced. The moving platform causes minor deflection to the object. Furthermore, note that no interaction torque is considered in the simulations, and this is confirmed by relatively uniform tracking error in the orientation ϕ, prior to and after collision.

The Cartesian wrench $\mathcal{F} = [F_x, F_y, \tau_\phi]^T$ generated by the impedance controller is illustrated in Figure 7.52, in which the total force is plotted in a solid line while the feedback linearizing term is plotted in a dash–dotted line. As seen in Figure 7.52, the Cartesian forces in x and y directions rapidly increase after a transient at collision time, while no interaction torques are observed in ϕ direction. Furthermore, it is observed that the feedback linearizing term produces only a small portion of the total force before collision, while after collision the interaction force $\mathcal{F}_e$ increases rapidly and this immediately affects the feedback linearizing term. This is due to the fact that the feedback linearizing wrench $\mathcal{F}_{fl}$, given in Equation 7.49, includes the measurement force $\mathcal{F}_m$.

Similarly, two sets of actuator forces of the manipulator are illustrated in Figure 7.53. The dash–dotted lines illustrate the base solution of the optimization problem introduced

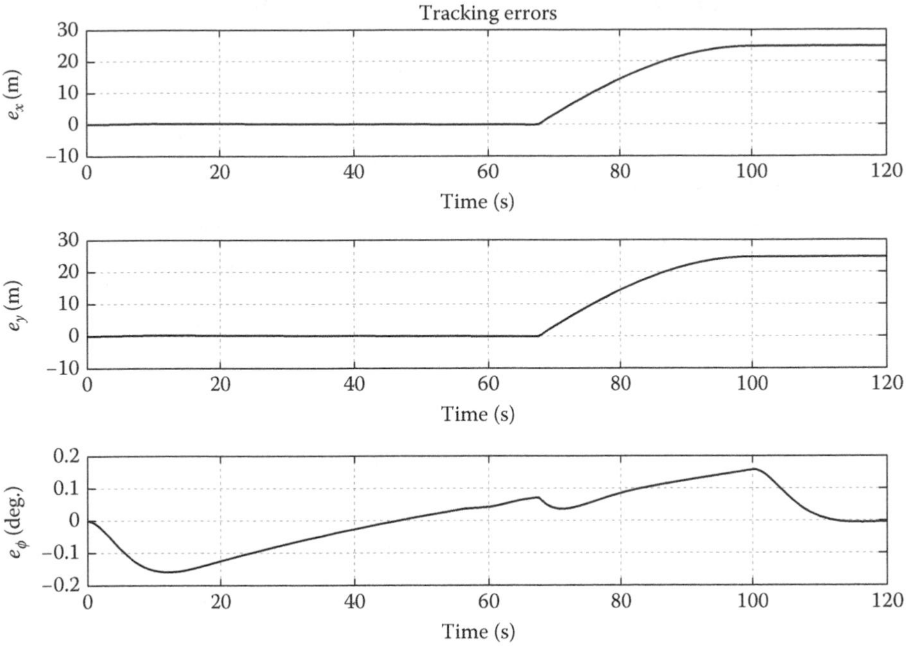

FIGURE 7.51
The closed-loop tracking performance of the planar $4R\underline{P}R$ manipulator colliding with a flat object; impedance control.

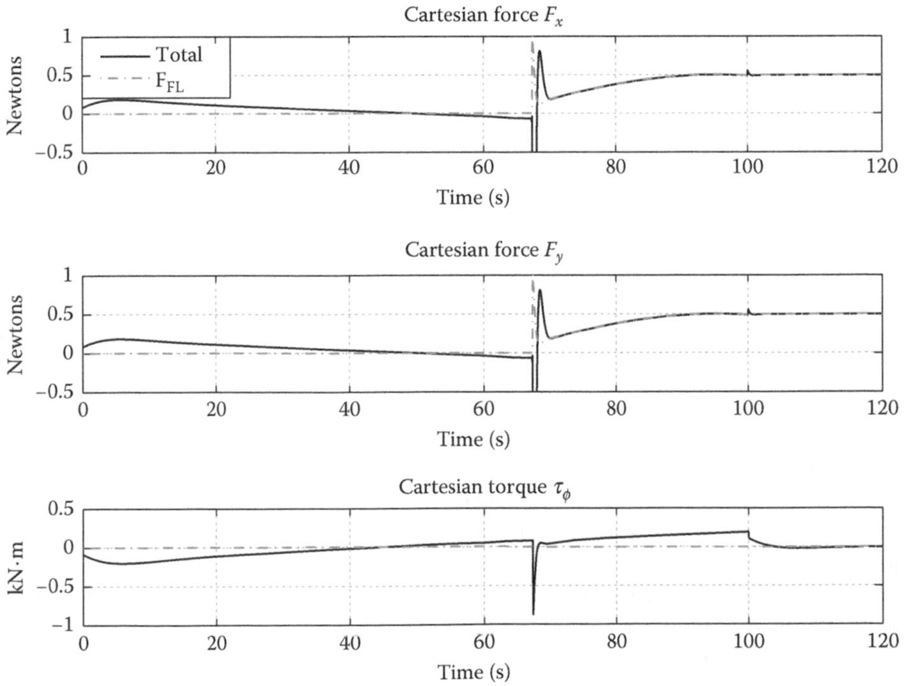

FIGURE 7.52
The Cartesian wrench of the planar $4R\underline{P}R$ manipulator colliding with a flat object; impedance control.

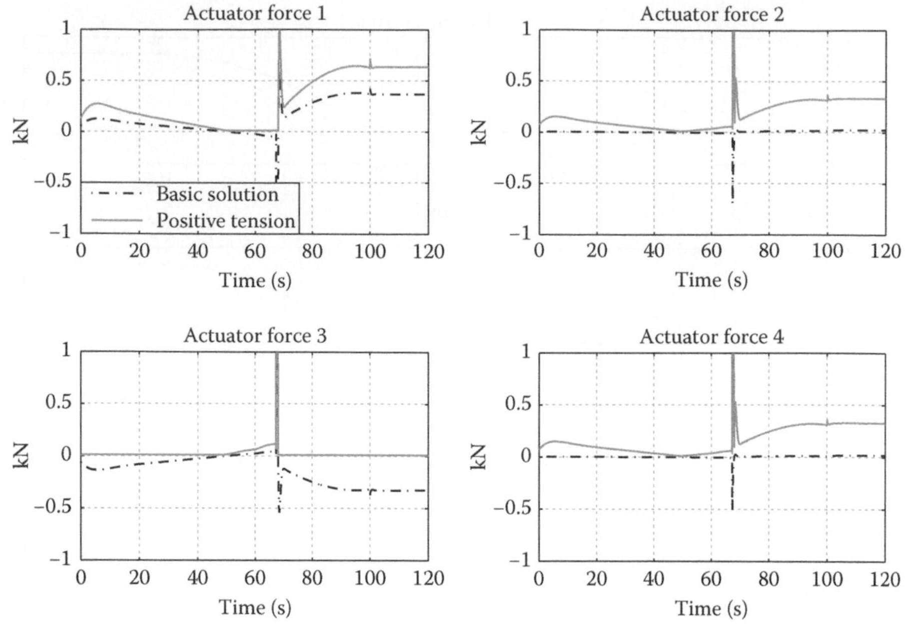

FIGURE 7.53
The actuator forces of the planar $4R\underline{P}R$ manipulator colliding with a flat object; impedance control.

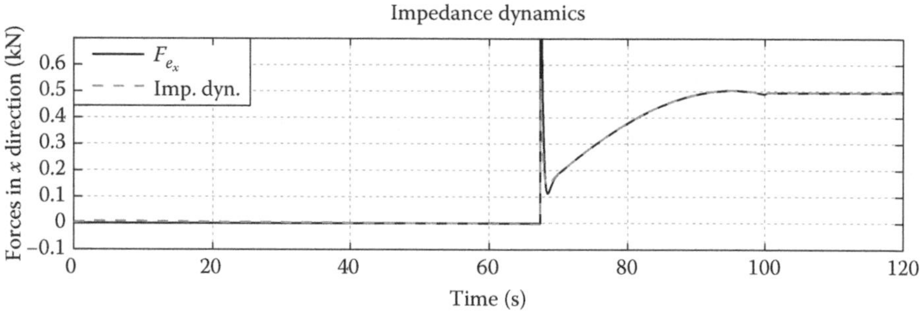

FIGURE 7.54
The interaction force F_x versus the impedance dynamics of the planar $4R\underline{P}R$ manipulator colliding with a flat object; impedance control.

in redundancy resolution scheme by Equation 6.76, while the solid line is the final optimization solution of the redundancy resolution problem detailed in Equation 6.79. The redundancy resolution is solved by the iterative numerical method. As seen in Figure 7.53, the actuator forces increase significantly after collision, while the tension forces derived from the redundancy resolution scheme remain positive at all times.

To examine the behavior of impedance controller, and to analyze how effective it can adjust the closed-loop dynamics to a desired impedance, the x component of the interacting force $\mathcal{F}_e$ is plotted in Figure 7.54 together with the impedance dynamics of the closed-loop system represented by the left-hand side of Equation 7.51. If impedance dynamics is ideally enforced in the closed-loop system, the two sides of Equation 7.54 would be identical, and therefore, the two plots would be the same. As seen in Figure 7.54,

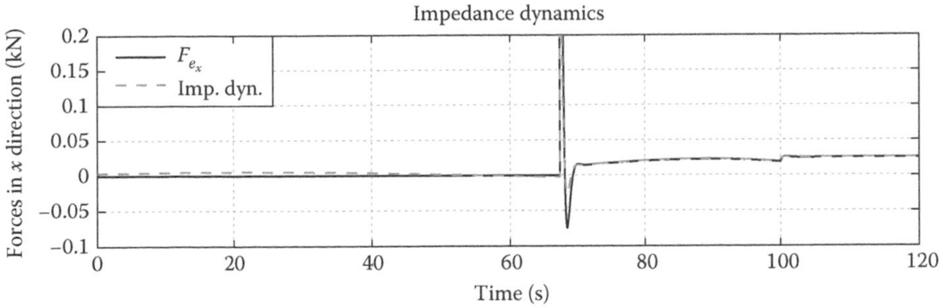

FIGURE 7.55
The interaction force F_{e_x} versus the impedance dynamics of the planar $4R\underline{P}R$ manipulator colliding with a flat object; impedance control with a different impedance requirement.

the two plots are not identical at all times, but they suitably follow each other with a minor difference. This confirms the objective of the impedance control and verifies how well the force–motion relation has been adjusted by a desired impedance dynamics.

To show how a designer can suitably assign the impedance requirement, consider another simulation in which all the parameters are the same as before, while the impedance coefficients are changed to $\boldsymbol{M}_d = 100 \cdot \boldsymbol{I}_{3\times 3}, \boldsymbol{C}_d = \boldsymbol{I}_{3\times 3}$, and $\boldsymbol{K}_d = 10 \cdot \boldsymbol{I}_{3\times 3}$. In this design, the impedance stiffness coefficient is reduced by a factor of 10, while impedance damping coefficient is reduced 20 times. Such impedance parameters are not suitable for a real implementation; however, an exaggerated change in parameters are considered to verify the effectiveness of impedance controller to regulate the performance toward a desired impedance. For such a case, the x component of the interacting force $\mathcal{F}_e$ is plotted in Figure 7.55 together with the impedance dynamics of the closed-loop system represented by the left-hand side of Equation 7.51. As seen in Figure 7.55, although the impedance parameters are significantly changed, the two plots suitably follow each other with a minor difference. This confirms that the force–motion relation can be suitably adjusted to a required impedance dynamics through the structure of impedance control.

To examine the performance of the impedance controller in the presence of external disturbance, another simulation is executed for the closed-loop system. In this simulation, a step disturbance force is applied to the system under the original trajectory shown in Figure 7.50. In this simulation, a 1 kN step disturbance force in x and y directions is applied at time 50 s, while at the same time, a 1 kN · m disturbance torque is applied in the ϕ direction. The desired impedance matrices are set at $\boldsymbol{M}_d = 100 \cdot \boldsymbol{I}_{3\times 3}, \boldsymbol{C}_d = 20 \cdot \boldsymbol{I}_{3\times 3}$, and $\boldsymbol{K}_d = 100 \cdot \boldsymbol{I}_{3\times 3}$ as the first simulation.

The closed-loop tracking error of the system is shown in Figure 7.56. By comparing this figure with Figure 7.51, it is seen that the effect of external disturbance is not completely attenuated by impedance controller, and the abrupt change in the tracking error at the instances when the disturbance wrench is applied is clearly seen in the response. Furthermore, as a result of nonvanishing disturbances applied to the system, static errors are accumulated in the closed-loop response of the system before collision. Figure 7.57 illustrates the Cartesian wrench generated by PD controller to carry out the required maneuver in the presence of disturbance. The effect of step disturbance wrench on the required controller forces is also clearly seen after time 50 s, in which the disturbance signature is seen in all the generated forces.

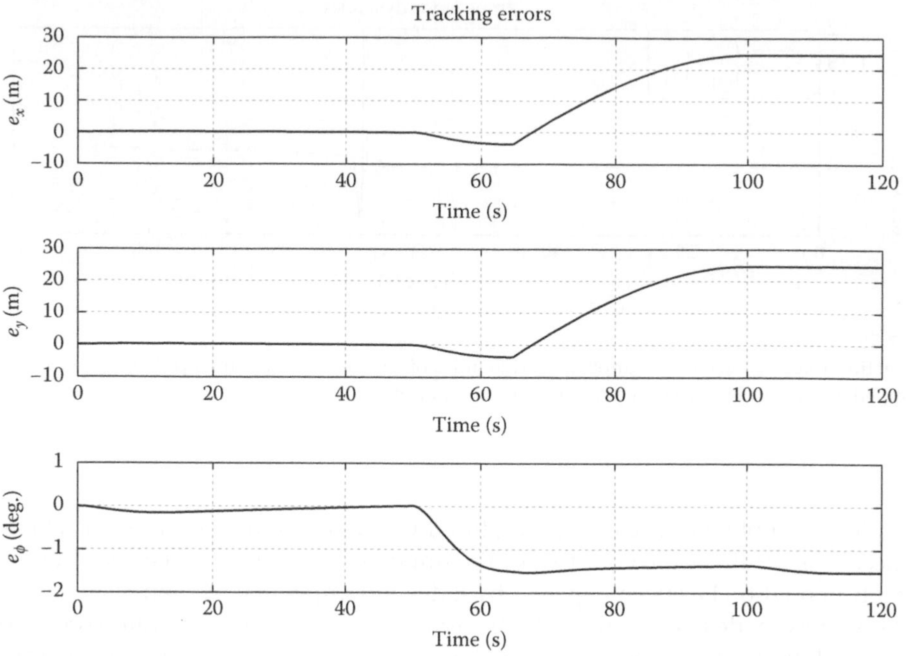

FIGURE 7.56
The closed-loop tracking performance of the planar $4R\underline{P}R$ manipulator colliding with a flat object; impedance control with disturbance.

Finally, the behavior of impedance controller on how effectively it can adjust the closed-loop dynamics to a desired impedance in the presence of disturbance has been examined in Figure 7.58. In this figure, the x component of the interacting force $\mathcal{F}_e$ is plotted together with the impedance dynamics of the closed-loop system represented by the left-hand side of Equation 7.51. As seen in Figure 7.58, the two plots differ by an offset corresponding to a nonvanishing disturbance acting on the system. It is assumed that the force measurement is unable to measure the disturbance force, and therefore, the impedance controller is unable to attenuate the effect of disturbance. However, if the disturbance force vanishes by time, or can be measured through by the force sensors, its effect on impedance assignment vanishes.

7.5.5 Impedance Control of the Stewart–Gough Platform

In this section, impedance control of the Stewart–Gough platform is studied in detail. The architecture of this manipulator is described in detail in Section 3.5.1 and shown in Figure 3.16. As explained earlier, in this manipulator the spatial motion of the moving platform is generated by six piston–cylinder actuators. Complete kinematic analysis of this manipulator is given in Section 3.5 and its Jacobian analysis is presented in Section 4.8. The dynamic analysis of this manipulator using Newton–Euler and virtual work approaches is given in Sections 5.3.3 and 5.4.5, respectively. The closed-form dynamic formulation being used in this section for the simulations is derived in Section 5.5.6. Different motion control strategies are implemented in Section 6.9. Stiffness control and direct force control of this manipulator have been studied in Sections 7.3.4 and 7.4.2, respectively. In what follows, the impedance control strategy is implemented on this manipulator.

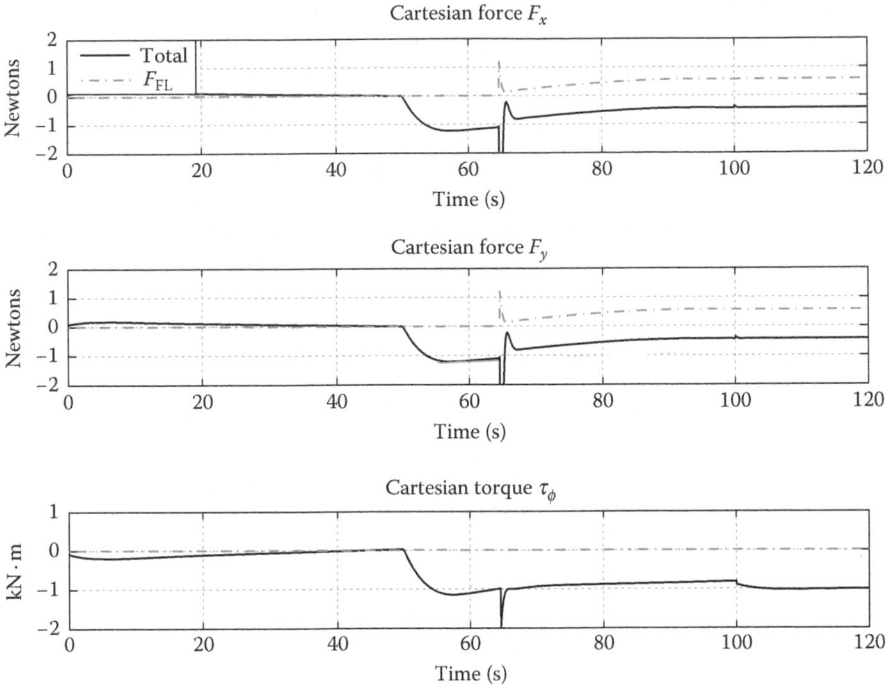

FIGURE 7.57
The Cartesian wrench of the planar $4R\underline{P}R$ manipulator colliding with a flat object; impedance control with disturbance.

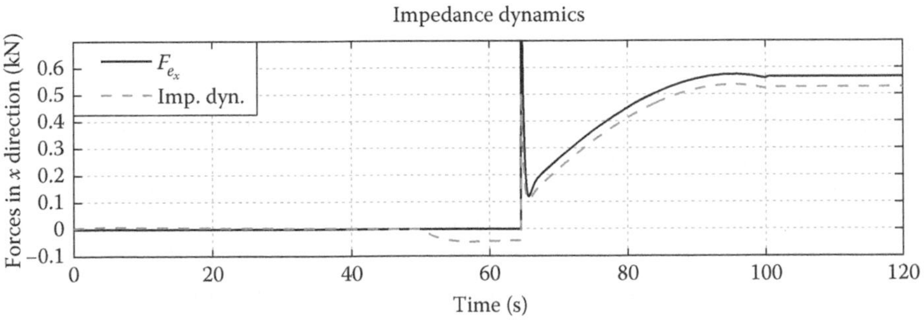

FIGURE 7.58
The interaction force F_{e_x} versus the impedance dynamics of the planar $4R\underline{P}R$ manipulator colliding with a flat object; impedance control with disturbance.

The impedance control strategy applied on the system has the structure shown in Figure 7.49, in which position feedback is considered in the outer loop, while force feedback is used in the inner loop. As shown in Figure 7.49, the motion-tracking error is directly determined from motion measurement by $e_x = \mathcal{X}_d - \mathcal{X}$ in the outer loop, and the motion controller is designed to satisfy the required impedance structure depicted in Equation 7.44. The required wrench $\mathcal{F}$ in impedance control scheme is based on inverse dynamics control, and consists of three main parts. In the inner loop, the impedance control scheme is based on a feedback linearization part in addition to a mass matrix adjustment, while in the outer loop, a linear motion controller is considered based on the desired impedance requirement such as that in Equation 7.44.

A linear impedance relation between the force and motion variables is generated through this structure. The desired impedance consists of a desired inductive $\boldsymbol{M}_d$, resistive $\boldsymbol{C}_d$, and capacitive $\boldsymbol{K}_d$ matrices, as given in Equation 7.44. The control effort applied to the system is given in Equation 7.46, in which $\boldsymbol{e}_F, \mathcal{F}_a$, and $\mathcal{F}_{fl}$ are given in Equations 7.47 through 7.49, respectively. Note that to generate these terms, dynamic formulation of the robot, and its kinematic and dynamic parameters are needed. In practice, exact information on dynamic matrices is not available, and therefore, a %10 perturbation in all kinematic and inertial parameters is considered in the following simulations.

As schematically shown in Figure 7.19, it is considered that a flat surface is placed in the workspace of the robot such that the robot-moving platform experiences contact with it in a segment of its maneuver. As shown in Figure 7.20, the desired trajectory in the following simulations is considered the same as in the stiffness control scheme depicted in Figure 7.20, while the position of the object is also considered the same as that in the stiffness control scheme. Thus, for the simulations, it is considered that $\boldsymbol{x}_l = [0.2, 0.2, 1.2]^T$ and the normal vector to the plane is $\boldsymbol{n}_l = [1, 1, 1]^T$. By this means, in the prescribed trajectory shown in Figure 7.20, the moving platform collides with the flat object at about 0.68 s, while the contact is held until the end of the simulation time.

The stiffness of the object is modeled by linear stiffness gain $\boldsymbol{K}_e = 10^5 \cdot \boldsymbol{I}_{6\times 6}$ (N/m), with a linear structural damping of $\boldsymbol{C}_e = 10^4 \cdot \boldsymbol{I}_{6\times 6}$ (N · m/s). The geometric and inertial parameters used in these simulations are given in Table 5.2, and the initial conditions for the states are all set at zero, except for the initial state of the moving platform height, which is set at $z_o = 1$ m. For the following simulations, the disturbance wrench applied to the moving platform is considered to be zero, while the gravity vector is considered as $\boldsymbol{g} = [0, 0, -9.81]^T$. The control effort is calculated from Equation 7.46, in which $\boldsymbol{e}_F, \mathcal{F}_a$, and $\mathcal{F}_{fl}$ are given in Equations 7.47 through 7.49, respectively. The desired impedance matrices are set at $\boldsymbol{M}_d = 100 \cdot \boldsymbol{I}_{6\times 6}, \boldsymbol{C}_d = 2 \times 10^3 \cdot \boldsymbol{I}_{6\times 6}$, and $\boldsymbol{K}_d = 10^4 \cdot \boldsymbol{I}_{6\times 6}$.

The final motion of the moving platform and the tracking errors have been illustrated in Figures 7.59 and 7.60. As seen in Figure 7.59, the desired and final closed-loop motions of the system are almost identical prior to collision. At almost 0.68 s, collision occurs, and the motion of the moving platform remains almost constant on the surface of the object. Greater details of the moving platform motion at this time are given in Figure 7.60, in which it is observed that after collision, a smooth motion of the moving platform on the object surface is produced. The moving platform causes minor deflection to the object. Furthermore, note that the error in the z direction is more than that in x and y directions. This is because of incomplete gravity compensation, which has significant influence in the z direction. Moreover, no interaction torque is considered in the simulations, the effect of which can be seen by relatively uniform tracking error in the orientation angles, prior to and after collision.

The Cartesian wrench $\mathcal{F}$ generated by the impedance controller is illustrated in Figure 7.61, in which the total force is plotted in a solid line while the PD term is plotted in a dashed line. As seen, the Cartesian forces in x, y, and z directions have a linear shape corresponding to the required acceleration of the moving platform prior to collision. These forces change rapidly and followed the interaction forces with a transient after collision time. Since there are no interaction torques considered in the simulation, the Cartesian torques have almost a linear shape corresponding to the required angular acceleration of the moving platform, although a vanishing transient is observed at the collision time. Furthermore, it is observed that the PD term produces only a small portion of the total force

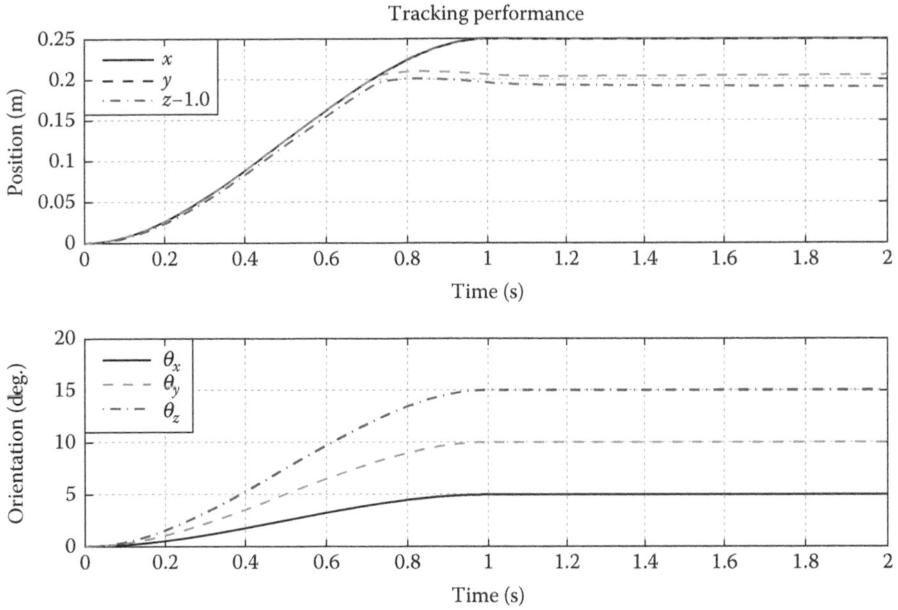

FIGURE 7.59
The desired and output trajectories of the Stewart–Gough platform colliding with a flat object; impedance control.

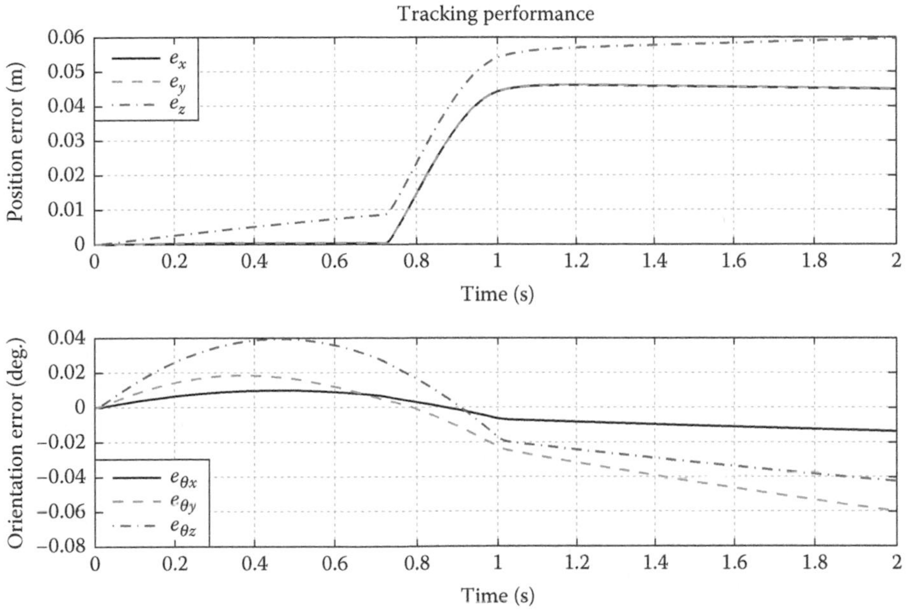

FIGURE 7.60
The closed-loop tracking performance of the Stewart–Gough platform colliding with a flat object; impedance control.

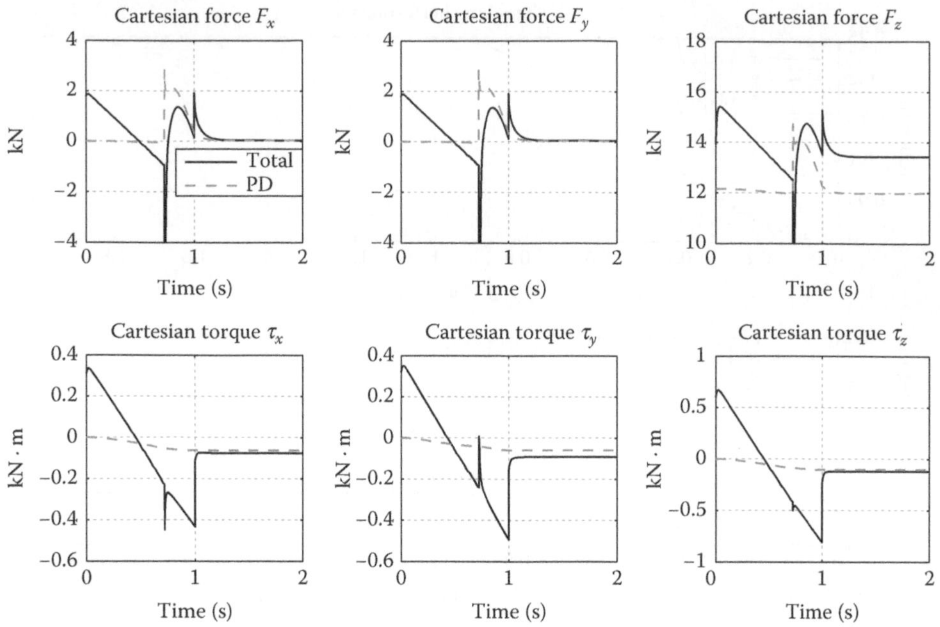

FIGURE 7.61
The Cartesian wrench of the Stewart–Gough platform colliding with a flat object; impedance control.

prior to collision, while after collision the interaction force $\mathcal{F}_e$ increases rapidly and this immediately affects the PD term. Similarly, the actuator forces of the manipulator are illustrated in Figure 7.62. As seen in Figure 7.62, the actuator forces have similar behavior as the Cartesian wrench prior to and after collision.

To examine the behavior of impedance controller on how effectively it adjusts the closed-loop dynamics to a desired impedance dynamics, the x and z components of the interacting force $\mathcal{F}_e$ are plotted in the top and bottom plots of Figure 7.63. In each figure, the interacting force is plotted in a solid line while the impedance dynamics of the closed-loop system, represented by the left-hand side of Equation 7.51, is plotted in a dashed line. If impedance dynamics is ideally enforced in the closed-loop system, the two sides of Equation 7.63 would be identical, and therefore, the two plots depicted by a solid line and a dashed line would be the same. As seen in the top figure, the two plots are identical in the x direction. However, in the bottom plot, the two plots are not identical and they have a small offset deviation from each other. This is because of incomplete gravity compensation, which has significant influence in the z direction. However, the difference is minimal and the plots follow each other suitably. From this figure it can be concluded how well the force–motion relation has been adjusted by desired impedance dynamics.

To show how a designer can suitably assign the required impedance requirements, consider another simulation in which all the parameters are the same as before, while the impedance coefficients are changed to $\boldsymbol{M}_d = 100 \cdot \boldsymbol{I}_{6\times6}$, $C_d = 10^4 \cdot \boldsymbol{I}_{6\times6}$, and $\boldsymbol{K}_d = 5 \times 10^4 \cdot \boldsymbol{I}_{6\times6}$. In this design, the impedance stiffness and damping coefficients are multiplied by a factor of 5, while impedance mass matrix remains unchanged. For such a case, x and z components of the interacting force $\mathcal{F}_e$ are plotted in the top and bottom part of Figure 7.64, respectively, together with the impedance dynamics of the closed-loop system represented by the left-hand side of Equation 7.51. As seen in Figure 7.63, although the impedance

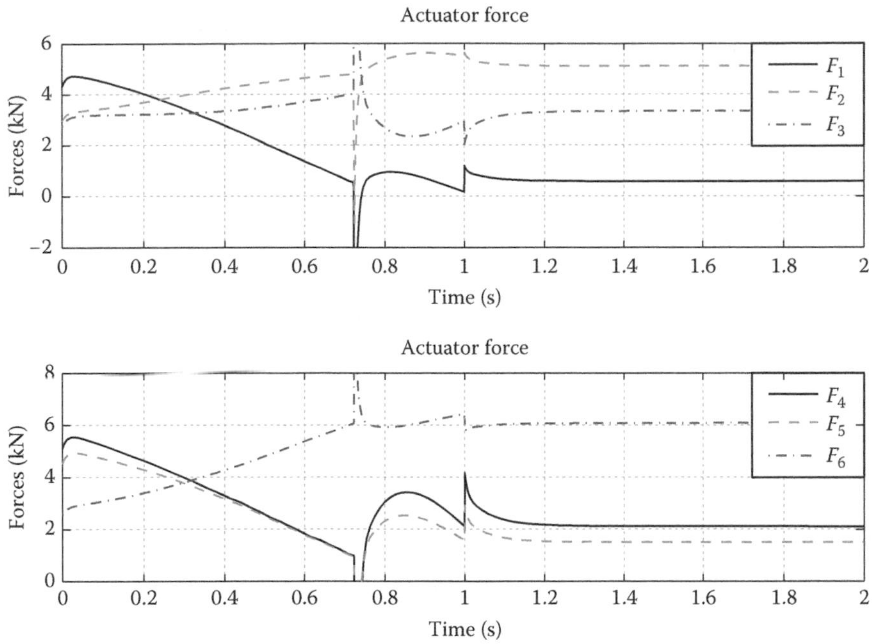

FIGURE 7.62
The actuator forces of the Stewart–Gough platform colliding with a flat object; impedance control.

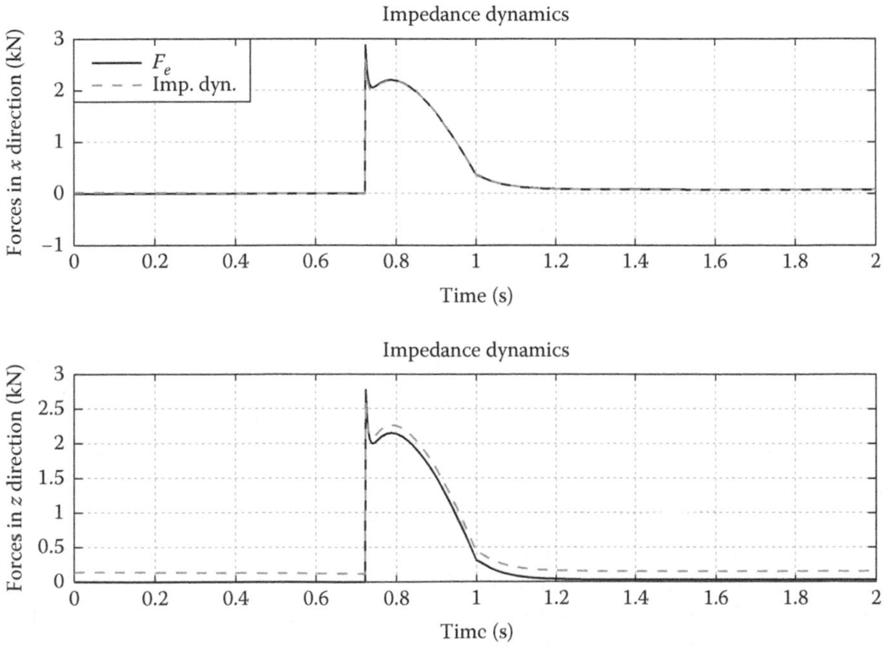

FIGURE 7.63
The interaction force F_{e_x} versus the impedance dynamics of the Stewart–Gough platform colliding with a flat object; impedance control.

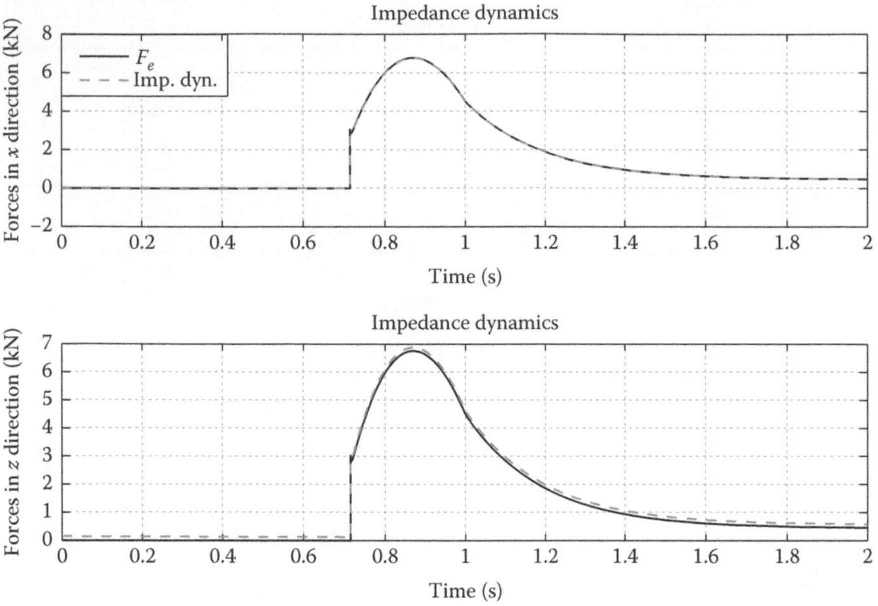

FIGURE 7.64
The interaction force F_{e_x} versus the impedance dynamics of the Stewart–Gough platform colliding with a flat object; impedance control with a different impedance requirement.

parameters change significantly, the solid and dashed lines in both plots follow suitably each other with a minor difference in the z direction. This confirms that the force–motion relation can be suitably adjusted to a required impedance dynamics through the structure of impedance control.

To examine the performance of the impedance controller in the presence of external disturbance, another simulation is performed for the closed-loop system. In this simulation, a unit step disturbance force is applied to the system under the original trajectory shown in Figure 7.59. A 10 kN step disturbance force in x, y, and z directions is applied at 50 s, while no disturbance torques are considered in this simulation. The desired impedance matrices are set at $\boldsymbol{M}_d = 100 \cdot \boldsymbol{I}_{6\times6}$, $\boldsymbol{C}_d = 2 \times 10^3 \cdot \boldsymbol{I}_{6\times6}$, and $\boldsymbol{K}_d = 10^4 \cdot \boldsymbol{I}_{6\times6}$, as in the first simulation.

The closed-loop tracking error of the system is shown in Figure 7.65. By comparing Figure 7.65 with Figure 7.60, it is seen that the effect of external disturbance is not completely attenuated by the impedance controller, and the abrupt change in the tracking error is clearly seen in the response at the time when the disturbance force is applied. Figure 7.66 illustrates the Cartesian wrench generated by the PD controller to perform the required maneuver in the presence of disturbance. The effect of step disturbance on the required controller forces is also clearly seen at the times after 0.5 s, at which the disturbance signature is seen in the generated forces in x, y, and z directions.

Finally, in the presence of disturbance, how effective the impedance controller can adjust the closed-loop dynamics to a desired impedance dynamics has been examined in Figure 7.67. In this figure, x and z components of the interacting force $\mathcal{F}_e$ are plotted in the top and bottom plots, respectively, together with the impedance dynamics of the closed-loop system, represented by the left-hand side of Equation 7.51. As seen in Figure 7.67, the solid and dashed plots differ by an offset corresponding to a nonvanishing disturbance acting on the system. It is assumed that the force measurement is unable to measure the

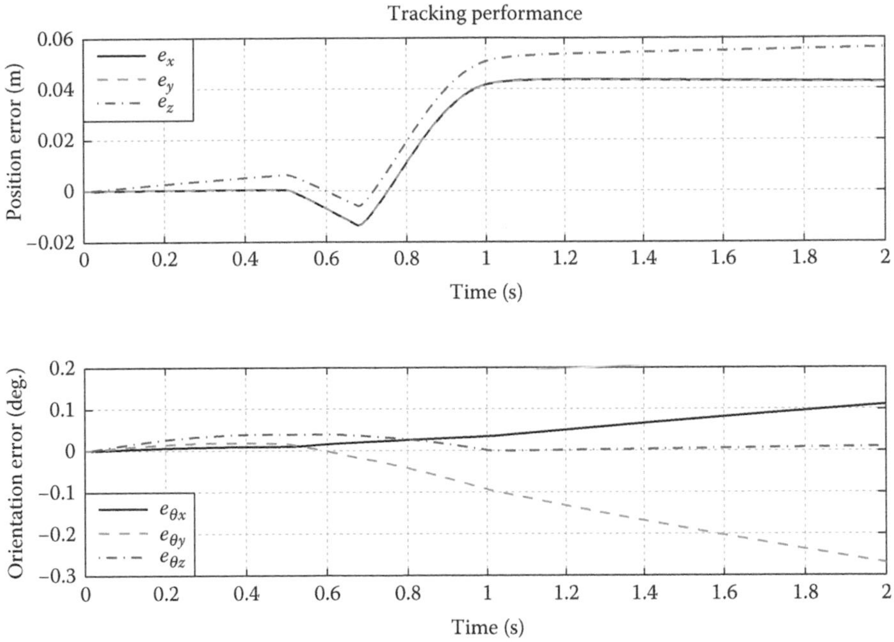

FIGURE 7.65
The closed-loop tracking performance of the Stewart–Gough platform colliding with a flat object; impedance control with disturbance.

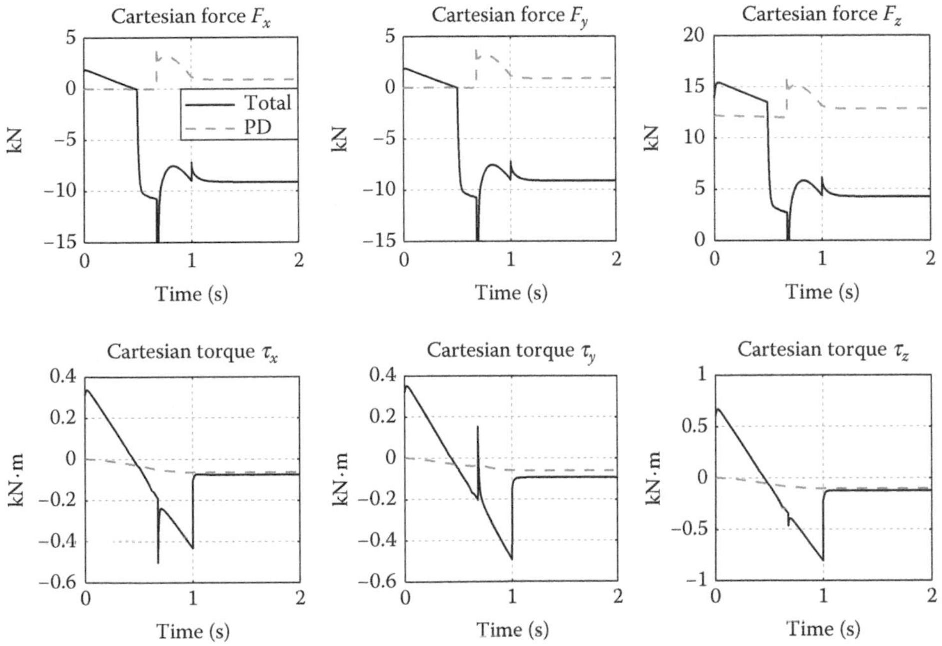

FIGURE 7.66
The Cartesian wrench of the Stewart–Gough platform colliding with a flat object; impedance control with disturbance.

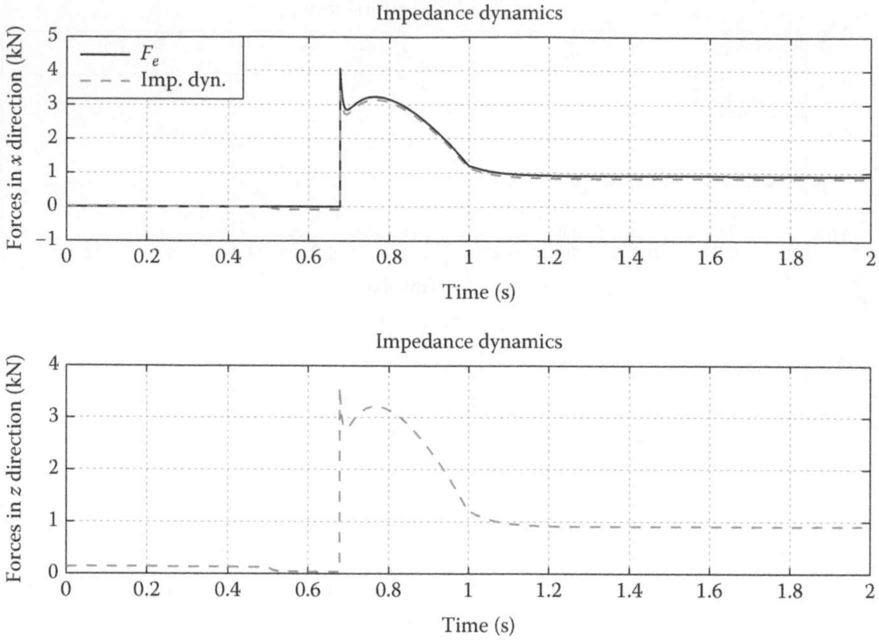

FIGURE 7.67
The interaction force F_{e_x} versus the impedance dynamics of the Stewart–Gough platform colliding with a flat object; impedance control with disturbance.

disturbance force, and therefore, the impedance controller is unable to attenuate the effect of disturbance. However, if the disturbance force vanishes by time or can be measured by the force sensors, its effect on impedance assignment vanishes.

PROBLEMS

1. Consider the five-bar mechanism shown in Figure 3.28. The dynamic formulation of this mechanism and its kinematic and inertial parameters, and a typical motion trajectory is given in Problem 5.1. Consider a flat surface within the workspace of the mechanism such as that modeled in detail in Section 7.3.3, and used in Equation 7.26 with $\boldsymbol{K}_e = 100 \cdot \boldsymbol{I}_{3\times 3}$ (N/m) and $\boldsymbol{C}_e = 100 \cdot \boldsymbol{I}_{3\times 3}$ (N · m/s) to represent the contact forces.
   a. For this mechanism, design a stiffness controller with complete inverse dynamics, and consider %10 perturbation in all geometric and inertial parameters. Set the object location such that a collision occurs during the motion. Simulate the closed-loop performance and tune the stiffness controller gains to reach a suitable stiffness during the contact.
   b. Plot and analyze closed-loop motion trajectories, tracking errors, Cartesian, and actuator forces versus time.
   c. Plot and analyze tracking errors in x and y directions versus the corresponding contact forces, and analyze the stiffness behavior of the closed-loop system during contact transient and at steady state.

d. Change the desired stiffness requirement and repeat part (c) to verify the performance of the controller.

2. Consider the five-bar mechanism shown in Figure 3.29. The dynamic formulation of this mechanism and its kinematic and inertia parameters, and a typical motion trajectory is given in Problem 5.2. Consider a flat surface within the workspace of the mechanism such as that modeled in detail in Section 7.3.3, and used in Equation 7.26 with $\boldsymbol{K}_e = 100 \cdot \boldsymbol{I}_{3\times3}$ (N/m) and $\boldsymbol{C}_e = 100 \cdot \boldsymbol{I}_{3\times3}$ (N · m/s) to represent the contact forces.

   a. For this mechanism, design a stiffness controller with complete inverse dynamics, and consider %10 perturbation in all geometric and inertial parameters. Set the object location such that a collision occurs during the motion. Simulate the closed-loop performance and tune the stiffness controller gains to reach a suitable stiffness during the contact.

   b. Plot and analyze closed-loop motion trajectories, tracking errors, Cartesian, and actuator forces versus time.

   c. Plot and analyze tracking errors in x and y directions versus the corresponding contact forces, and analyze the stiffness behavior of the closed-loop system during contact transient and at steady state.

   d. Change the desired stiffness requirement and repeat part (c) to verify the performance of the controller.

3. Consider a planar pantograph mechanism shown in Figure 3.30. The dynamic formulation of this mechanism and its kinematic and inertial parameters, and a typical motion trajectory is given in Problem 5.3. Consider a flat surface within the workspace of the mechanism such as that modeled in detail in Section 7.3.3, and used in Equation 7.26 with $\boldsymbol{K}_e = 100 \cdot \boldsymbol{I}_{3\times3}$ (N/m) and $\boldsymbol{C}_e = 100 \cdot \boldsymbol{I}_{3\times3}$ (N · m/s) to represent the contact forces.

   a. For this mechanism, design a stiffness controller with complete inverse dynamics, and consider %10 perturbation in all geometric and inertial parameters. Set the object location such that a collision occurs during the motion. Simulate the closed-loop performance and tune the stiffness controller gains to reach a suitable stiffness during the contact.

   b. Plot and analyze closed-loop motion trajectories, tracking errors, Cartesian, and actuator forces versus time.

   c. Plot and analyze tracking errors in x and y directions versus the corresponding contact forces, and analyze the stiffness behavior of the closed-loop system during contact transient and at steady state.

   d. Change the desired stiffness requirement and repeat part (c) to verify the performance of the controller.

4. Consider the planar three-degrees-of-freedom $3\underline{R}RR$ parallel manipulator shown in Figure 3.31. The dynamic formulation of this manipulator and its kinematic and inertial parameters, and a typical motion trajectory is given in Problem 5.4. Consider a flat surface within the workspace of the mechanism such as that modeled in detail in Section 7.3.3, and used in Equation 7.26 with $\boldsymbol{K}_e = 100 \cdot \boldsymbol{I}_{3\times3}$ (N/m) and $\boldsymbol{C}_e = 100 \cdot \boldsymbol{I}_{3\times3}$ (N · m/s) to represent the contact forces.

   a. For this manipulator, design a stiffness controller with completely inverse dynamics, and consider %10 perturbation in all geometric and inertial parameters. Set the object location such that a collision occurs during the motion. Simulate the closed-loop performance and tune the stiffness controller gains to reach a suitable stiffness during the contact.

b. Plot and analyze closed-loop motion trajectories, tracking errors, Cartesian, and actuator forces versus time.

c. Plot and analyze tracking errors in x and y directions versus the corresponding contact forces, and analyze the stiffness behavior of the closed-loop system during contact transient and at steady state.

d. Change the desired stiffness requirement and repeat part (c) to verify the performance of the controller.

5. Consider the planar three-degrees-of-freedom $4\underline{R}RR$ parallel manipulator shown in Figure 3.32. The dynamic formulation of this manipulator and its kinematic and inertial parameters, and a typical motion trajectory is given in Problem 5.5. Consider a flat surface within the workspace of the mechanism such as that modeled in detail in Section 7.3.3, and used in Equation 7.26 with $\boldsymbol{K}_e = 100 \cdot \boldsymbol{I}_{3\times3}$ (N/m) and $\boldsymbol{C}_e = 100 \cdot \boldsymbol{I}_{3\times3}$ (N · m/s) to represent the contact forces.

   a. For this manipulator, design a stiffness controller with complete inverse dynamics, and consider %10 perturbation in all geometric and inertial parameters. Set the object location such that a collision occurs during the motion. Simulate the closed-loop performance and tune the stiffness controller gains to reach a suitable stiffness during the contact.

   b. Plot and analyze closed-loop motion trajectories, tracking errors, Cartesian, and actuator forces versus time.

   c. Plot and analyze tracking errors in x and y directions versus the corresponding contact forces, and analyze the stiffness behavior of the closed-loop system during contact transient and at steady state.

   d. Change the desired stiffness requirement and repeat part (c) to verify the performance of the controller.

6. Consider the $3U\underline{P}U$ parallel manipulator shown in Figure 3.34. The dynamic formulation of this manipulator and its kinematic and inertial parameters, and a typical motion trajectory is given in Problem 5.7. Consider a flat object within the workspace of the mechanism such as that modeled in detail in Section 7.3.4, and used in Equation 7.29 with $\boldsymbol{K}_e = 100 \cdot \boldsymbol{I}_{3\times3}$ (N/m) and $\boldsymbol{C}_e = 100 \cdot \boldsymbol{I}_{3\times3}$ (N · m/s) to represent the contact forces.

   a. For this manipulator, design a stiffness controller with complete inverse dynamics, and consider %10 perturbation in all geometric and inertial parameters. Set the object location such that a collision occurs during the motion. Simulate the closed-loop performance and tune the stiffness controller gains to reach a suitable stiffness during the contact.

   b. Plot and analyze closed-loop motion trajectories, tracking errors, Cartesian, and actuator forces versus time.

   c. Plot and analyze tracking errors in x, y, and z directions versus the corresponding contact forces, and analyze the stiffness behavior of the closed-loop system during contact transient and at steady state.

   d. Change the desired stiffness requirement and repeat part (c) to verify the performance of the controller.

7. Consider the three-degrees-of-freedom $3R\underline{P}S$ manipulator represented in Figure 3.36. The dynamic formulation of this manipulator and its kinematic and inertial parameters, and a typical motion trajectory is given in Problem 5.9. Consider a flat object within the workspace of the mechanism such as that modeled in detail in Section 7.3.4, and used in Equation 7.29 with $\boldsymbol{K}_e = 100 \cdot \boldsymbol{I}_{3\times3}$ (N/m) and $\boldsymbol{C}_e = 100 \cdot \boldsymbol{I}_{3\times3}$ (N · m/s) to represent the contact forces.
   a. For this manipulator, design a stiffness controller with complete inverse dynamics, and consider %10 perturbation in all geometric and inertial parameters. Set the object location such that a collision occurs during the motion. Simulate the closed-loop performance and tune the stiffness controller gains to reach a suitable stiffness during the contact.
   b. Plot and analyze closed-loop motion trajectories, tracking errors, Cartesian, and actuator forces versus time.
   c. Plot and analyze tracking errors in x, y, and z directions versus the corresponding contact forces, and analyze the stiffness behavior of the closed-loop system during contact transient and at steady state.
   d. Change the desired stiffness requirement and repeat part (c) to verify the performance of the controller.
8. Consider the three-degrees-of-freedom $3\underline{P}RC$ translational parallel manipulator represented in Figure 3.36. The dynamic formulation of this manipulator and its kinematic and inertial parameters, and a typical motion trajectory is given in Problem 5.10. Consider a flat surface within the workspace of the mechanism such as that modeled in detail in Section 7.3.3, and used in Equation 7.26 with $\boldsymbol{K}_e = 100 \cdot \boldsymbol{I}_{3\times3}$ (N/m) and $\boldsymbol{C}_e = 100 \cdot \boldsymbol{I}_{3\times3}$ (N · m/s) to represent the contact forces.
   a. For this manipulator, design a stiffness controller with complete inverse dynamics, and consider %10 perturbation in all geometric and inertial parameters. Set the object location such that a collision occurs during the motion. Simulate the closed-loop performance and tune the stiffness controller gains to reach a suitable stiffness during the contact.
   b. Plot and analyze closed-loop motion trajectories, tracking errors, Cartesian, and actuator forces versus time.
   c. Plot and analyze tracking errors in x and y directions versus the corresponding contact forces, and analyze the stiffness behavior of the closed-loop system during contact transient and at steady state.
   d. Change the desired stiffness requirement and repeat part (c) to verify the performance of the controller.
9. Consider the Delta robot represented in Figure 3.39. The dynamic formulation of this manipulator and its kinematic and inertial parameters, and a typical motion trajectory is given in Problem 5.11. Consider a flat object within the workspace of the mechanism such as that modeled in detail in Section 7.3.4, and used in Equation 7.29 with $\boldsymbol{K}_e = 100 \cdot \boldsymbol{I}_{3\times3}$ (N/m) and $\boldsymbol{C}_e = 100 \cdot \boldsymbol{I}_{3\times3}$ (N · m/s) to represent the contact forces.
   a. For this manipulator, design a stiffness controller with complete inverse dynamics, and consider %10 perturbation in all geometric and inertial parameters. Set the object location such that a collision occurs during the motion. Simulate

the closed-loop performance and tune the stiffness controller gains to reach a suitable stiffness during the contact.

b. Plot and analyze closed-loop motion trajectories, tracking errors, Cartesian, and actuator forces versus time.

c. Plot and analyze tracking errors in x, y, and z directions versus the corresponding contact forces, and analyze the stiffness behavior of the closed-loop system during contact transient and at steady state.

d. Change the desired stiffness requirement and repeat part (c) to verify the performance of the controller.

10. Consider the Quadrupteron robot represented in Figure 3.40. The dynamic formulation of this manipulator and its kinematic and inertial parameters, and a typical motion trajectory is given in Problem 5.12. Consider a flat object within the workspace of the mechanism such as that modeled in detail in Section 7.3.4, and used in Equation 7.29 with $\boldsymbol{K}_e = 100 \cdot \boldsymbol{I}_{4\times4}$ (N/m) and $\boldsymbol{C}_e = 100 \cdot \boldsymbol{I}_{4\times4}$ (N · m/s) to represent the contact forces.

    a. For this manipulator, design a stiffness controller with complete inverse dynamics, and consider %10 perturbation in all geometric and inertial parameters. Set the object location such that a collision occurs during the motion. Simulate the closed-loop performance and tune the stiffness controller gains to reach a suitable stiffness during the contact.

    b. Plot and analyze closed-loop motion trajectories, tracking errors, Cartesian and actuator forces versus time.

    c. Plot and analyze tracking errors in x, y, and z directions versus the corresponding contact forces, and analyze the stiffness behavior of the closed-loop system during contact transient and at steady state.

    d. Change the desired stiffness requirement and repeat part (c) to verify the performance of the controller.

11. Consider the five-bar mechanism shown in Figure 3.28. The dynamic formulation of this mechanism and its kinematic and inertia parameters, and a typical motion trajectory is given in Problem 5.1. Consider a flat surface within the workspace of the mechanism such as that modeled in detail in Section 7.3.3, and used in Equation 7.26 with $\boldsymbol{K}_e = 100 \cdot \boldsymbol{I}_{3\times3}$ (N/m) and $\boldsymbol{C}_e = 100 \cdot \boldsymbol{I}_{3\times3}$ (N · m/s) to represent the contact forces.

    a. For this mechanism, assign a desired force trajectory using cubic polynomials, design a direct force controller with complete inverse dynamics, and consider %10 perturbation in all geometric and inertial parameters. Set the object location such that a collision occurs during the motion. Simulate the closed-loop performance and tune the controller gains to reach a suitable force tracking after the collision.

    b. Plot and analyze closed-loop force trajectories, force tracking errors, Cartesian, and actuator forces versus time.

    c. Plot and analyze the resulting motion trajectories versus the auxiliary motion variables generated by force feedback law and analyze the behavior of the force controller by intuition.

12. Consider the five-bar mechanism shown in Figure 3.29. The dynamic formulation of this mechanism and its kinematic and inertia parameters, and a typical motion

trajectory is given in Problem 5.2. Consider a flat surface within the workspace of the mechanism such as that modeled in detail in Section 7.3.3, and used in Equation 7.26 with $\boldsymbol{K}_e = 100 \cdot \boldsymbol{I}_{3\times3}$ (N/m) and $\boldsymbol{C}_e = 100 \cdot \boldsymbol{I}_{3\times3}$ (N · m/s) to represent the contact forces.

a. For this mechanism, assign a desired force trajectory using cubic polynomials, design a direct force controller with complete inverse dynamics, and consider %10 perturbation in all geometric and inertial parameters. Set the object location such that a collision occurs during the motion. Simulate the closed-loop performance and tune the controller gains to reach a suitable force tracking after the collision.

b. Plot and analyze closed-loop force trajectories, force tracking errors, Cartesian, and actuator forces versus time.

c. Plot and analyze the resulting motion trajectories versus the auxiliary motion variables generated by force feedback law and analyze the behavior of the force controller by intuition.

13. Consider a planar pantograph mechanism shown in Figure 3.30. The dynamic formulation of this mechanism and its kinematic and inertial parameters, and a typical motion trajectory is given in Problem 5.3. Consider a flat surface within the workspace of the mechanism such as that modeled in detail in Section 7.3.3, and used in Equation 7.26 with $\boldsymbol{K}_e = 100 \cdot \boldsymbol{I}_{3\times3}$ (N/m) and $\boldsymbol{C}_e = 100 \cdot \boldsymbol{I}_{3\times3}$ (N · m/s) to represent the contact forces.

    a. For this mechanism, assign a desired force trajectory using cubic polynomials, design a direct force controller with complete inverse dynamics, and consider %10 perturbation in all geometric and inertial parameters. Set the object location such that a collision occurs during the motion. Simulate the closed-loop performance and tune the controller gains to reach a suitable force tracking after the collision.

    b. Plot and analyze closed-loop force trajectories, force tracking errors, Cartesian, and actuator forces versus time.

    c. Plot and analyze the resulting motion trajectories versus the auxiliary motion variables generated by force feedback law and analyze the behavior of the force controller by intuition.

14. Consider the planar three-degrees-of-freedom 3$\underline{R}RR$ parallel manipulator shown in Figure 3.31. The dynamic formulation of this manipulator and its kinematic and inertial parameters, and a typical motion trajectory is given in Problem 5.4. Consider a flat surface within the workspace of the mechanism such as that modeled in detail in Section 7.3.3, and used in Equation 7.26 with $\boldsymbol{K}_e = 100 \cdot \boldsymbol{I}_{3\times3}$ (N/m) and $\boldsymbol{C}_e = 100 \cdot \boldsymbol{I}_{3\times3}$ (N · m/s) to represent the contact forces.

    a. For this manipulator, assign a desired force trajectory using cubic polynomials, design a direct force controller with complete inverse dynamics, and consider %10 perturbation in all geometric and inertial parameters. Set the object location such that a collision occurs during the motion. Simulate the closed-loop performance and tune the controller gains to reach a suitable force tracking after the collision.

    b. Plot and analyze closed-loop force trajectories, force tracking errors, Cartesian, and actuator forces versus time.

c. Plot and analyze the resulting motion trajectories versus the auxiliary motion variables generated by force feedback law and analyze the behavior of the force controller by intuition.

15. Consider the planar three-degrees-of-freedom $4\underline{R}RR$ parallel manipulator shown in Figure 3.32. The dynamic formulation of this manipulator and its kinematic and inertial parameters, and a typical motion trajectory is given in Problem 5.5. Consider a flat surface within the workspace of the mechanism such as that modeled in detail in Section 7.3.3, and used in Equation 7.26 with $\boldsymbol{K}_e = 100 \cdot \boldsymbol{I}_{3\times3}$ (N/m) and $\boldsymbol{C}_e = 100 \cdot \boldsymbol{I}_{3\times3}$ (N · m/s) to represent the contact forces.
    a. For this manipulator, assign a desired force trajectory using cubic polynomials, design a direct force controller with complete inverse dynamics, and consider %10 perturbation in all geometric and inertial parameters. Set the object location such that a collision occurs during the motion. Simulate the closed-loop performance and tune the controller gains to reach a suitable force tracking after the collision.
    b. Plot and analyze closed-loop force trajectories, force tracking errors, Cartesian, and actuator forces versus time.
    c. Plot and analyze the resulting motion trajectories versus the auxiliary motion variables generated by force feedback law and analyze the behavior of the force controller by intuition.

16. Consider the $3U\underline{P}U$ parallel manipulator shown in Figure 3.34. The dynamic formulation of this manipulator and its kinematic and inertial parameters, and a typical motion trajectory is given in Problem 5.7. Consider a flat object within the workspace of the mechanism such as that modeled in detail in Section 7.3.4, and used in Equation 7.29 with $\boldsymbol{K}_e = 100 \cdot \boldsymbol{I}_{3\times3}$ (N/m) and $\boldsymbol{C}_e = 100 \cdot \boldsymbol{I}_{3\times3}$ (N · m/s) to represent the contact forces.
    a. For this manipulator, assign a desired force trajectory using cubic polynomials, design a direct force controller with complete inverse dynamics, and consider %10 perturbation in all geometric and inertial parameters. Set the object location such that a collision occurs during the motion. Simulate the closed-loop performance and tune the controller gains to reach a suitable force tracking after the collision.
    b. Plot and analyze closed-loop force trajectories, force tracking errors, Cartesian, and actuator forces versus time.
    c. Plot and analyze the resulting motion trajectories versus the auxiliary motion variables generated by force feedback law and analyze the behavior of the force controller by intuition.

17. Consider the three-degrees-of-freedom $3R\underline{P}S$ manipulator represented in Figure 3.36. The dynamic formulation of this manipulator and its kinematic and inertial parameters, and a typical motion trajectory is given in Problem 5.9. Consider a flat object within the workspace of the mechanism such as that modeled in detail in Section 7.3.4, and used in Equation 7.29 with $\boldsymbol{K}_e = 100 \cdot \boldsymbol{I}_{3\times3}$ (N/m) and $\boldsymbol{C}_e = 100 \cdot \boldsymbol{I}_{3\times3}$ (N · m/s) to represent the contact forces.
    a. For this manipulator, assign a desired force trajectory using cubic polynomials, design a direct force controller with complete inverse dynamics, and consider %10 perturbation in all geometric and inertial parameters. Set the object location

such that a collision occurs during the motion. Simulate the closed-loop performance and tune the controller gains to reach a suitable force tracking after the collision.

b. Plot and analyze closed-loop force trajectories, force tracking errors, Cartesian, and actuator forces versus time.

c. Plot and analyze the resulting motion trajectories versus the auxiliary motion variables generated by force feedback law and analyze the behavior of the force controller by intuition.

18. Consider the three-degrees-of-freedom $3\underline{P}RC$ translational parallel manipulator represented in Figure 3.36. The dynamic formulation of this manipulator and its kinematic and inertial parameters, and a typical motion trajectory is given in Problem 5.10. Consider a flat object within the workspace of the mechanism such as that modeled in detail in Section 7.3.4 and used in Equation 7.29 with $\boldsymbol{K}_e = 100 \cdot \boldsymbol{I}_{3\times3}$ (N/m) and $\boldsymbol{C}_e = 100 \cdot \boldsymbol{I}_{3\times3}$ (N · m/s) to represent the contact forces.

    a. For this manipulator, assign a desired force trajectory using cubic polynomials, design a direct force controller with complete inverse dynamics, and consider %10 perturbation in all geometric and inertial parameters. Set the object location such that a collision occurs during the motion. Simulate the closed-loop performance and tune the controller gains to reach a suitable force tracking after the collision.

    b. Plot and analyze closed-loop force trajectories, force tracking errors, Cartesian, and actuator forces versus time.

    c. Plot and analyze the resulting motion trajectories versus the auxiliary motion variables generated by force feedback law and analyze the behavior of the force controller by intuition.

19. Consider the Delta robot represented in Figure 3.39. The dynamic formulation of this manipulator and its kinematic and inertial parameters, and a typical motion trajectory is given in Problem 5.11. Consider a flat object within the workspace of the mechanism such as that modeled in detail in Section 7.3.4, and used in Equation 7.29 with $\boldsymbol{K}_e = 100 \cdot \boldsymbol{I}_{3\times3}$ (N/m) and $\boldsymbol{C}_e = 100 \cdot \boldsymbol{I}_{3\times3}$ (N · m/s) to represent the contact forces.

    a. For this manipulator, assign a desired force trajectory using cubic polynomials, design a direct force controller with complete inverse dynamics, and consider %10 perturbation in all geometric and inertial parameters. Set the object location such that a collision occurs during the motion. Simulate the closed-loop performance and tune the controller gains to reach a suitable force tracking after the collision.

    b. Plot and analyze closed-loop force trajectories, force tracking errors, Cartesian, and actuator forces versus time.

    c. Plot and analyze the resulting motion trajectories versus the auxiliary motion variables generated by force feedback law and analyze the behavior of the force controller by intuition.

20. Consider the Quadrupteron robot represented in Figure 3.40. The dynamic formulation of this manipulator and its kinematic and inertial parameters, and a typical motion trajectory is given in Problem 5.12. Consider a flat object within the workspace of the mechanism such as that modeled in detail in Section 7.3.4, and

used in Equation 7.29 with $K_e = 100 \cdot I_{4\times4}$ (N/m) and $C_e = 100 \cdot I_{4\times4}$ (N · m/s) to represent the contact forces.

a. For this manipulator, assign a desired force trajectory using cubic polynomials, design a direct force controller with complete inverse dynamics, and consider %10 perturbation in all geometric and inertial parameters. Set the object location such that a collision occurs during the motion. Simulate the closed-loop performance and tune the controller gains to reach a suitable force tracking after the collision.

b. Plot and analyze closed-loop force trajectories, force tracking errors, Cartesian, and actuator forces versus time.

c. Plot and analyze the resulting motion trajectories versus the auxiliary motion variables generated by force feedback law and analyze the behavior of the force controller by intuition.

21. Consider the five-bar mechanism shown in Figure 3.28. The dynamic formulation of this mechanism and its kinematic and inertia parameters, and a typical motion trajectory is given in Problem 5.1. Consider a flat surface within the workspace of the mechanism such as that modeled in detail in Section 7.3.3, and used in Equation 7.26 with $K_e = 100 \cdot I_{3\times3}$ (N/m) and $C_e = 100 \cdot I_{3\times3}$ (N · m/s) to represent the contact forces.

    a. For this mechanism, design an impedance controller with complete inverse dynamics, and consider %10 perturbation in all geometric and inertial parameters. Set the object location such that a collision occurs during the motion, and assign suitable impedance matrices M_d, C_d, and K_d for the closed-loop system. Simulate the closed-loop performance and tune the impedance controller gains to reach a suitable impedance dynamics during the contact.

    b. Plot and analyze closed-loop motion trajectories, tracking errors, Cartesian, and actuator forces versus time.

    c. Plot and analyze the interaction forces in x and y directions versus the corresponding impedance dynamics, and analyze the impedance behavior of the closed, loop system before and after the contact.

    d. Change the desired impedance matrices and repeat part (c) to verify the performance of the controller.

22. Consider the five-bar mechanism shown in Figure 3.29. The dynamic formulation of this mechanism and its kinematic and inertia parameters, and a typical motion trajectory is given in Problem 5.2. Consider a flat surface within the workspace of the mechanism such as that modeled in detail in Section 7.3.3, and used in Equation 7.26 with $K_e = 100 \cdot I_{3\times3}$ (N/m) and $C_e = 100 \cdot I_{3\times3}$ (N · m/s) to represent the contact forces.

    a. For this mechanism, design an impedance controller with complete inverse dynamics, and consider %10 perturbation in all geometric and inertial parameters. Set the object location such that a collision occurs during the motion, and assign suitable impedance matrices M_d, C_d, and K_d for the closed-loop system. Simulate the closed-loop performance and tune the impedance controller gains to reach a suitable impedance dynamics during the contact.

    b. Plot and analyze closed-loop motion trajectories, tracking errors, Cartesian, and actuator forces versus time.

c. Plot and analyze the interaction forces in x and y directions versus the corresponding impedance dynamics, and analyze the impedance behavior of the closed-loop system before and after the contact.

d. Change the desired impedance matrices and repeat part (c) to verify the performance of the controller.

23. Consider a planar pantograph mechanism shown in Figure 3.30. The dynamic formulation of this mechanism and its kinematic and inertial parameters, and a typical motion trajectory is given in Problem 5.3. Consider a flat surface within the workspace of the mechanism such as that modeled in detail in Section 7.3.3, and used in Equation 7.26 with $\boldsymbol{K}_e = 100 \cdot \boldsymbol{I}_{3\times3}$ (N/m) and $\boldsymbol{C}_e = 100 \cdot \boldsymbol{I}_{3\times3}$ (N · m/s) to represent the contact forces.

    a. For this mechanism, design an impedance controller with complete inverse dynamics, and consider %10 perturbation in all geometric and inertial parameters. Set the object location such that a collision occurs during the motion, and assign suitable impedance matrices $\boldsymbol{M}_d, \boldsymbol{C}_d$, and $\boldsymbol{K}_d$ for the closed-loop system. Simulate the closed-loop performance and tune the impedance controller gains to reach a suitable impedance dynamics during the contact.

    b. Plot and analyze closed-loop motion trajectories, tracking errors, Cartesian, and actuator forces versus time.

    c. Plot and analyze the interaction forces in x and y directions versus the corresponding impedance dynamics, and analyze the impedance behavior of the closed-loop system before and after the contact.

    d. Change the desired impedance matrices and repeat part (c) to verify the performance of the controller.

24. Consider the planar three-degrees-of-freedom $3\underline{R}RR$ parallel manipulator shown in Figure 3.31. The dynamic formulation of this manipulator and its kinematic and inertial parameters, and a typical motion trajectory is given in Problem 5.4. Consider a flat surface within the workspace of the mechanism such as that modeled in detail in Section 7.3.3, and used in Equation 7.26 with $\boldsymbol{K}_e = 100 \cdot \boldsymbol{I}_{3\times3}$ (N/m) and $\boldsymbol{C}_e = 100 \cdot \boldsymbol{I}_{3\times3}$ (N · m/s) to represent the contact forces.

    a. For this manipulator, design an impedance controller with complete inverse dynamics, and consider %10 perturbation in all geometric and inertial parameters. Set the object location such that a collision occurs during the motion, and assign suitable impedance matrices $\boldsymbol{M}_d, \boldsymbol{C}_d$, and $\boldsymbol{K}_d$ for the closed-loop system. Simulate the closed-loop performance and tune the impedance controller gains to reach a suitable impedance dynamics during the contact.

    b. Plot and analyze closed-loop motion trajectories, tracking errors, Cartesian, and actuator forces versus time.

    c. Plot and analyze the interaction forces in x and y directions versus the corresponding impedance dynamics, and analyze the impedance behavior of the closed-loop system before and after the contact.

    d. Change the desired impedance matrices and repeat part (c) to verify the performance of the controller.

25. Consider the planar three-degrees-of-freedom $4\underline{R}RR$ parallel manipulator shown in Figure 3.32. The dynamic formulation of this manipulator and its kinematic and

inertial parameters, and a typical motion trajectory is given in Problem 5.5. Consider a flat surface within the workspace of the mechanism such as that modeled in detail in Section 7.3.3, and used in Equation 7.26 with $\boldsymbol{K}_e = 100 \cdot \boldsymbol{I}_{3\times3}$ (N/m) and $\boldsymbol{C}_e = 100 \cdot \boldsymbol{I}_{3\times3}$ (N · m/s) to represent the contact forces.

a. For this manipulator, design an impedance controller with complete inverse dynamics, and consider %10 perturbation in all geometric and inertial parameters. Set the object location such that a collision occurs during the motion, and assign suitable impedance matrices $\boldsymbol{M}_d, \boldsymbol{C}_d$, and $\boldsymbol{K}_d$ for the closed-loop system. Simulate the closed-loop performance and tune the impedance controller gains to reach a suitable impedance dynamics during the contact.
b. Plot and analyze closed-loop motion trajectories, tracking errors, Cartesian, and actuator forces versus time.
c. Plot and analyze the interaction forces in x and y directions versus the corresponding impedance dynamics, and analyze the impedance behavior of the closed-loop system before and after the contact.
d. Change the desired impedance matrices and repeat part (c) to verify the performance of the controller.

26. Consider the $3U\underline{P}U$ parallel manipulator shown in Figure 3.34. The dynamic formulation of this manipulator and its kinematic and inertial parameters, and a typical motion trajectory is given in Problem 5.7. Consider a flat object within the workspace of the mechanism such as that modeled in detail in Section 7.3.4, and used in Equation 7.29 with $\boldsymbol{K}_e = 100 \cdot \boldsymbol{I}_{3\times3}$ (N/m) and $\boldsymbol{C}_e = 100 \cdot \boldsymbol{I}_{3\times3}$ (N · m/s) to represent the contact forces.
    a. For this manipulator, design an impedance controller with complete inverse dynamics, and consider %10 perturbation in all geometric and inertial parameters. Set the object location such that a collision occurs during the motion, and assign suitable impedance matrices $\boldsymbol{M}_d, \boldsymbol{C}_d$, and $\boldsymbol{K}_d$ for the closed-loop system. Simulate the closed-loop performance and tune the impedance controller gains to reach a suitable impedance dynamics during the contact.
    b. Plot and analyze closed-loop motion trajectories, tracking errors, Cartesian, and actuator forces versus time.
    c. Plot and analyze the interaction forces in x, y, and z directions versus the corresponding impedance dynamics, and analyze the impedance behavior of the closed-loop system before and after the contact.
    d. Change the desired impedance matrices and repeat part (c) to verify the performance of the controller.

27. Consider the three-degrees-of-freedom $3R\underline{P}S$ manipulator represented in Figure 3.36. The dynamic formulation of this manipulator and its kinematic and inertial parameters, and a typical motion trajectory is given in Problem 5.9. Consider a flat object within the workspace of the mechanism such as that modeled in detail in Section 7.3.4, and used in Equation 7.29 with $\boldsymbol{K}_e = 100 \cdot \boldsymbol{I}_{3\times3}$ (N/m) and $\boldsymbol{C}_e = 100 \cdot \boldsymbol{I}_{3\times3}$ (N · m/s) to represent the contact forces.
    a. For this manipulator, design an impedance controller with complete inverse dynamics, and consider %10 perturbation in all geometric and inertial parameters. Set the object location such that a collision occurs during the motion, and assign suitable impedance matrices $\boldsymbol{M}_d, \boldsymbol{C}_d$, and $\boldsymbol{K}_d$ for

the closed-loop system. Simulate the closed-loop performance and tune the impedance controller gains to reach a suitable impedance dynamics during the contact.

b. Plot and analyze closed-loop motion trajectories, tracking errors, Cartesian, and actuator forces versus time.

c. Plot and analyze the interaction forces in x, y, and z directions versus the corresponding impedance dynamics, and analyze the impedance behavior of the closed-loop system before and after the contact.

d. Change the desired impedance matrices and repeat part (c) to verify the performance of the controller.

28. Consider the three-degrees-of-freedom $3\underline{P}RC$ translational parallel manipulator represented in Figure 3.36. The dynamic formulation of this manipulator and its kinematic and inertial parameters, and a typical motion trajectory is given in Problem 5.10. Consider a flat object within the workspace of the mechanism such as that modeled in detail in Section 7.3.4 and used in Equation 7.29 with $\boldsymbol{K}_e = 100 \cdot \boldsymbol{I}_{3\times3}$ (N/m) and $\boldsymbol{C}_e = 100 \cdot \boldsymbol{I}_{3\times3}$ (N · m/s) to represent the contact forces.

    a. For this manipulator, design an impedance controller with complete inverse dynamics, and consider %10 perturbation in all geometric and inertial parameters. Set the object location such that a collision occurs during the motion, and assign suitable impedance matrices $\boldsymbol{M}_d, \boldsymbol{C}_d$, and $\boldsymbol{K}_d$ for the closed-loop system. Simulate the closed-loop performance and tune the impedance controller gains to reach a suitable impedance dynamics during the contact.

    b. Plot and analyze closed-loop motion trajectories, tracking errors, Cartesian, and actuator forces versus time.

    c. Plot and analyze the interaction forces in x, y, and z directions versus the corresponding impedance dynamics, and analyze the impedance behavior of the closed-loop system before and after the contact.

    d. Change the desired impedance matrices and repeat part (c) to verify the performance of the controller.

29. Consider the Delta robot represented in Figure 3.39. The dynamic formulation of this manipulator and its kinematic and inertial parameters, and a typical motion trajectory is given in Problem 5.11. Consider a flat object within the workspace of the mechanism such as that modeled in detail in Section 7.3.4, and used in Equation 7.29 with $\boldsymbol{K}_e = 100 \cdot \boldsymbol{I}_{3\times3}$ (N/m) and $\boldsymbol{C}_e = 100 \cdot \boldsymbol{I}_{3\times3}$ (N · m/s) to represent the contact forces.

    a. For this manipulator, design an impedance controller with complete inverse dynamics, and consider %10 perturbation in all geometric and inertial parameters. Set the object location such that a collision occurs during the motion, and assign suitable impedance matrices $\boldsymbol{M}_d, \boldsymbol{C}_d$, and $\boldsymbol{K}_d$ for the closed-loop system. Simulate the closed-loop performance and tune the impedance controller gains to reach a suitable impedance dynamics during the contact.

    b. Plot and analyze closed-loop motion trajectories, tracking errors, Cartesian, and actuator forces versus time.

    c. Plot and analyze the interaction forces in x, y, and z directions versus the corresponding impedance dynamics, and analyze the impedance behavior of the closed-loop system before and after the contact.

d. Change the desired impedance matrices and repeat part (c) to verify the performance of the controller.

30. Consider the Quadrupteron robot represented in Figure 3.40. The dynamic formulation of this manipulator and its kinematic and inertial parameters, and a typical motion trajectory is given in Problem 5.12. Consider a flat object within the workspace of the mechanism such as that modeled in detail in Section 7.3.4, and used in Equation 7.29 with $\boldsymbol{K}_e = 100 \cdot \boldsymbol{I}_{4\times4}$ (N/m) and $\boldsymbol{C}_e = 100 \cdot \boldsymbol{I}_{4\times4}$ $(\mathrm{N} \cdot \mathrm{m/s})$ to represent the contact forces.

    a. For this manipulator, design an impedance controller with complete inverse dynamics, and consider %10 perturbation in all geometric and inertial parameters. Set the object location such that a collision occurs during the motion, and assign suitable impedance matrices $\boldsymbol{M}_d, \boldsymbol{C}_d$, and $\boldsymbol{K}_d$ for the closed-loop system. Simulate the closed-loop performance and tune the impedance controller gains to reach a suitable impedance dynamics during the contact.

    b. Plot and analyze closed-loop motion trajectories, tracking errors, Cartesian, and actuator forces versus time.

    c. Plot and analyze the interaction forces in x, y, and z directions versus the corresponding impedance dynamics, and analyze the impedance behavior of the closed-loop system before and after the contact.

    d. Change the desired impedance matrices and repeat part (c) to verify the performance of the controller.

Appendix A: Linear Algebra

In this book, it is assumed that the reader is familiar with the basic concepts and properties of vectors and matrices. These basics are not defined here. However, some advanced properties in linear algebra which are used in the book are briefly reviewed in this chapter. For additional information refer to [171].

A.1 Vectors and Matrices

A vector is defined as an array of n-tuple elements arranged into n rows. Usually lowercase letters with bold fonts like $\boldsymbol{x}$, $\boldsymbol{y}$, $\boldsymbol{z}$ are used to denote vectors. An n-tuple array $\boldsymbol{x}$ is said to belong to the space of $\mathbb{R}^n$, if all its elements x_i's, belong to the space of real numbers $\mathbb{R}$. Thus, $\boldsymbol{x} \in \mathbb{R}^n$ is defined as

$$\boldsymbol{x} = \begin{bmatrix} x_1 \\ x_2 \\ \vdots \\ x_n \end{bmatrix}, \tag{A.1}$$

in which $x_i \in \mathbb{R}, i = 1, 2, \ldots, n$. A row vector is defined as the transpose of a vector:

$$\boldsymbol{x}^T = \begin{bmatrix} x_1 & x_2 & \ldots & x_n \end{bmatrix}, \tag{A.2}$$

in which the superscript T denotes transposition operator. A matrix of dimension $m \times n$ is defined as an array of m rows and n columns. Usually uppercase letters with bold fonts like $\boldsymbol{A}, \boldsymbol{B}, \boldsymbol{C}$ are used to denote matrices. An $m \times n$ matrix $\boldsymbol{A}$ is said to belong to the space of $\mathbb{R}^{m \times n}$, if all its elements a_{ij}'s, belong to the space of real numbers $\mathbb{R}$. Thus, $\boldsymbol{A} \in \mathbb{R}^{m \times n}$ is defined as

$$\boldsymbol{A} = \begin{bmatrix} a_{11} & a_{12} & \cdots & a_{1n} \\ a_{21} & a_{22} & \cdots & a_{2n} \\ \vdots & & & \vdots \\ a_{m1} & a_{m2} & \cdots & a_{mn} \end{bmatrix}, \tag{A.3}$$

in which $a_{ij} \in \mathbb{R}, i = 1, 2, \ldots, m, j = 1, 2, \ldots, n$. A square $n \times n$ matrix $\boldsymbol{A}$ is said to be *upper triangular*, if $a_{ij} = 0$ for $i > j$; it is called *lower triangular*, if $a_{ij} = 0$ for $i < j$; and it is called diagonal, if $a_{ij} = 0$ for $i \neq j$. An $n \times n$ *diagonal* matrix the diagonal elements of which are all equal to 1 is called *identity* matrix and is denoted by $\boldsymbol{I}$. An $n \times n$ matrix whose elements are all equal to zero is called *null* matrix, and is denoted by $\boldsymbol{0}$. The transpose of an $m \times n$ matrix $\boldsymbol{A}$ is obtained by interchanging its rows and columns, and is denoted by $\boldsymbol{A}^T \in \mathbb{R}^{n \times m}$. A square matrix $\boldsymbol{A}$ is called *symmetric* if $\boldsymbol{A}^T = \boldsymbol{A}$, *skew-symmetric* if $\boldsymbol{A}^T = -\boldsymbol{A}$, and *orthogonal* if $\boldsymbol{A}^T\boldsymbol{A} = \boldsymbol{A}\boldsymbol{A}^T = \boldsymbol{I}$.

A.2 Vector and Matrix Operations

The *scalar* or *dot* product of vectors $\boldsymbol{x}$ and $\boldsymbol{y}$ belonging to $\mathbb{R}^n$ results in a scalar real number defined by

$$\boldsymbol{x} \cdot \boldsymbol{y} = \boldsymbol{x}^T \boldsymbol{y} = x_1 y_1 + x_2 y_2 + \cdots + x_n y_n. \tag{A.4}$$

The dot product of two vectors is commutative, $\boldsymbol{x} \cdot \boldsymbol{y} = \boldsymbol{y} \cdot \boldsymbol{x}$. The two vectors $\boldsymbol{x}$ and $\boldsymbol{y}$ are called *orthogonal*, if their dot product is zero, $\boldsymbol{x} \cdot \boldsymbol{y} = 0$. The *Euclidean norm* of a vector $\boldsymbol{x} \in \mathbb{R}^n$ denoted by $\|\boldsymbol{x}\|$ represents the geometric length of the vector and is obtained from

$$\|\boldsymbol{x}\| = \sqrt{\boldsymbol{x} \cdot \boldsymbol{x}} = \sqrt{x_1^2 + x_2^2 + \cdots + x_n^2}. \tag{A.5}$$

A unit vector denoted by $\hat{x}$, may be obtained from the division of the vector to its norm:

$$\hat{x} = \frac{\boldsymbol{x}}{\|\boldsymbol{x}\|}. \tag{A.6}$$

The following useful inequalities apply to Euclidean norm of vectors,

$$|\boldsymbol{x} \cdot \boldsymbol{y}| \leq \|\boldsymbol{x}\| \cdot \|\boldsymbol{y}\|, \tag{A.7}$$

$$\|\boldsymbol{x} + \boldsymbol{y}\| \leq \|\boldsymbol{x}\| + \|\boldsymbol{y}\|, \tag{A.8}$$

where A.7 is called Cauchy–Schwartz inequality, and A.8 is called triangle inequality.

The *vector* or *cross* product of vectors $\boldsymbol{x}$ and $\boldsymbol{y}$ belonging to $\mathbb{R}^n$ results in another vector $\boldsymbol{z} \in \mathbb{R}^n$ defined as

$$\boldsymbol{z} = \boldsymbol{x} \times \boldsymbol{y} = \begin{bmatrix} x_2 y_3 - x_3 y_2 \\ x_3 y_1 - x_1 y_3 \\ x_1 y_2 - x_2 y_1 \end{bmatrix}. \tag{A.9}$$

The dot and cross products have the following properties:

$$\boldsymbol{x} \times \boldsymbol{x} = \boldsymbol{0}, \tag{A.10}$$

$$\boldsymbol{x} \times \boldsymbol{y} = -\boldsymbol{y} \times \boldsymbol{x}, \tag{A.11}$$

$$\boldsymbol{x} \times (\boldsymbol{y} + \boldsymbol{z}) = \boldsymbol{x} \times \boldsymbol{y} + \boldsymbol{x} \times \boldsymbol{z}, \tag{A.12}$$

$$\alpha(\boldsymbol{x} \times \boldsymbol{y}) = (\alpha \boldsymbol{x}) \times \boldsymbol{y} = \boldsymbol{x} \times (\alpha \boldsymbol{y}), \tag{A.13}$$

$$\boldsymbol{x} \cdot (\boldsymbol{y} \times \boldsymbol{z}) = \boldsymbol{y} \cdot (\boldsymbol{z} \times \boldsymbol{x}) = \boldsymbol{z} \cdot (\boldsymbol{x} \times \boldsymbol{y}), \tag{A.14}$$

$$(\boldsymbol{x} \cdot \boldsymbol{y})\boldsymbol{z} = (\boldsymbol{x}^T \boldsymbol{y})\boldsymbol{z} = \boldsymbol{z}(\boldsymbol{x}^T \boldsymbol{y}) = (\boldsymbol{z}\boldsymbol{x}^T)\boldsymbol{y} = \boldsymbol{z}\boldsymbol{x}^T \boldsymbol{y}. \tag{A.15}$$

Equation A.14 is called the property of *triple* product. Note that the last equality in Equation A.15 is converting the vector product to matrix operation which is very useful in the simplification of dynamic equations. The vector product of the two vectors $\boldsymbol{x}$ and $\boldsymbol{y}$ may be expressed as the product of a skew-symmetric matrix obtained from the components of the vector by

$$\boldsymbol{x}_\times \text{ or } \boldsymbol{x}^\times = \begin{bmatrix} 0 & -x_3 & x_2 \\ x_3 & 0 & -x_1 \\ -x_2 & x_1 & 0 \end{bmatrix}, \tag{A.16}$$

to the other vector as

$$x \times y = x^{\times} y = -y^{\times} x. \tag{A.17}$$

A set of vectors $\{x_1, x_2, \ldots, x_n\}$ is called linearly independent, if and only if, linear combination of the vectors is zero only when all the combination coefficients are zero, that is,

$$\sum_{i=1}^{n} \alpha_i x_i = \mathbf{0} \Rightarrow \alpha_i = 0 \quad \forall i. \tag{A.18}$$

The rank of a matrix A denoted by rank(A) is the largest number of linearly independent columns (or rows) of A. Therefore, the rank of an $n \times m$ matrix cannot be greater than $\min(m, n)$. The inverse of a square $n \times n$ matrix is a matrix denoted by A^{-1} belonging to $\mathbb{R}^{n \times n}$ satisfying the following equation:

$$AA^{-1} = A^{-1}A = I, \tag{A.19}$$

where I denotes the identity matrix. The inverse of a matrix exists and is uniquely derived, if and only if, rank(A) $= n$, or equivalently det(A) $\neq 0$. Furthermore, the inverse of a matrix satisfies the following properties:

$$(A^{-1})^{-1} = A, \tag{A.20}$$

$$(AB)^{-1} = B^{-1}A^{-1}. \tag{A.21}$$

A matrix is called orthogonal, if consist of orthogonal column vectors. The inverse of an orthogonal matrix is its transpose, that is, $A^{-1} = A^T$.

If A and C are invertible square matrices of proper dimensions, the following equation holds:

$$(A + BCD)^{-1} = A^{-1} - A^{-1}B(DA^{-1}B + C^{-1})^{-1}DA^{-1}, \tag{A.22}$$

where the matrix product $(DA^{-1}B + C^{-1})$ shall be invertible. For a block partitioned matrix, the inverse can be determined from

$$\begin{bmatrix} A & B \\ C & D \end{bmatrix}^{-1} = \begin{bmatrix} A^{-1} + A^{-1}B\Delta^{-1}CA^{-1} & -A^{-1}B\Delta^{-1} \\ -\Delta^{-1}CA^{-1} & \Delta^{-1} \end{bmatrix} \tag{A.23}$$

$$= \begin{bmatrix} \Omega^{-1} & -\Omega^{-1}BD^{-1} \\ -D^{-1}C\Omega^{-1} & D^{-1} + D^{-1}C\Omega^{-1}BD^{-1} \end{bmatrix} \tag{A.24}$$

in which $\Delta = D - CA^{-1}B$ and $\Omega = A - BD^{-1}C$ shall be invertible. These relations are known as *matrix inversion Lemma*. For simple cases, where either B or C is equal to zero, these relations simplify to

$$\begin{bmatrix} A & 0 \\ C & D \end{bmatrix}^{-1} = \begin{bmatrix} A^{-1} & 0 \\ -D^{-1}CA^{-1} & D^{-1} \end{bmatrix}, \tag{A.25}$$

$$\begin{bmatrix} A & B \\ 0 & D \end{bmatrix}^{-1} = \begin{bmatrix} A^{-1} & -A^{-1}BD^{-1} \\ 0 & D^{-1} \end{bmatrix}. \tag{A.26}$$

The *null space* $\mathcal{N}$ of a matrix $\boldsymbol{A}$ is defined as

$$\mathcal{N}(\boldsymbol{A}) = \{\boldsymbol{x} \in \mathbb{R}^n : \boldsymbol{A}\boldsymbol{x} = \boldsymbol{0}\} \tag{A.27}$$

if $\boldsymbol{A}$ is an $n \times n$ matrix, then

$$\text{rank}(\boldsymbol{A}) + \dim\mathcal{N}(\boldsymbol{A}) = n. \tag{A.28}$$

Thus, a matrix is invertible, if and only if, the null space consists of only the zero vector, or $\dim\mathcal{N}(\boldsymbol{A}) = 0$.

Finally, the *induced norm* of a square matrix is defined as

$$\|\boldsymbol{A}\| = \sup_{\|\boldsymbol{x}\| \neq 0} \frac{\|\boldsymbol{A}\boldsymbol{x}\|}{\|\boldsymbol{x}\|}. \tag{A.29}$$

Therefore, it can be shown that

$$\|\boldsymbol{A}\boldsymbol{x}\| \leq \|\boldsymbol{A}\| \, \|\boldsymbol{x}\|, \tag{A.30}$$

$$\|\boldsymbol{A}\boldsymbol{B}\| \leq \|\boldsymbol{A}\| \, \|\boldsymbol{B}\|. \tag{A.31}$$

A.3 Eigenvalues and Singular Values

The *eigenvalues* of a square matrix $\boldsymbol{A}$ are the λ solutions of the following equation:

$$\det(\lambda \boldsymbol{I} - \boldsymbol{A}) = 0. \tag{A.32}$$

This equation is termed as *characteristics equation* of matrix $\boldsymbol{A}$, and its solutions $\lambda_1, \lambda_2, \ldots, \lambda_n$ are the eigenvalues of matrix $\boldsymbol{A}$. The eigenvalues of matrix $\boldsymbol{A}$ are the same as the eigenvalues of matrix $\boldsymbol{A}^T$. The vectors $\boldsymbol{v}_i$ satisfying the following systems of linear equations:

$$(\lambda_i \boldsymbol{I} - \boldsymbol{A})\boldsymbol{v}_i = \boldsymbol{0} \tag{A.33}$$

are called the corresponding *eigenvectors* of matrix $\boldsymbol{A}$.

If eigenvalues of an $n \times n$ matrix $\boldsymbol{A}$ are distinct, n linearly independent eigenvectors can be found to satisfy Equation A.33. The matrix $\boldsymbol{T}$ formed by column vectors $\boldsymbol{v}_i$, is then invertible and can be used as a *similarity transform* to diagonalize matrix $\boldsymbol{A}$. Matrix $\boldsymbol{\Lambda} = \boldsymbol{T}^{-1}\boldsymbol{A}\boldsymbol{T}$ is a transformed diagonal matrix, the diagonal elements of which are the eigenvalues of matrix $\boldsymbol{A}$.

Eigenvalues and eigenvectors are not defined for nonsquare matrices. Consider a nonsquare matrix $\boldsymbol{A} \in \mathbb{R}^{m \times n}$, in which $m < n$. For such a case, a square matrix with a dimension $m \times m$ may be generated by $\boldsymbol{A}\boldsymbol{A}^T$. The *singular values* of matrix $\boldsymbol{A}$ denoted by σ_i are defined as the square root of eigenvalues of $\boldsymbol{A}\boldsymbol{A}^T$,

$$\sigma_i(\boldsymbol{A}) = \sqrt{\lambda_i(\boldsymbol{A}\boldsymbol{A}^T)}. \tag{A.34}$$

Many tractable routines are developed to obtain singular values of a matrix. One of the most important techniques is called *singular value decomposition* (SVD), given by

$$A = \boldsymbol{U}\boldsymbol{\Sigma}\boldsymbol{V}^T \tag{A.35}$$

in which $\boldsymbol{U} \in \mathbb{R}^{m\times m}$ and $\boldsymbol{V} \in \mathbb{R}^{n\times n}$ are orthogonal matrices defined by

$$\boldsymbol{U} = \begin{bmatrix} u_1 & u_2 & \cdots & u_m \end{bmatrix}, \quad \boldsymbol{V} = \begin{bmatrix} v_1 & v_2 & \cdots & v_n \end{bmatrix} \tag{A.36}$$

and furthermore, $\boldsymbol{\Sigma} \in \mathbb{R}^{m\times n}$ is given by

$$\boldsymbol{\Sigma} = \left[\begin{array}{cccc|ccc} \sigma_1 & 0 & \cdots & 0 & 0 & \cdots & 0 \\ 0 & \sigma_2 & & 0 & 0 & \cdots & 0 \\ \vdots & & & \vdots & \vdots & & \vdots \\ 0 & 0 & \cdots & \sigma_m & 0 & \cdots & 0 \end{array}\right]. \tag{A.37}$$

The SVD of $\boldsymbol{A}$ may be computed as follows. First, the singular values σ_i of $\boldsymbol{A}$ are computed from the eigenvalues λ_i of $\boldsymbol{A}\boldsymbol{A}^T$ by the following equation:

$$\det(\boldsymbol{A}\boldsymbol{A}^T - \lambda_i \boldsymbol{I}) = 0; \quad \sigma_i = \sqrt{\lambda_i}. \tag{A.38}$$

These singular values can be then used to derive eigenvectors $u_1, u_2, \ldots, u_m$ that satisfy

$$(\boldsymbol{A}\boldsymbol{A}^T - \sigma_i^2 \boldsymbol{I})\boldsymbol{u}_i = \boldsymbol{0}. \tag{A.39}$$

The orthogonal matrix $\boldsymbol{U} = \begin{bmatrix} u_1 & u_2 & \cdots & u_m \end{bmatrix}$ is formed by these eigenvectors. Equation A.39 may be written as

$$\boldsymbol{A}\boldsymbol{A}^T\boldsymbol{U} = \boldsymbol{U}\boldsymbol{\Sigma}_m^2 \tag{A.40}$$

in which $\boldsymbol{\Sigma}_m$ is the diagonal part of matrix $\boldsymbol{\Sigma}$ defined as

$$\boldsymbol{\Sigma}_m = \begin{bmatrix} \sigma_1 & 0 & \cdots & 0 \\ 0 & \sigma_2 & & 0 \\ \vdots & & & \vdots \\ 0 & 0 & \cdots & \sigma_m \end{bmatrix}. \tag{A.41}$$

By this equation, one may find the first m columns of matrix $\boldsymbol{V}$, denoted by $\boldsymbol{V}_m$ as

$$\boldsymbol{V}_m = \boldsymbol{A}^T\boldsymbol{U}\boldsymbol{\Sigma}_m^{-1}. \tag{A.42}$$

The remaining $n-m$ columns of matrix $\boldsymbol{V}$, denoted by $\boldsymbol{V}_{n-m}$ may be generated to form an orthogonal matrix $\boldsymbol{V} = [\boldsymbol{V}_m \mid \boldsymbol{V}_{n-m}]$. $\boldsymbol{V}_{n-m}$ does not contribute directly to singular-value decomposition. This can be shown by

$$\begin{aligned} \boldsymbol{U}\boldsymbol{\Sigma}\boldsymbol{V}^T &= \boldsymbol{U}[\boldsymbol{\Sigma}_m \mid \boldsymbol{0}]\begin{bmatrix} \boldsymbol{V}_m^T \\ \boldsymbol{V}_{n-m}^T \end{bmatrix} \\ &= \boldsymbol{U}\boldsymbol{\Sigma}_m\boldsymbol{V}_m^T. \end{aligned} \tag{A.43}$$

Substitute V_m from Equation A.42, use orthogonality property of matrix U, and symmetricity of matrix Σ_m to show that

$$U\Sigma V^T = A. \tag{A.44}$$

Singular-value decomposition is very useful in many disciplines, it is applied especially to the study of linear inverse problems, and calculation of pseudo-inverse of matrix A.

A.4 Pseudo-Inverse

Inverse of matrices plays an important role in the study of linear inverse problems. The solution of a linear equation $Ax = y$ is simply found by inverse of matrix A as $x = A^{-1}y$. However, a matrix has an inverse only if it is square and nonsingular. The pseudo-inverse is a generalization of the inverse to the case of singular, or nonsquare matrices. In the field of robotics, pseudo-inverse is commonly used in the identification of the manipulator dynamics and in the redundancy resolution techniques developed for the control of redundant robots.

Theorem A.1

Consider a finite and real matrix $A \in \mathbb{R}^{m\times n}$, the pseudo-inverse of matrix A denoted by $A^\dagger$ is a unique $n \times m$ matrix that satisfies the following four conditions:

$$AA^\dagger A = A, \tag{A.45}$$

$$A^\dagger AA^\dagger = A^\dagger, \tag{A.46}$$

$$(AA^\dagger)^T = AA^\dagger, \tag{A.47}$$

$$(A^\dagger A)^T = A^\dagger A. \tag{A.48}$$

These four equations are called *Penrose conditions.*

Proof. First, prove existence of $A^\dagger$ that satisfies conditions A.45 through A.48. If $A = \mathbf{0}$, then the trivial solution $A^\dagger = \mathbf{0}$ satisfies all the four conditions. Assume $A \neq \mathbf{0}$ and let $r = \text{rank}(A)$, then A may be expressed as the product of an $m \times r$ matrix B and an $r \times n$ matrix C, such that

$$A = BC. \tag{A.49}$$

Using this decomposition of matrix A, the pseudo-inverse is given by

$$A^\dagger = C^T(CC^T)^{-1}(B^TB)^{-1}B^T. \tag{A.50}$$

It is easy to show that this formulation will satisfy all the above four conditions, and this completes the proof of existence of $A^\dagger$.

Next, prove the uniqueness of $A^{\dagger}$. Denote any two matrices satisfying conditions A.45 through A.48 by $A_1^{\dagger}$ and $A_2^{\dagger}$. Then

$$\begin{aligned}
A_1^{\dagger} - A_2^{\dagger} &= A_1^{\dagger}AA_1^{\dagger} - A_2^{\dagger}AA_2^{\dagger} \\
&= A^T A_1^{\dagger T} A_1^{\dagger} - A_2^{\dagger}A_2^{\dagger T} A^T \\
&= (AA_2^{\dagger}A)^T A_1^{\dagger T} A_1^{\dagger} - A_2^{\dagger}A_2^{\dagger T}(AA_1^{\dagger}A)^T \\
&= A^T A_2^{\dagger T} A_1^{\dagger} - A_2^{\dagger}A_1^{\dagger T} A^T \\
&= A_2^{\dagger}AA_1^{\dagger} - A_2^{\dagger}AA_1^{\dagger} \\
&= \mathbf{0}.
\end{aligned}$$

This completes the proof of the theorem. ■

A.4.1 Pseudo-Inverse Properties

The pseudo-inverse has the following properties:

1. $(A^{\dagger})^{\dagger} = A$.
2. $(A^T)^{\dagger} = A^{\dagger T}, (AA^T)^{\dagger} = A^{\dagger T}A^{\dagger}$.
3. $A^{\dagger} = (A^T A)^{\dagger} A^T = A^T (AA^T)^{\dagger}$.
4. $A^T AA^{\dagger} = A^{\dagger}AA^T = A^T$
5. Let A be an $m \times n$ matrix and B an $n \times p$ matrix. Then $(AB)^{\dagger}$ is not generally equal to $B^{\dagger}A^{\dagger}$. However, if rank(A) = rank(B) = n, then $(AB)^{\dagger} = B^{\dagger}A^{\dagger}$.
6. For an $m \times n$ matrix A, if $m < n$ and rank(A) = m, then AA^T is a nonsingular $m \times m$ matrix, and the pseudo-inverse $A^{\dagger} \in \mathbb{R}^{n \times m}$ is derived from

$$A^{\dagger} = A^T (AA^T)^{-1}. \tag{A.51}$$

   This is called *right inverse*.
7. For an $m \times n$ matrix A, if $m > n$ and rank(A) = n, then $A^T A$ is a nonsingular $n \times n$ matrix, and the pseudo-inverse $A^{\dagger} \in \mathbb{R}^{n \times m}$, is derived from

$$A^{\dagger} = (A^T A)^{-1} A^T. \tag{A.52}$$

   This is called *left inverse*.
8. For an $m \times n$ matrix A, if $m < n$ and rank(A) = m, consider the singular-value decomposition of matrix A as $A = U\Sigma V^T$, in which U, V are corresponding orthogonal matrices, and $\Sigma = [\Sigma_m | \mathbf{0}]$ as defined in Equations A.37 and A.41. Then the pseudo-inverse $A^{\dagger} \in \mathbb{R}^{m \times m}$ is derived from

$$A^{\dagger} = V\,[\Sigma_m^{-1} \mid \mathbf{0}]\,U^T. \tag{A.53}$$

A.4.2 Linear Inverse Problems

In this section, the pseudo-inverse is used to solve two distinct cases of linear inverse problems. Given a set of linear equations $Ax = y$, in which matrix A is a known $m \times n$ matrix, y is a known $m \times 1$ vector, and x is an unknown $n \times 1$ vector, when A is square $m = n$, and nonsingular, that is, rank$(A) = n$, then the solution is simply represented by inverse of matrix A as $x = A^{-1}y$, or equivalently $x = A^{\dagger}y$.

If matrix A is nonsquare $n \neq m$ and $m > n$, then rank$(A) = n$ and there exists no direct inverse to matrix A. In this case, the number of equations m, is more than the number of unknowns n, and the inverse problem consists of an overdetermined set of linear equations. In general, there exists no solution to this problem that can satisfy all the equations simultaneously. However, the pseudo-inverse may result in a solution that minimizes the Euclidean norm of the error ($e = Ax - y$):

$$\|Ax - y\| = \|e\| = \sqrt{e^T e}. \tag{A.54}$$

In order to prove this, note that for an optimal solution, the gradient of the cost function with respect to the unknown variable must be zero. Derive this gradient by

$$\begin{aligned}\frac{\partial(e^T e)}{\partial x} &= 2A^T e \\ &= 2A^T(Ax - y).\end{aligned} \tag{A.55}$$

To minimize $V = e^T e$, this gradient must be zero. Thus,

$$A^T A x = A^T y. \tag{A.56}$$

Since $A^T A$ is a full rank $n \times n$ square matrix, it is invertible. Therefore, the resulting least-squares optimal solution is derived by

$$x_{opt} = (A^T A)^{-1} A^T y. \tag{A.57}$$

The optimal solution is obtained from the left inverse of matrix A defined in Equation A.52.

Now consider the case where matrix A is nonsquare $n \neq m$ and $m < n$, then rank$(A) = m$ and there exists no direct inverse to matrix A. In this case, the number of equations m is less than the number of unknowns n, and the inverse problem consists of an underdetermined set of linear equations. In general, there exist many solutions to this problem that can satisfy all the equations simultaneously. For such a case, one may search among the many existing solutions for the one that optimizes a particular performance index. One general optimal solution may be obtained from the solution that minimizes its own norm $\|x\|$. For this case, it can be shown that the general solution may be obtained from the following equation:

$$x = A^{\dagger}y + (I - A^{\dagger}A)z \tag{A.58}$$

in which z is any arbitrary $n \times 1$ vector. In what follows, it is shown that Equation A.58 satisfies the linear problem $Ax = y$, and hence, it is a general solution to this problem.

$$\begin{aligned} Ax &= AA^{\dagger}y + A(I - A^{\dagger}A)z \\ &= AA^T(AA^T)^{-1}y + (A - AA^T(AA^T)^{-1}A)z \\ &= y + (A - A)z \\ &= y. \end{aligned}$$

It can be further shown that the base solution obtained by $x_o = A^{\dagger}y$ is the minimum norm solution for the problem that minimizes $\|x\|$. This can be proven by the fact that all other solutions have an added term to this solution which is perpendicular to the base solution.

$$\begin{aligned} (A^{\dagger}y)^T(I - A^{\dagger}A) &= y^T A^{\dagger^T}(I - A^{\dagger}A) \\ &= y^T\left[(I - A^{\dagger}A)A^{\dagger}\right]^T z \\ &= y^T(A^{\dagger} - A^{\dagger}AA^{\dagger})^T z \\ &= 0. \end{aligned}$$

In this derivation, first symmetricity of $I - A^{\dagger}A$, which is resulted from Equation A.47, is used to simplify the first equation, and Equation A.46 is used to simplify the last equation.

A.5 Kronecker Product

Kronecker product, denoted by $\otimes$, is an operation on two matrices of an arbitrary size resulting in a block matrix. If A is an $m \times n$ matrix and B is a $p \times q$, then the Kronecker product $A \otimes B$ is an $mp \times nq$ block matrix calculated by

$$A \otimes B = \begin{bmatrix} a_{11}B & a_{12}B & \cdots & a_{1n}B \\ a_{21}B & a_{22}B & \cdots & a_{2n}B \\ \vdots & \vdots & & \vdots \\ a_{m1}B & a_{m2}B & \cdots & a_{mn}B \end{bmatrix}. \tag{A.59}$$

Kronecker product has the following bilinear and associative properties:

$$\alpha(A \otimes B) = (\alpha A) \otimes B = A \otimes (\alpha B), \tag{A.60}$$

$$A \otimes (B + C) = A \otimes B + A \otimes C, \tag{A.61}$$

$$(A + B) \otimes C = A \otimes C + B \otimes C, \tag{A.62}$$

$$(A \otimes B) \otimes C = A \otimes (B \otimes C). \tag{A.63}$$

Furthermore, for matrices with appropriate size, the *mixed-product* property is as follows:

$$(A \otimes B)(C \otimes D) = AC \otimes BD. \tag{A.64}$$

The inverse and transpose on Kronecker product is calculated by the following equation. $(A \otimes B)$ is invertible, if and only if A and B are invertible. The inverse is given by

$$(A \otimes B)^{-1} = A^{-1} \otimes B^{-1}. \tag{A.65}$$

Moreover, transposition operator is distributive over Kronecker product.

$$(A \otimes B)^T = A^T \otimes B^T. \tag{A.66}$$

There is another property of Kronecker product, which is used in the simplification of dynamic equations. Consider an n-dimensional vector $x \in \mathbb{R}^n$, since the Kronecker product is not commutative, in general $I_{n\times n} \otimes x \neq x \otimes I_{n\times n}$, in which $I_{n\times n}$ denotes the $n \times n$ identity matrix. However, it can be shown that the following equality holds for any arbitrary vector $x \in \mathbb{R}^n$:

$$(I_{n\times n} \otimes x)\, x = (x \otimes I_{n\times n})\, x. \tag{A.67}$$

Finally, if partial derivative of a matrix $A(x)$, with respect to the vector $x \in \mathbb{R}^n$, is defined by the following block–column matrix:

$$\frac{\partial A}{\partial x} = \begin{bmatrix} \frac{\partial A}{\partial x_1} & \frac{\partial A}{\partial x_2} & \cdots & \frac{\partial A}{\partial x_n} \end{bmatrix}^T ; \tag{A.68}$$

then, it can be proven that the partial derivative of matrix product AB with respect to a vector $x \in \mathbb{R}^n$ can be determined by the following Kronecker product:

$$\frac{\partial}{\partial x}(A(x)B(x)) = (I_{n\times n} \otimes A)\frac{\partial B}{\partial x} + \frac{\partial A}{\partial x}B, \tag{A.69}$$

where $I_{n\times n}$ denotes the $n \times n$ identity matrix.

Appendix B: Trajectory Planning

In this appendix, some methods for trajectory generation in multivariable space are given. The computed trajectories may describe the desired motion of a manipulator for each degree-of-freedom in task space or joint space. Trajectory refers to a time history of position, velocity, and acceleration for each degree-of-freedom.

In order to command the desired motion of the manipulator by a human operator, the capability of specifying trajectories with simple descriptions has originated human–machine interface of a robot. Teaching pendant is a facility designed for the robot manipulators, by which the operator manually moves the robot to a desired goal position and orientation. Teaching pendant enables the operator to log the desired goal motion information in the robot processor by moving it to the required configuration. By this means, the initial and final motion information is logged on the robot processor. The robot shall provide the exact shape of the path for such maneuver, and its corresponding time trajectory. From the generated motion trajectory, the motion duration, the velocity profile, and other details are determined. Different methods are developed for robot manipulators to generate the desired trajectories from point-to-point information. In what follows, a review of the most important methods is given.

B.1 Point-to-Point Motion

Consider a point-to-point trajectory planning in which the task is to generate a trajectory from an initial configuration $q(t_o)$ to the desired final configuration $q(t_f)$. Some motion constraints may be considered for the trajectory as well. For example, the initial and final positions may be required to have zero velocity or the maximum velocity, or acceleration toward the trajectory may be limited. Nevertheless, there are infinite trajectories that satisfy a finite number of interpolation and constraint conditions.

The typical trajectories used in practice are from a simple and parameterizable family, such as polynomials of degree n, in which the $n+1$ parameters are found to satisfy the required interpolations and constraints. This is the approach considered in this section, and further extension of this approach to include a multiple number of via points will be described in the next section. Without loss of generality, consider trajectory planning for a single degree-of-freedom, since the trajectories for the remaining degrees-of-freedom will be created independently and in exactly the same way. Suppose that at time t_o the initial motion variable satisfies $q(t_o) = q_o$ and $\dot{q}(t_o) = \omega_o$, and it is desired to attain the following final values for the motion variable $q(t_f) = q_f$, $\dot{q}(t_f) = \omega_f$. In what follows, two different polynomial curve-fitting methods are given to generate the required motion trajectory based on these requirements.

B.1.1 Cubic Polynomials

Consider the case where the initial and final positions and velocity values are given. These four constraints may be satisfied by at least a third-order polynomial. Thus, consider a cubic polynomial with respect to time for the required trajectory

$$q(t) = a_o + a_1 t + a_2 t^2 + a_3 t^3. \tag{B.1}$$

The trajectory velocity is given as

$$\dot{q}(t) = a_1 + 2a_2 t + 3a_3 t^2. \tag{B.2}$$

Substitute the initial and final positions and velocity constraints into Equations B.1 and B.2, and write the thus obtained four equations into a matrix form. It yields

$$\begin{bmatrix} 1 & t_o & t_o^2 & t_o^3 \\ 1 & t_f & t_f^2 & t_f^3 \\ 0 & 1 & 2t_o & 3t_o^2 \\ 0 & 1 & 2t_f & 3t_f^2 \end{bmatrix} \begin{bmatrix} a_o \\ a_1 \\ a_2 \\ a_3 \end{bmatrix} = \begin{bmatrix} q_o \\ q_f \\ \omega_o \\ \omega_f \end{bmatrix}. \tag{B.3}$$

The determinant of the above 4×4 matrix is equal to $(t_f - t_o)^4$, and hence, this matrix is invertible provided $t_o \neq t_f$. Therefore, in general, a unique solution for the cubic polynomial coefficients may be determined from the inverse of Equation B.3. As a particular case, these cubic polynomial parameters can be simplified to the following equation, if the initial and final motion velocities are equal to zero $\omega_o = \omega_f = 0$. In such a case,

$$q(t) = q_o + 3(q_f - q_o)t^2 - 2(q_f - q_o)t^3, \tag{B.4}$$

and the corresponding velocity and acceleration trajectories are given as follows:

$$\dot{q}(t) = 6(q_f - q_o)t - 6(q_f - q_o)t^2, \tag{B.5}$$

$$\ddot{q}(t) = 6(q_f - q_o) - 12(q_f - q_o)t. \tag{B.6}$$

This trajectory is of particular interest for many applications, where the robot remains at rest at the initial and final times.

B.1.2 Quintic Polynomials

A cubic polynomial trajectory provides continuous positions and velocities at the initial and final times, but there are discontinuities in acceleration. In order to have continuity in accelerations, one may specify constraints on the acceleration as well. Assume that the initial and final accelerations of motion is specified by $\ddot{q}(t_o) = \alpha_o$, and $\ddot{q}(t_f) = \alpha_f$. In this case, a fifth-order polynomial may be considered to satisfy the required six constraints. Thus, consider a quintic polynomial with respect to time for the required trajectory.

$$q(t) = a_o + a_1 t + a_2 t^2 + a_3 t^3 + a_4 t^4 + a_5 t^5. \tag{B.7}$$

Substituting the initial and final constraints in Equations B.7 and its derivatives, the following matrix equation is obtained for the polynomial parameters:

$$\begin{bmatrix} a_o \\ a_1 \\ a_2 \\ a_3 \\ a_4 \\ a_5 \end{bmatrix} = \begin{bmatrix} 1 & t_o & t_o^2 & t_o^3 & t_o^4 & t_o^5 \\ 1 & t_f & t_f^2 & t_f^3 & t_f^4 & t_f^5 \\ 0 & 1 & 2t_o & 3t_o^2 & 4t_o^3 & 5t_o^4 \\ 0 & 1 & 2t_f & 3t_f^2 & 4t_f^3 & 5t_f^4 \\ 0 & 0 & 2 & 6t_o & 12t_o^3 & 20t_o^3 \\ 0 & 0 & 2 & 6t_f & 12t_f^3 & 20t_f^3 \end{bmatrix}^{-1} \begin{bmatrix} q_o \\ q_f \\ \omega_o \\ \omega_f \\ \alpha_o \\ \alpha_f \end{bmatrix}. \tag{B.8}$$

B.1.3 Linear Segments with Parabolic Blends

Another choice of path shape is linear, that is, a constant velocity trajectory for the motion. This may be obtained by linear interpolation from the present position q_o to the final position q_f. However, direct linear interpolation would cause the velocity to be discontinuous at the beginning and end of motion. To create a smooth path with continuous position and velocity, start with the linear function but add a parabolic blend region at each path point. During the blend portion of the trajectory, constant acceleration is used to change velocity smoothly. The linear function and the two parabolic functions are merged together such that the entire path is continuous in position and velocity.

In order to achieve this, specify the desired trajectory in three parts. The first part from time t_o to time t_b is a quadratic polynomial. This results in a linear velocity. At time t_b, called the blend time, the trajectory switches to a linear function. This corresponds to a constant velocity. Finally, at time $t_f - t_b$, the trajectory switches once again, this time to a quadratic polynomial so that the velocity is linear. Choose the blend time t_b such that the position curve remains symmetric.

Suppose that the initial time $t_o = 0$, and the motion is required to be at rest at the initial and final times, that is, $\dot{q}(0) = \dot{q}(t_f) = 0$. Furthermore, consider that the velocity at blend time and in the linear segment is given by a constant value denoted by Ω. Then, between the times 0 and t_b, the position and velocity curves are given by

$$q(t) = q_o + \frac{\Omega}{2t_b}t^2, \tag{B.9}$$

$$\dot{q}(t) = \frac{\Omega}{t_b}t = \alpha t, \tag{B.10}$$

in which the constant acceleration of the parabolic motion denoted by α is given by $\alpha = \Omega/t_b$.

The motion between time t_b and $t_f - t_b$ is given by the following linear segment:

$$q(t) = q(t_b) + \Omega(t - t_b) \tag{B.11}$$

in which $q(t_b)$ is found from Equation B.9 as

$$q(t_b) = \frac{1}{2}(q_o + q_f) - \Omega\left(\frac{1}{2}t_f - t_b\right). \tag{B.12}$$

Since the two segments must gently blend at time t_b, if the motion velocity is set by Ω, then we require to assign t_b to satisfy the following equation:

$$t_b = \frac{q_o - q_f + \Omega t_f}{\Omega}. \tag{B.13}$$

Furthermore, we have the constraint $0 < t_b < t_f/2$. This will immediately lead to the following constraints on t_f if Ω is given or on Ω if t_f is given:

$$\frac{q_f - q_o}{\Omega} < t_f < 2\frac{q_f - q_o}{\Omega}, \tag{B.14}$$

$$\frac{q_f - q_o}{t_f} < \Omega < 2\frac{q_f - q_o}{t_f}. \tag{B.15}$$

The last portion of the trajectory between $t_f - t_b$ and t_f is found by symmetry of the motion. The final linear segment trajectory with parabolic blend is given by

$$q(t) = \begin{cases} q_o + \frac{\alpha}{2}t^2 & 0 \le t \le t_b, \\ \frac{1}{2}(q_f + q_o - \Omega t_f) + \Omega t & t_b \le t \le t_f - t_b, \\ q_f - \frac{\alpha}{2}t_f^2 + \alpha t_f t - \frac{\alpha}{2}t^2 & t_f - t_b \le t \le t_f. \end{cases} \tag{B.16}$$

B.1.4 Minimum Time Trajectory

An important variation of linear segment trajectory with parabolic blend is obtained by minimizing the final time t_f for a given constant acceleration α. This is sometimes called a bang–bang trajectory since the optimal solution is achieved with a maximum acceleration $+\alpha$, until an appropriate switching time t_s. At this time, the trajectory abruptly switches to the minimum deceleration $-\alpha$ from t_s to t_f. Consider the case where the motion has zero initial and final velocities. Symmetry considerations would suggest that the switching time t_s is just half of the final time $t_s = t_f/2$. In such a case, denote Ω_s as the velocity at time t_s, and set $t_b = t_s$, then from Equation B.13, obtain

$$t_s = \frac{q_o - q_f + \Omega_s t_f}{\Omega_s}. \tag{B.17}$$

Furthermore, the symmetry condition implies that $t_s = t_f/2$, and $\Omega_s = (q_f - q_o)/t_s$. Hence

$$\alpha t_s = \frac{q_f - q_o}{t_s}, \tag{B.18}$$

which implies that

$$t_s = \left(\frac{q_f - q_o}{\alpha}\right)^{1/2}. \tag{B.19}$$

B.2 Specified Path with Via Points

In this section, the problem of planning a trajectory that passes through a sequence of configurations called *via points* is briefly examined. First, consider the simple example of a path specified by three points, q_o, q_1, and q_2, at times t_o, t_1, and t_2, respectively. If, in addition to these three constraints, four constraints on the initial and final velocities and accelerations are considered, a sixth-order polynomial may be considered to satisfy all the required constraints.

$$q(t) = a_o + a_1 t + a_2 t^2 + a_3 t^3 + a_4 t^4 + a_5 t^5 + a_6 t^6. \tag{B.20}$$

One advantage to this approach is, since $q(t)$ is continuously differentiable, there are no discontinuities in either velocity or acceleration at the via point, q_1. However, to determine the coefficients for this polynomial, a linear system of the seventh order shall be solved. The clear disadvantage to this approach is that as the number of via points increases, the dimension of the corresponding linear system increases and makes the method intractable.

An alternative is to use low-order polynomials for trajectory segments between adjacent via points. These polynomials are sometimes referred to as interpolating or blending polynomials. With this approach, it should be noted that continuity constraints such as that in velocity and acceleration are satisfied at via points. By this means, continuous motion is generated by switching from one polynomial to another. Given the initial and final times, t_o and t_f, respectively, with $q(t_o) = q_o$, $q(t_f) = q_f$, $\dot{q}(t_o) = \omega_o$, and $\dot{q}(t_f) = \omega_f$, the parameters of the cubic polynomial

$$q(t) = a_o + a_1(t - t_o) + a_2(t - t_o)^2 + a_3(t - t_o)^3 \tag{B.21}$$

can be computed from

$$\begin{aligned}
a_o &= q_o, \\
a_1 &= \omega_o, \\
a_2 &= \frac{3(q_1 - q_o) - (2\omega_o + \omega_1)(t_f - t_o)}{(t_f - t_o)^2}, \\
a_3 &= \frac{2(q_o - q_1) - (\omega_o + \omega_1)(t_f - t_o)}{(t_f - t_o)^3}.
\end{aligned}$$

A sequence of moves can be planned using this formulation by using the end conditions q_f, ω_f of the ith move as initial conditions for the subsequent move.

Now consider the case of linear segments with parabolic blends for the case in which there are an arbitrary number of via points specified. In order to generate such a trajectory, consider linear functions connect the via points, and parabolic blend regions are added around each via point. Furthermore, consider three neighboring path points which we will call points j, k, and l. The duration of the blend region at path point k is denoted by t_k. The duration of the linear portion between points j and k is denoted by t_{jk}. The overall duration of the segment connecting points j and k is denoted by $t_{d_{jk}}$. The velocity during the linear portion is denoted by ω_{jk} and the acceleration during the blend at point j is denoted by α_j.

As with the single-segment case, there are many possible solutions depending on the value of acceleration used at each blend. Given the values for all the path points q_k, and the desired durations $t_{d_{jk}}$, and the magnitude of acceleration to be used at each path point $|\alpha_k|$, one may compute the blend times t_k. For interior path points, these values are found from the following equations:

$$\begin{aligned}
\omega_{jk} &= \frac{q_k - q_j}{t_{d_{jk}}}, \\
\alpha_k &= \text{sign}(\omega_{kl} - \omega_{jk})|\alpha_k|, \\
t_k &= \frac{\omega_{kl} - \omega_{jk}}{\alpha_k}, \quad t_{jk} = t_{d_{jk}} - \frac{1}{2}t_j - \frac{1}{2}t_k.
\end{aligned} \tag{B.22}$$

The first and last segments must be handled slightly differently, since an entire blend region at one end of the segment must be counted in the total segment's time duration. For the first segment, solve for t_1 by equating two expressions for the velocity during the linear phase of the segment:

$$\alpha_1 t_1 = \frac{q_2 - q_1}{t_{d_{12}} - \frac{1}{2}t_1}. \tag{B.23}$$

This can be solved for t_1, the blend time at the initial point, and then ω_{12} and t_{12} are easily computed as follows:

$$\begin{aligned}
\alpha_1 &= \text{sign}(q_2 - q_1)|\alpha_1|, \\
t_1 &= t_{d_{12}} - \left(t_{d_{12}}^2 - \frac{2(q_2 - q_1)}{\alpha_1}\right)^{1/2}, \\
\omega_{12} &= \frac{q_2 - q_1}{t_{d_{12}} - \frac{1}{2}t_1}, \\
t_{12} &= t_{d_{12}} - t_1 - \frac{1}{2}t_2.
\end{aligned} \tag{B.24}$$

Likewise, for the last segment which connects the points $n-1$ and n, we have

$$\begin{aligned}
\alpha_n &= \text{sign}(q_{n-1} - q_n)|\alpha_n|, \\
t_n &= t_{d_{(n-1)n}} - \left(t_{d_{(n-1)n}}^2 - \frac{2(q_n - q_{n-1})}{\alpha_n}\right)^{1/2}, \\
\omega_{(n-1)n} &= \frac{q_n - q_{n-1}}{t_{d_{(n-1)n}} - \frac{1}{2}t_n}, \\
t_{(n-1)n} &= t_{d_{(n-1)n}} - t_n - \frac{1}{2}t_{n-1}.
\end{aligned} \tag{B.25}$$

These three sets of equations may be used to solve the blend times and velocities for a multi-segment path. Usually, the operator specifies only the via points and the desired duration of the segments. In this case, the system uses default values for acceleration for each degree-of-freedom. Sometimes, to make things even simpler for the operator, the system calculates durations based on default velocities.

Appendix C: Nonlinear Control Review

A brief introduction to state-space representation of dynamical systems is given here, and basic definitions and theorems for the stability of a class of nonlinear dynamical systems is reviewed. In this review, only autonomous dynamical systems are considered, and for a more general treatment of the subject the reader is referred to References 86,96.

C.1 Dynamical Systems

As seen in Chapter 5, the governing dynamic formulation for a general manipulator may be written as a set of second-order differential Equations 5.297. Furthermore, it is shown that, in general, the corresponding mass matrix is positive definite at all configurations, and hence, it is invertible. Therefore, by composing a suitable state vector consisting of the motion variables, and their time derivatives denoted by $\boldsymbol{x} = [x_1, x_2, \ldots, x_n]^T$, one may represent the general dynamic formulation of a robot manipulator in state space as following:

$$\dot{\boldsymbol{x}} = \boldsymbol{f}(\boldsymbol{x}, \boldsymbol{u}), \tag{C.1}$$

where $\boldsymbol{x}$ denotes the vector of state variable, $\boldsymbol{u}$ denotes the vector of control inputs, and $\boldsymbol{f}$ is a continuous function $\boldsymbol{f} : \mathbb{R}^n \to \mathbb{R}^n$, which is called a *vector field*. Equation C.1 represents open-loop dynamic formulation of the robot manipulator by a set of first-order differential equations. If the behavior of the system in the absence of the actuator forces has to be analyzed, then the dynamic formulation is simplified to

$$\dot{\boldsymbol{x}} = \boldsymbol{f}(\boldsymbol{x}). \tag{C.2}$$

This is generally not of interest for a motion tracking objective. In such a case, the actuator forces are designed to perform the required maneuvers. In such a case though, the control input $\boldsymbol{u}$ is found by the proposed control strategy as a function of state feedback, and in general, it can be determined by a nonlinear function

$$u = g(x).$$

Hence, the closed-loop dynamics may also be represented by Equation C.2:

$$\dot{\boldsymbol{x}} = \boldsymbol{f}(\boldsymbol{x}, \boldsymbol{g}(\boldsymbol{x})) = \boldsymbol{f}(\boldsymbol{x}).$$

Since the dynamic formulation of the system is a nonlinear function of $\boldsymbol{x}$, in order to analyze the stability of the closed-loop system, the definitions and theorems of the nonlinear systems apply. In what follows, a brief review on these concepts is given.

C.2 Stability Definitions

Definition C.1

Consider a nonlinear system given by the state-space representation C.2 on $\mathbb{R}^n$, and furthermore, suppose $f(\mathbf{0}) = \mathbf{0}$. Then, the origin of $\mathbb{R}^n$ is said to be the **equilibrium point** of the system.

Note that, if the equilibrium point of function f is not at the origin of state space, it can certainly be moved to the origin by a state transformation. Therefore, the assumption of $f(\mathbf{0}) = \mathbf{0}$ will not reduce the generality of definition. Furthermore, consider that the system represented by Equation C.2 has a null initial condition $x(0) = \mathbf{0}$. Then the trajectory of the states remain at zero for all the positive times, that is, $x(t) \equiv \mathbf{0}, \ \forall t > 0$. This trajectory is called the *trivial trajectory* for the system. Stability analysis of the system deals with the analysis of the behavior of nontrivial trajectories $x(t)$. Intuitively, the equilibrium point is called stable if, for arbitrary small initial conditions, the time trajectories of the solutions $x(t)$ remain close to the equilibrium point in some sense. Stability in sense of *Lyapunov* may be formally defined as follows:

Definition C.2

The system equilibrium point at origin is called

- **Stable**: If and only if, for any $\epsilon > 0$, there exists a $\delta(\epsilon) > 0$, such that

$$\|x(0)\| < \delta \Rightarrow \|x(t)\| < \epsilon, \quad \forall t > 0. \tag{C.3}$$

- **Asymptotic stable**: If it is stable, and furthermore, there exists $\delta > 0$ such that

$$\|x(0)\| < \delta \Rightarrow \|x(t)\| \to 0, \quad \text{as } t \to \infty. \tag{C.4}$$

- **Unstable**: If it is not stable.

Physical interpretation of these three definitions are as follows. An equilibrium point is called stable if for an arbitrary small initial condition from the equilibrium point, the trajectory is not diverging from it, and remain in a hyper-ball of radius ϵ around it. If furthermore, the trajectories are asymptotically converging toward the equilibrium point, then the equilibrium point is called asymptotically stable. In other words, asymptotic stability means that if the system is perturbed away from the equilibrium it will return asymptotically to it. If the equilibrium point is unstable, then for any neighborhood of the equilibrium point, no matter how small, there always exist at least one initial condition such that the time trajectory of the states leave this neighborhood as time evolves.

The above definitions of stability are local in nature. They may hold for initial conditions sufficiently close to the equilibrium point, but may fail for initial conditions farther away from it. All variations of stability may become *global*, if it holds for any arbitrary initial conditions.

Another important notion of stability may be defined by the boundedness of the trajectories. This definition is very useful in cases that the nonlinear system formulation is explicitly a function of time. In such a case, the initial time t_o, from which the system trajectories start becomes important in the stability analysis. Usually, the notion of uniformity is added to the stability definition to suggest that the stability of the equilibrium point is not directly dependent on the initial time t_o.

Definition C.3

A system trajectory $\boldsymbol{x}(t) : [t_o, \infty) \to \mathbb{R}^n$ with initial condition $\boldsymbol{x}(t_o) = x_o$ is said to be **uniformly ultimately bounded** (UUB) with respect to a set **S**, if there is a nonnegative constant $T(\boldsymbol{x}_o, \boldsymbol{S})$ such that

$$\boldsymbol{x}(t) \in \boldsymbol{S}, \quad \forall t \geq t_o + T. \tag{C.5}$$

Ultimate boundedness reveals the fact that the time trajectories $\boldsymbol{x}(t)$ starting from $\boldsymbol{x}_o$ at time t_o will ultimately enter and remain in the set **S**. Furthermore, uniformity adds the notion of initial time independence to the boundedness property. If the set **S** is a small region about the equilibrium point, then UUB is a practical notion of stability that is looser from asymptotic stability, but still very useful in control system design.

C.3 Lyapunov Stability

Lyapunov is a renowned mathematician who has formulated the notion of stability in dynamical systems in a mathematical framework. To introduce his direct method for stability analysis briefly, a physical interpretation of his works is given here. The philosophy of Lyapunov direct method for stability analysis of nonlinear systems may be seen as a mathematical extension of a fundamental physical observation. This observation is that, if the total energy of a system is continuously dissipating, then the system trajectories shall eventually settle down at an equilibrium point. To illustrate this concept let us examine an example.

Consider a body–spring–damper system illustrated in Figure 7.48b, in which both the spring and damper behave nonlinearly with respect to the state variable $\boldsymbol{x}$. Let us consider hardening spring modeled by $F_s = k_o x + k_1 x^3$, and the damper modeled by $F_d = b\dot{x}|\dot{x}|$. In these formulations, the spring and damping coefficients, k_o, k_1, and b are considered to be positive scalar constants. Furthermore, consider that no actuator force is applied to the system. The dynamic formulation of the system in absence of any disturbance forces may be written as

$$m\ddot{x} + b\dot{x}|\dot{x}| + k_o x + k_1 x^3 = 0, \tag{C.6}$$

in which m denotes the body mass and x denotes the body motion variable. The state variable for the system may be composed of the body motion and velocity as $\boldsymbol{x} = [x, \ \dot{x}]^T$. Stability analysis of this system is to examinc the resulting motion trajectories of the system $\boldsymbol{x}(t)$, with respect to relatively large initial conditions $\boldsymbol{x}(0)$. It is very hard to use the definition of stability to examine the behavior of the trajectories. However, it is relatively easy to derive and analyze the total energy of the system.

The total energy of the system may be derived from kinetic and potential energies as

$$V(x) = \frac{1}{2}m\,\dot{x}^2 + \int_0^x (k_o y + k_1 y^3)\,\mathrm{d}y$$

$$= \frac{1}{2}m\,\dot{x}^2 + \frac{1}{2}k_o x^2 + \frac{1}{4}k_1 x^4.$$

Let us examine the notions and definitions of stability corresponding to the total energy of the system.

- Zero energy corresponds to the equilibrium point at origin $x = \dot{x} = 0$
- Asymptotic stability implies the convergence of total energy toward zero
- Instability is related to the growth of energy

Therefore, the notion of stability is mostly related to the variation of energy rather than to the energy function itself. Let us examine the time variation of total energy of the system by

$$\dot{V}(\boldsymbol{x}) = m\,\ddot{x}\dot{x} + (k_o x + k_1 x^3)\dot{x}$$

$$= (m\,\ddot{x} + k_o x + k_1 x^3)\dot{x}. \tag{C.7}$$

In order to examine the behavior of the state trajectories $\boldsymbol{x}(t)$, it is essential to examine the solution of dynamic differential equations with respect to the initial conditions. A state trajectory shall satisfy the dynamic equation of motion represented by Equation C.6 in this example. Substitute C.6 into C.7 and simplify the time variation of total energy for this system as

$$\dot{V}(\boldsymbol{x}) = (-b\,\dot{x}\,|\dot{x}|)\dot{x} = -b\,|\dot{x}^3| \leq 0. \tag{C.8}$$

This equation implies that the total energy of the system is continuously dissipating until $\dot{x} = 0$, and this interpretation may be considered as a stable motion for the system.

In order to mathematically extend this observation, let us first examine the properties of the total energy function:

- Total energy function $V(\boldsymbol{x})$ is a scalar and a strictly positive definite function, that is, $V(\boldsymbol{x}) > 0, \quad \forall \boldsymbol{x} \neq \boldsymbol{0}$.
- Total energy function $V(\boldsymbol{x})$ is zero at equilibrium point, that is, $V(\boldsymbol{0}) = 0$.
- The time derivative of energy function $\dot{V}(\boldsymbol{x})$ is monotonically decreasing toward zero.

Definition C.4

A scalar continuous function $V(\boldsymbol{x})$ is locally positive definite if $V(\boldsymbol{0}) = 0$, and except at this point in a ball B_{R_o}, it is strictly positive $V(\boldsymbol{x}) > 0$.

Definition C.5

A locally positive-definite function $V(\boldsymbol{x})$ is a Lyapunov function candidate, if $V(\boldsymbol{x})$ is continuous and has continuous first partial derivatives in a neighborhood of the origin.

Note that, if the above condition holds for the whole state space, the scalar function is called globally positive definite. Furthermore, a negative-definite function $V(\boldsymbol{x})$ is defined if $-V(\boldsymbol{x})$ is a positive definite. The function $V(\boldsymbol{x})$ is called positive semidefinite, if $V(\boldsymbol{x})$ becomes zero at some points except the origin $V(\boldsymbol{x}) \geq 0$. By this introduction, the main theorem of Lyapunov for stability analysis can be stated as follows. The stability analysis based on this theorem is called the Lyapunov direct method.

Theorem C.1: Lyapunov Local Stability

Let $\boldsymbol{x} = \boldsymbol{0}$ be an equilibrium point for Equation C.2, and $D \subset \mathbb{R}^n$ be a domain containing equilibrium point. Let $V : D \rightarrow \mathbb{R}$ be a continuously differentiable positive-definite function, such that

$$V(\boldsymbol{0}) = 0 \quad \text{and} \quad V(\boldsymbol{x}) > 0 \quad \text{in} \quad D - \{\boldsymbol{0}\}$$
$$\dot{V}(\boldsymbol{x}) \leq 0 \quad \text{in} \quad D,$$

then $\boldsymbol{x} = \boldsymbol{0}$ is **stable**. Moreover, if

$$\dot{V}(\boldsymbol{x}) < 0 \quad \text{in} \quad D - \{\boldsymbol{0}\},$$

then $\boldsymbol{x} = \boldsymbol{0}$ is **asymptotically stable**.

Although the philosophy behind the Lyapunov stability analysis may be related to the behavior of the total energy in the system, the proof of Lyapunov direct method is not related to any physical interpretation of dynamical motion and energy. The proof is based on the Lyapunov level surfaces denoted by $\boldsymbol{S}(c)$ and defined by

$$\boldsymbol{S}(c) = \{\boldsymbol{x} \in \mathbb{R}^n \mid V(\boldsymbol{x}) = c\}. \tag{C.9}$$

The sketch of the proof is that if $V(\boldsymbol{x})$ satisfies the conditions given in the local stability theorem, then the Lyapunov level surfaces may be defined on the state space, which are not intersecting and are shrinking in size as the constant c decreases toward zero. By this means, the trajectories may never diverge from inside the level surfaces toward outside, and therefore, the system becomes stable. If furthermore, the time derivative of the Lyapunov function is strictly negative definite, the Lyapunov level surfaces will be monotonically shrinking toward zero, and hence asymptotic stability can be inferred.

Although this theorem is very strong in terms of mathematical formulation of stable motion, it has a number of drawbacks. First, the theorem gives only sufficient conditions, and is not bounded by the necessary conditions. Therefore, if the conditions are satisfied the stability is guaranteed. However, if a Lyapunov function candidate fails to satisfy the negative definiteness of its derivative, the inverse is not generally true and may not necessarily result in the instability of the equilibrium point. In such a case, the Lyapunov function candidate is not suitable for the analysis, and another candidate must be selected for the system.

The second drawback is lack of a general method to generate suitable Lyapunov functions for a system. Although, from physical interpretation given in this section, one may consider the total energy of the system as a first Lyapunov candidate; if it fails to satisfy the theorem conditions, no general method exists to search for a suitable function from this initial guess. Another limitation of this theorem is that, although in many systems the motion may be asymptotically stable, it is very hard to obtain suitable Lyapunov functions that satisfy negative definiteness of its time derivatives, and in many cases, the best function found just satisfies positive semi-definiteness. This drawback is well worked out and is remedied by the Krasovskii–Lasalle theorem given in the next section.

Many variations of the above theorem is given for different assumptions and conditions. Here only two variations are given, where the first one is for global stability condition and the second one is for UUB conditions.

Theorem C.2: Lyapunov Global Stability

Let $x = \mathbf{0}$ be an equilibrium point for Equation C.2, and $V : \mathbb{R}^n \to \mathbb{R}$ be a continuously differentiable positive-definite function, such that

$$\begin{aligned} V(\mathbf{0}) &= 0 \ \text{ and } \ V(x) > 0, \ \ \forall x \neq \mathbf{0}, \\ \text{As } \ \|x\| &\to \infty \Rightarrow V(x) \to \infty, \\ \dot{V}(x) &< 0, \ \ \forall x \neq \mathbf{0}, \end{aligned} \tag{C.10}$$

then $x = \mathbf{0}$ is **globally asymptotically stable**.

Theorem C.3: UUB Stability

Let $x = \mathbf{0}$ be an equilibrium point for Equation C.2, and $D \subset \mathbb{R}^n$ be a domain containing an equilibrium point. Let $V : D \to \mathbb{R}$ be a continuously differentiable positive-definite function, and $S(c)$ be any Lyapunov level surface of V defined by Equation C.9 for some positive constant $c > 0$. Then the trajectories $x(t)$ of the system are **uniformly ultimately bounded** with respect to S if

$$\dot{V}(x) < 0, \ \ \forall x \ \text{ outside of } \ S.$$

If $\dot{V}$ is strictly negative outside of S, then the trajectories outside S must be converging toward S. When the trajectory enters S, then the asymptotic behavior of the trajectories cannot be inferred. However, it can be seen that, the trajectories can never go outside of the S, if they enter this set, and therefore, they are trapped in that set. Therefore, UUB condition holds for all trajectories.

C.4 Krasovskii–Lasalle Theorem

As explained earlier, there are many cases where the motion is asymptotically stable, but it is very hard to obtain suitable Lyapunov functions that satisfy negative definiteness condition on $\dot{V}$. Invariant principle is discussed in this section to draw conclusions about asymptotic stability even when $\dot{V}$ is only a negative semidefinite. Other important conclusions may be drawn from the Krasovskii–Lasalle stability theorem in terms of region of convergence, and convergence toward periodic orbits and not the equilibrium point which are well treated in [96].

Definition C.6

A set $\boldsymbol{M}$ is an **invariant set** for a dynamical system represented by Equation C.2 if every trajectory of the system that starts from a point in the set remains in it for all future time.

$$\boldsymbol{x}(0) \in \boldsymbol{M} \Rightarrow \boldsymbol{x}(t) \in \boldsymbol{M}, \quad \forall t > 0.$$

Invariant set includes generalization of equilibrium point in state space. A set consisting of only an equilibrium point is an invariant set, since a trajectory starting at equilibrium point will remain on it at future time. However, a set of points producing a limit cycle is also an invariant set. Moreover, any trajectory of the system produces an invariant set, since the condition for invariant set holds for both these cases.

The Krasovskii–Lasalle theorem states conditions that guarantee convergence of the system trajectories to an invariant set. An extension of this theorem may be considered for asymptotic stability analysis of an equilibrium point, where the invariant set under study includes only the equilibrium point. Therefore, in such a case, asymptotic stability is guaranteed even when $\dot{V}$ is only negative semidefinite.

Theorem C.4: Krasovskii–Lasalle Theorem

For the system C.2 let $D \subset \mathbb{R}^n$ be a domain and $\Omega \subset D$ be a compact set that is invariant with respect to the system. Furthermore, let $V : D \rightarrow \mathbb{R}$ be a continuously differentiable function, such that $\dot{V} \leq 0$ in Ω.

Let E be the set of all points in Ω where $\dot{V} = 0$, and M be the largest invariant set in E. Then every solution starting in Ω converges to M as $t \rightarrow \infty$.

This theorem does not directly speak about asymptotic stability of the origin. However, it may be used to conclude on asymptotic stability provided the equilibrium point $\boldsymbol{x} = \boldsymbol{0}$ is the only point contained in the set M. In other words, if V does not vanish identically along any trajectory of the system other than the origin, then the origin is asymptotically stable, although $\dot{V} \leq 0$. This condition holds when the only solution that satisfies $\dot{V} = 0$, is the equilibrium point and other invariant set components are not included in the set M, in which $\dot{V} = 0$. This theorem is frequently used to prove asymptotic stability of nonlinear systems, where the Lyapunov function candidate can only satisfy $\dot{V} \leq 0$.

References

1. H. Abdellatif and B. Heimann. Computational efficient inverse dynamics of 6-DOF fully parallel manipulators by using the Lagrangian formalism. *Mechanism and Machine Theory*, 44(1):192–207, 2009.
2. M. Agahi and L. Notash. Redundancy resolution of wire-actuated parallel manipulators. *Transactions of CSME*, 33(4):561–573, 2009.
3. I.D. Akcali and H. Mutlu. A novel approach in the direct kinematics of stewart platform mechanisms with planar platforms. *Transactions of the ASME. Journal of Mechanical Design*, 128(1):252–263, 2006.
4. J. Albus, R. Bostelman, and N. Dagalakis. The NIST RoboCrane. *Journal of Robotic Systems*, 10(5):709–724, July 1993.
5. J. Angeles. *Fundamentals of Robotic Mechanical Systems: Theory, Methods and Algorithms*. Springer, New York, USA, 2003.
6. M. M. Aref, H. D. Taghirad, and S. Barissi. Optimal design of dexterous cable driven parallel manipulators. *International Journal of Robotics Research: Theory and Applications*, 1(1):29–47, 2009.
7. K. J. Åstrom and T. Hägglund. *Advanced PID Control*. ISA, Instrumentation Systems and Automation Society, Research Triangle Park, NC, 2006.
8. C. G. Atkeson, C. H. An, and J. M. Hollerbach. Estimation of inertial parameters of manipulator loads and links. *International Journal of Robotics Research*, 5(3):101–119, 1986.
9. R. S. Ball. *A Treatise on the Theory of Screws*. Cambridge University Press, Cambridge, 1900.
10. L. Baron and J. Angeles. The direct kinematics of parallel manipulators under joint-sensor redundancy. *IEEE Transactions Robotics and Automation*, 16(1):12–19, 2000.
11. G. Barrette and C.M. Gosselin. Determination of the dynamic workspace of cable-driven planar parallel mechanisms. *Transactions ASME Journal of Mechanical Design*, 127(2):242–248, 2005.
12. Y. B. Bedoustani, P. Bigras, H. D. Taghirad, and I. Bonev. Lagrangian dynamics of cable-driven parallel manipulators: A variable mass formulation. *Transactions of CSME*, 35(4):529–542, 2011.
13. Y. B. Bedoustani and H. D. Taghirad. Iterative-analytic redundancy resolution scheme for a cable-driven redundant parallel manipulator. In *IEEE/ASME International Conference on Advanced Intelligent Mechatronics (AIM)*, pp. 219–224, Montreal, Canada, July 2010.
14. I.A. Bonev. The true origins of parallel robots. http://www.parallemic.org/Reviews/Review007.html, January, 24, 2003.
15. I.A. Bonev, J. Ryu, S.-G. Kim, and S.-K. Lee. A closed-form solution to the direct kinematics of nearly general parallel manipulators with optimally located three linear extra sensors. *IEEE Transactions on Robotics and Automation*, 18:148–156, 2001.
16. P. Bosscher, A. Riechel, and I. Ebert-Uphoff. Wrench-feasible workspace generation for cable-driven robots. *IEEE Transaction on Robotics*, 22(5):890–902, Oct. 2006.
17. J.W. Brewer. Kronecker products and matrix calculus in system theory functions in dynamics of mechanical systems. *IEEE Transactions on Circuits Systems*, CAS-25(9):772–781, Sept. 1978.
18. H. Bruyninckx and J. De Schutter. Comments on closed form forward kinematics solution to a class of hexapod robots. *IEEE Transactions on Robotics and Automation*, 15:788–789, 1999.
19. K.L. Cappel. Motion simulator. US Patent No. 3,295,224, January 3, 1967.
20. B. Carlson et al. The large adaptive reflector: A 200-m diameter, wideband, cm-wave radio telescope. In *Radio Telescopes-Proc. of SPIE Meeting 4015*, pp. 33–44, Bellingham, WA, 2000.
21. M. Carricato and C. Gosselin. On the modeling of leg constraints in the dynamic analysis of Gough/Stewart-type platforms. *Journal of Computational and Nonlinear Dynamics*, 4(1):1–8, 2009.
22. D. H. Chaubert, A. O. Boryssenko, A. van Ardenne, J. G. Bij de Vaate, and C. Craeye. The square kilometer array (SKA) antenna. In *IEEE Int. Symp. Phased Array Systems and Technology*, pp. 351–358, Oct. 2003.

23. S. L. Chiu. Task compatibility of manipulator postures. *International Journal of Robotics Research*, 7(5):13–21, 1988.
24. Y. J. Chiu and M. H. Perng. Forward kinematics of a general fully parallelmanipulator with auxiliary sensors. *International Journal of Robotics Research*, 20(5):401–414, 2001.
25. T. S. Chung. An inherent stability problem in Cartesian compliance and an alternative structure of compliance control. *IEEE Transactions on Robotics and Automation*, 7(1):21–30, Feb 1991.
26. R. Clavel. Delta, a fast robot with parallel geometry. In *Proc. of the 18th International Symposium on Industrial Robots*, Lausanne, Switzerland, pp. 91–100, April 1988.
27. R. Clavel. Device for the movement and positioning of an element in space. US Patent No. 4,976,582, December 11, 1990.
28. R. Clavel. *Conception d'un robot paralléle rapide á 4 degrés de liberté*. PhD thesis, EPFL, Lausanne, Switzerland, 1991.
29. K. Cleary and M. Uebel. Jacobian formulation for a novel 6-DOF parallel manipulator. In *IEEE Conf. Robotics and Automation*, pp. 2377–2382, May 1994.
30. T. F. Coleman and Y. Zhang. *Optimization Toolbox 6, User's Guide*. The MathWorks, Inc., MA, USA, April 2011.
31. J. J. Craig. *Introduction to Robotics: Mechanics and Control*, 3rd edition. Prentice-Hall, Englewood Cliffs, NJ, 2005.
32. B. Dasgupta and P. Choudhury. A general strategy based on the Newton–Euler approach for the dynamic formulation of parallel manipulators. *Mechanism and Machine Theory*, 34(6):801–824, 1999.
33. B. Dasgupta and T.S. Mruthyunjaya. A Newton–Euler formulation for the inverse dynamics of the Stewart platform manipulator. *Mechanism and Machine Theory*, 33(8):1135–1152, 1998.
34. J. E. Dennis and R. B. Schnabel. *Numerical Methods for Unconstrained Optimization and Nonlinear Equations*. Prentice-Hall Series in Computational Mathematics. Prentice-Hall, Englewood Cliffs, NJ, 1983.
35. R. Di Gregorio. Analytic formulation of the 6-3 fully-parallel manipulators singularity determination. *Robotica*, 19:663–667, 2001.
36. R. Di Gregorio. Singularity-locus expression of a class of parallel mechanisms. *Robotica*, 20:323–328, 2002.
37. P. Dietmaier. The Stewart—Gough platform of general geometry can have 40 real postures. In *Proceedings of ARK*, pp. 7–16, Strobl, Austria, June 1998.
38. R. C. Dorf and R. H. Bishop. *Modern Control Systems*, 12th edition. Prentice-Hall, Upper Saddle River, NJ, 2010.
39. K. L. Doty, C. Melchiorri, E. M. Schwartz, and C. Bonivento. Robot manipulability. *IEEE Transactions on Robotics and Automation*, 11(3):462–468, 1995.
40. R. Dubey and J. Y. S. Luh. Redundant robot control using task based performance measures. *Journal of Robotic Systems*, 5(5):409–432, 1988.
41. J. Duffy. The fallacy of modern hybrid control theory that is based on orthogonal complements of twists and wrenches spaces. *Journal of Robotic Systems*, 7:139–144, 1990.
42. R. Dvorak, F. Freistetter, and J. Kurths. *Chaos and Stability in Planetary Systems*. Springer, New York, 2005.
43. B. S. El-Khasawneh and P. M. Ferreira. Computation of stiffness and stiffness bounds for parallel link manipulators. *International Journal of Machine Tools and Manufacture*, 39(2):321–342, 1999.
44. J. F. Engelberger. *Robotics in Practice*. AMACOM, New York, 1980.
45. L. Euler. Nova methodus motum corporum rigidorum determinandi. *Novii Comentarii Academiæ Scientiarum Petropolitanæ*, 20:208–238, 1776.
46. S. Fang, D. Franitza, M. Torlo, F. Bekes, and M. Hiller. Motion control of a tendon-based parallel manipulator using optimal tension distribution. *IEEE/ASME Transactions Mechatronics*, 9(3):561–568, 2004.

47. A. Fattah and G. Kasaei. Kinematics and dynamics of a parallel manipulator with a new architecture. *Robotica*, 18(5):535–543, 2000.
48. E. F. Fichter. A Stewart platform-based manipulator: General theory and practical construction. *International Journal of Robotics Research*, 5(2):157–182, 1986.
49. R. Fletcher. *Practical Methods of Optimization*. John Wiley and Sons, New York, 1987.
50. C. W. French et al. Multi-axial subassemblage testing (MAST) system: Description and capabilities. In *13th World Conf. on Earthquake Engineering*, pages 2146–51, Vancouver, Canada, August 2004.
51. J. Gallardo, J. M. Rico, A. Frisoli, D. Checcacci, and M. Bergamasco. Dynamics of parallel manipulators by means of screw theory. *Mechanism and Machine Theory*, 38(11):1113–1131, 2003.
52. X. S. Gao, D. Lei, Q. Liao, and G.-F. Zhang. Generalized Stewart–Gough platforms and their direct kinematics. *IEEE Transactions Robotics*, 21(2):141–151, 2005.
53. Z. Geng and L. S. Haynes. Neural network solution for the forward kinematics problem of a Stewart platform. *Robotics Computation Integration Manufacturing*, 9(6):485–495, 1992.
54. R. Ghorbani and Q. Wu. Adjustable stiffness artificial tendons: Conceptual design and energetics study in bipedal walking robots. *Mechanism and Machine Theory*, 44(1):140–161, 2009.
55. P. E. Gill, W. Murray, and M. H. Wright. *Numerical Linear Algebra and Optimization*. Addison-Wesley, Reading, MA, 1991.
56. R. Giovanelli et al. The Arecibo Legacy Fast ALFA survey I. Science Goals, Survey Design, and Strategy. *Astronomical Journal*, 130(1):2598–2612, 2005.
57. K. Glass, R. Colbaugh, D. Lim, and H. Seraji. Real-time collision avoidance for redundant manipulators. *IEEE Transactions on Robotics and Automation*, 11(3):448–457, 1995.
58. C. Gosselin. *Kinematic analysis optimization and programming of parallel robotic manipulators*. PhD thesis, McGill University, Montreal, June 15, 1988.
59. C. Gosselin. Stiffness mapping for parallel manipulators. *IEEE Transactions Robotics and Automation*, 6(3):377–382, 1990.
60. C. Gosselin. Parallel computational algorithms for the kinematics and dynamics of planar and spatial parallel manipulators. *Transactions of the ASME Journal of Dynamic Systems, Measurement and Control*, 118(1):22–28, 1996.
61. C. Gosselin and J. Angeles. Singularity analysis of closed-loop kinematic chains. *IEEE Transactions Robotics and Automation*, 6(3):281–290, 1990.
62. C. M. Gosselin and J. Angeles. A global performance index for the kinematic optimization of robotic manipulators. *ASME Journal of Mechanical Design*, 113(3):220–226, 1991.
63. V. E. Gough. Contribution to discussion to papers on research in automobile stability on control and in tyre performance. *Proceedings of the Automotive Division Instrument Engineering*, pp. 392–394, 1956.
64. Agur, A. M. R. *Grant's Atlas of Anatomy*. Lippincott Williams & Wilkins, Philadelphia, 2012.
65. D. T. Greenwood. *Advanced Dynamics*. Cambridge University Press, Cambridge, 2003.
66. M. Griffis and J. Duffy. A forward displacement analysis of a class of Stewart platforms. *Journal of Robotic Systems*, 6(6):703–720, 1989.
67. M. Grübler. *Getriebelehre*. Springer-Verlag, Berlin, 1917.
68. K. R. Han, W. K. Chung, and Y. Youm. New resolution scheme of the forward kinematics of parallel manipulators using extra sensors. *ASME Journal of Mechanical Design*, 118(2):214–219, 1996.
69. H.R. Harrison and T. Nettleton. *Advanced Engineering Dynamics*. Wiley, New York, 1997.
70. M. Hassan and A. Khajepour. Optimization of actuator forces in cable-based parallel manipulators using convex analysis. *IEEE Transactions on Robotics*, 24(3):736–740, 2008.
71. V. Hayward. Borrowing some design ideas from biological manipulators to design an artificial one. In *robots and biological system* Vol. 4, pp. 135–148. NATO Series. Springer-Verlag, New York, 1993.
72. V. Hayward. Design of a Hydraulic Robot Shoulder Based on a Combinatorial Mechanism. Chapter 6, pp. 297–310. *Lecture Notes in Control & Information Sciences*. Springer-Verlag, *The 3rd Int. Symp. Experimental Robotics III*, Japan, Oct. 1994.

73. V. Hayward. Design of Hydraulic Robot Shoulder Based on Combinatorial Mechanism. Chapter 3, pp. 297–310. *Experimental Robotics 3, Lecture Notes in Control and Information Sciences* 200. Springer-Verlag, New York, 1994.
74. V. Hayward and R. Kurtz. Modeling of a parallel wrist mechanism with actuator redundancy. *International Journal of Laboratory Robotics and Automation*, 4(2):69–76, 1992.
75. N. Hogan. Impedance control: An approach to manipulation: Part 1 theory. *Journal of Dynamics Systems, Measurements, and Control*, 107:1–16, 1985.
76. N. Hogan. Impedance control: An approach to manipulation: Part 2 implementation. *Journal of Dynamics Systems, Measurements, and Control*, 107:17–24, 1985.
77. J. Hollerbach and K. Suh. Redundancy resolution of manipulators through torque optimization. *IEEE Journal of Robotics and Automation*, 3(4):308–316, 1988.
78. M. A. Hosseini, H. R. M. Daniali, and H. D. Taghirad. Dexterous workspace optimization of a tricept parallel manipulator. *Advanced Robotics*, 25(3):1697–1721, 2011.
79. Z. Huang, Y. Cao, Y. W. Li, and L. H. Chen. Structure and property of the singularity loci of the 3/6–Stewart–Gough platform for general orientations. *Robotica*, 24(1):75–84, 2006.
80. Z. Huang, L. H. Chen, and Y. W. Li. The singularity principle and property of stewart parallel manipulator. *Journal of Robotics Systems*, 20(4):163–176, 2003.
81. K. H. Hunt. *Kinematic Geometry of Mechanisms*. Oxford University Press, Cambridge, 1978.
82. M. L. Husty. An algorithm for solving the direct kinematics of general Stewart–Gough platforms. *Mechanics and Machine Theory*, 31(4):365–379, 1996.
83. Adept Technology Inc. Adept robotics for packaging. Production Brochure, http://www.adept.com, 2008.
84. C. Innocenti. A novel numerical approach to the closure of the 6-6 Stewart platform mechanism. In *Proc. of the Fifth International Conference on Advanced Robotics: Robots in Unstructured Environments*, Pisa, Italy, pp. 852–855, 1991.
85. C. Innocenti. Forward kinematics in polynomial form of the general Stewart platform. *Journal of Mechanism and Design*, 123(2):254–260, 2001.
86. A. Isidori. *Nonlinear Control Systems*. Springer-Verlag, New York, 1999.
87. M. V. Ivashina, A. Van Ardenne, J.D. Bregman, J.G.B. de Vaate, and M. van Veelen. Activities for the square kilometer array (SKA) in Europe. In *Proc. of the Int. Conf. Antenna Theory and Techniques*, pp. 633–636, Columbus, OH, Sept. 2003.
88. P. Ji and Wu H. Algebraic solution to forward kinematics of a 3–DOF spherical parallel manipulator. *Journal of Robotic Systems*, 18(5):251–257, 2001.
89. P. Ji. and H. Wu. A closed-form forward kinematics solution for the 6-6P stewart platform. *IEEE Transactions on Robotics and Automation*, 17(4):522–526, 2001.
90. P. Ji and H. Wu. An efficient approach to the forward kinematics of a planar parallel manipulator with similar platforms. *IEEE Transactions Robotics and Automation*, 18:647–649, Aug. 2002.
91. H. Z. Jiang, J.F. He, and Z.Z. Tong. Characteristics analysis of joint space inverse mass matrix for the optimal design of a 6-DOF parallel manipulator. *Mechanism and Machine Theory*, 45(5): 722–739, 2010.
92. C. J. Jina, R. D. Nana, and H. Q. Gana. The fast telescope and its possible contribution to high precision astrometry. In *International Astronomical Union*, Vol. 3, pp. 178–181. Cambridge University Press, Cambridge, 2007.
93. S. Joshi and L. W. Tsai. A comparison study of two 3-DOF parallel manipulators: One with three and the other with four supporting legs. *IEEE Transactions Robotics and Automation*, 19:200–209, 2002.
94. W. Karush. *Minima of functions of several variables with inequalities as side constraints*. Masters dissertation, Department of Mathematics, University of Chicago, Chicago, Illinois, 1939.
95. S. Kawamura, H. Kino, and Ch. Won. High-speed manipulation by using parallel wire-driven robots. *Robotica*, 18(1):13–21, 2000.
96. H. K. Khalil. *Nonlinear Systems*, 3rd edition. Prentice-Hall, Upper Saddle River, NJ, 2002.

97. W. Khalil and O. Ibrahim. General solution for the dynamic modeling of parallel robots. *Journal of Intelligent and Robotic Systems*, 49(1):19–37, 2007.
98. O. Khatib. Unified approach for motion and force control of robot manipulators: The operational space formulation. *IEEE Transactions on Robotics and Automation*, 3(1):43–53, 1987.
99. M. A. Khosravi and H. D. Taghirad. Dynamic analysis and control of cable driven robots with elastic cables. *Transactions of CSME*, 35(4):543–577, 2011.
100. M. A. Khosravi and H. D. Taghirad. Experimental performance of robust PID controller on a planar cable robot. In *First International Conference on Cable–Driven Parallel Robots*, Stuttgart, Germany, September 2012.
101. D. Kim, W. Chung, and Y. Youm. Analytic Jacobian of in-parallel manipulators. In *IEEE Conf. Robotics and Automation*, San Francisco, CA, pp. 2376–2381, April 2000.
102. D. Kim and W. Y. Chung. Analytic singularity equation and analysis of six-dof parallel manipulators using local structurization method. *IEEE Transactions on Robotics and Automation*, 5(4):612–622, 1999.
103. C.A. Klein and B.E. Blaho. Dexterity measures for the design and control of kinematically redundant manipulators. *The International Journal of Robotics Research*, 6(2):72–83, 1987.
104. C.A. Klein and C.H. Huang. Review of pseudo-inverse control for use with kinematically redundant manipulators. *IEEE Transactions on Systems, Man and Cybernetics*, 13(2):245–250, 1983.
105. X. Kong and C. Gosselin. *Type Synthesis of Parallel Mechanisms*. Springer, New York, 2007.
106. D. M. Ku. Direct displacement analysis of a Stewart platform mechanism. *Mechanisms and Machine Theory*, 34:453–465, 1999.
107. H. W. Kuhn and A. W. Tucker. Nonlinear programming. In *The Proceedings of 2nd Berkeley Symposium*, Berkeley, pp. 481–492, 1951.
108. C. Lambert, A. Saunders, C. Crawford, and M. Nahon. Design of a one-third scale multi-tethered aerostat system for precise positioning of a radio telescope receiver. In *CASI Flight Mechanics and Operations Symposium*, Montreal, Canada, 2003.
109. G. Lebret, K. Liu, and F.L. Lewis. Dynamic analysis and control of a Stewart platform manipulator. *Journal of Robotic Systems*, 10(5):629–655, 1993.
110. T. Y. Lee and J. K. Shim. Forward kinematics of the general 6-6 Stewart platform using algebraic elimination. *Mechanism and Machine Theory*, 36(9):1073–1085, 2001.
111. Y. T. Lee, H. R. Choi, W. K. Chung, and Y. Youm. Stiffness control of a coupled tendon-driven robot hand. *IEEE Transactions on Control Systems Technology*, 14(5):10–19, 1994.
112. K. Levenberg. A method for the solution of certain problems in least-squares. *Quarterly Applied Mathematics*, 2:164–168, 1944.
113. F. L. Lewis, D. M. Dawson, and C. T. Abdallah. *Robot Manipulator Control: Theory and Practice*, 2nd edition. Marcel Dekker, New York, 2004.
114. H. Li, C. M. Gosselin, M.J. Richard, and B. Mayer St-Onge. Analytic form of the six-dimensional singularity locus of the general Gough–Stewart platform. *ASME Journal of Mechanical Design*, 128:279–287, 2006.
115. Y. Y. Li and Q. Xu. Kinematic analysis and design of a new 3–DOF translational parallel manipulator. *Transactions of the ASME. Journal of Mechanical Design*, 128(4):729–737, 2006.
116. A. Liégeois. Automatic supervisory control of the configuration and behavior of multibody mechanisms. *IEEE Transactions on Systems, Man, and Cybernetics*, 7(12):868–871, 1977.
117. H. C. Lin, T. C. Lin, and K. H. Yae. On the skew-symmetric property of the Newton–Euler formulation for open chain robot manipulators. In *Proc. of American Control Conference*, Seattle, Washington, USA, pp. 2322–2326, 1995.
118. K. Liu, J.M. Fitzgerald, and F.L. Lewis. Kinematic analysis of a Stewart platform manipulator. *IEEE Transactions on Industrial Electronics*, 40(2):282–293, 1993.
119. W. S. Lu and Q. H. Meng. Regressor formulation of robot dynamics: Computation and applications. *IEEE Transactions on Robotics and Automation*, 9(3):323–333, 1993.
120. O. Ma and J. Angeles. Architecture singularities of platform manipulators. In *IEEE International Conference on Robotics and Automation*, pp. 1542–1547, Sacramento, CA, 1991.

121. D. P. Marin, J. Baillieul, and J. Hollerbach. Resolution of kinematic redundancy using optimization techniques. *IEEE Transactions on Robotics and Automation*, 5(4):529–533, 1989.
122. D. Marquardt. An algorithm for least-squares estimation of nonlinear parameters. *SIAM Journal of Applied Mathematics*, 11:431–441, 1963.
123. M. T. Masouleh, C. Gosselin, M. H. Saadatzi, X. Kong, and H. D. Taghirad. Kinematic analysis of 5–RPUR (3T2R) parallel mechanisms. *Meccanica*, 46(2):131–146, 2011.
124. M. T. Masouleh, M. H. Saadatzi, and H. D. Taghirad. Workspace analysis of 5–PRUR parallel mechanisms (3T2R). *Journal of Robotics and Computer Integrated Manufacturing*, 28(1):437–448, 2012.
125. J. D. Mathewsa, D.D. Meiselb, K.P. Huntera, V.S. Getmana, and Q. Zhouc. Very high resolution studies of micrometeors using the Arecibo 430 MHz radar. *Icarus*, 126(1):157–169, 1997.
126. H. Mayeda, K. Yoshida, and K. Osuka. Base parameters of manipulator dynamic models. *IEEE Transactions on Robotics and Automation*, 6(3):312–321, 1990.
127. B. Mayer St-Onge and C. M. Gosselin. Singularity analysis and representation of the general Gough–Stewart platform. *International Journal of Robotics Research*, 19(3):271–288, 2000.
128. J. McPhee, P. Shi, and J.C. Piedboeuf. Dynamics of multibody systems using virtual work and symbolic programming. *Mathematical and Computer Modelling of Dynamical Systems*, 8(2):137–155, 2002.
129. J. L. Meriam. *Dynamics*. J. Wiley and Sons, New York, 1966.
130. J. L. Meriam and L.G. Kraige. *Engineering Mechanics: Dynamics*, 5th edition. Wiley, New York, 2007.
131. J. P. Merlet. Direct kinematics of parallel manipulators. *IEEE Transactions on Robotics and Automation*, 9(6):842–845, 1993.
132. J. P. Merlet. Solving the forward kinematics of a Gough-type parallel manipulator with interval analysis. *International Journal of Robotics Research*, 23(3):221–235, 2004.
133. J. P. Merlet. *Parallel Robots*, second edition. Kluwer Academic Publishers, Boston, MA, 2006.
134. J.P. Merlet. Singular configurations of parallel manipulators and grassmann geometry. *The International Journal of Robotics Research*, 8(5):45–56, 1989.
135. M.G. Mohamed and J. Duffy. A direct determination of the instantaneous kinematics of fully parallel robot manipulators. *ASME Journal of Mechanisms, Transmissions, and Automation in Design*, 107(2):226–229, 1985.
136. U. Mohrlen, D. Weber, R. Gonzalez, D. Max Schmid, T. Sulser, and R. Gobet. Robot-assisted minimal invasive pediatric urology. *Journal of Pediatric Urology*, 5(1):S45–S46, 2009.
137. J.J. Moré and D.C. Sorensen. Computing a trust region step. *SIAM Journal on Scientific and Statistical Computing*, 3:553–572, 1983.
138. J. J. Moré. *Numerical Analysis*, pp. 105–116. Lecture Notes in Mathematics. Springer-Verlag, Berlin, 1977.
139. Y. Nakamura and H. Hanafusa. Inverse kinematic solutions with singularity robustness for robot manipulator control. *ASME Journal of Dynamic Systems, Measurement and Control*, 108(3):163–171, 1986.
140. Y. Nakamura and H. Hanafusa. Optimal redundancy control of robot manipulators. *International Journal of Robotics Research*, 6(1):32–42, 1987.
141. P. Nanua, K. J. Waldron, and V. Murthy. Direct kinematic solution of a Stewart platform. *IEEE Transactions on Robotics and Automation*, 6(4):438–444, 1990.
142. C. C. Nguyen and F. J Pooran. Dynamic analysis of a 6 DOF CKCM robot end-effector for dual-arm telerobot systems. *Robotics and Autonomous Systems*, 5(4):377–394, 1989.
143. V. K. Nguyen. Consistent definition of partial derivatives of matrix functions in dynamics of mechanical systems. *Mechanism and Machine Theory*, 45(7):981–988, 2010.
144. National Institute of Standard and Technology. Robocrane project, large scale manufacturing using cable control. http://www.isd.mel.nist.gov/projects/robocrane, March 2001.
145. R. Oftadeh, M. M. Aref, and H.D. Taghirad. Explicit dynamics formulation of Stewart–Gough platform: A Newton–Euler approach. In *Proc. of the IEEE Int. Conf. IROS*, pp. 2773–77, Taiwan, Oct. 2010.

146. S. R. Oh and S. K. Agrawal. Cable suspended planar robots with redundant cables: Controllers with positive tensions. *IEEE Transactions on Robotics*, 21(3):457–465, 2005.
147. M. Okada, Y. Nakamura, and S. I. Hoshino. Development of the cybernetic shoulder—A three DOF mechanism that imitates biological shoulder motion. In *Int. Conf. Intelligent Robots and Systems*, Kyongju, Korea, pp. 543–548, 1999.
148. A. Omran and M. Elshabasy. A note on the inverse dynamic control of parallel manipulators. *Proceedings of the Institution of Mechanical Engineers, Part C: Journal of Mechanical Engineering Science*, 224(1):25–32, 2010.
149. O. M. O'Reilly. *Intermediate Dynamics for Engineers: A Unified Treatment of Newton–Euler and Lagrangian Mechanics*. Cambridge University Press, Cambridge, 2008.
150. V. Parenti-Castelli and R. D. Gregorio. A new algorithm based on two extra-sensors for real-time computation for the actual configuration of the generalized Stewart–Gough manipulator. *ASME Journal of Mechanical Design*, 112:294–298, 2000.
151. P. J. Parikh and S. Y. Lam. A hybrid strategy to solve the forward kinematics problem in parallel manipulators. *IEEE Transactions Robotics*, 21(1):18–25, 2005.
152. D. L. Pieper. *The kinematics of manipulators under computer control*. PhD thesis, Department of Mechanical Engineering, Stanford University, 1968.
153. S. R. Ploen. A skew-symmetric form of the recursive Newton–Euler algorithm for the control of multibody systems. In *1999 American Control Conference, San Diego*, pp. 3770–3773, CA, USA, 1999.
154. M. J. D. Powell. *Variable Metric Methods for Constrained Optimization. Mathematical Programming: The State of the Art*. Springer-Verlag, Berlin, 1983.
155. M. Raghavan. The Stewart platform of general geometry has 40 configurations. In *Proc. of the ASME Design and Automation Conference*, pp. 397–402, Chicago, USA, Sept. 1991.
156. R. Rembala and C. Ower. Robotic assembly and maintenance of future space stations based on the ISS mission operations experience. *Acta Astronautica*, 65(7-8):912–920, 2009.
157. P. L. Richard and C. M. Gosselin. Kinematic analysis and prototyping of a partially decoupled 4–DOF 3T1R parallel manipulator. *ASME Journal of Mechanical Design*, 129(6):611–616, 2007.
158. B. Roth. Screws, motors, and wrenches that cannot be bought in a hardware store. In *Proc. of the 1st International Symposium in Robotics Research*, Pittsburgh, USA, pp. 979–693, 1984.
159. H. Sadjadian and H. D. Taghirad. Comparison of different methods for computing the forward kinematics of a redundant parallel manipulator. *Journal of Intelligent and Robotic Systems*, 44:225–246, 2005.
160. H. Sadjadian and H. D. Taghirad. Kinematic, singularity and stiffness analysis of the hydraulic shoulder: A 3–DOF redundant parallel manipulator. *Advanced Robotics*, 20(7):763–781, 2006.
161. H. Sadjadian, H. D. Taghirad, and A. Fatehi. Neural networks approaches for computing the forward kinematics of a redundant parallel manipulator. *International Journal of Computational Intelligence*, 2(1):40–47, 2005.
162. J. K. Salisbury and I. J. Craig. Articulated hands: Force control and kinematic issues. *International Journal of Robotics Research*, 1(1):4–17, 1982.
163. L. Sciavicco and B. Siciliano. *Modelling and Control of Robot Manipulators*. Springer-Verlag, New York, 2001.
164. H. Seraji. Configuration control of redundant manipulators: Theory and implementation. *IEEE Transactions on Robotics and Automation*, 5(4):472–490, 1989.
165. H. Seraji. Task-based configuration control of redundant manipulators. *Journal of Robotic Systems*, 9(3):411–451, 1992.
166. L. F. Shampine. *Numerical Solution of Ordinary Differential Equations*. Chapman & Hall, New York, 1994.
167. A. Sokolov and P. Xirouchakis. Kinematics analysis of a 3–DOF parallel manipulator with R-P-S joint structure. *Robotica*, 23(2):207–217, 2005.
168. M. W. Spong, S. Hutchinson, and M. Vidyasagar. *Robot Modeling and Control*. John Wiley & Sons, Hoboken, NJ, 2005.

169. S. V. Sreenivasan, K. J. Waldron, and P. Nanua. Closed-form direct displacement analysis of a 6–6 Stewart platform. *Mechanism and Machine Theory*, 29(6):855–864, 1994.
170. D. Stewart. A platform with six degrees of freedom. *Proceedings of the UK Institute of Mechanical Engineering*, 180(1):371–386, 1965.
171. G. Strang. *Introduction to Linear Algebra*, 3rd edition. Wellesley-Cambridge Press, MA, 2005.
172. S. Tadokoro and T. Matsushima. A parallel cable-driven motion base for virtual acceleration. In *Proc. of Int. Conference on Intelligent Robots and Systems*, Hawaii, USA, pp. 1700–05, November 2001.
173. H. D. Taghirad and Y. B. Bedoustani. An analytic-iterative redundancy resolution scheme for cable-driven redundant parallel manipulators. *IEEE Transactions on Robotics*, 27(6):1137–1143, 2011.
174. H. D. Taghirad and P.R. Belanger. Intelligent built-in torque sensor of harmonic drive systems. *IEEE Transactions on Instrumentation and Measurements*, 48(6):1201–1207, 1999.
175. H. D. Taghirad and P. R. Belanger. H-infinity based robust torque control of harmonic drive systems. *Journal of Dynamic Systems, Measurements, and Control, ASME Pub.*, 123(3):338–345, 2001.
176. H. D. Taghirad and E. Jamei. Robust performance verification of IDCARC controller for hard disk drives. *IEEE Transactions on Industrial Electronics*, 55(1):448–456, 2008.
177. H. D. Taghirad and M. Nahon. Dynamic analysis of a redundantly actuated parallel manipulator: A virtual work approach. In *Iranian Conference on Electrical Engineering*, pp. 54–60, Tehran, Iran, May 2007.
178. H. D. Taghirad and M. Nahon. Dynamic analysis of a macro–micro redundantly actuated parallel manipulator. *Advanced Robotics*, 22(3):949–981, 2008.
179. H. D. Taghirad and M. Nahon. Kinematic analysis of a macro–micro redundantly actuated parallel manipulator. *Advanced Robotics*, 22(2):657–687, 2008.
180. Y. Ting, Y.-S. Chen, and H.-C. Jar. Modeling and control for a Gough–Stewart platform CNC machine. *Journal of Robotic Systems*, 21(11):609–623, 2004.
181. L. W. Tsai. *Robot Analysis: The Mechanics of Serial and Parallel Manipulators*. John Wiley and Sons, Inc., New York, 1999.
182. L. W. Tsai and S. A. Joshi. Kinematics and optimization of a spatial 3-UPU parallel manipulator. *ASME Journal of Mechanical Design*, 122:439–446, 2000.
183. K. Čapek. *Rossum's Universal Robots*. Doubleday, New York, 1923. translated by P. Selver.
184. K. J. Waldron and K. H. Hunt. Series-parallel dualities in actively coordinated mechanisms. *Robotics Research*, 4:175–181, 1988.
185. C. W. Wampler. Forward displacement analysis of general six-in-parallel SPS (Stewart) platform manipulators using soma coordinates. *Mechanism Machine and Theory*, 31(3):331–337, 1996.
186. J. Wang and C.M. Gosselin. A new approach for the dynamic analysis of parallel manipulators. *Multibody System Dynamics*, 2(3):317–334, 1998.
187. J. Wang, J. Wu, L. Wang, and T. Li. Simplified strategy of the dynamic model of a 6-UPS parallel kinematic machine for real-time control. *Mechanism and Machine Theory*, 42(9):1119–1140, 2007.
188. L. C. T. Wang and C. C. Chen. On the numerical kinematic analysis of general parallel robotic manipulators. *IEEE Transactions on Robotics and Automation*, 9(3):272–285, 1993.
189. Y. Wang. A direct numerical solution to forward kinematics of general Stewart–Gough platforms. *Robotica*, 25(1):121–128, 2007.
190. F. Wen and C. Liang. Displacement analysis of the 6–6 Stewart platform mechanisms. *Mechanism and Machine Theory*, 29(4):547–557, 1994.
191. D. E. Whitney. Resolved motion rate control of manipulators and human prostheses. *IEEE Transactions on Man–Machine Systems*, 10(2):47–53, 1969.
192. A. Wolf and M. Shoham. Investigation of parallel manipultors using linear complex approximation. *Journal of Mechanical Design*, 125:564–572, 2003.
193. C. Yang et al. PD control with gravity compensation for hydraulic 6-DOF parallel manipulator. *Mechanism and Machine Theory*, 45(4):666–677, 2010.

194. J. Yang and Z. J. Geng. Closed form forward kinematics solution to a class of hexapod robots. *IEEE Transactions on Robotics and Automation*, 14:503–508, 1998.
195. C. S. Yee. Forward kinematics solution of Stewart platform using neural networks. *Neurocomputing*, 16(4):333–349, 1997.
196. T. Yoshikawa. The manipulability of robotic mechanisms. *International Journal of Robotics Research*, 4(2):3–9, 1985.
197. T. Yoshikawa. *Foundation of Robotics: Analysis and Control*. MIT Press, Cambridge, MA, 1990.
198. D. Zlatanov, I. A. Bonev, and C. M. Gosselin. Constraint singularities of parallel mechanisms. In *IEEE International Conference on Robotics and Automation*, pp. 496–502, Washington, DC, USA, 2002.
199. D. Zlatanov, R. G. Fenton, and B. Benhabib. Singularity analysis of mechanisms and robots via a velocity-equation model of the instantaneous kinematics. In *IEEE International Conference on Robotics and Automation*, pp. 986–991, San Diego, CA, USA, 1994.
200. L. Zollo, B. Siciliano, C. Laschi, G. Teti, and P. Dario. An experimental study on compliance control for a redundant personal robot arm. *Robotics and Autonomous Systems*, 44(2):101–129, 2003.

Index

Q

R

T

U

V